Electricity and Electronics for Renewable Energy Technology

AN INTRODUCTION

POWER ELECTRONICS AND APPLICATIONS SERIES

Muhammad H. Rashid, Series Editor
University of West Florida

PUBLISHED TITLES

Advanced DC/DC Converters
Fang Lin Luo and Hong Ye

Alternative Energy Systems: Design and Analysis with Induction Generators, Second Edition
M. Godoy Simões and Felix A. Farret

Complex Behavior of Switching Power Converters
Chi Kong Tse

DSP-Based Electromechanical Motion Control
Hamid A. Toliyat and Steven Campbell

Electric Energy: An Introduction, Third Edition
Mohamed A. El-Sharkawi

Electrical Machine Analysis Using Finite Elements
Nicola Bianchi

Electricity and Electronics for Renewable Energy Technology: An Introduction
Ahmad Hemami

Modern Electric, Hybrid Electric, and Fuel Cell Vehicles: Fundamentals, Theory, and Design
Mehrdad Eshani, Yimin Gao, Sebastien E. Gay, and Ali Emadi

Modeling and Analysis with Induction Generators, Third Edition
M. Godoy Simões and Felix A. Farret

Uninterruptible Power Supplies and Active Filters
Ali Emadi, Abdolhosein Nasiri, and Stoyan B. Bekiarov

Electricity and Electronics for Renewable Energy Technology

AN INTRODUCTION

Ahmad Hemami
ADJUNCT PROFESSOR, MCGILL UNIVERSITY, MONTREAL, CANADA

CRC Press is an imprint of the
Taylor & Francis Group, an **informa** business

CRC Press
Taylor & Francis Group
6000 Broken Sound Parkway NW, Suite 300
Boca Raton, FL 33487-2742

First issued in paperback 2017

ISBN-13: 978-1-4822-6176-9 (hbk)
ISBN-13: 978-1-138-89299-6 (pbk)

Library of Congress Cataloging-in-Publication Data

Hemami, A.
Electricity and electronics for renewable energy technology : an introduction / Ahmad Hemami.
pages cm
Includes bibliographical references and index.
ISBN 978-1-4822-6176-9 (hardcover : alk. paper) 1. Electric apparatus and appliances. 2. Electronic apparatus and appliances. 3. Photovoltaic power systems. I. Title.

TK452.H466 2016
621.042--dc23 2015017783

Visit the Taylor & Francis Web site at
http://www.taylorandfrancis.com

and the CRC Press Web site at
http://www.crcpress.com

This book is dedicated to my wife, Shahla,

and to the memory of my parents.

Contents

Preface

This book is not about wind, solar, or any other renewable energy. It is about general and particular subjects in electricity and electronics that a technical person working in the generation and distribution of electricity from renewable sources needs to know. The purpose is to bring together in one book a general knowledge, to a moderated level, about the ways and the devices that are used in power generation from some renewable energy sources.

Modern wind turbines and photovoltaic solar systems use power converters to generate commercial-level alternating current electricity. This is quite different from the conventional methods employed in hydroelectric and fossil fuel power plants. The difference lies in the fact that in photovoltaic solar systems electrical energy is initially in the form of direct current, and in wind turbines, depending on the modes of operation, either induction generators are used or synchronous generators rotating at varying speeds are employed. In both cases, conversion by power electronics is inevitable.

The content of this book has been selected using numerous recommendations from the advisory committee members from industry for a 2-year program intended for training technicians to work in the wind industry. On the basis of their comments about the program and their expectation from the graduates working in the field, in addition to a knowledge of electricity, a general knowledge of basic electronics is essential, particularly as a preliminary step toward understanding the more advanced materials of gated transistors and power converters. Moreover, because, in practice, they must deal with and are exposed to PLCs and SCADA and industrial control in general, they need to have a fundamental and brief understanding of digital systems and digital electronics.

This book is written as a textbook for college and university students whose major is not electrical engineering. As a result, the necessary material, which for electrical engineering students appears in several books and with more in-depth analysis, is covered in a single book. In this way, all material judged to be necessary and at the level considered to be sufficient is presented to the reader. The relatively more advanced materials for some chapters are put separately under "advanced learning," which can be skipped by those schools/programs that prefer to do so. The advanced sections cover mathematics pertinent to the subject that within the chapter is described without extended mathematical analysis.

Recommendations from the reviewers about the content of the book, starting from simple subjects but having a fast pace to advanced material, were not uniform. Nevertheless, to make the book self-sufficient for many programs so that it can serve as the sole book for a few courses within a program, the content is kept as initially designed. In this respect, the first five chapters, dealing with the basics, can be skipped by some readers. In particular, for those students with insufficient knowledge of basic mathematics and also for those people in industry who want to use the text as a convenient reference book and may need a review to refresh their forgotten but necessary math skills, a chapter is devoted to basic mathematics.

The book starts with the basics of electricity, no previous knowledge is necessary, but it has a rapid pace and gradually covers materials that are quite advanced. In terms of the coverage of material, this a unique book of this caliber that brings several subjects of necessity together in a single volume. In addition, the pedagogic approach for certain topics, considered relatively difficult to understand by many, such as reactive power and the operation of a transistor, involves analogy to some more tangible physical systems. In this respect, certain analogies introduced in the book are for the first time ever.

This book can be used for a number of courses in any program for training qualified people to work for industry in the areas of electricity generation and distribution, as well as maintenance of machinery and process control. Typical courses are DC and AC electricity, motors and generators, solid-state electronics, and introduction to digital electronics. Any particular set of chapters can be selected for a desired course. Certain chapters may contain more advanced topics, but an instructor can ignore the parts that have been included for those who like to have them available with the rest in one volume.

The structure of the book is as follows: After the first three chapters of introduction, mathematical fundamentals, and the review of atomic structure of materials, the rest of the book is divided into two parts: electricity, from Chapters 4 through 12, and electronics from Chapters 13 through 24. The electricity part covers the basics of electricity and relationships, motors and generators, transformers, and networks and distribution. The electronics part covers the basics of electronics, diodes and transistors, switching devices, and power converters, and the three final chapters discuss digital electronics.

Chapter 4 discusses the basics of electricity. This chapter covers the introduction of electrical components, direct and alternating current, resistance, capacitance and inductance, and their physical forms and their properties as related to the physical form. Chapter 5 describes electric current, voltage, their measurement and their relationships for resistive components, and finally electric power. All the relationships for resistive loads in series and parallel and their combination are discussed in Chapter 6, while the other matters, such as Norton and Thevenin's theorems and the principle of superposition, which sound more difficult to some, are left for Chapter 12.

Chapter 7 is devoted to DC electric machines, where a good understanding of how these machines work and their performance and the relationships between current, voltage, torque, speed, and power can be gained from it. All the matters corresponding to the behavior of circuits containing the three categories of components (resistors, capacitors, and inductors), their corresponding power consumption and power relationships for single phase, are the subjects of Chapter 8. Three-phase systems are discussed in Chapter 9, which extends the rules of single-phase AC to three-phase AC. Transformers, both single phase and three phase, are studied in Chapter 10. In addition to principles, the chapter discusses transformer tests and transformer modeling. An instructor can select any desired parts of the material for class discussion.

Similar to Chapter 7 for DC machines, Chapter 11 is devoted to the way AC machines work, their categories and their corresponding speed, torque, and power relationships. Chapter 12 is about electric power transmission and distribution. It involves calculations of capacitance and inductance of power lines. It also covers the analysis of networks where the simpler laws of Chapters 4 to 6 are inadequate.

The second part of the book starts with an introduction of electronic devices and their functions, some preliminary definitions, and the distinction between analog and digital signals in Chapter 13. Diodes and special diodes, and their functions and in-circuit behavior, are discussed in Chapters 14 and 15, followed by diode rectifiers and filters in Chapter 16.

Transistors, their major functions, and behavior are the subject of Chapter 17, whereas elaborations on transistor circuits and the various forms of their application for amplification are continued in Chapter 18. Switchable diodes and gated transistors are introduced in Chapter 19, which serves for the discussion of power converters in Chapter 20.

After a suitably brief explanation of wind turbine and solar renewable energy systems, the application of power converters in these technologies is described in Chapter 21.

Finally, the fundamentals of digital electronics are introduced by presenting the number systems in Chapter 22, logic circuits and applications in Chapter 23, and encoders and decoders in Chapter 24.

Acknowledgments

The author would like to thank the reviewers for their helpful and constructive remarks and recommendations. I am grateful to the companies and institutions that contributed to the figures or other materials used in the book. Also, I thank the editors of the manuscript for their professional work. Special thanks go to my wife for her continuous and the true support she has always given me. Thanks are due also to other individuals in various institutions whose part in many ways has made the completion of this book possible.

Author

Ahmad Hemami, PhD, is an adjunct professor at McGill University in Montreal. He earned a BSc in mechanical engineering. He earned a PhD in system dynamics and control from the Department of Aeronautical and Mechanical Engineering, University of Salford, UK. Dr. Hemami has many years of work experience in industry and academia in the form of engineering practice, teaching, and research. His expertise is in systems, robotics, control and automation, and wind energy. He has taught several undergraduate and graduate courses in engineering disciplines, supervised a number of MSc and PhD students in Canada and the United States, and has served as a consultant for industry. The results of his analytical and experimental research on pertinent subjects have been presented in more than 100 journal and conference publications. His current work is on renewable energy and care for the environment.

1

Energy and Electricity

OBJECTIVES: After studying this chapter, you will be able to

- Explain why concern about energy is so important
- Understand the energy need in the world
- Define the sources of energy
- Define the sources of renewable energy
- Understand the importance of electrical energy
- Evaluate your knowledge of electricity and electronics
- Recognize the importance of renewable energy
- Learn some terms in electricity
- Understand the importance of safety at work and some basic safety rules
- Explain some examples of static electricity
- Understand a number of applications of static electricity

New terms:
Breaker, fuse, safety rules, safety standards, static electricity

1.1 Introduction

Electricity is a type of energy. Although it is not a type of energy stored in fossil and nuclear fuels, it can be obtained from conversion of the chemical energy contained in fossil fuel or nuclear energy contained in certain metals. It can also be obtained by converting mechanical energy in the form of wind and hydro to electrical energy. In fact, electric generators only convert mechanical energy to electricity, while the mechanical energy itself can be the result of converting some other types of energy. This is because energy cannot be generated; it can only be converted from one type to the other.

In gas, oil, and coal power plants the chemical energy of fossil fuel is first converted to mechanical energy and then mechanical energy is converted to electrical energy. Similarly, in nuclear power plants, atomic energy is first converted to heat then to mechanical energy through steam turbines and finally to electrical energy.

In hydraulic turbines the mechanical energy from the flowing water, and in wind turbines the mechanical energy from wind is converted to electrical energy. A direct conversion of heat/light from sun to electrical energy takes place in solar cells. Solar panels are a large number of solar cells put together. Solar heat can be gathered also by solar collectors, which contain a number of tubes that absorb heat from sunshine and warm up water (or another liquid, in general). At a larger scale an array of reflective mirrors receive the sun's heat and concentrate it at a common collector to boil water and generate steam. Geothermal energy is also another source of available energy without greenhouse effect and nuclear waste.

As you know, both fossil fuel and nuclear power plants have residual, which is not environmentally friendly. In other words, it is a contaminant for the environment. Water, wind, sun, and some other (very limited in quantity at the present time) renewable sources of energy are those without any harmful residual or contamination.

Two main reasons are behind the emphasis for renewable energy. One is contamination and global warming, and the other is the fact that the fossil fuels eventually dry up, and there will be no more oil, gas, and coal. The same thing can be said about nuclear energy because the reserves of nuclear fuels are not unlimited.

We and the future generations of mankind need to understand that humanity can only survive if we live in harmony with nature. We should eliminate any waste, and all our energy must ultimately be provided by renewable sources.

1.2 Need for Energy

Electricity is only one part of the energy that mankind needs. We cannot imagine living without electricity. Equally, we cannot imagine living without transportation, buildings, and so forth; all of these consume energy. The need for energy cannot be overemphasized. It is an everlasting fact. This is partly due to the increase in the population of the world and partly because of conversion to modern living by many nations. If the world population stayed constant, we could imagine that one day all the people could have a moderate modern life, if we would act wisely. The unfortunate fact is the population increases, and the unwise actions by many, from small scale at an individual level to large scale of nations and countries.

According to a 2011 report by British Petroleum "World primary energy consumption grew by 5.6% in 2010, the largest increase in percentage terms since 1973." This percentage is for the entire energy consumption, which implies transportation, heating (and cooling), electricity, and so forth.

Electrical energy, by itself, has a faster pace of growth. "Electricity demand is increasing twice as fast as overall energy use, and is likely to rise 76% by 2030 (from 2010)."*

Figure 1.1 depicts the growth rate of electricity during 2000 to 2012 period. Table 1.1 illustrates the rate of growth of electric energy for some countries and regions.

Despite the remarkable growth of renewable energy (mainly the number of installed wind turbines and solar panels) in many industrialized countries, the renewable energy covers only a small fraction of the total electrical energy demand (less than 2 percent in the world, 2.3 percent in the United States). Since 2011, China ranks first in terms of installed wind turbines per year, followed by the United States. Nevertheless, many

* From International Energy Agency (IEA).

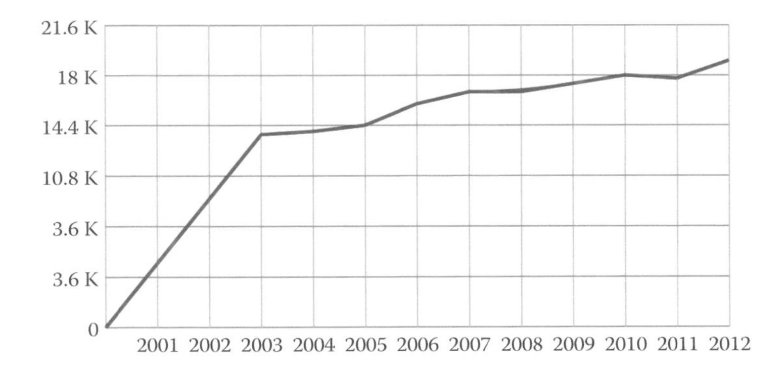

Figure 1.1 World electricity consumption (billion kWh). (From International Energy Agency.)

TABLE 1.1 Population and Energy Growth 1990–2008

	Population (Million)			Energy (Terawatt-Hour)		
	1990	**2008**	**Growth**	**1990**	**2008**	**Growth**
United States	250	305	22%	22.3	26.6	20%
EU-27	473	499	5%	19.0	20.4	7%
Middle East	132	199	51%	2.6	6.9	170%
China	1141	1333	17%	10.1	24.8	146%
Latin America	335	462	30%	4.0	6.7	66%
Africa	634	984	55%	4.5	7.7	70%
India	850	1140	34%	3.8	7.2	91%
Other	1430	1766	23%	36.1	42.2	17%
The World	5265	6688	27%	102.3	142.3	39%

regions have the potential and the capacity for much more wind and solar energy. China, for instance, has an estimated capacity of between 700 and 1200 gigawatts (GW). (Gigawatt is a measure of electric power. Giga stands for one billion. If you look at any electric device, even a simple one like a light bulb, the amount of its power is stamped on it [e.g., 100 W and 60 W].)

1.3 Renewable Energy

If we set aside the energy reserves of the Earth in the form of fossil fuel and metals that are used as nuclear reactor fuels, one can say that almost all other energy on the Earth comes either from the sun or from the heat stored under the Earth's crust. This is because except for tidal energy, which is attributed to the moon, the energy in the wind, waterfalls, and ocean waves all stem from the effect of sun on the Earth's atmosphere. The source of wind is the heat from the sun, and ocean waves are caused by the wind. Also, if biofuels are considered, the growth of plants is due to the sun's light and heat. Solar energy, as the name implies, involves direct energy from the sun.

Inside the Earth there exists a tremendous amount of thermal energy. This heat, called geothermal energy, can be used for heating buildings on a small scale, but it can also be used for generating steam and running steam turbines for generating electricity, as in power plants.

All the aforementioned categories of energy are called renewable because they do not create pollutants or dangerous residuals, which are harmful for the environment and all the living creatures on Earth. Also, they are regenerated, and, in general, they are abundant.

Presently, of the above-mentioned renewable energies, only wind and solar energy have gained momentum, because, in general, renewable energy is a low-grade energy (except hydro energy), meaning that although it is plentiful, it is not concentrated (the amount of energy per volume is low) or it is expensive to extract.

Despite all the advantages associated with renewable energy, a primary criterion is the cost. Compared to coal, oil, and gas, which determine the unit price for energy, renewable energy is still expensive. In particular, the initial cost for installations is high. Nevertheless, because the supply of all the fossil and similar fuels is limited, it is not too hard to understand that one day mankind must count only on those renewable sources of energy and nothing else.

Out of various renewable energy sources, the one that has been used frequently and for a long time is hydro energy. One reason for this is the large-sized hydraulic turbines. There is a significant difference between hydro energy and other renewable sources, which are of low grade. In the last 20 years the progress made in wind turbine technology and solar cells has made use of these devices more efficient and more cost effective, making them viable for large-scale and commercial use. However, tidal and sea wave energy are not yet candidates for commercially accepted production.

In general, we can say that renewable energy can be directly used, as appropriate to the usage, or it can be converted to electricity. For example, wind turbines in the past have been used for mechanical work, such as in grinding wheat and for pumping water from wells for irrigation. Although still this can be done, almost all wind turbines today change the wind energy into electricity. The obvious reason is that electrical energy can be relatively easily transmitted from the point of production to the point of consumption. Similarly, solar and geothermal energy can be directly used for heating or can be converted to electricity.

> Renewable energy can be defined as energy from the sources without pollutants of the environment, such as CO_2, radioactive waste, and harmful chemicals.

Converting wind energy or solar energy to electricity at the commercial level is performed in wind farms and solar farms with many turbines or solar panels. Their power output can be comparable to conventional power plants. For example, instead of one steam turbine with 500 MW, we may have 250 wind turbines of 2 MW. Nevertheless, the technology used for wind turbines and solar power plants is not the same as used in the conventional power generation, although both generate electricity.

This book is not intended to be about wind turbines or solar panels and how they work, though Chapter 21 has a short description of the main components and their specifications. The principal point is to describe how electricity is generated in these nonconventional power plants. This by itself is necessary knowledge for people in the field.

1.4 Electricity

The important and most significant feature of electrical energy is its transportability (see Figure 1.2). Compared to thermal and mechanical energy, one can say that electricity is relatively cheap to transport but rather expensive to store. For both transportation and storage the scale under consideration is very large. We are not referring to an ordinary pen size battery or a car battery. We are talking about transporting/storing electricity for a city.

As you progress through this book, step by step you will learn about electricity and many relevant topics. There is always more to learn, and you may go into more depth on many topics, but time and other factors do not allow one to grasp everything the first time.

Most likely, you have some basic knowledge of certain facts about electricity because you use it at home and in your cell phone and other consumer electrical/electronic equipment, even though you may be hearing about *direct current, alternating current*, and many other terms for the first time. You should not worry about it, as things become clearer in a timely manner and at a gradual pace.

Initially, electricity is divided into two categories: *electricity* and **static electricity**. There are similarities and differences between the two. This book is mainly about the first category, the type that we can use and our life nowadays much depends on it. That is to say that we are so used to it that we probably cannot live without it.

Static electricity: Type of electricity where charged bodies instantly discharge (when conditions allow). There is no long time duration for controlled flow of electrons, as in electric devices.

Figure 1.2 One important aspect of electricity is its transportability.

The next section describes what static electricity is. The proportion of the size of this paragraph to the size of the book is more or less the same as the ratio between the applications and the importance of static electricity to electricity.

1.5 Static Electricity

Although there are a few applications of static electricity, compared to those we call electric devices (because they need electricity to work), the number is so much smaller. At the large scale, lightning is an example of static electricity, which is accompanied by thunder. It is a large instantaneous flash of light, called the lightning bolt, which occurs as a result of a sudden discharge of stored electrical energy. There is tremendous amount of energy released this way, but the unfortunate fact is that we cannot use this energy for useful application. There is no way of using this energy, which is released in a fraction of a second. It is too much energy in too little time.

At the smaller scale, most of us have experienced a small spark accompanied by a shock sometimes when we open the door of a car or a room. Experiencing this phenomenon depends on the material of the sole of the shoe we are wearing and the carpet or ground we walk on, as well as the amount of humidity in the air. Usually, this happens in dry air, and if we have rubber soles. The phenomenon is as a result of accumulation of electrons (electrons are one of the fundamental particles of all the material around us). This shock is exactly a small-scale representation of what happens with lightning in clouds.

Accumulation of electrons takes place when certain materials rub together, for example rubber and some plastics. In the case of lightning it is the clouds (air with water vapor) that move relative to each other. If the electrons find a path to move through, then the discharge occurs. But, if they do not find a path, the effect will appear in a different form. For example, things stick together because there is some attraction force created between the materials. For instance, you have often seen that the thin plastic wraps stick together and that it is not easy to separate them. Also, some surfaces collect dust more than others. This is due to static electricity, again created by rubbing two dissimilar materials together.

This latter phenomenon is used for certain purposes in industry. One good example is the photocopier machine, in which the very small particles of toner (the dry ink, which is black carbon powder) stick to the parts of paper that have been charged through the process. A hot cylinder then melts them to penetrate into the paper in the form of ink. A similar usage is in the manufacturing of sand paper.

A very common application is used for painting large pieces such as automobile body parts. In order to have a uniform layer of paint all over a part to be painted, the process is automated. In passing through a chamber the charged tiny paint droplets stick to the body of the moving piece that is to be painted. Similarly, in industrial stacks, to prevent dust and smoke from going up and polluting the atmosphere, they are absorbed by a body that has received static electricity and attracts the small pieces of carbon and dust when passing nearby.

All the above examples, including stack cleaning and air filtering are among the industrial application of static electricity.

1.6 Safety at Work and Safety Rules

We work for our living. Therefore work must not endanger our lives nor must it harm the workers, but accidents always happen. As a professional individual when working with electricity, we must avoid any unsafe conditions by following the **safety rules**. Something that one must always care for is "safety." Every year all over the world in manufacturing sectors and service industries, a noticeable number of both fatal and nonfatal injuries happen for various reasons. Examples of these include falling, burning, getting crushed, and suffocating. The reason can be equipment failure, operator error, and negligence. Table 1.2 indicates the numbers of fatal electricity related injuries in all United States in 2010.

Safety rules: Sets of rules that deal with safe operation of a machine or completion of a task, thus minimizing the possible harm for an operator and others.

In all professions and in all industrial sites, there are many chances for something to go wrong, e.g. a machine breaks or some accident happens. But, a large percentage of accidents are due to lack of respect for safety rules and/or the lack of knowledge about what to do in case of an accident. It is everybody's responsibility to minimize the risk of anything that can be harmful to oneself or others. For all this, safety rules are set by the authorities in each country. To minimize the risks of something harmful or dangerous, one must always follow the safety rules.

At this time we will not mention those safety rules that require technical terms, which we have not yet introduced. Nevertheless, safety rules are often commonsense rules. In other words, in many cases one knows the rule by recognizing what way is safe to carry out a task and what way is not. The important thing to have always in mind is to *put safety first.*

Electricity, despite all its advantages, comes with certain dangers, and in working with electricity, one has the risk of facing the contained, but invisible, danger in it. Learning about electricity will make you aware of

TABLE 1.2 Fatal Occupational Injuries in 2010 Caused by Electricity

Event or Exposure	Total Number of Fatalities
Contact with electricity, unspecified	4
Contact with electricity of machine, tool, appliance, fixture	34
Contact with electricity wiring, transformer or other	62
Contact with overhead power lines	106
Contact with underground power lines	3
Struck by lightning	3
Total	212

Source: US Department of Labor.

these dangers. You will know about the risks and dangers, in the same way that you can visually judge the danger in a car coming toward you or a large rock that could start rolling down the hill.

As a first, common rule for safety, we can mention: *if there is a potential for a hazard or failure of equipment, never take a risk*. Working with faulty equipment is an example of taking a risk. One must always be sure about the reliability of the equipment. If we can put it as the next safety rule, *if you are not sure, or if you do not know, ask; do not assume.*

Live wires, or hot wires, can be dangerous. They can burn and kill. They may be quiet and what is inside them be invisible, but they can be lethal. Use precautions when working with electricity. Do not work on live wires unless you must and only if you are equipped with the knowledge and the proper equipment to do so.

Safety standards: Safety rules issued by authorities.

In the United States, **safety standards** are set and monitored by Occupational Safety and Health Administration (OSHA). In Canada a branch of Canadian Standards Association (CSA) takes care of occupational health and safety (OHS) matters. Both OSHA and CSA develop occupational health and safety guidelines and procedures. These are well documented and can be purchased in printed or electronic forms.

Safety standards by OSHA, CSA, and their equivalent organizations in other countries, describe the detailed property of the equipment to use, the steps and/or procedures for testing equipment, and so on.

Because we have not yet discussed much about electricity and its terms, we continue the discussion here with only a few items:

Fuse: Protective device for an electric device or circuit consisting of a piece of wire that melts and disconnects the circuit if current surpasses a specific value.

Breaker: Normally physical device that breaks an electric current and disconnects a device from electricity in order to protect it from damage.

1. Almost all electrical equipment is protected by a safety device that disconnects electricity. In the simplest form, this safety device is a **fuse**, which is a piece of wire that can melt upon getting very hot. A fuse is for the safety of equipment or apparatus. When something goes wrong, the wire in the fuse burns, or melts, and the device is no longer connected to electricity and in danger of being damaged. The piece of wire in a fuse is supposed to melt. A fuse can also protect an operator because if it does not disconnect a device from electricity, the danger of something going wrong is still there, no matter if the device is already damaged or not.

   Understand the importance of a fuse, or a **breaker**, which does the same thing in a different way. A breaker has a mechanism that disconnects electricity, without melting a junction wire. It can be put back to work by pressing a button after the cause of the problem is removed. One must always use a fuse or breaker of the correct size. This is not the physical size; it is the rating of the fuse or breaker that must be considered. It is normally written on it.
2. For any job you may need the proper tools and the proper clothing. Do not try to do a job without being prepared, which implies having the proper tools and the clothing. Ornamental attachments to clothes, particularly metallic ones, and pieces hanging from them can cause problems; they can get in the way or get

caught in a machine. Long hair is not an exception. Jewelry such as rings and sometimes even a watch should be taken off if they can cause a hazard.

If you are carrying tools, again, you must carry them correctly and properly. For example, if you are climbing a ladder, you should not have everything you need in your hands but rather in a secure bag, so that your hands are free and the tools do not slip or pop out of your pocket or bag.

3. If there is a danger of falling (you or others), or any other type of danger, you need to put sufficient signs to alert people of the danger. If you are working inside a hole, you must make sure that people do not knowingly or unknowingly drop something on top of you. If working on a machine, make sure that you put adequate signs so that others can know what is going on so that nobody turns on the machine.

1.7 Chapter Summary

- Mankind needs energy for living, in the form of heating and cooling, cooking, and making things.
- Essential energy comes from fossil fuel (coal, oil, and gas), atomic power, hydro, wind, sun, and a few other renewable sources.
- Fossil fuel and atomic energy come from mining coal, oil, and the radioactive metals from the ground, but, one day these reserves will come to an end. The only remaining sources of energy then will be the ones that are cyclically renewed in nature.
- The origin of all renewable energy (except geothermal energy) can be considered to be the sun and to some extent the moon.
- Consumption of energy by humans is increasing every day.
- Electricity is a type of energy. It is relatively easy to transfer from one place to another.
- Wind turbines convert mechanical energy from wind to electrical energy.
- Solar cells convert the sun's heat and light to electricity.
- Static electricity is a sort of electrical energy that cannot be stored and gradually consumed.
- Lightning is an example of static electricity.
- Static electricity can be put to useful applications in manufacturing sand paper, photocopy machines, filtering air, and industrial painting.
- Compared with other types of energy, electricity can be easily transported from one point to another.
- For any technical worker it is very important to understand the safety rules of his/her profession and follow those rules at work.
- A large number of accidents can be prevented if the safety rules are followed and if no risk is taken.

Review Questions

1. Why should we be concerned about energy?
2. Is electric energy the only type of energy we need?
3. If we have only electric energy (suppose sometime in the future), is it responsive to all our needs? Explain.
4. Presently, what are the sources of energy?
5. What is the advantage of electrical energy over other types?
6. How can we store electrical energy?
7. How can we store mechanical energy?
8. Why is safety so important?
9. Give a few examples of static electricity.
10. What are the main causes of accidents?
11. What is the difference between electricity and static electricity?
12. Is electricity dangerous? If so, in what ways?
13. Can static electricity be dangerous? Explain.
14. What does OSHA stand for?
15. What is OSHA?

2

Basic Mathematics and Systems of Measurement Units

OBJECTIVES: After studying this chapter, you will be able to

- Describe the applications of ratios and percentage
- Define the relationships between angles and sides in a right triangle
- Use trigonometric functions in finding angles and lengths in a geometric shape
- Explain the importance of measurement systems
- Define the principal units of measure in U.S. customary and metric systems
- Describe and determine the units of measure for most common entities in terms of the principal units of measure
- Convert units between imperial and metric systems for many common entities
- Extract formulas for calculating any variable from a relationship between several variables

New terms:
Cosine, cotangent, sine, tangent, trigonometric functions

2.1 Introduction

This chapter has been added for those readers who either need to learn or feel a need to refresh their knowledge on basic mathematical relationships, which is essential for later chapters. Those who are comfortable with the mathematics discussed here may skip all or parts of this chapter. Most of the mathematics discussed here are used in Chapter 8. Instructors may want to postpone this chapter until they reach Chapter 8, if necessary. Those readers who have not yet completed a course in trigonometry will find the contents of this chapter very helpful as well as necessary. The material is not extended beyond those topics essential for following the book chapters, and thus, this chapter is not a substitute for a course in mathematics.

Part of this chapter is devoted to measurement systems. A comprehension of other systems than the U.S. customary units, particularly the metric system of units, and a conversion between them is quite important for U.S. students. Metric system of units has become the industry standard in the field of wind turbines in the United States and elsewhere. For those who want to work in the field of renewable energy and wind turbine technology, it is a useful investment to learn this system.

For younger students it is important to know the relationships within the units of the either measurement system. This material, although trivial, often becomes challenging and causes problems in calculations for many students who do not have a previous exposure to it. The hardship becomes more pronounced for topics such as energy, work, power, and their relationships.

In this chapter, for better comprehension of the material and the examples, electrical entities and their measurement units have been avoided. Instead, more common and easy to understand entities such as length and weight have been employed, when necessary.

2.2 Ratio and Percentage

In practice, we do not need to know the exact value of an entity, but we need to compare it with something that we already know. For example, you may need a longer ladder to climb a wall, a longer stick to reach something on a tree, or a brighter lightbulb. In these cases you do not need to specify how long exactly a ladder or a stick must be, or how bright the lightbulb must be. Of course, as things become more critical and technical, one needs to specify the length, brightness, height, and other specifications with more precision. For now, however, we want to focus only on the ratio of two or more values.

The *ratio* can be defined for the same property of two or more items (e.g., length, weight, and height). If the wall in Figure 2.1 is 3.6 m (12 ft) and the ladder is 3.3 m (11 ft), then the ratio of the length of ladder to the

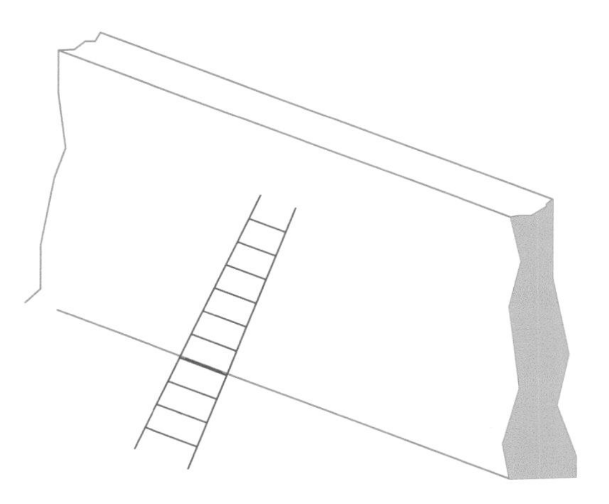

Figure 2.1 You need a longer ladder to reach to the top of the wall.

height of the wall can be defined. In defining a ratio one must pay attention to the following:

1. The ratio can be found by dividing the two measured values, one divided by the other.
2. Measurements must have been made using the same unit.
3. The ratio is just a number; it does not have any unit.

In the above example of the ladder and wall, we may divide 3.3 m by 3.6 m (or 11 ft by 12 ft) to get the ratio of the length of ladder to the height of the wall. The answer is 3.3/3.6, 11/12, or 0.9167. We get the same value no matter which measurement system is employed (as far as the same unit is used). The answer, however, may not be exact because we limit ourselves to a finite number of digits. We round up or down to the nearest value with the number of digits that we judge are sufficiently good.

Equally, we may get the ratio of the height of the wall to the ladder length. This gives the inverse of the previously found number. In this case, the number is 1.091 (1.0909 if we want to have four decimal digits). Remember that it is just 1.091 and not 1.091 ft, 1.091 m, or any other unit. If two measured values are equal, their ratio is 1, or unity.

The ratio of two values may be expressed in percentage. For this, the number obtained from calculation (division) is multiplied by 100. Then, using the values obtained previously, the length of the ladder is 91.67 percent of the height of the wall. This implies that if the height of the wall is divided by 100 and graduated lines are drawn on the wall, the ladder can reach (when it is parallel to the wall and not leaning at an angle) just over 91 of those divisions but cannot reach the 92nd of them.

More often, however, the percentage is used for the ratios smaller than unity, it is quite valid to have a percent number larger than 100. For example, the ratio of the height of the wall to the ladder length is 1.091. We may say that the wall height is 109 percent of the ladder length. The percentage sign is "%".

Example 2.1

How long is a ladder whose length is 140 percent of the wall height? The wall height is 3.6 m (12 ft).

Solution

The ratio of the length of the ladder to the wall height can be found by dividing the percent value by 100; that is 1.4 (140 ÷ 100). And the length of the ladder is

$$(1.4)(3.6) \approx 5 \text{ m} \quad [(1.4)(12) = 16.8 \text{ ft}]$$

Notice that for U.S. customary units you need to convert the 0.8 ft to inches and fraction of an inch. Therefore the ladder length is approximately 16 ft and 9 5/8 in. The beauty of the metric system is that you do not need to extend your calculation to the nearest fraction of an inch. All conversions are decimal. The exact number before rounding off was

$$\text{Ladder length} = (1.4)(3.6) = 5.04 \text{ m}$$

Using percentage or ratios often makes the study of a problem and making decisions easier. Particularly, for the variation of an entity it is easier to understand and more tangible to say a certain item "has been increased by 10 percent" rather than it "has been increased by 24 ft." Note that if something has an increase of 100 percent, its magnitude doubles. But, it is not possible to have a 100 percent decrease. If something has a decrease of 50 percent, it halves.

2.3 Area and Volume Ratios

We start this discussion with a simple example. Figure 2.2 depicts a general triangle. The measurement of all the three sides is shown in the figure. We want to see if the length of each side of this triangle increases by 20 percent how much the increase in the area (or weight) is. A direct practical application of this example is in wiring. For instance, if a thicker wire is used between two points (these points can be very far from each other), how much change can one expect in the weight of the wire?

The answer to the question is definitely not 20 percent. This is evident from Figure 2.3, in which the increase in size is 100 percent; that is, each side of the triangle has doubled in size.

For all geometric shapes, if the length size is increased by a given percentage, the area increases by the square of that percentage value. In this sense, the answer to the previous question is found as follows:

Increase in length = 20%
Ratio of the new length to the original length = $120 \div 100 = 1.20$
Ratio of the increase in area = $(1.20)^2 = 1.44$
Increase in area = $1.44 - 1.00 = 0.44$
Percentage of increase in area = $(100)(0.44) = 44\%$

This relationship can be extended to all applications in which the length and area come into picture. An important case is the relationship

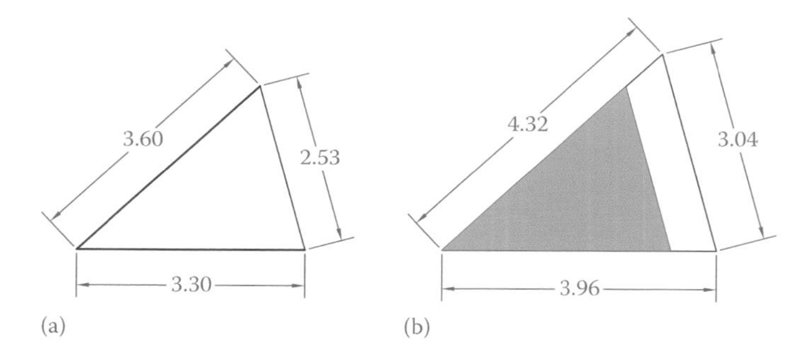

Figure 2.2 Each side of the triangle in (a) is 20 percent smaller than the triangle in (b).

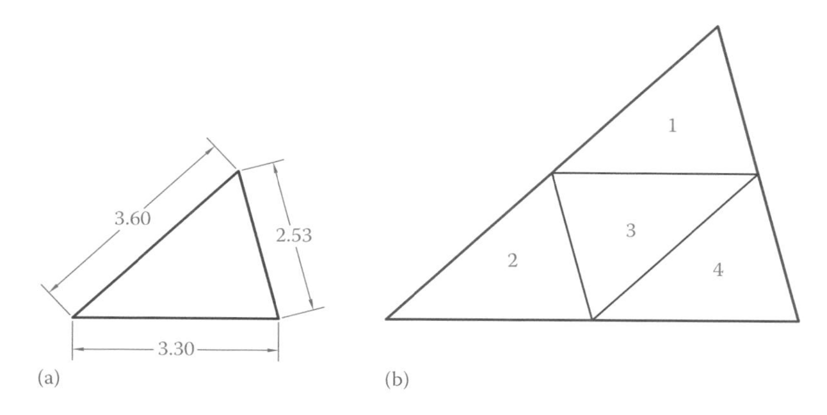

Figure 2.3 Each side of triangle (a) has doubled in (b), but the area has become 4× larger.

between the ratio of units of measure of length and the units of measure of area. For instance, 1 ft is 12 in. How many in^2 is 1 ft^2? The answer to this question is 12^2 or 144, as it is depicted in Figure 2.4. A common mistake is to consider 1 ft^2 to be 12 in^2. In the same way, 1 m is 1000 mm (millimeter) and, thus, 1 m^2 = 1,000,000 mm^2.

By the same token, conversion between the units for volume follows the same sort of relationship. This time the corresponding number must be cubed. The evidence is illustrated in Figure 2.5, which shows that if 1 yd is 3 ft, then 1 yd^3 is 27 ft^3.

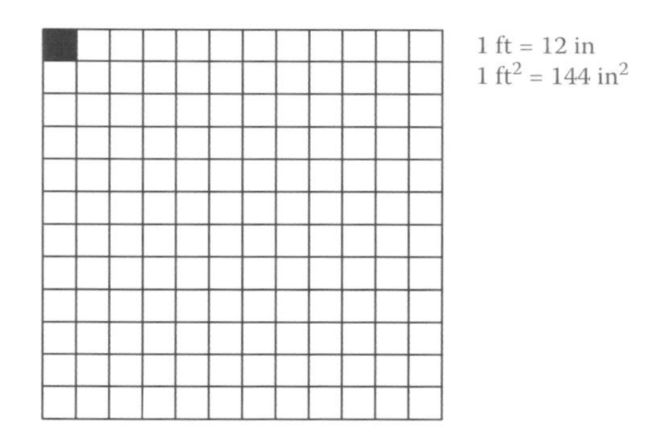

Figure 2.4 Relationship between 1 ft^2 and 1 in^2.

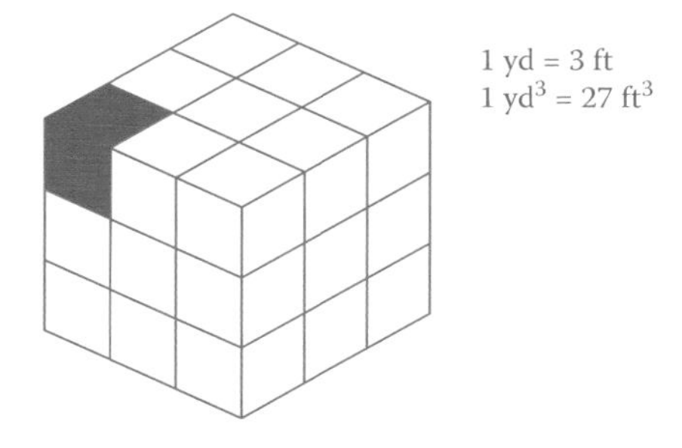

Figure 2.5 Relationship between 1 yd^3 and 1 ft^3.

Example 2.2

Find how many in^3 are in 1 ft^3.

Solution

$$1\ ft = 12\ in$$
$$1\ ft^3 = (12)^3\ in^3 = 1728\ in^3$$

Example 2.3

The area of a circle (square, triangle, etc.) is 1.5 times the area of another circle (square, triangle, etc.). What is the ratio between the radii (side lengths) of the two?

Solution

The ratio between the lengths of the radii or the sides is the square root of the given value. Thus,

$$\text{Ratio between the lengths} = \sqrt{1.5} = 1.225$$

Example 2.4

Containers A and B have similar shapes. (One is a smaller model of the other. You may assume two buckets of identical shapes.) The volume of container A is 50 times the volume of container B. Determine the ratio between the heights of the two containers.

Solution

The ratio between any two identical dimensions of the two containers is the cubic root of the given number. Thus,

$$\text{Ratio between the heights} = \sqrt[3]{50} = 3.384$$

2.4 Angles, Triangles, and Trigonometric Relationships

Trigonometric relationships are very important and have applications in various fields of engineering and physics, including electricity. They are a number of values and relationships corresponding to angles. The material described here is a prerequisite to the topics discussed in Chapter 8.

We start with a right angle triangle, as shown in Figure 2.6. Once you learned the fundamentals, then you see that they are not specific to right angle triangles and can be used for any angle in a triangle or any other geometric shape. First, you need to understand that in the few triangles shown in Figure 2.6b the following ratios are equal and are independent of any individual triangle and the lengths of their sides.

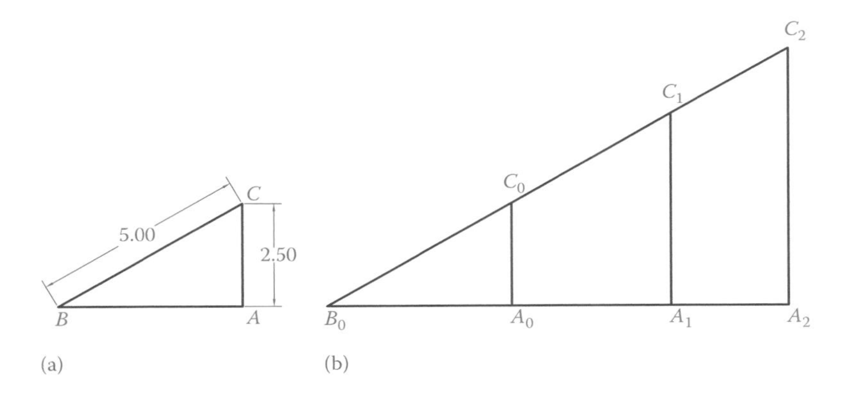

Figure 2.6 The triangle in (a) and those triangles in (b) are similar.

$$\frac{A_0C_0}{B_0C_0} = \frac{A_1C_1}{B_0C_1} = \frac{A_2C_2}{B_0C_2} \tag{2.1}$$

The above ratios and other ratios that can be written are independent of the lengths and reflect a property of the angles. The reason a right angle triangle is used is because it is very easy to understand different items in such a triangle. In Figure 2.6a the lengths of two sides, one opposite to the angle B and one the hypotenuse (the longest side, opposite to the right angle) are shown. On the basis of numbers shown, the above ratio is

$$\frac{A_0C_0}{B_0C_0} = \frac{AC}{BC} = \frac{2.5}{5} = 0.5 \tag{2.2}$$

because all the triangles shown are similar.

The important point to pay attention to is the fact that the above ratio has something to do with angle B, because if angle B changes, so do the lengths of AC, A_0C_0, A_1C_1, and A_2C_2. Thus, the above ratio can be used in various ways in order to find value of one of the three entities in a triangle when the other two are known. For instance, in the right angle triangle ABC, if the lengths AC and BC are known, we can determine the angle B (see below), and if the angle and one of the two sides are known, we can find the length of the other side.

In any right angle triangle, like ABC in Figure 2.6a, the following definitions are made:

1. **Sine** of angle B is the ratio of the opposite side to B to the hypotenuse, or $\frac{AC}{BC}$

   This is written as $\sin B = \frac{AC}{BC}$
2. **Cosine** of angle B is the ratio of the adjacent side to B to the hypotenuse or $\frac{AB}{BC}$

   This is written as $\cos B = \frac{AB}{BC}$
3. **Tangent** of angle B is the ratio of the opposite side to B to the adjacent side to $B = \frac{AC}{AB}$

   This is written as $\tan B = \frac{AC}{AB}$

Sine: Trigonometric function for an angle represented by the ratio of the length of the side opposite to the angle to the length of the hypotenuse in a right triangle setting.

Cosine: Trigonometric function for an angle represented by the ratio of the length of the side adjacent to the angle to the length of the hypotenuse in a right triangle.

Tangent: Trigonometric function for an angle represented by the ratio of the length of the side opposite to the angle to the length of the side adjacent to the angle in a right triangle representation.

Cotangent: Trigonometric function for an angle represented by the ratio of the length of the side adjacent to the angle to the length of the side opposite to the angle in a right triangle.

4. **Cotangent** of angle B is the ratio of the adjacent side to B to the opposite side to $B = \frac{AB}{AC}$

   This is written as cotan $B = \frac{AB}{AC}$

The same relationships can be written for angle C (with the appropriate changes).

The fourth relationship shows that

$$\text{Cotangent of any angle} = \frac{1}{\text{Tangent of the angle}}$$

This is why it is less used. In your scientific calculator you find buttons for *sin*, *cos*, and *tan*. These give the sine, cosine, and tangent of any angle, or conversely, they can be used to find the angle if its sine, cosine, or tangent is known.

In conjunction with the above relationships in a right triangle, the following relationship holds between the lengths of the three sides. Reference is given to Figure 2.6a. This relationship is called Pythagorean theorem.

$$a^2 = b^2 + c^2 \tag{2.3}$$

where

$a =$ length of the hypotenuse = length of the side opposite to angle $A = BC$

$b =$ length of the side opposite to angle $B = AC$

$c =$ length of the side opposite to angle $C = AB$

This relationship is visually observable in Figure 2.7 for $a = 5$, $b = 3$, and $c = 4$. The three numbers a, b and c are not always integral numbers. We may arbitrarily select two of them, but the value for the third one is determined through the mathematical relationship in Equation 2.3.

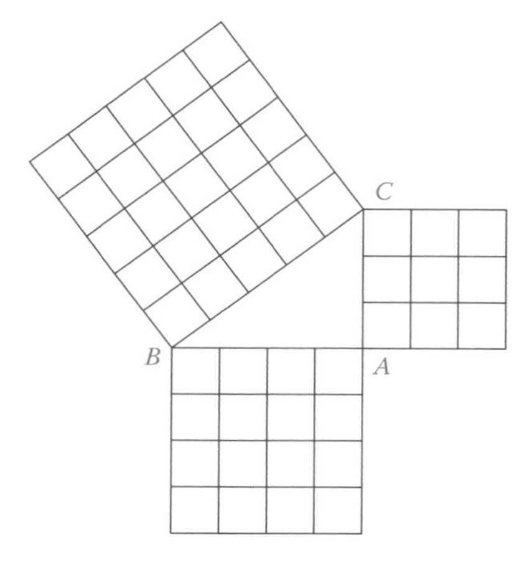

Figure 2.7 Observation of the Pythagorean theorem.

Notice that for the preceding definitions 1 to 4 a right angle triangle was employed only to define the ratios for the **trigonometric functions**. Otherwise, these functions are general, independent of the shape of a triangle. They correspond to an angle and can be used for any angle. Because it is customary and more common to indicate angles by the Greek letters, in the rest of this chapter we represent an arbitrary (general) angle by the Greek letter θ (pronounced theta). Throughout this book you will learn and use more Greek letters, as their use is common in electricity.

Trigonometric functions: Four functions sine (sin), cosine (cos), tangent (tan), and cotangent (cotan) for an angle.

Pay attention to the following points, which are pertinent to the use of trigonometric functions and can be observed from the aforementioned definition of these functions.

1. All the trigonometric functions (sin, cos, tan, and cotan) are ratios; therefore, they are just a number. They do *not* have a unit of measurement.
2. Sin and cos values are always less than 1. They can never be larger than 1.
3. If θ (theta) is an angle, then $(\sin \theta)^2 + (\cos \theta)^2 = 1$. (This expression is usually written as $\sin^2 \theta + \cos^2 \theta = 1$.)
4. $\text{Sin } 0 = 0$.
5. $\text{Cos } 0 = 1$.
6. $\text{Sin } 90° = 1$.
7. $\text{Cos } 90° = 0$.
8. $\text{Sin } 45° = \cos 45°$.
9. $\text{Tan } \theta = \dfrac{\sin \theta}{\cos \theta}$.
10. $\text{Tan } 45° = 1$.
11. $\text{Cotan } 45° = 1$.

As already noted, the applications of trigonometric functions are plenty. Some illustrations of common applications are given in Examples 2.9 through 2.11. Applications in electricity are shown in Chapter 8.

Example 2.5

Find the values of sin θ (sine theta), for θ = 10°, 30°, 45°, 60°, 180°, 225°, 270°, 300°, 400°, and 600°.

Solution

The values can be directly found from a calculator that has scientific buttons or the calculator on your computer (the scientific view) as follows. Note that the last two numbers are larger than 360°, but it does not matter because an angle is not restricted to be smaller than 360°. All numbers are rounded to three decimal points. (If you are checking these answers, make sure that your calculator shows the angles in degrees; otherwise, you will not get the correct answers.)

$\sin 10° = 0.174$
$\sin 30° = 0.500$
$\sin 45° = 0.707$

sin 60° = 0.866
sin 180° = 0.0
sin 225° = –0.707
sin 270° = –1.0
sin 300° = –0.866
sin 400° = 0.643
sin 600° = –0.866

Notice that some values are negative. For sine of an angle, if the angle is between 0 and 180° the sine value is positive. If the angle is larger than 360°, its value is reduced to a number between 0 and 360° by subtracting from it multiple times 360° (as many as necessary until it becomes between 0 and 360°). For example 1000° must be reduced to 1000° – 720° = 280°.

Example 2.6

Find the values of cos θ, for θ = –120°, –60°, –30°, 30°, 45°, 120°, 180°, 270°, 300°, and 360°.

Solution

Answers can be again found by using a calculator. Note that some angles are negative. It is always possible to have negative numbers.

cos –120° = –0.500
cos –60° = 0.500
cos –30° = 0.866
cos 30° = 0.866
cos 45° = 0.707
cos 120° = –0.500
cos 180° = –1.0
cos 270° = 0.0
cos 300° = 0.500
cos 360° = 1.0

Again, notice that some values are negative, but there is no direct correspondence between the sign of an angle and the sign of its cosine value. An angle can be negative, and its function (sin, cos, tan) can be negative. Also, an angle can be positive, but its function value can be negative. All the angles between –90° and +90° have a positive value for cosine function.

There is no direct correspondence between the sign of an angle and the sign of its trigonometric function value.

Example 2.7

Find the values of tan θ, for θ = –10°, 0°, 30°, 45°, 60°, 80°, 89°, 91°, 120°, and 300°.

Solution

Try to get the same values by using your calculator as follows. You may need to adjust the mode of your calculator to show angles in degrees.

$\tan -10° = -0.176$
$\tan 0° = 0.0$
$\tan 30° = 0.577$
$\tan 45° = 1.0$
$\tan 60° = 1.732$
$\tan 80° = 5.671$
$\tan 89° = 57.290$
$\tan 91° = -57.290$
$\tan 120° = -1.732$
$\tan 300° = -1.732$

Notice that the values for tangent of an angle can be smaller than 1 or larger than 1. The tangent value of 45° is 1; after that the values become larger until 90°. Watch the values for around 90° (and some corresponding angles, such as 270°), for which tangent of the angle is a very large.

The angle for which a trigonometric function is known is represented either by arc or by the inverse function. For example,

Arc sin (0.5) = $\sin^{-1} 0.5$ = the angle whose sine is 0.5 = 30°
Arc cos (0.5) = $\cos^{-1} 0.5$ = the angle whose cosine is 0.5 = 60°
Arc tan (0.5) = $\tan^{-1} 0.5$ = the angle whose tangent is 0.5 = 26.565°

Example 2.8

First, find the values for the following functions. Then use the inverse function (using the second key followed by the function key on the calculator) to return to the original values:

sin 150°, sin 270°, cos –60°, cos 240°, tan –100°, tan 234°

Solution

Although this action seems to be trivial, for none of the values one obtains the original number by a calculator or computer. This is shown below:

$\sin 150° = 0.5$, $\sin^{-1} 0.5 = 30°$
$\sin 270° = -1.0$, $\sin^{-1} -1.0 = -90°$
$\cos -60° = 0.5$, $\cos^{-1} 0.5 = 60°$
$\cos 240° = -0.5$, $\cos^{-1} -0.5 = 120°$
$\tan -100° = 5.671$, $\tan^{-1} 5.671 = 80°$
$\tan 234° = 1.376$, $\tan^{-1} 1.376 = 54°$

The reason for this is that the calculated values for inverse functions are always in a specific range. For $\sin^{-1}$ and $\tan^{-1}$ the angles are between –90 and 90 (sometimes it can change with the calculator model), and for $\cos^{-1}$ it is between 0 and 180.

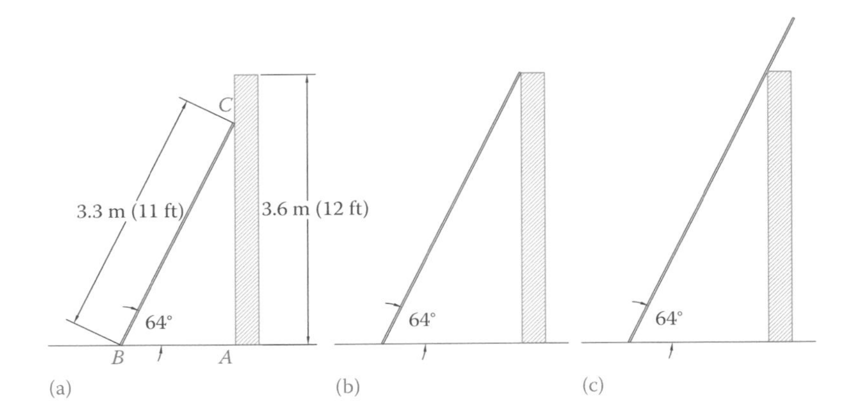

Figure 2.8 Wall and ladder. (a) Example 2.9, (b) Example 2.10, and (c) Example 2.11.

Example 2.9

Referring to Figure 2.8a, which shows the ladder in Figure 2.1 against a wall, if the wall is 3.6 m (12 ft) high and the ladder is 3.3 m (11 ft) long, at the angle shown at what height does the ladder touch the wall?

Solution

In the right triangle formed by the ladder, the ground and the wall, the length of the hypotenuse is 3.3 m (11 ft) and the angle B is 64°. By definition, the ratio of the side opposite to the given angle to the hypotenuse is equal to the sin of the angle.

Although not necessary, the assignment of letter names to the corners of the triangle in Figure 2.8a is the same as in Figure 2.6a. From the definition of sin of an angle it follows that

$$AC = BC \times \frac{AC}{BC} = BC \times \sin(B)$$

The length of the side opposite to the known angle (that the question is about) is therefore found from

(3.3)(sin 64°) = (3.3)(0.9) = 2.97 m or (11)(sin 64°) = (11)(0.9) = 9.9 ft

Example 2.10

Find the length of the ladder in Figure 2.8b. The ladder is just reaching the wall top.

Solution

In this case the length of one side and one angle is known and the hypotenuse is to be found. Using the same names for the corners, we have

$$BC = \frac{AC}{\sin B} = \frac{3.6}{\sin 64} = \frac{123.6}{0.9} = 4 \text{ m}$$

or

$$BC = \frac{12}{\sin 64} = \frac{12}{0.9} = 13.35 \text{ ft}$$

Example 2.11

From practicality and safety points of view, the length of the ladder must always extend from the reaching point, as shown in Figure 2.8c. If the ladder is 16 ft (4.8 m), how much does it horizontally extend toward the other side of the wall from the touching edge?

Solution

The extension of the ladder over the wall is

4.8 – 4 = 0.8 m (4 m was determined in Example 2.10)

Because the horizontal direction is adjacent to the known 64° angle, the extension in the horizontal direction can be found from the cos function. Therefore,

Horizontal extension = (0.8)(cos 64°) = (0.8)(0.44) = 0.35 m

In imperial units we have

Ladder extension over the wall = 16 – 13.35 = 2.65 ft

Ladder horizontal extension = (2.65)(cos 64°) = (2.65)(0.44) = 1.16 ft

2.5 Principal Measurement Entities and Systems of Measurement

Almost everything that we need for living is quantified by measurement, from all those items that we count to what we buy by weight, volume, and so on. Electricity is not an exception, and for various entities that we will discuss, measurement becomes essential. You are already aware of many of the measureable items such as weight, length, and time.

For measuring anything we need to define a specified quantity of that as the unit of measurement. For instance, one *foot* is a well-defined unit for length, based on which we can measure and perceive how much, for example, 2, 15, and 1000 ft are. Indeed, without a unit of measure, no measurement can have any meaning.

Among all the entities that we measure, some can be considered principal, from which other items can be derived. We have already seen examples of this in the measurement of length: If the unit for measuring length is feet (ft), then for area we can use ft^2 and for volume ft^3 may be employed. In the metric system we have meter (m) for length, and m^2 and m^3 for area and volume, respectively. Another example is the unit for measuring speed of something moving. If meter (foot) is the unit for measuring length (or distance) and second is the unit for measuring time, then meters per second (m/sec) or feet per second (ft/s) can be used for the unit of speed.

The most common entities that we use in our daily life and measure are length (distance, height), area, volume, speed, acceleration, force, weight (which has the same unit as force), mass, density, energy, time, and temperature. Other specialized measurements depend on the discipline; for example, in physics, mechanics, chemistry, and electricity, there are other entities that need to be measured, and each one has its unit of measure.

To measure what we need to measure, a system of measurement is defined, in which the units are established. There are a few major systems of measurement, some of which are becoming obsolete. Nowadays, the metric system is adopted by most of the industrialized countries. The imperial system (U.S. customary units) is another system of units that it is still used in the United States (and some other countries). One of the main objectives of this chapter is to facilitate understanding of the metric system units and the conversion from the U.S. customary units to their equivalent metric values for those who are not yet familiar with the metric system.

In most systems, three entities are considered as principal and the rest are derived from them, as in the case of area and speed. In all systems, one of these entities is time, the unit of which is the *second*. That is to say, the official unit for measuring time is 1 sec.

On the other hand, each unit can have its associated smaller or larger units for measurement of smaller or larger quantities. For example, for time in addition to second, one can use minute, hour, day, and year. Selection of an appropriate unit is by commonsense and adaptation to the quality involved. Within the framework of the application of units of measure we try to limit the discussion and examples to those that are of direct interest to implications in renewable energy.

In the metric system the three principal entities are length, mass, and time. The units of measure for these entities are meter (m), kilogram (kg), and second (s), respectively. The metric system is much easier to use than the imperial system. The metric system is particularly outstanding for deriving smaller and larger units for an entity, which (except for time) are obtained by dividing or multiplying a unit (like meter) by powers of 10.

In the imperial system the principal units are length, force, and time; the (standard) units are foot (ft), pound force (lb_F), and second (sec), respectively. The reason to accompany *force* here is because pound is also used for the measurement of mass; for the distinction of the two cases pound-mass (lb_m) is sometimes used when there can be an ambiguity. Of course, other larger and smaller units are also available/defined and used. For length, for instance, there are yards and miles, and some specialized units like nautical miles on the larger scale, and inches and mil (1/1000 in) on the smaller scale.

In all systems, special attention must be paid to the units of mass and force, because force and weight have the same nature and have the same unit; weight and mass are related to each other. The parameter relating them together is gravity acceleration. Mass is a property that depends on the amount and the nature of a material from which an object is made. Weight is the force exerted by Earth on the item. (If the same object is taken to the Moon, its mass does not change, whereas its weight is reduced.) Thus, gravity acceleration comes into view. The relationship between the formal unit of mass and the unit of force is

$$\text{unit of force} = \text{unit of mass} \times \text{unit of acceleration}$$

In the metric system the unit for measuring force is Newton (N). But, occasionally kilogram is used for measuring force (in the same fashion that pound is used for both mass and force), because in an older system, kilogram was the unit for force. Moreover, although lb_m is frequently used as a unit of mass, the official unit for mass in the imperial system is slug (not heard by many people). For this reason the following lines are very fundamental to the realization of mass and force units:

1. If the mass of an object is 1 lb_m, its weight on Earth is1 lb_F. Its official mass is 1/32.17 slug (32.17 ft/sec^2 is the gravity acceleration in the imperial system).
2. If the mass of an object is 1 kg, its weight on Earth is 1 kg, which is 9.81 N (9.81 m/sec^2 is the gravity acceleration in the metric system).

The three principal units in each system can be used to define and derive other units.

Conversion from one measuring system to the other is very important and often necessary. Not having a good understanding of the measuring units and the way to convert from one to the other can cause problems and inefficiencies. It is imperative to understand the relationships between different units in the same system and between the metric system and imperial system. In the next section the conversion between different units in the same system is discussed, and the conversion between metric and imperial systems is worked out in Section 2.7.

2.6 Other Measurement Entities and Their Derivation

In this section the focus is a conversion from one unit of measure to the other all corresponding to the same system. For example, the wind speed for a turbine can be mentioned in miles per hour (mph), but we may need to have it expressed in feet per second (ft/sec). In the metric system, speed can be expressed in km/hr or m/sec. Some students have difficulty doing this conversion. For this conversion, one needs to know the relationship between mile and feet and that between hour and second. A logical thinking leads to the correct numbers by which one unit must be divided or multiplies to obtain the other. Once the relationship between the units of measure is found, then that can be used for any quantity, because a quantity is a finite number of units.

We can show this better by a number of examples. In these examples, only the more common units are used. For the numbers corresponding to conversion between different units of measurement see the conversion tables in this chapter.

Example 2.12

If wind speed is 18 mph, what is it in ft/sec?

Solution

First, we see how much 1 mph is in terms of ft/sec. Then we multiply the answer by 18.

We notice that 1 mph implies one mile distance in 1 hour time. Knowing that 1 mile = 5280 ft instead of 1 mile, we replace it by 5280 ft. Thus, a multiplication is performed for converting the larger unit (mile) to the smaller unit (ft).

If wind, or anything else, moves 1 mph, then in 1 min it moves 60 times less, and in 1 sec it moves 60 × 60 = 3600 times less. Thus, a division is required here to express what happens during a shorter time. Therefore,

$$1\,\frac{\text{mile}}{\text{hour}} = \frac{5280\text{ ft}}{3600\text{ sec}} = 1.4667\,\frac{\text{ft}}{\text{sec}}$$

As noted, for 18 mph the previously obtained number must be multiplied by 18

$$18\text{ mph} = (18)(1.4667) = 26.4\text{ ft/sec}$$

Alternatively, you noticed that in the numerator and denominator we have substituted miles by its equivalent feet and hour by its equivalent number of seconds.

Example 2.13

If wind speed is 25 km/hr, what is it in m/sec?

Solution

In the same way as performed for Example 2.12, first, we find the equivalent of 1 km/hr in terms of m/sec Then we multiply the answer by 25. For this step we replace kilometer by meter and hour by second, noting that in the latter the distance moved in a second by anything that moves is less than the distance for an hour.

$$1\,\frac{\text{km}}{\text{hour}} = \frac{1000\text{ m}}{3600\text{ sec}} = 0.278\,\frac{\text{m}}{\text{sec}}$$

$$25\text{ km/hr} = (25)(0.278) = 6.94\text{ m/sec}$$

Example 2.14

The area of the blade in a wind turbine is 90 yd^2. What is its area in ft^2?

Solution

We observed in Section 2.2 that each square yard is 9 ft^2. Thus, for 90 yd^2 we must multiply 90 by 9.

$$(90)(9) = 810 \text{ ft}^2$$

Pay attention to the fact that when expressing a measured value in terms of two different units, the number corresponding to the smaller unit is always larger.

Example 2.15

Water pours into a reservoir at the rate of 45 L/sec. If the reservoir capacity is 125 m^3, how long does it take for an initially empty reservoir to be filled?

Solution

First, we should know that there are 1000 L in a cubic meter. We may either convert the capacity of the reservoir into liters or convert the flow rate to cubic meters per second. The necessary time, in seconds, then can be found directly.

$$\text{time duration} = \frac{\text{capacity}}{\text{flow rate}} = \frac{125}{0.045} = 2777.8 \text{ sec}$$

Normally, a given time like this must be converted to hours, minutes, and seconds, through consecutive dividing by 60.

$$2777.8 \text{ sec} = 46 \text{ min, } 17.8 \text{ sec}$$

Example 2.16

Water is poured out of a pipe at a rate of 30 gal/min How many cubic feet of water is poured out in 1.5 hour?

Solution

Although the unit for time is 1 sec, for this sort of problem it is not necessary to find the flow for 1 sec We first see that in whatever unit the water flow is, in 1.5 hour (that is 90 min) the water pours out 90 times of the amount in 1 min Thus, we need to multiply the flow rate by 90. Thus,

$$\text{water in 1.5 hr} = (30)(90) = 270 \text{ gal}$$

We must now express this in cubic feet. Thus, the relationship between cubic feet and gallon is necessary. If we do not have this in mind, it is always possible to find it in a handbook or on the Internet.

$$1\ \text{ft}^3 = 7.4805\ \text{gal}$$

Each 7.4804 gal make 1 ft^3 We should see how many times 7.4805 fits into 270. Division gives the answer:

$$270 \div 7.4805 = 36.1\ \text{ft}^3$$

Example 2.17

If 1 mile = 5280 ft, how much is 1 ft in miles? In other words, how many miles are in 1 ft?

Solution

This problem may look strange to some. But, it is completely valid and this sort of conversion is often found. It is, in fact, very simple and very important. You need only inverse the number.

$$1\ \text{ft} = \frac{1}{5280} = 0.0001894\ \text{mile}$$

Example 2.18

Find how many miles are in 18,000 ft (or how many miles is 18,000 ft).

Solution

This problem can be solved in two ways. First, you might think in terms associated with Example 2.17. If each foot is 0.0001894 mile, then 18,000 ft is

$$(18{,}000)(0.0001894) = 3.409\ \text{mile}$$

Alternatively, you can divide 18,000 by 5280: 18,000 ÷ 5280 = 3.409 miles.

From the operation viewpoint both operations are the same; however, because of round-off errors, the final answers may be slightly different.

2.7 Conversion between Systems of Measurement Units

In dealing with the conversion from imperial to metric system and vice versa, the good news is that not all the units are different. We have already seen that the unit of time in both systems is the same. Also, the units for all electrical entities are the same. Nevertheless, for those units that are not the same, a comprehension of both units is necessary for one who wants to work in the field. In this section the emphasis is more on those units that are regularly used in daily work associated with electricity and renewable energy technology. Examples of conversion between the two systems serve to illustrate this.

Table 2.1 summarizes those entities that are used more often, with their units in metric and U.S. customary units.

In the metric system, many smaller or larger units can be obtained by dividing or multiplying a unit by 10, 100, 1000, and so on. This is not done for all the units and numbers in the range. A number of prefixes are used to represent the smaller and larger units. You have seen some of these in Table 2.1 (e.g., kilo and mega). These prefixes are standard, and throughout this book you will use many of them. Table 2.2 shows the name and the corresponding multiplier for each. Pay attention to the symbols, which may be uppercase or lowercase. They should not be mixed up or substituted with other symbols. A new, Greek symbol is used here for "micro," as you see it in Table 2.2 (μ).

Table 2.3 shows the conversion between some units in the metric and U.S. customary systems. For converting values from metric to the corresponding U.S. unit, the given quantity must be multiplied by the number in the table.

Table 2.4 shows the conversion factors from U.S. units.

TABLE 2.1 More Common Measurement Units

	Metric		U.S. Customary	
	Main	Other/Remark	Main	Other/Remark
Length	meter (m)	micron, mm, cm, dm, km	foot (ft)	mil, inch, yard, mile
Force	newton (N)	kilo Newton (kN) = 1000 N	pound (lb)	(pound-force, lb_F)[a]
Time	second (s)	minute, hour, day, week, month, year, decade	second (sec)	minute, hour, day, week, month, year, decade
Area	sq. m	sq. km	sq. ft	acre, sq. mile
Volume	cubic meter	liter (L) = 1/1000 m^3, CC	cubic ft	gallon, ounce (oz.)
Speed	m/sec	km/hr	ft/sec	in/sec, mph, knot
Power	watt (W)-Nm/s	kW, MW, horsepower	lbft/sec	horsepower
Energy/work	joule (J), N.m	watt-hr, kW-hr, MW-hr	ftlb	
Pressure	pascal (Pa)	bar = 10^5 Pa	lb/in^2 (PSI)	bar, kips (1000 psi)
Mass	kilogram (kg)	ton = 1000 kg	slug	lb_m, ton (short) = 2000 lb_m
Density	kg/m^3	g/L	slug/ft^3	lb_m/ft^3
Torque[b]	Nm		ftlb[c]	ftin, oz-in
Heat	calorie	1 cal = 4.1868 joule	BTU	(British thermal unit)

[a] The unit lb can represent lb-force (lb_F), when used for force or lb-mass (lb_m), when used for mass.
[b] Note that work/energy has the same unit as torque, but they are two completely different entities and can never replace each other.
[c] The units lb-ft and ft-lb are both used interchangeably.

TABLE 2.2 Prefixes for Metric Units

Prefix	Symbol	Multiplier	Example
Peta	P	1,000,000,000,000,000	Used for very large numbers
Tera	T	1,000,000,000,000	Tera watt-hour = 10^{12} watt-hour: TW-hr
Giga	G	1,000,000,000	1.5 GW steam turbine
Mega	M	1,000,000	2 MW wind turbine
kilo	k	1000	5 kPa = 5000 Pa
hecto	h	100	Not much used
deca	da	10	20 daL
deci	d	1/10	4 dm = 0.4 L
centi	c	1/100	12 cm: 12 cm = 0.12 m
mili	m	1/1000	millimeter (mm), milliampere (mA)
micro	μ	1/1,000,000	microfarad (μF), micrometer (μm)
nano	n	1/1,000,000,000	15 nF capacitor
pico	p	1/1,000,000,000,000	50 pF
femto	f	1/1,000,000,000,000,000	For very small dimensions at atomic level

TABLE 2.3 Conversion from Metric to U.S. Units

1 m	3.28 ft	39.37 in	1.0936 yd	6.215×10^{-4} mile
1 km	3280 ft	39,370 in	1093.6 yd	0.6215 mile
1 m^2	10.764 ft^2	1550 in^2	1.196 yd^2	2.471×10^{-4} acre
1 m^3	35.306 ft^3	61,023.378 in^3	1.308 yd^3	264.173 gal
1 m/sec	3.28 ft/sec	39.37 in/sec	3.6 km/hr	2.237 mph
1 km/hr	0.911 ft/sec	10.936 in/sec	0.278 m/sec	0.6215 mph
1 N	0.2248 lb			
1 W	0.7376 lbft/sec	1.34×10^{-3} hp	3.412 BTU/hr	1×10^{-6} MW
1 joule = 1 Nm	0.7376 lbft	0.2388 cal	1/3,600,000 kW-hr	9.48×10^{-4} BTU
1 kg (mass)	2.2075 lb_m	0.0686 slug		
1 Pa	1.45×10^{-4} PSI	0.02088 lb/ft^2	1 N/m^2	1×10^{-5} bar
1 kg/m^3	1.94×10^{-3} $slug/ft^3$	0.0624 lb_m/ft^3		
1 cal	0.003968 BTU	3.088 lb-ft	4.1868 joule	1.163×10^{-3} Whr

TABLE 2.4 Conversion from U.S. Units

1 ft	0.3048 m	1/3 yd	12 in	12,000 mil
1 mile	1609.34 m	1.609 km	5280 ft	1760 yd
1 ft^2	0.0929 m^2	929 cm^2	144 in^2	0.1111 yd^2
1 ft^3	0.0283 m^3	7.4805 gallons	28.317 L	28,316.685 mm^3
1 ft/sec	0.3048 m/sec	1.0973 km/hr	0.6818 mph	1.6874 knot
1 mph	0.447 m/sec	1.609 km/hr	1.4667 ft/sec	2.475 knot
1 lb (force)	4.448 N			
1 lb-ft	1.3558 Nm	1.3558 joule	0.3238 cal	1.285×10^{-3} BTU
1 lb-ft/sec	1.3558 joule/s	1.3558 W	0.001817 hp	4.626 BTU/hr
1 slug (mass)	14.594 kg	32.17 lb_m		
1 psi	6894.757 Pa	0.00694 lb/ft^2	0.06894 bar	1 $pound/in^2$
1 U.S. gallon	0.003785 m^3	3.785 L	0.13368 ft^3	3785 CC (mL)
1 BTU	252.0 cal	1055.06 joule	778.1 lb-ft	1055.06 Nm

Example 2.19

If wind speed is 40 mph, what is it in meters per second?

Solution

Having Tables 2.2 and 2.3 make this type of conversion very easy. One needs to find in the table how much in meters per second, is each mile per hour. (Table 2.3). Then the corresponding number is multiplied by 40. Alternatively, the correspondence between meters per second and mph can be found (Table 2.2) and the quantity then must be divided by that number. For the logic behind it and to avoid mistakes, review the examples in Section 2.6.

$$\text{wind speed} = (40)(0.447) = 17.88 \text{ m/sec}$$

or

$$40 \div 2.237 = 17.88 \text{ m/sec}$$

Example 2.20

How many square inches are there in 24 m^2?

Solution

From Table 2.3, each m^2 is 1550 in^2. Thus,

$$24 \text{ m}^2 = (24)(1550) = 37{,}200 \text{ in}^2$$

Example 2.21

How many square miles are in 2.5 km^2?

Solution

Because 1 km = 0.6215 mile, then 1 km^2 = $(0.6215)^2$ = 0.3863 $mi.^2$ and 2.5 km^2 = (2.5)(0.3863) = 0.9657 $mi.^2$

2.8 Formulas, Relationships, and Equations

An equation is a mathematical relationship between a few variables. A variable is a measureable item that can assume values. Such a relationship is either found by logical reasoning and observation or is developed by experimental measurements. For example, the area of a circle (cross-section area of a circular cable, for instance) can be defined in terms of the radius of the circle as follows:

$$A = \pi r^2 \tag{2.4}$$

In this equation A and r are two variables, but π is a constant because its value is always 3.14159. When an equation exists that expresses the relationship between several variables, because of this relationship, it is not possible to have arbitrary values for all variables. As a result, any one of these variables can be defined in terms of the others. This gives a formula to determine that particular variable when the others are known. In other words, only one variable can be found if unknown.

We are going to use an equation with more variables to further this discussion. Consider Equation 2.5 with three variables. This is used in Chapter 8, but at this time, look at the characters in the equation as variables, without focusing on the application and the meaning of them.

$$X = 2\pi fL \tag{2.5}$$

X, f, and L are three variables that are related by Equation 2.5, and the 2π is the constant multiplier, which depends on the units that are used for X, f, and L. If the unit of measure for any of the variables changes, then, accordingly, the constant multiplier must be adjusted.

An important point we want to draw attention to is that Equation 2.5 is a relationship between three variables. This equation defines a formula to find X when f and L are known. Equally, we may write this relationship in other forms. Below are the formulae that define f or L.

$$f = \frac{X}{2\pi L} \tag{2.6}$$

$$L = \frac{X}{2\pi f} \tag{2.7}$$

Here f and L have been isolated from the rest by dividing both sides of Equation 2.5 by all other components. This is the general way to find any desired variable in terms of other parameters in an equation. Equations 2.5, 2.6, and 2.7 stem from a single equation. It is imperative that you understand this fact and easily extract the appropriate formula from an equation for calculation of a variable that you need to find.

As another example, consider the relationship in Equation 2.8 between A, B, and C, again without considering the application,

$$A = \sqrt{B^2 + C^2} \tag{2.8}$$

If we need to find B in terms of A and C, it is easy to understand that (by squaring each side of Equation 2.8)

$$A^2 = B^2 + C^2 \tag{2.9}$$

from which it follows that

$$B^2 = A^2 - C^2 \tag{2.10}$$

$$B = \sqrt{A^2 - C^2} \tag{2.11}$$

Example 2.22

The relationship between three variables x, y, and z is $x + y - z = 1$. Write the equation to find each variable in terms of the others.

Solution

To find each variable, it must be isolated from the others by mathematical operations. This is a general rule. For the current example the problem is easy. Think of transferring all other parameters to the other side of equation (in which the signs of all those items that are transferred must be reversed); alternatively, subtract from both sides all the other parameters, so that the one you need becomes the only one (it may become necessary to multiply both sides by –1).

The answers, thus, are

$$x = 1 - y + z$$

$$y = 1 - x + z$$

$$z = x + y - 1$$

2.9 Chapter Summary

- The ratio of two numbers is found by dividing them together. The result has no unit.
- Ratios can be converted to percentage if multiplied by 100.
- In a right triangle, sine of any of the two acute angles is the ratio of the length of opposite side to it to the length of the hypotenuse.
- In a right triangle, cosine of an angle is the ratio of the length of adjacent side to it to the length of the hypotenuse.
- In a right triangle, tangent of an angle is the ratio of the length of opposite side to it to the length of the adjacent side.
- Sin, cos, and tan are trigonometric functions and can be used for any, positive or negative, angle.
- Sin and cos functions for all angles are always a number between –1 and 1.
- Tan of an angle can be very large. It can be positive or negative.
- Tan of an angle can be found from dividing the angle's sin value by its cos value.
- A system of measurement units defines specific size of an entity as the unit of measure of that entity.
- Three basic (principal) entities for metric (imperial) measuring systems are the length, mass (force), and time.
- From a mathematical relationship (equation) between N variables, one can extract N formulas, each of which defines one variable in terms of the others.

Review Questions

1. If the ratio of a number A to a number B is larger than 1, is A larger or is B larger?
2. If the ratio of height of A to the height of B is 0.8, what is the percentage of height A compared to B?
3. If M and N are two quantities of the same entity, and M is 65 percent of N, what percentage is M of N?
4. What is the definition for the sine of an angle?
5. What is the definition for the cosine of an angle?
6. What is the definition for the tangent of an angle?
7. Can the sine of an angle be 1.5? Why or why not?
8. Can the tangent of an angle be 1.5? Why or why not?
9. Can the tangent of an angle be −100? Why or why not?
10. If the sine values of two angles are equal, can we say the two angles are equal?
11. What is a system of measure?
12. Define the principal units of measure in the U.S. customary system.
13. Define the principal units of measure in the metric system.
14. What is the principal unit of measuring speed in the U.S. system of measurement?

Problems

1. Find the difference between $\sin 10^\circ$ and $\tan 10^\circ$.
2. Find the difference between $\cos 5^\circ$ and $\sin 85^\circ$.
3. Calculate $\sin 25^\circ$ and $\sin 65^\circ$ and add them together. Compare the result with $\sin 90^\circ$. Which one is larger?

4. Make a table of the following values: sin 0, sin 10°, sin 20°, sin 30°, sin 40°, sin 50°, sin 60°, sin 70°, sin 80°, and sin 90°. Use three decimal places.

5. Find the angles between 0 and 90°, whose cos values are as follows: 0.25, 0.35, 0.45, 0.55, 0.65, 0.75, 0.85, and 0.95.

6. What is the angle that if its tangent is multiplied by 3 the result is 10?

7. What is the angle that if its tangent is multiplied by 3 the result is 100?

8. What is the angle that if its sin is multiplied by 3 the result is 2?

9. What is the angle that if its sin is multiplied by 2 the result is 3?

10. If A, B, C, and D are four variables and are related to each other by $A \cdot B \cdot C = D$, find each of the variables A, B, and C in terms of the others.

11. If A, B, C, and D are four variables and are related to each other by $A + B - C = D$, find each of the variables A, B, and C in terms of the others.

12. If A, B, C, and D are four variables and $\frac{A}{B} = \frac{C}{D}$, find each of the variables A, B, and C in terms of the others.

3

Atomic Structure of Materials

OBJECTIVES: After studying this chapter, you will be able to

- Define an atom and a molecule and the difference between them
- Describe the structure of an atom
- Explain the electrons, protons, and neutrons and identify their electric charges
- Describe the atomic number and its relationship with the numbers of electrons and protons
- Define the role of electrons in the properties of materials
- Define the arrangement of electrons orbiting around the nucleus
- Explain why electrons and protons do not attract each other in any atom or molecule
- Explain the role of neutrons in an atom and define the meaning of isotopes
- Categorize the elements based on their atomic structure
- Define valence electrons and their role in electrical properties of materials
- Define atomic weight and its relationship with weight of materials
- Define the similarities between different materials based on their valence electrons

New terms:
Atom, atomic mass, atomic number, atomic weight, compound, deuterium, electrons, elements, heavy water, isotope, molecule, neutron, nucleus, proton, valence, valence electrons

3.1 Introduction

All the physical materials around us are made up of a finite number of **elements**, in the same way that all the numbers we deal with can be represented only by 10 digits (0 to 9). The way that some or all of the ten digits are put together makes various numbers. We can make infinite numbers by ten digits. We may say elements are single materials that are not composed of separable substances. The number of known elements so far, as they appear in the Mendeleev table (periodic table of elements) is 118, and depending on the way and the magnitude they can combine together, they can compose millions of materials. The Mendeleev table as shown in Figure 3.1 contains various information about all elements. All other materials that we observe, i.e. those not in the Mendeleev table, are

Elements: Materials that cannot be decomposed to other simpler materials, like oxygen and copper. All elements are listed in the Mendeleev table.

Periodic Table of The Elements

Group 1	2	3	4	5	6	7	8	9	10	11	12	13	14	15	16	17	18	Shell
1 +1 −1 **H** [1.0078, 1.0082] 1																	2 0 **He** 4.0026 2	K
3 +1 **Li** [6.938, 6.997] 2-1	4 +2 **Be** 9.0122 2-2											5 +3 **B** [10.806, 10.821] 2-3	6 +2 +4 −4 **C** [12.009, 12.012] 2-4	7 +1 +2 +3 +4 +5 −1 −2 −3 **N** [14.006, 14.008] 2-5	8 −2 **O** [15.999, 16.000] 2-6	9 −1 **F** 18.998 2-7	10 0 **Ne** 20.180 2-8	K-L
11 +1 **Na** 22.990 2-8-1	12 +2 **Mg** [24.304, 24.307] 2-8-2											13 +3 **Al** 26.982 2-8-3	14 +2 +4 −4 **Si** [28.084, 28.086] 2-8-4	15 +3 +5 −3 **P** 30.974 2-8-5	16 +4 +6 −2 **S** [32.059, 32.076] 2-8-6	17 +1 +5 +7 −1 **Cl** [35.446, 35.457] 2-8-7	18 0 **Ar** 39.948 2-8-8	K-L-M
19 +1 **K** 39.098 -8-8-1	20 +2 **Ca** 40.078 -8-8-2	21 +3 **Sc** 44.956 -8-9-2	22 +2 +3 +4 **Ti** 47.867 -8-10-2	23 +2 +3 +4 +5 **V** 50.942 -8-11-2	24 +2 +3 +6 **Cr** 51.996 -8-13-1	25 +2 +3 +4 +7 **Mn** 54.938 -8-13-2	26 +2 +3 **Fe** 55.845 -8-14-2	27 +2 +3 **Co** 58.933 -8-15-2	28 +2 +3 **Ni** 58.693 -8-16-2	29 +1 +2 **Cu** 63.546 -8-18-1	30 +2 **Zn** 65.38 -8-18-2	31 +3 **Ga** 69.723 -8-18-3	32 +2 +4 **Ge** 72.630 -8-18-4	33 +3 +5 −3 **As** 74.922 -8-18-5	34 +4 +6 −2 **Se** 78.971 -8-18-6	35 +1 +5 −1 **Br** [79.901, 79.907] -8-18-7	36 0 **Kr** 83.798 -8-18-8	-L-M-N
37 +1 **Rb** 85.468 -18-8-1	38 +2 **Sr** 87.62 -18-8-2	39 +3 **Y** 88.906 -18-9-2	40 +4 **Zr** 91.224 -18-10-2	41 +3 +5 **Nb** 92.906 -18-12-1	42 +6 **Mo** 95.95 -18-13-1	43 +4 +6 +7 **Tc** -18-13-2	44 +3 **Ru** 101.07 -18-15-1	45 +3 **Rh** 102.91 -18-16-1	46 +2 +3 **Pd** 106.42 -18-18-0	47 +1 **Ag** 107.87 -18-18-1	48 +2 **Cd** 112.41 -18-18-2	49 +3 **In** 114.82 -18-18-3	50 +2 +4 **Sn** 118.71 -18-18-4	51 +3 +5 −3 **Sb** 121.76 -18-18-5	52 +4 +6 −2 **Te** 127.60 -18-18-6	53 +1 +5 +7 −1 **I** 126.90 -18-18-7	54 0 **Xe** 131.29 -18-18-8	-M-N-O
55 +1 **Cs** 132.91 -18-8-1	56 +2 **Ba** 137.33 -18-8-2	57-71 Lanthanoids	72 +4 **Hf** 178.49 -32-10-2	73 +5 **Ta** 180.95 -32-11-2	74 +6 **W** 183.84 -32-12-2	75 +4 +6 +7 **Re** 186.21 -32-13-2	76 +3 +4 **Os** 190.23 -32-14-2	77 +3 +4 **Ir** 192.22 -32-15-2	78 +2 +4 **Pt** 195.08 -32-17-1	79 +1 +3 **Au** 196.97 -32-18-1	80 +1 +2 **Hg** 200.59 -32-18-2	81 +1 +3 **Tl** [204.38, 204.39] -32-18-3	82 +2 +4 **Pb** 207.2 -32-18-4	83 +3 +5 **Bi** 208.98 -32-18-5	84 +2 +4 **Po** -32-18-6	85 **At** -32-18-7	86 0 **Rn** -32-18-8	-N-O-P
87 +1 **Fr** -18-8-1	88 +2 **Ra** -18-8-2	89-103 Actinoids	104 +4 **Rf** -32-10-2	105 **Db** -32-11-2	106 **Sg** -32-12-2	107 **Bh** -32-13-2	108 **Hs** -32-14-2	109 **Mt** -32-15-2	110 **Ds** -32-16-2	111 **Rg**	112 **Cn**	113 **Uut**	114 **Fl**	115 **Uup**	116 **Lv**	117 **Uus**	118 **Uuo**	-O-P-Q

																Shell
Lanthanoids	57 +3 **La** 138.91 -18-9-2	58 +3 +4 **Ce** 140.12 -19-9-2	59 +3 **Pr** 140.91 -21-8-2	60 +3 **Nd** 144.24 -22-8-2	61 +3 **Pm** -23-8-2	62 +2 +3 **Sm** 150.36 -24-8-2	63 +2 +3 **Eu** 151.96 -25-8-2	64 +3 **Gd** 157.25 -25-9-2	65 +3 **Tb** 158.93 -27-8-2	66 +3 **Dy** 162.50 -28-8-2	67 +3 **Ho** 164.93 -29-8-2	68 +3 **Er** 167.26 -30-8-2	69 +3 **Tm** 168.93 -31-8-2	70 +2 +3 **Yb** 173.05 -32-8-2	71 +3 **Lu** 174.97 -32-9-2	-N-O-P
Actinoids	89 +3 **Ac** -18-9-2	90 +4 **Th** 232.04 -18-10-2	91 +5 +4 **Pa** 231.04 -20-9-2	92 +3 +4 +5 +6 **U** 238.03 -21-9-2	93 +3 +4 +5 +6 **Np** -22-9-2	94 +3 +4 +5 +6 **Pu** -24-8-2	95 +3 +4 +5 +6 **Am** -25-8-2	96 +3 **Cm** -25-9-2	97 +3 +4 **Bk** -27-8-2	98 +3 **Cf** -28-8-2	99 +3 **Es** -29-8-2	100 +3 **Fm** -30-8-2	101 +2 +3 **Md** -31-8-2	102 +2 +3 **No** -32-8-2	103 +3 **Lr** -32-8-3	-O-P-Q

Key to Chart: Atomic Number → 50; Symbol → Sn; 2013 Atomic Weight → 118.71; Oxidation States → +2 +4; Electron Configuration → -18-18-4

Atomic weights are IUPAC 2013 values abridged to five significant digits.

Legend: Metallic solids; Non-metallic solids; Liquids; Gases

Figure 3.1 Mendeleev periodic table of the elements. (Copyright CRC Press.)

compounds and are made up of two or more elements chemically bonded together.

Going inside the elements has shown that they are all made out of only a few components. In this chapter, we are going to study this matter and see how electricity is related to the inside structure of the materials and how so many materials with various properties can be electrically categorized.

3.2 Material Properties

Before getting into the basic structure of all materials, we can easily recognize that there are noticeable differences between various materials as we look around us. Consider, soil, wood, plastic, cloth, a piece of iron, gold, and aluminum. We can define the similarities and the differences between them. All the metals can be categorized as shiny, solid (except mercury that at normal temperature is in the form of liquid), and relatively heavier than many other materials. Moreover, they are conductors of heat. That is, if one side of a reasonably long piece of metal is heated, the heat moves to the other side. This is not true for cloth, wood, or plastic.

One more difference between the above-named metals and the rest of the materials is that all the metals are among the elements, whereas others (e.g., cloth, wood, and plastic) are nonelements or **compound** materials. Elements are those seen in the Mendeleev table. Everything else is a compound, made out of some elements. Nevertheless, among the elements, conductivity of heat is a property of metals and the other elements do not have this property. For example, sulfur, phosphorus, and iodine are elements, but like soil and wood, they do not conduct heat. Whereas sulfur, phosphorus, iodine, and many elements are solid at normal temperature, some other nonmetallic elements are in the form of gas, like oxygen and hydrogen.

Compound: Any material that is not a simple element and is made of at least two elements.

As you may have noticed, all the metals are conductors of electricity in the same way that they are conductors of heat. You use a metallic wire to connect your car battery to the rest of the electric components in a car or to power your radio. So, as far as the electrical property of a material is concerned, the first step in their categorization is whether or not a material is a conductor of electricity.

3.3 Atom and Molecule

An **atom** is the smallest unit of an element having the properties of that element. If a piece of aluminum is broken into two pieces, each piece is still aluminum. Suppose that you can do this with the help of a microscope, and you continue to break the smaller piece into two. The very smallest piece that you obtain is still a piece of aluminum and has the properties of aluminum.

Atom: Smallest particle of an element having the same properties of the element, as comparable to a molecule that is the smallest particle of a compound material.

If you can physically repeat the same action millions of times, finally you end up with an atom of aluminum. You cannot further break it into

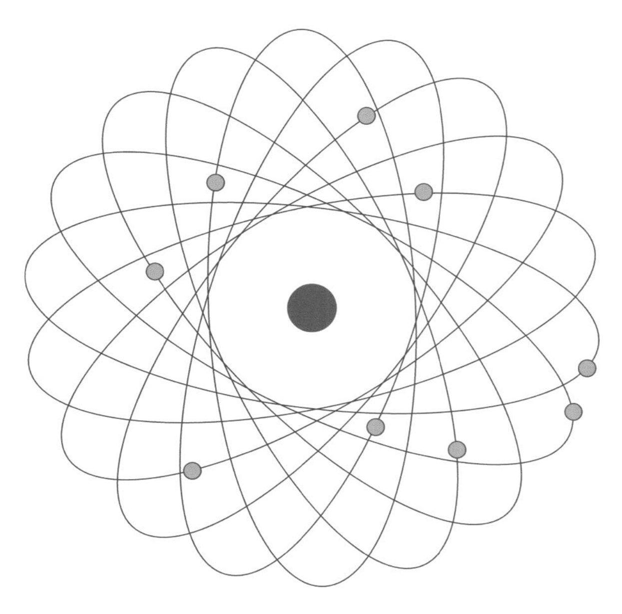

Figure 3.2 Basic structure of an atom.

smaller pieces and maintain its properties. That atom is the smallest piece that can still be called aluminum.

Nucleus: Part of an atom in the middle, consisting of protons (and neutrons when they exist), around which electrons orbit. The nucleus holds almost the entire mass of an atom.

Electrons: Main carrier of negative electric charge. One of the main components in the structure of an atom or molecule that orbits the nucleus at high speed.

Molecule: Smallest piece in a combined substance that still has the properties of the substance and cannot be broken into smaller pieces.

Nevertheless, if an atom is broken, a structure is observed in it. This structure consists of a **nucleus** in the middle and one or more flying **electrons** that orbit the nucleus at high speed. The electron(s) do not move in one plane and continuously alter their orbit, as shown schematically in Figure 3.2. We can assume they form a spherical boundary, layer, or shell around the nucleus. The speed of the electron motion is so high that one cannot find a free space in this shell, in the same way that when an electric fan is working you cannot identify a free space between the blades, and you do not dare check that by inserting your finger in it.

In the same manner that an atom is defined as the smallest possible fraction an element can have, a **molecule** is the smallest possible bit that a compound material can be divided into. A molecule has the same property of the material. For example, a molecule of salt is white, can dissolve in water, and tastes salty (salt is a compound made of chlorine and sodium).

At the atomic level a molecule consists of a number of atoms that are bonded together and compose the structure of a compound material. This bonding does not happen arbitrarily and is dictated by the nature of the forces between the atoms, based on the number of electrons and their arrangement as discussed in the next section.

3.4 Atomic Structure of Materials

In any atom, there is a relatively large space between the nucleus, the central part, and the shell(s) of electron orbits. The electrons are much smaller in size (and consequently much lighter) compared to the nucleus. This implies that the mass of an electron is concentrated in its nucleus. The nucleus is not a single piece and consists of one or more protons and neutrons. A **proton** is the smallest positively charged particle inside an atom, and a **neutron** is the smallest particle with no electric charge. Together they form the nucleus of an atom and determine its mass.

Proton: One of the two main particles in the nucleus of an atom. A proton holds positive charge.

Neutron: One of the two main particles in the nucleus of an atom. Existence of neutrons adds the atomic mass and defines isotopes of a chemical element.

The number of protons and electrons in any atom is the same, but that of the neutrons can be different or even zero. Electrons carry negative charges, and in normal conditions they counteract and neutralize the positive charges of protons, whereas neutrons can only affect the mass (or weight) of an atom because they are neutral and do not have any charge. Consequently, in equilibrium condition an atom is neutral.

Each element has a unique number of protons and electrons, which determine its physical and chemical properties. That is to say,

1. No two different elements have the same number of electrons (protons).
2. Each element can be identified by its number of electrons (protons).

The number of electrons and protons constitute the **atomic number** of an element.

Atomic number: Unique number for any element representing the number of protons in the element.

The gap between the nucleus and the orbits of electrons is relatively large (at the atomic scale). The approximate size of the nucleus of an atom is 10–100 fm (femtometer). Each femtometer is 10^{-15}m. The approximate size of an atom is 10,000 times more; that is 0.1 nm (nanometer). A nanometer is 10^{-9}m. You see that the empty space is quite large, about 10,000 times the nucleus.

We may expect that electrons that have negative charge will be attracted by protons that have positive charge. Because electrons move at very high speed around the nucleus, the centrifugal force (a force that wants to move them away from the center) neutralizes the attraction force between them. This is why they are not attracted to the center; otherwise, the world would shrink.

By the same token we expect that the protons that have the same electric charge will repel each other instead of staying in the center. Atomic theory suggests that there is an atomic strong force that keeps the protons and neutrons together. As far as this book is concerned, we are more interested in the application of atomic structure in electricity, and not in atomic science and further details of the composition of the nucleus, what causes the electrons to move so fast around the nucleus, and so on.

> Atomic number indicates the number of electrons or protons in the atom of an element.

For the elements with more than one electron, the motion of different electrons around the nucleus is not at the same radius. Motion takes place in various layers, each one at a different distance from the center. This distribution is not random and is based on a structure. Each layer, which is at a relatively fixed radius, can have a number of electrons, but this number is limited between a maximum and a minimum. Associated with each layer is a level of energy. Occasionally,

Atomic mass: The mass of an atom or molecule measured in atomic mass unit. The unit is 1/12 of the mass of carbon atom (atomic mass of carbon = 12). For the elements with isotopes, the atomic mass is the average value based on the atomic mass for each isotope and its abundance in nature. Atomic mass depends on the number of electrons, protons, and neutrons in an element.

Atomic weight: The ratio of the atomic mass of an atom or molecule to the atomic mass of carbon-12. Therefore, it does not have a unit since it is a ratio. The number for atomic weight is the same as that for atomic mass (without a unit).

Isotope: Variation of a chemical element in nature, having different numbers of neutrons in the nucleus.

an electron may jump from one layer to another layer because of a change in its energy level. This is why certain phenomena happen, because this can occur with billions of electrons. The energy level of an electron increases owing to an increase in temperature or by electrical stimulation.

The mass (and thus the weight for the material in the Earth's gravity field) of electrons is much smaller than those of the protons and neutrons. An electron has a mass of about 2000 times smaller than a proton. This is why the mass of any atom is almost all in the nucleus. The mass of a neutron is equal to the mass of a proton and an electron together. Thus, we may say, with good precision, the mass of protons and neutrons are the same. The mass of an atom, thus, depends on the numbers of its protons and neutrons. The more this number is, the more the mass of the atom (the **atomic mass**) is.

Atomic weight is used to compare the relative mass of all materials at the atomic and molecular scale. It is based on the number of protons and neutrons in an element. For a compound substance such as H_2O and CO_2, likewise, it depends on the total number of protons and neutrons in the molecule. In fact, atomic weight represents, and more recently has been replaced by, atomic mass (see Section 2.5 for mass and weight). It is a number based on which one can compare the mass (and the weight) of various substances and determine which one is heavier.

The basis (as the unit of measurement) for atomic weight has been revised. Formerly, it was the hydrogen atom (whose atomic weight was 1) before it was found out that an **isotope** (see Section 3.5) of hydrogen exists in nature.

> Atomic weight is a scale based on which the relative mass of substances can be determined. It is based on the total number of protons and neutrons in the atom of each element.

3.5 Differences between Materials

The atomic number is the signature for each element. It defines the number of protons and electrons, which is unique for each element. More important than the atomic number is the way the increase in the atomic number affects the properties of elements. The simplest atom, which is the first item in the Mendeleev table, is hydrogen. Hydrogen has only one proton and one electron (and no neutron), and, thus, its atomic number is 1. As you already know, hydrogen is a gas, it can burn (which implies chemically combining with oxygen) and produce heat, and the product is water. The next element after hydrogen is helium, which although a gas, is completely different from hydrogen and cannot combine with oxygen. The difference in the atomic structure with that of hydrogen is only that it has two protons (thus heavier than hydrogen) and two electrons.

Figure 3.3 depicts the fact that the number of electrons in elements increases one by one, but the electrons move in different orbits, with certain properties concerning the number of electrons in each orbit.

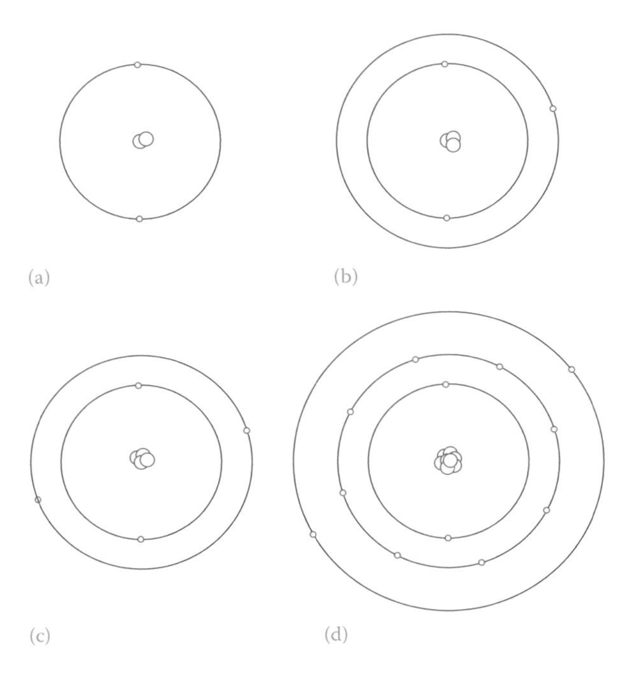

Figure 3.3 When each orbit contains its maximum number of electrons, any additional electrons move to a new orbit farther from the nucleus. (a) Maximum number for first orbit is 2. (b) A third electron moves to a new orbit. (c) The second orbit takes up to eight electrons. (d) After the second orbit fills, additional electrons move to a new (third) orbit.

The study of materials properties and their chemical behavior has led to the conclusion that the structure of the layers of orbits in which electrons move is such that

1. The first layer (nearest to the nucleus) can accept only 1 and 2 electrons.
2. The very last layer (farthest from the nucleus) can accept between 1 and 8 electrons.
3. The second layer also can accept only between 1 and 8 electrons.
4. The third layer can have up to 18 electrons.
5. The fourth and fifth layers when they exist can have up to 32 electrons.
6. The sixth layer (when it exists) can have up to 18 electrons.

Table 3.1 shows the number of electrons and their distribution in various shells for all elements. The above shows that electrons must not necessarily occupy each layer to its maximum capacity before they move to the next layer. As said, each layer of orbit of electrons is associated with a level of energy; thus, when referring to different orbital shells, the word level is sometimes used instead of layer.

The existence of neutrons in an atom does not have any effect on the atomic number. Therefore, the addition of neutrons does not change a substance. What is its significance then? A difference in the number of neutrons results in various isotopes of the same element. An isotope of an

TABLE 3.1 Distribution of Electrons in Shells for All Elements

Atomic Number	Element	No. of Electrons in Shells	Atomic Number	Element	No. of Electrons in Shells
1	Hydrogen	1	43	Technetium	2, 8, 18, 13, 2
2	Helium	2	44	Ruthenium	2, 8, 18, 15, 1
3	Lithium	2, 1	45	Rhodium	2, 8, 18, 16, 1
4	Beryllium	2, 2	46	Palladium	2, 8, 18, 18
5	Boron	2, 3	47	Silver	2, 8, 18, 18, 1
6	Carbon	2, 4	48	Cadmium	2, 8, 18, 18, 2
7	Nitrogen	2, 5	49	Indium	2, 8, 18, 18, 3
8	Oxygen	2, 6	50	Tin	2, 8, 18, 18, 4
9	Fluorine	2, 7	51	Antimony	2, 8, 18, 18, 5
10	Neon	2, 8	52	Tellurium	2, 8, 18, 18, 6
11	Sodium	2, 8, 1	53	Iodine	2, 8, 18, 18, 7
12	Magnesium	2, 8, 2	54	Xenon	2, 8, 18, 18, 8
13	Aluminum	2, 8, 3	55	Caesium	2, 8, 18, 18, 8, 1
14	Silicon	2, 8, 4	56	Barium	2, 8, 18, 18, 8, 2
15	Phosphorus	2, 8, 5	57	Lanthanum	2, 8, 18, 18, 9, 2
16	Sulfur	2, 8, 6	58	Cerium	2, 8, 18, 19, 9, 2
17	Chlorine	2, 8, 7	59	Praseodymium	2, 8, 18, 21, 8, 2
18	Argon	2, 8, 8	60	Neodymium	2, 8, 18, 22, 8, 2
19	Potassium	2, 8, 8, 1	61	Promethium	2, 8, 18, 23, 8, 2
20	Calcium	2, 8, 8, 2	62	Samarium	2, 8, 18, 24, 8, 2
21	Scandium	2, 8, 9, 2	63	Europium	2, 8, 18, 25, 8, 2
22	Titanium	2, 8, 10, 2	64	Gadolinium	2, 8, 18, 25, 9, 2
23	Vanadium	2, 8, 11, 2	65	Terbium	2, 8, 18, 27, 8, 2
24	Chromium	2, 8, 13, 1	66	Dysprosium	2, 8, 18, 28, 8, 2
25	Manganese	2, 8, 13, 2	67	Holmium	2, 8, 18, 29, 8, 2
26	Iron	2, 8, 14, 2	68	Erbium	2, 8, 18, 30, 8, 2
27	Cobalt	2, 8, 15, 2	69	Thulium	2, 8, 18, 31, 8, 2
28	Nickel	2, 8, 16, 2	70	Ytterbium	2, 8, 18, 32, 8, 2
29	Copper	2, 8, 18, 1	71	Lutetium	2, 8, 18, 32, 9, 2
30	Zinc	2, 8, 18, 2	72	Hafnium	2, 8, 18, 32, 10, 2
31	Gallium	2, 8, 18, 3	73	Tantalum	2, 8, 18, 32, 11, 2
32	Germanium	2, 8, 18, 4	74	Tungsten	2, 8, 18, 32, 12, 2
33	Arsenic	2, 8, 18, 5	75	Rhenium	2, 8, 18, 32, 13, 2
34	Selenium	2, 8, 18, 6	76	Osmium	2, 8, 18, 32, 14, 2
35	Bromine	2, 8, 18, 7	77	Iridium	2, 8, 18, 32, 15, 2
36	Krypton	2, 8, 18, 8	78	Platinum	2, 8, 18, 32, 17, 1
37	Rubidium	2, 8, 18, 8, 1	79	Gold	2, 8, 18, 32, 18, 1
38	Strontium	2, 8, 18, 8, 2	80	Mercury	2, 8, 18, 32, 18, 2
39	Yttrium	2, 8, 18, 9, 2	81	Thallium	2, 8, 18, 32, 18, 3
40	Zirconium	2, 8, 18, 10, 2	82	Lead	2, 8, 18, 32, 18, 4
41	Niobium	2, 8, 18, 12, 1	83	Bismuth	2, 8, 18, 32, 18, 5
42	Molybdenum	2, 8, 18, 13, 1	84	Polonium	2, 8, 18, 32, 18, 6

(*Continued*)

TABLE 3.1 (CONTINUED) Distribution of Electrons in Shells for All Elements

Atomic Number	Element	No. of Electrons in Shells	Atomic Number	Element	No. of Electrons in Shells
85	Astatine	2, 8, 18, 32, 18, 7	102	Nobelium	2, 8, 18, 32, 32, 8, 2
86	Radon	2, 8, 18, 32, 18, 8	103	Lawrencium	2, 8, 18, 32, 32, 8, 3 (?)
87	Francium	2, 8, 18, 32, 18, 8, 1	104	Rutherfordium	2, 8, 18, 32, 32, 10, 2 (?)
88	Radium	2, 8, 18, 32, 18, 8, 2	105	Dubnium	2, 8, 18, 32, 32, 11, 2 (?)
90	Thorium	2, 8, 18, 32, 18, 10, 2	106	Seaborgium	2, 8, 18, 32, 32, 12, 2 (?)
91	Protactinium	2, 8, 18, 32, 20, 9, 2	107	Bohrium	2, 8, 18, 32, 32, 13, 2 (?)
92	Uranium	2, 8, 18, 32, 21, 9, 2	108	Hassium	2, 8, 18, 32, 32, 14, 2 (?)
93	Neptunium	2, 8, 18, 32, 22, 9, 2	109	Meitnerium	2, 8, 18, 32, 32, 15, 2 (?)
94	Plutonium	2, 8, 18, 32, 24, 8, 2	110	Darmstadtium	2, 8, 18, 32, 32, 16, 2 (?)
95	Americium	2, 8, 18, 32, 25, 8, 2	111	Roentgenium	2, 8, 18, 32, 32, 17, 2 (?)
96	Curium	2, 8, 18, 32, 25, 9, 2	112	Copernicium	2, 8, 18, 32, 32, 18, 2 (?)
97	Berkelium	2, 8, 18, 32, 27, 8, 2	113	Ununtrium	2, 8, 18, 32, 32, 18, 3 (?)
98	Californium	2, 8, 18, 32, 28, 8, 2	114	Ununquadium	2, 8, 18, 32, 32, 18, 4 (?)
99	Einsteinium	2, 8, 18, 32, 29, 8, 2	115	Ununpentium	2, 8, 18, 32, 32, 18, 5 (?)
100	Fermium	2, 8, 18, 32, 30, 8, 2	116	Ununhexium	2, 8, 18, 32, 32, 18, 6 (?)
101	Mendelevium	2, 8, 18, 32, 31, 8, 2	117	Ununseptium	2, 8, 18, 32, 32, 18, 7 (?)

element has the same properties of that element but has a different atomic weight. Different isotopes of an element, when they exist, represent the varieties of the same element. For example, the normal hydrogen atom has no neutron, but an isotope of hydrogen, called **deuterium**, has 1 neutron. It is still hydrogen and bonds with oxygen. The product, however, is not ordinary water but **heavy water**. Heavy water is used in nuclear power plants for cooling the fuel rods. This water is called heavy water because the hydrogen in it is heavier than the ordinary hydrogen due to the extra neutron.

Deuterium: Isotope of hydrogen. When composed with oxygen, it produces heavy water.

Heavy water: Special water made out of deuterium (an isotope of hydrogen), thus with the chemical structure D_2O, which is used for storing fuel rods in a nuclear power plant.

The fact about isotopes is that they are normally very rare compared to the ordinary element. For instance, 99.98 percent (by weight) of all hydrogen available on Earth is regular and the rest is the other isotope. On an atom basis the ratio is 6420 to 1. For this reason, separating the rare isotopes from the rest needs a special process and is time-consuming and costly.

The existence of isotopes for some elements leads to a modification of the average mass for a substance (which is the criteria for allocation of the atomic weight). This translates to changes in the atomic weights of elements, which formerly were assumed to be integral numbers, but nowadays they carry the effect of the isotopes. For instance, the atomic weight of hydrogen used to be 1, whereas today this number is 1.00794.

Valence: Outermost orbit of electrons in the structure of an atom. Valence electrons move in this orbit.

3.6 Similarities between Materials Based on Valence Electrons

Valence electrons: The electrons on the outermost orbit around the nucleus of an atom.

The outermost shell of electrons in an atom is called the **valence** shell, and the electrons orbiting in that shell are called **valence electrons**. As said before, the valence shell can have between 1 and 8 electrons, with the exception of palladium (atomic number = 46). The valence electrons determine the properties of an element in relationship to bonding with the other elements. Moreover, elements can be classified based on their valence electrons, although this does not mean that all elements with the same number of valence electrons are similar. For instance, hydrogen with only 1 valence electron is quite different from sodium and potassium, also with 1 valence electron.

Furthermore, bonding of elements with each other greatly depends on the number of valence electrons. The total number of valence electrons cannot exceed 8. In this respect, elements whose valence electrons add up to 8 have a very strong tendency for bonding. Examples are hydrogen and chlorine (with 1 and 7 valence electrons, respectively) and many metals and oxygen (with 2 and 6 valence electrons). Nevertheless, there are always exceptions, in addition to the fact that the molecules can also be made of atoms of the same substance, such as H_2, which can be considered a molecule of hydrogen. In this regard, oxygen (6) can bond very well with the elements with 1 valence electron. The best example is water (H_2O), which is made up of one atom of oxygen and two atoms (or one molecule) of hydrogen. Figure 3.4 shows an element with 4 levels of orbit and with 1 valence electron. See Table 3.1 to determine what element it is.

When by either nature or bonding the number of valence electrons reaches a complete set, there is no more bonding tendency. For instance, helium has 2 valence electrons. (Therefore, its only orbital shell is complete.) It is an inert gas that does not bond with any other element. The same is true for argon and krypton, with 8 valence electrons. Figure 3.5 shows the bonding between hydrogen and oxygen, a strong bond in which two valence electrons from two hydrogen atoms share with the valence electrons of an oxygen atom to make a complete set of 8 valence electrons for the oxygen atom and 2 valence electrons for the hydrogen atom.

Electrons on the outermost orbit in an atom are called valence electrons. They determine the properties of materials.

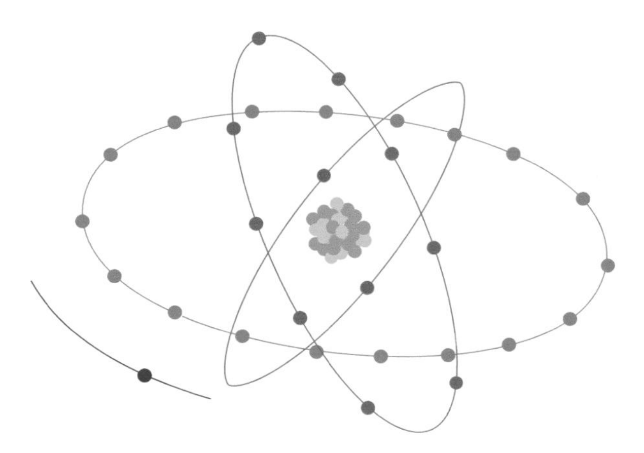

Figure 3.4 Element with four levels of orbits and one valence electron.

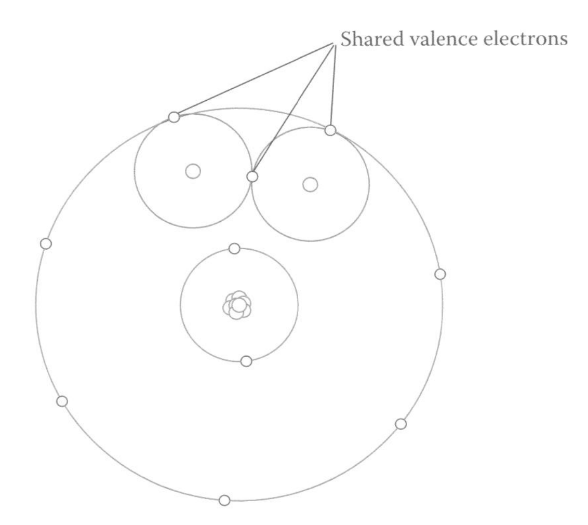

Figure 3.5 Schematic of bonding between hydrogen and oxygen.

3.7 Categorization of Materials: Conductors and Insulators

Study of the atomic structure of materials by physicists has shown that all metals have between 1 and 3 valence electrons. For instance, sodium, potassium and silver have 1 valence electron, iron and zinc have 2 valence electrons, and aluminum and gallium have 3 valence electrons. This does not mean that any element with 1, 2, or 3 valence electrons is a metal. Hydrogen, for example, has 1 valence electron, and helium has 2, both of them gases.

Moreover, all metals are conductors of both electricity and heat. This stems from the fact that the valence electrons of elements can jump or share more easily with other similar or same category atoms. In fact, flow of electricity is the movement of the valence electrons in a conductor. Receiving energy or being under stimulation is the force to move an electron from one atom to another. This happens in the form of a chain reaction with billions of electrons doing the same thing, simultaneously.

By the same token, elements having between 5 and 7 valence electrons are not conductors of either electricity or heat. Examples are sulfur and phosphorus. Observation of Table 3.1 reveals that carbon, tin, and lead have 4 valence electrons. These elements are conductors of electricity, also, though carbon cannot be considered a metal; and tin and lead are not among the best conductors (see Table 12.2). All the elements with their valence electrons completed (2 for helium and 8 for others) are inert gases. Inert implies that they do not bond with any other element.

In addition to categorizing elements to metals and insulators, as we will learn in Chapter 13, in electricity we have semiconductors, which are the basis for solid-state electronics, such as transistors. Semiconductor materials are made up of elements with 3 to 5 valence electrons (except carbon and some metals).

3.8 Electric Charge and Energy in Electrons

Electric charge, as the name implies, is a measure of electricity. Nevertheless, particularly in the beginning, it is difficult to have a feeling of the quantity of electricity because it is not tangible. In later chapters, we will discuss various electric-related quantities and their units of measure. At this time we define only what this entity is.

The smallest possible amount of electricity is the electric charge of an electron (no smaller amount can be found). Then, if we have *N* electrons the amount of electric charge is *N* times the charge of an electron. Think about *N* as a very large number, like 10^{20}. Consider lightning. It has a huge amount of electric charge. This is due to a huge number of electrons that are gathered together and make a cluster of electric charge.

There is also some energy associated with the electric charge. Again, think of the energy in lightning, which is the accumulation of the energy of many thousand billions of electrons. At an atomic level the energy associated with one electron is a small quantity. But this energy can be huge for a very large number of electrons together. The unit for energy at this very small scale is electron-volt. An electron-volt (eV) is the amount of energy that an electron gains (or loses) if moved between two points that have 1V potential difference. (For potential difference see Chapter 4.) Going from higher potential to lower potential point decreases energy (energy is released or given to other electrons). Consider the charged clouds, causing lightning and thunder. When they discharge, they release energy. Their potential difference is around 2 million volts (with reference to the ground, for instance). In this sense, each electron in a charged cloud has about 2 million electron-volt (2 MeV) energy.

As compared with the units of energy that we are more familiar with and are more tangible, 1 eV is about 1.602×10^{-19} joule. One joule is equal to 1 Newton-meter (Nm). 1 Nm is the energy that is required to lift an object that weighs 1 N by 1 m. That equals lifting an object that is 0.225 lb by 3.3 ft, which is equivalent to the energy required for lifting a 1 lb object by 9 in.

3.9 Chapter Summary

- Physical matters on Earth are made up of 118 elements.
- An atom is the smallest possible unit of an element that still holds all the properties of the element.
- A molecule is, likewise, the smallest reachable unit of a compound that still holds all the properties of that compound.
- A molecule is made up of two or more atoms.
- An atom is made up of a nucleus and a number of electrons orbiting around it at various radii.
- The nucleus contains protons and neutrons.
- Numbers of protons and electrons in an atom are the same.
- Electrons are much smaller than protons, and almost all the mass of an atom is in its nucleus.

- Electrons have negative charge, protons have positive charge, and neutrons have no charge.
- Charge of one electron is the smallest unit for measuring electric charge.
- In an element the number of neutrons can vary. This results in various isotopes of the same element. Isotopes of an element have the same properties of the element, but their weights are different.
- The number of electrons (or protons) in an atom defines its atomic number.
- Atomic weight is a number defining the relative mass (or weight) of various elements and compounds. It is based on the average number of protons and neutrons in the nucleus.
- Electrons move in various orbits. Each orbit can have a maximum number of electrons. These maximum numbers for electrons from innermost orbit are 2-8-18-32-32-18-8.
- The outermost orbit is called the valence shell, and its electrons are called valence electrons.
- Valence electrons determine the properties of materials.
- Most metals have between 1 and 3 valence electrons.
- Carbon and a few metals have 4 valence electrons.
- All elements with 5 to 7 valence electrons are nonconductors.
- Bonding tendency between elements depends on the number of valence electrons. When elements chemically bond to make a compound they fill their valence shell to contain 8 electrons.
- Electron-volt is a unit for measurement of energy at the atomic scale.

Review Questions

1. What is the difference between an atom and a molecule?
2. Does iron have an atom or a molecule?
3. Does hydrogen have an atom or a molecule?
4. What is in the nucleus of an atom?
5. Does a molecule have a nucleus?
6. Why can we assume that the mass of an atom is in its nucleus?
7. Can an atom have more electrons than protons? Why or why not?
8. Can an atom have no neutrons in the nucleus?

9. Can an atom have more neutrons than protons?

10. What defines the atomic number of an element?

11. What defines the atomic weight?

12. What is the atomic weight used for?

13. What is an isotope of an atom?

14. What is a valence electron?

15. Name an element with 9 valence electrons.

16. Can we say all elements that have between 1 and 3 valence electrons are metals? Explain.

17. How many valence electrons does carbon have?

18. Is carbon a metal or a nonmetal? Why?

19. Is carbon a conductor of electricity?

20. Is carbon a conductor of heat?

Problems

1. From Table 3.1, find all elements with 4 valence electrons.

2. From Table 3.1, find the lightest and the heaviest elements.

3. From Table 3.1, determine if selenium is a metal.

4. From Table 3.1, find all elements that have 8 valence electrons. What is the similarity between these elements?

5. From Table 3.1, find the electron structure of iron and copper. Use the Internet to find the properties of these two metals.

4

DC and AC Electricity

OBJECTIVES: After studying this chapter, you will be able to

- Differentiate between two main types of electricity
- Define direct current electricity
- Define alternative current electricity
- Explain what an electric circuit is
- Draw a schematic representation of an electric circuit
- Understand the basic components in any electric device or circuit
- Describe what a resistor in an electric circuit is and what effect it has on the circuit
- Describe what an inductor is and what it does
- Describe what a capacitor is and its role in electricity
- Describe resistance
- Explain what inductance is and how it is determined
- Explain what capacitance means and how it is calculated
- Understand two important properties of electricity, current and voltage
- Define electric current and electric voltage and make an analogy between electricity and water flow in a pipeline
- Name some of the units used to measure electricity and electric components

New terms:
AC circuit, alternating current, alternative current, ampere, capacitance, capacitor, circuit, coil, cycle, cyclic waveforms, DC circuit, direct current, electromotive force, Farad, ferromagnetic, frequency, Henries, Hertz, inductance, inductor, ohm, omega, potential difference, resistance, resistor, sinusoidal, specific resistance, volt, voltage, waveform

4.1 Introduction

Even if you do not know much about electricity yet, you use it every day. Unless you live in remote locations, or in undeveloped places, electrical networks can be seen and are in use. Even in those remote places, electricity can exist in other forms. You are familiar with batteries and know that they come in different sizes. Also, you are familiar with electricity that you have at home because you have lights, a refrigerator, a kitchen range (if electric), and so on. In all these devices, there is a switch to turn the device on and off, and there are wires between the device and the source of electricity. We are going to study what the source of electricity is and what is inside these devices, what happens when the switch is turned on, and how we can illustrate all of these in paper drawings. Furthermore, you learn what the difference is between the electricity from a battery and the one that is used for home lights.

4.2 Electricity Basics

We learned in Chapter 3 that in metals there are between 1 and 3 valence electrons and also that these electrons can be excited with heat (their level of energy increases). Electricity is the flow of electrons in any conductor. There are billions of these electrons even in a tiny piece of metal. When the electrons move together in one direction (even if for short periods of time), they form an electric current. Thus, electric flow implies simultaneous motions of billions of electrons.

Obviously, there must be a force to move all electrons in one direction; otherwise, there is no reason why they move in a particular direction, instead of moving in their orbits or in the case of free electrons (valence electrons with little bonding force that can easily travel from the orbit of one atom to another) moving randomly in all directions. Any source of electricity (such as a battery or a generator) provides the force to create this motion of electrons with a defined pattern.

An analogy can be made between electricity and water flow in a pipeline. This often helps to better understand the rules and facts of electric flow, because water flow can be observed and pictured in mind, whereas electricity and electric flow are obscure. You can imagine each drop of water as an electron. First, there has to be a force moving all the water in one direction. In the absence of such a force, the water stands still. Nevertheless, the following three points must be always kept in mind:

1. Even if the water is not flowing, drops of water can have local motion, like in waves generated by wind.
2. Not all the water drops have the same uniform motion (parallel with each other and with the same speed).
3. The path of water as a whole is defined by the waterway or the pipeline.

Electromotive force: Electrical potential difference causing electrical current between two points. A battery or an electric generator is a source of electromotive force, measured in volts.

In electricity, the force behind the motion of electrons is called **electromotive force**. This force pulls or pushes the electrons through a path along which they can move (e.g., through a conductor).

> Electricity is the flow of electrons. Motivation force causing their motion is called electromotive force.

In the flow of water in a pipe, or a stream, the current is defined in gallons per minute (or other units, such as liters per second in the metric system). This implies that if at some point the pipe is smaller in size, the water speed increases to maintain the flow rate. You can imagine the same thing for the flow of electrons. However, this analogy is not always true or perfect, and it will be combined with other facts.

If you consider the flow of electricity in a lightbulb (an electric device), you notice that flow of electrons is the same in the wire connecting the device (the lightbulb) to the electric source. While the device gets hot, the wire does not. The reason is that the electrons spend more energy in passing though the device. This becomes clearer for you in the following sections.

4.3 Electricity Effects and Electrical Devices

You have noticed the home electricity that is used for lighting, cooking, warming (heater), cooling (refrigerator), and motion (motors in home and garden devices). But, if you want to list the effect of electricity, you should not list the above. These devices use electricity as the source of energy to do their function, which depends on the device, its components, and the media used in them.

Electricity has three effects, and all we can do with electricity is based on one or more of these effects. These are thermal, magnetic, and chemical effects. The light in an incandescent lightbulb is due to thermal effect. The light in a fluorescent lightbulb is a result of thermal and chemical effects. All the motors work because of the magnetic effect of electricity.

> The main effects of electricity are thermal, magnetic, and chemical effects. All electric devices function based on one or more of these effects.

All the thermal devices (e.g., heaters and stove) use the thermal effect of the electricity. All the incandescent lightbulbs also work based on the thermal effect, which heats the element to a high temperature and results in emitting visible light.

The magnetic effect of electricity is the basis of the function of all the electric motors and generators. Electric motors and generators have wire windings in them that enhance the magnetic effect, as we will discuss in later chapters. Also, electric magnets (called electromagnets) are commonly used in many domestic and industrial applications.

In electric motors and generators, some electricity is converted to heat. Although this is unwanted because it is intrinsic to the use of electricity one cannot avoid this heat, which is a waste of energy. In other words, when using one effect of electricity, we may not be able to block the other effects.

You may have heard of using electricity to convert water into hydrogen and oxygen or for decomposing other compound materials, too. This is based on the chemical effect of electricity. This effect is frequently used in chemical and other industries. One good example is electroplating, which is covering the surface of a metal with another metal for protection or cosmetic reasons.

In a refrigerator (and other cooling devices), electricity is used in the compressor to compress the refrigerant, but the rest of the system is not electric and does not use electricity.

4.4 Main Electric Components

All the apparatuses that you can name as electric devices can be composed of one or more of the main electric components. An electric device is connected to electricity by means of a switch and some wires; it may have a fuse as well, and if it uses electricity from the wall outlet, it has a plug. Nevertheless, by *main components* we mean the following three basic elements: resistor, inductor, and capacitor.

We need to learn these components very well and understand their functions, before moving forward. Basically, these three elements have three different functions and their names when addressed imply that each one purely functions as expected, without crossing the border to functions of the other two. Any electric circuit can be composed of one or more devices and some wires. But, ultimately, it can be decomposed into a combination of these basic elements. Not every device should necessarily contain all three of these basic elements. For instance, a lightbulb has only a filament; that is, it consists of a resistor only. On the other hand, whenever we have a wire winding (like in a motor or an electromagnet), it can be considered a coil. A **coil** consists of a resistor and an inductor. A capacitor, however, is a storage device for electricity. It can come in various sizes. Fluorescent lights and flashes in cameras use capacitors. Capacitors are used in many other devices or circuits, such as in motor circuits.

Coil: Any wire wound in circles to form a helix, usually with a core to enhance the inductance.

In this section these basic components are introduced, but we will discuss more about them in the following sections and we will discuss direct current and alternating current electricity.

There are three main basic electric elements: resistor, inductor, and capacitor.

Resistor: (a) An electric component that only exhibits resistance (and no other reaction) to the flow of electricity. (b) Any of the standard components, made in different physical sizes, that are used in electrical and electronic circuits to absorb electrical energy.

4.4.1 Resistor

A **resistor** or a purely resistive device consumes electric power and converts it into heat. Examples of resistors are all electric lightbulbs filaments, electric kettle, heater, iron and stove elements, and any other heating element in a device such as electric water heater. Note that an incandescent lightbulb is not meant to convert electricity to heat like a heater, but to light up. However, in this process the filament is heated to a high temperature to emit light. Thus, the lightbulb gets hot, and it can be treated similar to an electric heater element.

The way a resistor works is that it introduces **resistance** to electric flow. In passing through such an element, electrons must consume more energy. This energy appears as heat in the element.

Resistance: (in electricity) (a) The property of resisting (but not blocking) the flow of electric current, leading to limiting the flow and absorbing (consuming) electric energy. Resistance is measured in ohms (Ω). (b) One of the three basic components that any electric load (a device using electricity) can be assumed to be composed of.

A heater element or lightbulb filaments are made up of a wire usually made out of tungsten or a tungsten alloy. The primary role of such an element is converting electricity to heat. Nevertheless, in many (electronic) circuits, such as in radio and television, the primary role of a resistor can be something else, but in the process a resistor gets hot as well. In such a case the heat must be removed by cooling. Figure 4.1 shows a wire-wound resistor. For protection and portability, such a resistor can be encapsulated in a ceramic casing.

Resistors of the second category (for electronic circuits) are available in various standard sizes and forms. A more detailed description of the resistors used in electronic devices is given in Appendix E. Figure 4.2 shows some examples of these resistors that are made of other material than a

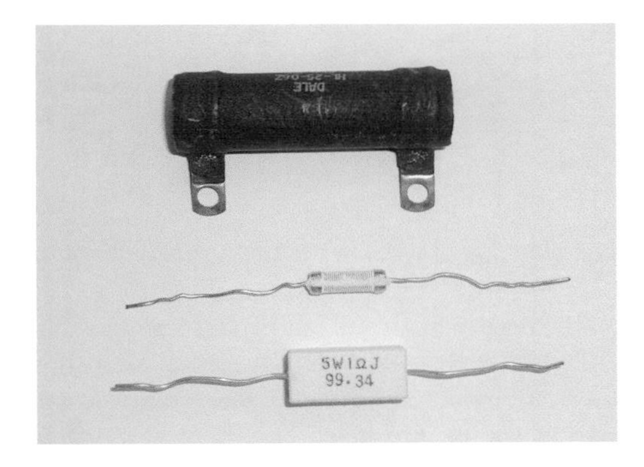

Figure 4.1 Wire-wound resistor.

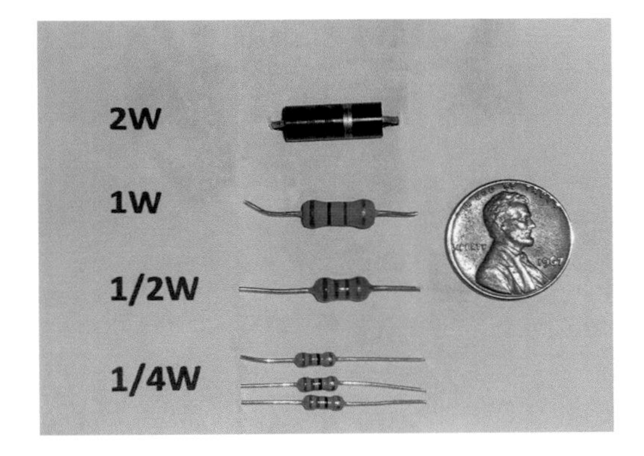

Figure 4.2 Examples of resistors used in electric and electronic devices.

Figure 4.3 Symbol for a resistor in an electric circuit.

metal and have a different structure than being wound. When used in an electric circuit, a special symbol is used to represent a resistor. Figure 4.3 shows the symbol for a resistor.

Inductor: A winding (a coiled wire) with only magnetizing property and without any electric resistance. Physically it is not possible to have a pure inductor, but at certain conditions, particularly high-frequency electric signals, the resistance of the coiled wire can be ignored in comparison with its inductance.

4.4.2 Inductor

An **inductor** is a storage device that can store electric energy by turning it into magnetism. This storage act is not similar to storing energy in a battery. Rather, it is a short-duration storage for a very small amount of electricity. Practically, a winding as shown in Figure 4.4 can do this job and behaves as an inductor. That is, a coil made of a wire wound around a support can be regarded as an inductor. An inductor can have a core. The core can be of any form such as a cylinder or prism, and it can be made out of any material such as paper, wood, plastic, or metal. As we discuss

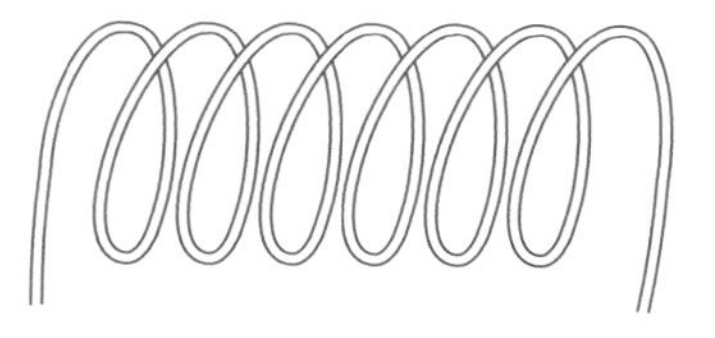

Figure 4.4 A coil (winding of wire) is the basis of an inductor.

Ferromagnetic: Type of material from the iron family that is suitable for magnetization.

later, for a metallic core to be useful (and not as a support) it must be a **ferromagnetic** (like iron and steel or certain special alloys) material.

Theoretically, an inductor has only the storage capability without converting any electricity into heat. This is to say that a coil of metal should ideally have no resistance to electric flow. But, in practice, a coil is made of wire, and the wire behaves as a resistor. The physical size of a coil (length, diameter, number of turns of the winding) and other factors define the property of a coil as an inductor. In practice, a coil can be very well accepted as an inductor if the resistor part of it is small (see Section 4.10).

Figure 4.5 shows the symbols for a coil with or without a (ferromagnetic) core. A ferromagnetic core enhances the inductor property of a coil. Figure 4.6 shows the physical shape of a coil.

> An inductor stores electric energy, but this storage action is in a small scale and with short duration.

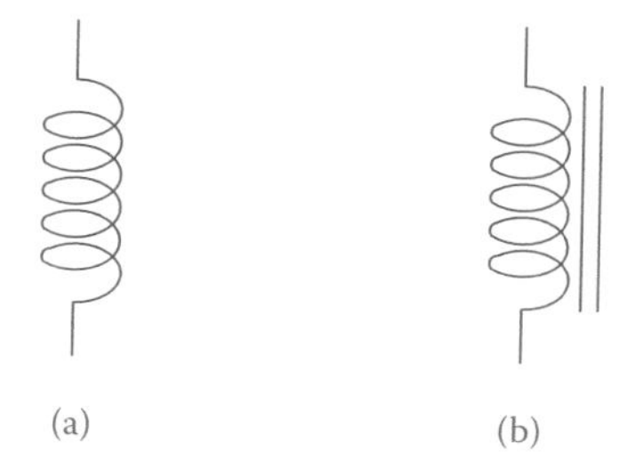

Figure 4.5 Symbol for inductors. (a) Coil without core and (b) coil with core.

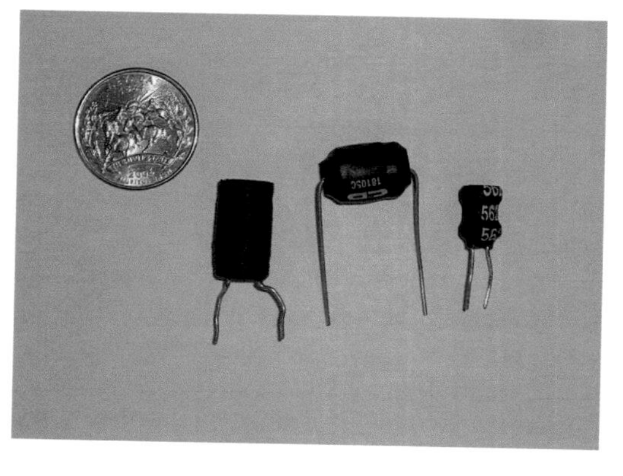

Figure 4.6 Examples of inductors.

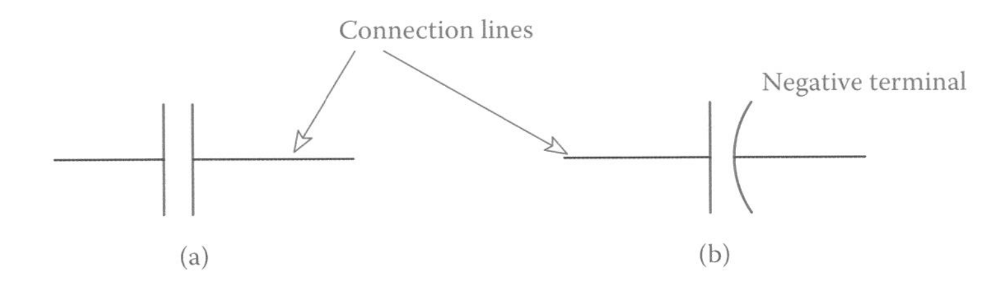

Figure 4.7 Symbol for a capacitor. (a) Capacitor without polarity and (b) capacitor with polarity.

4.4.3 Capacitor

A capacitor is another electricity storage device but completely different and based on a different structure and property than an inductor. A capacitor stores energy in the form of electric field, like a battery. Attention has to be paid to the fact that, although we use the term storing electric energy in an inductor and a capacitor, this storage is very small relative to what a battery does. In particular, the difference is more meaningful in terms of the time this energy can be reused. We may charge a battery, leave it charged for a long time, and then use the stored electricity for say one month. This is much beyond the storage in a capacitor or an inductor. For instance, if instead of a (rechargeable) battery, we use a capacitor, it gets charged in a few seconds or less (can be in microseconds); after some time, if the stored energy is not leaked out, we may use it for a period around the same that it took for charging. Think of the duration of a flash in photography.

You may question what type of storage such a function is useful for. Well, one example is storing numbers in a calculator or a computer during its number processing. We will discuss about the other applications of a capacitor in the later chapters.

Figure 4.7 shows the symbol of a capacitor in a **circuit**. As shown, capacitors are either polarized (i.e., one terminal must be always connected to the positive side and the other terminal to the negative side of the circuit) or nonpolarized (i.e., it does not matter which side is connected to positive). The basic structure of a capacitor is two metallic plates separated by an insulator. In practice, the two plates and the insulator material between them are rolled to form a cylinder, as shown in Figure 4.8. In this way, a more compact and portable package is obtained. Very small capacitors are made differently, and come in a different shape than cylindrical form. Figure 4.9 shows some examples of capacitors.

Circuit: Any combination of electric and electronic components connected together by wires to be connected to an electric source.

> A capacitor stores electric energy, like an inductor, but in a different way. This storage action is not on the same scale as in batteries.

4.5 Electric Circuit

To use electric energy in any device, one needs to connect the device to the electric supply by wires. An electric circuit represents the connection of a device to the electric source. In the simplest form a circuit must have a source to which one or more devices are connected through two wires.

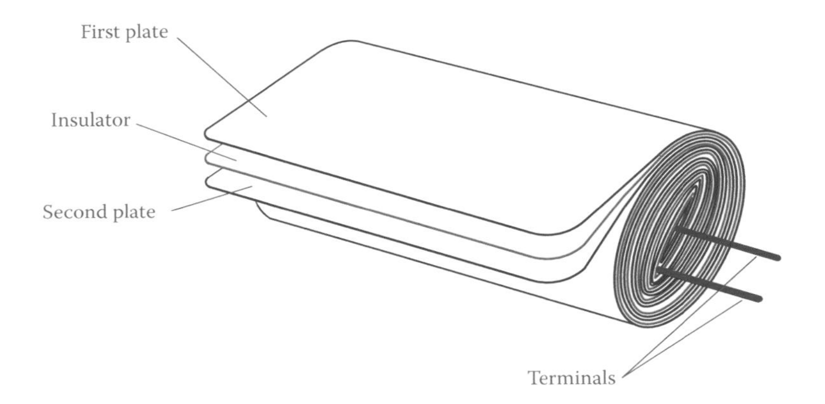

Figure 4.8 Basic construction of a capacitor.

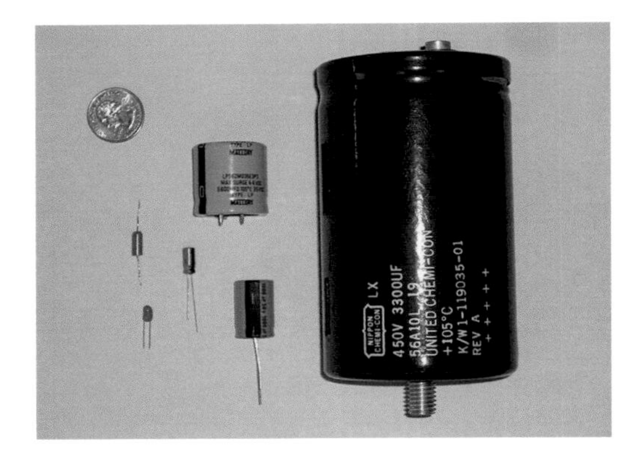

Figure 4.9 Capacitors with various sizes and shapes.

One may also include a switch and a fuse. A switch is a control device, and a fuse is a safety device.

The role of the switch is to control energizing the load when desired. The role of the fuse is to protect the load and the source from damage. If something goes wrong, a fuse disconnects a device from electricity, thus saving the device from getting damaged; also, it can prevent a potential fire in case a heating element is involved. We discuss more about this protection in later chapters. For now, know that a fuse is an important part of any electric circuit. Figure 4.10 shows such a simple circuit.

A fuse is a necessary and important part of any electric circuit.

Figure 4.11 depicts a circuit for a light connected to electricity. This is the physical arrangement for the wires and the light, switch locations, and so on. Whereas this diagram is good for wiring purposes, it is not so suitable for analysis that is needed before a wiring diagram can be produced. For analysis purposes it is customary to simplify this diagram and show all the components inside a circuit by their respective symbols and connect them together in the logical way and order. More importantly,

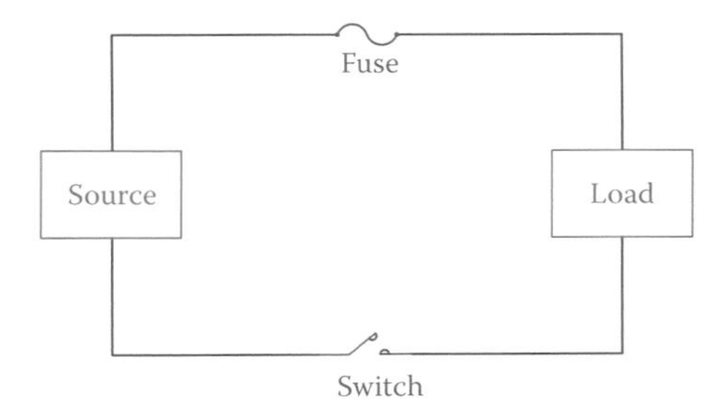

Figure 4.10 Simple electric circuit.

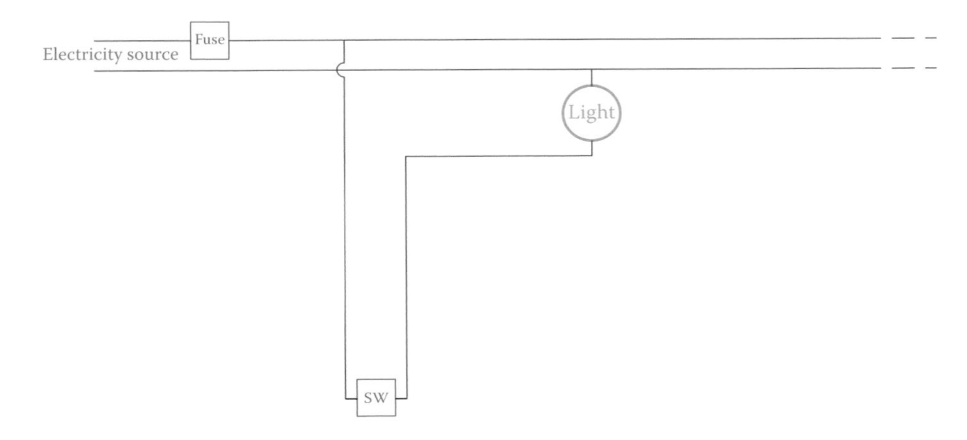

Figure 4.11 Sample of a circuit at home lighting.

because all devices are made up of the three basic aforementioned elements, any device is represented by its equivalent basic elements. In this sense, the circuit in Figure 4.11 is represented as illustrated in Figure 4.12.

Note that in Figure 4.12 for each component a standard symbol is used. A pair of parallel short and long lines denote a battery. (Here we used a battery as the electricity source.) Moreover, all the wires are represented by

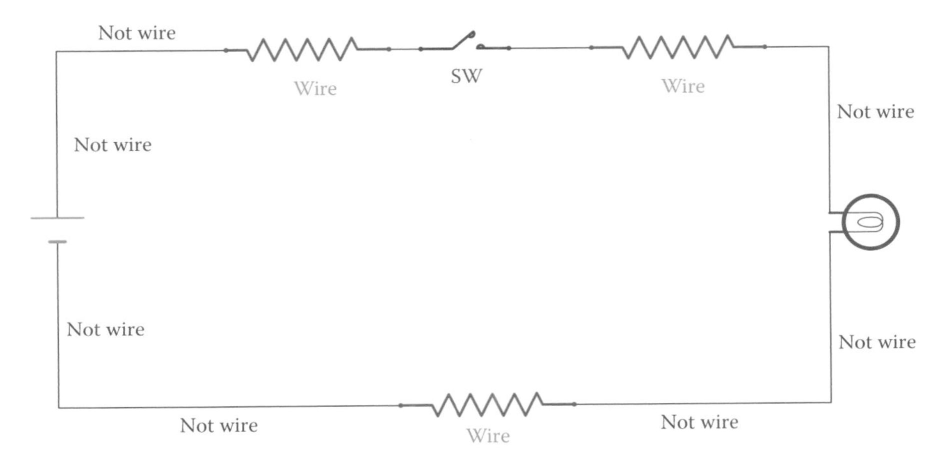

Figure 4.12 Representation of the electric circuit in Figure 4.11.

a resistor denoting the value of the resistor (see Section 4.10), and the lines connecting various elements are just drawing lines (showing the order of connections) and they do not represent wires. This is very important, and you must pay attention to it all the time.

In Chapter 6 we discuss how to combine all the resistors in the circuit of Figure 4.12 and show them by only one resistor. When the particular function of a device is not of primary importance to the study of the circuit, a device is shown as a load to the electric source in the circuit. In this sense, the electric circuit consists of a load and a source. The source must be capable of driving (energizing) the load.

4.6 Electric Flow

Electricity is the flow of electrons in a circuit, as we learned earlier. The electric source provides the supply of electrons and makes them move in a circuit. In this sense, if a circuit has no source, then no electricity, and thus no electric flow, can exist. Also, if a circuit is open (it does not form a loop), there is no path for electrons to flow. Thus, for electric flow we need a power supply and a closed loop. But, how fast is the electric flow and how fast do the electrons move and in what direction?

> For electric flow in a circuit, the circuit must be closed, forming a loop.

Electric flow in a circuit is very fast, almost as fast as the speed of light. But, electrons do not (and cannot) move that fast. Electrons move with a speed much slower, around a few inches per second (see Section 5.2). This implies that when we turn a light switch on, the transfer of electric energy to the light is immediate, and we do not need to wait until the electrons start moving and get to the light. That is to say, although electrons can only move a few inches per second, when they get energized, they transfer their energy to the neighboring electrons and this process continues in a chain reaction. As a result, the load receives energy immediately.

As it is universally accepted, electrons have negative charge. Also, as you know, a battery, as a power source, has two terminals, positive and negative. A power source pushes the electrons inside a circuit, and because the circuit must be closed, electrons must circulate. There are two conventions that are used to define the direction of the electric flow in a circuit. Both are valid and can be used in practice. Nevertheless, one should always follow one of them to avoid confusion. These two conventions are

1. Electric flow is in the same direction as the flow of electrons; that is, from negative side toward the positive side.
2. Electric flow is assumed to be from the positive side to the negative side. This latter is the conventional direction of electric flow.

> In this book we employ the conventional electric flow. That is, electric flow is from positive toward negative.

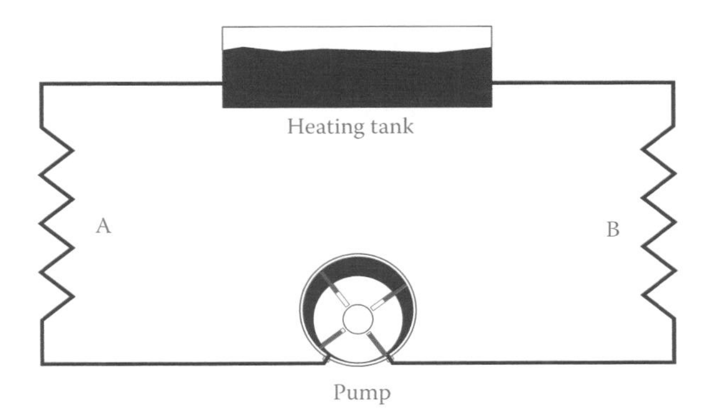

Figure 4.13 Analogy of a pump with an electric power source.

The role of a power supply in an electric circuit can be compared to the action of a water pump in a closed circuit (like in the hot water heating system that the pump is used for circulating water). A pump pushes water to flow. This analogy also can make understanding electric flow and its rules and relationships easier. Think about a pump. It pushes the water by pressurizing it. (So, there is a pressure difference between the two sides of the pump.) Also, there is a flow rate that depends on the pump pressure and the piping that water must flow through. See Figure 4.13.

4.7 DC and AC

We start making a distinction between the two types of electricity in use. If you have not already noticed the difference between the electricity you get from a battery and the electricity used at home, this is the starting point to realize their differences. There are two distinct types of electricity that they have similarities and differences. Their similarity is that they are both electricity, and certain definitions, rules, and formulations associated with their application are the same. Their difference is in the way they are generated, their effects, and most of the relationships associated with their use. The two categories are DC and AC.

DC stands for **direct current** and AC stands for **alternating current** or **alternative current** (both terms are used). In direct current, electrons move only in one direction, from the negative terminal of the power supply through the external circuit to the positive terminal of the power supply. That is, as long as the circuit is closed, there is a continuous motion of electrons all going in one direction. In alternative current, electrons have a back and forth motion. That is, they move in one direction, but after a short period they change direction and move in the opposite direction. They do this continuously as long as the circuit is closed.

Note that it is the motion of electrons that delivers electric energy to a device; thus, for alternating current, as long as the electrons are in motion, the electricity flow continues and its effect can be observed.

Direct current: Type of electricity in which there are positive and negative poles (or sides) and electric current is always in the same direction between the two poles, as opposed to alternative current in which the current direction continuously changes (at a fixed frequency).

Alternating current: Type of electricity in which current continuously and regularly changes direction (i.e., electrons in the wires move back and forth, as opposed to direct current in which electrons move only in one direction).

Alternative current: Same as alternating current.

In the analogy between a pump circulating a liquid and electricity, you can imagine that for direct current the pump always rotates in the same direction and sends the fluid inside the pipes in one direction, whereas for alternating current it alternately switches direction and sends the fluid into the pipes in two directions. For example, as shown in Figure 4.13, for one minute it circulates the fluid clockwise and in the following minute it circulates the fluid counterclockwise.

Normally, in a fluid circuit this is not done. However, to better understand the meaning of alternating current, this assumption is made. Note that the pump used in this circuit is a vane pump, which, in principle, can pump in both directions (other pumps can usually work in one direction). In the circuit illustrated in Figure 4.13, assume that water is heated in the tank and the purpose of circulation is to send hot water to the coils at A and B. We want to show that for the purpose of heating sites A and B, it does not matter in what direction the water circulates. It can also circulate first in the clockwise direction and after a while in the counterclockwise direction. As far as heating is concerned, the direction of circulation and its switching are immaterial. This is exactly what happens in the alternating current. For many (but not all) applications the direction of current does not matter, and more importantly, if the current continuously changes direction, as far as a current exists, we can benefit from the electric energy. For instance, in an electric heater or a lightbulb, the energy in the moving electrons changes to heat. The direction of the motion of the electrons does not play any role.

In the alternating current the interval between switching direction must be constant; otherwise, there will be a chaos. This interval is not long and is much smaller than a second. The number of switches of direction per second is a property of alternating current electricity and plays a significant role. The electric supply at home is alternating current. In North America the number of switching directions is 120 times per second. In European and many other countries it is 100 times per second.

4.8 Electric Potential and Electric Current

Similar to a fluid flow, there are two parameters associated with electric flow in an electric circuit. These two parameters are of paramount importance and must be well understood. Figure 4.14 helps to better understand these two parameters. Four scenarios are shown in Figure 4.14 for a water stream. In these scenarios, water falls from a higher level to a lower level. This level difference is necessary; otherwise, there will be no flow. As can be seen, in two cases the level difference is large, and in the other two it is small. Also, the amount of water flowing in two cases is large and in the other two cases is small. For electricity we refer to the level difference as **potential difference**, and it is always this relative difference that causes an electric flow. As depicted in the four scenarios, water can have a large level difference but a small flow rate, or a small level difference but a large flow rate, and so on. The flow rate of electricity is called **current intensity** or just **current**. Current determines the rate of electric charge that moves in a circuit. Because electric charge is associated with

Potential difference: The difference in intensity in an electric field or various points in an electric circuit, also called voltage. The potential difference between the + and – terminals in a car battery is around 12 V.

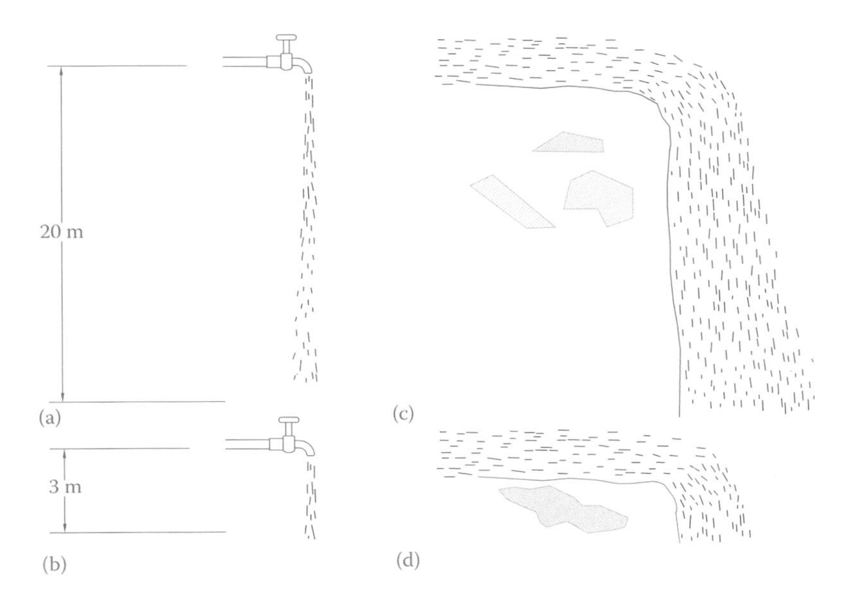

Figure 4.14 Analogy of water level and current with electric voltage and current. (a) High potential difference, low current. (b) Low potential difference, low current. (c) High potential difference, high current. (d) Low potential difference, high current. (From Hemami, A., *Wind Turbine Technology*, 1E ©2012 Delmar Learning, a part of cengage Learning Inc. Reproduced with permission from http://www.cengage.com/permissions.)

electrons, current is proportional to the number of electrons that move in one second.

A simpler and more widely used term for potential difference is **voltage**. The voltage (or potential difference) determines how much electromotive force is behind the electrons to push them in a circuit. Thus, it depends on the electricity source. For example, the voltage of a small battery (e.g., AA or AAA size) is small, and the voltage of electricity at the wall outlet is much larger. The current, however, depends on how easy or difficult the electrons are allowed to move in a circuit; thus, it depends on the circuit.

Voltage: A main property of electricity representing the intensity of electric charges based on accumulation of electrons. Normally a voltage difference between two points defines the potential for a discharge. Voltage is measured in volts.

> Electric current is a measure of the electricity flow rate; it is proportional to the number of electrons that move in 1 sec. It is not the speed of electrons.

To measure the voltage and current, we need to have units for their measurement. Voltage is measured in **volt**. On a small battery you see the writing "1.5 V," where V represents volt. Also, if you look at a lightbulb that you use at home, you will see "120 V." You probably know that one cannot light up this lightbulb with a battery. It is because of this voltage incompatibility. There is one more reason for not being able to light up the above lightbulb with a 1.5 V battery, which we will discuss in Chapter 5.

Volt: The unit for measurement of voltage (or potential difference) in an electric source or load or between two points in an electric circuit. The voltage in a pen size battery is 1.5 V.

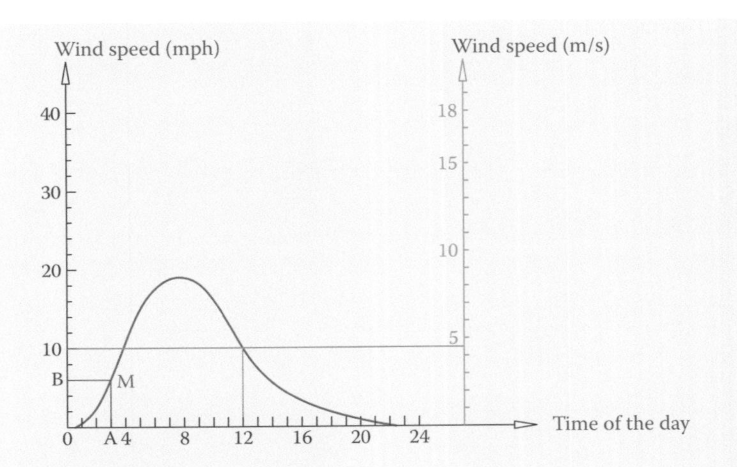

Figure 4.15 Example of a graph.

Graphs and Graphic Presentation of Variables

A graph is a convenient way of illustrating the variation of the values of an entity in terms of one (or more) parameters. For example, the variation of the temperature versus time for a day can be represented by a graph, or the area of a circle versus its diameter can be illustrated by a graph.

In the simplest form a graph has two axes: a horizontal axis and a vertical axis. The values of the changing variable (for example, time in the case of temperature) are shown on the horizontal axis and the values of the entity under study are shown on the vertical axis. Often the parameter on the horizontal axis is called the *independent variable*, whereas that on the vertical axis represents the *dependent variable*. Each pair of values of the horizontal and vertical axes define a point, which can be found from drawing two lines from those two points parallel to the axes. This is shown in Figure 4.15, where point M is obtained from point A on the horizontal axis and point B on the vertical axis.

The above graph shows the variation of wind speed during a period of 24 hours for a particular region on a specific day. As can be understood, wind speed is not the same when it blows, and can be zero, too (when there is no wind).*

By this graph one can see the wind speed for a given hour as well as the time for a given wind speed. The values for each can be read from the graduations on the axes. For example, the wind speed at 12 o'clock is 10 mph. Furthermore, this graph has two scales for the vertical axis. Not all graphs need to have this. Here, since wind speed can be measured in mile per hour (mph) or in m/s (meters per second) two scales is very useful. You can see that at 12 o'clock the wind speed is about 4.5 m/s. Notice also that a wind speed of 10 mph corresponds to two points on the horizontal axis. That is, at around 4:00 and at 12:00 the wind speed is 10 mph.

* This graph should not be confused with another one looking similar to this, in which the horizontal axis is the wind speed and the vertical axis shows the probability of wind at various speeds.

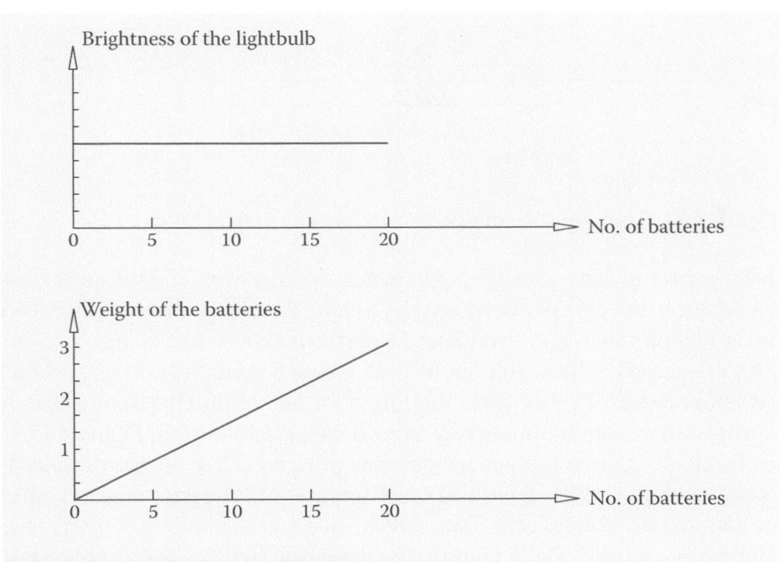

Figure 4.16 Examples of linear graphs.

Note that many graphs are obtained as a result of some calculation. This makes it faster for a user, since instead of performing the calculation one can read the answer from the graph. However, there will be some approximation and inaccuracy in this process.

Sometimes a graph is used only to see the pattern of variation of a variable. In Figure 4.16, for instance, the first graph shows that there is no change in the variable. The second graph (which is a sloped line) depicts that the variable of interest has a linear relationship with the independent variable.

In this example we have assumed that you have 20 similar batteries that you want to carry with you to a remote place for lighting purposes. Suppose that you put them in a box and connect all the positive terminals together and all the negative terminals together. The first graph shows that the brightness of a light connected to these batteries does not change, whatever the number of batteries. The second graph shows that as you increase the number of batteries the weight of the box increases. You can decide how many batteries you take if you cannot carry all.

Although, as mentioned earlier the electric current is proportional to the number of electrons that pass in a circuit during a one second period, this is not a convenient unit for measuring the electric current. This is, first, because of the extremely large number of electrons that move and, second, because there is no way of counting those electrons. A better unit for measuring current is **ampere**, abbreviated **amp** and denoted by the capital letter A, e.g., 2A (2 amperes, or 2 amps), 3A, and so on.

Ampere: Unit to measure electric current.

Both voltage and current can be measured by the appropriate measuring devices. (See Sections 5.5 and 5.6 for how to measure the current and voltage in a circuit.)

Now that we have learned about voltage, we may graphically illustrate the difference between DC and AC. Graphical illustration of an entity is a very useful and convenient method for understanding and picturing its variation

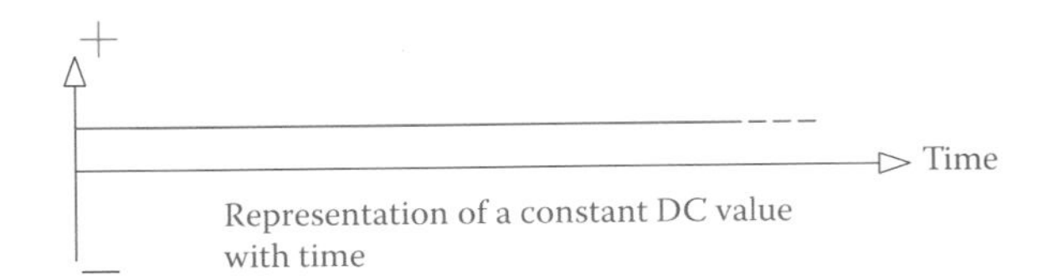

Figure 4.17 Graph of the voltage in a DC source versus time.

Waveform: The form (shape) of variation of a signal (electric or other) that varies with time. The shape of a rectangular waveform, for example, is made up of succeeding rectangles.

Sinusoidal: The most common and preferred waveform for alternating current electricity. The variation of a sinusoidal waveform is according to the values of sine (or cosine) of an angle for 360° (one revolution) change.

Cycle: Pattern of all values that will be continuously repeated in a cyclic variable. For example, in alternative current electricity a cycle is the complete set of positive and negative values based on a sine function of an angle when it varies from zero to 360°.

with respect to some changing parameter, or comparing it with some other variables. In the case of electricity, it is usually the *time* with respect to which we study the values and variation of a particular entity like voltage and current in a circuit. The beginning of time for such a graph is usually when a switch is closed. In this sense the graph for DC electricity from a battery during a short period of time (say a few hours) is as shown in Figure 4.17.

In an AC source the voltage changes polarity. This change of polarity is not random and has a desired order. Figure 4.18 depicts some examples of patterns for voltage variation, which, in general, can be categorized as alternating current. These patterns are, however, more used for special purposes in electronic devices such as in radio and TV. Normally, when alternating current electricity is addressed, particularly when comparing DC with AC, a **waveform** with a **sinusoidal** pattern, as shown in Figure 4.19, is referenced. The commercial AC electricity has this waveform. In other words, for AC electricity a sinusoidal waveform is desired, and it is always assumed that a source generates electricity with this pattern (see the text on "sinusoidal waveform" in this chapter). If the waveform of electricity is far from a sinusoidal pattern, corrections may become necessary.

As can be seen from all the illustrated waveforms, a pattern is repeated continuously with respect to time. Each repetition is called a **cycle**, and

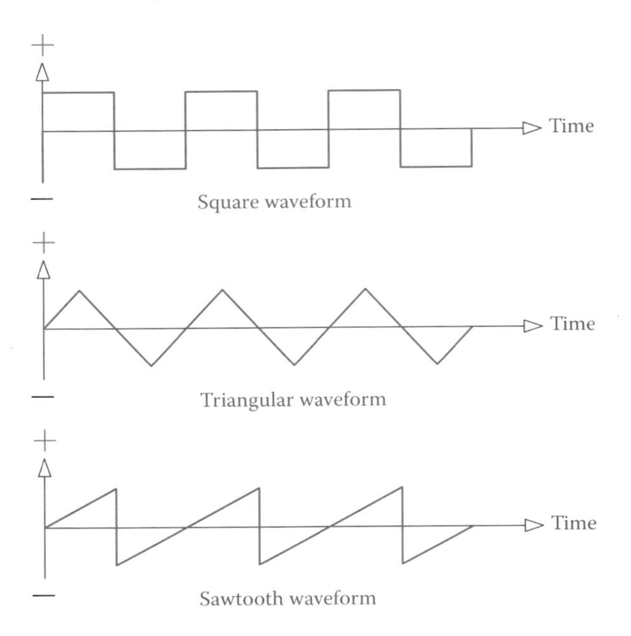

Figure 4.18 Various alternating current waveforms.

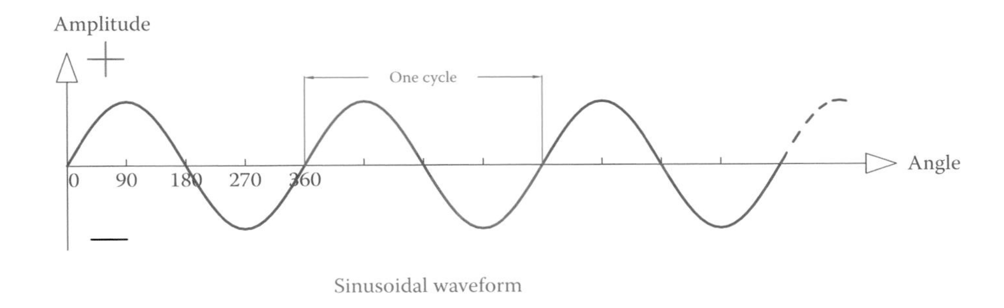

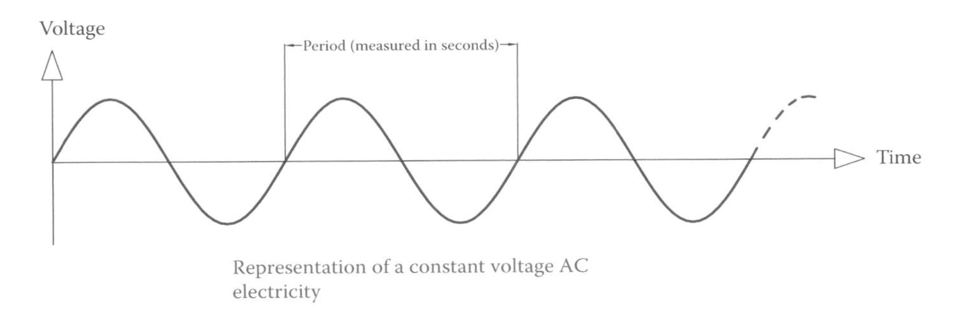

Figure 4.19 Sinusoidal alternating current waveform.

all the patterns shown in Figures 4.18 and 4.19 can be called **cyclic waveforms**. The duration for a cycle is called a **period**, measured in seconds. The magnitude of a cyclic waveform is called **amplitude**. Here only three repetitions are shown, but the pattern is assumed to continue infinitely. Because the same pattern is repeated, in order to denote or study the details of a waveform, only one cycle is sufficient.

Cyclic waveforms: Waveform corresponding to a variable entity whose variation has a repeat of the same pattern over and over.

Sinusoidal Waveform

A sinusoidal waveform is as shown in Figure 4.19. The waveform has alternative positive and negative values that form two equal areas between the curve and the horizontal axis. The important fact about the curve in this waveform is that the values at each point on the curve are obtained from the sine function [check your calculator for the sin(e) function]. Sine of an angle is a value corresponding to that angle. Each cycle of a sinusoidal waveform corresponds to 360°, which is one revolution. If you enter all angles between zero and 360° and plot a graph of the sine values of these angles, you get a curve similar to one cycle of what you see in Figure 4.19.

In order to understand better the meaning and the application of sinusoidal function consider a frame in the stream of wind, as shown in Figure 4.20. If you are interested to see how much air passes through the frame at any angle, this is where the effect of sine function comes into play. For example, when the angle between the direction of wind and the frame is zero (case B), no air passes through the frame. Case A (frame perpendicular to the air flow) corresponds to maximum amount of air passing through the frame. Then between zero and maximum, the flow of air depends on the angle. In case A the angle is 90°, and in case B it is zero. Now from the value of the sine function you can see that at an angle of 30° the amount of air flow is half of the maximum flow (sin 30° = 0.5).

There are many other applications similar to this example. For instance, if the frame is a solid plate, the amount of pressure from air on the plate is maximum in position A, zero in position B, and so on.

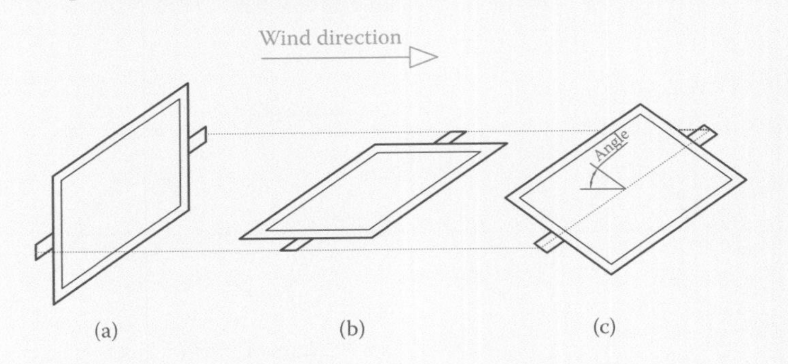

Figure 4.20 Physical example representing sinusoidal function: the amount of a fluid (for instance, air) passing through a frame. (a) Maximum flow. (b) Zero flow. (c) Some amount between zero and maximum, based on the angle.

Frequency: The number of repetitions per second of any cyclic phenomenon. In AC electricity, the number of cycles per second for alternating current.

Hertz: Cycles per second. The unit for measurement of frequency.

AC circuit: Any electrical circuit with alternating current as the energy source.

DC circuit: Electric circuit with only DC source(s).

In alternating current the number of cycles that happen in one second is called **frequency**. Frequency plays an important role in AC electricity and all the apparatus that work with alternating current. Frequency is measured in cycles per second or **Hertz** (Hz); for instance 50 Hz and 60 Hz. Frequency implies how many times per second the direction of current changes. For example, in 60 Hz electricity (North American standard) the current direction changes 120 times in each second.

As mentioned earlier, AC and DC devices cannot be mixed up. In this sense, when the power supply in a circuit is DC the circuit is called a **DC circuit** and when it is AC the circuit is called **AC circuit**.

4.9 Resistance to Electric Flow

In an electric circuit a wire is the path for the flow of electrons. This path can be an easy way or it can exhibit some form of difficulty for the electrons to move, and, thus, the electrons must struggle to pass and spend a lot of energy. The more difficult an electric path is, the more resistance the path has. This can be regarded as analogous to a pumping circuit, as shown in Figure 4.13. In Figure 4.13 the resistance of the straight section of pipe is smaller than the resistance of the section in the form of a zigzag. In electric circuits the wires between the source and the electric device or between the switch and the electric device are equivalent to the straight part of the pipes shown in the pumping circuit. They represent the easy path. An electric device, however, is designed to extract the electrical energy, and, thus, it has much more resistance for the electrons to pass. In this way the electrons must give away a lot of their energy. The resistance to electric flow is a measure of the difficulty in a path encountered by electrons to pass.

Resistance to electric flow, called **electric resistance**, or just **resistance**, is represented by the letter R. A device can have a large resistance or a small resistance. Also, a small length of a wire has a very small resistance that can be ignored, but the resistance of a few miles of the same wire cannot be ignored.

Resistance is measured in **ohm**. In other words, ohm is the unit of measurement for resistance; e.g., 1 ohm, 10 ohms, and 5000 ohms. If in a circuit the resistance is 300 ohm, then we may write $R = 300$ ohms. Instead of writing "ohm," a symbol is used for this unit of measurement. This symbol is the Greek letter Ω (equivalent to capital W), called **omega**. For instance, we may write $R = 300\ \Omega$. The lowercase ω (equivalent to w) is not used for this purpose. We discuss this and how much 1 Ω is in Chapter 5.

Ohm: The unit for measurement of the electric resistance and electric reactance.

Omega: The name for the Greek letter Ω/ω (corresponding to W/w). The upper case omega (Ω) is used to represent ohm (for example, 5 Ω, 20 Ω, 1000 Ω, etc.).

Normally, a lightbulb filament and a heater or stove element has resistance, which does not matter whether it is connected to AC or DC and at the operating temperature has a constant value (its resistance is constant).

Higher resistance in a circuit implies higher obstruction to electric current. Thus, in general, in conjunction with higher resistance we have lower current. This relationship will be further studied in Chapter 5.

As we will see later, all wires have resistance and can behave like (small or large) resistors. Because all electric circuits have wires, it is not possible to have an electric circuit with no resistance.

4.9.1 Resistance of a Wire

The resistance of a wire is a function of three parameters: the length, cross-section area, and material. The resistance of a wire is directly proportional to its length. That is, if two wires are made of the same material and have equal thicknesses, but the length of the first wire is twice as much as the length of the second, the former has twice as much resistance. The resistance is inversely proportional to the thickness; thus, if two wires have the same length and are made of the same material but one has twice as much cross-section area as the other, the latter has twice as much resistance as the first one.

The above facts, together with the effect of the material, can be put in a formula as follows:

$$R = \frac{\rho l}{A} \tag{4.1}$$

where R is the resistance in ohm, l is the length, and A is the cross-section area. The Greek letter ρ (pronounced rho) is the representation of the material. It is called **specific resistance** and is the resistance of a piece of material having a certain specified dimension. Because in the metric and imperial system the units are different, ρ has different value in each system of measurement. See the next section, before moving on.

Specific resistance: Same as resistivity the electric resistance of a specific size (based on the measurement system) of a metal or material.

Specific Resistance of Materials

To find the electric resistance of a wire or the resistance that a nonmetallic part introduces to electric current, we should know the specific resistance of the material for the wire (part). Specific resistance is the electric resistance (inside a circuit) of a piece of the material with specified dimensions at a specified temperature. Formally, the dimension must be one unit length with a cross-sectional area of one unit. But, practically, some other values are used, as described below. Thus, its value and measurement unit depends on what system of measurement is employed. The specific resistance is shown by the Greek letter rho (ρ), which is equivalent to the English letter r.

Metric system: In the metric system, specific resistance is defined as the resistance value (in ohm) of a piece of the material whose length is 1 cm and whose cross-sectional area is 1 cm^2 at a temperature of 20°C. (See Chapter 2 for the definition of measurement units.) Figure 4.21 shows a piece of metal with these dimensions. According to Equation 4.1, thus l = 1 cm and cross section A = 1 cm^2.

Because resistance is measured in ohm (Ω), the specific resistance has a unit such that when multiplied by length and cross-section area, with their respective units, the resultant is ohm. Thus, its unit in metric system is a quantity that if multiplied by centimeters and divided by centimeters squared, the resultant is ohm. That is,

$$\text{Unit for specific resistance} \times \frac{\text{cm}}{\text{cm}^2} = \text{ohm} \rightarrow \text{specific resistance unit} = \text{ohm.cm}$$

Alternatively, the value is frequently given in ohm.m (Ω.m), which is 1/100th of the value in ohm.cm because 1 cm is replaced by 1/100 m. The values of specific resistance for a number of substances are given in Table 4.1.

Imperial system: In the imperial system, specific resistance is defined as the resistance of a piece of a material whose length is 1 ft and whose cross-sectional area is 1 **circular mil**. A circular mil (CM) is a unit for measurement of (small) area, specifically, wire thickness (see Appendix D for circular mil). It is the area of a circle

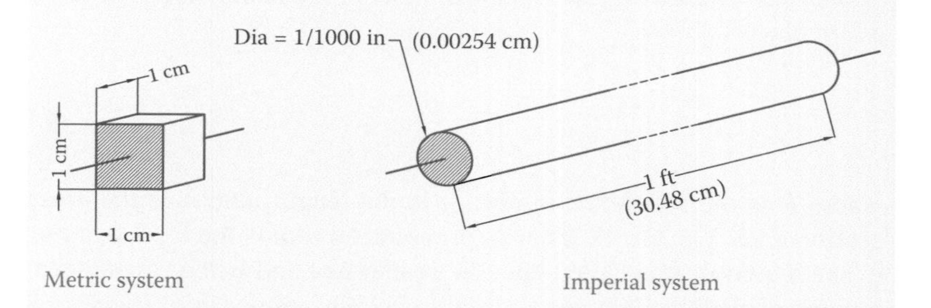

Figure 4.21 Physical dimensions of a wire for definition of specific resistance in metric and imperial systems.

TABLE 4.1 Comparison of Specific Resistance of Some Metals and Insulators

Material	Ohm-CM/ft	Ohm-cm
Aluminum	15.94	2.65×10^{-6}
Constantan	272.97	45.38×10^{-6}
Copper	10.09	1.678×10^{-6}
Gold	13.32	2.214×10^{-6}
Iron	57.81	9.61×10^{-6}
Lead	13.23	2.2×10^{-6}
Molybdenum	32.12	5.34×10^{-6}
Nickel	41.69	6.93×10^{-6}
Platinum	63.16	10.5×10^{-6}
Silver	9.546	1.587×10^{-6}
Tungsten	31.76	5.28×10^{-6}
Zinc	35.49	5.90×10^{-6}
Carbon (graphite)	15–30	2.5×10^{-6}–5×10^{-6}
Air	7.82×10^{22}–19.85×10^{22}	1.3×10^{16}–3.3×10^{16}
Glass	6×10^{16}–6×10^{20}	10^{10}–10^{14}
Rubber	6×10^{19}	10^{13}
Teflon	6×10^{28}–6×10^{30}	10^{22}–10^{24}

whose diameter is 1 mil. 1 mil is 1/1000th of an inch. Thus, 1 circular mil is

$$1/4 \times \pi \times (1 \div 1000)^2 = 3.14/4 \times 1 \div 1{,}000{,}000 = 0.785 \times 10^{-6}\ \text{in}^2$$

The unit for the specific resistance in the imperial system of measurement is

$$\text{Unit for specific resistance} \times \frac{\text{ft}}{\text{CM}} = \text{ohm} \rightarrow \text{specific resistance unit} = \text{ohm.CM/ft}$$

Make sure you do not mix up circular mil CM (uppercase) with centimeter cm (lowercase).

Example 4.1

The specific resistance of a tungsten heating element at its operating temperature is 492 ohm.CM/ft. What is the resistance of the element if its length is 2 ft and its thickness is 100 CM?

Solution

Substituting for ρ, l, and A in Equation 4.1 leads to

$$\frac{492 \times 2}{100} = 98.4\ \Omega$$

Example 4.2

The specific resistance of copper is 1.678×10^{-6} ohm.cm. Find the resistance of 250 m of a copper wire if its thickness is 1.31 mm^2.

Solution

Because the specific resistance is specified in ohm.cm, we need to convert both the length and the thickness in cm and cm^2, respectively.

$$l = 250 \times 100 = 25{,}000 \text{ cm}$$

$$A = 1.31 \div 10^2 = 0.0131 \text{ cm}^2$$

$$R = 1.678 \times 10^{-6} \times \frac{25{,}000}{0.0131} = 3.2\ \Omega$$

4.10 Inductance

Inductance: A property associated with the magnetic effect of electricity, exhibited by a winding carrying electric current. The magnetic field generated by a winding is proportional to the winding inductance.

Henries: Plural of Henry. Henry is the unit of measurement of inductance of an inductor.

Inductance is a property observed in a wire only when it is in the form of a winding. A long piece of wire has no inductance if straight, only resistance. However, if it is in the form of a winding, then, in addition to resistance, it has inductance. Inductance deals with the magnetic property of electricity. In a wire winding, each turn creates a magnetic field and these magnetic fields are added together, creating a larger magnetic field (see Section 7.2). In order to generate such a magnetic field, some electric energy must be spent. Inductance defines a parameter that relates the energy necessary for generation of a magnetic field and physical properties of a winding. Energy consumed for magnetization is, in fact, stored in the winding in the form of a magnetic field.

The symbol for representing inductance of an inductor is L (always capital), and the unit for its measurement is **henries**, which is denoted by H (capital). For instance, we may write $L = 0.01$ H, $L = 5$ mH (millihenry), etc.

Factors that influence inductance are the length of a winding (not the length of the wire), the cross-sectional area of the winding, the number of turns in the winding and the material of the core (if there is a core). These can be put in a formula:

$$L = \frac{\mu N^2 A}{l} \tag{4.2}$$

where l is the common length of the winding and its core, A is the (mean) cross-sectional area of the winding, N is the number of turns, and the Greek letter mu (μ) introduces the effect of the core (see "Permeability"). Because a wire has thickness or a winding may consist of several layers of wire on top of each other, it is necessary to use the average (mean) cross section for A. Figure 4.22 shows an inductor with only five turns.

We learn more about the effect of inductance in a circuit in Section 8.5.

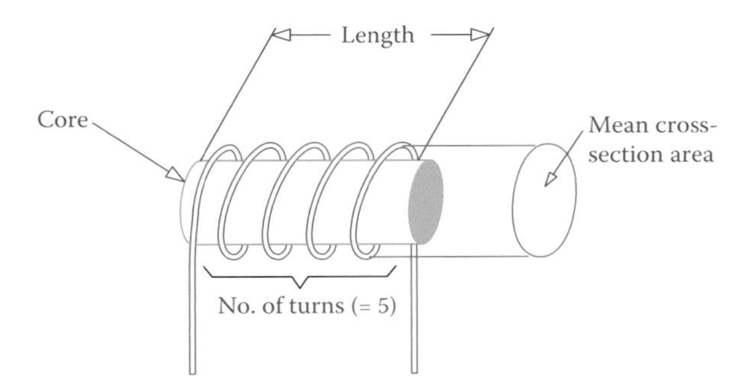

Figure 4.22 Definition of parameters in an inductor.

Example 4.3

The mean (average) diameter of a cylindrical 100 mm long winding of 500 turns is 10 mm. It has a high permeability core for which $\mu = 2.5 \times 10^{-2}$ H/m. Find its inductance.

Solution

The given values can be substituted in Equation 4.1 after conversion

$$\text{Length } l = 0.1 \text{ m}$$

$$\text{Cross section } A = 1/4 \times \pi \times (10 \div 1000)^2 = 0.785 \times 10^{-4} \text{ m}^2$$

$$L = \frac{(2.5 \times 10^{-2})(500)^2(0.785 \times 10^{-4})}{0.100} = 4.9 \text{ H}$$

Example 4.4

The mean diameter of a cylindrical winding is 1 in, and its length is 2 in. If the winding has 300 turns, find the inductance of the winding. Winding has a steel core for which $\mu = 8.75 \times 10^{-4}$ H/m.

Solution

Note that μ has a very small value. Also, pay attention to the fact that its value is in H/m (henry/meter). Thus, the dimensions of the winding must be converted to metric system. Finally, notice that the unit for μ is such that if multiplied by area and divided by length the resultant will be in henries.

The cross section and the length of the winding in metric system of units are

$$l = 2 \div 39.37 = 0.0508 \text{ m} \qquad (1 \text{ in} = 1 \div 39.37 \text{ m})$$

$$A = 1/4 \times \pi \times (1 \div 39.37)^2 = 5.067 \times 10^{-4} \text{ m}^2$$

From Equation 4.2

$$L = \frac{(8.75\times10^{-4})(300)^2\ (5.067\times10^{-4})}{0.0508} = 0.785\ \text{H}$$

Permeability

Many electrical devices such as motors and transformers that we will discuss later have windings. The role of a winding is concentrating the magnetic effect of electricity. A winding is a wire in the form of a coil with many loops (turns).

Almost all windings, especially those in large sizes, have a metallic core. It is *very important* that this core be made from the category of metals that are **ferromagnetic**. Otherwise, the core has no or little effect. The ferromagnetic class of metals is those in the family of iron and steel that have magnetic effect (can be magnetized). Metals such as gold, copper, and aluminum have no such effect and in this respect cannot be used. The effect of a ferromagnetic core is to increase (by thousands of times) the magnetic lines generated by a winding when electric current passes through it. This property of a winding is called inductance. The more the inductance of a winding, the greater the magnetic effect.

This property of metals to enhance the magnetic effect is called permeability. Permeability of a core inside a winding directly affects the inductance of the winding. Permeability of different materials is compared to each other by relative permeability, which is a nondimensional number and specifies the permeability of various materials with respect to vacuum. Table 4.2 shows the relative permeability of some materials.

TABLE 4.2 Comparison of Permeability of Some Materials

Material	Relative Permeability
Mu-metal (a highly magnetic alloy)	20,000
Permalloy	8000
Electrical steel (special steel for motors rotors, etc.)	4000
Regular steel	100
Nickel	100–600
Aluminum	1.000022
Platinum	1.000265
Wood	1.00000043
Air	1.00000037
Vacuum	1
Copper	0.999994
Superconductor	0

4.11 Capacitance

Capacitance is a property of a **capacitor**. A capacitor can store energy, but "how much electricity can be stored in a capacitor?" depends on its capacitance. A capacitor stores energy in the form of an electric field (an inductor stores it in the form of a magnetic field). In other words, capacitance is a measure of the electric capacity of a capacitor.

The uppercase C is used to represent capacitance, and the unit of measurement for capacitance is **farad** (denoted by F). For instance, for a particular capacitor we may write $C = 51 \times 10^{-6}$ F. One farad is a large value, and usually its fractions are used instead, like 51 microfarad (51 μF).

Basically, a capacitor is made of two conducting surfaces (metals) separated by a layer of an insulated material between them (it could also be air or another gas). The capacitance of such a capacitor is directly proportional to the common area of the two surfaces and inversely proportional to the distance between them. Also, the insulator material adds its own effect because some materials are better insulators than the others. Thus, capacitance can be expressed by

$$C = \frac{kA}{d} \tag{4.3}$$

where C is the capacitance in Farad, A is the shared area of the two conductors, and d is the distance between the surfaces of the two plates, as shown in Figure 4.23. The effect of the insulator is represented in k, called the dielectric constant of an insulator, the value of which also depends on the units of measurement (metric or imperial). The value of the constant k is measured in F/in or F/m. It can be seen that varying the common area by moving one plate with respect to the other changes the capacitance. This is used in the older radios, where a tuning knob was turned for selecting a station.

Capacitance: Measurable property of a capacitor for storing electricity, measured in Farad.

Capacitor: Electric device mainly made of two conductors separated by an insulator. It has the property of storing electric energy in the form of an electric field. Energy stored in a capacitor can only be discharged instantly, not like a battery in which the stored energy can be gradually discharged.

Farad: Unit for measurement of capacitance of capacitors. Farad is a relatively large unit and fractions of it such as microfarad and picofarad are commonly used.

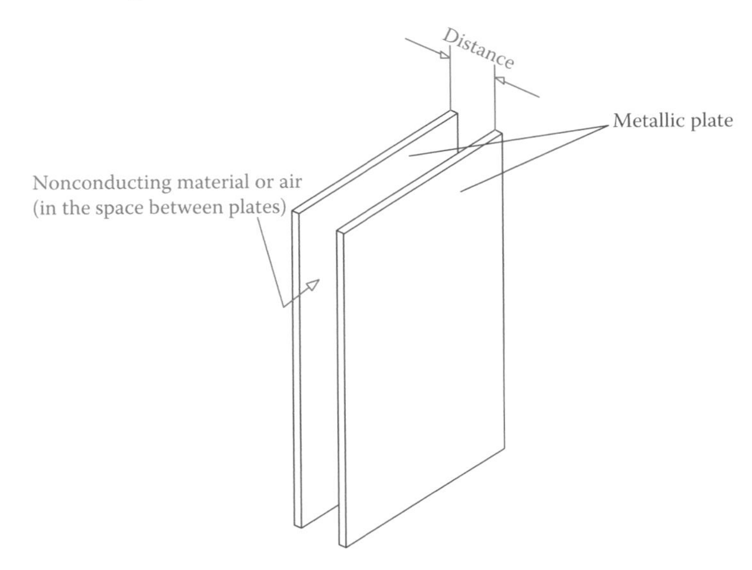

Figure 4.23 Basic parameters of a capacitor.

TABLE 4.3 Dielectric Constant of Air = $k_0 = 8.85 \times 10^{-12}$ F/m

Material	k/k_0
Air	1
Liquid air	1.5
Glass	3.8–14.5
Nylon	3.5–22.4
Oil (cotton seed)	3
Oil (heavy oil)	3
Oil (mineral)	2.1
Paint	5–8
Paper (dry)	2
Porcelain	5–7
Quartz	4.7–5
Rubber	2–4
Rubber (hard)	2.8
Sand (dry)	5
Silicon	11–12
Soil (dry)	2.4–2.9
Tantalum oxide	11.6
Teflon	2.1
Water	4–88
Water (pure)	88
Wax (mineral)	2.2–2.3
Wood	2–6

For various materials the constant k is defined in terms of the relative (to air) dielectric constant. For air this value is approximately

$$k_{air} = 8.85 \times 10^{-12} \text{ F/m}$$

Table 4.3 shows the relative dielectric constants for a number of selected materials.

Example 4.5

A capacitor is made up of two aluminum foils separated by a piece of paper, all three rolled together. The paper acts as an insulator between the aluminum sheets. In some capacitors the paper is wetted by oil for longer life. For dry paper (not oiled) the value of k is 0.356×10^{-12} F/in. If the dimensions of the two aluminum foils are 1 inch wide and 10 inches long (before rolling), and the thickness of the paper is 0.002 inches, what is the capacitance of the capacitor?

Solution

All the values are given in imperial system. Thus, we may directly plug the values into Equation 4.3.

$$C = 0.356 \times 10^{-12} \times \frac{10.0}{0.002} = 1.78 \times 10^{-9}\text{F} = 1.78\ \text{nF (nanofarad)}$$

Recall from the conversion table in Chapter 1 that nano is the prefix in metric system for 10^{-9}.

Example 4.6

In the metric system the constant in Equation 4.3 is 17.7×10^{-12} F/m. If two plates each having an area of $25 \times 250\ \text{mm}^2$ are separated by a piece of dry paper 0.1 mm thick, what is the capacitance of the capacitor made out of these plates?

Solution

First, we need to convert the dimensions into meters. Thus,

$$d = 0.1\ \text{mm} = 0.0001\ \text{m}$$

$$A = 25 \times 250\ \text{mm}^2 = 0.025 \times 0.250\ \text{m}^2 = 0.00625\ \text{m}^2$$

$$C = 17.7 \times 10^{-12} \times \frac{0.00625}{0.0001} = 1106.25 \times 10^{-12} = 1.10625 \times 10^{-9}$$

$$\text{F} \approx 1.1 \times 10^{-9}\,\text{F} = 1.1\text{nF}$$

4.12 Chapter Summary

- Electricity is the flow of electrons.
- For a continuous flow of electrons it is necessary to have a closed circuit.
- A pressure difference (potential difference), called electromotive force, is required to cause the flow of electrons in a circuit.
- Potential difference is also called voltage.
- Current is a measure of how many electrons move in one second. However, for measuring current a different unit, an ampere, is used.
- Electric current is not the speed of electrons in motion.
- All electric devices function based on one or more of the three main properties of electricity: thermal, magnetic, and chemical effects.
- Potential difference is measured in volts.
- Electric energy can exist in two forms: direct current and alternating current.
- In direct current the flow of electrons is always in the same direction.

- In alternating current, electrons continuously alter their direction of movement.
- Direction of current is a convention (arbitrarily selectable). In this book, the positive direction of current is considered to be from positive side to negative side (in a circuit). This is from higher voltage to the lower voltage.
- An electric circuit is made up of a source and some loads connected to the source. A source provides electricity and a load consumes electricity.
- In general, loads can be of three basic types: resistor, inductor, and capacitor. A resistor converts electric energy into heat. Inductors and capacitors store electric energy momentarily.
- A resistor has resistance, an inductor has inductance, and a capacitor has capacitance. Resistance, inductance, and capacitance can be calculated from their corresponding formulas. In physical circuits they can be measured.
- All wires have resistance and behave as resistors. Because in an electric circuit all parts are interconnected by wires, there is no electric circuit without resistance.

Review Questions

1. What is electricity?
2. What causes electric flow?
3. How many types of electricity can you name?
4. What is direct current electricity?
5. What is alternative current electricity?
6. What is alternating current electricity?
7. What is the simplest electric circuit composed of?
8. What is a source in an electric circuit?
9. What are the basic components in any electric device or circuit?
10. Does an electric circuit contain all the basic components? Explain.
11. Describe resistance in an electric circuit.

12. Describe a resistor.

13. What is an inductor and what does it do to electricity?

14. What is a capacitor and what role does it play in electricity?

15. What is the physical form of an inductor? What is it made of?

16. What is the construction of a simple capacitor? What are the parameters that determine the capacitance of a capacitor?

17. What is voltage and what is the unit for measuring it?

18. Using the analogy of water flow, what represents voltage?

19. What is the unit for measuring electric current?

20. Name the units of measure for resistance, inductance, and capacitance.

Problems

1. A piece of copper wire is 40 ft long. If the resistance of this piece of wire is 0.2 Ω what is the resistance of a piece of the same wire that is 100 ft long?

2. The resistance of a piece of wire is 1 Ω. The cross-section area of the metallic core of the wire is 0.05 mm^2. What is the resistance of the same length of a wire of the same material, but with a cross-section area of 0.2 mm^2?

3. You have a 100 ft wire. The resistance of this wire is 1 Ω. You cut this wire from the middle and twist the two parts together along the entire length (so that you have 50 ft of double wire). What is the resistance of the double wire?

4. In Problem 3, if you divide the 100 ft wire into four equal segments and wrap all four pieces together with their ends twisted together, what is the resistance of the quadruple wire?

5. A coil has 500 turns and its inductance is 0.4 H. If you cut the coil from the middle length, so that each half has 250 turns, what is the inductance of each half coil?

6. The coil in Problem 5 must be rewound so that it has half-length for the same wire. In this process the cross-section area increases by 10%. What is the new value for its inductance? Also, what is the change in its resistance?

7. A winding with 400 turns has a steel core for which $\mu = 8.75 \times 10^{-4}$ H/m. The mean cross section of the winding is 0.25 in^2 and its inductance is 200 mH (millihenries). What is the length of the winding in mm?

8. You put an 8 × 11.5 inch sheet of paper between two aluminum foils of the same size and roll them together to form an 8 inch height cylinder. Then you cut that into four pieces and solder two wires to each piece of aluminum foil in each piece. What is the capacitance of each home-made capacitor, if the paper thickness is 2 mil? (Hint: See Example 4.5.)

9. What is the capacitance if in Problem 8 you use two sheets of paper between the aluminum foils?

10. In Example 4.3, what is the resistance of the winding if it is made out of the wire in the Example 4.2?

11. If the wire in Example 4.2 cracks and a gap of 0.01 mm is created between the two wire segments, what is the capacitance of the capacitor formed this way?

5

Voltage, Current, and Power

New terms:
Ammeter, coulomb, digital multimeter (DMM), electric power, galvanometer, multimeter, Ohm's law, ohmmeter, voltmeter

OBJECTIVES: After studying this chapter, you will be able to

- Describe what the main entities of electric flow are
- Explain how current is measured
- Explain how electric voltage is measured
- Define the units of measure for current and voltage
- Know what an ammeter is
- Know what an ohmmeter is
- Know what a voltmeter is
- Describe a multimeter and its use
- Measure the voltage and current in an electric circuit
- Use an ohmmeter to measure the resistance of a resistor or part of a circuit
- Use an ohmmeter for testing capacitors and inductors
- Explain the factors defining electric power
- Explain the importance of each factor
- Perform calculations for basic circuits
- Calculate energy consumption of simple devices

5.1 Introduction

Practical understanding of electricity and its usage starts with a good comprehension of the meaning of electric current, electric voltage, and electric power, and the relationships between them. In this chapter these terms are defined, and before going further on, make sure that you have a clear understanding of the meaning of these terms and what they represent. At a practical level, one does not measure electric flow by either counting electrons or caring about what happens to the electrons. The tools used to measure electricity entities are graduated and calibrated in terms of units that are more appropriate for use rather than dealing with individual electrons and their electric charge.

In Chapter 4 we discussed electric current, voltage, and their unit of measure. In this chapter we will discuss how they must be measured in a circuit. As well, we will discuss their relationships with electric power. Understanding of power is quite important, because in many devices it is the *power* that determines or dictates other quantities.

5.2 Electric Current

In Chapter 4 we discussed that the electric current is a measure of the rate of flow (i.e., how many per second) of electrons. In fact, electric current is the rate of the electric charge of those electrons, because we are concerned about the electric charge, not the numbers, corresponding to electrons. If 6.241×10^{18} electrons move through a wire in 1 sec (i.e., if these many electrons pass a given cross section of a wire during a 1 sec period), the electric current is 1 amp (1 A). The amount of electric charge corresponding to this number (6.241×10^{18}) of electrons is called 1 **coulomb**. Thus, we may say a flow of 1 coulomb electricity in 1 sec is 1 A.

Coulomb: Measure of the amount of electricity equal to the electric charge of 6.241×10^{18} number of electrons.

For these many electrons to move, it is not necessary that their speed be high. It is the volume that counts more, because it is the amount of electric charge that is important. The speed of electrons does not play any role in electrical current, although electricity travels fast, close to the speed of light. The reason for electricity to move very fast is the simultaneous transfer of electric charge along a conductor. In fact, from a mechanics view, electrons cannot go very fast, because although very tiny, they still have mass and follow the rules of motion.

Yet you may not get a tangible feeling for how much 1 A of current is. This will gradually become clearer for you as we continue this discussion. Consider a lightbulb at home; on these, "110 V, 100 W" is written. See how bright (and hot) the filament is when it is connected to electricity. One amp current is around the electric flow rate giving that much heat and intensity to that lightbulb. To see how much it affects a human body, refer to Table 5.1, which shows that even a current of 0.5 A kills a person in less than a second if it passes through the heart. Note that the numbers in the table are in mA (1/1000th of amp).

The fuse box at many homes in North America has a capacity of 100 A. That means it is possible to have a maximum current of 100 A at home. Each regular switch at home is capable of carrying 15 A. From these numbers you can imagine what can happen if you touch the wires.

TABLE 5.1 Effect of Electric Current through Human Body

Current (mA)	Effect
Less than 1	No sensation
Less than 3	Mild sensation and possible sudden shock
3–10	Painful shock, let go current for most of people
10–15	Local muscle contraction, hands freeze to the conductor for some people
15–50	Loss of muscle control, freezing to the conductor, burns
50–100	Difficulty breathing, collapse, and unconsciousness; death for prolonged contact
100–200	Heart problem (ventricular fibrillation), and death if more than 1/4 sec
Over 200	Clamping action of the heart, respiratory paralysis, death if current is not stopped

5.3 Electric Voltage

Voltage is the electric potential that causes electrons to move around a closed circuit. Volt is the unit of measure for voltage. Volt is defined as the value of the potential difference for which the energy of one coulomb of electric charge (i.e., the charge of 6.241×10^{18} electrons) is one joule. Joule is a unit for measuring energy. This official definition of volt may not be much help to understand how much 1 V is. A better understating is possible by considering that each small dry battery you use in your battery-operated devices is 1.5 V, the car battery is 12 V, and the electricity at home is around 115 V. Also, lightning during a thunder storm has millions of volts.

You can touch the two sides of a small battery (1.5 V) without any fear, while you might be cautious about doing the same for a car battery. You should not touch the wires (if bare) at home, because if the voltage there does not kill, it definitely causes injuries and gives a disturbing shock. Similarly, higher voltages are more dangerous; lightning is a high-intensity voltage that if directly hits someone, there is no hope for survival.

Although Table 5.1 shows that the intensity of an injury from electricity depends on the current, as we will see in Section 5.4, the more voltage there is, the more the corresponding current. As a result, the intensity of an electric shock depends on its voltage, and therefore the higher the voltage, the more dangerous it is.

5.4 Ohm's Law: Relationship between Voltage, Current, and Load Resistance

Ohm's law is probably the most fundamental as well as important relationship that defines the relationship between voltage and current in a circuit. Try to master the meaning of Ohm's law before continuing any further.

Ohm's law: One of the most important laws of electric circuits: relationship between voltage across a component, the current in the component and the electric resistance exhibited by the component to the flow of electricity. For a simple resistor it is $V = RI$.

Ohm's law states that if the current in a resistor with a resistance R is I, then the voltage across the resistor (the voltage between the two ends of the resistor) is V, such that

$$V = RI \tag{5.1}$$

where R is in ohm, I is in amp, and V is in volt.

This law also implies that if a voltage of V volt is applied to a resistance of R ohm, then the current is I ampere; that is, the current, voltage, and resistance between two points are always related to each other.

Example 5.1

A lightbulb filament and the wires connecting it to a 12 V battery altogether have a resistance of 5 Ω. Find the current is in the lightbulb filament?

Solution

Substituting for the voltage and the resistance in Equation 5.1 leads to

$$12 = 5 \times I$$

$$I = 12 \div 5 = 2.4 \text{ A}$$

Example 5.2

If the same lightbulb as in Example 5.1 is connected to a 1.5 V battery, what is the current?

Solution

Resistance of the lightbulb does not change, because it is the physical property of the metallic wires involved. Thus,

$$I = 1.5 \div 5 = 0.3 \text{ A}$$

Note, however, that when a filament is warmed and its temperature has changed, its resistance also changes. Here, for simplicity, we have assumed that the change in temperature is not high enough to affect the resistance.

There are other meanings embedded in Ohm's law, which we need to pay attention to.

1. The relationship between the voltage across a resistor and the current through that resistor is linear. That is, if the voltage doubles, the current doubles, too.
2. By the same token, if the resistance of the resistor does not change, then, if the voltage drops in value (decreases), the current also decreases. Similarly, if the voltage increases, the current increases.
3. For a constant resistor, if the voltage across it remains unchanged, the current through it remains unchanged. Alternatively, if the current through the resistor does not change, it implies that the voltage across it has not changed.

Note that it is always the voltage applied to a resistor that determines how much the current through the resistor is.

In conjunction with Equation 5.1 we have the following equations that determine current in terms of the voltage and resistance and the resistance in terms of the voltage and the current:

$$I = \frac{V}{R} \tag{5.2}$$

$$R = \frac{V}{I} \tag{5.3}$$

> It is always the voltage applied to a resistor that determines how much the current through the resistor is.

Example 5.3

A resistive element (has only resistance) has a resistance of 50 Ω and is connected to 120 V. If as a result of the generated heat the resistance of the element increases by 10 percent, what current is in the element?

Solution

Initial current in the element is

$$120 \div 50 = 2.4 \text{ A}$$

After the element is heated its resistance increases by 10 percent and changes to

$$50 \times \frac{110}{100} = 55\ \Omega$$

$$I = 120 \div 55 = 2.18\ \Omega$$

Example 5.4

While the resistive element in the previous example is connected to the 120 V, the voltage changes to 130 V; determine the new current in the element.

Solution

Change in the voltage is relatively small, and it does not affect the resistance of the element. Thus, the new current is

$$130 \div 55 = 2.36\ \Omega$$

Example 5.5

When a lightbulb is connected to 120 V supply, it lights up and the current is 0.5 A. If the applied voltage is 220 V instead, what is the current?

Solution

Although for this problem one can numerically find a value for the new current, because the voltage is almost doubled, the physical lightbulb cannot withstand the higher current and its filament will blow.

5.5 Measuring Electric Current

Ammeter: Device to measure electric current.

Any electric circuit has a current in it based on the components in the circuit and based on the voltage of its source. Often, it is necessary to measure the current in a circuit for diagnosing problems and repairs. For measuring current we use an **ammeter**, a device directly graduated in amps and decimal fractions of amp. To measure current in a circuit, an ammeter must be inserted inside the circuit; that is, it must become part of the loop forming the circuit. Figure 5.1 shows that for measuring the current in a circuit you need to open the circuit at one (appropriate) point. Then you connect the two leads of the meter to the open ends of the circuit. In this way the ammeter integrates to the loop and becomes part of the circuit. See Figure 5.2.

> To measure current in a circuit, an ammeter must be inserted inside the circuit. The circuit must be opened for this purpose.

Multimeter: Device for electrical measurements with selectable switches to function as voltmeter, ohmmeter, and ammeter, and some more capabilities (all in the same unit).

For measuring current one can use an ammeter, which measures the electric current only, or use a **multimeter**. A multimeter is a multipurpose device that can measure current in addition to voltage and resistance. It has the capability to measure additional entities, such as capacitance and frequency. In circuit schematics a circle with a letter "A" in it represents an ammeter, as shown in Figure 5.1. Similarly, a circle with a letter "V" in it represents a voltmeter, which measures voltage. Note that all the components (including the source) and wires in a single circuit (one loop only) have the same current. In Chapter 6 we will discuss multiloop circuits.

Because in DC electricity current has one direction and in AC electricity current direction constantly changes, measuring current in AC and DC is not done by the same ammeter. For DC a DC meter must be used. In multimeters switching from AC meter to DC and from current to voltage and so on can be done using a selector switch with which one selects the desired choice. In measuring DC current the red lead of the meter must be connected to the positive side and the black lead to the negative side. If the leads are switched, the reading will be negative. (The needle is forced to

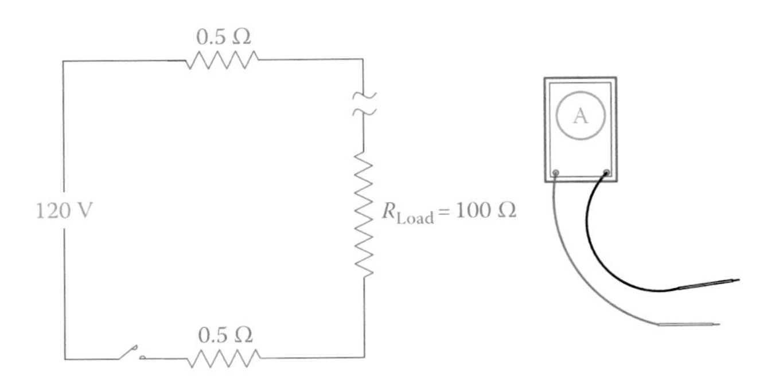

Figure 5.1 Step 1 for measuring the current in a circuit.

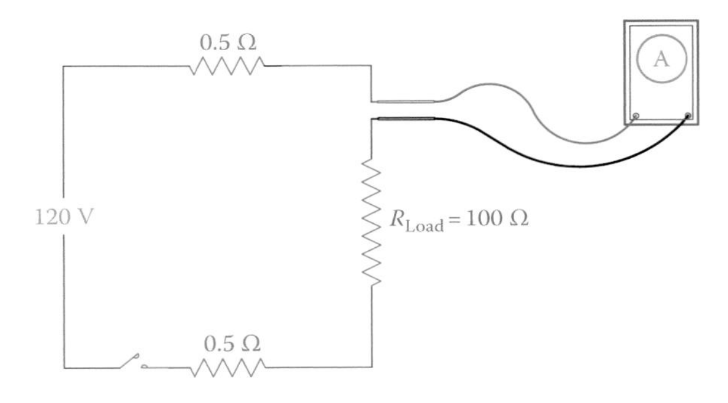

Figure 5.2 Step 2 for measuring the current in a circuit.

turn to left in an analog device.) In some ammeters (not multimeters) with a needle the zero point is in the middle and the motion of the needle indicates both positive and negative readings. This is helpful for the circuits in which current can be either positive or negative.

5.6 Measuring Electric Voltage

Because voltage is the potential difference *between* two points, to measure voltage, the two leads of a **voltmeter** must be connected to those points. Pay attention for measuring voltage; you should not open the circuit. Whereas for measuring current, one must open the circuit. In Figure 5.3 we need to measure the voltage *across* the load. Thus, the voltmeter is connected at points A and B so that the load is between points A and B (Figure 5.4). The measured value is the voltage applied to the load. A voltmeter, in fact, measures the voltage difference between two points.

Voltmeter: Electrical instrument to measure electric voltage.

Note that, whereas in a single (one loop) circuit there is only one current, there are various voltages depending on the number of components in the circuit and where the measurement is made. For instance, in Figure 5.5 there is a 100 Ω load and two 0.5 Ω wires connecting the load to the 120 V power supply. We may measure the voltage between each pair of points A, B, C, D, and E; for example, A-B, A-D, B-C, B-E, and so on. The measurement across the source shows the source voltage. Note that in

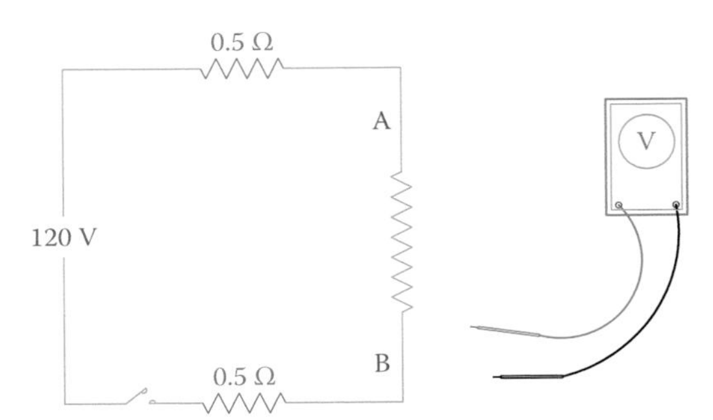

Figure 5.3 Use a voltmeter to measure voltage between two points.

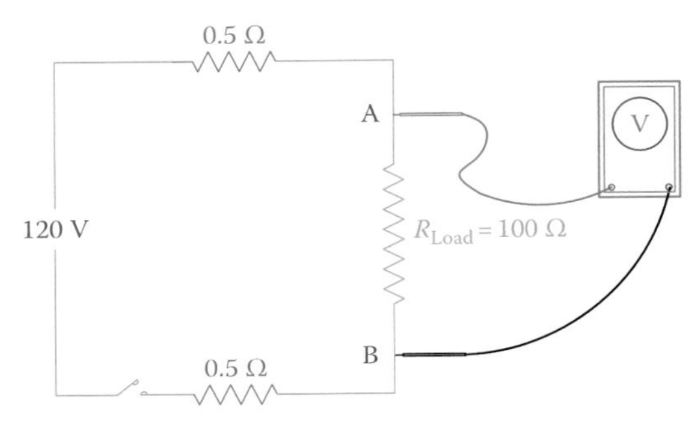

Figure 5.4 Measurement of voltage across two points.

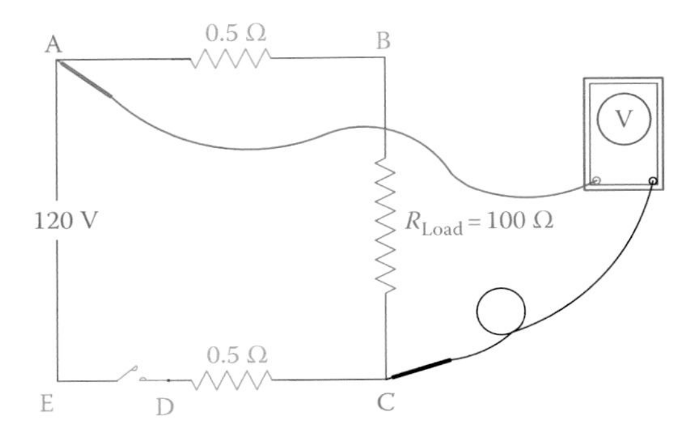

Figure 5.5 There are numerous voltages between various points in any circuit.

Figure 5.5 all the points A to E are selected at a graphically suitable point in the line connecting two elements together. Any other point on each line denotes the same point of the circuit.

In DC electricity, voltage measurement shows the polarity, too. So, if the positions of the leads of a meter are swapped, in a digital meter the reading will appear with a negative sign, but in an analog meter the reading cannot be done because the needle is forced to move to the left.

A voltmeter measures the voltage difference between two points by connecting the meter leads to those points.

Example 5.6

In the circuit shown in Figure 5.6, measurement of the voltage across R_2 shows a value of 75 V and a measure of the current in the circuit shows a value of 0.68 A. What is the value of the resistor R_2?

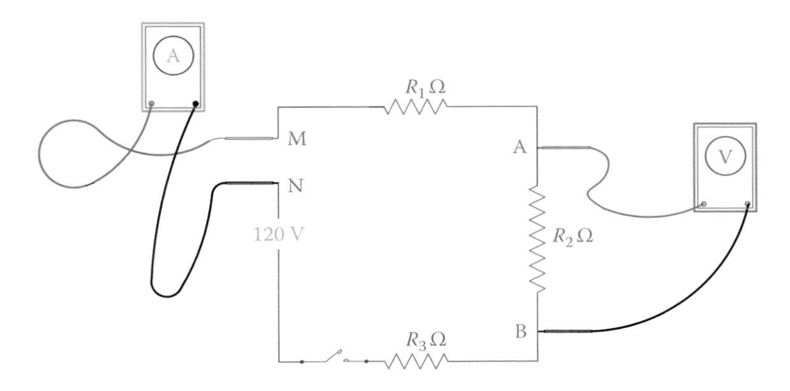

Figure 5.6 Circuit of Example 5.6.

Solution

Value of the resistance can be found from Equation 5.3 as

$$75 \div 0.68 = 110\ \Omega$$

Pay special attention to Figure 5.6 and the way the voltmeter and the ammeter are used for measuring voltage and current. The voltage for R_2 must be measured across R_2, say between points A and B, but the current can be measured at any point where the circuit is opened and the ammeter is inserted in it.

In any measurement, care must be taken that all the connections are clean and tight. This is especially true for the leads of a handheld meter. Make sure that you firmly hold the leads against the contact points. Otherwise, mistakes in readings are possible.

> For any measurement, make sure that the measuring leads are firmly held at the contact points.

5.7 Multimeter

A multimeter is a must-have tool for someone working with electricity and electronics. It is always necessary to verify values, to measure voltage, current, and resistance values in a circuit for diagnostic purposes, or to identify a damaged component. A basic multimeter can measure the three basic electric values (i.e., voltage, current, and resistance), but more functions can also be included in a multimeter.

Initially, there are two types of multimeters, analog and digital. Analog multimeters work based on the deflection of a needle from its zero position point. During a measurement one needs to read the correct value just under the needle. The main part of an analog meter is a **galvanometer** that consists of a small winding. The winding can rotate about a pin, and a needle is attached to it. It also has a spring to return the needle to the home (zero) position. Depending on the intensity of the current through the winding, it deflects from its rest position. Thus, a galvanometer is a

Galvanometer: A device consisting of a needle attached to a coil that can rotate around a pin shaft as a result of an electric current flowing through the coil. The coil behaves also as a spring, limiting the motion of the needle. A galvanometer is used to measure electric current, but the needle position can be graduated for other electric entities, like voltage.

current-sensitive device. By making some changes to a galvanometer and adding more components, it can be used as an ammeter, a voltmeter, an **ohmmeter**, and other measurement devices. The deflection of the needle in an analog meter is always from left to right.

Ohmmeter: Device for measurement of electric resistance.

Figure 5.7 shows a typical analog multimeter. It has a selector rotary switch by which one selects the category and range of values to be measured. Not all the meters are built the same way and have the same range of values and the same number of terminals. More details of the switches and terminals in the multimeter of Figure 5.7 are illustrated in Figure 5.8. In addition to a main selector switch for changing from volt to ohm or amp and selecting their range of values (e.g., 100 mA, 500 mA, 25 V, and 500 V), there is another switch for AC, –DC, and +DC. There are two main terminals for connecting the black (common) and red leads, but additional terminals are used for 10 A current (current larger than 500 mA and up to 10 A), 500 V and 1000 V (higher voltages). For measuring these relatively high values the black lead still goes to the common terminal, but the red lead must be inserted to the appropriate hole.

Figure 5.7 Typical analog multimeter.

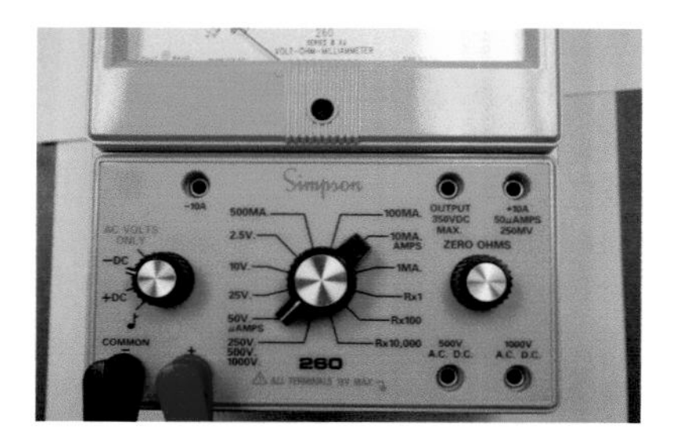

Figure 5.8 Switches and terminals on an analog multimeter.

Figure 5.9 Various graduations on a typical analog multimeter.

Figure 5.9 illustrates the set of various readings for the multimeter shown in Figure 5.7. As can be understood, these graduations are different, and for one position of the needle, there are various numbers. Each graduation corresponds to one or more selector switch positions. For instance, for a selector switch range of 2.5, 25, and 250 (see Figure 5.8) one must use the graduations ending in one of these numbers, whereas for other selections the numbers are between 0 and 10 (which must be multiplied by a power of 10, accordingly). Also, in the meter shown, the AC values are shown in red, whereas the DC values are in black.

Note that the zero value for ohm measurement (top scale) is to the extreme right (all the other zero values are on the extreme left). This is because a higher resistance value leads to a lower current value (see Ohm's law in this chapter) and vice versa.

A typical **digital multimeter (DMM)** is depicted in Figure 5.10. It has a rubber casing that protects it against shocks and scratches. Normally, it is easier to work with a digital multimeter because it directly gives the measured values in numbers. A DMM does not have any moving parts and works based on converting an analog reading to a digital value (analog to digital conversion; see Chapter 22).

Digital multimeter (DMM): A device to measure resistance, current, voltage, and other electrical parameters, in which the reading is automatically adjusted and displayed by 3 or 4 digits, as compared to analog multimeters in which the position of a rotating needle represents a measured value.

Similar to an analog multimeter, a DMM has a common terminal to which the black lead is connected and the red lead goes into one of the other terminals, depending on the design of the DMM. In the DMM shown in Figures 5.10 and 5.11 the red terminal is inserted in the hole to the right of the common (com) terminal, but for measuring current (based on its value if it is in the order of amps or much smaller) the other two holes are used. It is always safer to use the higher current terminal first

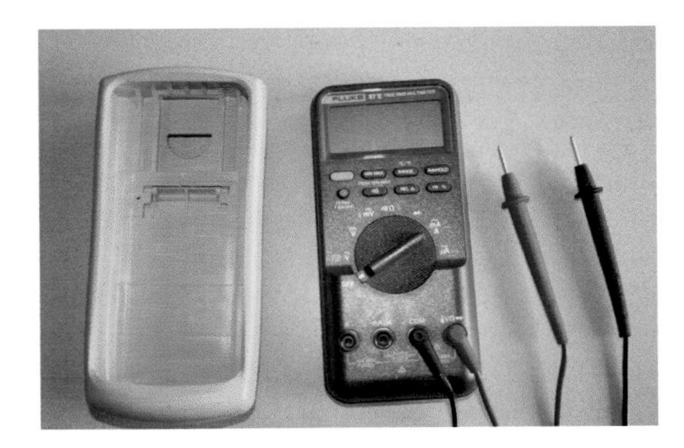

Figure 5.10 Typical digital multimeter (DMM).

Figure 5.11 Selector switch and the display of a typical DMM.

(the one on the left) before using the middle terminal for higher precision. In this way, you do not subject the fine meter to a current much larger than its capacity, which can damage the meter (or blow its fuse).

In the multimeter of Figure 5.10 the switch can be positioned to off, when the meter is not in use, or it can be set for AC voltage, DC voltage, small DC voltage (mV), resistance (ohm, Ω), and current (Ampere A, milliamp mA, and μA micro-amp). There is one more selection for diode (test); we discuss this in Section 14.6. Better DMMs have four digits, as shown in Figure 5.11. Some cheaper DMMs have only three digits. Normally, on the basis of the measured values, the scale is automatically adjusted; for instance, in Figure 5.11 the scale is automatically set to kΩ (appearing on the screen).

5.8 Measuring Resistance

A resistor's resistance or the resistance of part of a circuit can be measured by an ohmmeter or a multimeter. An ohmmeter or the ohmmeter part of a multimeter works based on measuring the current in a circuit where the item to be measured constitutes the main load of that circuit. Nevertheless, the graduations are made in terms of resistance values, not current values. In this sense, the following differences exist between an ohmmeter and a voltmeter or ammeter:

1. An ohmmeter needs a battery to power its circuit, whereas for measuring current or voltage, no battery is needed because there is already a current (when switch is closed).
2. Because current is measured, using Ohm's law, for a smaller resistance, there is a large current and for a larger resistance the current is smaller. This necessitates that the zero reading for resistance (in an analog meter) be at the extreme right side of the scale (the highest current) and the larger resistance values be on the left side.

3. Because the ohmmeter battery loses its strength with time, an ohmmeter has a zero adjustment knob that is used to bring the needle to zero reading. It is always necessary to adjust the zero reading before any measurement is made with an analog ohmmeter. This can be done by directly putting together the ends of the two leads, then turn the adjustment knob to bring the needle to zero.
4. Measuring resistance must always be performed when the power to a circuit (the resistance of a part of which is to be measured) is turned off. Otherwise, the reading will be erroneous.

An ohmmeter can be used for measuring and verifying the value of any resistor or resistive element. In addition, it can be used for checking the continuity of a circuit to see if any part of a circuit is open or there is any short (two points unnecessarily contacting) in the circuit.

> When measuring the resistance of part of a circuit, the power to the circuit must be off.

5.8.1 Checking a Capacitor and an Inductor with an Ohmmeter

An ohmmeter (usually, only analog meters) can also be used to check if a capacitor is good or damaged. If a capacitor is damaged, it either becomes short (when the two plates of the capacitor contact each other) or open (contacts are lost). To check a capacitor, it is connected to an ohmmeter. If it is short, then it shows a high current (near zero resistance) and the meter needle stays at the same point. If it is open, then it shows a very high resistance (the needle stays in the very left side of the meter). If the capacitor is good, the needle of the meter quickly moves to the right and slowly goes back to the left.

In a similar way we can use an ohmmeter to see if an inductor is fine or if it is damaged. A digital or an analog multimeter can be used. Note that in a DMM when the value to be measured is beyond the range of values selected an *overload* sign will appear on the screen by the letters "OL" as shown in Figure 5.12. Specifically, when measuring resistance, an open circuit (equivalent to a very high resistance measured) is realized if the meter shows OL. An inductor can be shorted or can be open. If after connecting its two ends to an ohmmeter high resistance is noticed, then it

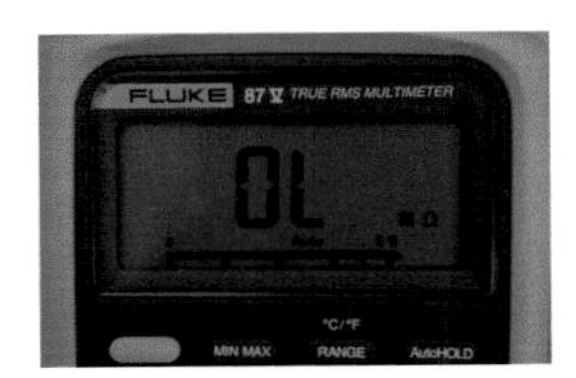

Figure 5.12 Very high value resistance or an open circuit.

is open. A short inductor shows zero resistance. A good inductor shows a small value of resistance but not zero.

5.9 Electric Power

Electric power: Power in the form of electricity and measured by electrical units (power is the amount of work in 1 sec).

It is always the **power** in an electric circuit that determines how much energy is going to a device, a system, or a place, and how much that energy costs. A good comprehension of the relationships for **electric power**, unit of power measurement, and the rules that determine power is very important. You must always bear in mind that for a power source (e.g., generator and battery), power rating implies the maximum power that the device can deliver. For a consumer, (a load) power rating implies the power requirement by the device so that within the desired operating conditions (voltage and current) it can function with good efficiency and with minimum risk of damage. The required power must be available to the device so that it functions properly. If that much power is not provided (by the source, the line, or the power supply), some shortcoming can happen that can lead to damage and failures.

Before continuing further, be sure to understand the meaning of power and the difference between power and energy. See the text in this chapter on Energy and Power.

In electricity, power is always obtained from the product of voltage and current. In later chapters we will discuss other factors that come into play and that power (especially in AC) has different categories. For the time being you need to understand that power is proportional to voltage and current. The analogy given in Chapter 4 between electricity and water flow can help you to understand this fact better.

Energy in water depends on both the height and the volume of water. Likewise, the power (energy in one second) in the water depends on the height and the water flow rate. It is obvious that out of the four scenarios shown in Figure 4.14, case c has the largest power. For this case both the current and the voltage have the larger values. As can be seen, both the height (corresponding to voltage in an electric system) and the flow rate of water (corresponding to current) influence the power. If the current does not change but the voltage doubles, the power doubles, and if the voltage does not change but the current doubles, the power doubles. This relationship is

$$P = VI \tag{5.4}$$

where P is in watts, V is in volts, and I is in amps.

Equation 5.4 clearly shows that for the power to be constant (this situation often occurs in practice), if the voltage decreases, then the current must increase or if the current increases, the voltage must decrease. Current increase beyond a rated value is equivalent to overload, which is not desirable.

Example 5.7

Measured current in the filament of a lightbulb when connected to a 12 V battery is 5 A. What is the power consumption of the lightbulb?

Solution

Power can be directly found by the product of the voltage and the current:

$$P = VI = (12)(5) = 60 \text{ W}$$

Energy and Power

"Energy" is the potential to do *work*, and it can be in different forms, such as electrical energy, nuclear energy, thermal energy (heat), wind energy, and solar energy. "Work" here implies mechanical or other type of work. Mechanical work is, for instance, when a weight is lifted or when an engine drives a car. Mechanical work is more tangible compared to other types of work. For instance, a motor can do mechanical work, but a battery does not directly do mechanical work. However, it can run a motor that does mechanical work; thus, a battery has potential to work. It has electrical energy. Energy can be converted from one form to another. For instance, consider a steam turbine that can do mechanical work. That is, conversion of heat to mechanical energy. In this sense, when an electric kettle heats water, it performs work. It consumes energy and converts it to work.

Energy can be measured, like any other entity, in its appropriate unit(s). Suppose that a machine or device has energy. How much energy does it have? Energy can be measured in terms of heat units like calorie and BTU (British thermal unit), work units, or energy units. The unit for energy is joule, and units for work can be foot-pound and newton-meter. One joule is, in fact, one newton-meter.

In conjunction with energy we have *power*. We may always ask the question: "If a machine can do a certain amount of work, how long does it take to do it?" For example, how long does it take for a kettle to boil the water in it? The answer to this question stands in *power*, which is also used for comparison between various energy sources. Power is the amount of work done in 1 sec by a device that can do work. This is the measure of strength of energy sources. For instance, a smaller motor has less power than a larger motor. That is, it can do less work than the larger motor in the same duration of time or, in doing the same work for the smaller motor it takes more time. Similarly, it takes more time for a smaller (less powerful) kettle to boil the same amount of water compared to a larger kettle.

Thus, power is determined from energy divided by time.

$$\text{Power} = \frac{\text{Energy}}{\text{Time}}$$

Electrical power is normally measured in watt and kilowatt. Energy is sometimes (especially in the case of electrical energy) measured in terms of the unit of power multiplied by the unit of time; that is, watt-second and kilo-watt-hour are used as units of energy.

For almost any device, including lightbulbs, the power is written on the device itself or on a nameplate. In addition, the operating voltage is always shown. For example, on a lightbulb you may see "100 W, 120 V." This implies that

1. Operating voltage of the lightbulb is 120 V. You should not connect this bulb to a voltage that is considerably higher than 120 V; whereas it is generally acceptable if you connect it to 125 V, but if you apply 180 V to this bulb, you will definitely burn it out.
2. If connected to 120 V, the power consumed by the lightbulb is 100 W. Moreover, if you connect it to a higher voltage, the power will be more than 100 W, and if the voltage used for the bulb is lower, then the power is lower than 100 W.

Note that physically the lightbulb does not change and it has only a filament in it with certain resistance. The resistance of the filament when it is cold is less than when it is lighted.

Example 5.8

What is the resistance of the filament in a 100 W, 120 V lightbulb?

Solution

The current in the lightbulb filament can be found from Equation 5.4:

$$100 = (120)(I) \quad \longrightarrow \quad I = 100 \div 120 \text{ A}$$

From Equation 5.1, however, we have

$$(R)(100 \div 120) = 120 \text{ V}$$

This leads to $R = 144\ \Omega$.

Example 5.9

What is the power consumption of a 100 W, 120 V lightbulb if it is connected to 110 V?

Solution

No matter to what voltage the lightbulb is connected, its filament does not change. When the lightbulb is connected to 110 V, its filament still gets hot and with good precision we can assume that its resistance stays at 144 Ω (see Example 5.8).

The current due to connecting this lightbulb to 110 V is

$$I = 110 \div 144 = 0.764 \text{ A}$$

Thus, the power of the lightbulb is

$$P = (110)(0.764) = 84 \text{ W}$$

It can be clearly seen that if the voltage applied to a lightbulb (or any other device with resistive element) is not the same as its rated voltage, then one cannot expect to obtain the rated power.

Note that if the applied voltage is much smaller than the rated value, then the lightbulb filament is not heated enough and its resistance cannot be assumed to stay the same. Its resistance becomes slightly smaller.

As can be seen from the Ohm's law and from the relationship for power, there are four interrelated electrical entities for any resistive load: resistance, voltage, current, and power. When any two of these values are known, the other two can be obtained. Equations 5.1 through 5.4 can be combined together to give rise to the formulas for determination of any of the four entities in terms of two other ones. Four of these equations are already given by Equations 5.1 through 5.4:

a. Power

$$P = VI = (RI)(I) = RI^2 \tag{5.5}$$

$$P = VI = (V)(V/R) = \frac{V^2}{R} \tag{5.6}$$

b. Current

$$I = P \div V \quad \text{(from Equation 5.4)} \tag{5.7}$$

$$I = \sqrt{\frac{P}{R}} \quad \text{(from Equation 5.5)} \tag{5.8}$$

c. Voltage

$$V = P \div I \quad \text{(from Equation 5.4)} \tag{5.9}$$

$$V = \sqrt{PR} \quad \text{(from Equation 5.6)} \tag{5.10}$$

d. Resistance

$$R = \frac{P}{I^2} \quad \text{(from Equation 5.5)} \tag{5.11}$$

$$R = \frac{V^2}{P} \quad \text{(from Equation 5.6)} \tag{5.12}$$

All of the Equations 5.1 through 5.12 can be put together in the form of charts shown in Figure 5.13.

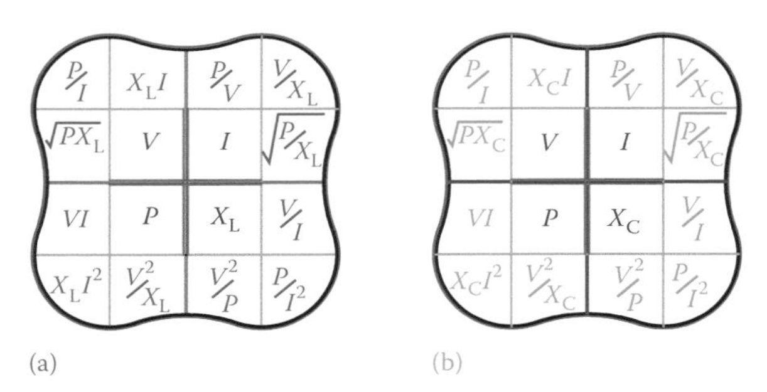

Figure 5.13 Summary for the relationships among voltage, current, resistance, and power for (a) inductor and (b) capacitor.

Example 5.10

In a resistive element the current is 4 A and the power consumed by the element is 160 W. What is the resistance of the element? What is the voltage across the element?

Solution

Resistance of the element is found in terms of the power and current. From Equation 5.11 we have

$$R = \frac{160}{4^2} = \frac{160}{16} = 10\ \Omega$$

Voltage across the element can be found from either of Equations 5.1, 5.9, or 5.10:

$$V = (10)(4) = 40\ \text{V}$$

Example 5.11

When a voltage of 110 V is applied to a resistor, the power generated in the form of heat is measured to be 60 W. What is the resistance of the resistor?

Solution

We may directly use Equation 5.12:

$$R = \frac{110^2}{60} = \frac{1210}{6} = 201.7\ \Omega$$

Note that the power consumed by the resistor converts to heat.

Example 5.12

If the voltage applied to the resistor in Example 5.11 increases by 10 percent, what is the percentage of increase in power?

Solution

Increase in voltage is

$$110 \times 0.10 = 11 \text{ V}$$

Thus, the new voltage is 121 V. The new power can be found, having the voltage and the resistance, from Equation 5.6:

$$P_{\text{New}} = \frac{121^2}{201.7} = 72.59\,\text{W}$$

and the percentage of power increase is

$$\frac{72.59 - 60}{60} = \frac{12.59}{60} = 0.21 \text{ or } 21\%$$

5.10 Measuring Electric Power and Energy

From the fact that power is the product of voltage and current, it is obvious that a watt-meter to measure power must work based on a needle (if analog) or a display (if digital) whose displacement or value is proportional to the values for both the voltage and the current in a circuit. While a watt-meter can be used in panels of electric generators to indicate the instantaneous power, it is normally not used by a technician, and it is not included in the functions of a multimeter.

However, measuring energy is very common for the calculation of electric energy cost. Each house and building is equipped with an electricity meter that measures the consumed electric energy. The basis of such a meter is a circular disk that rotates and causes a series of gears to rotate as in a clock. The speed of rotation of this disk is proportional to the power consumed in a circuit, and the number of revolutions during a period represents the energy consumed during that period.

The unit for measuring power is watt (see the table of units and conversions in Chapter 2). A kilowatt (1000 watts) is a larger unit, and for even larger measures of energy production, megawatt (1,000,000 watt) is a more common unit of power.

Because energy is the product of power and time (see the text on Energy and Power in this chapter), we may write

$$E = PT \tag{5.13}$$

If P is the power in watt and T is the time in second, then E determines energy in joules. It is very common in electricity that a larger unit of energy be employed. If the power is given in kW and time in hr (3600 sec), then the unit of energy is kW-hr (kilowatt-hour). Kilowatt-hour is a unit of energy equivalent to (1000)(3600) = 3,600,000 joule.

A better understanding of kW-hr is possible in the following manner. One kW-hr is the energy consumption of one 100 W lightbulb in 10 hours, or of two 100 W lightbulbs in 5 hours, or ten 100 W lightbulbs in 1 hour.

Example 5.13

A 100 W lightbulb is turned on 5 hours per day. Calculate the amount of energy consumed by the lightbulb in a year.

Solution

Number of hours the light is on in a year = (365)(5) = 1825 hr

$$\text{Energy} = (100)(1825) = 182{,}500 \text{ W-hr} = 182.5 \text{ kW-hr}$$

Example 5.14

If the cost of electricity is 8 cents per kW-hr, what is the cost of electricity used by the lightbulb in Example 5.13?

Solution

$$\text{Annual cost} = (182.5)(0.08) = \$14.60$$

Example 5.15

If the average daily electricity consumption in a household is 4 kW-hr during a given season, what is the monthly cost of electricity if the rate of electricity is 7.5 cents per kW-hr?

Solution

$$\text{Average monthly consumption} = (30)(4) = 120 \text{ kW-hr}$$

$$\text{Cost} = (120)(0.075) = \$9$$

Example 5.16

You have a walkway light that remains on during night and turns off during daytime. It uses a 60 W lightbulb. If you change the light bulb with an energy saving bulb, which uses 10 W, how much electric energy would you save if you consider the yearly average night hours to be 11 hr. If the rate of electricity is 8.5 cents per kW-hr, how much money do you save in one year?

Solution

$$\text{Difference in power} = 60 - 10 = 50 \text{ W}$$

$$\text{Energy saving per night} = (50)(11) = 550 \text{ W-hr} = 0.55 \text{ kW-hr}$$

Annual energy saving = (365)(0.55) = 200.75 kW-hr

Annual amount of saving = (200.75)(0.085) = $17.06

5.11 Chapter Summary

- Voltage applied to a circuit and the current in the circuit are not independent.
- If a circuit has resistive load, the higher the load is, the smaller the current is.
- Voltage difference between two points equals the product of the resistance between those points and the current.
- To measure current, an ammeter is used. The circuit must be opened at one point and the ammeter be inserted in the circuit.
- There is only one current value for all components in a single circuit.
- The unit to measure current is ampere (amp), and its symbol is A.
- To measure the voltage difference between two points, a voltmeter must be connected between those points without opening the circuit.
- Voltage difference depends on the points to which the voltmeter is connected.
- The unit to measure voltage is volt, and its symbol is V.
- To measure current and voltage, a circuit must be energized.
- The unit to measure resistance is ohm, and its symbol is the Greek letter Ω (omega).
- For measuring the resistance between two points in a circuit the power must be off.
- A multimeter is a device containing a voltmeter, ammeter, and ohmmeter in one unit. It can have other capabilities (e.g., frequency measurement).
- All meters can be made analog (with a moving needle), or digital, which directly gives the measured value in numbers. Most digital multimeters have four digits. Others have three digits.
- An ohmmeter can be used also for testing a capacitor or an inductor to see if it is damaged.
- Any electric device has a nominal voltage. This voltage is written on the device or its nameplate, together with the device power. This is the voltage that best suits the operation of that device.
- Power of a device depends on the voltage applied to it. If the nominal voltage is applied, power is the nominal power written on the device.
- If the voltage applied to a device is much different from the nominal voltage, the device will be damaged or will not deliver the expected power.
- Power is determined from the product of the applied voltage and current in a device.

- Applied voltage to a device determines its current and, thus, its power. For a load this is the power that the device absorbs from the circuit or the power supply.
- In a source the rated power is the maximum power that it can deliver. For a load the rated power is the amount of power required for functioning of the device.
- Energy is power multiplied by time.
- In electricity the most common unit of energy is kW-hr (kilowatt-hour).
- We pay for electricity based on the amount of energy, normally expressed in kW-hr.

Review Questions

1. What are the two main entities associated with electricity?
2. What is the unit for measuring current called?
3. How is electric current measured, in terms of connecting a meter?
4. What is the name of the device with which current is measured?
5. When measuring current in a circuit, should the power be on or off?
6. What is the name of the device for measuring voltage?
7. Explain how you measure voltage between two points.
8. What is the name of the unit for measuring voltage?
9. What is the difference between potential difference and voltage difference?
10. How much do you read/measure if you connect both leads of a voltmeter to the same point in a circuit? Explain why.
11. If you have a circuit without any source, how much can you read the voltage between any two points? Explain why.
12. What do you use to measure the resistance of a resistor?
13. What is the unit for measuring resistance?

14. If you have only a voltmeter and an ammeter and you need to measure the value of resistance between two points of a circuit, describe how you can do this.

15. If you have an ohmmeter and you need to measure the value of resistance between two points of a circuit, should the power be on or off?

16. What is a multimeter?

17. What color lead do you connect to the terminal marked "com"?

18. In an electric circuit, *power* is proportional to which two entities?

19. If the voltage applied to a lightbulb is lowered, does its power go up or down? Explain.

20. If the current in a lightbulb filament goes up, does its brightness go up or down?

21. Can you use a multimeter to test if a capacitor is damaged or it is good? If yes, explain with which of the devices in the multimeter.

22. What are the two main faults in a damaged capacitor?

23. What are the two main faults in a damaged inductor?

Problems

1. The heating element of a stove is marked 240 V, 1500 W. How much current do you expect to be flowing through the element when it is connected to 240 V electricity?

2. Find the resistance of the heating element in Problem 1.

3. If the resistance of a heating element is 40 Ω and it is connected to 240 V, what is the current in the element?

4. If the element in Problem 3 carries a current of 5.5 A, what voltage is applied to it?

5. The heating element in Problem 1 is connected to 220 V (instead of 240 V). What power does it takes from electric source it is connected to?

6. When the element of Problem 1 is connected to 220 V, what is the current in the element?

7. If a lightbulb is 100 W, 120 V. When you measure the voltage at home it shows 110 V. What is the power consumption of the lightbulb?

8. When you connect the lightbulb in Problem 7 to 110 V, you will not get the full brightness of the lightbulb as when you connect it to 120 V. Approximately, how much less percent is the brightness?

9. You have bought a kettle in Europe. On its nameplate it says "1500 W, 220 V." Suppose that when you used it in Europe, it could boil water in 5 min. At your home the electricity voltage is 110 V. At home you connect the kettle to electricity and it does not boil the same amount of water in 5 min. Explain why.

10. Approximately how long do you expect that it should take for the kettle in Problem 9 to bring the water to boil at your home?

11. What is the resistance of the element in the kettle of Problem 9?

12. If you want to change the element of your kettle (in Problem 9) so that it boils water in 5 min, what size resistance should you look for (assuming it is available)?

Project

Find an electric device at home or at a store. Write down the information provided on its case, on its nameplate, or on the device itself. How much of this information do you understand? Discuss the information in the class.

6

DC Circuits Relationships

OBJECTIVES: After studying this chapter, you will be able to

- Explain series circuits
- Understand the equivalent resistor in series resistors
- Explain parallel circuits
- Understand the equivalent resistor in parallel resistors
- Explain combined series and parallel circuits
- Understand the equivalent resistor for a set of resistors combined together in series and parallel
- Use a voltage divider for providing a desired voltage
- Perform power calculations for all DC circuits
- Describe the behavior of a capacitor when in a DC circuit
- Describe the behavior of an inductor in a DC circuit
- Understand what "time constant" in an electric circuit is

New terms:
Combined circuit, equivalent resistor, parallel circuit, series circuit, time constant, voltage divider

6.1 Introduction

Not all electric circuits are composed of a simple device and a single loop. Circuits can have more of the same component put together or they can have any number of the basic components put together in various combinations. This could be because of different parts of the same device or when a source is feeding various devices. For instance, at home there is only one source of electricity, whereas many devices are powered by the same source. This is true for all the electric devices, say in a factory, which are ultimately powered by one source of electric power.

If we learn how to deal with a few components when put together, then we will be able to extend that to the case when thousands of components are put together. The application of what you learn here is widespread. At small scale, consider few components, like in a stove that has a few heating elements, a light and so on, or in a battery charger (Chapter 16). At large scale, consider a city where each house has several devices (e.g., lights, radio, TV, kitchen appliances). Today, electricity in cities is provided by more than one source. The power network, or grid, is fed at different points. So, we need to study the cases where there is more than one power source.

In this chapter we discuss how to deal with such circuits when connected to a DC source. In Chapter 8 we discuss these combinations in AC circuits.

6.2 Series Circuit

Series circuit: Type of electric circuit in which the loads form a single loop between the two terminals of the power source.

When a number of components are put together in such a way that when connected to a source they form a single loop, we say these components are in series (with each other), and the circuit formed this way is called a **series circuit**. A series circuit is shown in Figure 6.1. In this circuit, there are three resistors (or resistive components) and two inductors in series and connected to 110 V. A switch is added for turning the power on and off. Note that it is not defined if the 110 V source is DC or AC (when no other information is included). In the case it was DC we could show the source by a symbol for battery, and in the case it was an AC source we could write 110 VAC (when an AC source is used, normally the frequency should also be given).

Because there is only one loop, there is only one current for all the components. This is a property of a series circuit.

To analyze this circuit (finding the current, power, etc.), it is first necessary to reduce this circuit. The first step will be to substitute all the components of the same type by a single component equivalent to all those. This is what is always pursued (for parallel and combination circuits, as well).

In this chapter we discuss only finding the equivalent circuit for resistors. To determine the equivalent component for inductors and capacitors, refer to Chapter 8.

> In a series circuit, all components in series with each other have the same current.

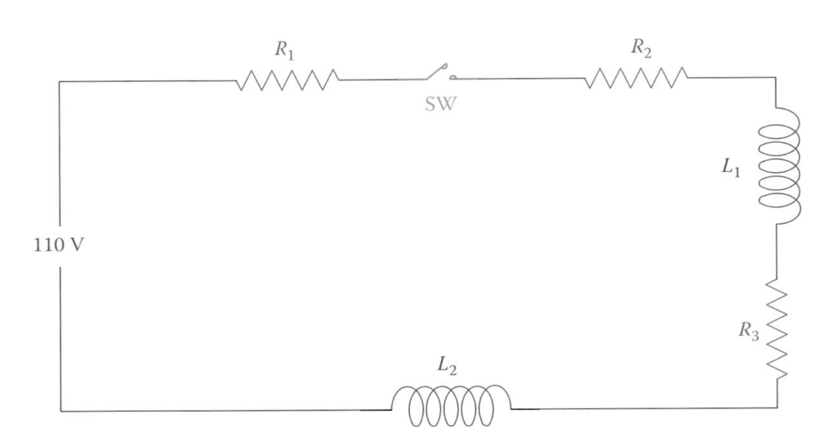

Figure 6.1 A general series circuit.

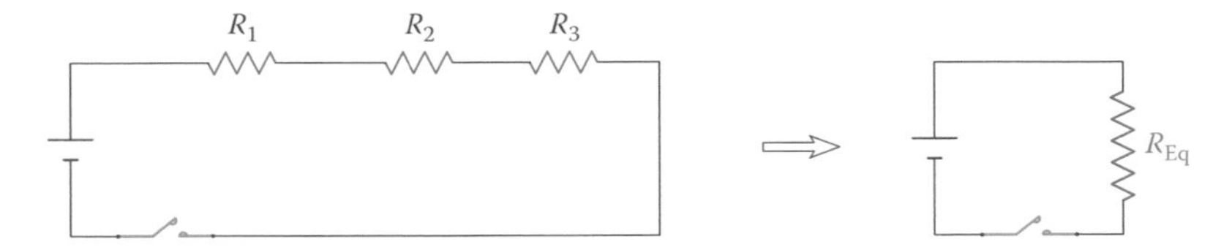

Figure 6.2 Resistors in series must be reduced to their equivalent resistor.

6.3 Resistors in Series

Resistors in series are shown in Figure 6.2, which contains only three resistors, but, in general, it could be any number. Any such circuit must be simplified to the equivalent circuit, which contains only one resistor. The **equivalent resistor**, denoted by R_{Eq}, or often just R, for a series circuit is

Equivalent resistor: Resistor that can replace two or more resistors in series, parallel, or in any combination, that is, having the same effect on the circuit.

$$R_{Eq} = R = R_1 + R_2 + R_3 + \ldots \quad (6.1)$$

Example 6.1

What current is in a circuit containing three resistors connected to a 12 V battery if two of the resistors are 5.1 Ω each and the third one is 51 Ω?

Solution

First, we find the equivalent resistor, as

$$R = 5.1 + 5.1 + 51 = 61.2\ \Omega$$

The current can be determined (from Equation 5.2)

$$I = 12 \div 61.2 = 0.196\ \text{A} \approx 0.2\ \text{A}$$

Example 6.2

In the following circuit, what current is in the 800 Ω resistor (see Figure 6.3)?

Solution

The current in the 800 Ω resistor is the same as in the other resistors. First, find the equivalent resistor:

$$R = 100 + 300 + 400 + 800 = 1600\ \Omega$$

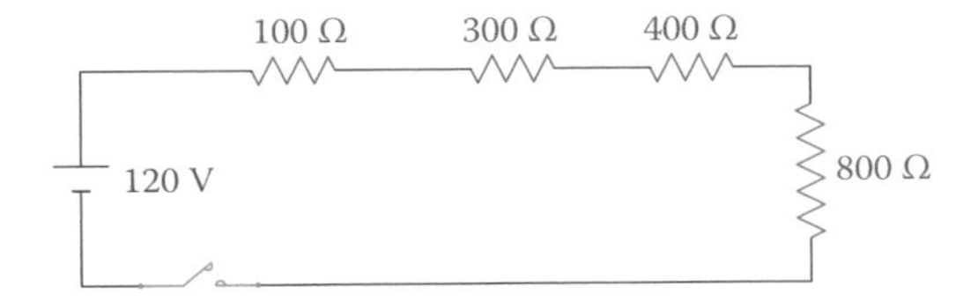

Figure 6.3 Resistors in Example 6.2.

and the current is calculated as

$$I = 120 \div 1600 = 0.075 \text{ A}$$

Alternatively, you may directly find the current from

$$I = \frac{120}{(100 + 300 + 400 + 8000)} = 0.075 \text{ A} = 75 \text{ mA}$$

Which way is easier for you depends on your habit and how familiar you are with your calculator.

Example 6.3

In the circuit of Example 6.2, what voltage is across the 800 Ω resistor?

Solution

Using Equation 5.1, the voltage across this resistor is

$$V_{800} = (800)(0.075) = 60 \text{ V}$$

Note the subscript 800 is used for the voltage in order to denote the voltage across the 800 Ω resistor.

Note also that we may say the voltage difference across the 800 Ω resistor is 60 V. Either term is correct to use and implies that if a voltmeter is connected to the two sides of the resistor, it should read 60 V. Furthermore, if the two ends of the resistor are called A and B, we may say the voltage difference between A and B is 60 V.

Example 6.4

Current in a device must not exceed 4 A. This device has a resistance of 24 Ω. If the available voltage is 115 V, determine the resistance of a resistor to be put in series with this device, so that the current is 4 A.

Solution

Voltage that causes a current of 4 A to flow in the device is

$$V = RI = (24)(4) = 96 \text{ V}$$

The remainder of the voltage from the source is

$$115 - 96 = 19 \text{ V}$$

Because this resistor must also carry 4 A current, its resistance is

$$19 \div 4 = 4.75 \ \Omega$$

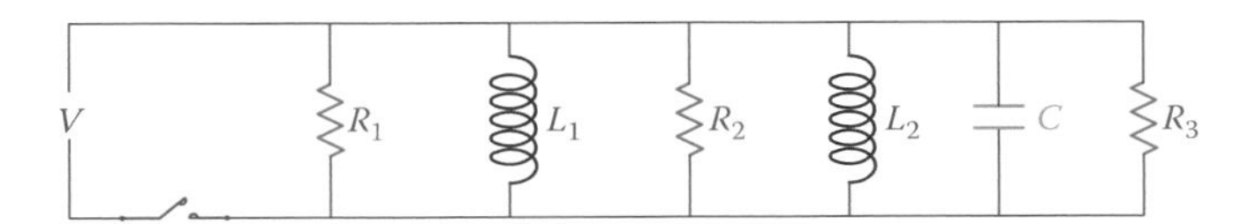

Figure 6.4 Example of a parallel circuit.

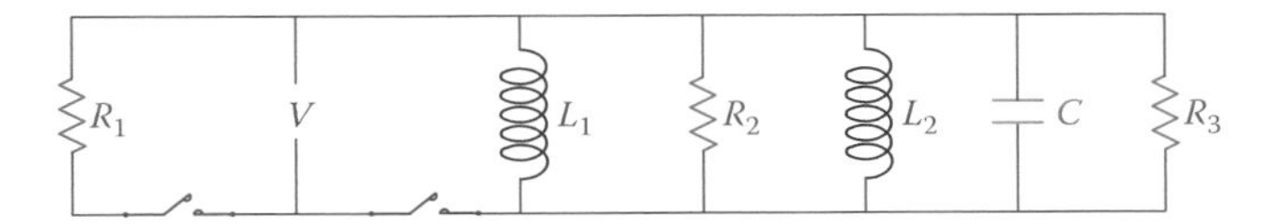

Figure 6.5 Although this is equivalent to Figure 6.4, the circuit is almost never shown this way.

6.4 Parallel Circuit

In a **parallel circuit** all the components are directly connected to the electric source. In this way, the full voltage of the source is applied to all the components in the circuit. A parallel circuit is normally shown as in Figure 6.4. You see that each component in the circuit makes a separate loop containing the source. You may show the parallel circuit of Figure 6.4 as shown in Figure 6.5 (i.e., the components are shown both sides of the source), but this is almost never done, although the two circuits represent the same thing.

Parallel circuit: Electric circuit having two or more electric components with their terminals connected to the same point in the circuit, so that current divides between the branches formed by those components.

In a parallel circuit each loop containing the source has a current, which is independent of the currents in the other loops. In this sense, in a parallel circuit there is not a common current for all the components, but the applied voltage to all the components in parallel is the same. Moreover, as can be observed, the current through the source is the sum of all the currents through the various components in the circuit. A source must be capable of delivering the total current required by all the loads.

Here again in order to find the current and determine the power requirement from the source all components of the same type are substituted by their equivalent component. We study parallel resistors in this chapter but leave the study of equivalent inductor and capacitor for Chapter 8.

> In a parallel circuit all the components in parallel with each other have the same voltage.

6.5 Resistors in Parallel

Figure 6.6 illustrates three resistors R_1 to R_3 in parallel. These resistors can be substituted by a single resistor that is equivalent to all those resistors. The formulation for finding the equivalent resistor for three parallel resistors can be generalized and extended to embrace as many that can be in parallel.

Because each branch (resistor) in Figure 6.6 has a separate current, we can identify the currents for R_1 to R_3 as I_1, I_2, and I_3, respectively. Then the

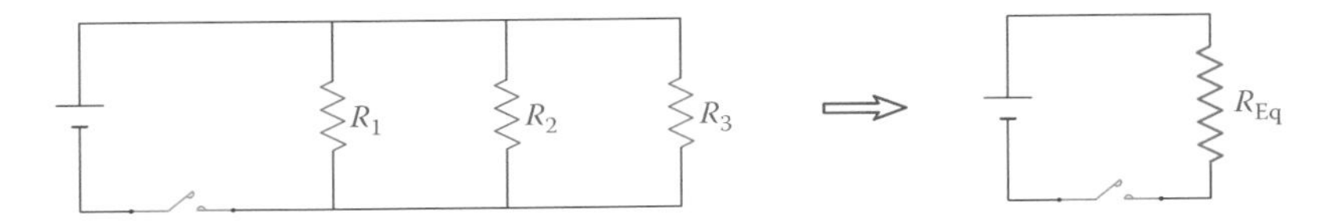

Figure 6.6 Resistors in parallel are substituted by a single equivalent resistor.

total current in the circuit, that is, the current that must be provided by the power supply, is

$$I = I_1 + I_2 + I_3 \tag{6.2}$$

According to Ohm's law, the values of the various currents in Equation 6.2 are

$$I_1 = \frac{V}{R_1}, \quad I_2 = \frac{V}{R_2}, \quad I_3 \frac{V}{R_3}, \quad I = \frac{V}{R} = \frac{V}{R_{Eq}}$$

because the voltage V is the same for all branches and for the equivalent resistance. Note that for simplicity the subscript Eq is dropped from R_{Eq}, and only R is used for the equivalent resistor.

Substituting for I_1, I_2, I_3, and I in Equation 6.2 leads to

$$\frac{V}{R} = \frac{V}{R_1} + \frac{V}{R_2} + \frac{V}{R_3}$$

which leads to

$$\frac{1}{R} = \frac{1}{R_1} + \frac{1}{R_2} + \frac{1}{R_3} \tag{6.3}$$

This is the relationship between parallel resistors. Note that in Equation 6.3 the value of R obtained this way is smaller than any of the R_1 to R_3 values. Equation 6.3 gives the inverse value of R.

Thus, R can be found from (inverting both sides of Equation 6.3)

$$R = \frac{1}{\frac{1}{R_1} + \frac{1}{R_2} + \frac{1}{R_3}} \tag{6.4}$$

In the case there are only two resistors, Equation 6.4 can be simplified to

$$R = \frac{R_1 R_2}{R_1 + R_2} \tag{6.5}$$

Example 6.5

If four 1000 Ω resistors are connected in parallel, what is the resistance value of the equivalent resistor?

Solution

Employing Equation 6.4, we have

$$R = \frac{1}{\frac{1}{1000}+\frac{1}{1000}+\frac{1}{1000}+\frac{1}{1000}} = \frac{1}{\frac{4}{1000}} = \frac{1}{0.004} = 250\ \Omega$$

Example 6.6

A 510 Ω and a 51 Ω resistor are put in parallel. What is the resulting resistor?

Solution

Because there are only two resistors, one can directly use Equation 6.5

$$R = \frac{510\times 51}{510+51} = \frac{26{,}010}{561} = 46.36\ \Omega$$

Note that the equivalent resistor is smaller than the 51 Ω. If in calculation you ever came up with a number larger than the smallest resistor in the parallel resistors, check your calculation because a mistake has happened.

Example 6.7

Resistance of the equivalent resistor for a 47 Ω resistor parallel with another resistor is 15 Ω. What is the resistance of the second resistor?

Solution

We may employ Equation 6.3 and write

$$\frac{1}{15} = \frac{1}{47} + \frac{1}{R_2}$$

From which we may find R_2

$$\frac{1}{R_2} = \frac{1}{15} - \frac{1}{47} = 0.04539 \rightarrow R_2 = 22\ \Omega$$

Note that the values for resistors in the above examples are among those of the standard resistors (see Appendix E for standard resistors).

It is very helpful to notice that when similar resistors are put in parallel, the resistance of the ensemble is obtained from dividing their common resistance value by the number of units. For instance, if four 1000 Ω resistors are in parallel, they are equivalent to a 250 Ω resistor (see Example 6.5), and the resultant of two 51 Ω resistors is 25.5 Ω. Because 25.5 Ω is not a standard resistor, if one needs to use a resistance of this value in a circuit, one way is to use two 51 Ω resistors in parallel with each other.

6.6 Combined Series and Parallel Resistors

Arrangement of the components in a circuit can be in such a way that it may not be modeled either as a series circuit, or as a parallel circuit. The resistors in the circuit of Figure 6.7 illustrate a simple example. This circuit has components that are in series (have a common current) and components that are in parallel (a few branches start from some points). In other words, all components neither have the same current nor voltage across all components. Note that the way various resistors (and, in general, various components) are shown in Figure 6.7 is not the only way to represent this circuit. For example, the same circuit can be shown as in Figure 6.8. It is better to reduce the representation of a circuit to its simplest form, before anything else, because it helps to understand and analyze the circuit better. Often, it is an art to bring a complicated circuit to its simplest representation. Always we have to bear in mind that two components are in parallel if the same voltage is applied to them, and two components are in series if they are part of the same loop and share the same current.

Combined circuit: Circuit having a number of the same electric components in both series and parallel at the same time.

The way to tackle a **combined circuit** (find the current, power, etc.) depends on the way the components are laid out. In the circuits of Figure 6.8 it is easy to understand that first we may replace the resistors R_1 and R_2 by their equivalent resistor and also resistors R_3 and R_4 by their equivalent resistor. Then we end up with three resistors that are in series with each other. The rules of resistors in series can, then, be applied.

> Two components are in parallel if the same voltage is applied to them, and two components are in series if they are part of the same loop and share the same current.

Figure 6.9 shows another example of a combined circuit. It is a little more involved than the one in Figures 6.7 and 6.8. Dealing with a combined circuit of resistors requires, in general, that we find the single

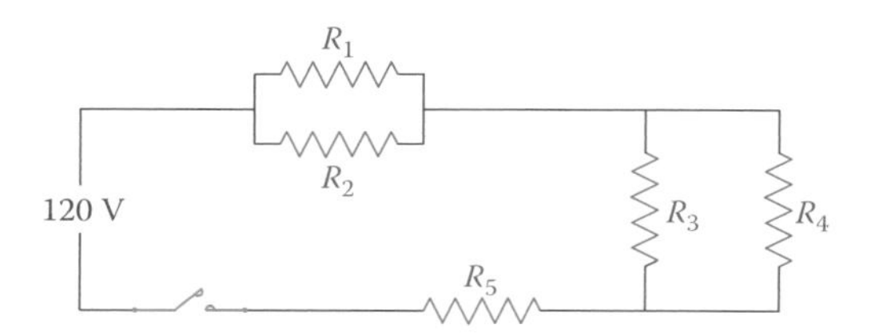

Figure 6.7 Example of a combined circuit.

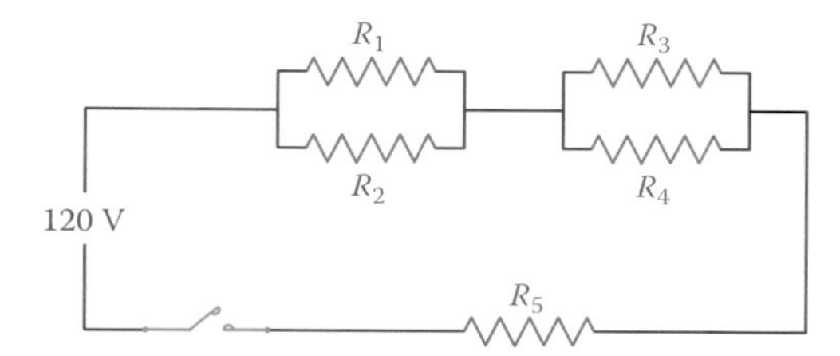

Figure 6.8 Alternative representation of the same circuit shown in Figure 6.7.

resistor equivalent to all those in the circuit. Nevertheless, before doing so, we need to know what the purpose of this analysis is and avoid unnecessary computations.

As said before, reducing combined resistors to their single equivalent resistor depends on the circuit. Step by step and from inside out we need to replace the parallel resistors by their equivalent resistor and the series resistors by their equivalent resistor. For instance, to reduce the circuit of Figure 6.9, the following steps are necessary:

1. Substitute the parallel resistors R_2 and R_3 by their equivalent resistor R_{23}.
2. Substitute the series resistors R_1 and R_{23} by their equivalent resistor R_{123}.
3. Substitute the parallel resistors R_5 and R_6 by their equivalent resistor R_{56}.

   The result of these operations is shown in Figure 6.10.

   The circuit of Figure 6.10 is still a combined circuit. Thus we need to further do the following.
4. Substitute the parallel resistors R_{123} and R_4 by their equivalent resistor R_{1234}.

Now we have only three resistors R_{1234}, R_{56}, and R_7 that are in series with each other, as shown in Figure 6.11. The equivalent resistor for these can be found now by adding the values of these resistors (Equation 6.1).

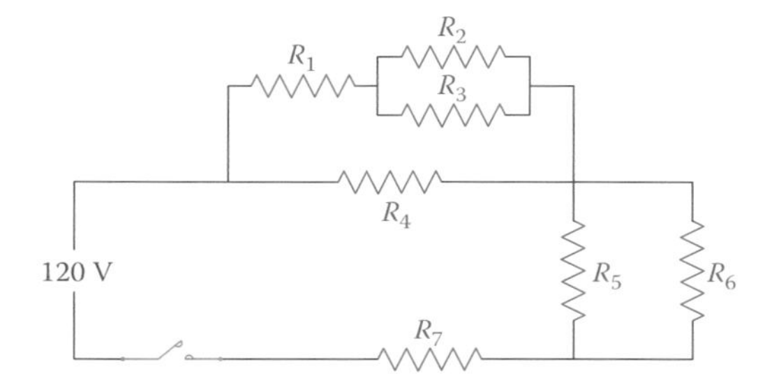

Figure 6.9 Another example of a combined circuit (more involved than that in Figure 6.7).

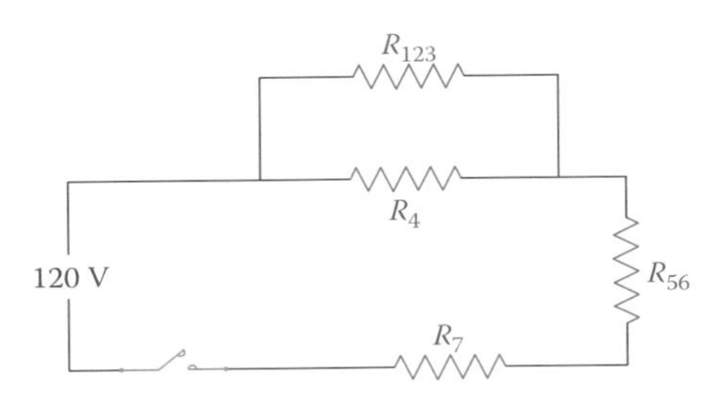

Figure 6.10 Result of operations 1 to 3 to the circuit of Figure 6.9.

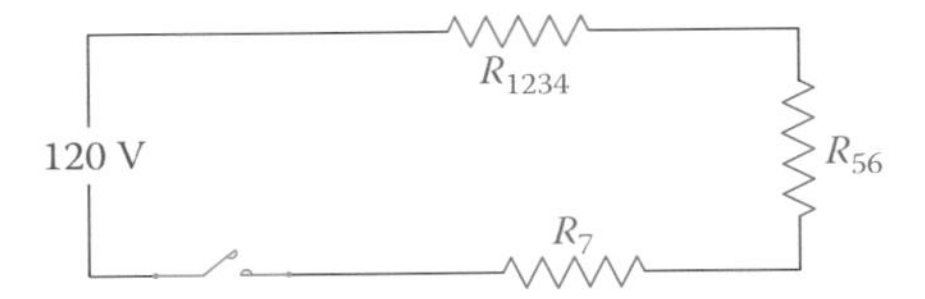

Figure 6.11 Simplification after steps 1 to 4 for the circuit in Figure 6.9.

Example 6.8

Find the equivalent resistance for the circuit of Figure 6.9, if $R_1 = 240\ \Omega$, $R_2 = 300\ \Omega$, $R_3 = 470\ \Omega$, $R_4 = 510\ \Omega$, $R_5 = 300\ \Omega$, $R_6 = 300\ \Omega$, and $R_7 = 2400\ \Omega$.

Solution

Following steps 1 to 4 as mentioned above leads to

1. Using Equation 6.5,

$$R_{23} = \frac{300 \times 470}{300 + 470} = 183.12\ \Omega$$

2. R_{23} is in series with R_1

$$R_{123} = 240 + 183.12 = 423.12\ \Omega$$

3. R_{56} is the resultant of R_5 and R_6 in parallel. However, since both are 300 Ω their equivalent resistance is 150 Ω.
4. R_{123} is in parallel with R_4; thus,

$$R_{1234} = \frac{423.12 \times 510}{423.12 + 510} = 231.26\ \Omega$$

Finally, the equivalent resistance of R_{1234}, R_{56}, and R_7 is

$$R = 231.26 + 150 + 2400 = 2781.26\ \Omega \approx 2781\ \Omega$$

Example 6.9

In Example 6.8, how much total current is in the circuit that the power supply must provide?

Solution

Because all the resistors can be substituted by a single resistance of 2781 Ω, the total current that the power supply must be able to provide is

$$120 \div 2781 = 0.043 \text{ A} = 43 \text{ mA}$$

Often in a circuit, like that under study, it is necessary to find the current in all or some of the resistors. For this task, an additional calculation is necessary. In the problem of Example 6.8, the current in resistor R_7 is the same as the total current as calculated, because R_7 is a part of the resistors in series with each other in the final calculation. However, if the current in any other resistor is required, it is not readily available without some more analysis. The general way to do so is employing Ohm's law for various parts of the circuit, as required. In this sense, if the current in a resistor is required, we need to have the voltage across the resistor (and divide by the value of the resistor). For example, if we need to know the current in resistor R_4 in the Example 6.8, we must determine the voltage across it, first.

Before we can go ahead with how to perform this task, we must discuss two rules, one for the series circuits and one for the parallel circuits.

6.7 Rule of Series Circuits

As we have discussed, all the components in a series circuit have the same current, but there are different voltages associated with a number of components in series. Figure 6.12 shows three resistors in series with each other and the voltage measured between their ends is V (any number of items can considered; three is randomly selected).

Because there are three resistors, three voltages can be defined denoted by V_1, V_2, and V_3 in Figure 6.12; if the circuit current is I, then

$$V_1 = R_1 I, \quad V_2 = R_2 I, \quad V_3 = R_3 I$$

The rule of series circuits states that

$$V = V_1 + V_2 + V_3 \tag{6.6}$$

That is, the sum of all the voltages across the components is equal to the supply voltage.

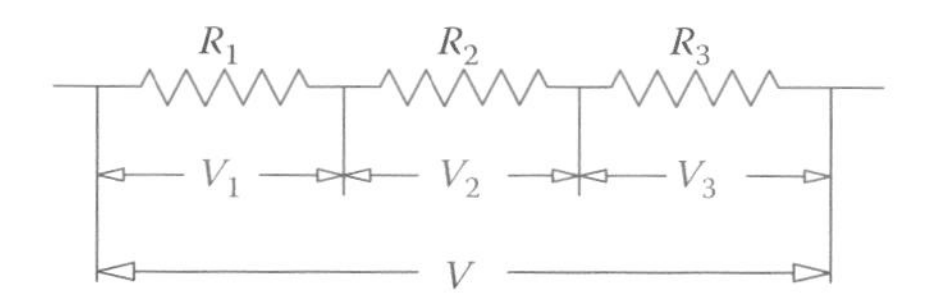

Figure 6.12 Three resistors in series under an applied voltage *V*.

Example 6.10

If a 240 Ω resistor and a 510 Ω resistor are in series and are connected to a 12 V battery with wires of negligible resistance (say, 0.5 Ω, which is negligible compared to 240 Ω), what voltage is across the 510 Ω resistor?

Solution

We need to find the current first. The common current in both resistors is

$$I = \frac{12}{240 + 510} = 0.016\ \text{A}$$

The voltage across the 510 Ω resistor is

$$V_{510} = (510)(0.016) = 8.16\ \text{V}$$

Example 6.11

If the current through three resistors in series with each other is 0.06 A and the resistance of the resistors are 470 Ω, 680 Ω, and 750 Ω, what is the applied voltage to the three resistors?

Solution

Applied voltage can be found from

$$V = (470 + 680 + 750)(0.06) = (1900)(90.06) = 114\ \text{V}$$

Alternatively, we may find the individual voltages and add them together

$$V_1 = (470)(0.06) = 28.2\ \text{V}$$

$$V_2 = (680)(0.06) = 40.8\ \text{V}$$

$$V_3 = (750)(0.06) = 45\ \text{V}$$

$$V = 28.2 + 40.8 + 45 = 114\ \text{V}$$

> In a series circuit the sum of all the voltages across the components (or voltage drops in the components) is equal to the total circuit voltage.

Example 6.12

In the circuit shown in Figure 6.13, find the value of R_3.

Solution

From Ohm's law the value for resistance R_3 can be found if the voltage across it and its current are known. The voltage across R_3 is found to be

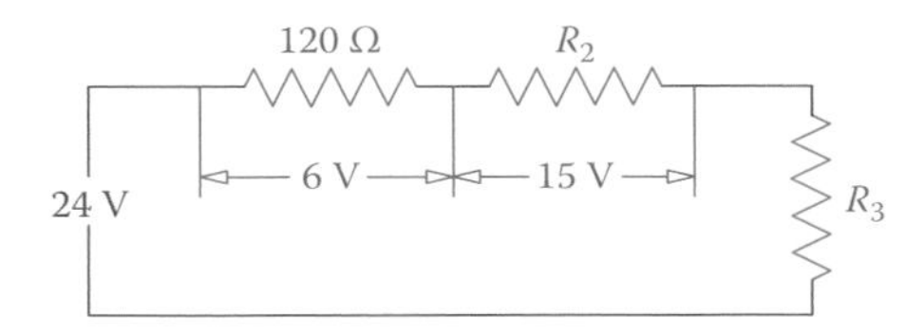

Figure 6.13 Example 6.12.

$$V_3 = 24 - (15 + 6) = 3 \text{ V}$$

and the current through R_3 is the same as that through the 120 Ω resistor, which is

$$I = 6 \text{ V} \div 120\ \Omega = 0.05 \text{ A}$$

Thus,

$$R_3 = 3 \div 0.05 = 60\ \Omega$$

Note that, alternatively, you could deduce that since the voltage across the 120 Ω resistor is 6 V, the resistance of R_3 must be 60 Ω, because the voltage across it is 3 V. This proportionality is always true for the voltage across series components. In general, we may say that for two resistors in series the voltage across the larger resistor is larger than the voltage across the smaller resistor.

For two series resistors the voltage across the larger resistor is larger than the voltage across the smaller resistor.

6.8 Rule of Parallel Circuits

In a parallel circuit, there are as many independent currents as there are branches, in the same way that in a series circuit, there are as many independent voltages as the number of components. Independent is used, because there are also currents that are dependent on the other currents and can be obtained by adding or subtracting from each other, as shown in Figure 6.14.

As marked in Figure 6.14, there are various currents in any circuit containing parallel components. There exist, however, some relationships between these currents. These relationships can be found from the rule of

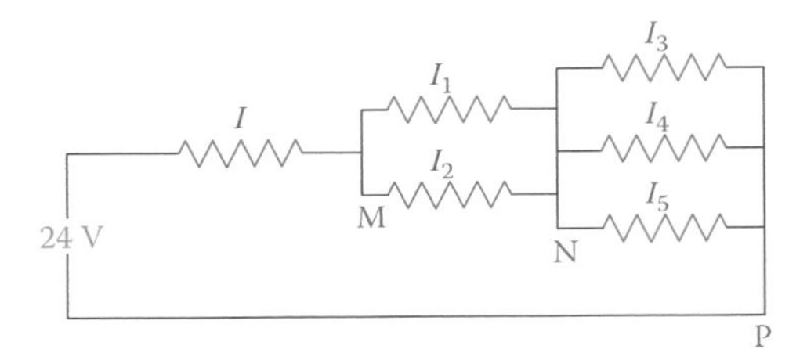

Figure 6.14 Different currents in branching (parallel) resistors.

parallel circuits. For more complicated circuits containing more than one source, other rules such as Kirchhoff's current law must be employed (see Section 12.6.2).

The simple rule of parallel circuits that you need to understand is that when a current carrying line divides into two or more parallel branches the current in that line divides between the branches. This is well illustrated in Figure 6.14. Current I divides into I_1 and I_2. Thus,

$$I_1 + I_2 = I \tag{6.7}$$

Also, the two currents I_1 and I_2 branch into three divisions carrying I_3, I_4, and I_5. In this sense,

$$I_3 + I_4 + I_5 = I_1 + I_2 = I \tag{6.8}$$

Example 6.13

In the circuit of Figure 6.15, $I = 1.5$ A and the current in R_2 is 0.5 A. What current is in R_1?

Solution

Because the sum of the currents I_1 and I_2 is 1.5 A and I_2 is 0.5 A, thus,

$$I_1 = 1 \text{ A}$$

Example 6.14

In the circuit shown in Figure 6.16 the voltage difference between points A and B is 9 V. What is the current in $R_1 = 90\ \Omega$ and $R_2 = 180\ \Omega$?

Solution

$$\text{Current in } R_1 = 9 \div 90 = 0.1 \text{ A}$$

$$\text{Current in } R_2 = 9 \div 180 = 0.05 \text{ A}$$

Note the important fact that because the voltage across the two resistors is the same (being in parallel), the current in the 90 Ω resistor is twice as

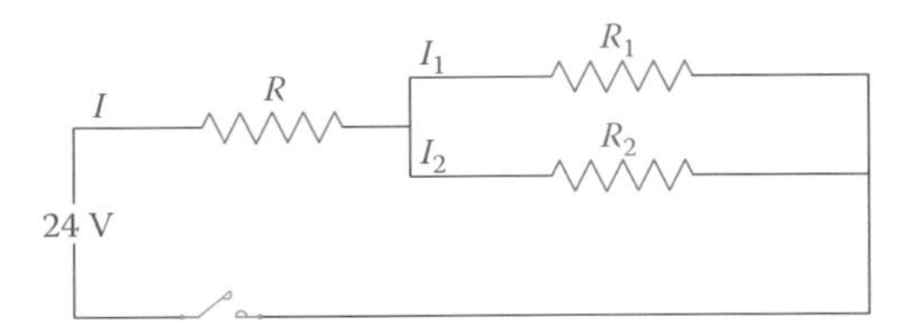

Figure 6.15 Circuit of Example 6.13.

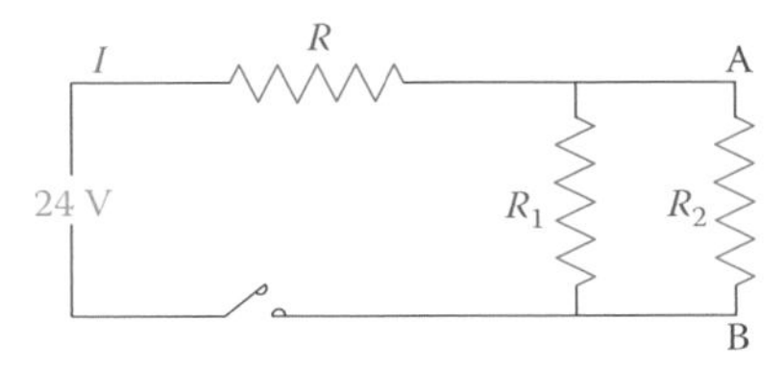

Figure 6.16 Circuit for Example 6.14.

much as the current through the 180 Ω resistor. In general, we may say that when two resistors are parallel, the current in the smaller resistor is larger than the current in the larger resistor.

> In two parallel resistors the current in the smaller resistor is larger than the current in the larger resistor.

6.9 Examples of Combined Circuits

In dealing with combined circuits, there is no unique method of analysis. Depending on the circuit and what information is required, one must determine the simplest way. Also, more than one method of solution might be found for certain problems. The general approach of reducing the parallel and series circuits to their respective equivalents must be successively followed. In this section a number of examples are solved to illustrate the matter.

Example 6.15

In the problem of Example 6.13, what is the resistance of resistor R if the voltage across $R_2 = 9$ V and the circuit current is 0.15 A?

Solution

Current through R is given (0.15 A), and the voltage across R can be determined from the other information given:

$$\text{Voltage across } R = V_R = 24 - 9 = 15 \text{ V (from series circuit rule)}$$

and the resistance of R is

$$R = 15 \div 0.15 = 100\ \Omega$$

Example 6.16

In the circuit of Figure 6.15, $R = 10\ \Omega$, $R_1 = 9\ \Omega$, and $R_2 = 18\ \Omega$. What is the current in R_1 if the circuit current is 1.5 A?

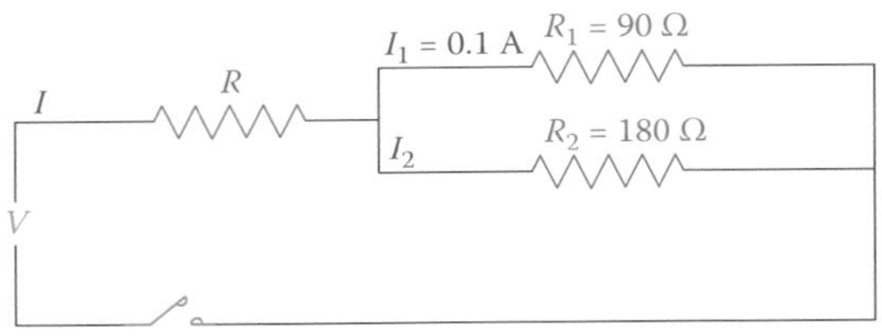

Figure 6.17 Circuit of Example 6.16.

Solution

$$V_R = RI_R = (10)(1.5) = 15 \text{ V}$$

$$V_{R1} = V_{R2} = 24 - 15 = 9 \text{ V}$$

$$I_{R1} = V_{R1} \div R_1 = 9 \div 9 = 1 \text{ A}$$

Example 6.17

In the circuit of Figure 6.17, if $R = 36\ \Omega$, what is the supply voltage?

Solution

Using Figure 6.17, the current in R_1 is 0.1 A. Because R_1 and R_2 are in parallel, the current through R_2 is 0.05 A (R_2 is twice in size as R_1). Thus, the total current in the circuit is 0.15 A.

Supply voltage is the sum of the voltage across R and the voltage across the parallel resistors.

$$\text{Voltage across } R = (36)(0.15) = 5.4 \text{ V}$$

$$\text{Voltage across } R_1 \text{ and } R_2 = (90)(0.1) = 9 \text{ V}$$

$$\text{Supply voltage} = 5.4 + 9 = 14.4 \text{ V}$$

Example 6.18

The circuit in Figure 6.18 is the same as that in Figure 6.9, shown differently. (It might look different, but it is the same). If $R_1 = 240\ \Omega$, $R_2 = 300\ \Omega$, $R_3 = 470\ \Omega$, $R_4 = 510\ \Omega$, $R_5 = 300\ \Omega$, $R_6 = 300\ \Omega$, and $R_7 = 2400\ \Omega$, the total current was found (see Examples 6.8 and 6.9) to be approximately 43 mA. How much current is in R_4?

Solution

To find the current in R_4, we need to determine the voltage across it, first. If the voltage across R_7 and the voltage across

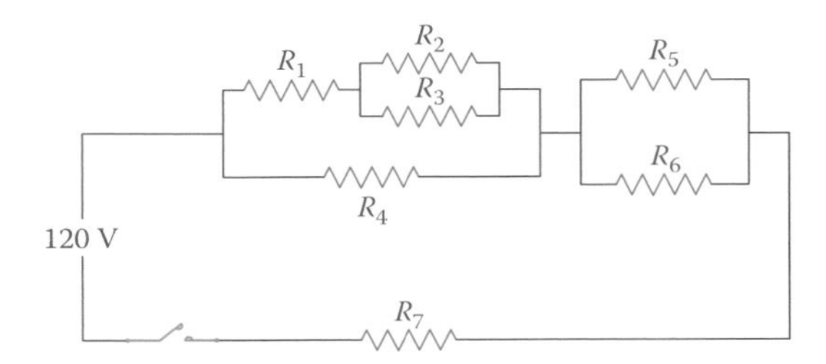

Figure 6.18 Circuit of Example 6.18.

R_5 (or R_6) can be subtracted from the supply voltage, then the voltage across R_4 can be determined.

$$V_{R7} = R_7 I = (2400)(0.043) = 103.2 \text{ V}$$

In finding the voltage across R_5 and R_6 we need to use their equivalent value R_{56} if the 43 mA current is to be used.

$$V_{R5} = V_{R6} = R_{56} I = (150)(0.043) = 6.45 \text{ V} \quad \text{(see Example 6.8)}$$

$$V_{R4} = 120 - (103.2 + 6.45) = 120 - 109.65 = 10.35 \text{ V}$$

$$I_{R4} = V_{R4} \div R_4 = 10.35 \div 510 = 0.020 \text{ A} = 20 \text{ mA}$$

Example 6.19

In the problem of Example 6.18, what is the power consumption of R_2?

Solution

To find the power in R_2, either the voltage across it or the current in it must be known (in addition to its resistance). If we concentrate on the part of the circuit containing R_1, R_2, R_3, and R_4, we recall from Example 6.8 that the circuit was reduced to that of Figure 6.10. Because the total current is 43 mA and the current in R_4 is 20 mA (from Example 6.18), the current in R_{123} is 23 mA (43 – 20).

From the fact that the voltage across R_4 (and R_{123}) is 10.35 V (see Example 6.18) and the current in R_{123} is 23 mA, we can conclude that the voltage across R_2 and R_3 in the original circuit is

$$\begin{aligned} 10.35 - (R_1)(0.023) &= 10.35 - (240)(0.023) \\ &= 10.35 - 5.52 = 4.83 \text{ V} \end{aligned}$$

Power consumption of the resistor R_2 can now be directly found from Equation 5.6:

$$P = \frac{V^2}{R} = \frac{4.83^2}{300} = 0.07776 \text{ W} = 77.76 \text{ mW}$$

6.10 Voltage Divider

Voltage divider: Electric circuit made up of components in series that can deliver two or more smaller voltages from a voltage applied to the circuit.

A **voltage divider** is a circuit composed of series components that is used to provide one or more voltages from a single voltage supply. These voltages are lower than that of the power supply. This is not the most efficient way of obtaining a desired voltage from a power supply, but it is often the simplest or the easiest way. Suppose that you need 7.5 V for an application, but you have 12 V power supply. In such a case you may use a voltage divider to obtain 7.5 V from 12 V.

Figure 6.19 depicts the arrangement of a voltage divider. This voltage divider is made out of three resistors. On the basis of the voltage of the power supply and the values of the resistors, a current passes through all the resistors. Then various voltages can be observed between points A-B, A-C, B-C, and B-D in addition to the supply voltage V across A-D. The load is connected between the appropriate points across a resistor, based on the desired voltage. One important fact about the use of voltage dividers is that the load resistance cannot be small compared to the resistor across which it is connected. Otherwise, the voltage relations will not remain as desired. As a rule of thumb, you may say the resistance of the load must be 10 times or more larger than the resistance of the resistor to which it is connected.

Various voltages that are available from a voltage divider can be directly found from the values of the resistors and the supply voltage. For instance, for the voltage divider in Figure 6.19, we may write the following relations. This formulation is general and can be used for any voltage divider.

$$V_{AB} = V \times \frac{R_1}{R_1 + R_2 + R_3}$$

$$V_{BC} = V \times \frac{R_2}{R_1 + R_2 + R_3}$$

$$V_{CD} = V \times \frac{R_3}{R_1 + R_2 + R_3} \tag{6.9}$$

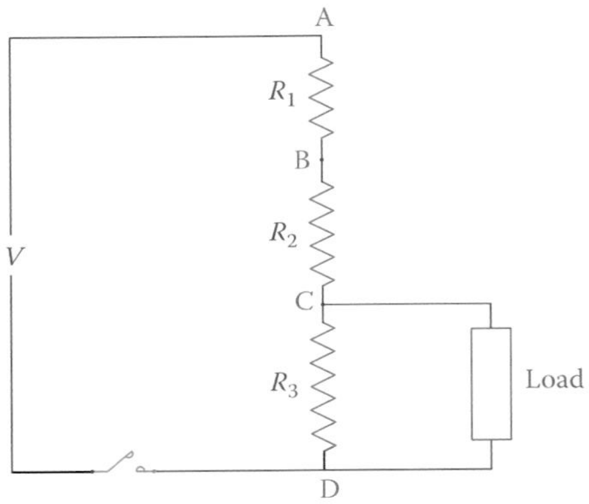

Figure 6.19 Voltage divider.

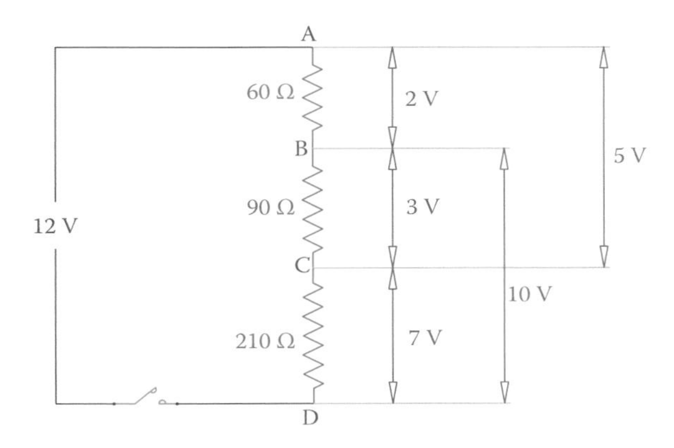

Figure 6.20 Getting different voltages from one source with a voltage divider.

$$V_{AC} = V \times \frac{R_1 + R_2}{R_1 + R_2 + R_3}$$

$$V_{BD} = V \times \frac{R_2 + R_3}{R_1 + R_2 + R_3}$$

As can be seen from this formulation, to find the voltage between any two points, the supply voltage must be multiplied by the ratio of the resistance between those points to the total resistance in series.

In Figure 6.20 a 12 V battery is used as a power supply, but with this arrangement various voltages can be obtained as shown.

Example 6.20

You need to make a voltage divider consisting of two resistors to obtain 7.5 V from a 12 V battery. If you already have a 51 Ω resistor, what other resistor do you need?

Solution

This problem has two solutions, based on if you use the 51 Ω resistor as the smaller or the larger of the two resistors. Suppose that you will use it as the larger resistor. If the value of the resistor to be found is represented by *R*, from the relationships given in Equation 6.9,

$$\frac{51}{51 + R} = \frac{7.5}{12} = 0.625$$

This is a simple equation that can be solved. In case you are not yet familiar with this type of equation, you may want to write it differently, which can look easier, that is (invert both sides),

$$\frac{51+R}{51} = \frac{12}{7.5} = 1.6$$

The value of resistor R can be found to be

$$R = 30.6\ \Omega \approx 30\ \Omega$$

30 Ω is a standard resistor and can be used for this purpose.

If, instead, you want the 51 Ω resistor as the smaller resistor, then

$$\frac{51}{51+R} = \frac{4.5}{12} = 0.375$$

or

$$\frac{51+R}{51} = \frac{12}{4.5} = 2.67$$

from which the value of R can be found to be

$$R = (51)(2.67 - 1) = (51)(1.67) = 85\ \Omega$$

6.11 Capacitors in DC Circuits

As we learned earlier, a capacitor acts as a storage device to electricity. The amount of storage depends on the capacity of the capacitor. We want to study what happens if a capacitor is included with resistors connected to a DC circuit. Figure 6.21a is a simple DC circuit with a resistive load, to which a capacitor is added, as in Figure 6.21b. An analogous equivalent hydraulic circuit, for better tangibility, is shown in Figure 6.22. Adding a capacitor to an electric circuit is equivalent to adding a closed tank with limited capacity or a tall tank (with walls that are tall enough and comparable with the main reservoir) at the end of a pipeline.

Water flows in the pipeline in Figure 6.22a as the electricity flows in the circuit of Figure 6.21a, as long as the reservoir (battery) is not emptied. Nevertheless, in the case of Figure 6.22b a current of water flows as long as the small tank is not full. That is, for a short period of time, there is a current in the pipeline, but after a while water stops flowing. Moreover, this flow of water is not uniform during the period. In the beginning (when

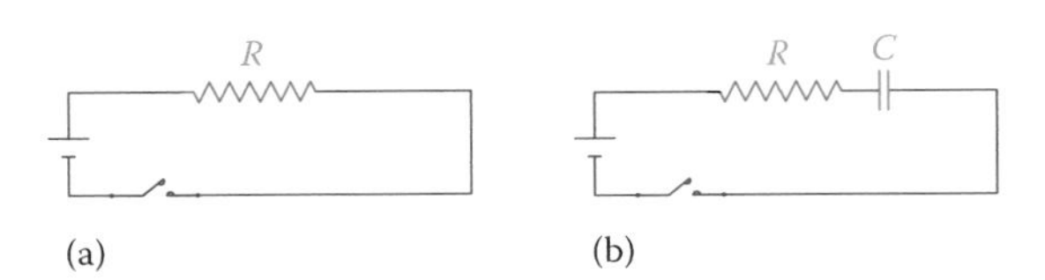

Figure 6.21 Adding a capacitor to a DC circuit. (a) Circuit without capacitor; normal current after switch is closed. (b) Circuit with a capacitor; current only for a short time no current afterwards.

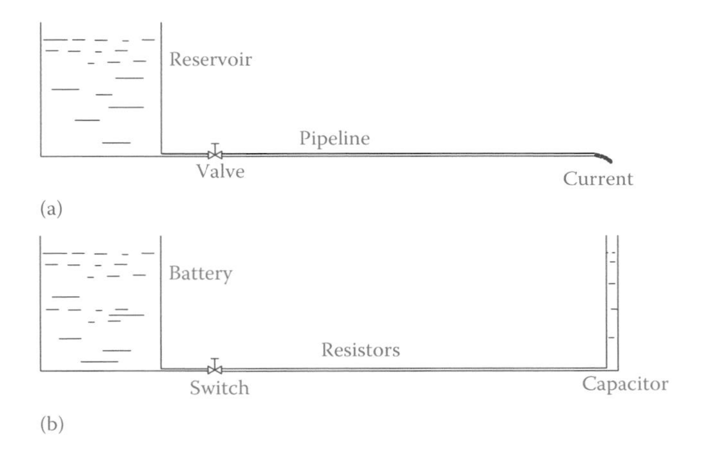

Figure 6.22 Adding a small tank to the end of a pipeline. (a) Hydraulic system analogous to Figure 6.21a. (b) Hydraulic system analogous to the DC circuit with a capacitor, in Figure 6.21b.

the tank is empty), more water flows but toward the end the flow of water becomes less and less, approaching zero. This is exactly what happens in the electric circuit containing a capacitor, too.

When a capacitor is inserted inside a DC circuit, for a short period of time after the switch is turned on, current flows in the circuit. In the beginning this current is higher but gradually becomes smaller and smaller until it diminishes. This is when the capacitor has charged, and it does not accept electric charge anymore. At this time and afterward, there is no current flowing in the circuit. Thus, except for a short period in the beginning a capacitor in a DC circuit blocks the circuit and does not allow any current.

A charged capacitor contains electricity and behaves like live electricity. It can be dangerous if the voltage is sufficiently high. The voltage level to which a capacitor charges is the same as that of the battery (power supply) if the capacitor becomes fully charged. If the circuit is broken (the switch is turned off) while the capacitor is charging, it only partially charges and to a voltage less than the applied voltage.

> A capacitor in a DC circuit blocks the current, except for only a short period following a change such as after a switch is closed (or opened if already closed).

It is interesting to know how long it takes for a capacitor to charge. In fact, in many timing devices, including electronic watches, this time period is used for time keeping. The time for charging or discharging a capacitor has a direct correspondence with the problem for a water tank to fill (under a given water pressure) or empty if already full (based on the height of water and the resistance of the pipeline from which water discharges).

Here a simple rule is given for the calculation of the time for charging a capacitor. At the end of this chapter we provide additional material on this subject. Referring to Figure 6.21b the circuit consists of a resistor

Time constant: Duration of time for a circuit containing a capacitor or inductor and resistors to reach 63 percent of a new value after a change has happened to the circuit (e.g., its power has been turned on or off).

with resistance R and a capacitor with capacitance C in series with each other connected to battery (power supply) of voltage V. We define a **time constant** τ (the Greek letter corresponding to English t, and pronounced tow).

$$\tau = RC \tag{6.10}$$

When R is defined in ohm and C in farad, τ is automatically expressed in seconds. As can be seen, the time constant τ depends on the values of R and C. It takes approximately 5 times the time constant τ for the capacitor to fully charge to voltage V. This time starts from the instant that the switch is closed. But, if during this time interval the switch is opened, the capacitor does not fully charge. The time constant is sometimes shown by T (uppercase) instead of τ.

Example 6.21

A 51 Ω resistor and a 10 mF (millifarad) capacitor in series are connected to a 12 V battery. How long does it take for the capacitor to charge?

Solution

The time constant is calculated first:

$$\tau = RC = (51)(10 \div 1000) = 0.51 \text{ sec}$$

The required time for charging the capacitor = $5\tau = (5)(0.51) =$ 2.55 sec

Example 6.22

How much current is in the circuit of Example 6.21 after 5 sec following closing the switch?

Solution

Current diminishes to zero at the end of the charging time (2.55 sec). Thus, the current at 5 sec after the switch is closed is 0.

Because this time period of charging is relatively very small, it can be ignored compared with longer periods of time i. Thus, in general, we consider a DC circuit to be open when it contains a capacitor.

> Except for a very short period in the beginning, a capacitor in a DC circuit behaves as an open circuit and does not allow any current.

> It takes approximately 5 times the time constant for a capacitor to either charge or discharge.

6.12 Inductors in DC Circuits

Figure 6.23 illustrates a simple DC circuit consisting of a resistor with which an inductor is added in series. We are interested to see what the effect of this inductor to the circuit is. The effect of an inductor in an electric circuit is always to oppose a change in the circuit current. For instance, at the instant the switch is closed, current tends to increase from zero to a value that depends on the voltage and resistance in the circuit. An inductor wants to keep the current unchanged. Eventually, after a short period of change, the current settles at a value and remains constant for this DC circuit.

Adding an inductor in an electric circuit is analogous to adding a flexible (stretchable) tank (reservoir) to a hydraulic pipeline. When the tank stretches, its volume changes. When the valve in the pipeline is opened to let water flow, the flowing water must fill the flexible reservoir before it can continue its normal flow. Thus, part of the flow is used to fill the reservoir. Once the reservoir is filled, water flow continues in the pipeline at its normal rate. In other words, the flow of water is initially resisted and limited for a short period of time but afterward will have a continuous and constant current.

> The effect of an inductor in an electric circuit is to oppose a change in the circuit current (keeping the current unchanged).

Here again it is interesting to know how long it takes for the electric current in Figure 6.23b to reach its constant value or for the water in Figure 6.24b to reach its normal flow. Associated with the circuit resistance R and inductance L values, we may define a time constant (similar to the case for a capacitor) as follows:

$$\tau = \frac{L}{R} \tag{6.11}$$

When L is defined in henries and R in ohms, τ will be in seconds. This equation implies that for smaller values of R the duration of the effect of inductor is longer. It takes approximately 5 times the time constant for the current to reach its constant value.

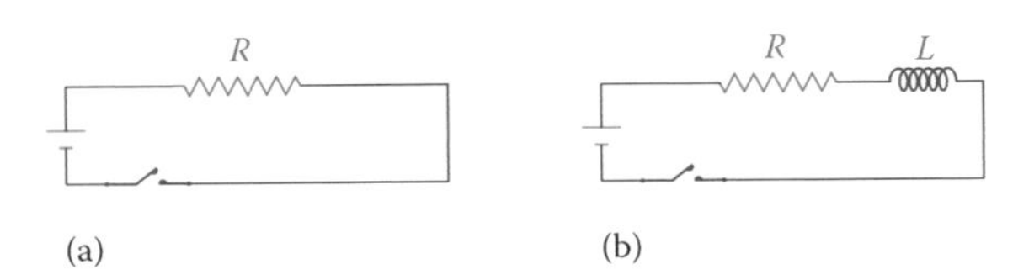

Figure 6.23 Adding an inductor in series with a resistor in a DC circuit. (a) Original circuit. (b) Circuit with an inductor.

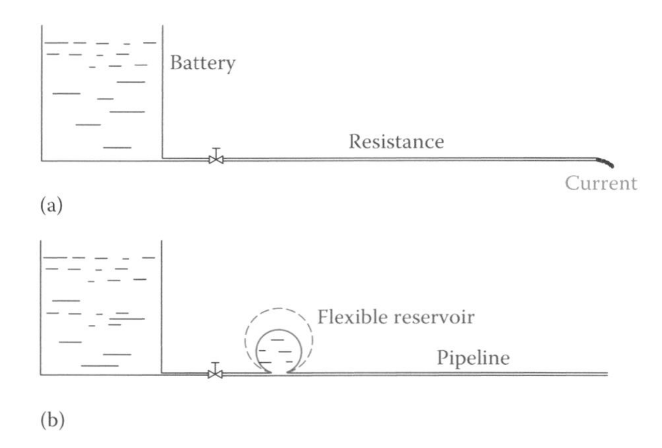

Figure 6.24 Analogy of a hydraulic system to an electric circuit when an inductor is added. (a) Circuit without an inductor. (b) Circuit with an inductor.

Example 6.23

What is the time constant in a circuit containing a resistor and a coil in series? The resistor has a value of 51 Ω, and the coil has an inductance of 11 mH and a resistance of 4 Ω.

Solution

Substituting in Equation 6.11 gives

$$\tau = \frac{L}{R} = \frac{0.011}{51+4} = 0.0002 \text{ sec} = 0.2 \text{ msec}$$

Example 6.24

What is the final current in the circuit of Example 6.23 if the battery voltage is 6 V? And how long after the switch is closed does the current reach this final value?

Solution

The final current is established after the effect of the inductor on current ends. The final current is found from Ohm's law. The total resistance of the circuit is 51 + 4 = 55 Ω, and on the basis of Ohm's law,

$$I = \frac{V}{R} = \frac{6}{55} = 0.109 \text{ A} = 109 \text{ mA}$$

It takes approximately 5 times the time constant before the current reaches this value; thus, the current reaches 109 mA after

$$(5)(0.2) = 1 \text{ msec}$$

It takes approximately 5 times the time constant for the effect of an inductor in a DC electric circuit to disappear.

Time Constant in DC Circuits

CAPACITOR

A capacitor has a storage capability for electricity. When it is part of a DC circuit, it exhibits an apparent opposition to a change in the circuit voltage. When a switch turns on or off the power to an electric circuit, it introduces a change of voltage to the circuit. During turning the power on in a DC circuit containing a capacitor, first, the capacitor is charged. Similarly, at power off, first, a capacitor in the circuit releases its charge to the circuit. Consequently, a delay is associated with both turning on and turning off a DC circuit. As a result of this delay, it takes a short while before the circuit containing a capacitor reaches its equilibrium and comes to its final condition when a change in voltage, either increasing or decreasing, occurs. This delay depends on the capacitance of the capacitor and the resistances in the circuit.

The delay just mentioned is approximately 5 times the time constant of the circuit.

The time constant for a circuit containing capacitors with an equivalent capacitance C and resistors with an equivalent resistance R is

$$\text{Time constant} = RC$$

When R is in ohm and C is in Farad, the time constant is determined in second.

The figure below shows the two cases of increasing and decreasing the voltage in a DC circuit. Because this change in voltage can be any value, it is normalized and its magnitude is shown as 1, which percentagewise represents 100 percent. The figure illustrates the gradual change of voltage during the time it takes for this change. The time constant is depicted as T, and the percentage change after the elapse of each time constant T is also shown.

For an increase in voltage (like turning on a switch), part (a) in the figure illustrates that subsequent to the initiation of an increase in voltage, after T seconds the voltage increases by 63 percent of difference between the new and the old values. In the figure, the old value is zero, but it can be a nonzero value. After $2T$ sec the voltage is augmented to 86 percent of the change. After $5T$, 99 percent of the change is reached, which is almost the total change to occur.

Part (b) of the figure illustrates a decrease in voltage. If the change is caused by turning the power off, so that the final voltage is zero, 63 percent of decrease in voltage takes place after one time constant. That is, the circuit voltage drops to 37 percent of its initial value. Accordingly, after $2T$ the voltage drops to 14 percent and after $5T$ the decrease is

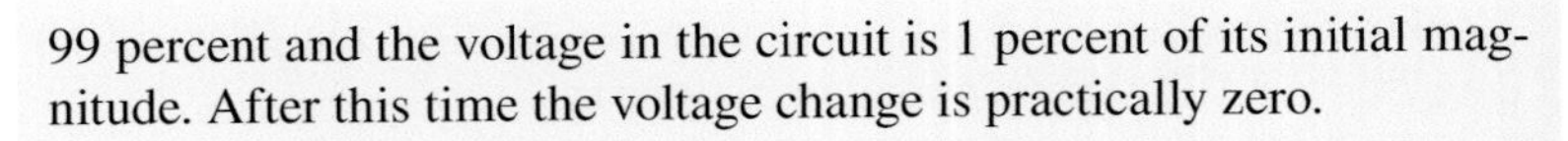

99 percent and the voltage in the circuit is 1 percent of its initial magnitude. After this time the voltage change is practically zero.

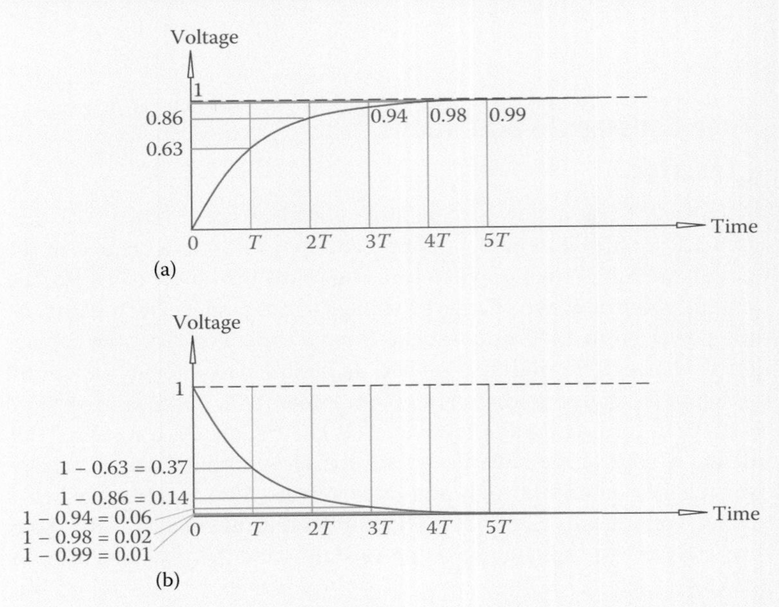

The delay introduced by a capacitor has many applications, particularly in electronic circuits used in radio, TV, communications, and so on, as well as in electrical applications that need control and regulation. By changing the time constant, through R and C, a desired delay can be obtained for any application.

INDUCTOR

In the same way that a capacitor causes a delay for the voltage change in a circuit, because of its intrinsic electricity storage property, an inductor in a circuit causes a delay. However, the delay caused by an inductor is because of the property of an inductor to resists a change in the current in the circuit it is a part.

The time constant for a circuit containing an inductor and a resistor is defined by

$$\text{Time constant} = \frac{L}{R}$$

When L is henries and R is in ohms the calculated value of time constant is in seconds.

All what was said about the time delay for changing voltage in a circuit containing capacitors and resistors and the corresponding graphs is true for a circuit having inductor in it. However, in this case, the values and the percentage of change shown at the end of periods T, $2T$, $3T$, etc., corresponds to changes of current in a circuit and not voltage.

6.13 Power in DC Circuits

In the previous two sections, we studied the effect of a capacitor and an inductor in a DC circuit. On the basis of this study we know that this effect is only for a short period of time after a switch is turned on or off. Turning a switch on and off is, in fact, a change of power to a circuit. In many electric circuits when a switch is turned on, no other condition changes for a relatively long time. However, in electronic circuits many factors can change with time and new conditions are introduced repeatedly and on a continuous basis. For example, in a radio circuit the signals (voltages created by a microphone) corresponding to someone's talking, or music, continuously vary.

In this respect, the time scale of electric circuits and electronic circuits are completely different. Without loss of generality we can assume that for DC electric circuits we deal only with resistors (or resistive components), because the existence of a capacitor introduces an open in a circuit, and an inductor only adds its resistance to a circuit. The matter is different for AC circuits, which we will discuss in Chapter 8. Moreover, electronic circuits are dealt with later in Chapters 13 to 20. As a result, from now on when we refer to a DC electric circuit it is assumed that we have only resistive elements in the circuit (no capacitor and no inductor).

In this section we consider the power associated with a DC circuit. As noted before, the resistors are the consumers of power and the electric power supply is the provider of that power. A circuit can be a series circuit, a parallel circuit, and a series-parallel combination circuit. In all these cases the total power consumption by the circuit is the sum of all the individual power consumption of all the resistors in the circuit. To find power consumption for a circuit, thus, we may (a) find and add together all the individual powers, (b) find the power in the equivalent resistor of the circuit, or (c) find the power consumption in various parts of the circuit. The following example illustrates the matter.

Example 6.25

The circuit shown below is the same circuit as shown in Figure 6.14. It is illustrated this way for your convenience (the physical circuit could have been not so straightforward). Values of resistors are shown in the figure. We want to find the total power (consumption) of this circuit. Pay attention to the numbering of the resistors. This way, the same number subscripts are used for the currents and the resistors; that is, the current in R_1 is I_1 and so on, except for R_6 (whose current is the total current in the circuit). See Figure 6.25.

Solution

To find the individual power consumptions, the current for each resistor should be known. Also, for calculation of the power using the total current, the equivalent resistance of the whole circuit is required. Although the second way needs less

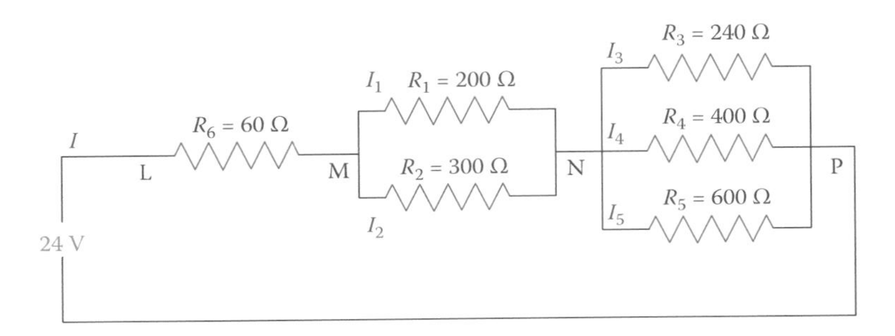

Figure 6.25 Circuit and values for Example 6.25.

calculation, here we are going to find the power in both ways as a demonstration that the two values are equal.

There are two sets of parallel resistors in this circuit, which when substituted by their equivalent resistors, respectively, the whole circuit reduces to three resistors in series. For ease of reference, we numerate each step as we proceed:

1. Find the equivalent values of each parallel set of resistors:

$$\text{Equivalent resistance of } R_1 \text{ and } R_2 = R_{12} = \frac{(200)(300)}{500} = 120\ \Omega$$

$$\text{Equivalent resistance of } R_4 \text{ and } R_5 = R_{45} = \frac{(400)(600)}{1000} = 240\ \Omega$$

$$\text{Equivalent resistance of } R_3 \text{ and } R_{45} = R_{345} = \frac{(240)(240)}{480} = 120\ \Omega$$

2. Find the equivalent resistance of series resistors $R_6 + R_{12} + R_{345}$:

   $R_{Total} = R_6 + R_{12} + R_{345} = 60 + 120 + 120 = 300\ \Omega$

   At this point we may use the rule of a voltage divider to determine the voltages across different points as follows:

$$\text{Voltage between points L and M} = V_{LM} = 24 \times \frac{60}{300} = 4.8\text{ V}$$

$$\text{Voltage between points M and N} = V_{MN} = 24 \times \frac{120}{300} = 9.6\text{ V}$$

$$\text{Voltage between points N and P} = V_{NP} = 24 \times \frac{120}{300} = 9.6\text{ V}$$

3. Find the total current:

$$I = \frac{24}{300} = 0.08\text{ A} = 80\text{ mA}$$

4. Find the currents in R_1 and R_2:

$$I_1 = \frac{9.6}{200} = 0.048\ \text{A} = 48\ \text{mA}$$

$$I_2 = \frac{9.6}{300} = 0.032\ \text{A} = 32\ \text{mA}$$

5. Find the currents in R_3, R_4, and R_5:

$$I_3 = \frac{9.6}{240} = 0.040\ \text{A} = 40\ \text{mA}$$

$$I_4 = \frac{9.6}{400} = 0.024\ \text{A} = 24\ \text{mA}$$

$$I_5 = \frac{9.6}{600} = 0.016\ \text{A} = 16\ \text{mA}$$

6. Verify the currents in R_1 to R_6:

$$I_1 + I_2 = 48 + 32 = 80\ \text{mA}$$

$$I_3 + I_4 + I_5 = 40 + 24 + 16 = 80\ \text{mA}$$

7. Find the sum of the powers in all resistors:

$$\begin{aligned} P &= (60)(0.08)^2 + (200)(0.048)^2 + (300)(0.032)^2 \\ &\quad + (240)(0.04)^2 + (400)(0.024)^2 + (600)(0.016)^2 \\ &= 0.384 + 0.4608 + 0.3072 + 0.384 + 0.2304 + 0.1536 \\ &= 1.920\ \text{W} \end{aligned}$$

8. Find the power from equivalent resistance:

$$P = (300)(0.08)^2 = 1.920\ \text{W}$$

9. Also, you may want to verify the following:

$(4.8)(0.08) + (9.6)(0.08) + (9.6)(0.08) = 0.384 + 0.768 + 0.768 = 1.92\ \text{W}$ (using $P = VI$)

6.14 Chapter Summary

- In a series circuit all components are connected one after another, so that they form a single loop.
- In a series circuit the current is the same for all components in series with each other.
- In a series circuit, applied voltage divides between the components. The component with larger resistance has a larger voltage across it.
- In a parallel circuit the same voltage is applied to all components in parallel. Each component has a direct connection to the supply voltage.

- In a parallel circuit there are as many currents as there are circuit branches.
- In a parallel circuit, total current is obtained by adding all the individual currents (the currents in all branches). This is the current through the power supply, and the power supply must be able (have enough power) to provide this current.
- In a parallel circuit the component with the largest resistance has the least current.
- In either a series circuit or a parallel circuit composed of resistors, one can find one single resistor to replace all the resistors. This resistor is called the equivalent resistor.
- In a series circuit the equivalent resistor is larger than all the resistors. In a parallel circuit the equivalent resistor is smaller than all the resistors.
- A combined circuit has both series and parallel parts.
- A combined circuit consisting of resistors can be represented only by one single *equivalent resistor.*
- All the rules of series and parallel circuits apply to series parts and parallel parts of a combined circuit, respectively.
- A voltage divider is an arrangement that allows various voltages from a single power source with one single voltage. Various voltages obtained by a voltage divider are always smaller than the power supply voltage.
- A capacitor in a DC circuit acts as an open circuit and does block the current.
- An inductor in a DC circuit adds its resistance to the circuit.
- Associated with a capacitor (or an inductor) in an electric circuit, there is a time constant that depends on the circuit resistance and the capacitor capacitance (inductor inductance).
- For only a very short period of time a capacitor or an inductor in a DC circuit affects the current in the circuit. This period is approximately 5 times the time constant.
- Power in any DC circuit can be found from the total current and the equivalent circuit or from adding the individual powers of all the resistors.

Review Questions

1. What is a series circuit? How many loops does a series circuit have?

2. What is the equivalent resistor for a set of series resistors? Is it larger or smaller than the largest resistor? Is it larger or smaller than the smallest resistor?

3. In series resistors, how many switches do you need to turn off the power from all the resistors?

4. How do you define parallel circuits?

5. Are the four elements on the top of an electric oven in series or in parallel?

6. How do you determine the equivalent resistor in parallel resistors?

7. What is the relationship for voltages in parallel resistors?

8. What is the relationship for currents in parallel resistors?

9. What is meant by combined series and parallel circuit?

10. Can you substitute a set of resistors combined together in series and parallel by a single resistor?

11. What is a voltage divider used for?

12. Can one obtain any desired voltage from a voltage divider?

13. Can you get 12 V from a voltage divider if the power supply is 9 V?

14. Which way is better to find power in a DC circuit, adding all the individual powers or using the equivalent circuit and its current?

15. In a DC circuit, is the power consumption of all the loads larger or the power to be delivered by the power supply? If so, why?

16. Does a capacitor conduct electricity when in a DC circuit?

17. Does an inductor conduct electricity when in a DC circuit?

18. What is "time constant" in a DC electric circuit? Does it mean that the time is constant?

19. When dealing with an electric circuit, when/what does normally define the start of time?

Problems

1. A lightbulb is connected to 120 V electricity through a switch, using three pieces of wire (one between the + and the switch, one between the switch and the lightbulb and one between the lightbulb and the –) of the required

lengths. If the resistance of the lightbulb filament is 70 Ω, the resistance of each shorter wire is 0.5 Ω and the resistance of the wire between the light and the switch is 1 Ω, draw a schematic of the circuit. What is the total resistance in the circuit?

2. What is the current in the circuit of Problem 1?

3. In Problem 1, what is the power that the source must supply?

4. In Problem 1, how much of the power provided by the power supply is used in the filament of the lightbulb?

5. Where does the rest of the power provided by the power supply go?

6. What is the equivalent resistance of the circuit in Figure P6.1?

7. If the current in the circuit of Figure P6.2 is 12 mA, what is the resistance R_2?

8. In the following circuit (Figure P6.3), what is the voltage across R_3?

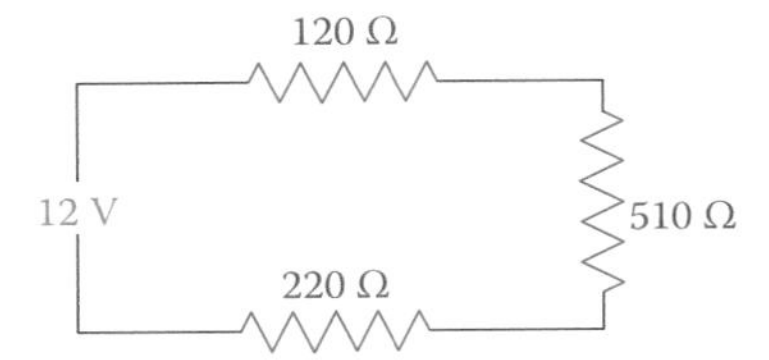

Figure P6.1 Circuit of Problem 6.

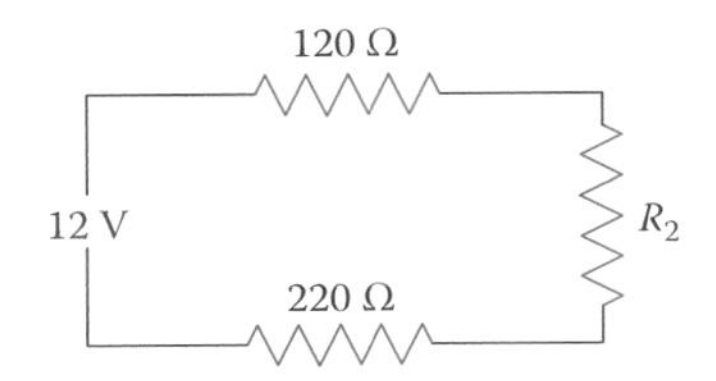

Figure P6.2 Circuit of Problem 7.

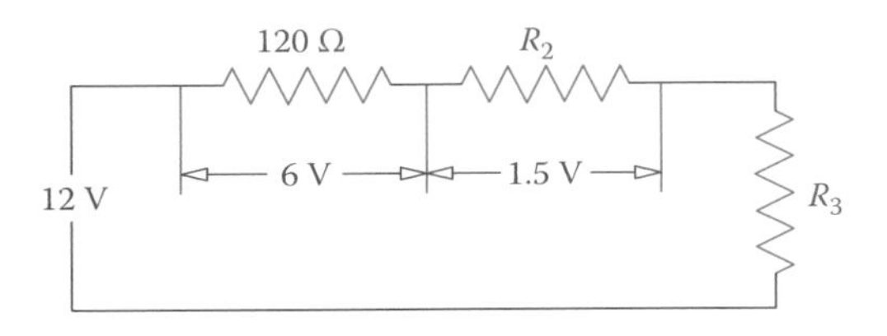

Figure P6.3 Circuit of Problem 8.

9. In the circuit of Problem 8, find the value of R_3.

10. What is the total power used in the circuit of Figure P6.3?

11. Find the total current in the circuit of Figure P6.4.

12. Explain why the circuit in Figure P6.5 is a parallel circuit.

13. Find the current in the circuit shown in Figure P6.5.

14. Find the equivalent resistance of the circuit shown in Figure P6.6.

15. Find the current in each of the 51 Ω resistors in the circuit of Problem 14.

16. In the circuit of Figure P6.6 determine, without calculation, if the current in the left 100 Ω resistor is higher or lower than that on the right?

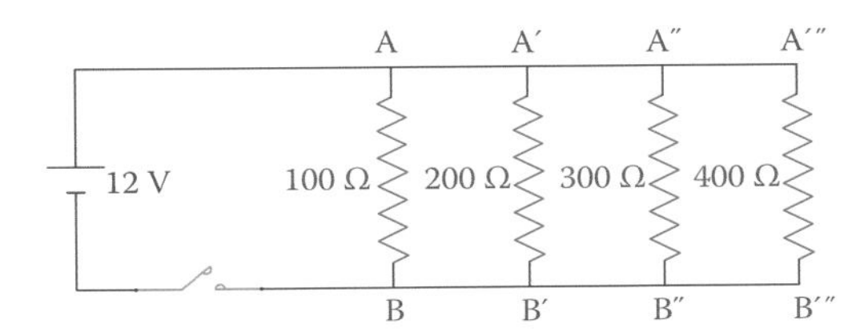

Figure P6.4 Circuit of Problem 11.

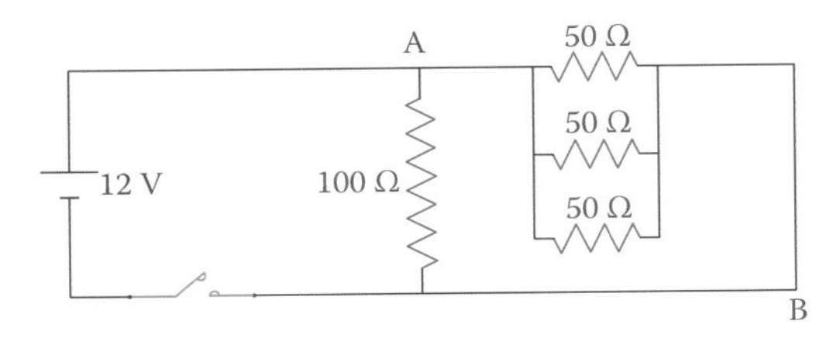

Figure P6.5 Circuit of Problems 12 and 13.

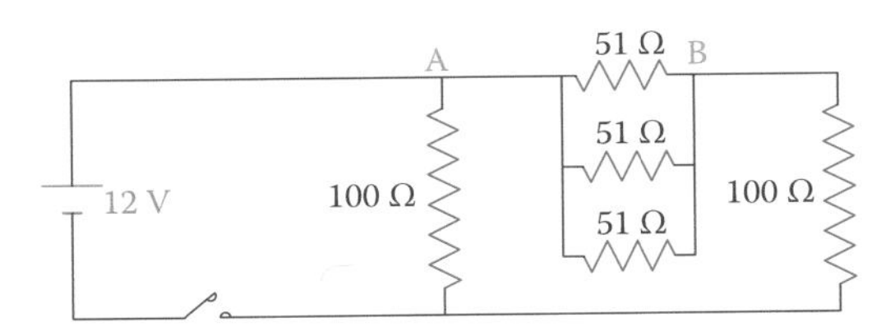

Figure P6.6 Circuit of Problem 14.

17. Find the equivalent resistor of the circuit shown in Figure P6.7.

18. Find the power consumption of the circuit of Figure P6.7.

19. In the circuit of Figure P6.8, if all the resistors have similar power capacity, what is wrong with the design of this circuit?

20. How much is the total power consumption of the circuit in Figure P6.8?

21. Using the conversion tables in Chapter 2, determine how many calories are generated in 1 sec in the circuit of Problem 20.

22. Determine how many calories are generated in the circuit of Problem 20 within 24 hr.

23. In the circuit shown in Figure P6.9 the coil has 10 Ω resistance. What is the time constant of the R-L circuit?

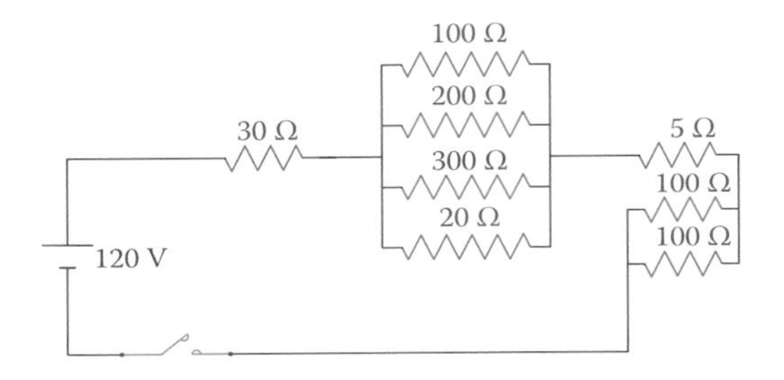

Figure P6.7 Circuit of Problem 17.

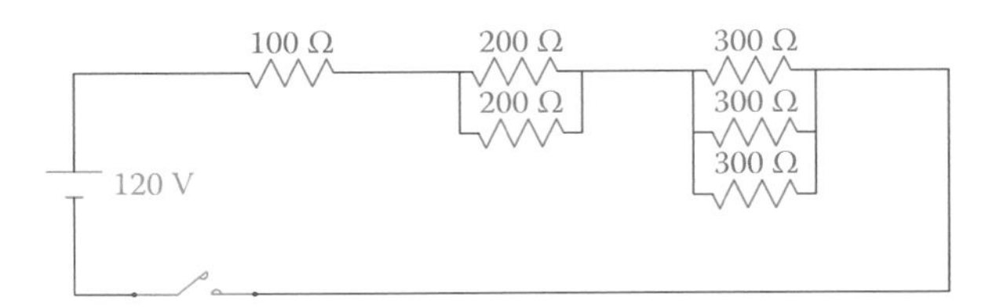

Figure P6.8 Circuit of Problem 19.

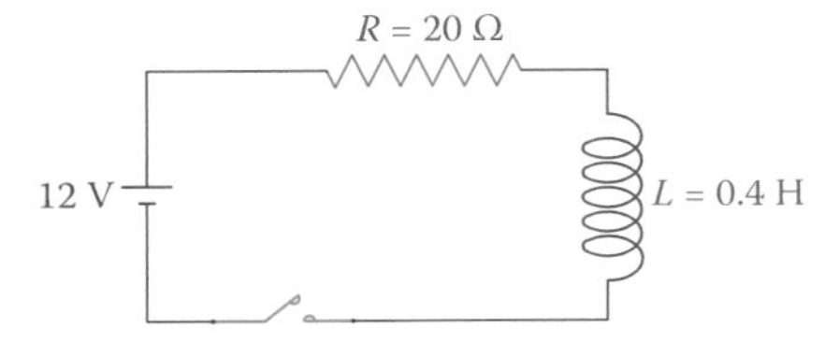

Figure P6.9 Circuit of Problem 23.

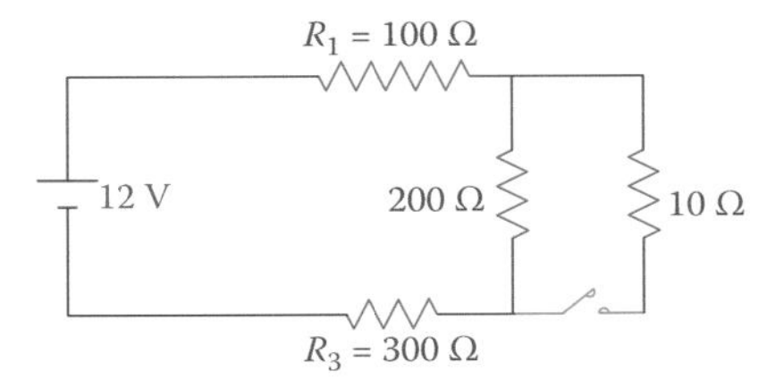

Figure P6.10 Circuit of Problem 25.

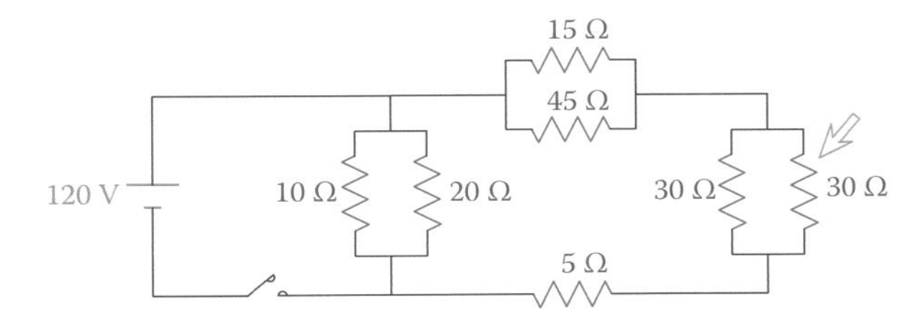

Figure P6.11 Circuit of Problem 26.

24. What current is in the circuit of Figure P6.9 after the effect of the coil ends?

25. In the circuit shown in Figure P6.10, find the power to be provided by the battery (a) when the switch is open and (b) when the switch is closed.

26. In the circuit shown in Figure P6.11, find the power consumed in the resistor marked by an arrow.

7

DC Motors and Generators

OBJECTIVES: After studying this chapter, you will be able to

- Define a magnetic field
- Differentiate between a magnet and an electromagnet
- Determine poles of an electromagnet
- Express fundamentals of DC electric motors
- Explain Lorentz force
- Understand components of a DC machine and the importance of each one
- Define the characteristic curve for the performance of a machine
- Explain similarity between a DC motor and a DC generator
- Describe Faraday's law in electricity
- Describe the difference between various types of DC motors
- Understand the use of different DC motors for different applications
- Understand the rules of batteries in series and in parallel
- Explain the conditions to put DC generators in parallel together
- Define rated value(s) for an electric machine
- Describe what efficiency is
- Explain why efficiency can never be 100 percent
- Calculate the efficiency of a motor

New terms:
Armature, back electromotive force, brush, characteristic curve, commutator, counter electromotive force (EMF), cumulative compound wound, diamagnetic, differential compound wound, electromagnet, ferromagnetic, long shunt compound, magnetic field, paramagnetic, performance curve, permanent magnet, prime mover, rated value, servomotor, short shunt compound, slip rings

7.1 Introduction

One of the properties of electricity is magnetism. All motors work based on this property of electricity and its effects, as you will see in this chapter and Chapter 11. This chapter covers the fundamentals of DC machines. The reason why the term "machine" is used here is that the basic construction of a motor and a generator, particularly for DC electricity, is the same. A machine can be used as a generator (converting mechanical energy to electrical energy) or a motor (converting electrical energy to mechanical energy).

There are many different types of motors, particularly in AC electricity, but in this book we discuss only the construction of major motors, because the variety in structure and physical differences can give different properties for the operation of a motor. Also, the discussion of magnetism, by itself, is very involved and falls beyond the scope of this book. We just discuss the matter very briefly.

7.2 Magnetism

Certain metals of the iron family can become magnetized if their atoms align such that this property is enhanced. This alignment can be due to contact with another magnet, electricity influence, being placed in a magnetic field (being in the vicinity of a magnet), or by natural events in nature. As a result of magnetism, magnetic media can attract or reject other media by inserting attraction force or rejection forces on them.

Ferromagnetic: Type of material from the iron family that is suitable for magnetization.

Diamagnetic: Materials without any significant magnetism property, such as wood and clay. They can never become a magnet.

Paramagnetic: Metals with very little (negligible) magnetic property that never can retain any magnetism after being in a magnetic field, as opposed to ferromagnetic materials.

Magnetic field: Limited part of the space around a magnet where the magnetic effect exists.

Flux lines: Imaginary lines around a magnet indicating the direction of magnetic effect. Stronger magnetism effect implies more flux lines through the same area.

In nature the metallic form and salts (compounds) of the **ferromagnetic** materials can be found as magnets. This is the way magnetism was first recognized. Ferromagnetic materials are those that can relatively easily become magnets, and they can retain their magnetism. Some other materials, called **diamagnetic** materials, can never become a magnet. A third category is called **paramagnetic** materials, which can exhibit a very small degree of magnetism, but they can never retain it.

Examples of ferromagnetic material are iron, steel, chromium, and nickel. Examples of diamagnetic material are copper, lead, mercury, silver, tin, and graphite (carbon). Examples of paramagnetic material are tungsten, aluminum, magnesium, sodium, and titanium.

A **magnetic field** is a part of the space around a magnet that the magnetic influence can be felt and measured. For example, there is a magnetic field between the two poles of a horseshoe magnet, as shown in Figure 7.1. Also, the Earth has a magnetic field, between North Pole and South Pole. To elaborate on the meaning of field, consider the Earth also having a gravity field. As one moves away from the Earth, the gravity field becomes weaker until it becomes so small that it can be ignored. At that point, everything becomes weightless. A field can be assumed to consist of a number of imaginary lines that show the orientation of the field. For example, for a magnetic field we always consider the direction to be from North Pole to South Pole, and the gravity field around the Earth, it is always toward the center of Earth. For a magnetic field these lines are called **flux lines**.

As well as orientation, a field has strength. This strength is not constant. A magnetic field can be strong, or it can be weak at a certain point. As one moves away from the magnetic field, it becomes weaker. For instance,

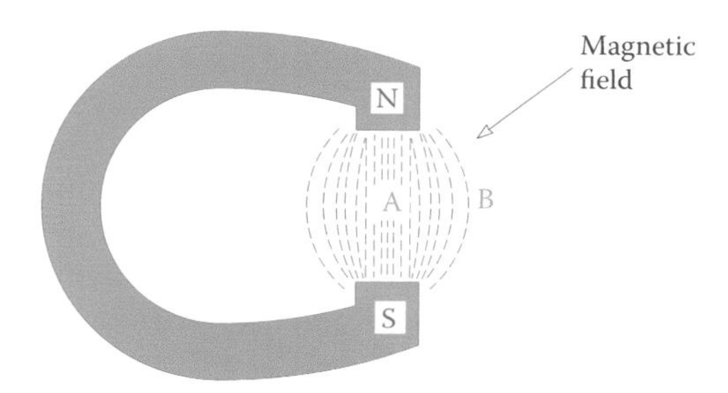

Figure 7.1 Magnetic field.

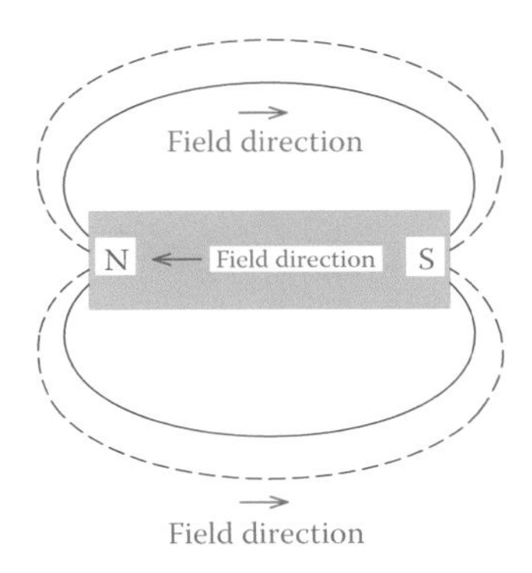

Figure 7.2 Direction of a magnetic field inside and outside the magnet body.

the magnetic field in Figure 7.1 is stronger at point A than at point B. A stronger field can be assumed to have more lines (magnetic flux lines) per area; that is, the flux lines are denser.

As you know, a magnet can be in various forms, a yoke or a bar, for instance. No matter what the shape of a magnet is, the magnetic field direction is always from North pole toward South pole outside the magnet body along the flux lines and from south to north inside the magnet. The magnetic flux lines form closed curves. This is as shown in Figure 7.2.

As we will see shortly, a magnetic field can be created by electricity. The core body of such a magnet loses its magnetic property when electricity is turned off. However, many magnets do retain their magnetism. These are called **permanent magnets**. Today, strong permanent magnets can be made from alloys of rare metals and iron. They are employed in many motors and recently in some small generators such as those in some wind turbines.

Permanent magnet: Magnet with permanent magnetic property that cannot be turned on and off or altered.

The magnetic property of a permanent magnet is due to its atomic structure. Thus, if a bar magnet is broken in the middle, each piece by itself becomes a perfect magnet having two poles, one north and one south. In the same way, if each of the two broken magnets is divided again into pieces, each piece becomes a magnet. Inversely, two magnets can be attached together only when their opposite poles are put together. This is why opposite poles attract and similar poles repel each other.

Heat is not good for a permanent magnet. A magnet can lose it magnetism by being heated. This changes the atomic structure of the metal. So, if you want to destroy a magnet, you can heat it. Alternatively, if a magnet is struck hard by a hammer, it will lose its magnetism.

A ferromagnetic material can easily become a magnet.

7.3 Electricity and Magnetism

One of the properties of electricity is magnetism. When electric current passes through a conductor, an electric field is generated around the conductor. This can be checked by placing a compass near the current carrying wire. The compass needle aligns itself with the field generated by the

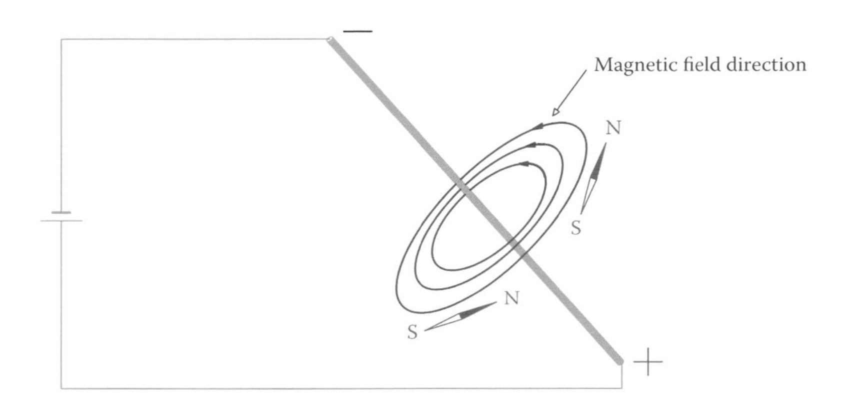

Figure 7.3 Magnetic field of a current carrying wire.

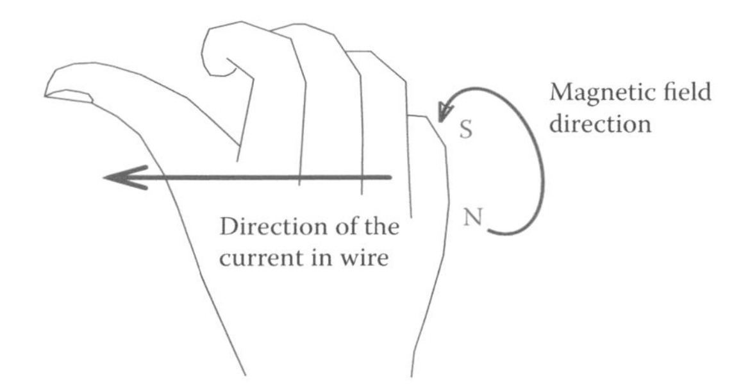

Figure 7.4 Right-hand rule to find the direction of the magnetic field around a wire. (From Hemami, A., *Wind Turbine Technology*, 1E ©2012 Delmar Learning, a part of Cengage Learning Inc. Reproduced with permission from http://www.cengage.com/permissions.)

electric current. This is shown in Figure 7.3. Each piece of the wire creates a magnetic field around it. The field lines are circles perpendicular to the wire segment; thus, depending on the point on this circle, the direction of the magnetic field changes. The needle of a compass always stays tangent to the magnetic lines. This magnetic field is all around the wire; so, for a straight wire the magnetic field forms a cylinder.

Note that in Figure 7.3 the current direction is from positive to negative, using the conventional direction of current. The direction of the magnetic field generated this way can always be found by the right-hand rule, as shown in Figure 7.4. That is, if the thumb in the right hand is aligned with the current direction, the fingers show the magnetic field direction.

The strength of this magnetic field depends on (1) the current in the wire and (2) the distance from the wire. The higher the current is, the stronger the field is. Also, the farther from the wire, the weaker the field is.

7.3.1 Magnetic Field of an Inductor

An inductor is a wound wire: when connected to electricity, it carries a current. In the previous section we learned that if a straight wire carries a current, then a cylindrical magnetic field is created around it. In the case of a coil (an inductor) the wire is circular, and normally there are a large

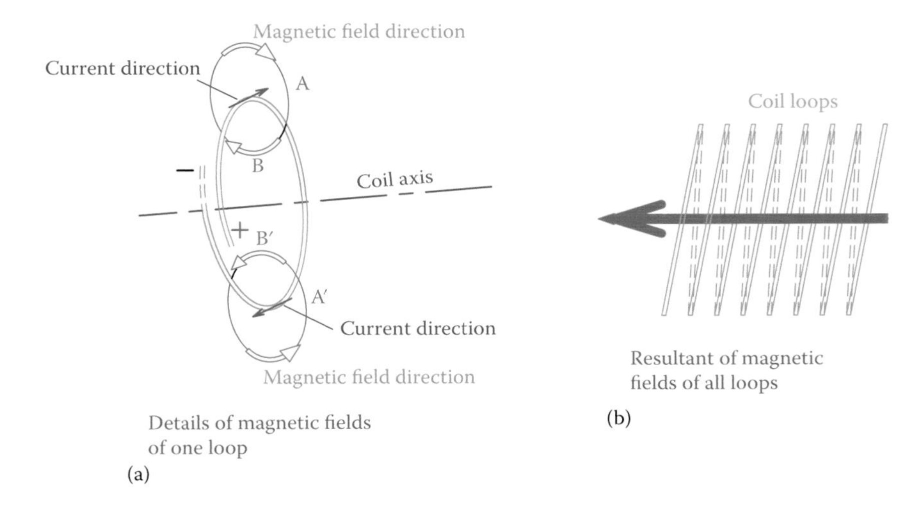

Figure 7.5 Magnetic field of a coil. (a) Field direction for each individual loop. (b) Resultant magnetic field of all loops.

number of loops that carry the same current. In this sense, there are many magnetic fields created because of the electric current, and they interact with each other owing to the loops in the wire. Figure 7.5 shows what happens in such a case.

For each loop in the coil the magnetic fields form a torus around the loop. Magnetic lines at two points are shown in Figure 7.5a. At some points (such as A and A′), magnetic lines are in the opposite direction and cancel each other, but at some other points (such as B and B′) they combine with each other and their effects add together. The resultant magnetic field for each loop is along a line perpendicular to the loop (along the loop axis). Magnetic fields of all the loops, then, add together to construct the magnetic field of the coil. This field is along the coil axis, as shown in Figure 7.5b.

In conclusion, when a coil carries electric current, a magnetic field is generated. The direction of this magnetic field depends on the direction of the current. Figure 7.6 shows a coil and the direction of its magnetic field based on the electric current direction. As noted before, the direction of the magnetic field is from north to south outside of the magnet generated by the current flowing in the coil (from + toward –).

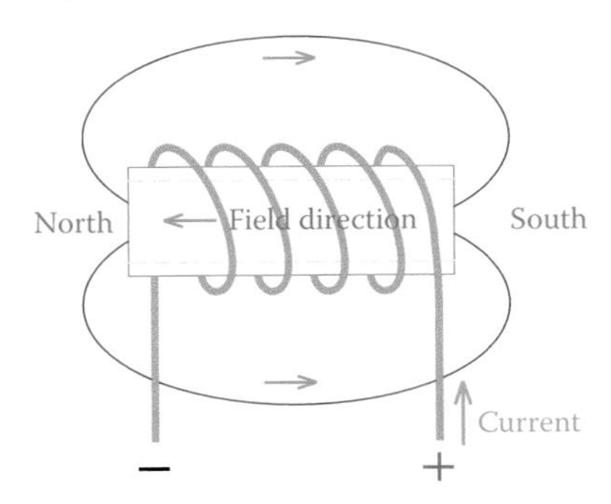

Figure 7.6 Direction of the magnetic field of a coil.

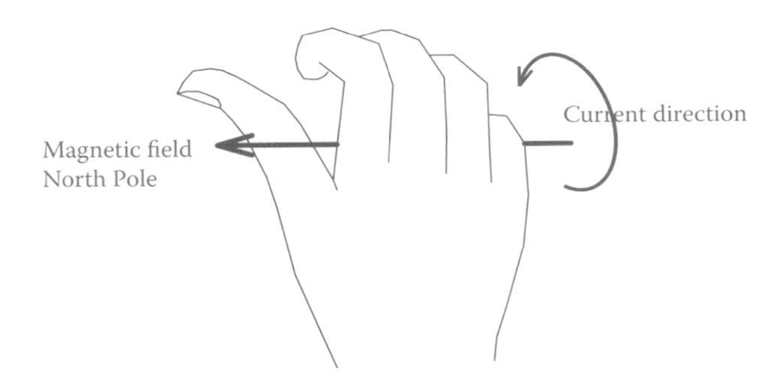

Figure 7.7 Right-hand rule to find the direction of the magnetic field of a coil. (From Hemami, A., *Wind Turbine Technology*, 1E ©2012 Delmar Learning, a part of Cengage Learning Inc. Reproduced with permission from http://www.cengage.com/permissions.)

Moreover, the field strength depends on the current as well as other factors, such as the physical dimensions of the coil and the material of its core. To enhance the magnetic field of a coil, a core made of a ferromagnetic material is inserted inside the coil.

The right-hand rule again can be helpful here to determine which side is the north and which side is the south, knowing the current direction. This is shown in Figure 7.7. This rule states that if your right-hand fingers are in the direction of current, then your thumb indicates the direction of the magnetic field (the thumb shows the north pole of the resulting magnetic field). Study this figure together with Figure 7.6. Also, note the difference between Figures 7.4 and 7.7.

> The right-hand rule states that if your right-hand fingers are in the direction of the current in a coil of wires (from + toward –), then your thumb indicates the north pole of the coil magnetic field.

7.3.2 Electromagnet

The magnet created by the coil in Figure 7.6 is not a permanent magnet. As soon as the current is turned off, the magnetic field vanishes and the magnetic effect disappears. This arrangement (a coil and a core in it) is called an **electromagnet**, and it has many industrial applications. As you might imagine, it can be used for collection of (ferromagnetic) metals and for exerting force, in addition to many other applications such as opening and turning a switch or a valve on and off.

Electromagnet: A (not permanent) magnet made by a wire coil wrapped around a ferromagnetic core when carrying an electric current. The magnet can be turned on and off or its strength can be adjusted.

7.4 Lorentz Force

Many motors work based on Lorentz force. It is a force that is created because of the interaction of a magnetic field and electric current. If a wire carrying electricity happens to be inside a magnetic field, then a force is exerted on the wire. This force has a direction that depends on both the direction of current and the direction of the field (see Figure 7.8). Also, as

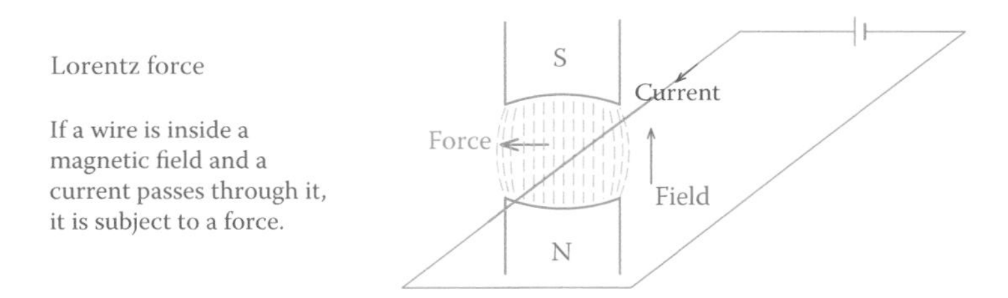

Figure 7.8 Definition of the Lorentz force.

can be expected, the magnitude of this force depends on the current and the strength of the magnetic field.

The direction of the Lorentz force can be determined from the right-hand rule, as shown in Figure 7.9. The right-hand rule for the Lorentz force states that if the right-hand fingers are along the magnetic field and the thumb is in the direction of the current, then the direction of the force exerted on the wire is along an arrow coming out of the palm.

> **Lorentz force law**: If a wire is inside a magnetic field and carries electricity, then a force is exerted on the wire.

A DC motor consists of a winding (many loops of wire), called an **armature**, that when connected to electricity an electric current will pass through all its loops. A motor also has a magnet that provides the necessary magnetic field. The armature is inside the magnetic field and can rotate about its axis. A force is exerted on each of the individual wires in the armature. All this force leads to a motion, and with proper arrangements the armature has a rotational motion about its axis. This is shown in Figure 7.10 for one loop. To provide electricity for a rotating component, use is made of a **brush** and **slip rings**, schematically illustrated

Armature: Rotating part in a DC machine, especially the windings of this rotating part.

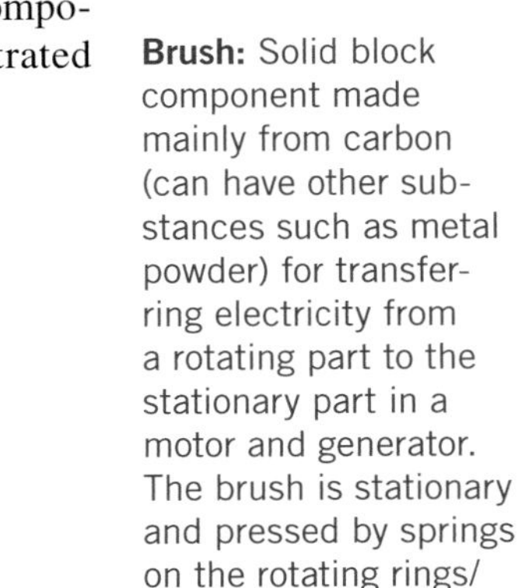
Brush: Solid block component made mainly from carbon (can have other substances such as metal powder) for transferring electricity from a rotating part to the stationary part in a motor and generator. The brush is stationary and pressed by springs on the rotating rings/segments.

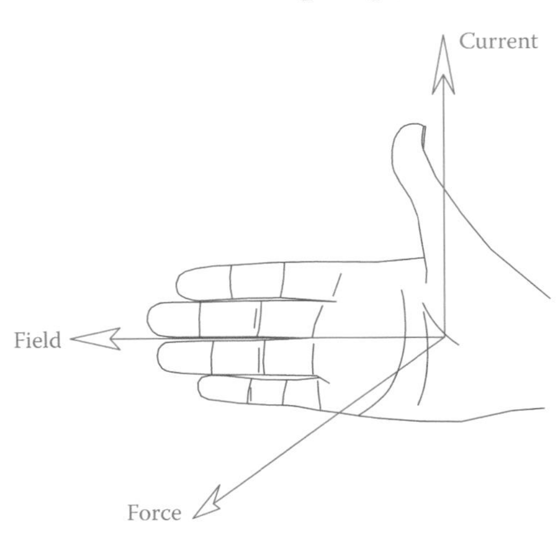

Figure 7.9 Right-hand rule for determination of the Lorentz force direction. (From Hemami, A., *Wind Turbine Technology*, 1E ©2012 Delmar Learning, a part of Cengage Learning Inc. Reproduced with permission from http://www.cengage.com/permissions.)

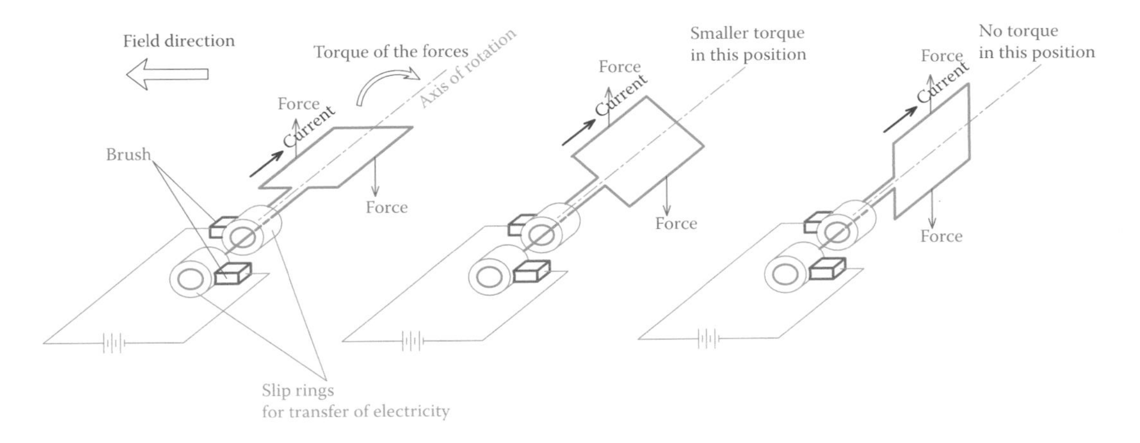

Figure 7.10 Principal structure of a DC motor.

in Figure 7.10. A brush having a fixed position can slide on a cylindrical surface that rotates.

In practice, an armature always has a metallic core. It is not a hollow frame of wires. An armature core is very important for the operation of any electric machine, because without it a large percentage of available power is lost. The armature core holds the windings; it is made up of laminated iron or steel (a ferromagnetic material) and is part of the structure of the entire rotating part of a machine. Figure 7.11 illustrates a typical armature assembly consisting of the armature windings, armature core, and a **commutator**, all mounted on a shaft.

As can be understood from the Lorentz force law, and as illustrated in Figure 7.10, for each loop of the armature, there are two forces exerted to the two sides of the loop. The relative directions of these forces vary with the loop angle with respect to the magnetic field, and, consequently, their resulting torque about the axis of rotation changes between zero and a maximum value. Moreover, after 180° of rotation, the directions of the two forces (and consequently the direction of their torque) change, which is not suitable for a continuous rotating motion. This is shown in Figure 7.12. To keep the torque in the same direction so that the armature shaft continues moving in the same direction, it is necessary to alter the direction of the current in the armature at a certain position. This alteration of current in a DC motor is achieved by a commutator. Instead of the two slip

Commutator: Part of the armature of a direct current motor or generator consisting of many small metallic (normally copper or brass) segments in the form of a ring which can slide under a brush. The brushes are spring pressed on the commutator and transfer the current from the rotating commutator to the outside of the armature. The commutator has the role of rectifying the otherwise AC current in the armature.

Slip rings: Set of two (in single-phase) and three (in three-phase) in some types of AC electric machines that are connected to windings that rotate with the rotor. Brushes pressed by springs on the slip rings can transfer electricity from or to the winding from the external circuit.

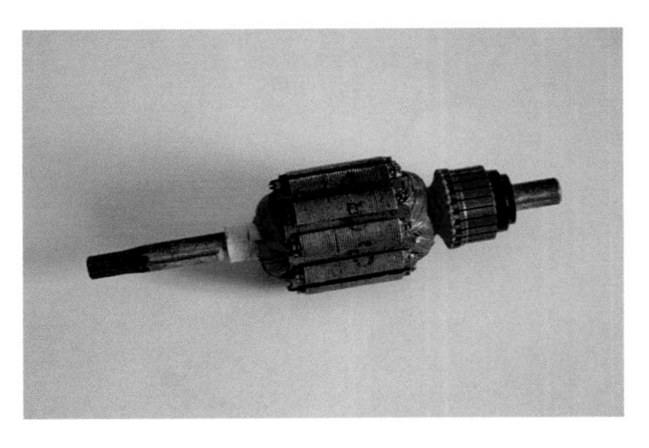

Figure 7.11 Example of an armature assembly.

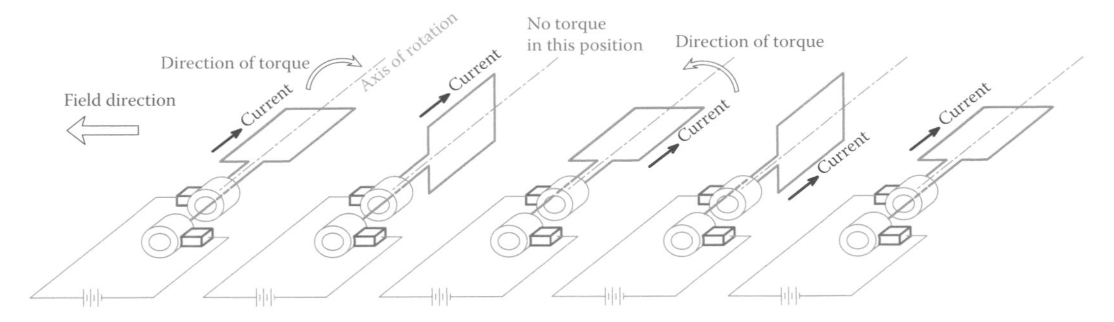

Figure 7.12 Change in direction of force (and torque) after 180° of rotation.

rings for transfer of electricity (in Figure 7.10), if two half rings are used (isolated from each other), as they rotate with the shaft, the brushes change contact from one to the other. In this way the connections to a power supply, and subsequently the current direction, switch. Each DC motor (and generator) has a commutator, which is a mechanical device to change the direction of current in its armature and maintain the torque direction. In practice, there is more than one loop, and, therefore, instead of two half cylinders, there are many smaller cylinder segments put together, each one isolated from the rest, and all isolated from the shaft body. The two ends of each loop of the armature are connected to two of these segments 180° apart. Each loop by itself can comprise a bundle of looped wires (in series with each other), the two ends of which are connected to the segments on the commutator.

A commutator is illustrated in Figure 7.13. It consists of a number of copper segments that are arranged around the periphery of a cylinder. Each opposite pair of these segments is connected to the two ends of one of the loops of the armature, and, therefore, a commutator has twice as many segments as the number of bundle loops in the armature. The concept of a commutator can be more clearly understood from Figure 7.14, which shows only two loops and four commutation segments. Copper segments in a commutator are connected to the power supply through brushes that can slide on the commutator when it rotates. Thus, the brushes are stationary and do not rotate. Brushes are made of carbon (or a similar manufactured electric conductive material, containing carbon) and are spring loaded (sufficiently pressed on the commutator by some springs) to have a good contact. Otherwise, they spark in operation. Normally, a

Figure 7.13 Commutator in a DC motor.

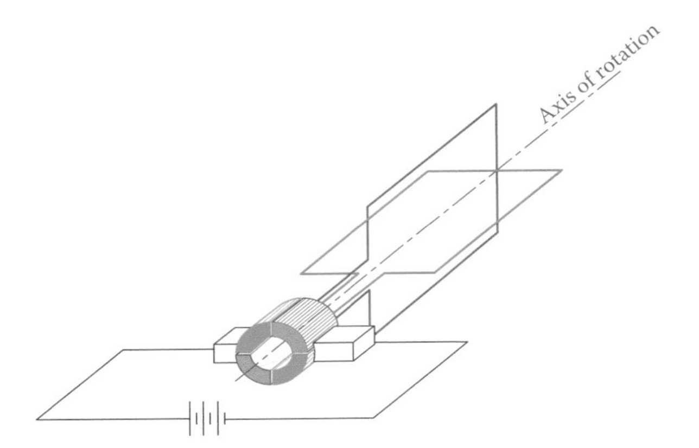

Figure 7.14 Four commutator segments for two-loop armature. (From Hemami, A., *Wind Turbine Technology*, 1E ©2012 Delmar Learning, a part of Cengage Learning Inc. Reproduced with permission from http://www.cengage.com/permissions.)

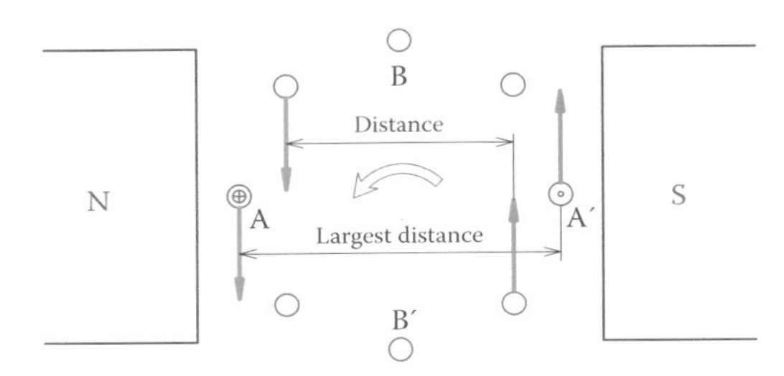

Figure 7.15 Loops with maximum and minimum torque.

brush can cover a few of the commutator segments. Therefore, more than one loop is connected to electricity at a time. These are the loops with maximum force and giving maximum torque to a motor shaft (see Figure 7.15). After a few degrees of rotation each loop disengages and another loop substitutes for it. Thus, during one revolution a motor always benefits from engaging the part of winding that gives the highest torque, while the rest of the loops carry no current.

A commutator is a mechanical device that by switching connection points while turning changes the direction of current in the armature of a DC machine.

7.5 Faraday's Law

Faraday's law is the opposite of Lorentz force law. It states that if a wire is moved inside a magnetic field such that it cuts the magnetic flux lines, then an electromagnetic force is generated in that wire (a voltage difference is created between the two ends of the wire). This is the principle of

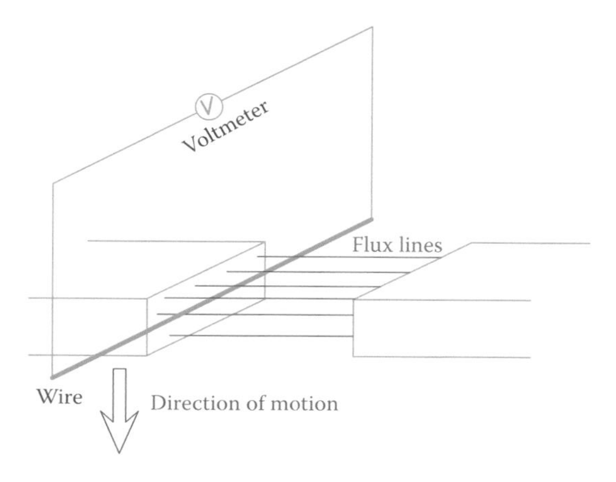

Figure 7.16 Faraday's law, the principle of working of a generator.

operation of a generator. Figure 7.16 shows this law for one single wire. As shown, the direction of motion is perpendicular to the direction of the magnetic field flux lines; it, therefore, cuts the flux lines.

> If a wire moves inside a magnetic field such that it cuts the magnetic flux lines, an electromotive force is generated in that wire.

Figure 7.17 illustrates the same wire and the same magnetic field as in Figure 7.16, except that the motion of the wire is in parallel with the direction of the flux lines, and therefore the lines are not cut by the wire. In such a case, no voltage difference is generated in the wire. In general, when a wire moves inside a magnetic field, its motion can vary between these two cases, thus, partly perpendicular and partly parallel to the magnetic field lines. The generated voltage is always proportional to the component of motion that is perpendicular to the field lines (any motion can be decomposed into two components in two perpendicular directions).

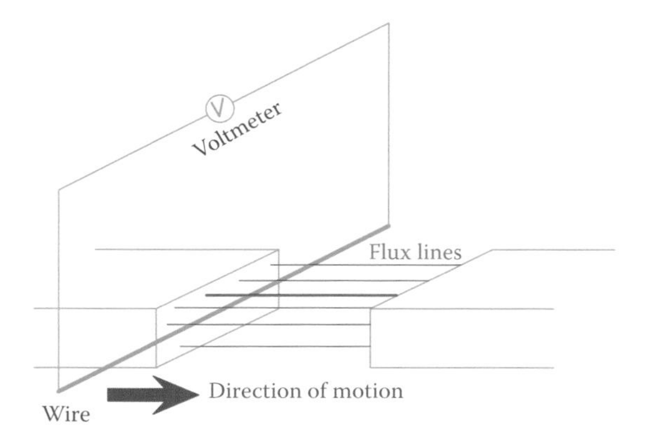

Figure 7.17 No voltage is generated in the wire because it moves parallel to flux lines.

In a generator, there are a large number of wires, and they are in the form of loops that during rotation of the generator shaft move through a magnetic field and cut the magnetic field lines. Figure 7.18 depicts one of these loops. Voltage generation takes place only in the parts of each loop of wire that are parallel to the axis of rotation. The other segments are to complete the loop and connect it to the outside. Moreover, when rotating, the two sides of a loop move in different directions. For example, for a counterclockwise rotation of the loop in Figure 7.18, when the segment AB of the loop moves downward, the segment CD moves upward. As a result, the voltages generated in the two opposite sides of a loop of wire are in the opposite directions and therefore add together (if they were in the same direction, they would oppose and cancel each other). The direction of the electromotive force generated in a wire in relation to the direction of the magnetic field and motion direction is determined from the right-hand rule. This rule is different from that for the Lorentz force, and it is shown in Figure 7.19. Compare this figure with Figure 7.9 to notice the difference.

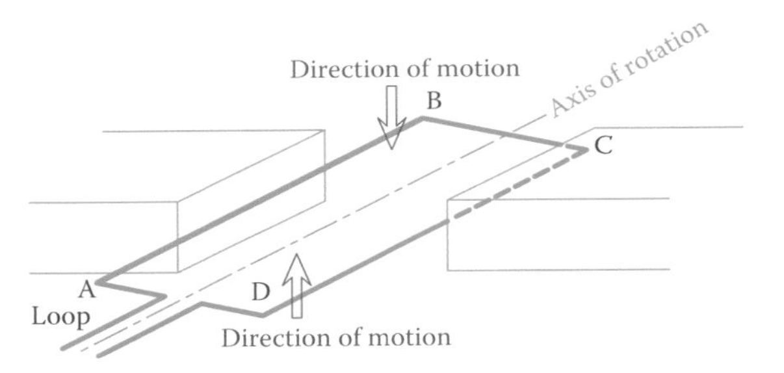

Figure 7.18 Counterclockwise rotation of one loop of the winding in a generator.

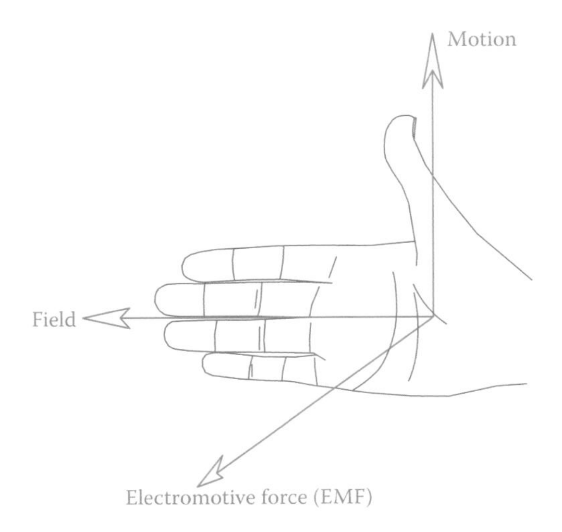

Figure 7.19 Right-hand rule for a generator. (From Hemami, A., *Wind Turbine Technology*, 1E ©2012 Delmar Learning, a part of Cengage Learning Inc. Reproduced with permission from http://www.cengage.com/permissions.)

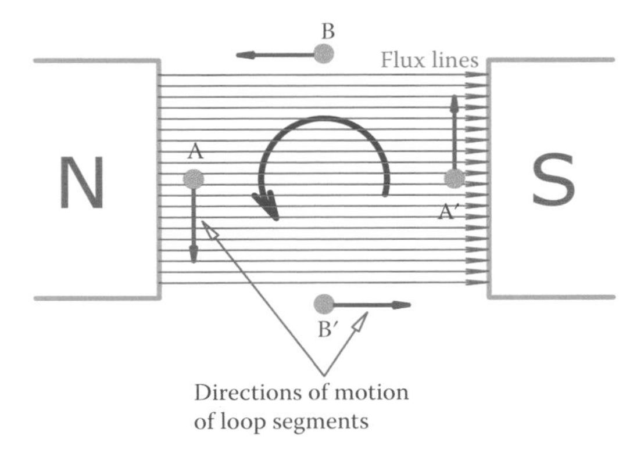

Figure 7.20 Maximum voltage corresponds to position AA′ and zero voltage at BB′.

Voltage generated in a loop depends on the position of the loop with respect to the magnetic field. This electromotive force is not the same in all positions within a 360° range. The cross section of one loop is shown in Figure 7.20. The direction of motion is shown for the wire segments in the loop. When the loop assumes the position shown by points B and B′, direction of motion of the wires in the loop is parallel to the flux lines. Consequently, no flux lines are cut, and no voltage is generated. As the rotation takes place and the angle of the loop changes, more and more flux lines are cut by the wires, and in the position defined by points A and A′ the number of wires cut by the wire is the maximum. Thus, generated voltage is at a maximum in this position.

7.6 DC Machines

In electricity the word machine is used for both a motor and a generator. When an electric machine is turned by a **prime mover**, it converts mechanical energy to electrical energy, thus behaving as a generator. When electricity is applied to the machine, it converts electrical energy to mechanical energy, thus functioning as a motor. Therefore, a DC machine is an electrical machine that works with DC electricity as a motor or produces DC electricity as a generator. A basic DC machine can behave as a motor or a generator. (Today's machines are sometimes composed of a basic machine and additional circuitry for specialized functions, like embedded electronic circuits. Thus, they may not readily function two ways.) Construction and electrical components of a DC motor and a DC generator are, therefore, the same, and it is not necessary to repeat the common features, particularly those associated with their structure.

Prime mover: Mechanical power source (e.g., steam turbine, gas turbine, diesel engine) that turns an electric generator or any other device requiring mechanical energy.

As it was mentioned in Section 7.4 for a DC motor, in order for a DC machine to work, a magnetic field is necessary, inside which the armature rotates. The armature is mounted on a shaft, which is attached to the stationary body by bearings at its ends. Connection of the armature (rotating part) to the outside (stationary) is through the commutator. Here we consider a number of scenarios according to which a DC machine can

Performance curve: Same as characteristic curve: A line or curve that shows the relationship between two or more parameters related to the performance of a device.

Characteristic curve: Line or curve that shows the relationship between two or more parameters affecting the functionality of a device, like the relationship between voltage and speed in an electric motor.

be constructed. The difference between these scenarios lies in the way the magnetic field is created. Because the performance of a DC machine depends on the interaction between the magnetic field and the current in the armature, the difference between these various scenarios affects the performance of the machine. Performance is normally the relationship between the torque (load) and speed or between the speed and the applied voltage (for generator the speed and the generated voltage). This relationship is depicted by a curve referred to as the **performance curve** or the **characteristic curve**.

Permanent magnet: The (magnetic) field in such a machine is provided by a permanent magnet. The magnetic strength of this field is constant in all conditions. Figure 7.21 shows the schematic of a permanent magnet (PM) DC machine. The geometrical arrangement of the magnetic field and the armature can vary. Normally, magnetic poles in a DC machine are salient, as opposed to some other machines that the poles are not physically distinguishable. Operation of a permanent magnet machine is smooth, and compared with other classes, its power loss is smaller owing to not having any magnetizing coil. Nevertheless, there is a limitation to the power capacity of such a machine.

Separately excited machine: In this configuration, shown in Figure 7.22, the magnetic field is provided through an electric current passing through the field winding. The magnetic field, in fact, is provided by an electromagnet. Two different DC voltages are applied to the machine, one for generation of the magnetic field and one for the armature. Because

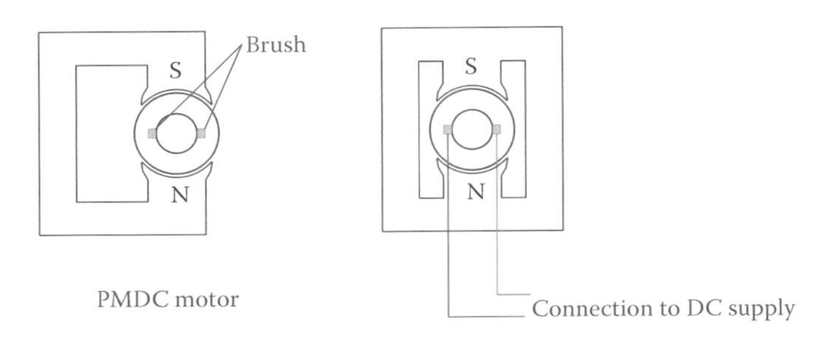

Figure 7.21 Schematic structure and pole arrangement of a permanent magnet DC machine.

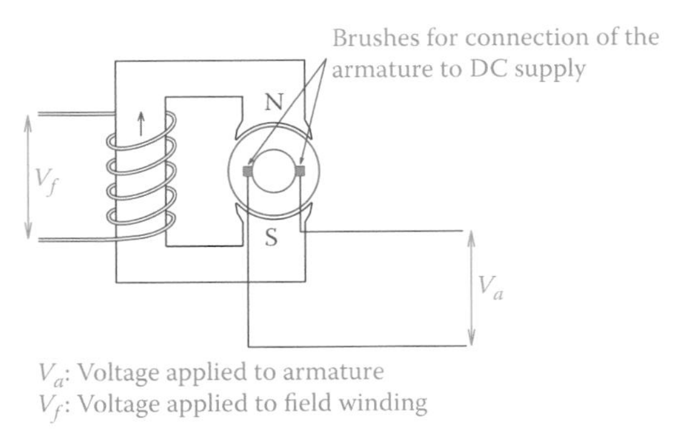

Figure 7.22 Separately excited machine.

there are two voltages, the currents in the field winding and the armature are independent and can be separately controlled. This gives the highest degree of control of the machine. A motor made this way is called a **servomotor**.

Servomotor: Motor whose speed or displacement can be controlled (must be capable of being controlled).

To show the principle of construction of an electric machine more easily, normally, the body is not shown and only the windings and the terminals are depicted. Thus, instead of Figure 7.22, the simpler Figure 7.23 is used. Accordingly, for the rest of the figures the simpler representation is employed.

Series wound DC machine: In this configuration the field winding and the armature winding are put in series with each other, as depicted in Figure 7.24. A single voltage is applied to the open ends of the two windings (terminal voltage), and this voltage is divided between the field winding and the armature winding. In this way, the current in the field winding is the same as the current in the armature. The only thing that can be changed to control and vary the behavior of the machine is the applied voltage.

Shunt wound DC machine: In this case the two windings are in parallel with each other, and, thus, the same voltage is applied to both the field winding and the armature, but they have different currents. Figure 7.25 shows the winding arrangement of a shunt wound DC machine.

Compound-wound DC machine: It is possible to combine the construction of a series wound and a shunt wound DC machine for performance

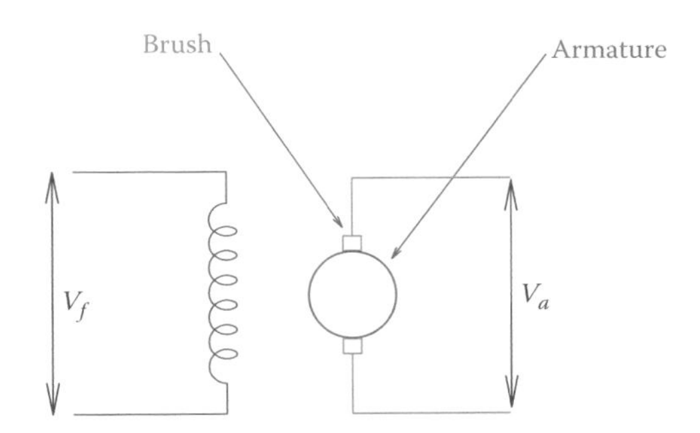

Figure 7.23 Simple schematic of a separately excited DC machine.

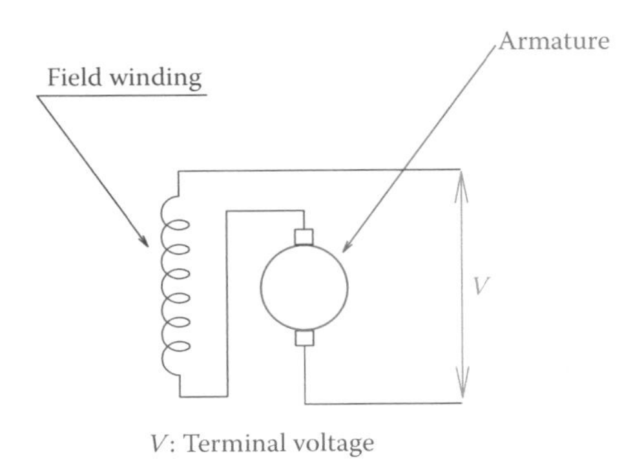

Figure 7.24 Series wound DC machine.

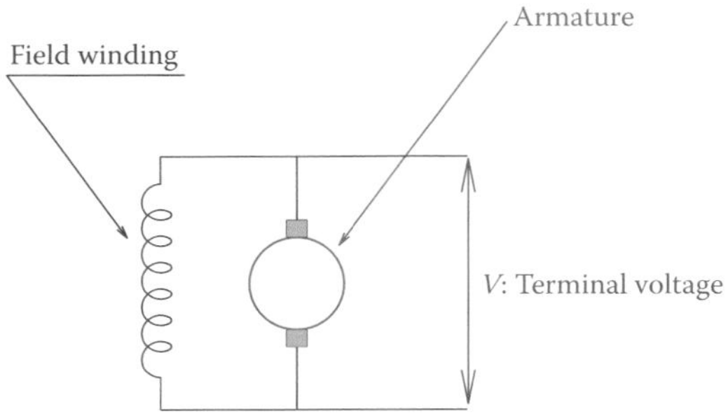

Figure 7.25 Shunt wound DC machine.

improvement. Such a machine has two field windings, one in parallel with the armature winding and the other one is in series with it. There are various possible ways to have this arrangement, as shown in Figure 7.26. The two arrangements shown are called **long shunt compound** and **short shunt compound** machines, respectively.

Note that whereas in the preceding sections a magnetic field was represented by one pair of (north and south) poles, in practice, it is possible to have more poles for the magnetic field. This affects the speed of rotation of a machine. Figure 7.27 depicts when two pairs of magnetic poles are used in a machine.

Long shunt compound: A kind of compound wound DC electric machine in which the shunt winding receives the entire terminal voltage, as opposed to the short shunt compound machine. The shunt winding current is independent of the series winding current.

Short shunt compound: A kind of compound wound DC electric machine in which the field shunt winding receives only a part of the terminal voltage (The field has a shunt winding plus a series winding). The shunt winding is parallel with the armature winding, only, and the current in the shunt winding is a fraction of the field series winding current.

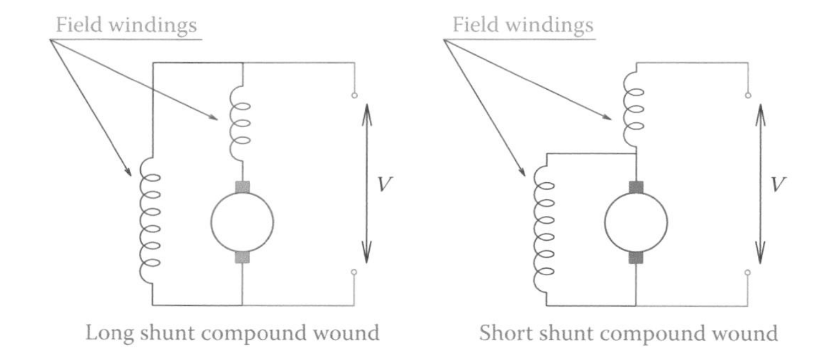

Figure 7.26 Long shunt compound and short shunt compound DC machine.

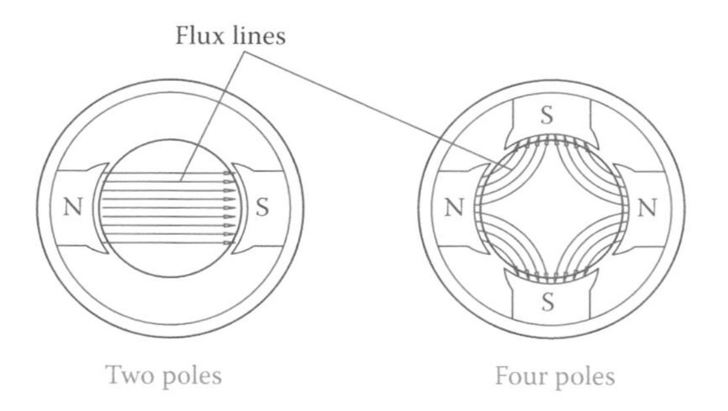

Figure 7.27 Comparison of magnetic field flux lines in two and four magnetic poles.

7.6.1 DC Machines Characteristic Curves

As far as the performance of a machine and its characteristic curves are concerned, for motors we normally are interested in the speed versus applied voltage relationship and torque versus speed relationship. The first category shows the variation of the speed of a motor with the increase in the applied voltage. This is important in applications where the speed of rotation of a load connected to the motor is to be controlled. A hypothetical case (because each case depends on the aforementioned types of a machine) is shown in Figure 7.28. This curve implies that as the voltage is increased, the speed of the motor is also increased, and this increment in speed is proportional to the voltage increase. In other words, the relationship is linear (the curve is, indeed, a line in this case). However, this relationship is true for values above an initial minimum voltage below which a motor does not move owing to friction.

The second category exhibits the relationship between speed and torque. This curve is very important when matching a load to a motor (or vice versa). When running a load by a motor, the power requirement of the load must be satisfied by the motor. That is, if at any instant the motor cannot provide the power demanded by the load, it can slow down, or even stall. Because power in a rotating system depends on the product of the rotational speed and the torque on the shaft, the operating point is always a point when the torque-speed curve of the load and the torque-speed curve of the drive meet, as shown in Figure 7.29 for a typical case (the load curve corresponds to a fan or blower). Point A in this figure is the operating point for a particular situation. If the speed of the load is to be increased, say to that of point B, the demand for torque also increases (based on the load characteristic curve). In such a case, either the applied voltage to the motor must be increased (so that a new curve passing through point B defines the motor characteristics), or the motor cannot run the load at the new speed (if either the maximum allowed voltage or the motor maximum power is reached). A motor and its load always come to a balance (equilibrium) with each other. This equilibrium point defines the operating point.

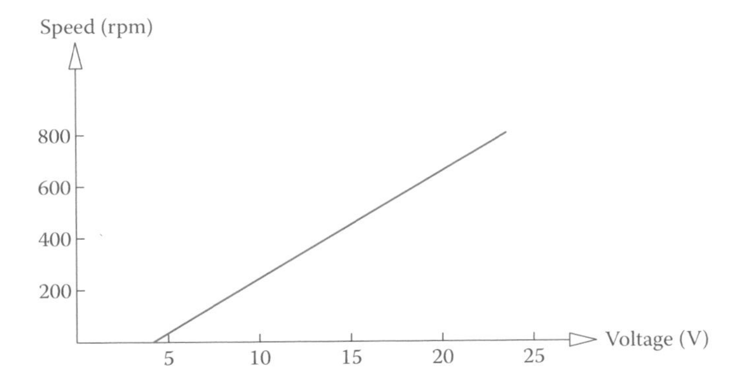

Figure 7.28 A characteristic curve indicating a linear relationship between applied voltage and motor speed (load torque is constant).

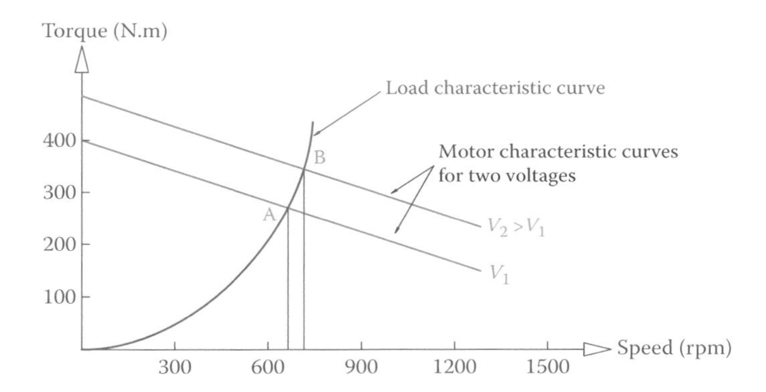

Figure 7.29 A typical torque-speed curve for a motor and for its load.

In Figure 7.29 the curve for the load signifies that the required torque increases as the speed increases. This is a typical situation for many devices such as pumps, fans, compressors, and blowers.

Power demand in a rotating load equals the product of the rotational speed and the torque on the shaft.

A motor and its load always run at an equilibrium speed, which is the point of intersection of their torque-speed characteristic curves.

The speed-torque characteristic curve can be drawn differently, showing speed on the vertical axis and torque on the horizontal axis. This does not change the property of a machine under study. Either curve can serve the purpose of studying the variation of one parameter in terms of the changes in the other. Usually, the variable that is susceptible to variation is shown on the horizontal axis. In many loads the torque is the variable subject to change. For example, in a lift depending on the number of people, and in lifting loads depending on the weight of the load, torque on the shaft changes. Figure 7.30 shows two sets of graphs showing the

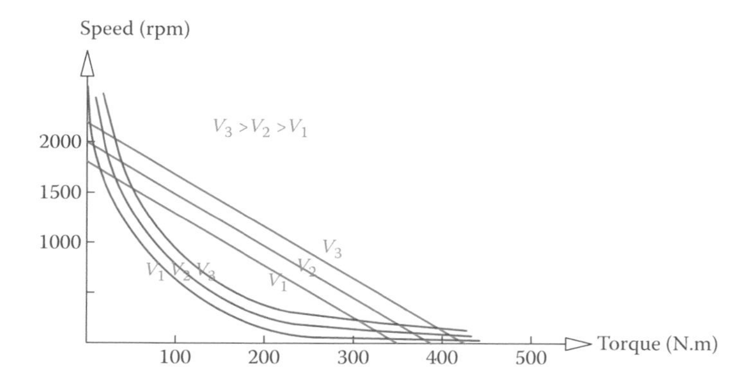

Figure 7.30 Two sets of speed-torque characteristic curves for two different types of DC machines.

speed-torque relationships but with the two axes interchanged compared to those in Figure 7.29. Three separate performance curves between (angular) speed and torque are shown for each set. Each one corresponds to a different applied voltage. It is a common practice that a set of curves are shown on the same graph, which, in fact, brings the effect of changing the applied voltage into view.

Another class of characteristic curve that can be useful is the relationship between the torque and the current that a motor takes to run (usually the current in the armature). We show that later for particular motors.

For a generator, normally, the voltage that it delivers is of more concern. Thus, the two types of curves of interest are voltage-speed graph, indicating the output voltage variation with the speed at which the generator is turned and the voltage-current curve that determines the output voltage at various loads (electric loads connected to a generator). Two typical curves are depicted in Figures 7.31 and 7.32, respectively. In both cases the third parameter is assumed to be constant; otherwise, one cannot represent the corresponding variations on two axes. This is why often a series of curves are shown for different values of the third parameter (like in Figure 7.30).

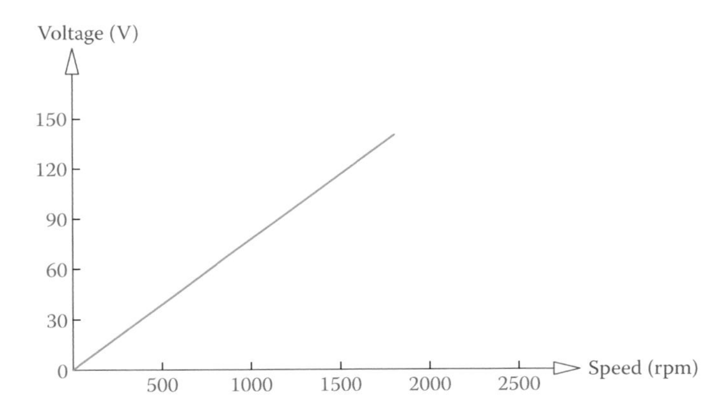

Figure 7.31 Voltage versus speed of rotation in a DC generator (load kept constant).

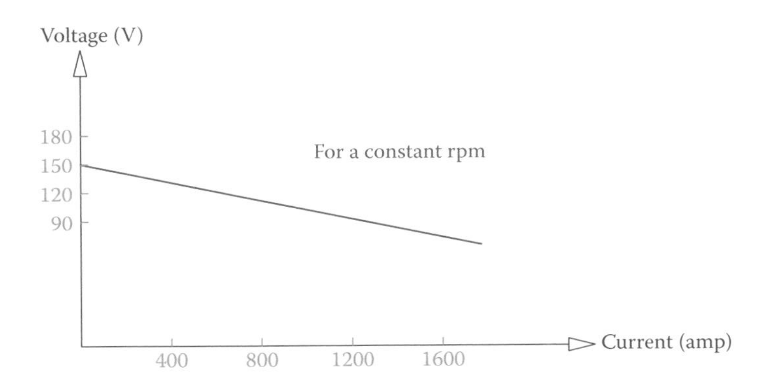

Figure 7.32 Voltage versus current (load) in a DC generator (speed kept constant).

It is to be noted that the characteristic curve in Figure 7.31 does not imply that for all the values of speed there is a corresponding value for voltage. For instance, if one runs the generator at 25 rpm, then a very low voltage can be generated. Any machine must be operated in the range of values for which it is designed. A generator, for instance, may have been designed to run at around 1500 rpm. This speed is the **rated value** for the machine. The rated value does not need to be a single figure; for example, a machine can have a rated speed between 1000 and 1750 rpm. The fact that the curve in Figure 7.31 passes through the origin implies that theoretically if the rpm is zero then zero voltage is generated.

Rated value: Value(s) affecting the operation of a device that the device is designed for and must operate under conditions within the neighborhood of those values. For example, an electric device has a rated voltage and current. Operating the device outside of the rated conditions is either inefficient or harmful to the device.

Performance of DC machines is based on two general equations. The first equation determines the torque developed in a DC machine. This torque is proportional to the strength of the magnetic field and to the armature current. Thus,

$$T = K\phi I_a \tag{7.1}$$

where T is the torque, ϕ is the field strength, I_a is the armature current, and K is a constant that depends on the physical size and design of a machine as well as the units that are used for T, ϕ, and I_a.

The second equation is based on the fact for all electrical motors that if a motor is prevented from turning a very high current passes through its winding and as the speed increases, this current decreases. This is why a motor can be damaged if is it not turning, either because of overloading (a large load that is beyond the capacity of the motor) or some braking action. In particular, for a DC motor, if the motor is stopped, a high voltage is applied to its armature winding (thus, a high current through it), but as the motor speeds up, a **counter electromotive force** (CEMF) is developed in the armature that opposes the applied voltage (thus, reduces the effect of the applied voltage). This counter electromotive force (or **back electromotive force** [BEMF]) depends on the speed of rotation and is directly proportional to it.

Counter electromotive force (CEMF): Same as back electromotive force (BEMF).

Back electromotive force: (BEMF) Electromotive force that is generated in an electric motor due to its rotation (as happens in a generator) and is in the opposite direction to the applied voltage and tends to stop the motor from rotation. It acts as a brake to the operation of a motor.

The second equation can be written in the form of

$$V - V_a = V_{\mathrm{CEMF}} = K_N\phi N \tag{7.2}$$

where V is the voltage applied to the armature winding by the line (which depends also on the machine type), V_a is the voltage difference causing a current in the armature, V_{CEMF} is the amount of counter electromotive force, N is the speed of rotation, and K_N is a constant value that depends on the physical size and design of a machine as well as the units used for the voltage and speed. The value ϕ is the field strength, as in Equation 7.1. It depends on the type of a DC machine and can be constant or variable. Equation 7.2 clearly shows that if the rotational speed is zero (the first moments when a machine starts), the entire voltage V will be applied to the armature, whereas after a motor starts the voltage leading to the armature current is

$$V_a = V - V_{\mathrm{CEMF}} = V - K_N\phi N \tag{7.3}$$

which depends on the rotational speed (note that Equation 7.3 is the same as Equation 7.2). V_{CEMF} is the generated electromotive force in a winding.

Equation 7.3 is employed when a machine works as a motor. For a generator the term counter electromotive force is normally not used, because it represents the generated voltage and it is larger than the terminal voltage V. Thus, for a generator, Equation 7.2 assumes the form

$$V = E - V_a = K_N \phi N - V_a \quad (7.4)$$

where $E = K_N \phi N$ is the generated voltage, V_a represents the voltage drop in the armature winding, and V is the terminal voltage at the generator output.

7.6.2 DC Motors

The general structure of DC motors and the principle of their operation were explained in the previous two sections. In this section we concentrate on the characteristic curves and performance differences between various types of DC motors.

7.6.2.1 Permanent Magnet DC Motor

This motor has a permanent magnet, and the magnetic field strength ϕ is constant (or it is better to say, its variation is negligible). In this case, the only variable in Equation 7.1 is the current I_a. This implies that the variation of motor torque is proportional to I_a or vice versa; that is, the variation of I_a is proportional to the required torque. The latter makes more sense, because as the motor load increases (i.e., increase of torque requirement by a load that must be driven), current in the motor increases. Theoretically, when there is no load, then the current is zero. However, in practice, because of the losses (friction and dissipated heat), even for zero load (zero torque) there is a small current. The theoretical and actual cases are represented by the lines shown in Figure 7.33. Note that this diagram illustrates the variation of a motor torque versus the motor current; it is an additional curve compared to those described in Figures 7.28 and 7.29. Note also that the armature current is, in fact, proportional to the voltage applied to the armature. Consequently, in a version of this diagram instead of current armature voltage could be used.

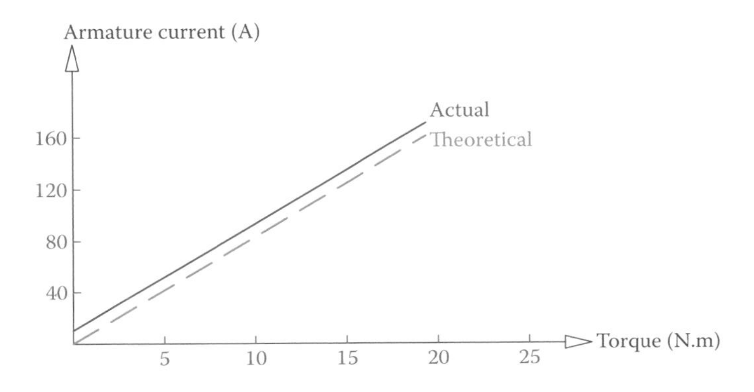

Figure 7.33 Armature current-torque relationship in a permanent magnet DC motor.

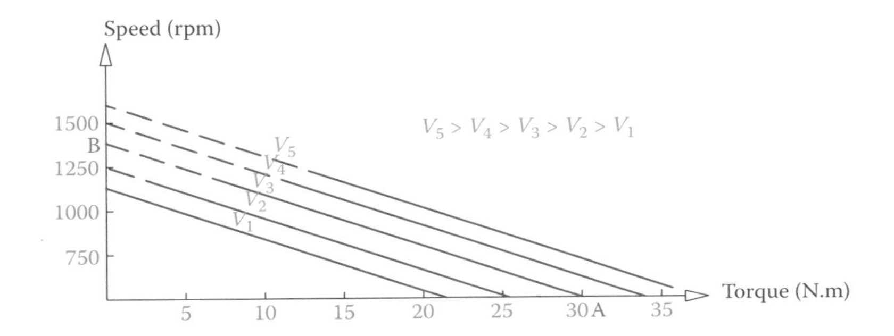

Figure 7.34 Speed-torque relationship in a permanent magnet DC motor.

Figure 7.34 illustrates the speed-torque relationship for a typical permanent magnet DC motor. This relationship is also linear, and the corresponding curve is a straight line. Note that there are a series of lines indicated in Figure 7.34, each corresponding to a different applied voltage (e.g., V_1 to V_5), where $V_1 < V_2 < V_3 < V_4 < V_5$. The torque-speed relationship directly follows from Equations 7.1 and 7.3, taking into account Ohm's law in the armature winding ($V_a = I_a R_a$) and noting that the magnetic field is constant.

For each characteristic curve (such as the middle one associated with V_3), point A corresponds to zero rotational speed. For the applied voltage, this is the highest torque that a motor can deliver. Moreover, point B corresponds to no load rotational speed. This is the highest speed that the motor can turn when voltage V_3 is applied to it, but no load can be connected to the shaft. This part of the curves is shown by a dashed line, because it corresponds to a situation that is not normally realistic, or it is outside the range of operation of a motor.

For a constant load (constant torque), the relationship between the applied voltage and speed of a permanent magnet DC motor is as shown in Figure 7.28, and it is not repeated here.

7.6.2.2 Series Wound DC Motor

In a series wound motor, current in the field winding and the current in the armature are the same. Because the magnetic field strength is directly proportional to the current, Equation 7.1, as a result, can be written as

$$T = K\phi I_a = K_a I_a^2 \tag{7.5}$$

where K_a is constant (the product of K and another constant value). The characteristic curve denoting the relationship between the motor torque and the armature current is not linear. For example, for a double value for torque the current increases by 41 percent (based on Equation 7.5: $1.41^2 = 2$).

From the construction viewpoint, in a series motor, wires for the field winding are thicker than their counterparts in the other wound-field DC motors. This is because the same current flows through both the armature and the field windings. In a motor, wire thickness and the number of loops in various windings are among the parameters that determine the constants such as K_a in Equation 7.4.

One of the distinguishing features of a series wound motor is its very high starting torque. For this reason, series motors are used for loads that need a high torque to start, such as locomotion and lifting equipment, like in trains and cranes. A typical curve is illustrated in Figure 7.35, representing the speed-torque relationship. When such a motor is powered, load must already be connected to it. Otherwise, the motor having a low torque load may develop incremental speed, leading to runaway speed, which is quite dangerous and can cause severe damage of the motor (see the speed corresponding to the no load end of the graph in Figure 7.35).

As discussed earlier, the operating point (which lies within the range of the rated values of a motor) is the point of balance between a motor and its load, found by the intersection of the performance curve for the applied voltage to a motor and its load characteristic curve. For a constant torque (i.e., a load whose torque has negligible variation) any necessary alteration of the motor speed must be done through changing the applied voltage. Figure 7.36 shows a sample of the approximate relationship between voltage and speed.

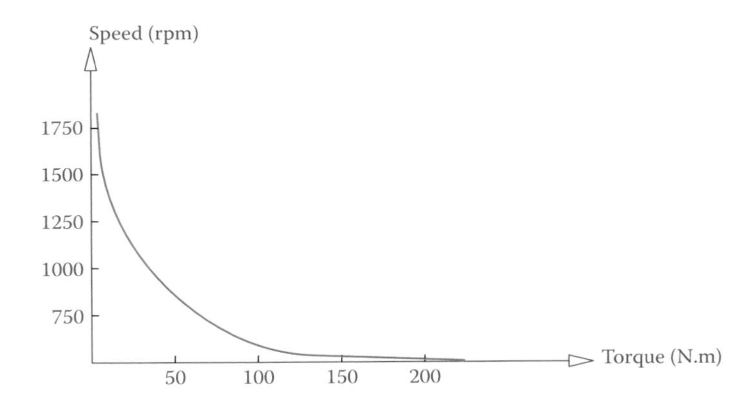

Figure 7.35 Speed-torque relationship in a series wound DC motor.

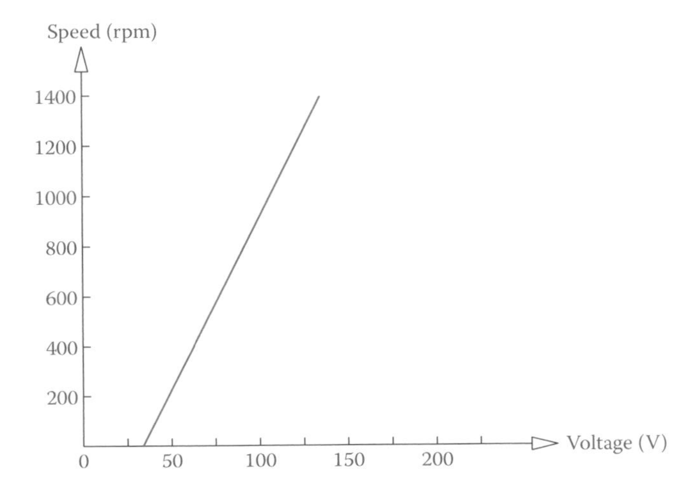

Figure 7.36 Speed-voltage relationship for a constant torque.

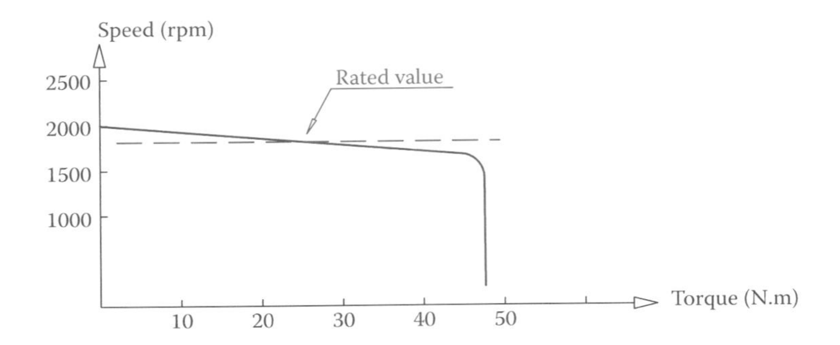

Figure 7.37 Speed-torque curve for a typical shunt wound DC motor.

7.6.2.3 Shunt Wound DC Motor

Not all loads require the characteristics of a series wound motor. If the speed of operation of a device must be almost constant irrespective of the load, then a better choice is a shunt wound motor. The speed-torque characteristic curve of a shunt wound motor is a straight line, within the operational region, with a slope better than that of a permanent magnet motor (see Figure 7.34). This is shown in Figure 7.37.

7.6.2.4 Compound Wound DC Motor

Compound wound motors have been invented to improve the performance of the machine in terms of torque-speed and torque-current relationships. Performance of a motor depends on many design parameters; the field arrangement is only one of them. Two methods for combining the series and parallel field windings have been presented in Figure 7.26 (long and short shunt compound). For each arrangement, nevertheless, there are two ways that the field windings can be inserted (connected) inside the machine. Consequently, the two magnetic fields generated by the two windings can be helping each other (added effect), or they can be opposing each other. For the long shunt compound arrangement these two ways are shown in Figure 7.38. As illustrated, in Figure 7.38a the two fields are adding together (see the polarity of the winding denoted by a plus sign),

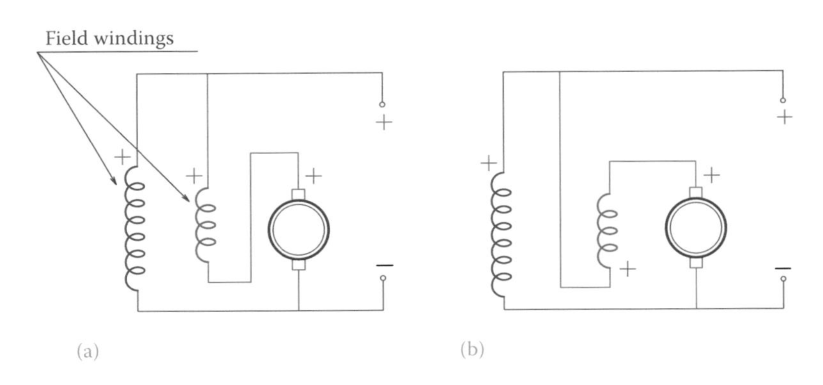

Figure 7.38 (a) Cumulative and (b) differential compound wound long shunt DC motor.

whereas the windings in Figure 7.38b have the reverse polarity. The resulting magnetic field, thus, is equal to the difference between the two magnetic fields. For this reason the first category is called **cumulative compound wound** motor, and the second category is called **differential compound wound** motor.

In general, no matter which arrangement is used for a compound wound DC motor, the performance falls between that of a series wound motor and a shunt wound motor. A typical scenario for the speed-torque relationship is shown in Figure 7.39.

To compare the performances of the preceding five types of DC motors side by side, Figure 7.40 depicts the speed-torque performance of them for a common operating point. In application, any motor operates around an operating point within the operating region (rated values) that it is designed for (from the maximum efficiency and power points of view). For any variation of torque around the operating point the variation of speed can be smaller or larger depending on the type of motor. As can be understood from Figure 7.40, a separately excited motor has a much better capability of running at a constant speed, whereas for higher starting torques a series wound motor is preferable (as mentioned before).

Cumulative compound wound: Type of DC electric machine with both parallel and series field winding where the magnetic effects of the parallel and series windings add together, as opposed to the differential compound wound.

Differential compound wound: Type of DC electric machine with both parallel and series field winding where the magnetic effects of the parallel and series windings are opposite to each other (the magnetic field from the parallel winding is reduced by the smaller magnetic field of the serial winding).

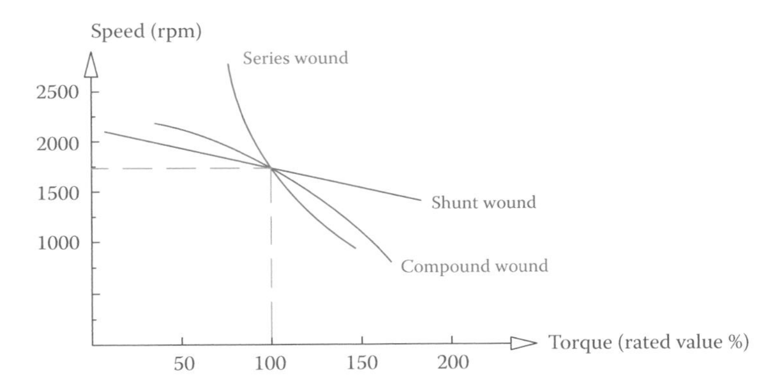

Figure 7.39 Speed-torque curve for a compound wound motor with respect to a series wound and a shunt wound motor.

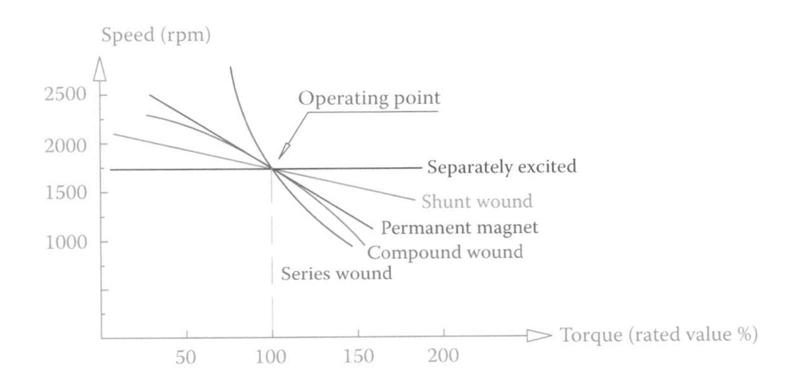

Figure 7.40 Comparison of the speed-torque curve for all types of DC motors.

7.6.3 DC Generators

In the same way that there are different types of DC motors, one can say, in general, that each one can perform as a generator if rotated by a driver. Nevertheless, the reason for having various types of motor is to obtain a desirable performance of the motor. Except for the separately excited machine and permanent magnet machine, the (magnetic) field must be developed within the machine based on some current flowing through the field winding, which depends on the applied voltage. In this sense, the operation of a machine as a generator must build up from the small amount of magnetism in the core of its armature. This regenerative build up action must continue until a generator reaches its nominal voltage. For certain arrangement of field and armature windings (such as in differential compound winding), where the two fields oppose each other, this build up process may never take place and the machine may never reach the operational condition.

The voltage developed in the armature winding is according to Equation 7.4. Thus, the performance of a DC generator varies with its design. In general, the faster a generator is turned, the higher its voltage.

As stated before, for a generator, one should see the variation of the produced voltage with the load current. This is shown for various types of generators in Figure 7.41, which shows the change in voltage versus the load current around an operating point; it is exaggerated for better clarity. As can be observed from Figure 7.41, except for a series wound generator, the generated voltage decreases with an increase in the line (load) current. This implies that in a series wound generator, if by any reason the current goes up, it further increases the voltage, which, in turn, increases the current. Such an unstable behavior is undesirable and may lead to destruction of a generator. For this reason, series wound generators are not very common. All the other generators have a self-regulatory behavior, and any increase in the load current reduces the voltage (and thus the current), and the machine reaches a stable new working condition.

A series wound DC machine should never be used as a generator.

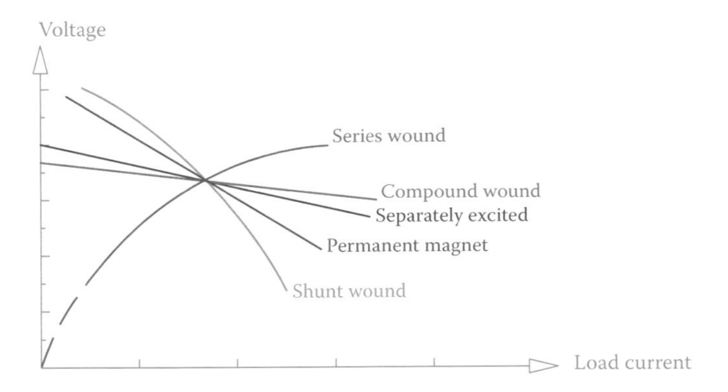

Figure 7.41 Variation of voltage with the load current in DC generators.

7.7 DC Sources in Parallel

Putting electricity sources in parallel with each other is to provide the required electric power from more than one source. This is out of necessity when expansions occur, and more loads are to be powered by electricity. Remember that the power capacity of power supplies must always be greater than the power demand by the loads.

In DC electricity the condition to be met for electric sources to be put in parallel is that their voltages be equal. In no way can one put a 12 V battery together with a 24 V battery and expect that they both provide electricity to some loads. When the voltages of the sources are the same, the current demanded by the loads will be shared among all the sources. In putting sources in parallel, in fact, the current is distributed among them. If two sources are exactly identical, then their currents are equal.

For connecting sources in parallel the positive sides of all must be connected together and the negative sides together. Then the common positive point is the + terminal and the common negative point is the – terminal. Mistakes in connecting sources together in the right manner lead to damage to the equipment and injury to people.

> For putting DC sources in parallel the important necessary condition is that their voltages be equal.

7.7.1 Batteries in Parallel

Figure 7.42 shows three batteries in parallel with each other. A simple example of putting two batteries in parallel is in boosting a weak battery by jumper wires for starting a car on cold winter days. Note that the positive sides must be connected together and the negative sides together. If this is not respected, the weaker battery becomes a load to the good battery and high current flows through the two batteries. This can cause sparks, temperature rise, and even explosion of the batteries.

In connecting batteries in parallel, for the purpose of more power, it is always better to have similar batteries of the same condition, so that the current is equally distributed among them.

7.7.2 DC Generators in Parallel

Generators are not exactly like batteries, though both are the sources of DC electricity. Batteries are for low power, while generators can provide

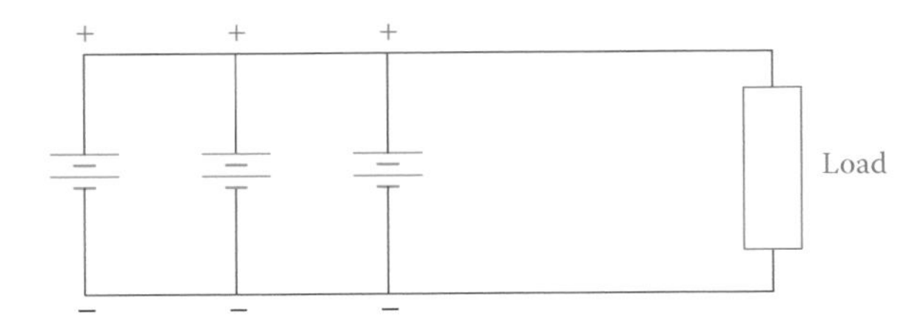

Figure 7.42 Batteries in parallel.

large amounts of electric power. Generators can be of different power capacity and different type, while the difference between batteries is not so wide. Combining generators of various types may not always be possible, but different sizes of generators of the same type can be combined. When generators of different sizes are in parallel, the larger generator takes a higher portion of the load current and the smaller generator provides smaller power. For two batteries when the similar terminals are connected together, there is no reason for polarity change. In the case of DC generators, if the voltage of one of the generators decreases (especially if it is smaller and has a lower current), it is possible that eventually it changes state and becomes a motor getting power from the larger generator. Thus, voltage regulation becomes very important in order to prevent this from happening. Voltage regulation for each machine is performed through speed control.

7.8 Efficiency

For a motor to work (produce mechanical energy) it needs to use electrical energy. Also, for a generator to produce electrical energy, it must be fed with mechanical energy. In general, any machine needs energy to work. If the energy that is produced by a machine is called the output energy and the energy that is given to it is called the input energy, the output energy is always smaller than the input energy. The ratio of the output to input energy defines the efficiency of a machine.

To determine the efficiency, it is more reasonable to consider the energy for one second, which is the *power*. If the output power is denoted by P_{out} and the input power is represented by P_{in}, efficiency is represented by the ratio

$$\text{Efficiency} = \frac{P_{out}}{P_{in}} \times 100 \tag{7.6}$$

The difference between the output power and the input power is the power that is lost in a machine in the form of heat in the electrical parts (wire windings) and mechanical parts (moving parts). This heat is not recoverable; thus, it is considered a loss of energy.

$$P_{out} = P_{in} - P_{lost} \tag{7.7}$$

Efficiency is always stated as a percentage. This is the reason for multiplying by 100 in Equation 7.6. Because P_{out} is always less than P_{in}, efficiency is always a number less than 100. There is no machine or device with an efficiency of 100 percent. However, it is possible to find machines and devices with a high efficiency, for example, up to 98 percent. But, efficiency can be much lower; for example, in machines with an efficiency of 35 percent (e.g., a combustion engine). Efficiency of motors and generators can reach high values of around 90 percent for well-designed large machines, and it can be as low as 60 percent for small poorly designed machines.

For any machine—electrical, mechanical, or thermal—efficiency is not always constant; it depends on the operating conditions and the output. For some value of the output power the efficiency of a machine is maximum, and for other values it is less than that. Usually, within the rated value of any device its efficiency is near maximum. For example, if a motor is rated for 240 hp, its efficiency is highest at around this power output, but if it is operated at say 120 hp, then its efficiency can be much lower.

> Usually, within the rated value of a machine or device its efficiency is the highest.

If a machine or device is made up of several components with known efficiencies, then the entire efficiency of the device depends on the individual efficiencies of its components. In such a case we can calculate the total efficiency as

$$\text{Efficiency}_{\text{total}} = \text{eff}_1 \times \text{eff}_2 \times \text{eff}_3 \times \cdots, \tag{7.8}$$

where eff_1, eff_2, etc. are the efficiencies of the components.

Example 7.1

An electric motor is used for driving a lift. If on one occasion (because the power demand depends on the number of people in the lift) the power consumption is 7.2 kW and the power output (as required by the lift) on the motor is 5 kW, what is the motor efficiency?

Solution

The efficiency of the motor for the case can be determined from Equation 7.6 as

$$\text{Efficiency} = \frac{5}{7.2} \times 100 = 0.69444 \times 100 = 69.4\%$$

Example 7.2

Suppose that the motor in Example 7.1 drives a gearbox whose efficiency is 90% and there is also a pulley system with 88% efficiency. What is the total efficiency of the whole system?

Solution

From Equation 7.8 we get

$$\text{Efficiency}_{\text{total}} = \left(\frac{69.4}{100}\right)\left(\frac{90}{100}\right)\left(\frac{88}{100}\right) = 0.55 \times 100 = 55\%$$

Example 7.3

Suppose that when the lift in the previous examples is at full load, its total efficiency is 75%. What is the efficiency of the motor in this case if the efficiencies of the gearbox and the cable system are assumed to be the same as before?

Solution

For this case we have (M_{eff} is motor efficiency)

$$\left(\frac{M_{eff}}{100}\right)\left(\frac{90}{100}\right)\left(\frac{88}{100}\right) = 0.75$$

which leads to

$$M_{eff} = \frac{0.75}{(0.90)(0.88)} \times 100 = 94.7\%$$

Example 7.4

If the output power by the motor in Example 7.3 is 8.2 kW, what is the power that the motor takes from electricity?

Solution

Input power (which must be larger than the output power) is a value that if multiplied by the motor efficiency the resultant is 8.2 kW. Thus,

$$P_{in} = \frac{P_{out}}{0.947} = \frac{8.2}{0.947} = 8.66 \text{ kW}$$

7.9 Chapter Summary

- A magnetic field is the space around a magnet where a magnetic effect exists.
- Only ferromagnetic materials are good for magnetism and magnetic applications.
- An electromagnet is a magnet made up of a coil wound around a ferromagnetic core; its magnetism can be turned on and off.
- Poles of an electromagnet can be determined from the right-hand rule: If the fingers show the direction of current in the winding (from + toward –), then the thumb shows the north pole.
- A DC electric motor works based on Lorentz force, as a result of the interaction between a magnetic field and electric current.
- Lorentz force law states that if a current carrying wire is placed in a magnetic field, then a force is exerted on the wire. Direction and magnitude of this force depends on the magnetic field and the current.
- The right-hand rule for motors states that if the fingers of the right hand are aligned in the direction of the magnetic field and

the thumb shows the direction of current, then the direction of the Lorentz force is perpendicular to those two and coming out of the palm.

- Main components of a DC machine are a magnetic field, an armature winding, and a commutator. Armature windings have a ferromagnetic core, and together with the commutator are mounted on a shaft.
- Performance of any machine can be observed from its characteristic curve, which shows the variation of a parameter like the developed torque in a motor versus another parameter such as the voltage applied to the motor.
- For an electric motor one should know the variation of speed versus the applied voltage and the variation of speed versus the torque on the shaft.
- The same DC machine can operate as a DC motor and as a DC generator; if turned, it generates electricity, and if given electricity, it turns.
- Faraday's law is the basis for electric generators. It states that if a wire moves inside a magnetic field, then an electromotive force (voltage difference) is generated in the wire.
- There are various types of DC motors. The difference between them stems from how the magnetic field in a motor is provided. Similarly, there are various types of generators.
- Series wound motors have a very high starting torque; they are good for locomotion. Shunt wound (parallel wound) motors have a more constant speed of operation. Compound wound motors have two field windings, and their performance characteristic stands between series and shunt wound motors.
- To provide more electric power, generators are put in parallel together. The principal condition to put DC generators in parallel is that they have the same voltage.
- Rated value in an electric machine is the range of values of voltage, power, speed, and torque for which the machine is designed. A machine has the highest efficiency at its rated value(s).
- Efficiency is the ratio of output power to the input power in a machine or device. Normally, efficiency can never be 100 percent.

Review Questions

1. What is the field of a magnet?
2. What are three categories of materials in term of magnetism?
3. What is the difference between a magnet and an electromagnet?

4. What is a ferromagnetic material?

5. What is the difference between a permanent magnet and an electromagnet?

6. What is the direction of a magnetic field inside a magnet?

7. What are the two main ingredients for an electric motor?

8. What does the right-hand rule state about the magnetic field of an electromagnet?

9. What is the structural difference between a DC motor and a DC generator?

10. Briefly explain the Lorentz force law.

11. What is an armature?

12. How many different types of DC motors do you know?

13. What is the main element of difference between various types of motors?

14. What is the reason to have more types of motors if they are all DC?

15. Can a motor work without a magnetic field? Why or why not?

16. Describe the right-hand rule for motor operation.

17. Describe the right-hand rule for generator operation.

18. Which one is simpler, a shunt wound DC motor or a permanent magnet DC motor? Why?

19. Explain the similarity between a DC motor and a DC generator.

20. Briefly explain Faraday's law.

21. Why are batteries put in series?

22. Why are batteries put in parallel?

23. What is the condition to connect batteries in parallel?

24. Can one connect a battery and a DC generator in parallel?

25. Can a 10 kW motor be powered by a 10 kW generator?

26. Does a machine have always the same efficiency?

27. If the efficiency of a motor is 90 percent and we use this motor as a generator, can we say its efficiency is 110 percent?

28. What are the losses in a generator?

29. Explain why efficiency is always less than 100 percent.

Problems

1. In Figure P7.1 the characteristic curve of a motor is shown.
   a. What type of motor is it?
   b. If the motor runs at 750 rpm, what is the torque provided by the motor?

2. The torque on the shaft of a DC motor is 150 N.m and remains constant at this value. If the characteristic curve of this motor is as shown in Figure P7.2, determine
   a. Speed of the motor if it is powered by 110 V supply
   b. New speed if the voltage of the power supply goes up to 120 V
   c. Speed if the voltage is 115 V

3. Average efficiency of a 250 hp motor for the range of operating conditions is 90 percent. If the motor is used 20 hours per day, how much energy is lost in the motor per day?

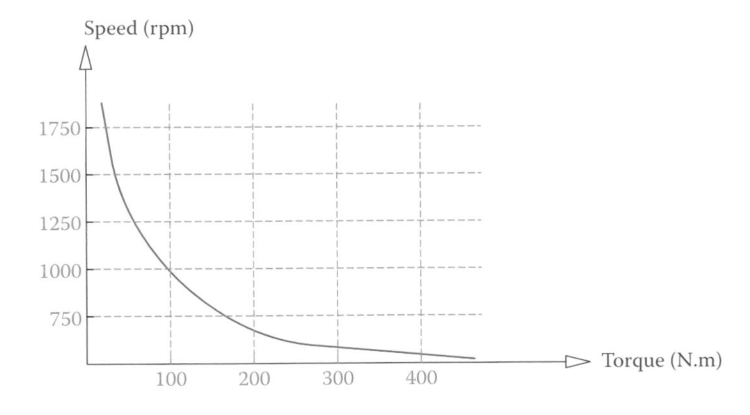

Figure P7.1 Problem 1.

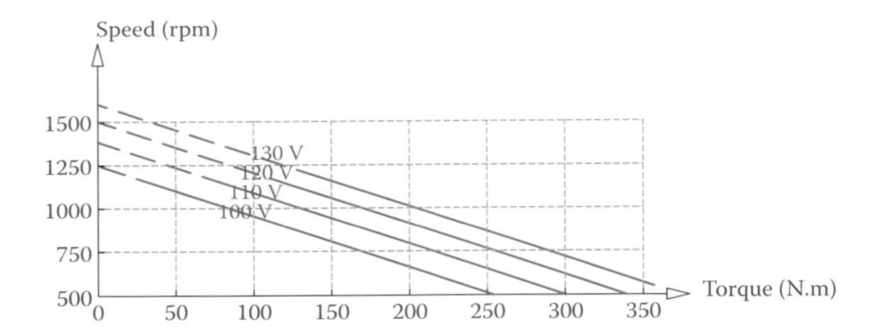

Figure P7.2 Problem 2.

4. For running a conveyor a DC motor is employed through a gearbox. For the maximum load on the conveyor, 100 hp of power is required. The gearbox has an efficiency of 90 percent and the efficiency of the conveyor cannot go below 80 percent. What is the required output power from the motor?

5. What must the rated power of a motor in kW be in order to run the conveyor in Problem 4 if its efficiency is 92 percent?

6. What is the total efficiency of the conveyor system in Problem 4?

7. A lift in a wind turbine is used at an average rate of 55 hours per year. There are two options to replace the damaged original motor that provided 2 hp to the lift. One has an efficiency of 70%, and the other has an efficiency of 92%. Calculate the difference in the annual energy consumption of the two motors.

8

AC Circuits Relationships

OBJECTIVES: After studying this chapter, you will be able to

- Find current and power in almost all AC circuits
- Use vectors for representing voltage, current, and their relationships in AC electricity
- Describe differences between DC and AC circuits
- Explain three types of power in AC and the relationship between them
- Find the equivalent of inductors and capacitors in series and in parallel with each other
- Describe phase difference
- Define reactance, and understand inductive and capacitive reactance and how to measure them
- Explain when inductors and capacitors are put in an AC circuit, what happens, and why there will be phase difference between current and voltage
- Solve problems involving resistors, inductors, and capacitors connected to AC electricity
- Solve problems when resistive loads, inductive loads, and capacitive loads are combined together, either in series or in parallel, in AC circuits
- Note the meaning of leading and lagging
- Design voltage dividers for AC circuits
- Define resonance and its effects in series and parallel *RLC* circuits
- Understand the importance of power factor and its effect on power in a circuit
- Describe power factor correction and how to perform it in practice

New terms:
Active power, apparent power, capacitive reactance, effective value, effective voltage, impedance, in phase, induction, inductive reactance, lagging, leading, out of phase, peak voltage, peak-to-peak, reactance, reactive power, real power, resonance, resonance frequency, resonant frequency, true power, Volt-Ampere-Reactive (VAR)

8.1 Introduction

We learned in Chapter 6 how a load in a DC electrical circuit can comprise various components arranged in different combinations. Alternatively, we may say a load could be decomposed in various components. The same is true when dealing with AC circuits and AC loads.

In addition, because of the nature of alternative current, its effect on electrical components is not the same as DC. There are quite a number of new features and terms that one needs to know when dealing with alternating current.

This chapter deals with all these new features and the behavior of various components when subject to AC. Although the rules as we have learned are the same, the calculation of current and power are quite different because of the features of AC electricity and the corresponding relationships.

In dealing with AC circuits it is absolutely necessary to learn about vectors, which contain the additional dimension one needs to define and present AC relationships. Thus, this chapter starts with vectors.

8.2 Vectors

In studying and dealing with AC circuits the use of vectors is very helpful and facilitates the comprehension of additional matters that do not appear in DC electricity. Understanding of vectors is as important (and perhaps as easy) as numbers. Vectors have many applications, and the mathematics of dealing with them is more involved than discussed here. We only consider them as much as necessary to understand AC electricity relationships.

Numbers like 5, 34, and 5000 give you certain information about quantity. You can describe anything such as the number of houses, size of an angle, and weight of a car provided that a unit is associated to it, such as 5 lb of apples. A negative number is also possible to describe certain entities such as temperature; for example, –20°. But, negative numbers cannot be used for everything. For instance, you do not say –34 houses.

In addition to quantity, some entities have a direction, and mentioning a quantity is not sufficient to describe them. For instance, wind can be blowing at 22 mph, but at what direction? Other examples of those entities that have both quantity and direction are force and velocity (speed). A vector is used to represent an entity that has both direction and quantity.

A vector is represented by an arrow, the direction of which indicates the direction and its length represents the quantity. Figure 8.1 illustrates a few examples of vectors. To address vectors, they may be given a name like V_1, V_2, etc., which are only names referring to two different vectors.

Figure 8.1 shows various examples of how vectors are used in representation of force, speed, and so on. Attention must be paid to the following points:

1. V_2 and $-V_1$ show how V_2 is related to V_1 and the relationship between a vector V_1 and its negative. V_2 is 2 times longer in total length (including the arrow head) compared to V_1, and $-V_1$ has exactly the same length as V_1, but it points to the opposite direction.
2. V_3 is a vertical vector, while V_1 is horizontal, and V_4 is inclined.
3. V_4 can also be denoted by AB.
4. The points of action (supposing a force) of V_5 and V_6 are the same (point C).

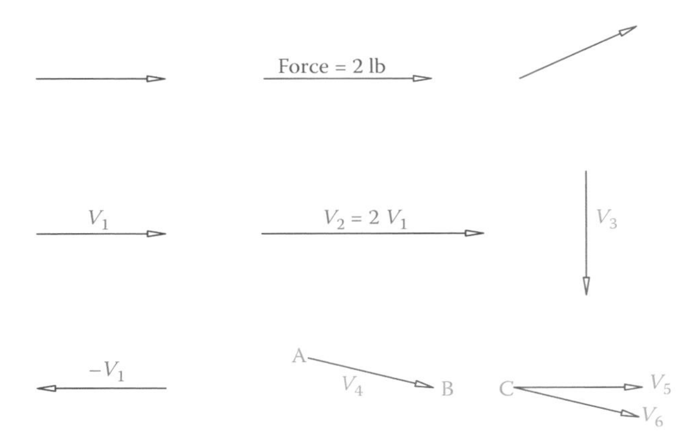

Figure 8.1 Representation of vectors.

When two or more vectors have the same point of action, they can be replaced by their resultant vector. In this sense, vectors can be added together. Depending on the angle between vectors, their sum has its own direction and length. The sum of two vectors can be smaller than each one of them. Figure 8.2 illustrates a number of cases for the sum of two vectors A and B. It also depicts how one vector can be subtracted from another. The resultant (sum) of two vectors is the diagonal of the parallelogram formed by the vectors. Alternatively, it can be found by the process shown in Figure 8.3, where one of the vectors is moved to the end of the other vector, and the sum is represented by the line connecting the starting point of the first vector to the end of the moved vector. For this process, it

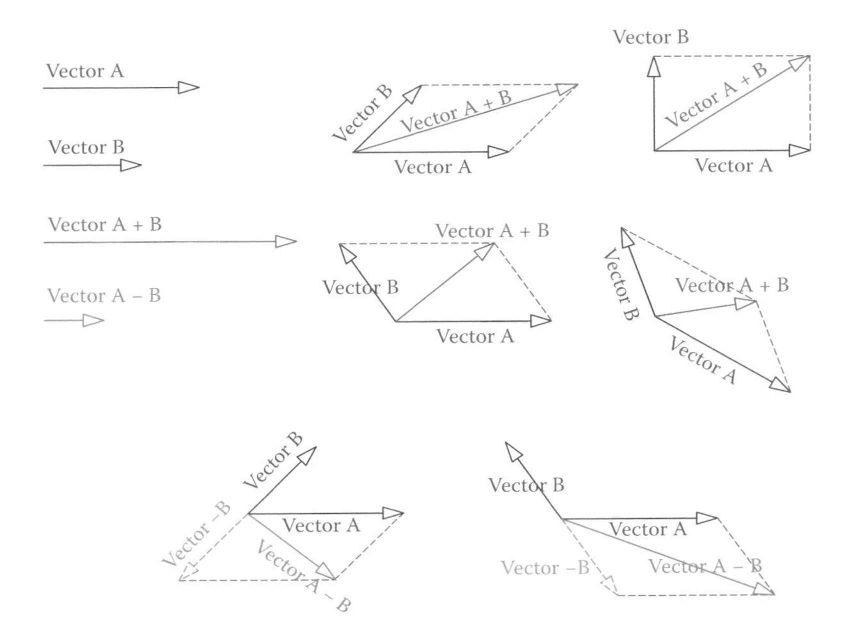

Figure 8.2 Adding two vectors together.

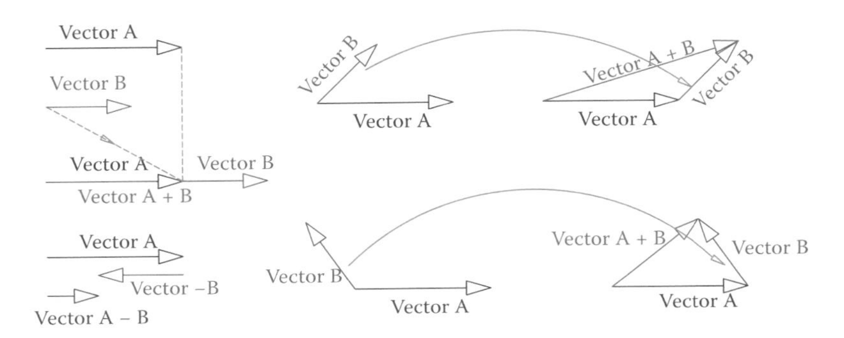

Figure 8.3 Adding two vectors together, alternative method.

does not matter which vector is selected first. As can be understood from Figures 8.2 and 8.3, when two vectors have the same direction, their sum is also in the same direction and its length is equal to the total length of the two vectors. Also, when two vectors have the opposite directions, their sum is in the direction of the larger vector and its length is equal to the difference between the two vectors.

In adding three or more vectors, one can add two of them at a time and add the resultant to the third, and continue the same process for the rest.

8.3 AC Circuit Measurements and Values

So far, you have learned a lot about DC circuits, how to measure voltage current, and so on. The same can be done for AC circuits, but, first, you need to learn some new definitions. In DC circuits the voltage or current can be positive or negative, and that depends on the definition of positive and negative direction of current. In AC, however, because current continuously switches direction, the notion of positive and negative for current becomes meaningless. Nevertheless, because voltage difference is a relative entity, it will still make sense to determine the voltage between two points in a circuit being positive or a negative. It is always true that if the voltage at a point A is higher than the voltage at point B, then V_{AB} is positive, whereas V_{BA} is negative. Furthermore, the current direction is from A toward B.

Peak voltage: Highest magnitude of a sinusoidal waveform (like in AC electricity).

Peak-to-peak: Magnitude between the minimum and the maximum values in a sinusoidal waveform. Frequently used for voltage and current in AC electricity.

First, let's consider the meaning of voltage in AC. Consider Figure 8.4, which shows the amplitude variation for a few cycles of a sinusoidal waveform (see the text box for sinusoidal waveform in Chapter 4). For this waveform the **peak voltage** is 5 V and the **peak-to-peak** voltage is 10 V.

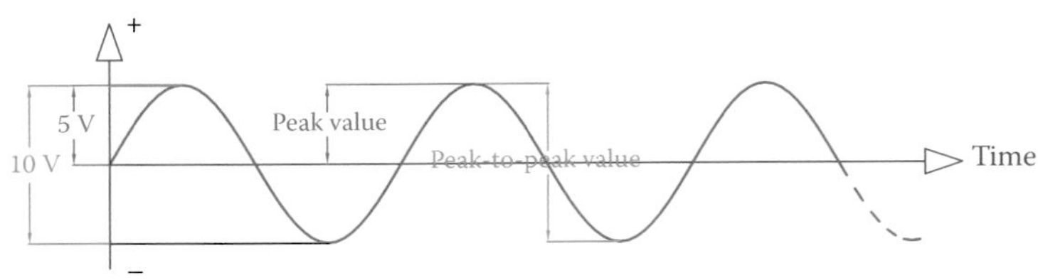

Figure 8.4 Definitions for an AC waveform.

Both these voltages refer to the same waveform; thus, they are equivalent. But, how does this voltage correspond to a DC voltage of 5 V?

In answering this question, let's consider a circuit made up of a lightbulb connected to a 5 V DC source and compare it with another circuit in which the same lightbulb is connected to an AC source with a peak voltage of 5 V. Which lightbulb is brighter (or is the brightness the same for the two)?

> A peak voltage of 5 V is equivalent to a peak-to-peak voltage of 10 V, and vice versa.

Effective voltage: Because the voltage continuously fluctuates in alternating electricity, the effective voltage is the value that if a DC system was used instead, it produced the same power. For sinusoidal waveform the effective voltage is 0.707 of the peak voltage.

Effective value: Equivalent DC value of an AC value based on producing/consuming the same power. This is mathematically equal to the root mean square (RMS) of the AC waveform. In AC electricity, effective value can be used for voltage and current.

The answer is that the two lightbulbs *do not* have the same brightness. The brightness of the one connected to DC is higher. The fact that electricity fluctuates between its maximum and minimum peaks reduces the effect of providing energy to the lightbulb. The amount of electricity given to the lightbulbs in this scenario is defined by the power delivered to them. Suppose now that the AC electric source is chosen such that it does provide the same power to the lightbulb as the 5 V DC source and in both cases the brightness of the lightbulb is the same. In such a case we say the **effective voltage** of the AC source is equal to the voltage of the DC source (5 V in this example). In general, one can define **effective value** for a cyclic waveform. The following relationships exist between the peak value, the peak-to-peak value and the effective value in a sinusoidal waveform (not for any other waveform).

$$\text{Effective value} = \frac{\sqrt{2}}{2}\,\text{peak value} = 0.707\ \text{peak value} \tag{8.1}$$

$$\text{Effective value} = \frac{\sqrt{2}}{4}\,\text{peak-to-peak value} = 0.3536\ \text{peak-to-peak value} \tag{8.2}$$

If the effective value is known, then the peak value is found from

$$\text{Peak value} = \sqrt{2}\ \text{effective value} = 1.41\ \text{effective value} \tag{8.3}$$

Figure 8.5 shows the plots for a 5 V DC and a 5 V AC electricity for a short period of time. It also shows the symbol used for an AC source

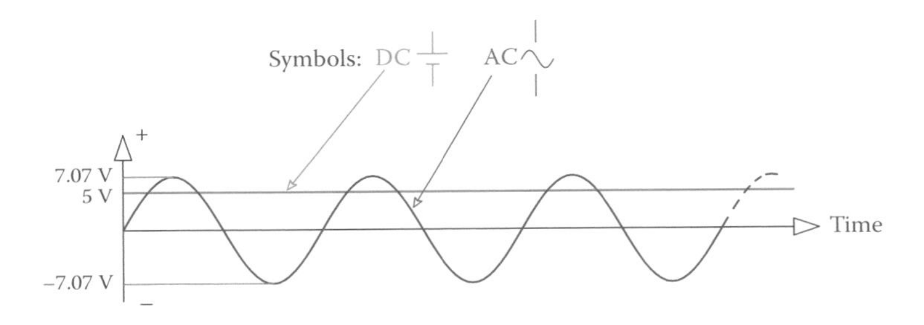

Figure 8.5 Comparison of a 5 V DC source and a 5 V AC source.

(compared to the two, long and short, parallel lines for a DC source) when inside a circuit.

Example 8.1

What is the effective voltage of an AC power supply with a peak voltage is 18 V?

Solution

$$V_{\text{eff}} = (18)(0.707) = 12.73 \text{ V}$$

Example 8.2

What is the peak-to-peak voltage for a power source with an effective voltage of 9 V?

Solution

$$\text{Peak-to-peak voltage} = (9)\left(\frac{4}{\sqrt{2}}\right) = (9)(2.83) = 25.46 \text{ V}$$

From the above example one can understand that there is a large difference between the effective voltage in an AC source and the peak-to-peak value of the same voltage. Normally, when an AC voltage is defined, the given value is the effective voltage, unless otherwise stated. For instance, in North America the electricity at home has a nominal voltage between 110 and 120 V. This implies that the effective voltage has such a value. In most of Europe where the mains voltage is 220 V, the peak voltage is 310 V. Often when an AC voltage is given, its frequency must also be defined; for example, 120 V, 60 Hz. If the frequency is not given, then, to denote that, a given voltage is AC the term VAC can be used; for example, 115 VAC.

Example 8.3

What are the peak and the peak-to-peak values of 115 V AC, 60 Hz voltage?

Solution

Frequency is independent of the voltage; thus, it does not have any effect on the voltage values.

$$\text{Peak voltage} = \left(\sqrt{2}\right)(115) = 162.63 \text{ V}$$

$$\text{Peak-to-peak voltage} = (2)(162.63) = 325.27 \text{ V}$$

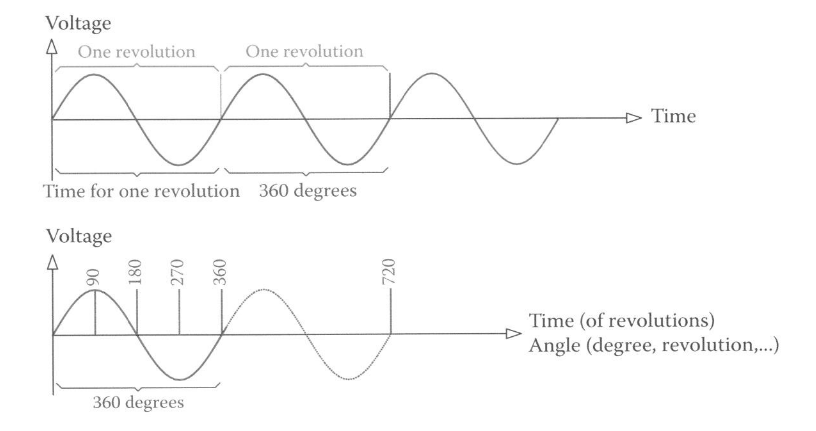

Figure 8.6 Equivalence of time and angle in periodic signals.

> Unless otherwise stated, a given AC voltage value refers to the effective value of the voltage.

In conjunction with the representation of an AC waveform, as shown in Figure 4.17, each cycle of a periodic signal corresponds to 360° of rotation. In this respect, the variation of a waveform can be expressed either in terms of time or in terms of angle, as shown in Figure 8.6. This is especially useful when one cycle is considered. The relationship between time and angle is established based on equating the duration of one cycle to 360°. For example, for 60 Hz AC the duration of each full cycle is 1/60th of a second, i.e., for 360°. The time duration for 90° is, thus, one quarter of that, i.e., 1/240th of a second.

8.4 Resistors in AC Circuits

When one or more resistors is connected to an AC source a circuit is formed, and, as in the case of resistors connected in a DC circuit, each resistor carries a current. We may determine the current(s), voltage drop(s), and power for the entire or parts of the circuit. Resistors may be in series and in parallel or may form a combination of the two, as discussed in Chapter 6.

All the relationships discussed in Chapter 6 for DC circuits can be directly used for AC circuits containing only resistors. This also includes the rules for series and parallel resistors and combination circuits.

When an AC voltage is applied to a resistor, AC current flows through the resistor. Ohm's law must be used in order to find the current. The important point to take into account is that whichever value of the voltage (effective, peak, or peak-to-peak) is used, the corresponding value for the current is obtained. That is to say, if the peak value of the voltage is given, then the value obtained for the current based on Ohm's law will be the peak value of the current, and so on.

For calculation of power, however, it is always the effective values that must be considered. If other values are given, the effective values must be determined first. Corresponding to Equation 5.4 for the power in a DC circuit, the power consumption of a resistor in an AC circuit is

$$P_{resistor} = V_{eff} I_{eff} \quad (8.4)$$

where P is in watts, V_{eff} is the effective voltage across the resistor in volts, and I_{eff} is the effective current in the resistor in amps. This power turns into heat in the resistor (part of that turns into light in an incandescent lightbulb).

Similarly, all the corresponding relationships between current, voltage, and power, as given in Equations 5.5 through 5.12 can be directly applied to AC electricity when a circuit involves only resistors. In all cases, the effective values of voltage and current are to be used. These equations are not repeated here.

Example 8.4

An AC power supply is connected to a 16 Ω resistor. If the power consumed is 36 W, what is the peak voltage of the power supply?

Solution

Having given the power and the resistance, the (effective) voltage can be found from Equation 5.10:

$$V_{eff} = 24 \text{ V}$$

$$V_{peak} = (1.4)(V_{eff}) = 33.6 \text{ V}$$

Example 8.5

In an experimental lab, when an AC power source whose peak-to-peak voltage is 140 V is connected to a heating element in a vessel, it boils water in 150 sec. What should be the voltage of a DC power supply to replace the AC power source so that it takes the same time to boil water?

Solution

Boiling water is due to the energy delivered by the heating element. Because energy is the product of power and time in order to bring water to boiling within the same amount of time the power must be the same in both cases. Because the heating element does not change, the only condition is that the DC voltage be equal to the effective value of the AC voltage.

$$V = \frac{140}{2.82} = (0.3536)(140) = 49.5 \text{ V} \approx 50 \text{ V}$$

8.5 Inductors in AC Circuits

In Chapter 4 you learned about inductors and inductance. It is advisable that you review that material before continuing this section for a better grasp of the subject (see Section 4.10). You also learned that the effect of an inductor in a DC circuit is that it resists any change in the current (see Section 6.12). This happens at both the time that a current is to develop in a circuit (switching on a circuit) and when the electric connection to a circuit is turned off. The phenomenon takes place for only a very short period of time. It can be before a magnetic field is developed in the inductor (when a switch is turned on) or before the circuit current has completely dropped to zero (when a switch is turned off). In fact, a change in current implies any reduction or increase in current from an initial value to a new value, none of them necessarily zero.

The reason for this phenomenon stems from the Faraday's law (see Chapter 7), which was the basis for generators. Recall that if a wire moved in a magnetic field, then a voltage was generated in it. Now we want to extend this observation. First, instead of moving the wire in the magnetic field, we may move the magnetic field with respect to the wire (see Figure 8.7). The effect is the same because the two of them have a relative motion with respect to each other. Further, the relative motion can be substituted by varying the strength of the magnetic field, i.e., keeping both the wire and the magnet stationary but changing the field strength. This is what happens when a magnetic field is developing, i.e., coming into existence from zero or disappearing (reducing to zero). In both cases, a voltage is generated in the wire.

When a coil of wire (an inductor) is connected to DC electricity, a current is building up from zero, making a magnetic field. The wire, itself, is in that field, and therefore, a voltage is generated in it. This generated voltage is in the opposite direction of the voltage making the field; thus, it is opposing the current. The same thing happens when an inductor is disconnected from a DC source.

The phenomenon just discussed is called Lenz's law. Before a statement for Lenz's law is made, let's define the term **induction**. When an electromagnetic force (emf) is generated in an inductor, a more common statement employed is that "an emf is induced in the inductor." Thus, induction is used instead of generation, which goes well with inductor and inductance. Lenz's law states that an induced emf always opposes its cause, meaning that the direction of the current owing to the induced

Induction: Generation of electricity in a wire when the magnetic flux is cut by the wire (e.g., when wire moves with respect to a magnetic field or the strength and/or direction of the magnetic field varies).

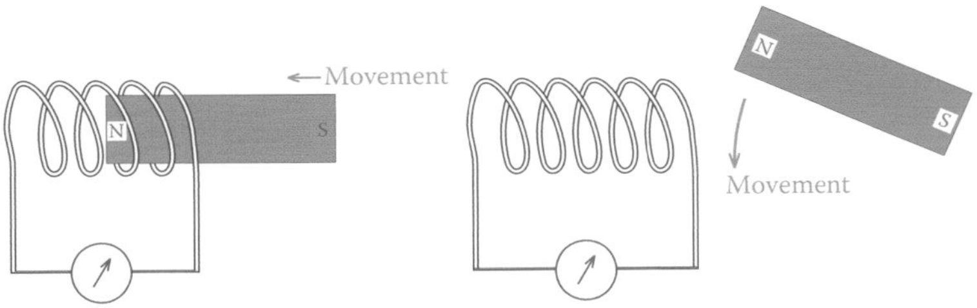

Figure 8.7 Moving a magnet near a coiled wire induces electricity in the coil.

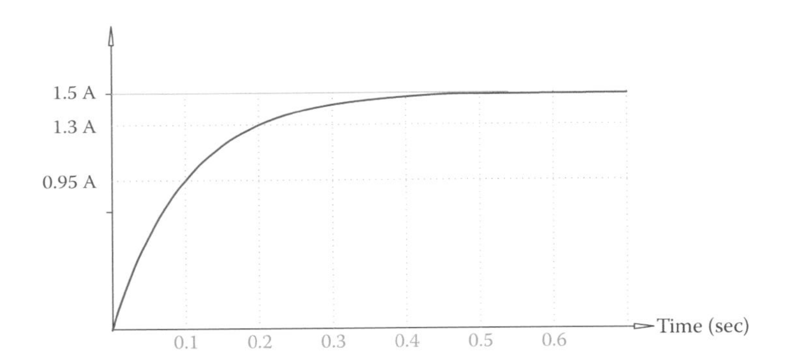

Figure 8.8 Gradual change of current from zero to 1.5 A in a circuit containing a coil.

voltage is opposite to the direction of the current that made the induction. As a result of this opposition, the current in a DC circuit containing an inductor after a switch is turned on is not an abrupt change and, instead, is a gradual growth from zero to a steady value. This is illustrated in Figure 8.8.

As can be observed, it takes a short time for the current to settle to its steady value. The steady value of the current depends on the total resistance in the circuit. The length of time that takes for the current to reach its final value depends on (1) the value of inductance of the coil (measured in henries) and (2) the total resistance of the circuit (measured in ohm).

8.5.1 Inductor Connected to AC

For the sake of better understanding, first, let's assume a square wave AC signal provided by a source and connected to an inductor, shown in Figure 8.9. Furthermore, we assume that the resistance of the coil is negligible and can be ignored.

When the circuit containing an inductor is turned on, a current starts to develop in the circuit, starting from zero and reaching a high value, according to the curve as shown in Figure 8.8. But, before the current reaches its maximum value the voltage has a sudden change and the polarity changes. Consequently, a current in the opposite direction starts to develop in the same manner. On the basis of the gradual development of current and the fact that the AC voltage switches back and forth from negative to positive and vice versa, it is easy to understand that for the circuit of Figure 8.9 the current looks as illustrated in Figure 8.10. It can be further understood from Figure 8.10 that

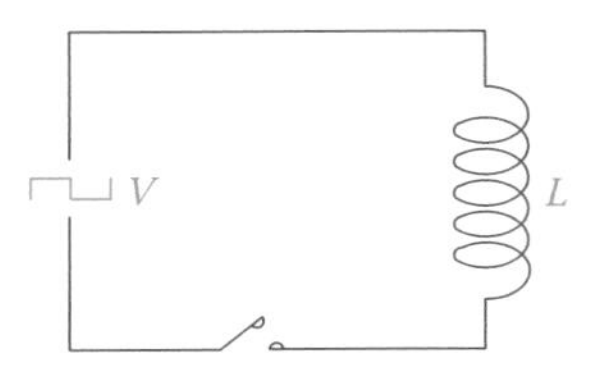

Figure 8.9 Inductor connected to an AC source with square waveform.

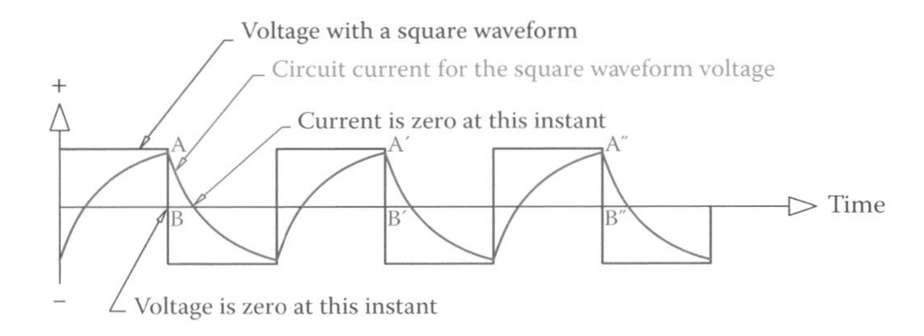

Figure 8.10 Current in an inductor connected to an AC source with square waveform.

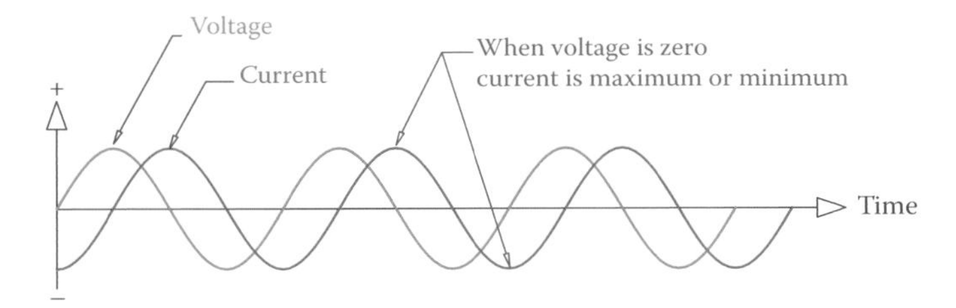

Figure 8.11 Current in an inductor connected to an AC source with sinusoidal waveform.

1. There is always a current flowing in the circuit.
2. Current changes between positive and negative, i.e., it is AC.
3. Current has its maximum (and minimum) value at the time that the voltage is zero (points A, A′, A″, etc., and points B, B′, B″, etc.).

If instead of a square wave, we have a sinusoidal AC waveform, which is the common case, the current is also sinusoidal and looks like Figure 8.11. Its peaks correspond to the instants the voltage value is zero.

Note that, as depicted in Figure 8.11, in an AC circuit containing an inductor (only) the current is 90° behind the applied voltage. This means that the current reaches its maximum (minimum) after the voltage has reached its maximum (minimum), and the difference between the time that voltage reaches its peak and the time that current reaches its peak is 1/4 of a cycle, or 90°. This is always the case for such a circuit.

8.5.2 AC Current in an Inductor

We established that when an inductor is connected to an AC source, a current exists in the circuit. This implies that, on the basis of Ohm's law, there is some resistance (to electric flow) in the circuit associated with this current. This so-called resistance (it is not really resistance because there is zero ohm value in the circuit, as assumed) can be determined. But, first let's establish another fact that the current in this circuit changes with frequency of the AC electricity. This can be easily understood from the comparison of currents shown for two cases in Figure 8.12. Again, for better understanding, a square waveform is assumed. The waveform

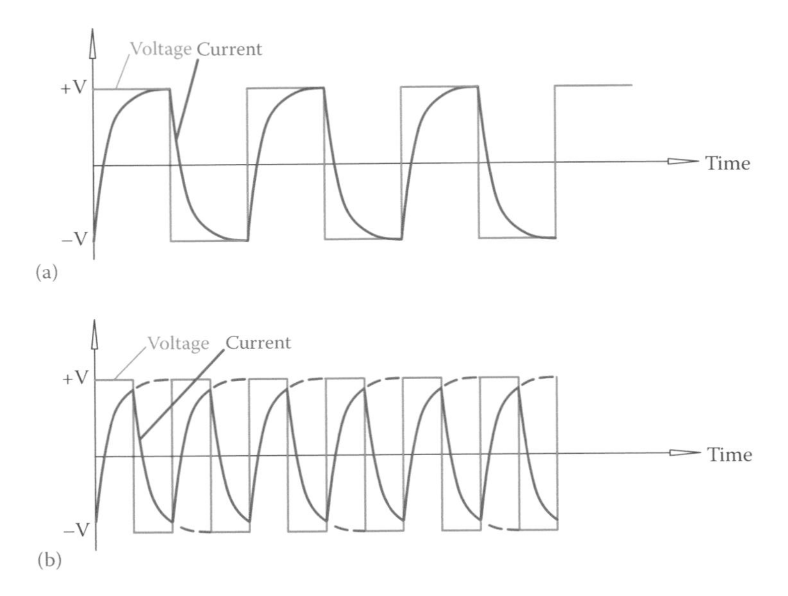

Figure 8.12 Comparison of currents in two AC square wave signals when frequency doubles. (a) Lower frequency: larger peak current. (b) Higher frequency: smaller peak current.

in Figure 8.12b has 2 times the frequency as in Figure 8.12a. The current in Figure 8.12b is smaller than the current in Figure 8.12a. The reason for the difference in the value of current is that in Figure 8.12b there is not sufficient time for reaching the peak value before the polarity switching occurs. The same is true for a sinusoidal waveform.

Reactance: The apparent resistance (measured in ohms) that a capacitor or an inductor when connected in an AC circuit exhibits to the flow of electricity. Reactance depends on AC frequency and unlike a resistor the energy involved does not convert to heat.

It can be followed from Figure 8.12 that the current through an inductor has an inverse effect with the frequency of the AC line. The effect of the insertion of an inductor in an AC circuit is exhibited in the form of impedance to the current, but because it is not a resistance (with ohm value that can be measured), it is called **reactance**. So, reactance in an AC circuit is what exhibits a resistance to the flow of current, but it is not because of a resistive element that converts the electrical energy to heat. Reactance is measured in ohms. To specify that reactance is due to an inductor, when necessary, it is more particularly addressed as **inductive reactance**. The reactance of an inductor, denoted by X_L, depends on the frequency and the inductance, and can be found from

Inductive reactance: The apparent resistance to flow of electricity exhibited by an inductor in an AC circuit. It is measured in ohm and determines the current in the inductor based on the applied voltage.

$$X_L = 2\pi fL \quad (8.5)$$

where π is a constant (π = **3.14**159265), f is the frequency of the AC electricity measured in Hz, and L is the inductance measured in henries (see Section 4.10).

Example 8.6

What is the inductive reactance of an inductor when connected to 60 Hz electricity? The inductor has an inductance of 0.05 H.

Solution

It follows directly from Equation 8.5 that

$$X_L = (2)(3.14)(60)(0.05) = 18.85\ \Omega$$

Example 8.7

Find the current for an inductor when connected to a 9 V, 50 Hz AC power source, if its inductance is 20 mH.

Solution

$$X_L = (2)(3.14)(50)(0.020) = 6.28\ \Omega$$

$$I = 9 \div 6.28 = 1.43\ \text{A}$$

Example 8.8

What is the current in the inductor in Example 8.7 if the frequency of the power source is 60 Hz?

Solution

$$X_L = (2)(3.14)(60)(0.020) = 7.64\ \Omega$$

$$I = 9 \div 7.54 = 1.19\ \text{A}$$

You can observe from the above examples that when the frequency goes up, the reactance of an inductor goes up proportionally to it and the current in the inductor decreases.

> When the frequency of electricity connected to an inductor goes up, the reactance of the inductor increases proportionally to the frequency and the current through the inductor decreases.

8.5.3 Inductors in Series with Each Other

In the same way that resistors could be combined in series and in parallel with each other, inductors can be put in series or in parallel, when necessary. Figure 8.13 illustrates three inductors in series. The relationships for a number of inductors put in series with each other are

$$L = L_1 + L_2 + L_3 + \ldots \tag{8.6}$$

$$X_L = X_{L1} + X_{L2} + X_{L3} + \ldots \tag{8.7}$$

L is the equivalent inductor for all the inductors in series, and X_L is the equivalent inductive reactance of them.

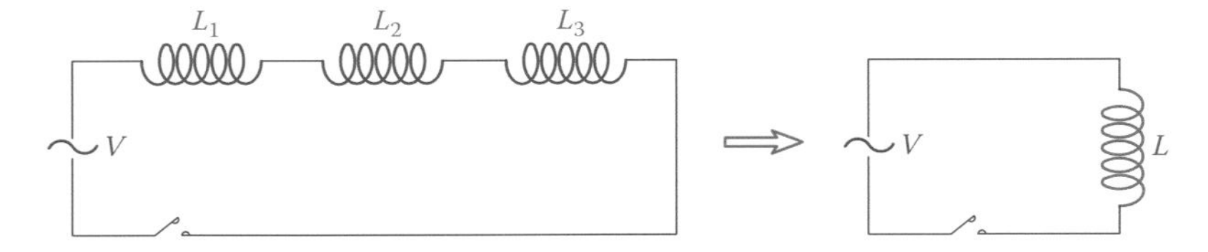

Figure 8.13 Inductors in series.

When inductors are connected in series, all of them share the same current. Accordingly, similar to the case for resistors, the total voltage divides among them, proportional to their reactance values.

Example 8.9

What is the current in three inductors in series, when the applied voltage is 48 V? The frequency of the power supply is 60 Hz, and the inductors are 5, 10, and 15 mH.

Solution

For this problem, either Equation 8.6 or 8.7 can be used. Using the former takes less calculation because similar calculations must be repeated for each inductor to find its reactance.

$$L = 5 + 10 + 15 = 30 \text{ mH}$$

$$X_L = (2)(3.14)(60)(0.030) = 11.31\ \Omega$$

$$I = 48 \div 11.31 = 4.24 \text{ A}$$

Example 8.10

At what frequency does the current in the circuit of Example 8.9 reduce to 3 A?

Solution

For the current to reduce to 3 A the total inductive reactance must be

$$X_L = (48) \div (3) = 16\ \Omega$$

then

$$(2)(\pi)(f)(0.030) = 16$$

$$f = \frac{16}{(2)(3.14)(0.03)} = 85 \text{ Hz}$$

(The answer is rounded to an integral number.)

Note that you could shortcut the solution and find the answer by the ratio of the reactance values in the two cases. That is,

$$\frac{f}{f'} = \frac{X_L}{X_L'}$$

which stems from Equation 8.5. Thus,

$$\frac{f}{60} = \frac{16}{11.31}$$

$$f = (60)(16) \div (11.31) = 85 \text{ Hz}$$

Example 8.11

In Example 8.9, find the voltage difference across each inductor.

Solution

This problem resembles exactly the problem of finding voltages for resistors in series (see Section 6.10). We first find the individual reactance values for the three inductors:

$$X_{L1} = (2)(3.14)(60)(0.005) = 1.885\ \Omega$$

$$X_{L2} = (2)(3.14)(60)(0.010) = 3.770\ \Omega$$

$$X_{L3} = (2)(3.14)(60)(0.015) = 5.655\ \Omega$$

$$X_L = 1.885 + 3.770 + 5.655 = 11.31\ \Omega$$

According to Equation 6.9, the voltages across the various resistors (here inductors) can be found from the ratio of the reactance value of each inductor to the total reactance value multiplied by the applied voltage (48 V). Thus,

$$V_1 = (48)\left(\frac{1.886}{11.31}\right) = 8 \text{ V}$$

$$V_2 = (48)\left(\frac{3.770}{11.31}\right) = 16 \text{ V}$$

$$V_3 = (48)\left(\frac{5.655}{11.31}\right) = 24 \text{ V}$$

However, pay attention to the fact that the ratio between the reactance values is the same as the ratio between the values of the inductances. That is, all X_L values are obtained from multiplying the inductance values by $(2\pi f)$. In this sense, the voltage values across each inductor can be found directly by using the inductance values rather than the reactance values.

Then

$$V_1 = (48)\left(\frac{5}{5+10+15}\right) = 8\text{ V}$$

$$V_2 = (48)\left(\frac{10}{5+10+15}\right) = 16\text{ V}$$

$$V_3 = (48)\left(\frac{15}{5+10+15}\right) = 24\text{ V}$$

Example 8.12

In Example 8.9, find the voltage across each inductor if the frequency of the AC line is 100 Hz, instead of 60 Hz.

Solution

On the basis of the discussion in Example 8.11, understand that the voltage distribution between the inductors is independent of the frequency. Therefore, the voltages across L_1, L_2, and L_3 will remain 8, 16, and 24 V at all frequencies.

8.5.4 Inductors in Parallel with Each Other

If one needs to connect inductors in parallel with each other, as shown in Figure 8.14, then the rule of parallel inductors can be used. You have noticed that the rule for inductors in series is the same as that used for resistors in series. The same is true for inductors in parallel with each other. Either of the following equations can be used.

$$\frac{1}{L} = \frac{1}{L_1} + \frac{1}{L_2} + \frac{1}{L_3} + \ldots \tag{8.8}$$

$$\frac{1}{X_L} = \frac{1}{X_{L1}} + \frac{1}{X_{L2}} + \frac{1}{X_{L3}} + \ldots \tag{8.9}$$

Example 8.13

Three inductors with inductance 5, 10, and 15 mH are put in parallel with each other and connected to 120 V line. If the frequency of the line is 60 Hz, what is the current in the line?

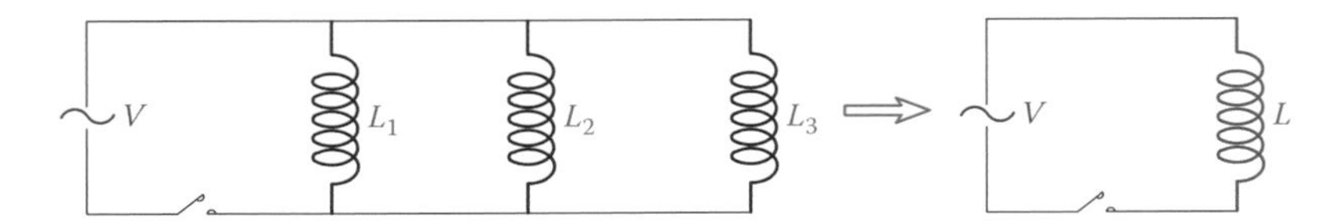

Figure 8.14 Inductors in parallel.

Solution

For this problem it is easier to use Equation 8.8.

$$\frac{1}{L} = \frac{1}{5} + \frac{1}{10} + \frac{1}{15} = \frac{11}{30}$$

The equivalent inductance for the three inductors is

$$L = 2.727\ \text{mH}$$

and the reactance of them for 60 Hz frequency is

$$X_L = (2)(3.14)(60)(0.002727) = 1.028 \approx 1\ \Omega$$

Thus, the current in the line is about 120 A (120 V ÷ 1 Ω).

Note that the resulting inductance is smaller than that of the smallest (5 mH) inductor.

Example 8.14

Understand that the current in the line when the three inductors with inductance 5, 10, and 15 mH are connected in parallel is very high (120 A). Suppose that this is a case in reality when you had to replace a damaged inductor but that you did not have the exact value and you had decided to make the equivalent inductor by combining the available ones.

Suppose that you had made a mistake and instead of connecting these inductors in series you had connected them in parallel. How many times the current calculated in Example 8.13 is compared with the expected current? (The line is 120 V and 60 Hz.)

Solution

When these three inductors are in series, their equivalent inductance is

$$L = 5 + 10 + 15 = 30\ \text{mH}$$

and their reactance is

$$(2)(3.14)(60)(30) = 11.3\ \Omega$$

and

$$I = 120 \div 11.3 = 10.6\ \text{A}$$

$$120\ \text{A} \div 10.6\ \text{A} = 11.32$$

Thus, the current is 11.32 times larger than expected. This can be very damaging in a real situation.

Example 8.15

The three parallel inductors in Example 8.13 (with inductance 5, 10, and 15 mH) were, in fact, taken from a device that worked with 400 Hz. Find what the current was through each individual inductor and what the total current was for 48 V applied voltage.

Solution

First, we need to find the reactance for each inductor.

$$X_{L1} = (2)(3.14)(400)(0.005) = 12.566\ \Omega$$

$$X_{L2} = (2)(3.14)(400)(0.010) = 25.133\ \Omega$$

$$X_{L3} = (2)(3.14)(400)(0.015) = 37.699\ \Omega$$

The three currents for each inductor are, respectively,

$$I_1 = 48 \div 12.566 = 3.82\ \text{A}$$

$$I_2 = 48 \div 25.133 = 1.91\ \text{A}$$

$$I_3 = 48 \div 37.699 = 1.27\ \text{A}$$

And the total current is the sum of the three currents

$$I = 3.82 + 1.91 + 1.27 = 7\ \text{A}$$

Example 8.16

The three inductors in the previous example (5, 10, and 15 mH) are to be replaced by one inductor. How much must the reactance of this inductor be?

Solution

We can either find the equivalent reactance value from Equation 8.9, as

$$\frac{1}{X_L} = \frac{1}{12.566} + \frac{1}{25.133} + \frac{1}{37.699} = 0.146 = \frac{1}{6.854}$$

$$X_L = 6.854\ \Omega$$

or

$$X_L = \frac{V}{I} = \frac{48}{7} = 6.857\ \Omega$$

(The small difference is because of round-off errors.)

8.6 Capacitors in AC Circuits

As we learned in Chapter 6 (see Section 6.11), when a capacitor is subject to a voltage across its terminals, it starts charging until its charge becomes at the level of the applied voltage. During the time that charging takes place a current flows in the circuit (wires connecting the capacitor to the power source). This current is due to the electrons moving in the wires and not in the capacitor itself, as shown in Figure 8.15. This current is not long lasting and decays to zero. When the voltage across the capacitor is the same as that of the power supply, the current is zero. The change in the current in the circuit containing the capacitor is shown in Figure 8.16. At the first instance the connection is established the current has its maximum value.

The same argument is true when a charged capacitor discharges through a circuit. A current in the circuit flows and continues until the capacitor is fully discharged. At the first instant the capacitor is connected to the circuit, the current has its highest magnitude, which gradually decays to zero. Figure 8.17 shows the current change in a discharging capacitor. (The change pattern is based on a decaying exponential curve.) Consider a circuit as shown in Figure 8.17a in which a capacitor is in a DC circuit with two resistors R_1 and R_2. When the switch is turned on and the resistors are powered, the capacitor charges and the voltage across it is the same as that of the battery. In Figure 8.17b the charge is shown as 1, which implies 100 percent of the battery voltage. Suppose now that at this time the switch is turned off. This instant is the beginning of discharge for the capacitor. The discharge takes place through the resistors R_1 and R_2, which form a closed loop with the capacitor. The voltage across the capacitor changes from the initial value to zero in a smooth way and depending on the time constant defined by C and resistors R_1 and R_2 (see Chapter 6 for time constant). In Figure 8.17b, T represents the time constant (as you remember, it takes approximately 5 times the time constant for the circuit to settle to a new value after a change in the voltage has occurred).

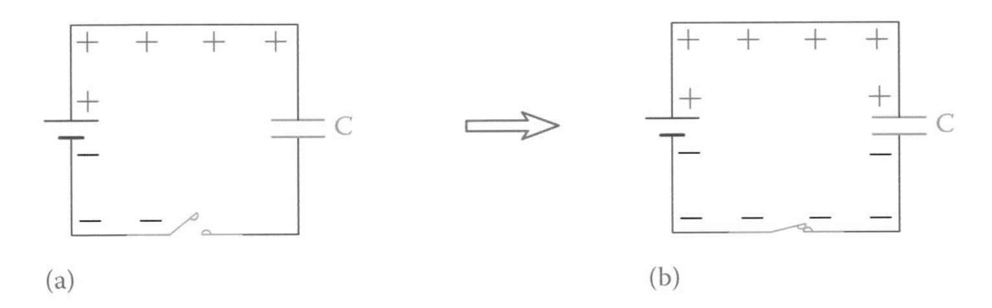

Figure 8.15 Capacitor charging instants (a) before and after (b) switch is closed.

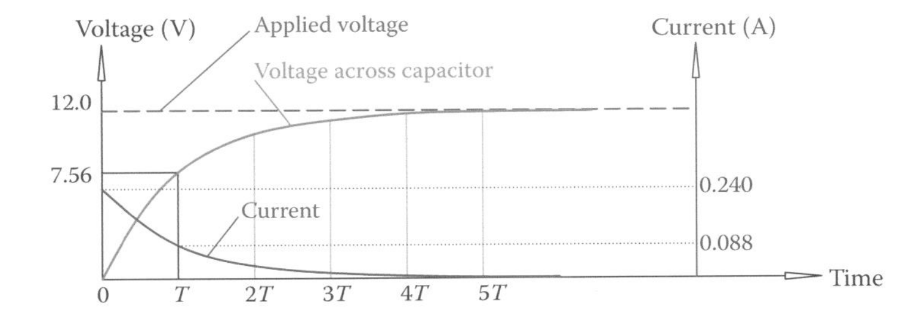

Figure 8.16 Current change in the circuit containing a charging capacitor.

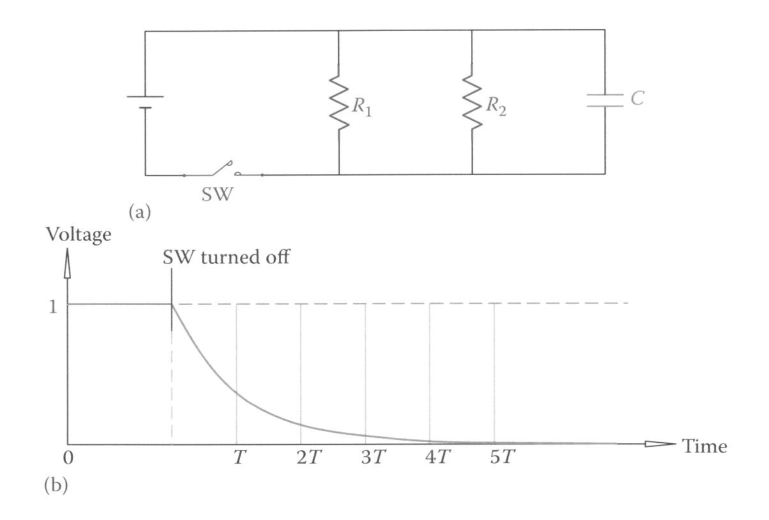

Figure 8.17 Current change in the circuit containing a discharging capacitor. (a) The DC circuit containing a capacitor. (b) After the switch is turned off following being on, the discharge current starts from some value and decays to zero.

8.6.1 Capacitor Connected to AC Electricity

For better understanding of what happens in an AC circuit containing a capacitor, we first assume a square wave AC signal. When the connection is made, the capacitor starts charging, but after it is charged (or before it is fully charged, depending on the capacitance), the half cycle terminates and the polarity changes. The charged capacitor now must discharge and start charging in the opposite polarity. This process is shown in Figure 8.18. Because this charging and discharging happens continuously while the circuit is switched on, there is always some current flowing in the wires connecting the source to the capacitor, but not inside the capacitor. Thus, the circuit has a current, which can be measured. Associated with this current, according to Ohm's law, an ohmic value (that is measured in ohms) can be determined. The latter represents a current limiting entity, while it is not a resistance as in a resistor. It does not turn the electricity into heat. Similar to what we have seen for inductors, the associated current limiting entity is called reactance, and because it stems from a capacitor, it is called **capacitive reactance**.

Capacitive reactance: Reactive effect of a capacitor when connected to AC electricity, measured in ohms. Capacitive reactance determines the current in a circuit containing a capacitor.

Figure 8.19 shows two different possible cases for the current in a circuit with a square wave power source and a capacitor. In Figure 8.19a the

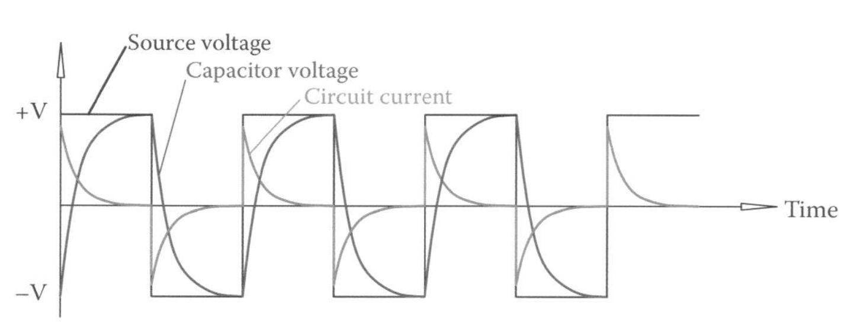

Figure 8.18 Current due to a capacitor in a circuit with square wave voltage.

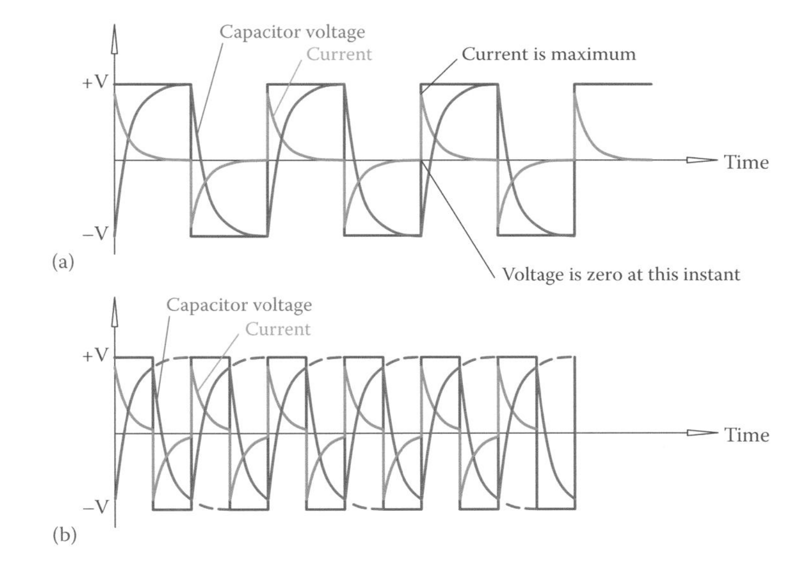

Figure 8.19 Effect of capacitance and frequency on the current in AC circuits with capacitor. (a) Lower frequency: there is enough time for charging and discharging currents to fall to zero. (b) Higher frequency: there is not enough time for current during charging and discharging to become zero; result: more current in the case of higher frequency.

capacitor is fully charged before it starts to discharge; in Figure 8.19b the capacitor does not get sufficient time to charge during a half cycle. In Figure 8.19a, by the end of each half cycle the current has dropped to zero, while in Figure 8.19b the current has a nonzero value. We can say that the average magnitude of the current in Figure 8.19b is larger than the current in Figure 8.19a because there is less duration of zero value current.

The difference between the current values is due to the capacitance of a capacitor (see Section 6.11). It can be seen that a larger capacitance leads to a larger current. Figure 8.19 also reflects the effect of frequency. The larger the frequency is, the less a capacitor has time to completely charge and discharge. Thus, the average current is larger because there are no or smaller periods of zero current. Figure 8.19 also shows that the maximum current (peak value of current) corresponds to the instance where the applied voltage is zero (the voltage changes sign; it passes through zero). This is more evident for a sinusoidal signal, as shown in Figure 8.20,

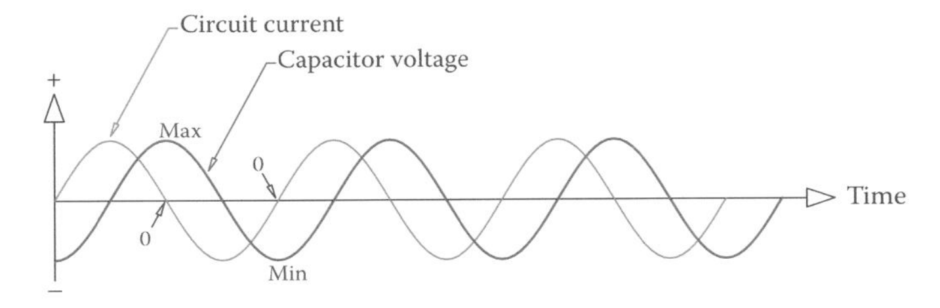

Figure 8.20 Current due to a capacitor in a sinusoidal wave AC circuit.

because the changeover from charge to discharge (and this is when the current has its maximum absolute value) starts at the time the voltage changes sign. Moreover, for the case of a pure capacitor connected to an AC source (sinusoidal waveform) the circuit current reaches its maximum (minimum) value at 90° before the applied voltage reaches its maximum (minimum). This is the opposite of the case for an inductor, as shown in Figure 8.20.

8.6.2 Current in AC Circuits Containing Capacitor

An observation of the current in Figure 8.19 reveals that the line frequency and the capacitance of a capacitor in an AC circuit have direct effect on the magnitude of the current. A larger capacitor (higher capacitance) takes more time to charge (or discharge), and a smaller capacitor takes less time to charge (or discharge). More time for charging and discharging implies higher current because current does drop to zero after charging. A higher frequency implies smaller cycle time and, thus, less available time for charging and discharging. Less available time implies that the current does not get sufficient time to reach zero; thus, more nonzero current (higher current). Therefore, the current in a circuit containing a capacitor is proportional to both the capacitor capacitance and the source frequency. In consonance with what was said about an inductor in an AC circuit, the effect of a capacitor is exhibited by its capacitive reactance.

Capacitive reactance of a capacitor is determined from

$$X_C = \frac{1}{2\pi f C} \tag{8.10}$$

and is measured in ohms. In Equation 8.10, f is the line frequency in Hz and C is the capacitance in farads. It implies that as the frequency increases the value of X_C decreases. Also, as C increases, the value of X_C decreases (which implies a higher current). Thus, larger capacitors are leading to higher currents (for the same frequency), in consonance with what was said before.

> When the frequency of an AC source connected to a circuit containing a capacitor increases, capacitive reactance of the circuit decreases and circuit current increases.

Example 8.17

What is the circuit current when a 12 V, 60 Hz electricity source is connected to a 51 μF capacitor?

Solution

Capacitive reactance is calculated from Equation 8.10 as

$$X_C = \frac{1}{(2)(3.14)(60)(0.000051)} = 52\ \Omega$$

$$I = \frac{12}{52} = 0.231\,\text{A} = 231\,\text{mA}$$

Example 8.18

What is the reactance of the capacitor in Example 8.17 if it is connected to a signal with 10,000 Hz frequency?

Solution

Using the new value of frequency in Equation 8.10 gives

$$X_C = \frac{1}{(2)(3.14)(10,000)(0.000051)} = 0.312\,\Omega$$

From this example you notice better a fact about capacitors in AC circuits. Whereas for 60 Hz frequency the reactance of a capacitor is 52 Ω, this value reduces to 0.3 Ω when the frequency is 10 KHz. For radio and TV signals that have a much higher frequency this value is almost zero. This means that a capacitor behaves like a solid connection.

At very high frequencies a capacitor behaves as a solid connection.

Example 8.19

When a capacitor is connected to a 6 V, 50 Hz power source, the current is 500 mA. What is the capacitance of the capacitor?

Solution

$$X_C = \frac{6}{0.5} = 12\,\Omega$$

Thus,

$$\frac{1}{(2)(3.14)(50)(C)} = 12$$

$$C = \frac{1}{(2)(3.14)(50)(12)} = 0.000265\,\text{F} = 265\,\mu\text{F}$$

8.6.3 Capacitors in Series

Capacitors in series are shown in Figure 8.21. We need to find the one equivalent capacitor that replaces those in series. The rule for series

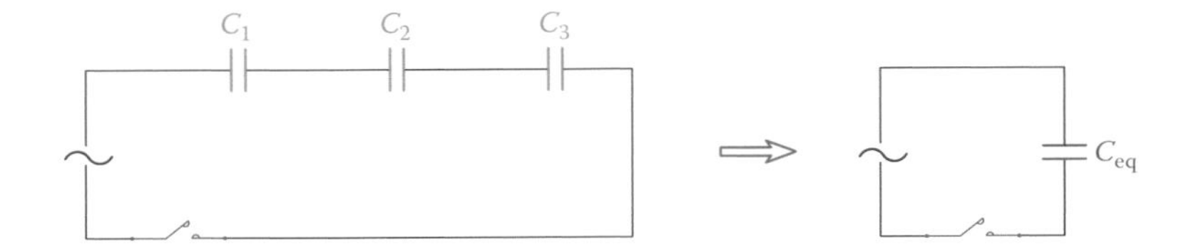

Figure 8.21 Capacitors in series.

capacitors is the opposite of the rule for resistors and inductors. For series capacitors,

$$\frac{1}{C_{eq}} = \frac{1}{C_1} + \frac{1}{C_2} + \frac{1}{C_3} + \ldots \tag{8.11}$$

If both sides of the above equation are multiplied by $1/2\pi f$, then we get

$$\frac{1}{2\pi f C_{eq}} = \frac{1}{2\pi f C_1} + \frac{1}{2\pi f C_2} + \frac{1}{2\pi f C_3} + \ldots$$

which is

$$X_{C_{eq}} = X_{C1} + X_{C2} + X_{C3} + \ldots \tag{8.12}$$

Thus, for finding the equivalent capacitor for series capacitors, Equation 8.11 must be used; however, if the reactance values are employed, they add together.

Example 8.20

In a previous repair job in a circuit, three capacitors had been put in series with each other instead of the damaged capacitor. If these capacitors are 47, 68, and 100 μF, find the value of the capacitor in the original circuit.

Solution

Using Equation 8.11, we find

$$\frac{1}{C_{eq}} = \frac{1}{47} + \frac{1}{68} + \frac{1}{100} = 0.045825 = \frac{1}{21.74}$$

Thus, the original capacitor has been a 22 μF (the nearest standard value capacitor).

Example 8.21

In part of a circuit there is a 68 μF capacitor, which must be replaced. You see that if it is urgent, you can use a 100 μF and a 150 μF in series with each other. If the voltage across the 68 μF capacitor has to be 4.8 V at 60 Hz, what is the difference in current if you do this substitution?

Solution

Reactance of the 68 μF capacitor is

$$X_{68} = \frac{1}{(2)(3.14)(60)(0.000068)} = 39\ \Omega$$

and the circuit current is

$$4.8 \div 39 = 0.123\ \text{A} = 123\ \text{mA}$$

Reactance of substitution circuit is

$$X_C = \frac{1}{(2)(3.14)(60)(0.000100)} + \frac{1}{(2)(3.14)(60)(0.000150)}$$

$$= 26.5 + 17.7 = 44.2\ \Omega$$

The circuit current then is

$$4.8 \div 44.2 = 0.109\ \text{A} = 109\ \text{mA}$$

Current difference = 123 – 109 = 14 mA.

8.6.4 Capacitors in Parallel

We may need to put capacitors in parallel to obtain a capacitance value that cannot be found among standard capacitors. Capacitors in parallel are shown in Figure 8.22.

The rule for capacitors in parallel is easier. You add their capacitances together. That is,

$$C_{eq} = C_1 + C_2 + C_3 + \ldots \tag{8.13}$$

To find the equivalent reactance, however, they must be treated as resistors in parallel. This stems from multiplying the two sides of Equation 8.13 by $2\pi f$. Thus,

$$2\pi f C_{eq} = 2\pi f C_1 + 2\pi f C_2 + 2\pi f C_3 + \ldots$$

or

$$\frac{1}{X_C} = \frac{1}{X_{C1}} + \frac{1}{X_{C2}} + \frac{1}{X_{C3}} + \ldots \tag{8.14}$$

Example 8.22

A 33, 47, and 10 μF capacitor are put in parallel together. How much is the resulting capacitor?

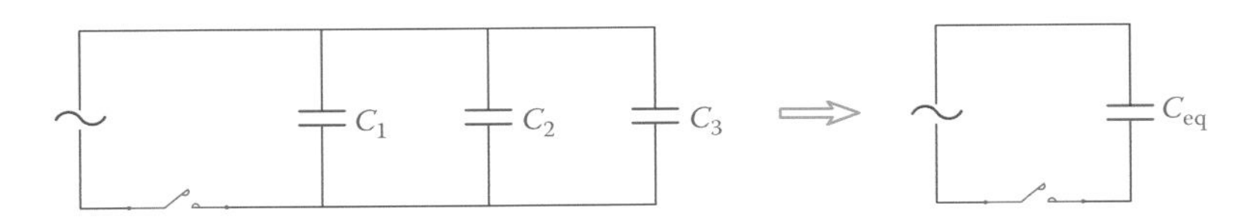

Figure 8.22 Capacitors in parallel.

Solution

It is sufficient just to add the three values together

$$C = 33 + 47 + 10 = 90\ \mu F$$

Note that 90 μF can also be obtained from putting two standard 68 and 22 μF capacitors in parallel.

Example 8.23

The normal current in a circuit containing the three capacitors (in parallel) in Example 8.22 is around 0.4 A. One day you notice the current has decreased to 0.2 A. You suspect that one of the capacitors may have become disconnected or damaged. Which capacitor could have been damaged? The frequency of the line is 60 Hz.

Solution

You can answer this question without calculation. There are three branches in the circuit, each containing one of the capacitors. Because current values are proportional to the capacitance of the capacitors, the highest current corresponds to the 47 μF capacitor and the lowest current corresponds to the 10 μF capacitor. Also, the total current (0.4 A) corresponds to 90 μF capacitance. Because the current has approximately halved (in practice, we cannot read the precise values from measurement devices) and because 47 is approximately half of 90, we may conclude that the 47 μF capacitor is the problematic one.

8.7 Power in Inductors and Capacitors

In the same way that power could be found for resistors, we may find power for circuits containing inductors and capacitors. As stated in Section 8.4, as far as we use the effective values for AC voltage and current, all the relationships in Chapter 5 for resistors in DC circuits can be directly used for AC circuits. Hereafter the subscript "eff" is dropped and whenever AC variables are used, they represent the effective value. We want to extend the previous relationships to cover the circuits containing inductors and capacitors.

At this time we concentrate on the power relationships. More specifically,

$$P = VI = (RI)(I) = RI^2 \tag{5.5}$$

$$P = VI = (V)(V/R) = \frac{V^2}{R} \tag{5.6}$$

for the power consumption in a resistor. Here P is the power, V is the voltage across a resistor (effective voltage if AC), and I is the current through a resistor (effective current if AC).

Similarly, for an inductor we can define

$$P_{\text{inductor}} = VI = X_L I^2 = \frac{V^2}{X_L} \tag{8.15}$$

where V is the voltage across the inductor, I is the current through it, and X_L is the inductive reactance for the frequency of the circuit containing the inductor.

In a similar manner, for a capacitor (or the equivalent of a number of capacitors) in an AC circuit, we may define the power associated with the capacitor in the form of

$$P_{\text{capacitor}} = VI = X_C I^2 = \frac{V^2}{X_C} \tag{8.16}$$

where V is the voltage across the capacitor, I is the circuit current (due to the capacitor), and X_C is the capacitive reactance for the frequency of the circuit. (Remember that in reality there is no current flowing through a capacitor. The current flows in the wires connecting a capacitor in a circuit. Nevertheless, for simplicity, we may refer to I as the capacitor current because it is due to the capacitor).

The summary for the relationships among voltage, current, resistance, and power in Chapter 5 can be extended for inductors and capacitors. These relationships are shown in Figure 5.13 Extended.

Note the important fact that we did not use the term "consume" for the power in a capacitor or an inductor, and we have emphasized earlier that the reactance is not like a resistance generating heat. Where does the power of a capacitor or an inductor then go?

In fact, the power that an inductor or a capacitor takes from an AC circuit is stored in the component in the form of magnetic (for an inductor) or electrical (for a capacitor) energy. This energy-storing action is for one half of a cycle only. In the following half cycle, this stored energy is given back to the circuit. In this way, an inductor or a capacitor constantly exchanges energy with the circuit it is connected to, as long as the circuit is turned on and there is a power source powering it. The result of this exchange of energy is that the net energy taken from a circuit by a pure capacitor or a pure inductor is zero.

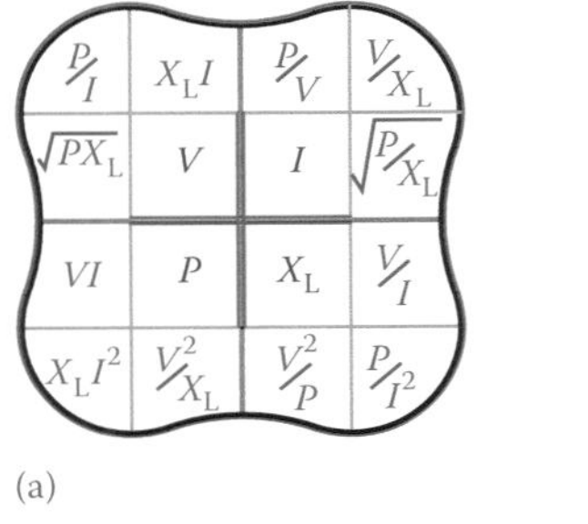

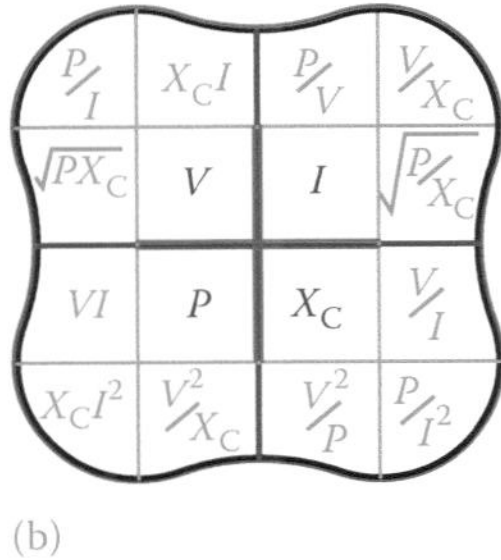

Figure 5.13 Extended Summary for the relationships among voltage, current, resistance, and power for (a) inductor and (b) capacitor.

> In an AC circuit an inductor or a capacitor continuously exchanges energy with the circuit power source by storing energy in half of each cycle and giving back that energy in the other half of the cycle. Thus, net energy consumption of a capacitor or a pure inductor is zero.

The above fact is more clearly understood from Figure 8.23, which shows the power consumed for each 1/4 cycle of an AC circuit containing an inductor or a capacitor. The figures shown are the repeats of Figures 8.11 and 8.20, put side by side. Power is shown for each 1/4 cycle. Power is obtained by the product of current and voltage, but, because these two parameters change sign, power is positive when both voltage and current are either positive or negative, and power is negative in the other cases. In this sense, for two of the 1/4 cycles the power consumption of an inductor or a capacitor is positive and for the other two 1/4 cycles the power consumption is negative (meaning that they provide power to their circuits). It can be mathematically shown that the positive values and the negative values are equal and they cancel each other. As a result the net power consumption of a pure inductor or a pure capacitor is zero.

It is important to note that, although we say the net value of the power is zero, the circuit power supply must be capable of providing power to the component. The values determined by Equations 8.14 and 8.15 are the power that a circuit containing these elements must be able to handle.

There is another important issue here that must be realized. The power described thus far is not in the form of heat, light, or work of a motor. In this sense, it *cannot* be measured in watts. To make a distinction between this type of power and the power that is tangible in the form of heat or work, the power associated with an inductor or a

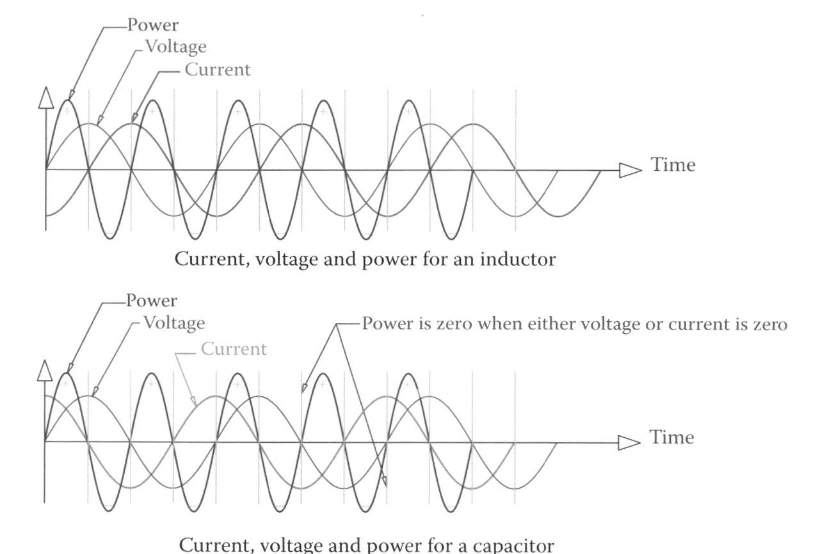

Figure 8.23 Relationship between current, voltage, and power in pure inductors and capacitors.

capacitor is called **reactive power** (because it corresponds to a reactance), and in this book we denote that by Q to be different from P, the power that turns into heat, work, etc. P in AC circuits is called **real power** or **active power**. It is also referred to as **true power** in some texts. Also, to distinguish between the unit for the measurement of active power and reactive power, its unit of measurement is **VAR**, which stands for **Volt-Ampere-Reactive**.

Reactive power: Power in an AC circuit corresponding to an inductor or a capacitor. Reactive power is not consumed but is momentarily stored in the component (inductor or a capacitor) and is returned to the circuit in the next half cycle (when the AC polarity alters).

Real power: Another name for active power.

Active power: The portion of electric power in AC circuits that converts to heat or mechanical work (like the work by a motor). It is also called true power and real power.

True power: Same as active power.

Volt-Ampere-Reactive (VAR): Unit of measurement for reactive power.

Example 8.24

What is the power in a 100 mH inductor when connected to 120 V, 50 Hz line? What is the power if it is connected to a 60 Hz line?

Solution

50 Hz:

$$X_L = (2)(3.14)(50)(0.100) = 31.4\ \Omega$$

$$I = 120 \div 31.4 = 3.82\ \text{A}$$

$$Q = (31.4)(3.82)^2 = 458.6\ \text{VAR}$$

60 Hz:

$$X_L = (2)(3.14)(60)(0.100) = 37.7\ \Omega$$

$$I = 120 \div 37.7 = 3.18\ \text{A}$$

$$Q = (37.7)(3.18)^2 = 382\ \text{VAR}$$

As can be seen, as the frequency goes up, the power decreases. This conclusion could be reached from Equation 8.15 because the voltage does not change, but as the frequency goes up, the reactance increases.

Example 8.25

A 68 μF capacitor is inserted in a part of the circuit where $V = 100$ V. If the circuit frequency is 50 Hz, what are the current and the power requirement of the capacitor? What is the power if the frequency increases to 60 Hz?

Solution

50 Hz:

$$X_C = \frac{1}{(2)(3.14)(50)(0.000068)} = 46.83\ \Omega$$

$$I = 100 \div 46.83 = 2.14\ \text{A}$$

$$Q = \frac{V^2}{X_C} = \frac{100^2}{46.83} = 214 \text{ VAR}$$

60 Hz:

$$X_C = \frac{1}{(2)(3.14)(60)(0.000068)} = 39\ \Omega$$

$$I = 100 \div 39 = 2.56 \text{ A}$$

$$Q = \frac{V^2}{X_C} = \frac{100^2}{39} = 256.4 \text{ VAR}$$

In phase: When two cyclic waveforms of the same shape and frequency have their pairwise maximum and minimum points occurring at the exact same time.

Out of phase: Fact of having a finite time difference between the instances that two waveforms of the same shape and frequency reach their maximum or minimum values. This can be extended for logic signals by comparing their rising or falling edges timing.

Leading: Opposite of lagging, implying that a cyclic waveform reaches its maximum (minimum) *before* another waveform of the same pattern and same frequency. Most commonly employed in AC electricity (see also lagging).

Lagging: Implying that a cyclic waveform reaches its maximum (minimum) *after* another cyclic waveform of the same frequency. More commonly used in AC circuits to indicate the timing relationship of current with respect to voltage.

For a capacitor, as the frequency goes up, the power goes up, too, as can be observed or as can be understood from the equations.

> In AC circuits, if the frequency increases, the power for an inductor decreases, but the power for a capacitor increases.

8.8 Phase Difference

When we have two sine waves with the same frequency, the duration of one cycle is the same for both. Nonetheless, irrespective of their peak values, there are two possibilities:

1. The two waveforms reach their maximum values (and accordingly their minimum values) at the same instant. In this case they are synchronized or **in phase** with each other.
2. It is possible that the two waveforms do not have their maximum (and minimum) values occurring exactly at the same time. In such a case they are said to be **out of phase** with each other.

8.8.1 Leading and Lagging

When two waveforms are out of phase, then the way to express the time difference between the two is by stating the angle difference for one cycle, i.e., the angle value of the first waveform when the other one has a zero value. This is shown in Figure 8.24, where there is a phase difference of 30° between the waveforms A and B.

In conjunction with the phase difference are two other terms: **leading** and **lagging**. When the waveform A is ahead of B (i.e., when it reaches its maximum value before B reaches its maximum value), it is said to be leading waveform B. At the same time, B is behind (following) A, and it is said to be lagging A. Note that when A is leading B, it also reaches its minimum value and zero value before B reaches those values. The amount of leading or lagging is expressed by the **phase angle**, which is the phase difference between two waveforms. Note that the phase angle must always

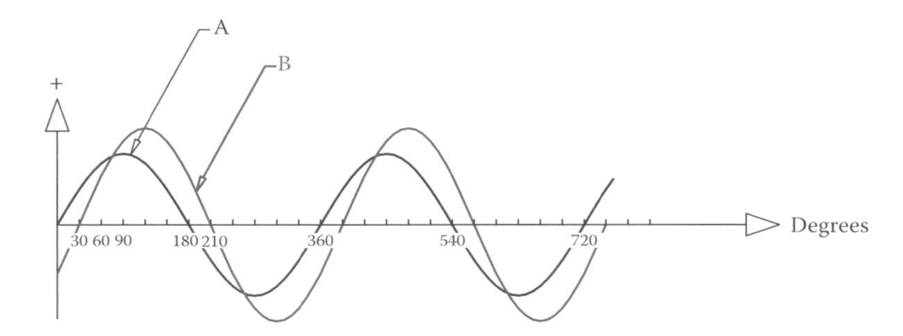

Figure 8.24 Definition of phase difference.

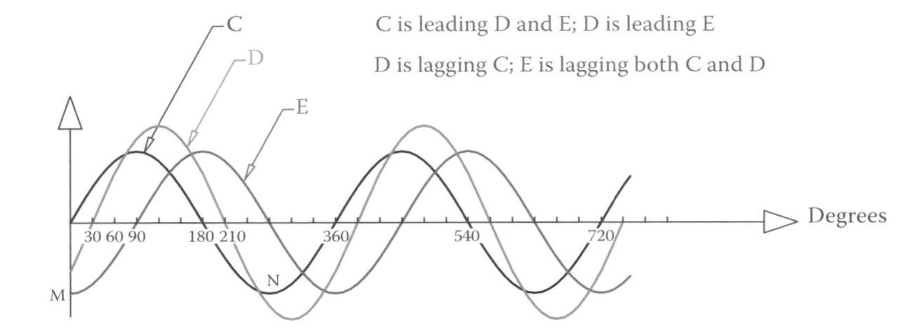

Figure 8.25 Definition of leading and lagging.

be less than 180°. It is not usual to say, for instance, that E leads C by 270°, as shown in Figure 8.25, by comparing points M and N (the minimum peak values of them); instead, we say C leads E by 90°. In AC electricity, more often the 180° range for phase angle is expressed between –90° and +90°, as will become evident later.

Example 8.26

In Figure 8.26, determine if the waveform A is leading B or lagging B and by how many degrees.

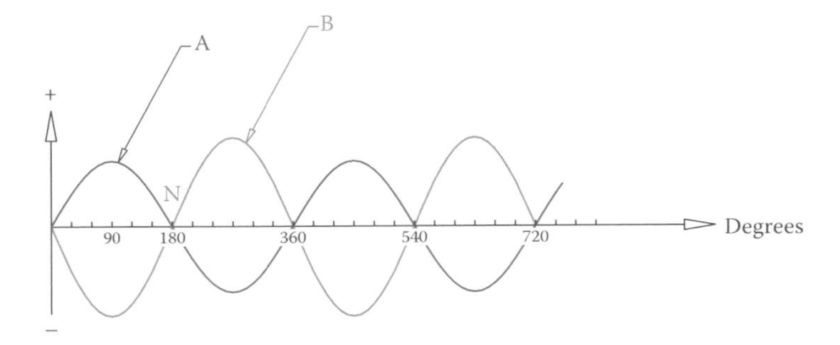

Figure 8.26 Waveforms A and B in Example 8.26.

Solution

In this example, A and B have all their zero values at the same time (points M and N); only, when A is increasing (to get to its maximum value), B is decreasing (to reach its minimum value). In this particular case, A and B have 180° phase difference. It is possible to say either A is leading B or B is leading A. (Note the fact that A reaches its maximum value before B is immaterial because if the waveforms are drawn from another point, such as N, then B reaches its maximum before A.)

8.8.2 Phase Difference between Voltage and Current in AC Circuits

When the load in an AC circuit consists of only resistors, then the current at each instant of time is proportional to the voltage. This implies that, in changing through its sinusoidal waveform, when voltage is zero, the current is zero, too; and when the voltage is at its positive or negative peak, the current is at its positive or negative peak, respectively. In this sense, we can say there is no phase difference between the voltage and the current.

This is not the case, however, when the load is inductive or capacitive. On the basis of the discussion in Section 8.5.1 and as shown in Figure 8.11, for an inductor in an AC circuit the current reaches its maximum (minimum) with a delay after the voltage has reached its maximum (minimum) and this delay is 90°. In other words, the current is lagging the voltage by 90°.

By the same token, and as shown in Figure 8.20, for a capacitor in an AC circuit the current leads the voltage by 90°. That is, for a capacitor the case is in the opposite way of that for an inductor. For better clarity the three cases for the three types of loads are depicted in Figure 8.27.

In an AC circuit

- For a resistive load there is no phase difference between current and voltage.
- For an inductive load the voltage leads the current by 90° (current lags voltage).
- For a capacitive load the current leads the voltage by 90° (voltage lags current).

Understanding the preceding concept is quite important in AC circuits. To quickly remember the correct state, the two words *ICE* and *ELI* can be helpful. Here E is used for electromotive force, which is the voltage, and I stands for current. The two words help to remember that for a C component (capacitor) in the circuit I is before E (current leads) and for an L (inductor), E is before I (voltage leads).

Later in this chapter we see that when dissimilar components are used in a circuit the phase angle is not 90° and can be any angle between –90° and +90°.

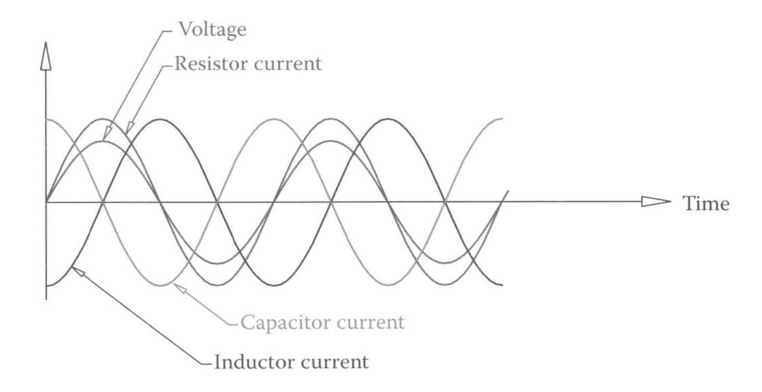

Figure 8.27 Phase relationships between current and voltage for the three types of loads in AC.

8.8.3 Use of Vectors to Show the Phase Difference

From this point the application of vectors in AC circuits starts and knowledge about vectors helps to tackle all the associated problems with ease and better understanding. The phase difference as described before is very important in AC circuits, but it becomes tedious if two sinusoidal waveforms must be drawn in order to show their phase difference, especially if we are not always dealing with the simple case of 90° phase difference. A simpler way of showing the phase difference between two waveforms *with the same frequency* is through representing each waveform by a vector. The length of this vector denotes the peak value of the waveform (employing a convenient scale), and its angle is decided by selecting an arbitrary direction (usually horizontal) for a reference waveform. For example, a horizontal (on the paper) vector of 6 in (15 cm) length can be drawn for a 120 V sinusoidal voltage. Then, a horizontal vector of 3 in (7.5 cm) can represent another AC voltage that is 60 V and is in phase with the former, and a 2 in (5 cm) vertical vector can represent an AC current value of 2.5 A, which is leading the reference voltage by 90°. Note the fact that it is not necessary to use the same scale for two different entities (here voltage and current). Otherwise, one or more vectors become so small in size (compared to the others) that their values cannot be seen or read. However, similar entities must have the same scale.

Figure 8.28 illustrates the above-mentioned voltage and current waveforms and other values by vectors. Because the AC waveforms with

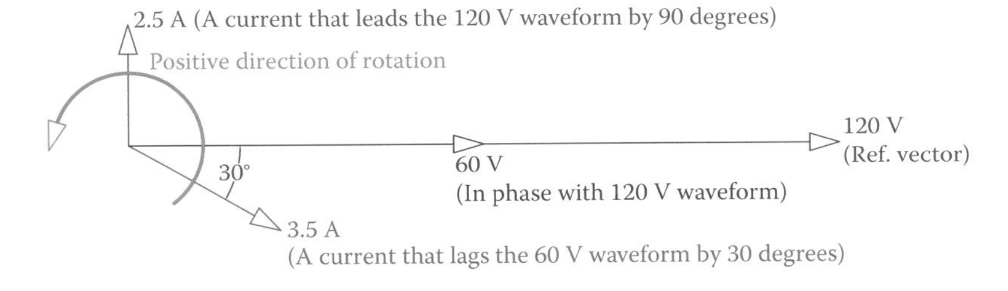

Figure 8.28 Representation of waveforms and phase relationships by vectors.

different frequencies cannot be mixed together, it is always understood that the waveforms shown together on a graph have the same frequency. The positive sense of direction to determine leading and lagging is counterclockwise (CCW). Thus, if a vector $\mathbf{V}_1$ must be rotated counterclockwise to coincide with a vector $\mathbf{V}_2$, it lags $\mathbf{V}_2$. For instance, in Figure 8.28 the 2.5 A vector leads the 60 V vector, but the 3.5 A vector lags it.

Vector representation of cyclic values is both very convenient and easy. It clearly shows the phase difference between two entities. In Figure 8.28 the angle of the 3.5 A vector with the reference vector is –30°. The minus sign is understood from the fact that this vector is below the reference line. Hence, its phase angle with the reference vector is 30° lagging.

8.9 Power in AC

Power in AC electricity is more complex than that in DC electricity. We have already learned about active and reactive power. It is imperative to understand the notion of power in AC circuits and understand the difference between various power terms.

8.9.1 Impedance

Impedance: Load in AC circuits consisting of resistive and reactive components. Impedance is measured in ohms and can be represented by a vector. It is normally denoted by *Z*.

To normalize the relationships between the voltage, load current, and power in AC circuits, irrespective of the type of load, the term **impedance** is employed. In accordance with the Ohm's law for DC circuits (see Section 5.4), in AC circuits one can write

$$V = ZI \tag{8.17}$$

where I is the current in a load, V is the voltage across the load, and Z is the load impedance. Z can be for a resistive load, a capacitive load, an inductive load, or a combination of them. If the load is resistive, Z reduces to R and the load impedance is equal to its resistance.

The relationship between R, X_C, X_L, and Z depends on how the three components appear in a circuit. The cases that only one of the three components exists are as discussed earlier, where Z is then equal to the resistance, capacitor reactance or inductor reactance. Complex combination of various components (R, L, and C) is possible, but at this time we confine our discussion to when a circuit can be reduced to R, L, and C in series or in parallel with each other.

8.9.2 AC Power

In consonance with Equation 8.17, load power in an AC circuit can be determined from

$$\text{Power} = ZI^2 = VI \tag{8.18}$$

Apparent power: Power that must be supplied by an AC source for powering its loads.

The power found this way is called the **apparent power**. This is the power that a power supply must provide for a load. The unit for measure of apparent power is VA (standing for Volt-Ampere). You may pronounce it Vee-A or say volt-amp. In conjunction with the two previously defined

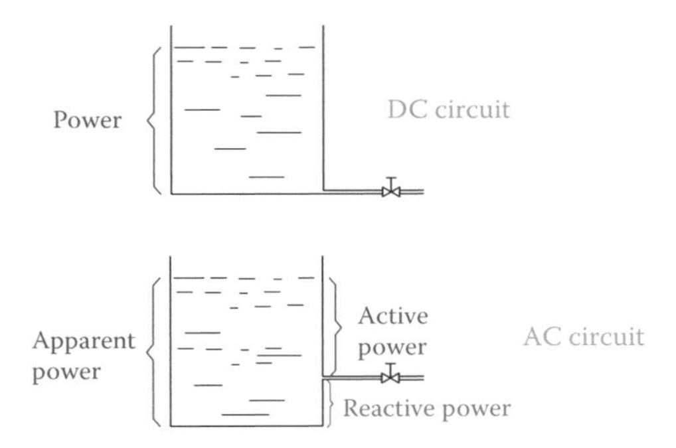

Figure 8.29 An analogy for comprehension of apparent, active, and reactive powers in AC. (From Hemami, A., *Wind Turbine Technology*, 1E ©2012 Delmar Learning, a part of Cengage Learning Inc. Reproduced with permission from www.cengage.com/permissions.)

terms, active power and reactive power, only a part of this apparent power converts to heat and work (active power). The rest of that (reactive power) exchanges between the reactive components (inductor and/or capacitor) and the electricity source. Although the reactive power is not consumed, it must be present; that is, the source must be able to maintain it. Figure 8.29 exhibits an analogy for a better understanding of this fact. (To have the desired volume/height of water at the outlet, the reservoir must be full; that is, the volume/height below the outlet must also be provided.)

To find the apparent power in a circuit, we cannot just add together the active and reactive parts. The relationship between the three types of AC power is

$$S = \sqrt{P^2 + Q^2} \tag{8.19}$$

where S is the apparent power, P is the active power, and Q is the reactive power of a circuit. Q is the sum of the reactive powers in inductive components and capacitive components, but because these two types of loads behave as in the opposite way of each other, their sum always is represented by the difference between their algebraic values. In other words, their sum is smaller than the larger value. Furthermore, power is always positive. Thus, we may say

$$Q = |\text{Capacitor power} - \text{Inductor power}| \tag{8.20}$$

(The two vertical lines imply the absolute value.)

In conjunction with Equation 8.19, the three power values can be represented in the vector form as shown in Figure 8.30 (see Section 8.2). The evidence resides in the Pythagorean theorem in the triangle formed. Note that the net reactive power found from Equation 8.20 can be capacitive (when the capacitive power is larger) or it can be inductive (when the inductive power is larger). Both cases are shown in Figure 8.30. The difference does not affect the calculated value for apparent power, but it leads to the angle between the vectors for apparent power and the active power to be positive or negative. In Figure 8.30a this angle is considered to be negative, whereas in Figure 8.30b it is positive. The angle between

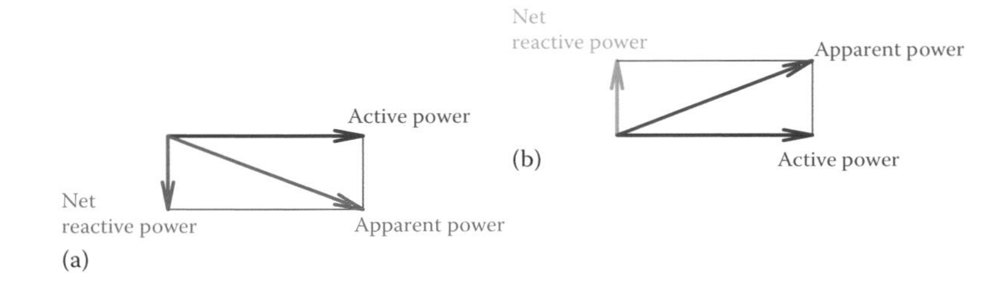

Figure 8.30 Relationship between the three power values in AC. (a) Negative phase angle. (b) Positive phase angle.

the apparent power vector and the active power vector turns out to be the phase angle for a circuit. You will see later why this angle represents the phase angle of a circuit.

For a purely resistive load, and for special cases that can occur and Q becomes zero, the value of apparent power is the same as the active power. For such a case, moreover, the impedance is the same as the resistance.

8.9.3 Power Factor

From the preceding discussion it follows that any AC circuit has a phase angle and that this angle can theoretically change between –90° and +90°. The cosine of this angle denotes the ratio of the active power to the apparent power. This ratio is a number smaller than unity and always positive. This number is called the power factor of a circuit or a load, and it is normally denoted by *pf*. Thus,

$$pf = \cos\text{(phase angle)} \tag{8.21}$$

On the basis of this definition, the relationships for power in an AC circuit are

$$\text{Apparent power} = S = VI \tag{8.22}$$

$$\text{Active power} = P = \text{Apparent power} \times pf = VI \times pf \tag{8.23}$$

$$\text{Reactive power} = Q = VI \times \sqrt{1 - pf^2} \tag{8.24}$$

For the above formulae the actual value of *pf* is used, which is smaller than 1. In expressing the power factor, this number can be multiplied by 100 and the *pf* is stated in percent; for example, 95 percent instead of 0.95. Both methods can be used.

Example 8.27

In an AC circuit the active power is 250 W and the reactive power is 100 VAR. What is the power factor and what is the phase angle?

Solution

From Equation 8.19 it follows that

$$S = \sqrt{250^2 + 100^2} = 269.26 \text{ VA}$$

$$pf = 250 \div 269.26 = 0.93 \text{ or } 93\%$$

$$\text{Phase angle} = \cos^{-1}(0.93) = 21.80° \cong 22°$$

For a purely resistive load, and when Q happens to be zero (the value of apparent power being the same as the active power), the power factor is 1.

8.10 Series *RLC* Circuits

Thus far we have learned about inductors and capacitors, as well as resistors in AC circuits. A general electric circuit can contain all these components. They can be in series, in parallel, or in combination. In this section and the next section we learn about the simpler cases, i.e., when components are in series or in parallel with each other. If a circuit has more than one component of the same type, first, we reduce the circuit by using the corresponding laws for series and parallel combination of each type (see Sections 8.5.3, 8.5.4, 8.6.3, and 8.6.4).

In a series *RLC* circuit the three basic elements are in series with each other, which means that they all have the same current. The formulation covers the general case of three types of load being present in a circuit. If any of the components is absent (usually, the inductor or the capacitor not the resistor), then in calculations the corresponding value for that element can be set to zero or the associated term deleted from the formulae.

Figure 8.31 illustrates an *RLC* series circuit. Note that the order the three components are shown is not important. Thus, in Figure 8.32 the two circuits are equivalent. In tackling the circuit at hand, we need to know the relationship between the applied voltage and the current and the power consumption of the circuit, using all the knowledge that has been gained so far.

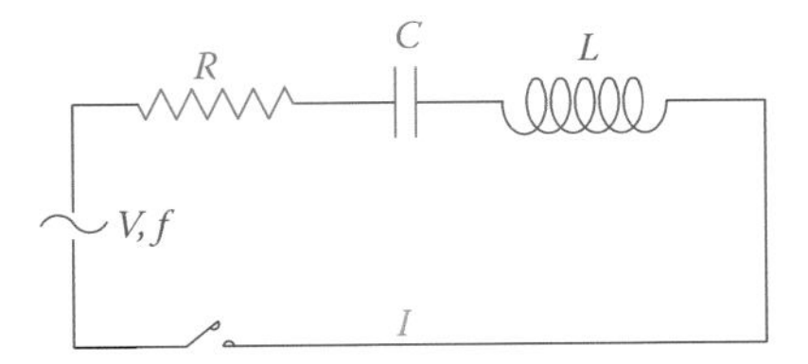

Figure 8.31 A series *RLC* circuit.

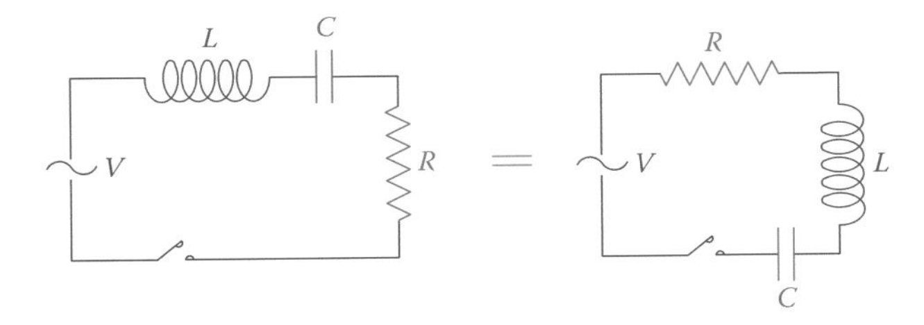

Figure 8.32 Two equivalent circuits.

The simplest question with a series *RLC* circuit is finding the current in the circuit if the particulars of the loads and the applied voltage are given. In the above circuit V is the applied voltage, I is the common current for all the three elements, f is the frequency, and R, L, and C represent the values for resistance, inductance, and capacitance, respectively, of the three components in the circuit.

First, referring to the previously discussed cases, we recall that the applied voltage in this circuit is divided between the three components. In this regard, the corresponding voltages across R, L, and C are denoted by V_R, V_L, and V_C, respectively. Recall that each of these voltages follows the rules that we learned about the relationship between current and voltage in each component. That is, the voltage across the resistor is in phase with the current, the voltage across the capacitor is lagging the current by 90° and the voltage across the inductor is leading the current by 90°. We now try to show these variables in the vector form. Because the three components have the same current, the most appropriate reference entity for showing the vectors is the circuit current I. Figure 8.33 shows these vectors, irrespective of their numerical values (because we do not have any number values given yet) but with correct orientation. Nevertheless, remember that the scale for current and voltage can be different, whereas the scale for all identical values (voltages here) must be the same.

As can be seen in Figure 8.33, the voltage across R is in phase with the current, the voltage across L is leading the current, and the voltage across C is lagging the current.

This implies that R, X_C, and X_L cannot be algebraically added together. Z is the vector sum of these three values. After their values have been determined, Z can be found. Because a capacitor and an inductor have opposite effects, their corresponding vectors are opposite to each other, and, thus, their sum is represented by a number smaller than the larger value ($X_C - X_L$ or $X_L - X_C$. Figure 8.34 illustrates two cases, one when

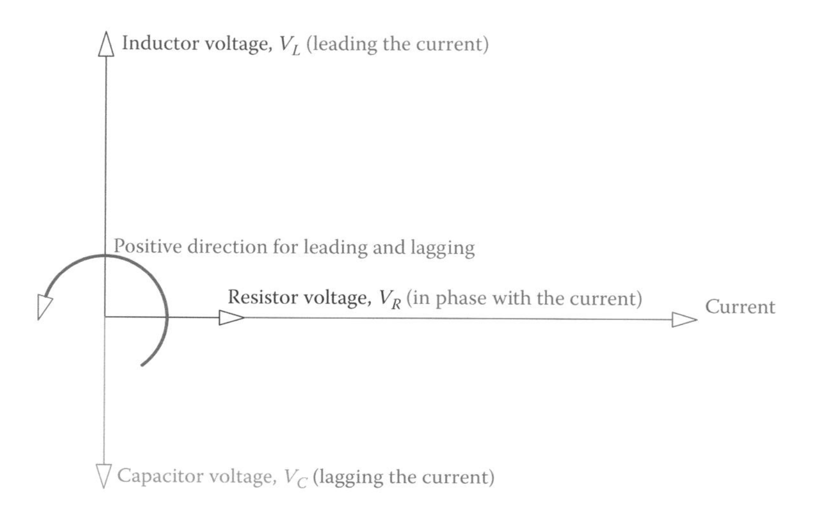

Figure 8.33 Vectors for the current and the three different voltages in the *RLC* series circuit.

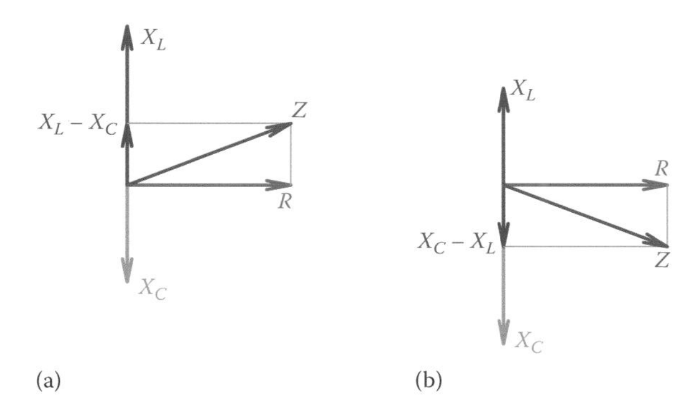

Figure 8.34 Relationship between R, X_C, X_L, and Z. (a) Positive phase angle. (b) Negative phase angle.

$X_C > X_L$ and one when $X_C < X_L$). For similar values the value of Z obtained is the same for both cases.

Using the Pythagorean theorem the value of the impedance Z can be written as

$$Z = \sqrt{R^2 + (X_L - X_C)^2} \tag{8.25}$$

(Note that it does not matter if one enters $X_L - X_C$ or $X_C - X_L$).

Example 8.28

In a series RLC, circuit $R = 30\ \Omega$, $L = 15$ mH, and $C = 51\ \mu$F. If the source voltage and frequency are 12 V and 60 Hz, respectively, what is the current in the circuit?

Solution

$$X_L = (2)(3.14)(60)(0.015) = 5.655\ \Omega$$

$$X_C = \frac{1}{(2)(3.14)(60)(0.000051)} = 52\ \Omega$$

$$Z = \sqrt{30^2 + (52 - 5.655)^2} = 55.21\ \Omega$$

$$I = 12 \div 55.21 = 0.217\ \text{A} = 217\ \text{mA}$$

Figure 8.35a is the same as Figure 8.34a. Because for the three components in series the current is the same, if all the values represented by the three sides of the triangle are multiplied by the current, then a similar triangle, as shown in Figure 8.35b, results. The sides of this triangle represent the voltages, since $RI = V_R$, $ZI = V$, and so on. Furthermore, if the sides of this new triangle are multiplied by I one more time, again a similar triangle results, which reflect the values for the various powers, as shown in Figure 8.35c.

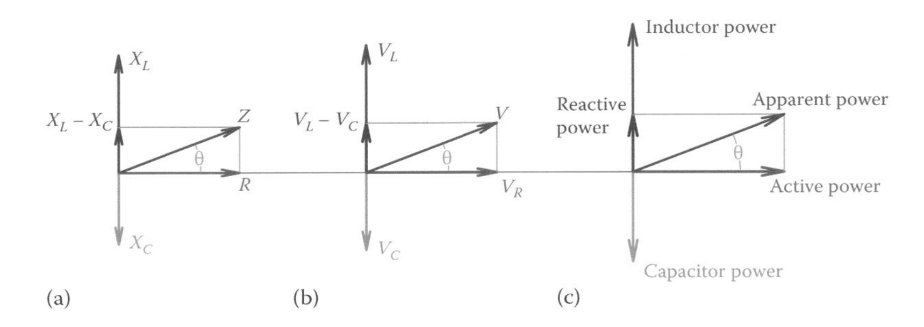

Figure 8.35 Similar triangles showing powers, voltages, and their relationships in a series *RLC* circuit. (a) Circuit resistance versus circuit impedance. (b) Voltage across resistance versus the applied voltage. (c) Circuit active power versus apparent power.

As can be understood from Figure 8.35b and as mentioned earlier (see Section 8.8), the three voltage values V_R, V_L, and V_C cannot be algebraically added together. These voltages are not in phase with each other and, hence, can only be added as vectors. In this respect, the relationship between V_R, V_L, and V_C is

$$V = \sqrt{V_R^2 + (V_L - V_C)^2} \tag{8.26}$$

where V is the applied voltage. The second term under the radical is in fact the sum of the voltage across the inductor and the voltage across the capacitor, but, because these two voltages are 180° out of phase, their sum appears as a difference in values.

8.10.1 Power Factor in Series *RLC* Circuits

Figure 8.35a–c clearly show that for a series *RLC* circuit the power factor can be found from one of the following relationships:

$$pf = \frac{R}{Z} = \frac{V_R}{V} = \frac{\text{Active power}}{\text{Apparent power}} \tag{8.27}$$

It is also easy to understand if the circuit is more inductive or more capacitive. This is reflected in the vector sum $X_L - X_C$ and $V_L - V_C$ (see Figure 8.34).

Example 8.29

A series *RLC* circuit consists of a 20 Ω resistor, a 51 μF capacitor, and a 25 mH inductor. If the source frequency is 50 Hz, and the circuit current is 350 mA, what is the applied voltage?

Solution

$$R = 20\ \Omega$$

$$X_L = (2)(3.14)(50)(0.025) = 7.85\ \Omega$$

$$X_C = \frac{1}{(2)(3.14)(50)(0.000051)} = 62.445\ \Omega$$

$$Z = \sqrt{20^2 + (7.85 - 62.445)^2} = 58.15\ \Omega$$

$$V = ZI = (58.15)(0.35) = 20.4 \approx 20\ \text{V}$$

In a series *RLC* circuit the voltages across the three components are not in phase with each other.

Example 8.30

If the applied voltage to the circuit of Example 8.29 is 12 V, what is the voltage across the capacitor?

Solution

In Example 8.29 the applied voltage was 20 V. The distribution of this voltage among the three components is as follows:

$$V_R = (20)(0.35) = 7\ \text{V}$$

$$V_C = (62.445)(0.35) = 21.85\ \text{V}$$

$$V_L = (7.85)(0.35) = 2.75\ \text{V}$$

In the current case, when the applied voltage is 12 V (i.e., 0.6× the previous case), because nothing has changed in the circuit, the current will be accordingly smaller by 0.6 times. As a result, all voltages will be 0.6 times smaller. The voltage across the capacitor, therefore, is

$$V_C = (21.85)(0.6) = 13.1\ \text{V}$$

You see that it is not necessary always to do repeat all the calculation. Also, note that the voltage across the capacitor is larger than the applied voltage of 12 V. This is always possible, and we do not expect all the voltages to be smaller than the applied voltage.

Example 8.31

In the circuit of Example 8.29, what is the phase angle between the voltage and the current?

Solution

The answer can be found by fist finding the power factor from any of the relationships in Equation 8.27. The values of *R* and *Z* are readily available and are the best choice.

$$pf = \frac{R}{Z} = \frac{20}{58.15} = 0.344$$

$$\text{Phase angle} = \cos^{-1}(0.344) = 70°$$

Knowing only the value of the phase angle to be 70° is not sufficient, and we need to express if the circuit is leading or lagging (i.e., if the current leads the voltage or it lags the voltage). This can be judged based on the values of X_C and X_L. If $X_C > X_L$, then the current leads the voltage and if $X_C < X_L$, the current lags the voltage.

Example 8.32

When a series *RLC* circuit is subject to 48 V, V_R is 15 V, and V_L is 22 V. What is the voltage across the capacitor?

Solution

According to Equation 8.26,

$$V = \sqrt{V_R^2 + (V_L - V_C)^2} = 48\text{ V}$$

Because $V_R = 15$ V, $(V_L - V_C)^2$ is determined as (directly from Equation 8.26)

$$(V_L - V_C)^2 = V^2 - V_R^2 = 48^2 - 15^2 = 2079 = 45.6^2$$

Attention must be paid here in calculating $(V_L - V_C)$ because there are two answers for it. This also reflects the fact that, when squared, both $(V_L - V_C)$ and $(V_C - V_L)$ lead to the same result. Thus,

$$\pm(V_L - V_C) = 45.6\text{ V}$$

Having $V_L = 22$ V leads to only one acceptable value. (The other value is negative and is not acceptable.) Hence, the value for V_C is

$$V_C = 67.6\text{ V}$$

In a series *RLC* circuit the voltages across *R*, *L*, and *C* must be added as vectors in order to find their sum.

Example 8.33

In Example 8.32, if the applied voltage was 24 V (instead of 48 V) and the other voltages are as given ($V_R = 15$ V and $V_L = 22$ V), what is the voltage across the capacitor?

Solution

$$V = \sqrt{V_R^2 + (V_L - V_C)^2} = 24\text{ V}$$

For $V_R = 15$ V, $(V_L - V_C)^2$ is determined as

$$(V_L - V_C)^2 = V^2 - V_R^2 = 24^2 - 15^2 = 351$$

As before, there are two answers for $(V_L - V_C)$. Thus,

$$\pm(V_L - V_C) = 18.73 \text{ V}$$

This time the value of $V_L = 22$ V leads to two answers for V_C, as follows V_C:

$$V_C = 22 + 18.83 = 40.73 \text{ V}$$

$$V_C = 22 - 18.73 = 3.27 \text{ V}$$

Note that in Example 8.32, one of the answers became negative, which was not acceptable.

Example 8.34

For the circuit shown in Figure 8.36, find current, voltage across the capacitor, power factor, and phase angle. The operating frequency is 60 Hz.

Solution

In this circuit, there is no inductor. Hence, in Equation 8.26 the value for X_L is set to zero.

$$R = 100 \ \Omega$$

$$X_C = \frac{1}{(2)(3.14)(60)(0.000045)} = 58.95 \approx 59 \ \Omega$$

$$Z = \sqrt{100^2 + (59)^2} = 116 \ \Omega$$

(Note that it is not necessary to put a minus sign for 59.)

$$I = 125 \div 116 = 1.08 \text{ A}$$

$$V_C = (59)(1.08) = 63.5 \text{ V}$$

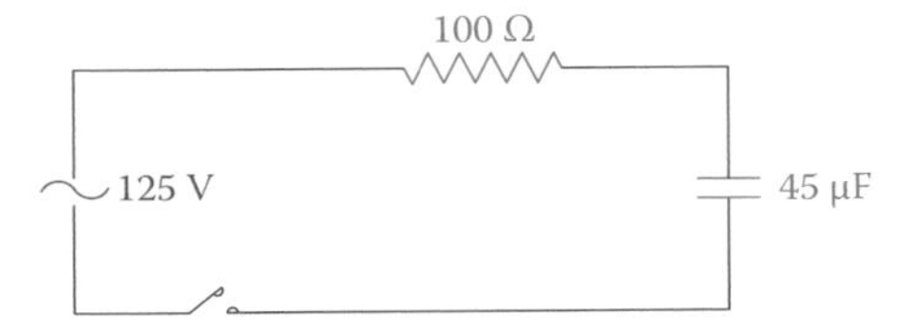

Figure 8.36 Figure for Example 8.34.

$$pf = \frac{R}{Z} = \frac{100}{116} = 0.86$$

$$\text{Phase angle} = \cos^{-1}(0.86) = 30.5°$$

One of the characteristics of a series *RLC* circuit is that the voltage between two points can be higher than the applied voltage.

8.11 Parallel *RLC* Circuits

In parallel *RLC* circuits the three basic components are in parallel with each other, and, therefore, all are subject to the same voltage. The current for each branch, however, depends on the impedance of the branch and can be individually determined by employing Ohm's law. For a parallel *RLC* circuit the voltage is common for all the three types of components because it is the same voltage that is applied to each component. Nevertheless, on the basis of the previous discussions the currents in the three branches are not in phase with each other. This means that the currents in the three branches do not simultaneously reach their peak values or zero values. Hence, the total current cannot be determined by algebraically adding the individual values of the currents in the resistor, inductor, and capacitor.

A parallel *RLC* circuit is shown in Figure 8.37. As in the case of series *RLC* circuits, we need to find the total current and the power consumption for the whole circuit or for each individual branch.

For this circuit the voltage applied to each component in each branch is the same. Therefore, the current in each component can be found from dividing the voltage by the branch impedance. Then the currents can be added together. However, on the basis of previous discussions, because the currents in the three components are not in phase with each other (they do not reach their maximum and minimum values at the same time), they cannot be algebraically added together and must be added in vector form. Figure 8.38 illustrates the vector representation of the three currents in a typical parallel *RLC* circuit. It shows that the current in the resistor is in phase with the applied voltage, the current in the capacitor leads the applied voltage (remember *ICE*) and the current in the inductor lags the voltage (remember *ELI*). Furthermore, note that for this vector representation of the currents and voltage in a parallel *RLC* circuit, because the voltage is the common variable for all branches, you start by drawing the vector for the voltage as the reference vector. (In series *RLC* circuit you started this process by drawing the vector for the current.)

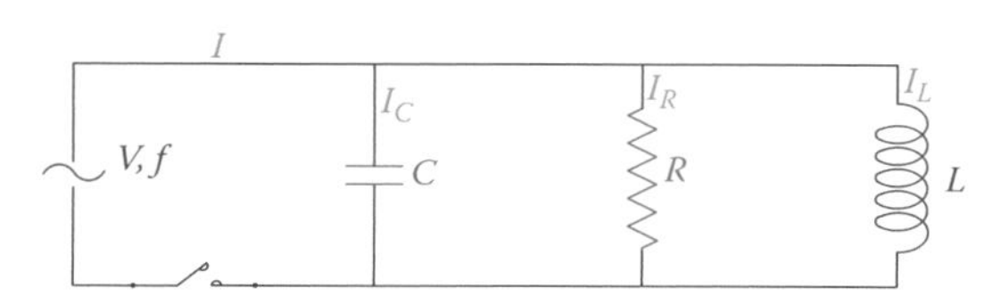

Figure 8.37 Schematic of parallel *RLC* circuits.

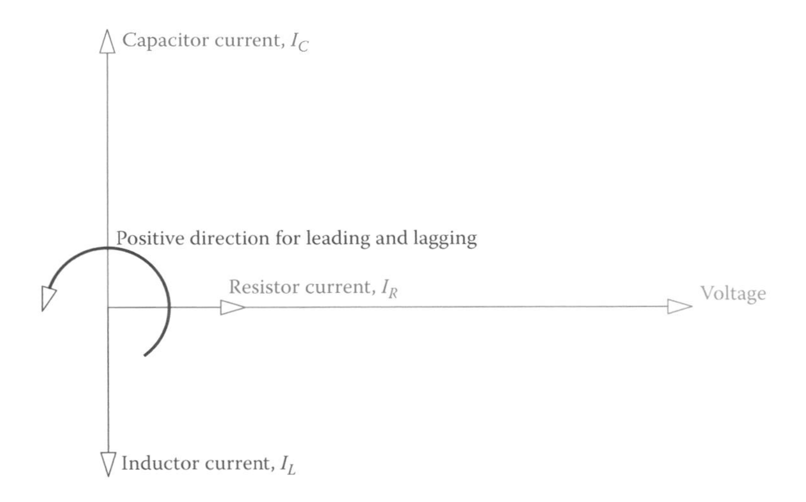

Figure 8.38 Vectors for the voltage and the three different currents in the *RLC* parallel circuit.

> To find the total current in a parallel *RLC* circuit, one needs to find the vector sum of the currents in *R*, *L*, and *C*.

Because the current in the inductor and the current in the capacitor are 180° out of phase, in adding them together their values are subtracted from each other. Thus, the relationship for the total current of the circuit, *I*, and the individual component currents I_R, I_L, and I_C is

$$I = \sqrt{I_R^2 + (I_L - I_C)^2} \tag{8.28}$$

Example 8.35

In the circuit shown in Figure 8.39 the current is 1.8 A. If the current through the capacitor is 1.5 A, find the applied voltage and the resistance of the resistor.

Solution

For 60 Hz frequency the reactance of the capacitor is

$$X_C = \frac{1}{(2)(3.14)(60)(0.000045)} = 58.94 \cong 59\ \Omega$$

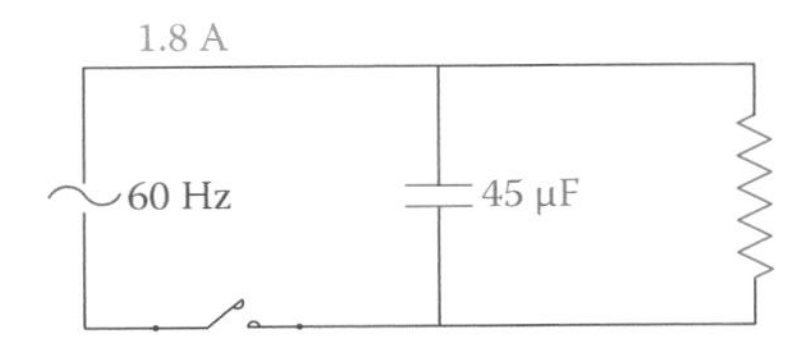

Figure 8.39 Circuit corresponding to Example 8.35.

Thus, the applied voltage is

$$(59)(1.5) = 88.5 \text{ V}$$

Because this circuit has no inductor, the value of L in Equation 8.28 is set to zero and the result is

$$I = \sqrt{I_R^2 + I_C^2}$$

which leads to

$$I_R = \sqrt{1.8^2 - 1.5^2} = 0.995\text{A} \cong 1 \text{ A}$$

and the resistance of the resistor is

$$88.5 \div 1 = 88.5 \ \Omega$$

If in Equation 8.28, the values for I_R, I_L, and I_C are replaced by $\frac{V}{R}$, $\frac{V}{X_L}$, and $\frac{V}{X_C}$, and I is written as the ratio of the applied voltage to the circuit impedance Z, we have

$$\frac{V}{Z} = \sqrt{\left(\frac{V}{R}\right)^2 + \left(\frac{V}{X_L} - \frac{V}{X_C}\right)^2}$$

By omitting V from both sides the relationship between Z and R, L, and C can be found then as

$$\frac{1}{Z} = \sqrt{\left(\frac{1}{R}\right)^2 + \left(\frac{1}{X_L} - \frac{1}{X_C}\right)^2} \quad (8.29)$$

Equation 8.29 can be used to find the equivalent impedance of the three components in parallel. The circuit current can also be found this way by dividing the applied voltage by Z or by directly multiplying $\frac{1}{Z}$ by the applied voltage.

Example 8.36

In the circuit shown in Figure 8.40, $R = 55\ \Omega$, $L = 0.08$ H, and $C = 1\ \mu\text{F}$, find the impedance of the circuit and the applied voltage.

Solution

$$X_L = (2)(3.14)(60)(0.08) = 30.16 \ \Omega$$

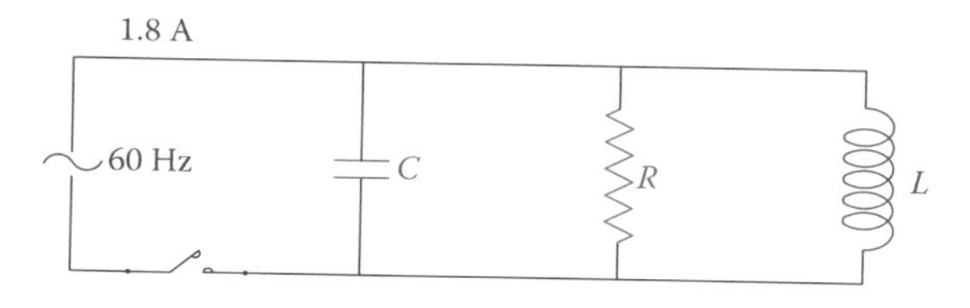

Figure 8.40 Circuit for Example 8.36.

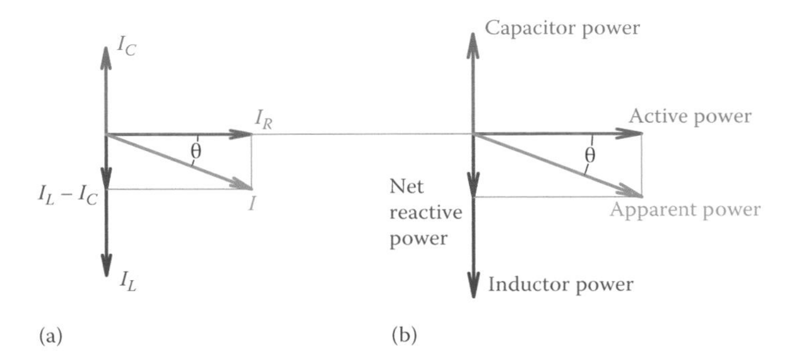

Figure 8.41 Vectors for (a) currents and (b) powers in parallel *RLC* circuits.

$$X_C = \frac{1}{(2)(3.14)(60)(0.000001)} = 26.5\ \Omega$$

$$\frac{1}{Z} = \sqrt{\left(\frac{1}{55}\right)^2 + \left(\frac{1}{30.16} - \frac{1}{26.5}\right)^2} = 0.01875 = \frac{1}{53.33} \quad \rightarrow$$

$Z = 53.33\ \Omega$

Applied voltage = $V = ZI = (53.33)(1.8) = 96$ V.

Equation 8.29 also implies that the value for Z is smaller than R for parallel *RLC* circuits. A vector representation of I_R, I_L, I_C, and I is shown in Figure 8.41, which also shows the powers in the three components and the apparent power. Reactive power is the vector sum of the inductive and capacitive powers. Depending on if inductive power (Q_L) or the capacitive power (Q_C) is larger the vectors for I and the apparent power S fall below or above the horizontal reference. The former implies that the current lags voltage and the latter denotes that the current leads voltage. Because in practice the majority of applications (including home and industrial circuits) are parallel circuits, any circuit is categorized to be leading or lagging. If in a circuit the current leads the voltage, the circuit is said to be leading; if the current lags the voltage, the circuit is said to be lagging.

8.11.1 Power Factor in Parallel *RLC* Circuits

Figure 8.41 shows a lagging circuit. In practice, most of circuits are lagging because of presence of electric motors, unless the effects of electric motors are compensated by inserting capacitors that introduce capacitive power to a circuit (see power factor correction). The power factor in a parallel *RLC* circuit is determined from

$$pf = \frac{Z}{R} = \frac{I_R}{I} = \frac{\text{Active power}}{\text{Apparent power}} \tag{8.30}$$

Note that the power factor by itself is not sufficient to describe a circuit. It has to be accompanied by the statement for leading or lagging. A circuit may have the same power factor in two cases, either leading or lagging.

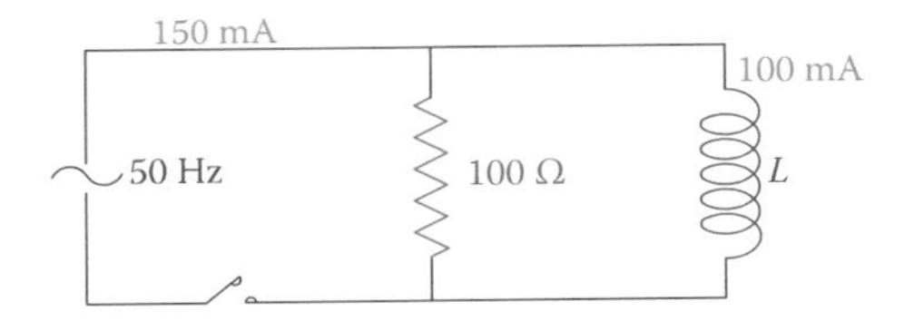

Figure 8.42 Circuit of Example 8.37.

Sometimes the leading or lagging is attributed to the power factor. For example, one may say a circuit has a leading power factor of 0.90.

Example 8.37

In the circuit shown in Figure 4.42 the total current is 150 mA and the current through the inductor is 100 mA. Determine what the applied voltage is. Also, knowing that the frequency is 50 Hz, find the value of L.

Solution

Applied voltage can be found by multiplying the resistor current by 100 Ω. Having only a resistor and an inductor in this circuit Equation 8.28 leads to

$$I_R = \sqrt{I^2 - I_L^2} = \sqrt{150^2 - 100^2} = 111.8\ \text{mA} = 0.1118\ \text{A}$$

$$V = (100)(0.1118) = 11.18\ \text{V}$$

$$X_L = 11.18 \div 0.100 = 111.8\ \Omega$$

$$L = \frac{X_L}{2\pi f} = \frac{111.8}{(2)(3.14)(50)} = 0.3559\ \text{H} = 35.6\ \text{mH}$$

In a parallel AC circuit, if the current leads the voltage, the circuit is said to be leading; if the current lags the voltage the circuit is said to be lagging.

8.12 Voltage Divider in AC Circuits

In the same way that a series DC circuit can be used as a voltage divider, in AC circuits a voltage divider can be devised by components in series. Inductors are not usually used for this purpose, because of their relative higher price, but capacitor use is common. A voltage divider can consist of a number of resistors and capacitors, but it could be only made of capacitors. The following examples show the applications.

Example 8.38

In the circuit shown in Figure 8.43, find the value of C for the capacitor, such that the voltage across R is twice the voltage across C.

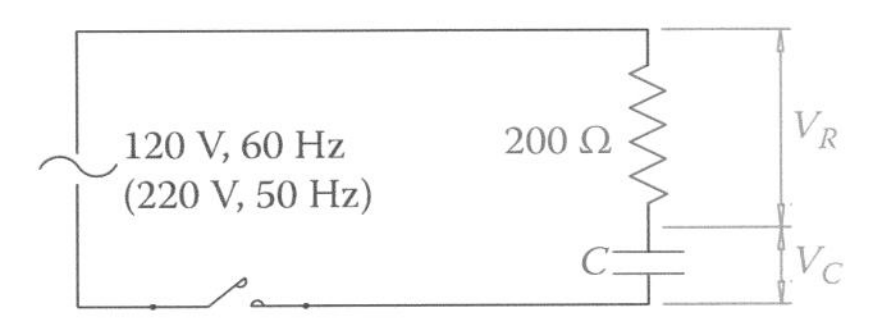

Figure 8.43 Circuit of Example 8.38.

Solution

Because the ratio of the voltages across R and C are given, in order to have $V_R = 2\ V_C$ it is necessary that $R = 2\ X_C$. Thus,

$$X_C = 200 \div 2 = 100\ \Omega$$

$$C = \frac{1}{(2)(\pi)(60)(100)} = 0.000026\ \text{F} = 26\ \mu\text{F}$$

Example 8.39

In the problem of Example 8.38, if we need the voltage across the capacitor to be 100 V, what is the size of the capacitor?

Solution

When the applied voltage is 120 V and the voltage across the capacitor is 100 V, the voltage across the resistor is

$$V_R = \sqrt{V^2 - V_C^2} = \sqrt{120^2 - 100^2} = 66.33\ \text{V}$$

$$I_C = I_R = \frac{66.3}{200} = 0.3315\ \text{A}$$

$$X_C = \frac{100}{0.3315} = 301\ \Omega$$

$$C = \frac{1}{(2)(\pi)(60)(301.5)} = 8.8 \times 10^{-6} F = 8.8\ \mu\text{F}$$

Note that an 8.8 μF capacitor can be made from standard size capacitors (2.2 and 3.3 μF).

Example 8.40

In the following circuit (Figure 8.44) the resistor in Example 8.39 is replaced by a capacitor. The requirement is that again with C_2 = 8.8 μF voltage V_2 = 100 V. If the applied voltage is 120 V, what is the capacitance of C_1?

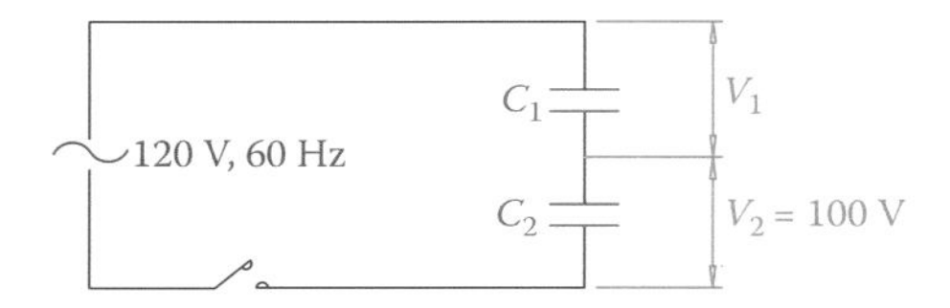

Figure 8.44 Circuit of Example 8.40.

Solution

In this voltage divider the components are similar; thus, voltages can be added together algebraically. That is, when $V_2 =$ 100 V, $V_1 = 120 - 100 = 20$ V. The capacitance of C_1 thus is 5 times of the capacitance of C_2. That is $C_1 = 8.8 \times 5 = 44$ μF.

8.13 Resonance

Resonance: Special condition in AC circuits where all the energy stored by inductive components is provided by capacitive components, and vice versa. This occurs in a particular frequency. This condition implies other facts such as the net reactive power to be zero, the power factor to be unity, and a number of other correlations based on the circuit layout.

In any AC circuit consisting of resistors, capacitors, and inductors, either in series or in parallel, a condition can happen in which the reactive power of the capacitors and of the inductors become equal. This condition is called **resonance**. Simultaneous with the capacitive reactive power and the inductive reactive power being equal, other features can reflect resonance. Remember that we always reduce a circuit to a single resistor, a capacitor, and an inductor. Thus, for this discussion we assume one of each component in the circuit.

In fact, when resonance happens, the inductive reactance and the capacitive reactance are equal to each other:

$$X_L = X_C \tag{8.31}$$

In a circuit with a fixed frequency, resonance can happen if the condition in Equation 8.31 is true. On the other hand, since both X_L and X_C are functions of frequency, if the frequency of a circuit changes, at a unique frequency these values can become equal. Figure 8.45 shows the variation of the impedance for the three basic types of loads in a circuit versus frequency. The horizontal axis implies frequency increase. A resistor is independent of frequency; thus, its impedance is constant, represented by a line parallel to the horizontal axis. The impedance of an inductor is

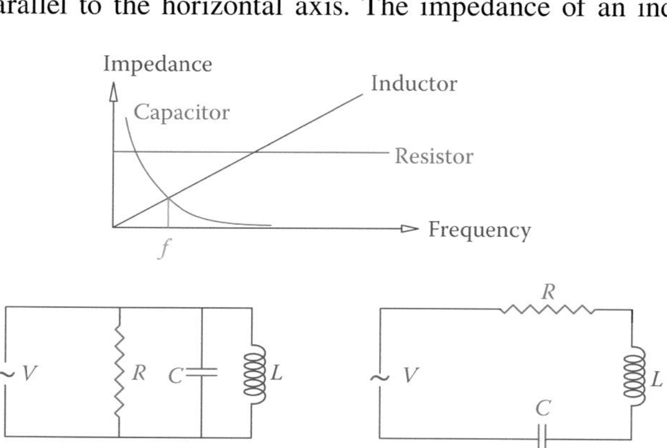

Figure 8.45 Resonance condition in AC circuits.

proportional to the frequency and augments as the frequency increases. For a capacitor the reverse happens and its impedance decreases (though not linearly) as the frequency increases. At the point of intersection of the two curves, $X_L = X_C$ and the frequency at that point is called the **resonant frequency** or **resonance frequency** and is denoted by f_R.

Resonance frequency can be found from equating X_L and X_C. This leads to

$$f_R = \frac{1}{2\pi\sqrt{LC}} \tag{8.32}$$

Resonant frequency: A unique frequency for each AC circuit containing both reactive components (inductors and capacitors) at which the resultant reactance of all capacitive components is equal to the resultant reactance of all the inductive components. As a result, the two types of components cancel the effect of each other, and the total reactive power of the circuit is zero.

Resonance frequency: Frequency at which resonance happens in an AC circuit.

Example 8.41

Find the resonance frequency of a 40 mH inductor and a 51 μF capacitor.

Solution

Values of the capacitance and inductance in Farad and Henry can directly be plugged in Equation 8.32. Thus,

$$f_R = \frac{1}{2\pi\sqrt{(0.040)(0.000051)}} = \frac{1}{(2)(3.14)(0.00143)}$$

$$= 111.43 \cong 112\ \text{Hz}$$

Example 8.42

Find the capacitance for a capacitor to become in resonance with a 40 mH inductor at 60 Hz frequency.

Solution

At 60 Hz the reactance of the capacitor must be the same as the reactance of the inductor. Thus,

$$X_C = X_L = (2)(\pi)(60)(0.040) = 15\ \Omega$$

$$C = \frac{1}{(2)(\pi)(60)(15)} = 0.000176\ \text{F} = 176\ \text{mF}$$

Note that if this value is not among the standard values for capacitors, one can make such a value by combining a number of standard capacitors in series and/or parallel.

8.13.1 Resonance in Series Circuits

When resonance occurs in a series *RLC* circuit, the resonance condition (Equation 8.31) leads to other relationships or properties. These are

1. Voltage across the inductor is equal to the voltage across the capacitor.
2. Voltage across the resistor is equal to the applied voltage (according to Equation 8.26).

3. Impedance of the circuit has its lowest value and is equal to R (see Equation 8.25).
4. Circuit current assumes its maximum value because the impedance is minimum.
5. Power factor for the circuit becomes equal to 1, and the phase angle is zero.
6. Apparent power has its lowest value and becomes equal to the active power because the power factor is 1.

8.13.2 Resonance in Parallel Circuits

Similar to the series circuits, when resonance occurs in a parallel *RLC* circuit the resonance condition (Equation 8.31) leads to other relationships or properties:

1. Current in the inductor is equal to the current in the capacitor.
2. Current in the resistor is equal to the total circuit current (according to Equation 8.28).
3. Impedance of the circuit has its highest value and is equal to R (see Equation 8.29).
4. Circuit current assumes its minimum value because the impedance has the highest value.
5. Power factor for the circuit becomes equal to 1, and the phase angle is zero.
6. Apparent power has its lowest value and becomes equal to the active power because the power factor is 1.

Items 5 and 6 are the same as for the series resonant circuits, but the rest are quite different.

When resonance occurs in a parallel *RLC* circuit, a local current circulates between the inductor and the capacitor. This current can be very high, while the circuit current as seen from the source can be low. This phenomenon is used in induction heaters (in industry for heating metals when necessary, e.g., heating bearings for mounting or dismounting) and in induction cookers (for domestic use). In such an application a high current is flowing through an inductor, whereas the current provided by the power line is small. This means that the rating of the wires and breakers are much smaller than the current in the inductor. The current in the inductor creates (induces) local currents in the piece to be warmed, without even touching it. In the case of an induction cooker the body of the cooking pan becomes hot owing to local currents created by induction. This is shown in Figure 8.46. The efficiency

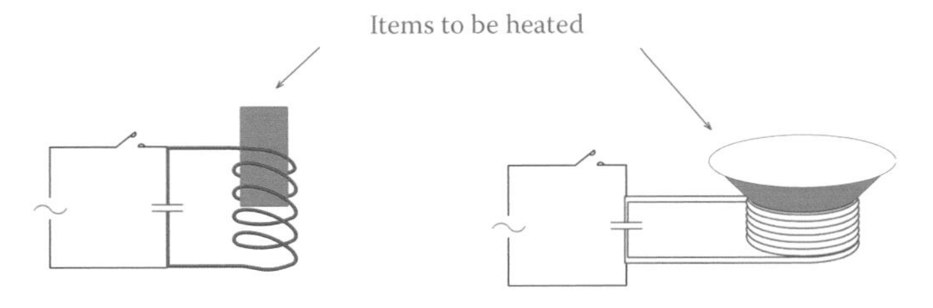

Figure 8.46 Principle of induction heater and induction cooker.

of induction heating is very high, and the process is very fast compared to conventional heating in which a great part of energy is used for heating air and the intermediate media between the source and the body to be heated.

8.14 Power Factor Correction

Power factor correction and power factor improvement are similar actions that are frequently employed in industry, when a lot of motors are used for industrial operations. A motor can be modeled by a resistor and an inductor. As a result, the power factor of the circuit containing those motors is lower than 1. There are two disadvantages corresponding to low power factors: one is the higher power that a power source (generator) must provide (the apparent power), and the other is the voltage drop in the lines because of higher current. Examples of this are given in Chapter 11 on motors and generators.

Power factor correction is performed by adding one or more capacitors in parallel with the loads. This will compensate for the reactive load of the inductor part of the load(s) and lowers the phase angle (thus, increases the power factor). Further discussion of this subject is left for Chapter 11 (see Section 11.8.5). Here only we show an example of how inserting a capacitor in a circuit can improve the power factor. This is, in fact, based on resonance condition in which the power factor is exactly equal to 1, but, in practice, this is hard to achieve and a power factor as close to one as possible is desirable.

Example 8.43

An electric load consists of a resistor of 80 Ω in parallel with an inductor the reactance of which at the operating frequency of 60 Hz is 70 Ω. The applied voltage is 120 V. Find the capacitor that if in parallel with the load changes the power factor to 95 percent.

Solution

Current power factor can be found by either of the terms in Equation 8.30. Any change in the power factor is due to the capacitor only. We use the currents for this purpose.

$$I_R = 120 \div 80 = 1.5 \text{ A}$$

$$I_L = 120 \div 70 = 1.714 \text{ A}$$

$$I = \sqrt{1.5^2 + 1.714^2} = 2.73 \text{ A}$$

$$pf = 1.5 \div 2.73 = 0.55$$

If the new power factor was to be 100 percent, the problem would have become easier because we need only to set the reactance of the capacitor to become equal to the reactance of the inductor. Although this is not in the question, we will find the proper capacitor to do this, for comparison. Thus,

Capacitor for 100 percent power factor:

$$C = \frac{1}{(2)(\pi)(60)(70)} = 0.000038\,\text{F} = 38\,\mu\text{F}$$

Capacitor for 95 percent power factor:

Figure 8.47 shows the current for the two cases: (1) the present setting and (2) after power factor correction. This helps to find the required current in the capacitor, based on which we can determine X_C and therefore the required C.

After the capacitor is added to the circuit, the current in the circuit changes from its old magnitude (2.37 A) to a new magnitude. This new current can be found from the desired power factor (0.95) and the resistor current (1.5 A)

$$I_{\text{New}} = 1.5 \div 0.95 = 1.58\ \text{A}$$

From this new current the difference between the current in the inductor and the current in the capacitor (see Equation 8.28) can be found. Thus,

$$I_L - I_C = \sqrt{1.58^2 - 1.5^2} = 0.3\ \text{A}$$

Therefore,

$$I_C = I_L - 0.3 = 1.714 - 0.3 = 1.414\ \text{A}$$

$$X_C = 120 \div 1.414 = 84.87\ \Omega$$

$$C = \frac{1}{(2)(\pi)(60)(84.87)} = 0.000031\,\text{F} = 31\,\mu\text{F}$$

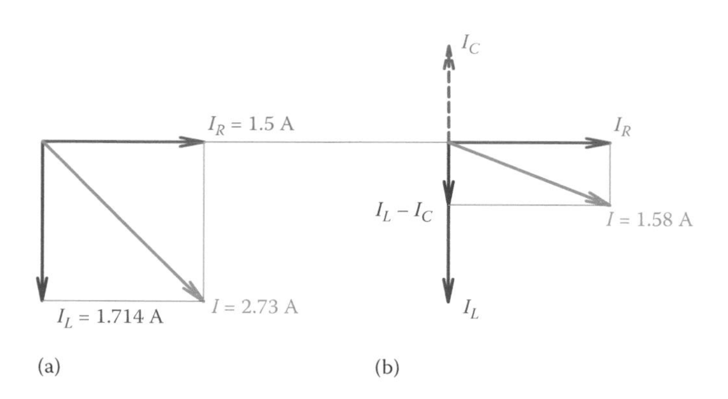

Figure 8.47 Currents (a) before and (b) after power factor correction for Example 8.43.

As expected the value obtained is smaller than 38 μF, which changes the power factor to 1.

Note that if the value of the capacitance to be added is more than 38 μF, the power factor decreases again, but this time with a leading power factor.

Example 8.43 clearly shows the advantage of power factor correction. The line current has been reduced from 2.73 to 1.58 A as a result of power factor correction, while there has not been any change in the current corresponding to the active power.

8.15 Chapter Summary

- A vector is an entity that has both a magnitude and a direction.
- When two sinusoidal waveforms with the same frequency do not reach their maximum and minimum values simultaneously, there is a phase difference between them.
- Phase difference implies how much the time delay between two sinusoidal waveforms of the same frequency is. It can be expressed in degrees because one cycle corresponds to 360°.
- Vectors are used in AC circuits to represent the phase relationship between current and voltage.
- If voltage in an AC waveform has a sinusoidal form, current also has a sinusoidal form.
- The highest value of an AC waveform is called peak value.
- The difference between maximum value and minimum (negative) value in a sinusoidal waveform is called peak-to-peak value.
- The cyclic waveform has an effective value. For a sinusoidal waveform the effective value is 0.707 of the peak value.
- When expressing an AC voltage or current value, like 12 V, it is the effective value that is expressed.
- When a meter measures voltage or current in an AC circuit, it is the effective value that is shown by the meter.
- Resistors in DC and AC circuits are treated the same way.
- When resistive loads are connected to AC, voltage and current are in phase (there is no phase difference).
- The rule of inductors in series and parallel is similar to the rule of resistors in series and parallel.
- The rule of capacitors in series and parallel is the opposite of the rule for resistors. In parallel capacitors their capacitances add together.
- When an inductor or a capacitor is inserted in an AC circuit, the effect is exhibited in the form of reducing the current. This effect is represented in the form of a resistance, measured in ohms, but it is called reactance; inductive reactance for an inductor and capacitive reactance for a capacitor.
- An AC circuit, in general, can have resistive loads, inductive loads, and capacitive loads.

- Resistive loads consume active (or real, or true) power.
- Inductive and capacitive loads consume reactive power.
- Reactive power is the power that is stored in an inductor or a capacitor in one half cycle of an AC waveform. It is given back to the circuit in the next half cycle. However, the source in the circuit must provide this power.
- Apparent power is the value obtained by multiplying the voltage and current in an AC circuit.
- In an AC circuit consisting of only inductive loads, current reaches its maximum value a quarter of cycle after the voltage reaches its maximum value. Current is 90° behind voltage. Being behind is expressed as "lagging."
- In an AC circuit consisting of only capacitive loads, current reaches its maximum value a quarter of cycle before the voltage reaches its maximum value. Current is 90° ahead of voltage. Being ahead is expressed as "leading."
- A general AC circuit can have resistive, inductive, and capacitive loads. Depending on the values of capacitive and inductive loads, current can be leading or lagging the voltage. The amount of phase difference is not 90°; it depends on the sizes of the three types of loads.
- The resistive (*R*), inductive (*L*), and capacitive (*C*) loads can be in parallel or in series, thus, parallel *RLC* and series *RLC* circuits.
- In all ordinary electricity usage, the loads are in parallel with each other.
- An AC circuit is said to be leading when current is leading voltage. It is said to be lagging if voltage is leading current (current is lagging voltage).
- In an AC circuit the ratio of voltage value over current is the circuit impedance. Impedance is measured in ohms and depends on the values of *R*, *L*, *C*, and frequency.
- Power factor is the ratio between the active power and the apparent power in a circuit.
- The formulae for power factor in series and parallel *RLC* circuits are not the same.
- Maximum and desired value for the power factor is unity (1).
- In a circuit that has a power factor less than 1, it is possible to change the power factor to around 1. This is called power factor correction.
- A power factor of 1 implies that the circuit is neither leading nor lagging (no phase difference between current and voltage).
- In an AC circuit, resonance occurs when the inductive reactance of the inductive loads and the capacitive reactance of the capacitive loads are equal to each other.
- When resonance happens, certain other conditions are in order. For example, in a series *RLC* circuit the current becomes maximum, whereas in a *RLC* parallel circuit the current becomes minimum.

Review Questions

1. What is effective value in AC electricity?
2. What is the relationship between effective value and peak-to-peak value?
3. What is the phase difference between current and voltage in a resistive load? How much is it in a purely inductive load?
4. When a capacitor, an inductor, and a resistor are connected in series, how many degrees out of phase are the voltage across the inductor and the voltage across the capacitor?
5. When a number of capacitors are put in series with each other, is the equivalent capacitor larger or smaller than all those capacitors?
6. When a number of inductors are put in series with each other, is the equivalent inductor larger or smaller than each one?
7. When a capacitor, inductor, and resistor are connected in parallel, how many degrees out of phase are the voltage across the inductor and the voltage across the capacitor?
8. In an AC circuit made out of some capacitors in parallel with each other, if the frequency goes up, does the current go up or down? Why?
9. In an AC circuit consisting of a number of resistors (only), what happens to the current if the frequency goes up?
10. A circuit is made up of a capacitor in parallel with an inductor. If the frequency goes down, does the current go up or down?
11. If a parallel *RLC* circuit is lagging, what happens to the current if the frequency goes up (does it increase or decrease)? Explain.
12. Is it possible in a series *RLC* circuit to have a voltage larger than the applied voltage between two points in the circuit? Why or why not?
13. When in a parallel *RLC* circuit resonance occurs, what is the relationship between current in the inductive load and current in the capacitive load?
14. Why can the power factor not be more than 1?

15. Why is a small power factor not desirable?

16. What is resonance frequency?

Problems

1. The peak-to-peak voltage of an AC line is 36.8 V. What is its effective voltage?

2. The consumed energy by a 10 Ω resistor when connected to an AC voltage source is 130 W. What is the peak voltage of the source?

3. An inductor is 265 mH. What is its reactance when connected to a 120 V, 60 Hz AC line?

4. Four inductors each having an inductance of 0.50 mH are connected in series. What is the total reactance when connected to a 50 Hz line?

5. If all the inductors (in series) in Problem 4 are connected to a 120 V, 60 Hz source, what is the circuit current?

6. When connected to a 60 Hz source, the inductor in the circuit shown in Figure P8.1 has 15 Ω reactance. Find the current in the circuit.

7. In the circuit shown in Figure P8.2 the voltage across *R* is 25 V. Find the current in the inductor.

8. A capacitor is 2.2 μF. What is its reactance when connected to a 60 Hz AC line?

9. In the circuit shown in Figure P8.3, what is the current?

Figure P8.1 Circuit of Problem 6.

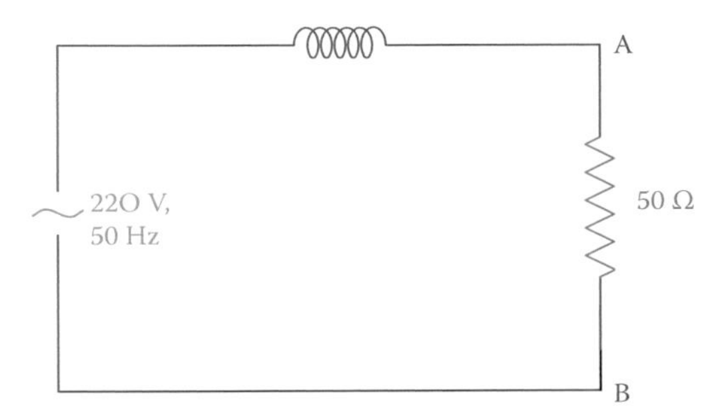

Figure P8.2 Circuit of Problem 7.

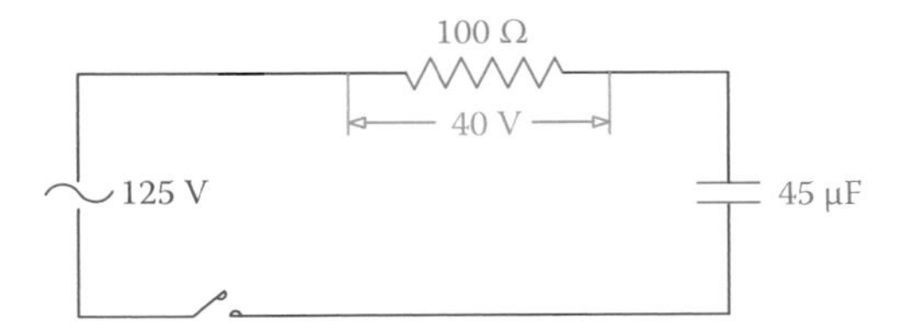

Figure P8.3 Circuit of Problem 9.

10. In the circuit of Problem 9 what is the frequency of the applied voltage?

11. A capacitor and a 40 Ω resistor are connected in series to a 60 Hz line and the current is 1.8 A. If the capacitive reactance of the capacitor is 50 Ω, what is applied voltage?

12. Referring to Figure P8.3, how much is the voltage across the resistor if the line frequency changes to 400 Hz?

13. A capacitor, a 40 Ω resistor and a 186 mH inductor are connected in series to a 60 Hz, 120 V source. If the capacitive reactance of the capacitor is 50 Ω, what is the total current flow in the circuit?

14. In Problem 13, what is the capacitance of the capacitor (the frequency is 60 Hz)?

15. In Problem 13, if the applied voltage was 48 V, what would voltage across the capacitor be?

16. In the circuit shown in Figure P8.4 the applied voltage is 48 V. If the capacitance is 96 μF, $R = 50\ \Omega$, and $L = 186$ mH, how much is the voltage between points C and E? Select one of the following answers:
    a. Smaller than 48 V
    b. 48 V
    c. Larger than 48 V
    d. Depends on the frequency

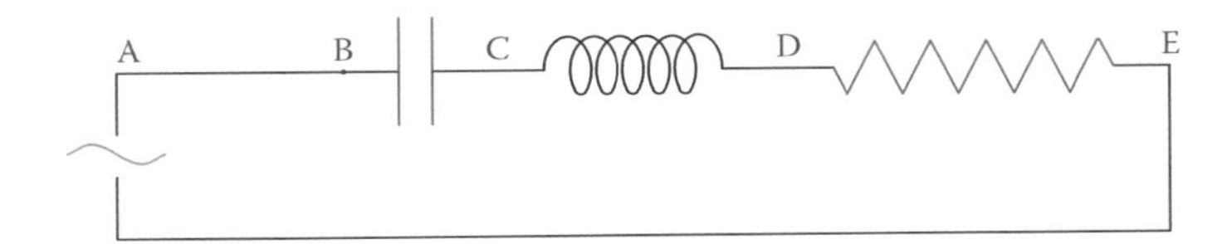

Figure P8.4 Circuit for Problems 16 through 21.

17. In the circuit of Problem 16 if the frequency is 50 Hz, what is voltage between C and E?

18. In a series *RLC* circuit (as shown in Figure P8.4) the applied voltage is 120 V, 60 Hz, and $R = 47\ \Omega$. If the voltage between points B and D is 80 V, what is the current?

19. In the circuit of Problem 16 the applied voltage is 12 V, what is the voltage across points A and B?

20. Find the frequency at which the circuit of Problem 16 goes to resonance.

21. What is the current of the circuit of Problem 16 at resonance frequency?

22. In the circuit shown in Figure P8.5, how much is the apparent power if the applied voltage is 120 V, $C = 22\ \mu F$, $L = 186$ mH, $R = 25\ \Omega$, and $f = 50$ Hz?

23. Compare the apparent power in the circuit of Problem 22 for $f = 50$ Hz and $f = 100$ Hz.

24. Find the total current in the circuit shown in Figure P8.6 when the applied voltage is 12 V, 50 Hz, $C_1 = 51\ \mu F$, and $C_2 = 22\ \mu F$.

25. In the circuit depicted in Figure P8.7, $R = 100\ \Omega$, $L =$ 200 mH, and $C = 51\ \mu F$. Find the resonance frequency.

26. In the circuit of Problem 25, if at resonance the current in the inductor is 3 A, determine the voltage across the resistor. What is the current in the capacitor branch?

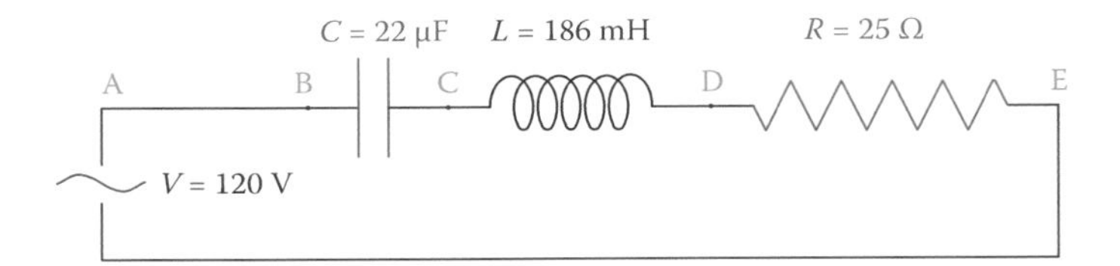

Figure P8.5 Circuit of Problem 22.

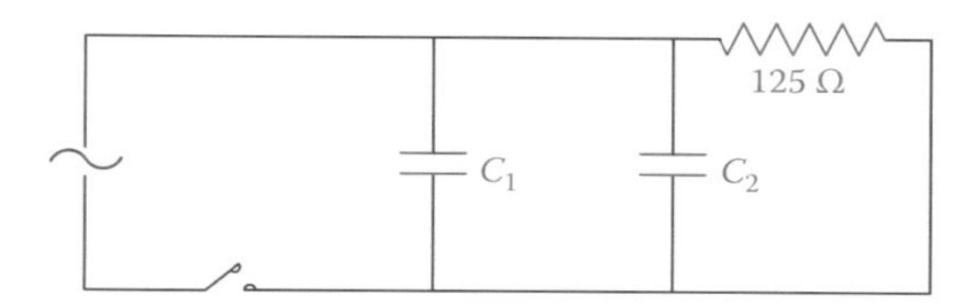

Figure P8.6 Circuit of Problem 24.

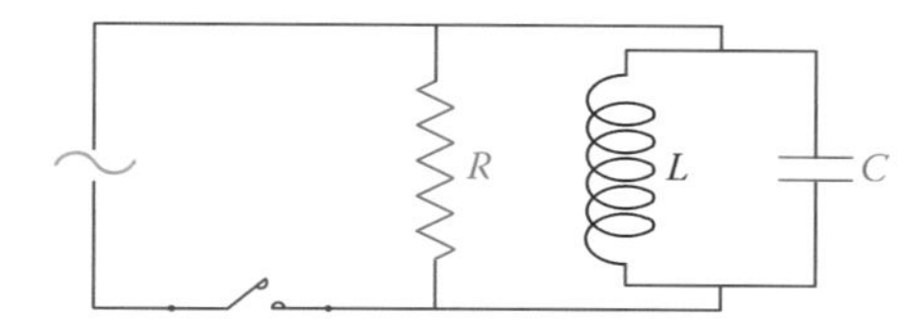

Figure P8.7 Circuit of Problem 25.

27. If in Problem 25 the resistor is 1 MΩ. What is the current in the resistor when the circuit is at resonance? The voltage is the same as obtained in Problem 26.

28. In Problem 27, how much is total current?

29. Supposing that for the circuit shown in Figure P8.7 the total current at the resonant frequency is 6 mA. What is the total current at the frequency twice the resonant frequency?

30. A coil has a resistance of 5 Ω, and its inductance is 0.2 H. If five of these coils are put in parallel with each other and connected to 100 V, 60 Hz, what is the active power, reactive power, and apparent power in the circuit?

31. If a capacitor with $C = 63$ μF is put in parallel with the five inductors in Problem 30, does the real (active) power go up or down?

32. If a 63 μF capacitor is put in series with the five inductors, what is the change in the real power?

33. In Problem 32, if you see the waveforms for current and voltage on an oscilloscope, which one reaches its maximum first?

34. For the circuit shown in Figure P8.8, find the capacitor value for resonance at 60 Hz. $L = 0.015$ H.

35. For the circuit in Problem 34, find the current passing through the inductor at resonance if $R = 12$ Ω and $V = 96$ V.

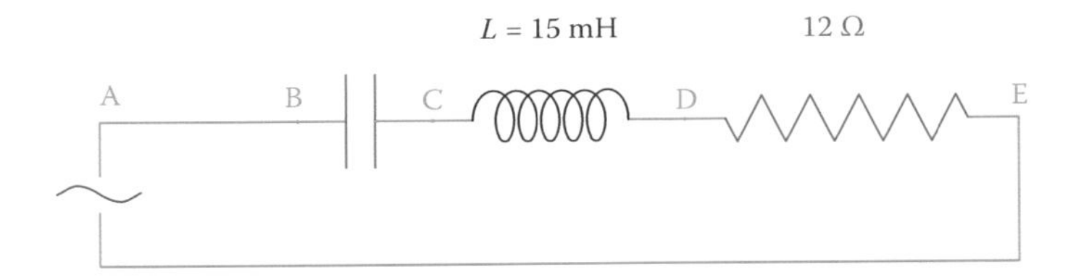

Figure P8.8 Circuit of Problem 34.

36. In the circuit in Problem 35, find the following voltages:

    V_{AB}, V_{BC}, V_{AC}, V_{BC}, V_{BD}, V_{CD}, V_{CE}, when at resonance

37. Find the same voltages as in Problem 36, for the frequency of 60 Hz, when $C = 1.1$ μF.

38. A 450 nF capacitor in parallel with a resistor has 16.25 mA current. Find the value of R if the total circuit current is 35 mA. The frequency is 50 Hz.

39. Find the capacitance C of the capacitor in the circuit of Figure P8.9.

40. Five inductors and capacitors with values as given or determined in Problem 34 are put in series as shown in Figure P8.10a, and together with a 120 Ω resistor are connected to 120 V, 60 Hz. What is the current?

41. If the arrangement in Problem 40 is changed as in Figure P8.10b, find the current.

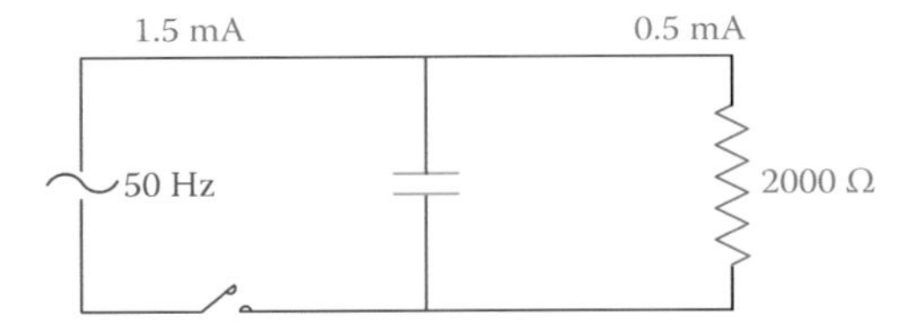

Figure P8.9 Circuit of Problem 39.

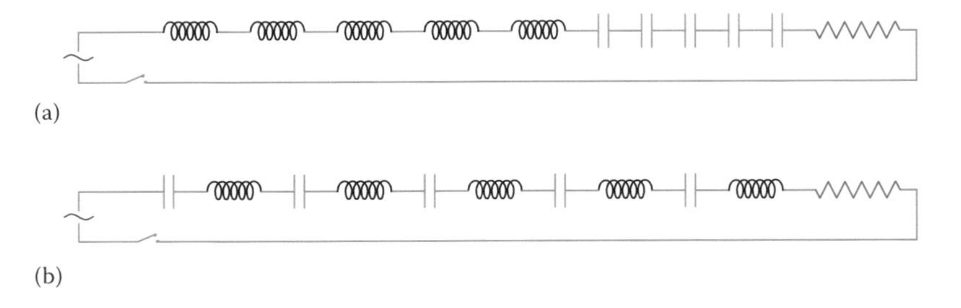

Figure P8.10 Circuit of Problem 40. (a) All similar components put together. (b) Capacitors and inductors put in pairs.

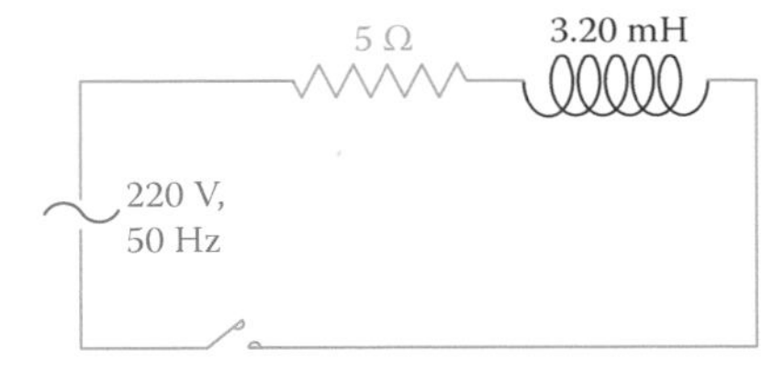

Figure P8.11 Circuit for Problem 43.

42. In Problem 40, put all elements in parallel with each other. Does the current go up or down? Calculate it.

43. The circuit in Figure P8.11 represents a motor connected to a 50 Hz voltage. Find the current in the motor and the (active) power.

44. In Problem 43, how much is the change in the apparent power if the frequency is 60 Hz? Does the power factor increase or decrease?

45. Find the voltage across R in the circuit shown in Figure P8.12.

46. Compare the active power, apparent power, and power factor in the circuit of Figure P8.13 with that of Figure P8.12, and explain the difference.

47. Find the power factor in Problem 46 for 50 Hz frequency. Conclude whether the frequency at which $pf = 1$ is larger or smaller than 60 Hz.

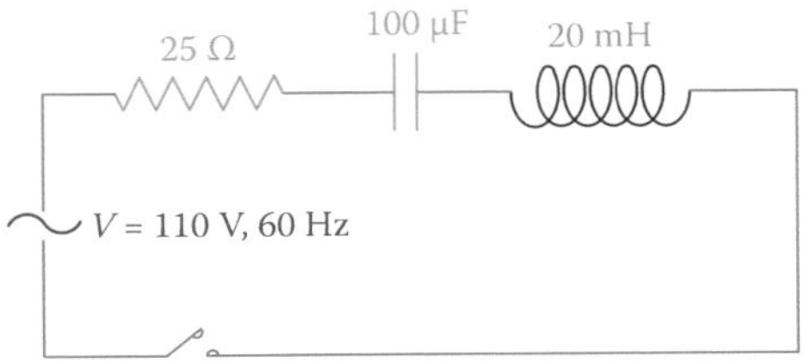

25 Ω
100 μF
20 mH
V = 110 V, 60 Hz

Figure P8.12 Circuit of Problem 45.

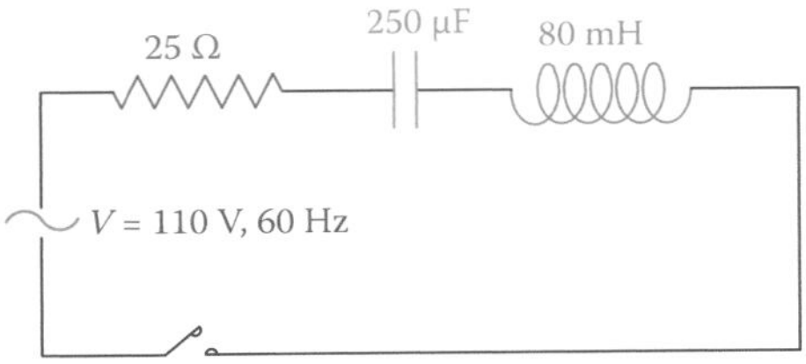

25 Ω
250 μF
80 mH
V = 110 V, 60 Hz

Figure P8.13 Circuit of Problem 46.

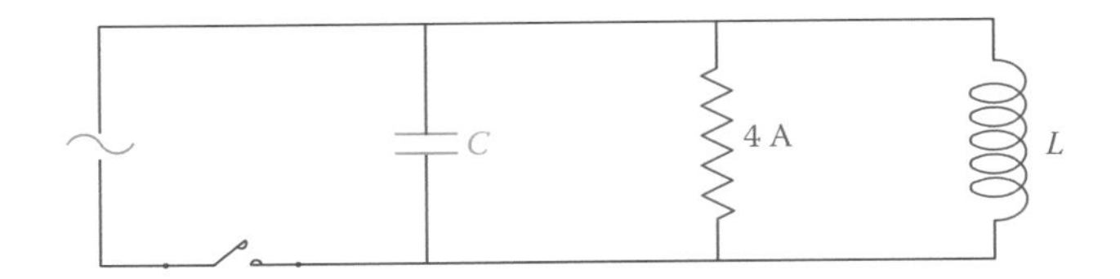

Figure P8.14 Circuit of Problem 48.

48. What is the total current in the circuit shown in Figure P8.14 if the power is 236 W, $C = 100\ \mu F$, $L = 136$ mH, and $f = 50$ Hz?

49. Find the reactive power, power factor, and the apparent power for the circuit of Problem 48.

50. If a 33 μF capacitor is added (in parallel) to the circuit of Problem 48, find the new circuit current, and the new power factor and apparent power.

Advanced Learning: AC Problem Solving Using Complex Numbers

This section shows how complex numbers are used in the analysis and solution of alternating current problems. Complex number operations can be regarded as a way of representing vectors. It is assumed that an advanced reader already knows about the concept of operator j, its mathematics, and the rules for addition, subtraction, multiplication, and division of complex numbers. Therefore, no explanation of the pertinent mathematics is made here.

Introduction

The material presented in this section is only an extension of what has already been discussed within the chapter in terms of the relationships between voltage, impedance, and current and the relationship between the various voltages in a series circuit and between various currents in a parallel circuit. Indeed, in using complex numbers an AC circuit is treated in the same way that DC problems were formulated according to Ohm's law. Hence, the voltages and currents add together linearly; the exception, however, is that instead of resistances (R) in DC, here we are dealing with impedances (Z). By using the approach in this section we will be able to tackle a number of problems that could not be easily treated before, as we will see from some examples.

The operator j is such that $j^2 = -1$. It has no tangible numerical meaning, but it represents a rotation by 90°. You may imagine a vector. Multiplying this vector by j implies rotating it 90° in the positive direction (CCW) and multiplying it by $-j$ implies rotating it by 90° clockwise.

In dealing with this approach, the impedances in a circuit must be defined first. Figure A8.1 illustrates the necessary definitions. After the impedances are defined it is a matter of employing Ohm's law and voltage-current relationship. This is shown in the examples that follow.

Examples

The examples below illustrate a few problems solved by using complex numbers. Most of these problems are more involved than the examples seen within the chapter but can be easily tackled by the new approach.

Example A8.1

Find the current and the phase angle in the circuit shown in Figure A8.2.

Solution

This is the same circuit as in Example 8.29 for *RLC* series circuits, solved in a different way here.

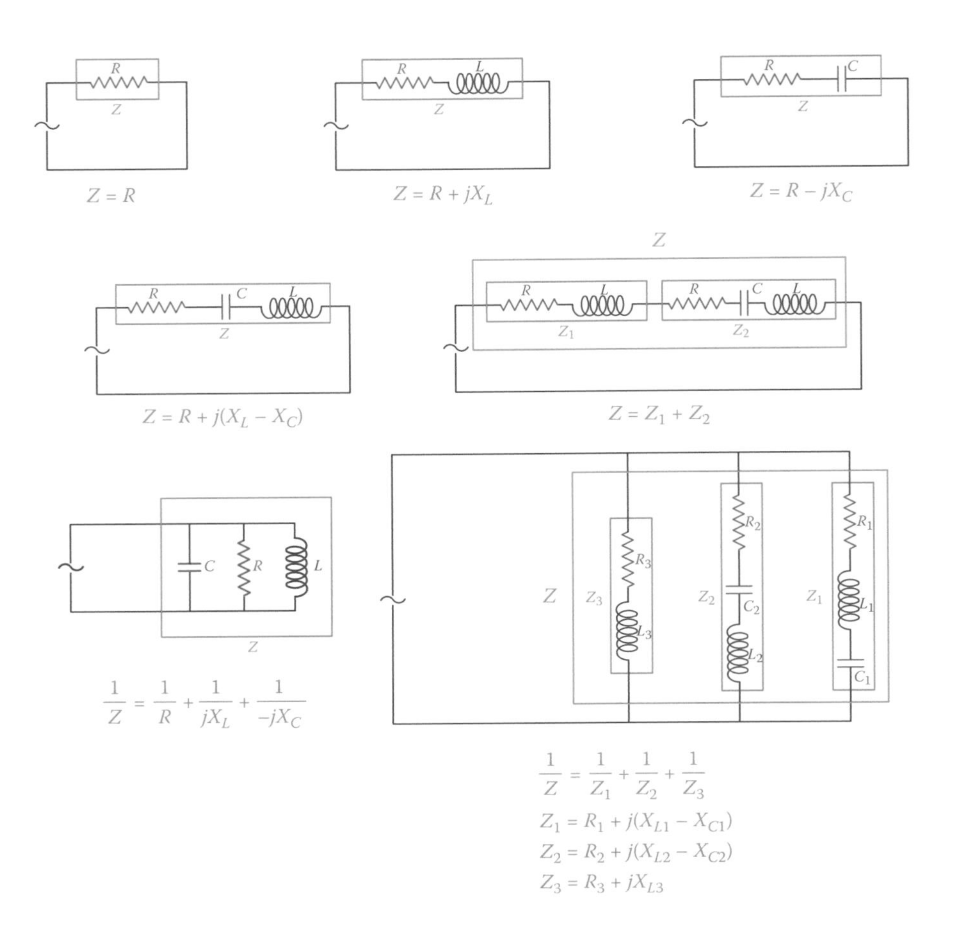

Figure A8.1 Expression of complex *Z* for some *RLC* combinations.

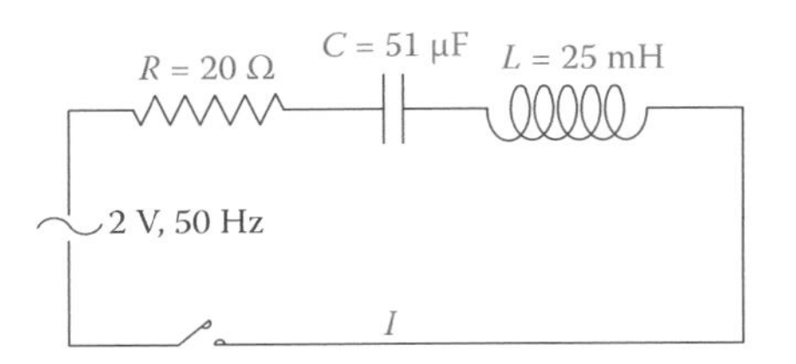

Figure A8.2 Circuit for the problem in Example A8.1.

$$X_L = 7.85\ \Omega$$

$$X_C = 62.445\ \Omega$$

$$Z = 20 + 7.85j - 62.45j = 20 - 54.60j$$

$$I = \frac{2}{20 - 54.6j} = \frac{2(20 + 54.6j)}{20^2 + 54.6^2} = \frac{40 + 109.2j}{3381.16}$$

$$= 0.0118 + 0.0323j$$

This form of representation for the circuit current (or for a voltage) does include the information about the value and the phase angle of current, although it does not directly show them. Hence, it needs conversion to a magnitude and an angle, exactly as dealing with a vector when its perpendicular components are known. Some calculation details are shown for this example, but not for the rest of examples.

$$I = 0.0118 + 0.0323j = \sqrt{0.0118^2 + 0.0223^2} \angle \tan^{-1}\left(\frac{0.0223}{0.0118}\right)$$

$$= 0.0344 \text{ A} \angle 70° = 34.4 \text{ mA} \angle 70°$$

Thus, the circuit current is 34.4 mA, and it is leading the voltage by 70°. If the phase angle had a negative value, then the current would be lagging the voltage.

Example A8.2

Find the current in the circuit shown in Figure A8.3.

Solution

$$X_{L1} = (2)(\pi)(50)(0.025) = 7.854\ \Omega$$

$$X_{L2} = (2)(\pi)(50)(0.033) = 10.367\ \Omega$$

$$X_C = \frac{1}{(2)(\pi)(50)(0.000068)} = 46.810\ \Omega$$

For this problem we denote the currents in the branches as I_1 and I_2. After finding each individual current we add them together for the circuit current.

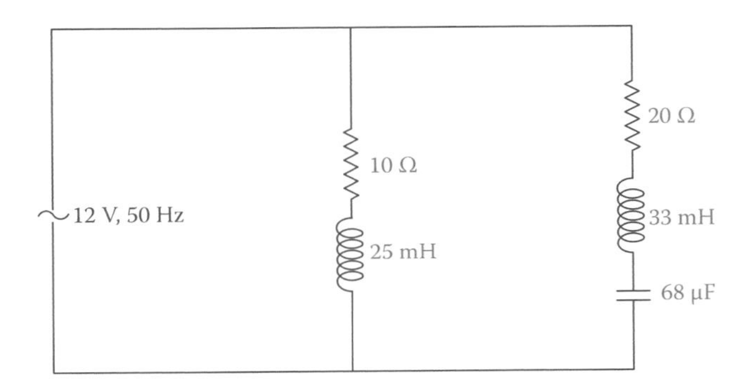

Figure A8.3 Combined circuit for the problem in Example A8.2.

$$I_1 = \frac{V}{Z_1} = \frac{12}{10 + 7.854j} = \frac{(12)(10 - 7.854j)}{161.685}$$
$$= 0.742 - 0.583j \rightarrow 0.9436\ \text{A}\angle - 38.2^\circ$$

$$I_2 = \frac{V}{Z_2} = \frac{12}{20 - 36.443j} = \frac{(12)(20 + 36.443j)}{1728.09}$$
$$= 0.139 + 0.253j \rightarrow 0.2887\ \text{A}\angle 61.2^\circ$$

$$I = I_1 + I_2 = (0.742 - 0.583j) + (0.139 + 0.253j)$$
$$= 0.881 - 0.330j \rightarrow 0.941\ \text{A} \angle\ 20.5^\circ$$

Thus, the circuit current is 941 mA lagging the voltage by 20.5°.

Example A8.3

Find the current in the circuit of Figure A8.4.

Solution

In this circuit the values for the inductor and capacitor have been chosen so that for the given frequency they are at resonance.

$$X_L = 8.29\ \Omega \cong 8.3\ \Omega$$

$$X_C = 8.0\ \Omega$$

$$I_1 = \frac{V}{Z_1} = \frac{12}{10 + 8.3j} = \frac{(12)(10 - 8.3j)}{168.79}$$
$$= 0.711 - 0.590j \rightarrow 0.923\ \text{A}\angle - 39.7^\circ$$

$$I_2 = \frac{V}{Z_2} = \frac{12}{10 - 8j} = \frac{(12)(10 + 8j)}{164}$$
$$= 0.732 + 0.585j \rightarrow 0.937\ \text{A}\angle\ 38.6^\circ$$

$$I = I_1 + I_2 = 1.443 - 0.005j \rightarrow 1.44\ \text{A}\angle\ 0.2^\circ$$

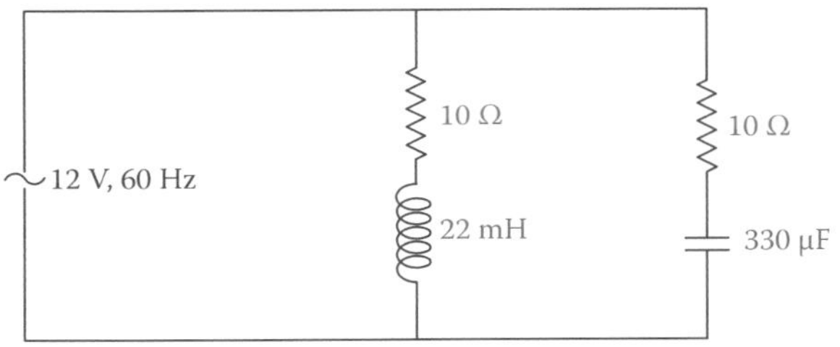

Figure A8.4 Circuit for the problem in Example A8.3.

It can be seen that the current in the main branch is practically in phase with the voltage. This current is mostly due to the two resistors.

Example A8.4

Repeat the problem for the circuit of Figure A8.4 if both resistors are 1 Ω.

Solution

Inductive and capacitive reactance values are the same as in the previous example. Thus,

$$X_L = 8.29\ \Omega \cong 8.3\ \Omega$$

$$X_C = 8.0\ \Omega$$

The branch currents are

$$I_1 = \frac{V}{Z_1} = \frac{12}{1+8.3j} = \frac{(12)(1-8.3j)}{69.79}$$

$$= 0.172 - 1.427j \rightarrow 1.437\ \text{A}\angle -83^\circ$$

$$I_2 = \frac{V}{Z_2} = \frac{12}{1-8j} = \frac{(12)(1+8j)}{65} = 0.185 + 1.477j \rightarrow 1.488\ \text{A}\angle\ 83^\circ$$

$$I = I_1 + I_2 = 0.357 - 0.05j \rightarrow 0.36\ \text{A}\angle\ 8^\circ$$

We can observe that while the branch currents are noticeable, current through the source is quite small.

Example A8.5

Find the current in the circuit of Figure A8.5. Also, find the voltage drop V_{AB} across the 2 Ω resistor. The supply voltage is 14 V.

Solution

For this problem we write the voltage relationships for both loops containing the power source. Then we solve the two simultaneous equations for the two unknowns I_1 and I_2. The important fact is that these equations are linear. Alternatively, we find I_1 in terms of I_2 (or vice versa) based on the fact that the same voltage is applied to both branches. Then substituting in one of the voltage relationships leads to a linear equation in one single unknown.

$$X_{L1} = (2)(\pi)(50)(0.025) = 7.854\ \Omega$$

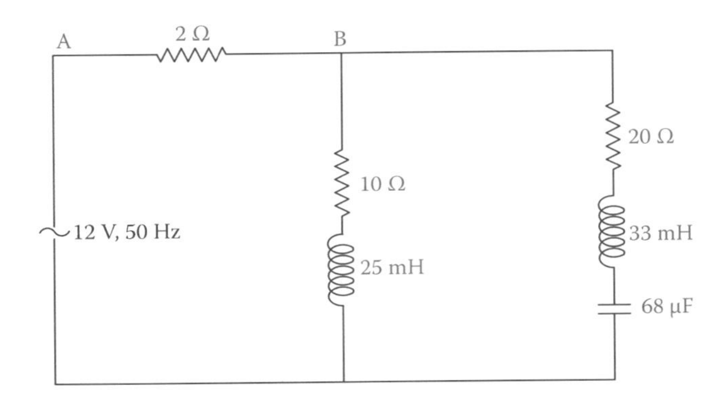

Figure A8.5 Figure for Example A8.5.

$$X_{L2} = (2)(\pi)(50)(0.033) = 10.367\ \Omega$$

$$X_C = \frac{1}{(2)(\pi)(50)(0.000068)} = 46.810\ \Omega$$

$$Z_1 = 10 + 7.854j$$

$$Z_2 = 20 + 10.367j - 46.810j = 20 - 36.443j$$

Voltage relationships:

$$(2)(I_1 + I_2) + Z_1 I_1 = 14$$

$$(2)(I_1 + I_2) + Z_2 I_2 = 14$$

Thus,

$$(10 + 7.854j)I_1 = (20 - 36.443j)I_2$$

which leads to

$$I_1 = \left(\frac{20 - 36.443j}{10 + 7.854j}\right)I_2 = \frac{(20 - 36.443j)(10 - 7.854j)}{100 + 61.685}I_2$$

$$= (-0.553 - 3.225j)I_2$$

$$I_1 + I_2 = (0.467 - 3.225j)I_2$$

Plugging into equation containing I_2

$$(2)(I_1 + I_2) + Z_2 I_2 = (0.9335 - 6.450j + 20 - 36.443j)I_2$$
$$= (20.9335 - 42.893j)I_2 = 14$$

leads to

$$I_2 = \frac{14(20.9335 + 42.893j)}{2278}$$

$$= 0.12865 + 0.2636j \rightarrow 0.2933\ A\angle\ 64°$$

Then

$$I_1 = (-0.533 - 3.225j)(0.12865 + 0.2636j)$$
$$= 0.7816 - 0.5554j \rightarrow 0.9585 \text{ A}\angle -35.4^\circ$$

Finally

$$I_1 + I_2 = (0.7816 - 0.5554j) + (0.12865 + 0.2636j)$$
$$= 0.910 - 0.292j \rightarrow 0.956 \text{ A}\angle -17.8^\circ$$

And the circuit current is

$$I = 0.956 \text{ A at } 18^\circ \text{ lagging.}$$

The voltage drop across the resistor is found from Ohm's law:

$$V_{AB} = RI = (2)(0.956) = 1.912 \text{ V}$$

9

Three-Phase Systems

OBJECTIVES: After studying this chapter, you will be able to

- Explain the difference and the similarity between single-phase and three-phase systems
- Explain the advantages of three-phase systems
- Understand why three-phase systems are more common in industry
- Extend the relationships in single-phase AC electricity to three-phase systems
- Understand delta and wye connections in three-phase electricity
- Describe the relationships and differences between wye and delta connections
- Make a distinction between balanced and unbalanced loads
- Calculate power and power factor in three-phase electricity
- Apply power factor correction for three-phase electricity

New terms:
Balanced load, delta connection, Δ connection, inrush current, line current, line voltage, neutral line, null line, phase current, phase voltage, star connection, unbalanced load, wye connection, Y connection

9.1 Introduction

All over the world, except in very small local generators, commercial electricity is generated and distributed using a three-phase AC system. Three-phase electricity is, in fact, three single-phase AC systems integrated together. Obviously, this is because of advantages that a three-phase system has over three separate single-phase systems. Moreover, all three systems are generated simultaneously and by one machine. This makes the generation process more efficient. Similarly, on the consumption side, because of the same reasons many industrial loads such as electric motors can be made three phase, which makes them more efficient and even better (see Chapter 11) than equivalent single-phase loads. Except for certain relationships, all that we have learned so far about single-phase AC electricity applies to a three-phase electricity system. In this chapter we study the extension from single phase to three phase and the differences in formulation when it applies.

9.2 Three-Phase Electricity

In Figure 9.1 a single load connected to a source constitutes a simple circuit, as we have seen in previous chapters. The only difference is that the

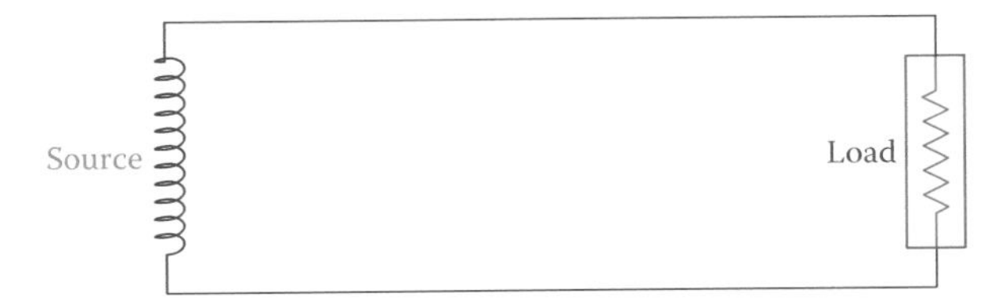

Figure 9.1 A single circuit consisting of a source and a load.

source is represented by a winding, because all generators have windings in them.

Figure 9.2 illustrates three of these assumed AC circuits near each other. Intentionally, the generators are shown forming a triangle, but at this time it is only a graphic layout. For ease of reference these systems are named A, B, and C. Supposing now that these three single-phase electric circuits are such that we can connect together the points that are near each other; that is, point A to B′, B to C′, and C to A′, as shown in Figure 9.3. If this can be done, then we may use only three wires for connection to the loads, instead of 6, provided that the loads permit this. This is shown in Figure 9.3, thus saving wire cost.

The arrangement on the right-hand side of Figure 9.3 can be rearranged to look as shown in Figure 9.4b, which is one of the ways a three-phase load can be connected to a three-phase source. The necessary condition to be able to connect the associated points in this circuit (A to B′, and so on) is that these points must have the same electric potential (the same voltage with respect to a reference) at any instant of time.

In Chapter 11 you learn about AC generators that work based on a magnetic field that rotates with respect to a winding. If there is one winding, the generated electricity is single phase. But, if there are three windings as shown in Figure 9.5, then the generated electricity is three phase. As

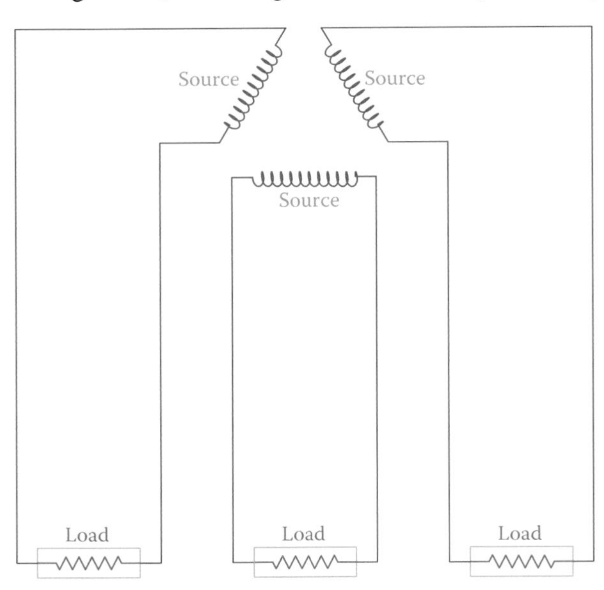

Figure 9.2 Three single-phase systems shown together.

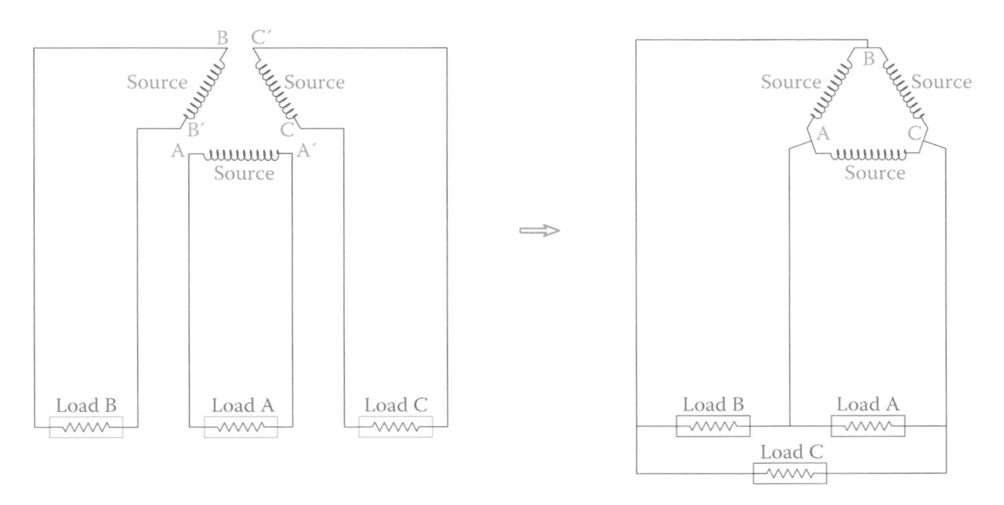

Figure 9.3 Saving in the number of wires used.

you see in Figure 9.5, the three windings are placed at 120° from each other. This causes a phase difference between the three voltages that are generated. The reason is obvious; one of the windings sees the magnetic field before the other two, and again the second winding sees it before the third, before everything is repeated. In other words, what happens to one winding in a given time happens to the other two windings with a delay that corresponds to 120° and 240° physical displacement. As a result of the rotation of the magnetic field, the generated voltage in each winding

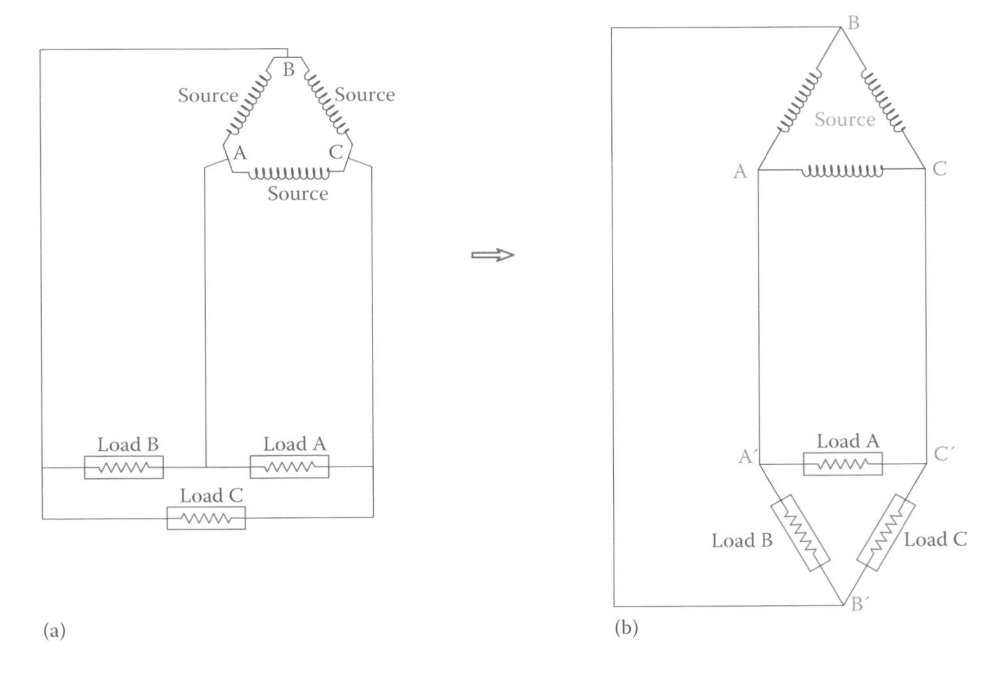

Figure 9.4 Common representation of a three-phase system. The three loads in (a) are shown as in (b).

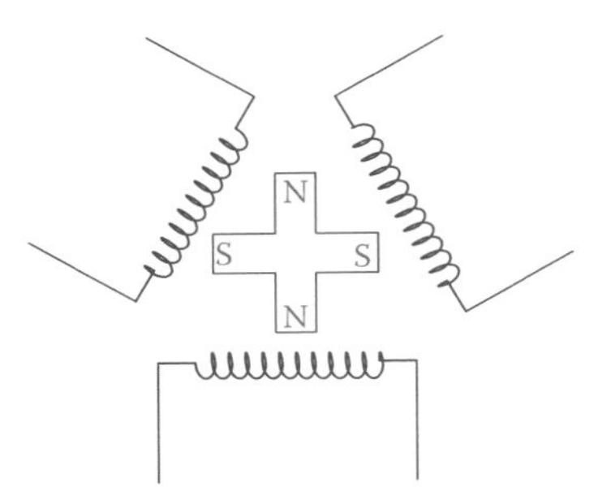

Figure 9.5 A simple representation of a three-phase generator, indicating the reason for phase difference between the voltages in the three windings.

has a sinusoidal form. For one cycle variation of the magnetic field the generated voltages in each winding are illustrated in Figure 9.6.

When put together, the three sinusoidal waveforms of a three-phase electricity system are shown in Figure 9.7. The three phases are addressed as phases A, B, and C, or phases 1, 2, and 3. No matter what the three

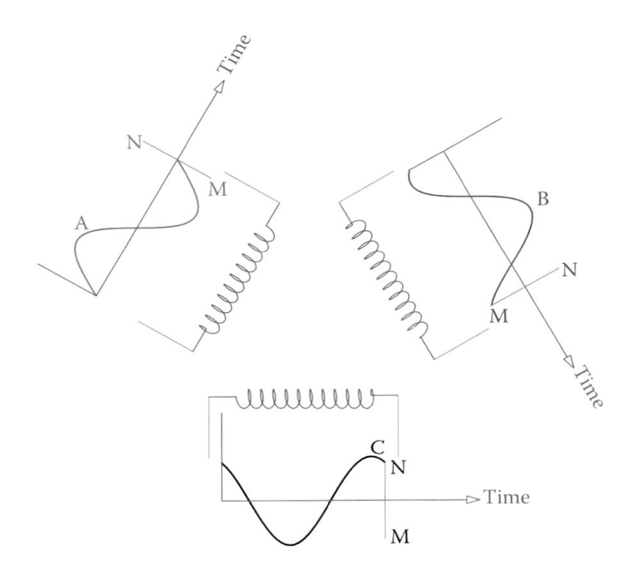

Figure 9.6 Individual waveform variation of the voltage in each winding for a one cycle period.

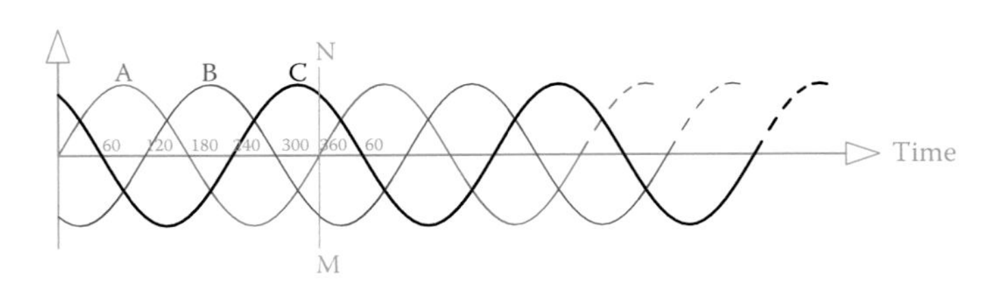

Figure 9.7 Waveforms of a three-phase system shown together on a common axis.

phases are called, their order is very important and must be respected. Observe that in Figure 9.7 phase A reaches its maximum before B and C. B reaches its maximum after 120°, and phase C reaches its maximum 120° after B (240° after A). Also, notice that the vertical axis has no title because it can represent voltage or current. In Figure 9.7, line MN corresponds to the time for one full cycle and the variations of each phase are those shown in Figure 9.6.

9.3 Properties of a Three-Phase System

The fact that the three waveforms of a three-phase system are 120° apart gives desirable properties to it that makes it attractive from a practical viewpoint. The very first property of the voltages in three-phase electricity is that at each instant of time the sum of all the voltages is zero. This can be mathematically shown, but here we can observe that from the graphics in Figure 9.8 for only a few points.

At any instant, such as those marked by lines 1, 2, 3, and 4, one can verify by measurement on the figure that the sum of the values of the voltages of the three phases is zero. For example, at the instant denoted by line 1, the value of phase B is zero and the other two phases A and C have equal values but with opposite signs (A is positive and C is negative). This is true for any other instant of time. If the voltages of the three phases are denoted by V_A, V_B, and V_C, then

$$V_A + V_B + V_C = 0 \quad (9.1)$$

The above fact leads to another way that the three voltage sources in a three-phase system can be connected together. This is shown in Figure 9.9. In this connection, one end of the three windings are connected together, and the voltage is taken from the other end. The common end can be used as a zero voltage reference point. This point is usually grounded, but there may also be a connection to this point, which is termed **neutral line** or **null line**. The neutral line goes to a similar point on the generation side.

Neutral line: Fourth line in a three-phase system, which ideally does not carry current and can be grounded; also called null line.

Null line: Same as neutral line in a three-phase system.

In a three-phase AC system at any instance the sum of the voltages in the three lines is zero. This is because these voltages are not in phase with each other.

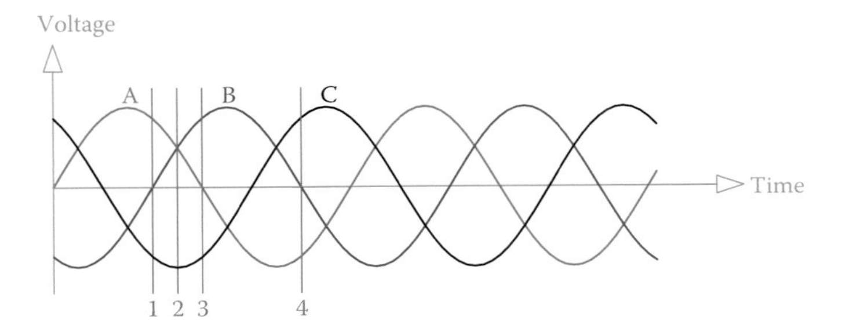

Figure 9.8 The sum of the voltages of the three phases are always zero.

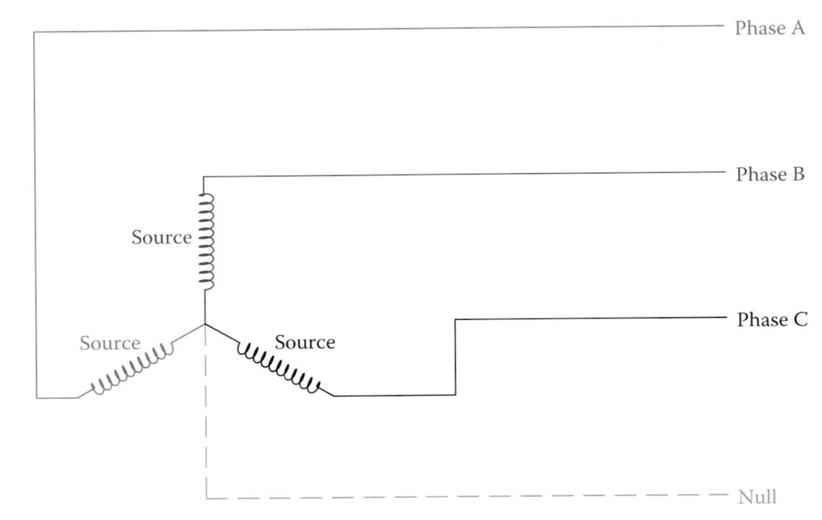

Figure 9.9 Alternative connection method in three-phase electricity.

Delta connection: Way of connecting loads and sources to a three-phase system, where the components of a load or a source form a triangle (or delta, Δ) each corner of which is connected to the three lines.

Δ connection: Same as delta connection.

Y connection: Same as wye connection.

Wye connection: A way of connecting loads and sources to a three-phase system, where the components of a load or a source form a letter Y (that is one side of each component are tied together) and the free sides of the Y are connected to the three lines.

Star connection: Same as wye connection. One of the two ways for connection of loads and sources in a three-phase system.

9.3.1 Wye Connection and Delta Connection

The former method of connecting the source voltages together is called **delta connection** (or **Δ connection**) because a Δ is formed. The latter form of connection is called **Y connection** (**wye connection**) or **star connection**. Note that in delta connection there are only three wires (there is no place for a fourth wire, but in Y-connection one can use three or four wires. The dashed (null) line in Figure 9.9 can be replaced by a short ground wire, which connects the common point to ground. It is also common practice to carry this wire while also grounding it, all depending on the case, as will be discussed in this chapter.

Note the difference between the voltages in the two cases. At this time the connection to the source is in focus. In a delta connection the voltage generated in the generator winding is between each of the two lines, whereas in the wye connection the voltage between each line and the null line is equal to that generated in the winding.

To distinguish the difference, the terms line voltage and phase voltage are employed. These definitions are depicted in Figure 9.10. When there

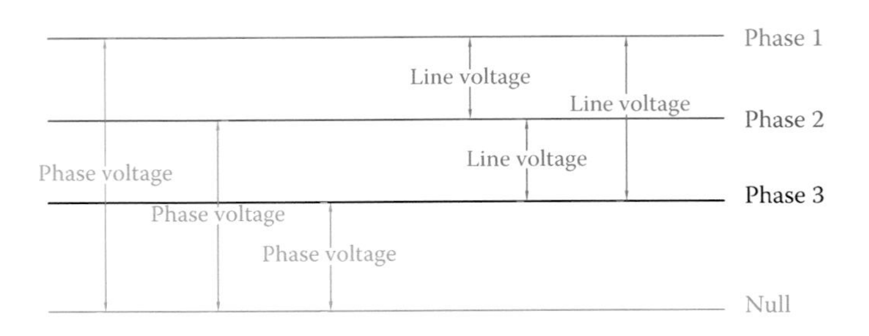

Figure 9.10 Definition of line voltage and phase voltage.

are three lines carrying three-phase electricity, voltage between each two lines is called line-to-line or simply **line voltage**. In the presence of a fourth line the voltage between each line and the common point (or the null line) is called **phase voltage**. This is irrespective of how the connection is at the source.

Line voltage: Voltage between each pair of (three) lines bringing electricity to the loads in a three-phase system.

Phase voltage: Voltage between each line and the common reference point (neutral, which is normally grounded) in a three-phase system.

9.3.2 Voltage Relationships

Another property of three-phase electricity is the relationship between line voltage and phase voltage. Unless the voltages are disturbed by various load conditions, line voltages for all three phases are the same. Also, all the three phases have the same phase voltages. Similar to the single-phase electricity, line voltage and phase voltage express the effective values of the voltages, unless mentioned otherwise. If line voltage is denoted by V_L and the phase voltage is represented by V_P, then

$$V_L = \sqrt{3}\, V_P \tag{9.2}$$

Example 9.1

Voltage across each winding of a generator when working is 120 V. Find the line voltage and the phase voltage for the circuit connected to this generator (1) when the generator windings are delta connected and (2) when the windings are wye connected.

Solution

1. Delta connection: Because the voltage across each winding is 120 V, then the line voltage is 120 V. The phase voltage thus can be found from Equation 9.2 as

$$V_P = \frac{V_L}{\sqrt{3}} = \frac{120}{1.73} = 69.28 \text{ V}$$

2. Y connection: In this case for each phase, voltage between the two ends of its winding is 120 V. Thus, the phase voltage is 120 V.

The line voltage is determined from Equation 9.2 as

$$V_L = \sqrt{3}\, V_P = (1.73)(120) = 207.85 \text{ V}$$

Note that, as this example shows, depending on how the wires are connected together at the source, the voltages can be different. This becomes very important when connecting a source and a load to a circuit. Often, the windings in a load are not initially connected. There are six terminals for either a three-phase source or a three-phase load (two terminals for the two ends of each winding). A technician must connect these together in the correct way for the desired connection.

In expressing the voltage in three-phase electricity, in order to make it clear that nobody doubts whether the given value is for the line voltage or the phase voltage, it is common to state both values. This is particularly done for standard voltage values. For example, in expressions like 120/208 V line and 220/380 line there is no doubt that the larger value is for the line voltage.

A three-phase source or load has six terminals. They must be connected together in the correct way to form delta connection, or wye connection, as desired.

9.4 Load Connection in Three-Phase Systems

As it was previously mentioned, a three-phase system is like three single-phase systems. In this sense the loads connected to three-phase electricity can be (and in many cases are, as we will see) independent single-phase loads. At this time, suppose that there are three similar loads connected to the three

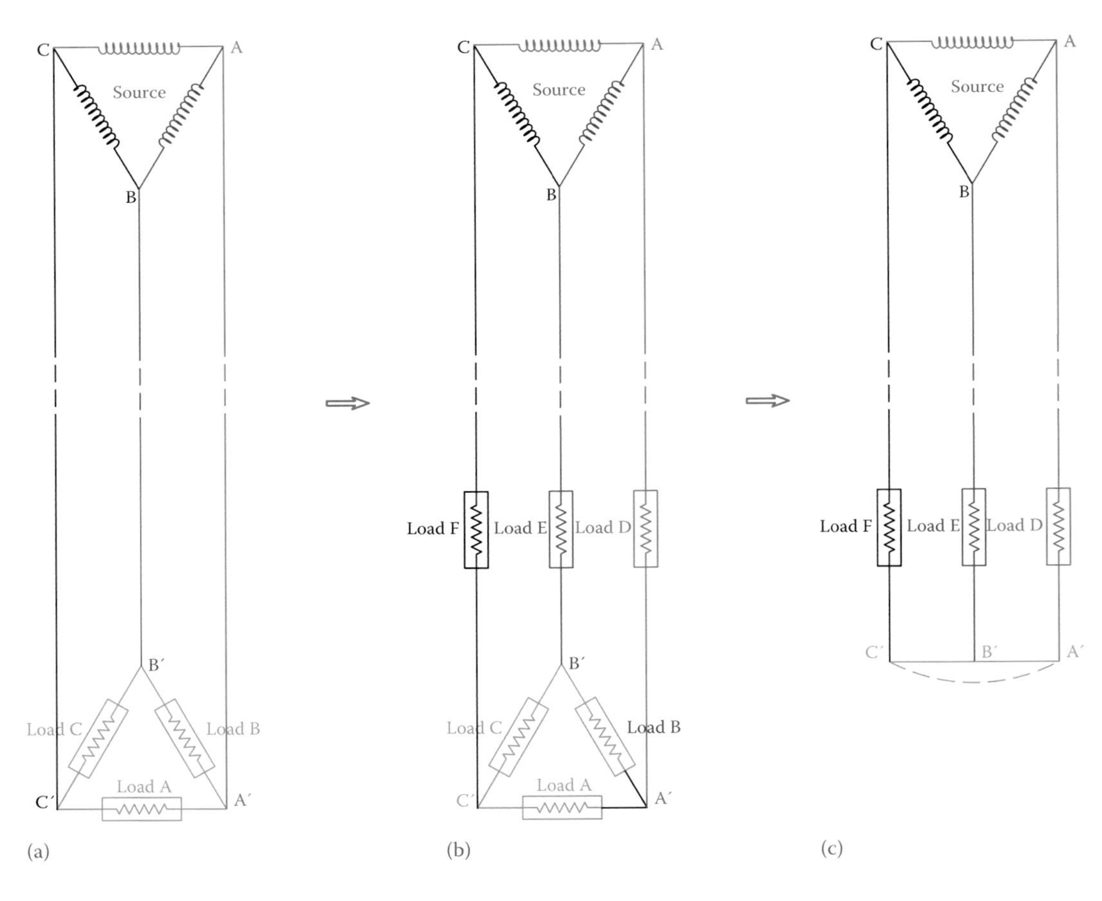

Figure 9.11 Connection of load to a three-phase system. (a) Three loads connected in delta form. (b) Three other loads can be added to the circuit. Then loads A, B, and C can have little or zero voltage across them. (c) Loads A, B, and C can be removed and points A′, B′, and C′ can be put together; thus, wye connection.

lines of a three-phase system, as shown in Figure 9.11a. Note that the source is shown with delta connection, but it could have a wye connection, too. The difference would have been in the line voltage received by the loads.

Current flows from the source to the loads and while the instantaneous direction and magnitudes of the currents in the three lines are not the same, their effective magnitudes are equal, because the loads are assumed to be similar.

Now suppose that three other similar loads are added to this circuit, as illustrated in Figure 9.11b. These new loads reduce the current in the three lines, thus reducing the voltages at points A′, B′, and C′. If these loads are large enough, then the voltage values at points A′, B′, and C′ can be so small that without overloading the circuit we may completely remove the three former loads and just connect these three points together, as shown in Figure 9.11c. The result, as can be observed in Figure 9.11c, is that the loads of a three-phase system can be connected by employing either delta connection, or wye connection. This is particularly true for three-phase loads that consist of three simultaneous similar sets of single loads.

On the basis of the above discussion, because connection at the source can be either delta or wye, and at the load can also be either delta or wye, there are four possible ways of connecting sources and loads in a three-phase circuit. These are depicted in Figures 9.12 and 9.13.

> In delta connection the ends of each (load or source) element are connected together, such that they form a triangle. Then the three corners of the triangle are connected to the three electricity lines.

> In wye connection one end of all the (load or source) elements are connected together; this point may be grounded. The other ends are connected to the electricity lines. A letter Y is formed by the elements.

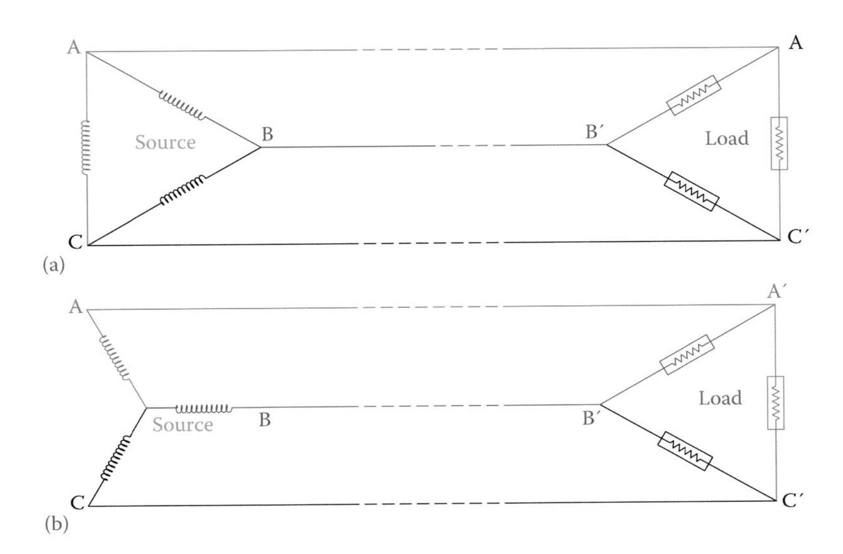

Figure 9.12 (a) Source and load are both delta connected and (b) source is wye connected and load is delta connected.

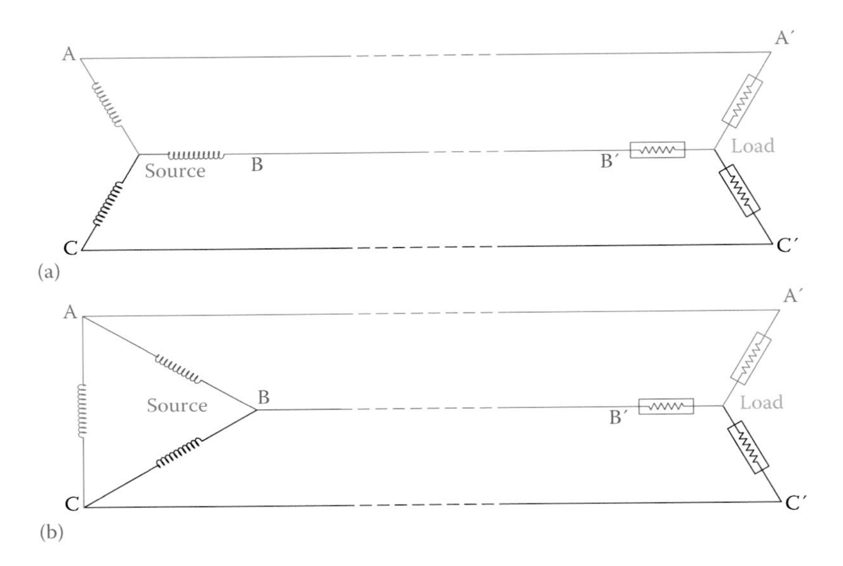

Figure 9.13 (a) Source and load are both wye connected and (b) source is delta connected and load is wye connected.

9.4.1 Balanced Loads and Unbalanced Loads

Similar to what we have already seen in Chapter 8 for the type of loads in single-phase electricity, the loads connected to a three-phase source can contain resistive, inductive, and capacitive components. No matter whichever component is present and if they are in series or parallel, if all the three loads connected to a three-phase circuit are exactly similar in arrangement and magnitudes, then the load is said to be balanced and to the three-phase circuit they form a **balanced load**.

Balanced load: Three identical loads in a three-phase system that come together as a three-phase load.

With the exception of three-phase loads (see Section 9.5) it is extremely difficult if not impossible to arrange all the independent loads connected to a three-phase circuit to be exactly identical. For example, in the electricity distribution in cities the generated power is three-phase; however, the electricity coming to individual houses is single phase. Each group of houses is fed by one phase of the three-phase electricity. Because the electricity consumption of each house is independent of others, there is no way to have a balanced load.

For a balanced load the currents in the three lines are exactly equal (the three lines have the same current magnitude and power factor). Moreover, if there is a 4th line, the current in this line is zero. This is not the case for **unbalanced loads**.

Unbalanced load: When in a three-phase system, the load consists of three separate and not identical single-phase devices.

When connecting loads to a three-phase circuit, normally, we are not concerned of how the connection at source is (delta or wye). The concern is about the line voltage and phase voltage that determine the appropriate connection for a load. More often, there are three lines that constitute the three phases of a three-phase circuit. The voltage between each pair is the line voltage. A load is connected either in delta or in wye, as depicted in Figure 9.14. If a wye connection is employed, the common point is usually grounded. This helps for protection and fault detection.

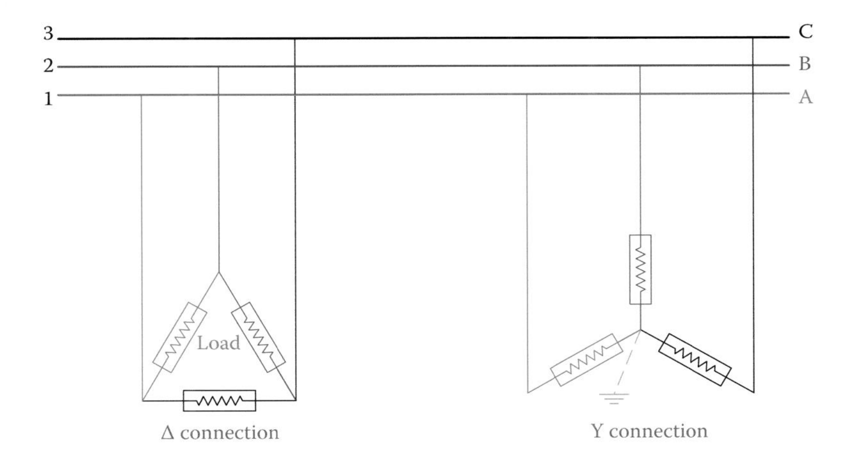

Figure 9.14 Delta and wye connection of a load to a three-phase circuit.

The voltage across each individual load is the phase voltage. If a wye connection is employed, phase voltage implies the voltage between each line and the common (grounded) point, but for a delta connection the phase voltage and the line voltage are the same. The definitions for line and phase voltages are also shown in Figure 9.15.

Example 9.2

If the line voltage in Figure 9.14 is 200 V, what is the voltage applied to each load element (1) when delta connection is used and (2) when the loads are wye connected to the line?

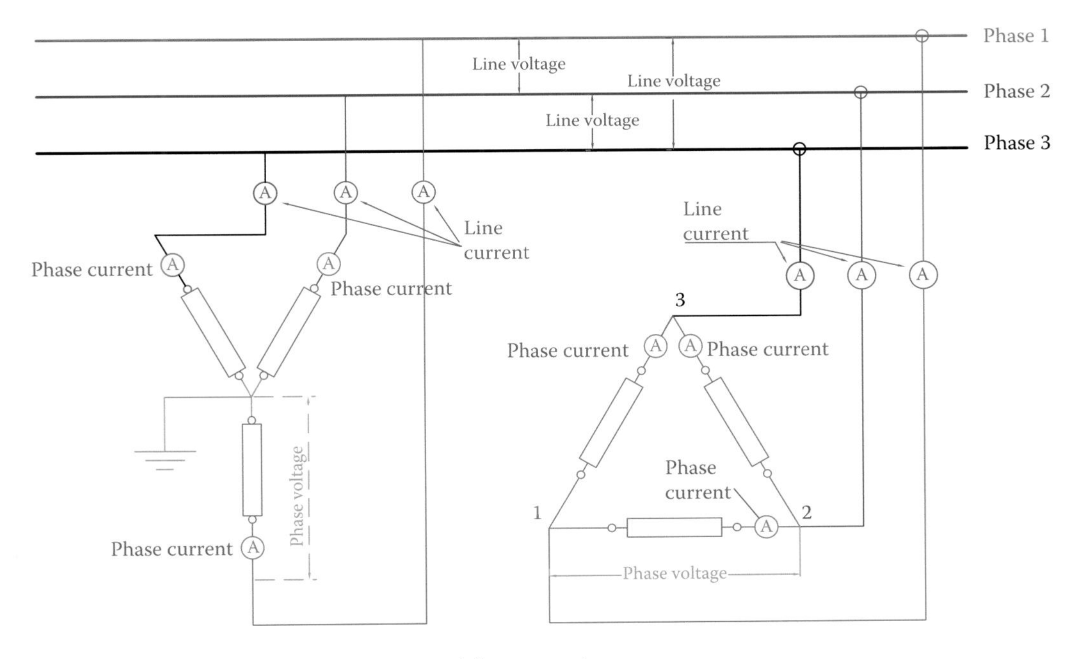

Figure 9.15 Definition of phase current and line current.

Solution

1. In this case the whole line voltage is applied to each load element. Thus, the voltage across the element is 200 V.
2. In this case the 200 V line voltage is distributed between two elements. The relationship in Equation 9.2 again applies, and hence the voltage across each element is

$$V_P = \frac{V_L}{\sqrt{3}} = \frac{200}{1.73} = 115.5 \text{ V}$$

A three-phase load is a balanced load.

9.4.2 Line Current and Phase Current

Phase current: Current inside each electric component connected to a three-phase system, as opposed to the line current representing the current in the lines carrying electricity to that component.

Line current (in three-phase systems): Current in lines bringing electricity to the loads in a three-phase system.

In conjunction with phase and line voltage terms, we also have **phase current** and **line current**. Line current is the current that one can measure in each of the phase lines to which loads are connected. Phase current is the current passing through the load. For two sets of loads, one with delta connection and one with wye connection, these currents are illustrated, also, in Figure 9.15, where the circle with a letter "A" inside implies an ammeter.

9.5 Three-Phase Loads

A three-phase load is a set of three exactly similar combinations of electric components arranged in exactly the same way (in series, parallel or other). The reason to have three-phase loads is the higher power demand. For instance, in industry the majority of motors are three phase, particularly the larger motors that deliver large magnitudes of power to conveyors, pumps, blowers, machine tools, and so on. As studied before, the three components in AC are resistors, inductors, and capacitors. In this sense, the three-phase loads can be purely resistive (consisting of heating elements only), resistive and inductive (like electric motors), and resistive, inductive, and capacitive (like many motors that are accompanied by capacitors for various reasons).

Because the three separate sets of loads are identical and are simultaneously connected to electricity, they are balanced. Thus, all three-phase loads automatically form a balanced load for a three-phase circuit.

Dealing with three-phase loads is much simpler than if independent single-phase loads are connected to a three-phase circuit. In fact, many of the relationships and calculations for three-phase loads resemble their counterparts for single-phase systems. This is because the individual parts of the loads on the three phases behave alike. For instance, in a load consisting of resistors and inductors, as we have seen before, there is a phase difference between the current and the voltage (see Chapter 8). For a three-phase load this phase difference is the same for all the three phases, and, as a result, we need to find it for only one phase.

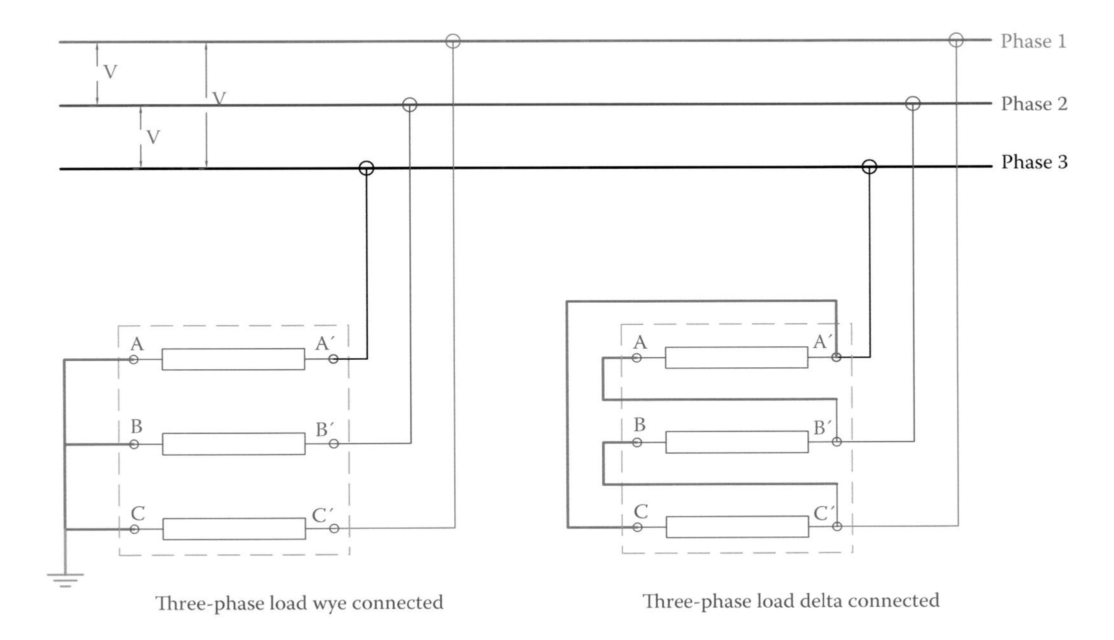

Figure 9.16 Wye and delta connection of the same load.

Analyzing three-phase loads is tackling the same sort of problems that we have seen in Chapter 8. Therefore, we need to know the relationships between voltage, impedance, current, power, and power factor, as well as power factor correction. Again, here we have three types of power: active power, reactive power, and apparent power. We need, though, to make a distinction between delta connection and wye connection. A three-phase load can initially be open, with six terminals to deal with. On the basis of how these terminals are connected together and to the external lines they form a wye or a delta connection. Figure 9.16 shows how the 6 terminals A, B, C, A′, B′, and C′ can be connected for wye and delta connection. The two methods of connection are not equivalent and affect the current and power taken from a circuit (see Section 9.5.1).

Dealing with unbalanced loads is more difficult, and each branch must be individually analyzed and the results be put together.

9.5.1 Three-Phase Relationships for Balanced Loads

Referring to Figure 9.15, which depicts the definition of line and phase currents as well as line and phase voltages for both delta and wye connection of a load, the following relationships always exist:

For wye-connected load,

$$\text{Line current} = I_{\text{L}} = \text{phase current} = I_{\text{Ph}} \tag{9.3}$$

$$\text{Line voltage} = V_{\text{L}} = \sqrt{3}\ \text{phase voltage} = \sqrt{3}\ V_{\text{Ph}} \tag{9.4}$$

For delta-connected load,

$$V_L = V_{Ph} \tag{9.5}$$

$$I_L = \sqrt{3}\, I_{Ph} \tag{9.6}$$

Current in a component can always be found by Ohm's law. If the impedance of each of the three elements of a three-phase load is denoted by Z, then

$$\text{Phase current} = I_{Ph} = V_{Ph} \div Z \tag{9.7}$$

Equations 9.3 to 9.7 can determine all values if two of them are known.

Note that for a balanced load all the phase currents are equal and all the line currents are also equal. Also, referring to Figure 9.17, although the line currents are equal in magnitude, they cannot be in phase with each other. As shown, for both the delta-connected load and the wye-connected load, at a given instant while the currents on two phases are toward the load, the current on the third phase is in the opposite direction (remember that in AC electricity the current direction continuously changes; here the direction for current implies that if at a given instant its magnitude is positive or negative). Because the AC current continuously changes direction, those shown in Figure 9.17 are momentary values and for one instant later the directions will be different. We want to say that for a balanced load the three currents have the same relationships as the voltages have. That is, for sinusoidal voltage the variation of currents is sinusoidal and with a phase difference, as shown in Figure 9.7, and moreover, their instantaneous sum is zero.

$$I_1 + I_2 + I_3 = 0 \tag{9.8}$$

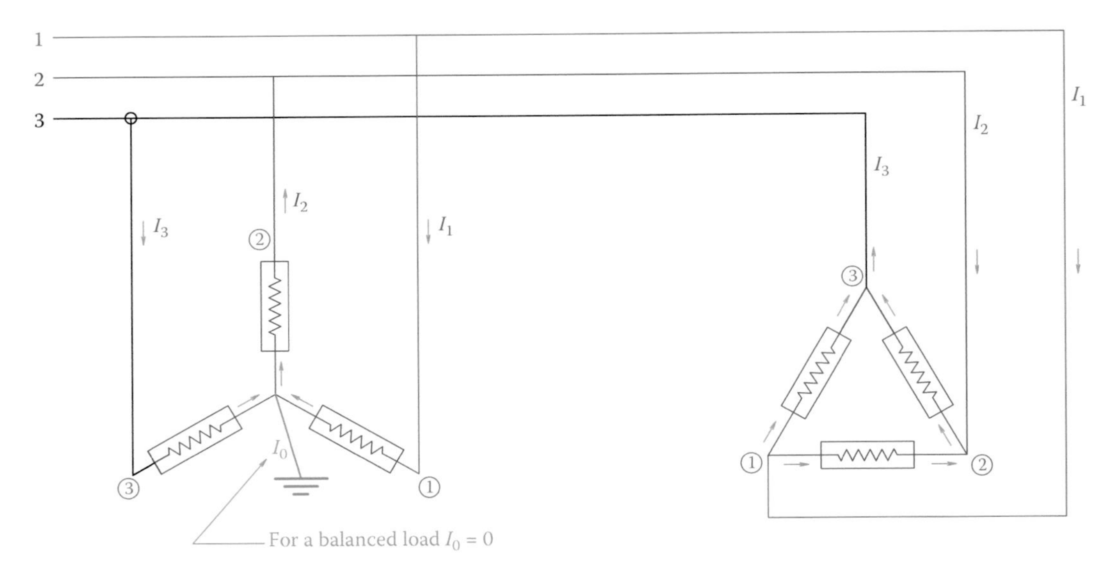

Figure 9.17 Instantaneous direction of currents in three-phase systems.

Equation 9.8 implies that at any instant the sum of the currents in the three lines of a three-phase system is zero for balanced loads. In this sense, current I_0, associated with the grounded wire or the neutral line (when it is present) is zero. This is the reason why the neutral line is smaller in size than the three phase lines.

For wye connection,

$$\text{Line current} = \text{phase current}$$

$$\text{Line voltage} = \sqrt{3}\ \text{Phase voltage}$$

For delta connection,

$$\text{Line voltage} = \text{Phase voltage}$$

$$\text{Line current} = \sqrt{3}\ \text{Phase current}$$

Example 9.3

A three-phase heater consists of three elements each having 20 Ω resistance. (1) If the heater is connected to 220/380 V line (that is, line voltage is 380 V) using delta connection, what is the current in each element? (2) What is the current in the lines bringing in the three-phase electricity? (3) Also, find the heating power of the device in kW.

Solution

1. When delta-connected, each element is subject to the line voltage of 380 V, as the line voltage and phase voltage are the same in this case. Since we are dealing with heating elements, all the loads are resistors. Thus,

$$I_{Ph} = \frac{380}{20} = 19\ \text{A}$$

2. The line current based on Equation 9.6 is

$$I_L = \sqrt{3}\ I_{Ph} = \left(\sqrt{3}\right)(19) = 32.9\ \text{A}$$

3. The heating power depends on the power consumption of the resistive elements. This can be found by any of the power formulae in Chapter 7. Because there are three elements, the power in one can be multiplied by 3.

$$\text{Power} = (3)(380)(19) = 21{,}660\ \text{W}$$

Example 9.4

Compare the line current and the power for the heater in Example 9.3 if it is wye connected to the same three-phase electricity.

Solution

In this case it follows from Equation 9.4 that

$$\text{Phase voltage} = \frac{1}{\sqrt{3}}V_L = \frac{380}{1.73} = 219.4 \text{ V}$$

$$I_{Ph} = \frac{219.4}{20} = 10.97 \text{ A} \approx 11 \text{ A}$$

And from Equation 9.3,

$$I_L = I_{Ph} \approx 11 \text{ A}$$

$$\text{Power} = (3)(219.4)(11) = 7220.23 \text{ W}$$

This example clearly shows that in this case the power is 3 times smaller than in the previous case.

Example 9.5

If in the problem of Example 9.3 one of the phases is disconnected by a fault, determine the currents in the other two lines and what the power is.

Solution

Refer to Figure 9.15 with the delta connection and suppose that one of the lines does not exist. In such a case we have two parallel single-phase loads connected to 380 V electricity. One of the parallel branches has a 20 Ω resistor, and the other one has two 20 Ω resistors in series (40 Ω). All the loads are resistive, and the current in the connecting lines is

$$I = \frac{380}{20} + \frac{380}{40} = 19 + 9.5 = 28.5 \text{ A}$$

The power is found by adding the individual powers in the two parallel branches

$$P = (20)(19)^2 + (40)(9.5)^2 = 10{,}830 \text{ W}$$

which also could be found from

$$P = (380)(28.5) = 10{,}830 \text{ W}$$

Example 9.6

If in the problem of Example 9.4 (wye-connected load) one of the wires is broken by an incident, determine the current in the remaining lines and the power supplied.

Solution

In the same manner that Example 9.5 was tackled, if one line is disconnected, we study the circuit of the remaining single-phase system. For this example the remaining circuit consists of 40 Ω resistor connected to 380 V. The current and power, therefore, are

$$I = \frac{380}{40} = 9.5 \text{ A}$$

$$P = (380)(9.5) = 3610 \text{ W}$$

Example 9.7

Figure 9.18 shows three identical coils that in series with three resistors are connected to a three-phase system. The line is 220/380 V and the frequency is 60 Hz. The resistors are 40 Ω each. The resistance of each coil is 4 Ω, and its inductance is 100 mH. Determine the current in the line and the power factor for this load.

Solution

Figure 9.18 shows that the load is wired for delta connection. Each coil is equivalent to a resistor and an inductor in series. The resistance of each individual load, thus, is 44 Ω. For each phase,

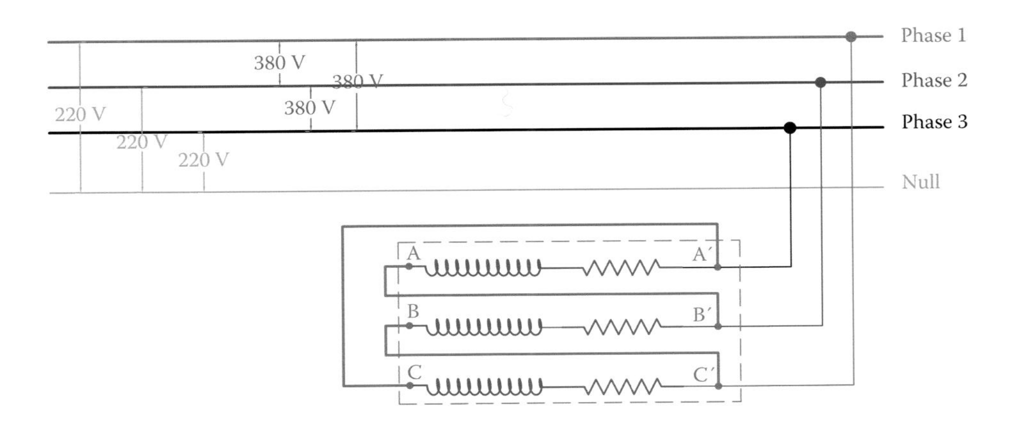

Figure 9.18 Circuit of Example 9.7.

$$X_L = (2)(\pi)(60)(0.100) = 37.7\ \Omega$$

$$Z = \sqrt{44^2 + 37.7^2} = 57.94\ \Omega$$

$$I_{Ph} = \frac{380}{57.94} = 6.56\ \text{A}$$

$$I_L = \sqrt{3}(6.56) = 11.36\ \text{A}$$

The power factor can be found from the ratio $\frac{R}{Z}$ as

$$pf = \frac{44}{57.94} = 0.76 \rightarrow 76\%$$

This power factor is the same for all loads, and therefore it represents the power factor of the three-phase load.

9.5.2 Power in Three-Phase Electricity

The relationships for power for three-phase systems must be separately stated. There are two aspects of power relationships, one is with regards to the three types of power (active, reactive, and apparent), and the other is with regards to the fact that we have phase voltage and phase current versus line voltage and line current. The relationships between active, reactive, and apparent power are the same that we have seen before (see Chapter 8), that is

$$pf = \cos\ (\text{phase angle}) \tag{9.9}$$

$$\text{Active power} = P = \text{Apparent power} \times \text{Power factor} = S \times pf \tag{9.10}$$

$$\text{Reactive power} = Q = \text{Apparent power} \times \sin(\text{phase angle})$$

$$= S \times \sqrt{1 - pf^2} \tag{9.11}$$

But, then the expression for apparent power S for both delta connection and wye connection of a load is

$$S = \sqrt{3} \times \text{Line voltage} \times \text{Line current} \tag{9.12}$$

or

$$S = 3 \times \text{Phase voltage} \times \text{Phase current} \tag{9.13}$$

The two expressions in Equations 9.12 and 9.13 are equivalent because in delta connection, where the phase and line voltage are the same, the line current is $\sqrt{3}$ times the phase current and in wye connection that the line current and the phase current are the same the line voltage is $\sqrt{3}$ times the phase voltage.

Equations 9.12 and 9.13 must be used together with Equations 9.10 and 9.11 in order to avoid any confusion when a load is entirely resistive or entirely reactive. When a load is entirely resistive, the power factor is 1 and the apparent power and active power are the same. Likewise, when a load is entirely reactive, for instance, when a set of capacitors are connected to a line, the power factor is zero and the apparent power and the reactive power are the same.

Example 9.8

For the load in Example 9.7, determine the active, reactive, and apparent powers.

Solution

As shown in Figure 9.18, the line voltage is 380 V, which is also the phase voltage for the load, because it is delta connected. We find the apparent power by both Equations 9.12 and 9.13 in order to show that both must lead to the same answer. The line and phase currents have been found in Example 9.7 to be 11.36 and 6.56 A, respectively.

$$S = \left(\sqrt{3}\right)(380)(11.36) = 7477 \text{ VA}$$

$$S = (3)(380)(6.56) = 7478 \text{ VA}$$

Note that the difference is because of the round-off error. Active and reactive powers are then found from Equations 9.10 and 9.11.

$$P = S \times pf = (7477)(0.76) = 5682 \text{ W}$$

$$Q = (7477)(\sqrt{1 - 0.76^2}) = 4860 \text{ VAR}$$

Example 9.9

A small three-phase generator provides power to a three-phase oven. The resistors in the oven are 100 Ω each. The line voltage is 173 V. If the windings in the generator are delta connected, find the current in the generator winding (1) when the heating elements are connected in delta and (2) when they are connected in wye.

Solution

The arrangement is shown in Figure 9.19. This problem is simplified in order to demonstrate the relationships between the values of the currents due to altering the way a load is connected to a source.

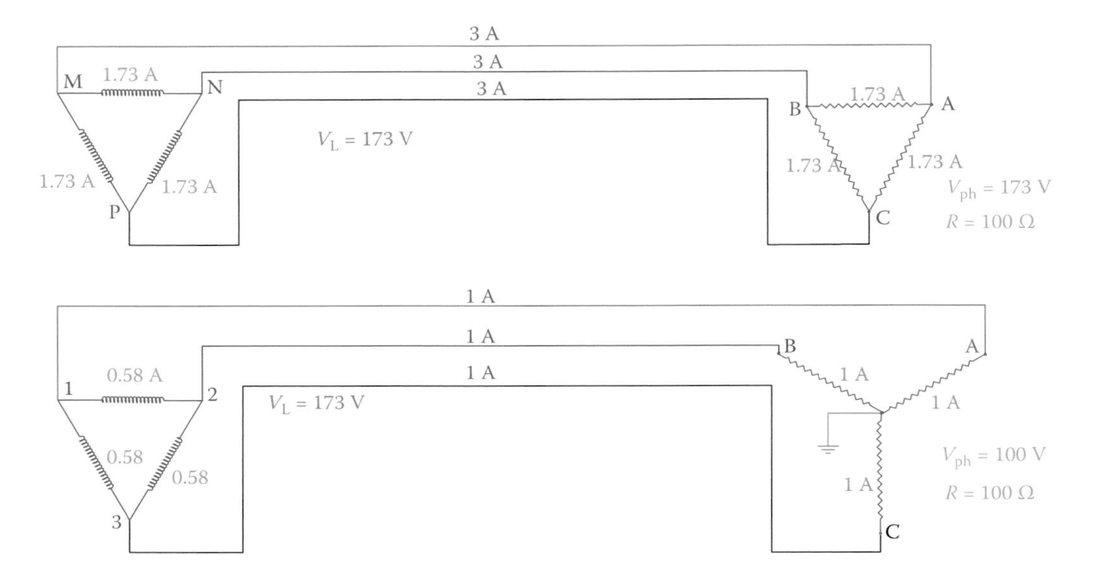

Figure 9.19 Circuit of Example 9.9.

1. When both the load and the source are connected in the same way, the currents in the source winding are the same as the current in the load branches. For this example,

$$I_{Ph} = \frac{173}{100} = 1.73 \text{ A}$$

$$I_L = \sqrt{3}\, I_{Ph} = (1.73)(1.73) = 3 \text{ A}$$

$$\text{Power} = (\sqrt{3})(173)(3) = 900 \text{ W} \quad \text{(using Equation 9.12)}$$

On the generator side

$$I_{winding} = I_{Ph} = \frac{3}{\sqrt{3}} = 1.73 \text{ A}$$

2. If the load is wye connected,

$$I_{Ph} = I_L = \frac{V_{Ph}}{R} = \frac{173/\sqrt{3}}{100} = 1 \text{ A}$$

$$\text{Power} = (\sqrt{3})(173)(1) = 300 \text{ W (using Equation 9.12)}$$

and on the generator side

$$I_{winding} = I_{Ph} = \frac{1}{\sqrt{3}} = 0.58 \text{ A}$$

All the results are shown in Figure 9.19.

9.5.3 Power Factor Correction for Three-Phase Circuits

In the same way that for single-phase electricity the power factor can be improved by adding appropriate capacitive load in parallel with other loads, in three-phase electricity a bank of capacitors can be added for power factor correction. Like any other three-phase load, the capacitors can be wye connected or delta connected, as shown in Figure 9.20. The principle of determining the capacitor size and the procedure is the same as discussed in Chapter 8 (see Section 8.14). Because this task is performed for balanced loads, the necessary calculations can be based on the information for one phase.

In what follows we see examples of power factor improvement, which depict the details of the necessary work.

Example 9.10

A three-phase load takes 10 A if delta connected to 120/208 V, 60 Hz electricity. If the power factor for this load is 66 percent, find the size of a set of capacitors, delta connected, to improve the power factor to 95 percent.

Solution

The line current is 10 A in each of the three wires. Because both the load and the capacitors are delta connected, there is a one-to-one relationship between the elements. We may directly work with the given current.

A part of the 10 A current goes toward the active power and part corresponds to the reactive power:

Current corresponding to the active power = (10)(0.66) = 6.6 A

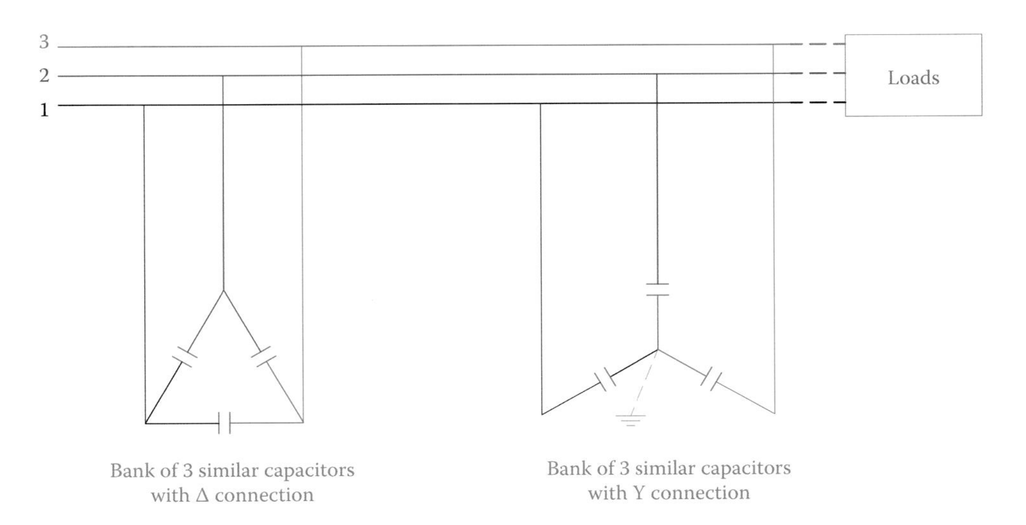

Figure 9.20 Connection of capacitors for power factor correction.

Current corresponding to the reactive power $= (10)\sqrt{1-0.66^2}$
$= (10)(0.75) = 7.5$ A

When capacitors are added, the part of the current for the active power does not change and is still 6.6 A. But the part corresponding to the reactive component(s) changes and so does the total current (not 10 A anymore, but smaller). We need to find the difference between the two values of the old and new current for reactive power. That is the current for the capacitor(s).

New line current after capacitors are added $= \dfrac{6.6}{0.95} = 6.95$ A

New current corresponding to the reactive load $=$
$(6.95)\,(\sqrt{1-0.95^2}) = (6.95)(0.31) = 2.15$ A

Difference between the old current and new current for reactive power = 7.5 – 2.15 = 5.35 A

Current in the line due to the capacitors = 5.35 A

On the basis of the voltage, the current due to a capacitor, and the frequency we may find the capacitance.

$$X_C = \frac{208}{5.35} = 38.9\ \Omega$$

$$C = \frac{1}{(2)(3.14)(60)(38.9)} = 68\times 10^{-6}\text{F} = 68\ \mu\text{F}$$

Note that not any 68 μF capacitor can be employed. The capacitors must be rated for more than 208 V, say 250 V.

Example 9.11

The power factor of a 12 kW three-phase load is 0.75 inductive (or lagging). Find the capacitance of a set of wye-connected capacitors to raise the power factor to 0.92. The line voltage and frequency are 400 V and 50 Hz, respectively.

Solution

In this problem, because we have the active power (kW), we base all the calculations on the power values. With a power factor of 0.75, the apparent and reactive powers are

$$S = \frac{12}{0.75} = 16\ \text{VA}$$

$$Q = (16)(\sqrt{1-075^2}) = 10.58\ \text{kVAR}$$

After the power factor is improved, the new apparent power is

$$S_{\text{New}} = \frac{12}{0.92} = 13\ \text{kVA}$$

and the new reactive power is

$$Q_{New} = (13)(\sqrt{1 - 0.92^2}) = 5.11 \text{ kVAR}$$

The difference between the new reactive power and the previous value is due to the reactive power of the capacitors. Thus, the size of the capacitors can be found from the relationship for capacitor power and three-phase power (see Section 8.7).

$$Q_{Cap} = 10.58 - 5.11 = 5.47 \text{ kVAR}$$

Because the capacitors are to be wye connected, the voltage across each capacitor is $\frac{400}{\sqrt{3}}$ V. This voltage gives rise to a phase current (also line current) of $\frac{400}{\sqrt{3}X_C}$, where X_C is the reactance for each capacitor. It follows from Equation 9.12 that

$$\text{Capacitor power} = Q_{Cap} = \left(\sqrt{3}\right)(400)\left(\frac{400}{\sqrt{3}X_C}\right) = 5.47 \text{ kVAR}$$

from which X_C can be determined. Thus,

$$X_C = \frac{400^2}{5470} = 29.25\ \Omega$$

and the capacitor size is

$$C = \frac{1}{(2)(3.14)(50)(29.25)} = 108.8 \times 10^{-6} \text{F} = 108.8\ \mu\text{F}$$

Capacitors with the above capacitance may not exist. In such a case the nearest standard value capacitor that is available must be used or it might be arranged by combination; the selected capacitor must have the rating for 231 V ($400/\sqrt{3}$). An available choice is a 120 μF, 450 V capacitor.

Example 9.12

A three-phase 12 kW motor connected to 400 V and 50 Hz line has a power factor of 0.75. Find the new power factor if a bank of three 100 μF capacitors is connected to the same line (1) when the capacitors are delta connected and (2) when they are wye connected.

Solution

A motor always introduces a lagging power factor in an AC circuit. When capacitors are connected to the same line, they introduce capacitive reactive power to the line. The effect of capacitors is to reduce/cancel the inductive effect of the motors. For this problem we find the (reactive) power introduced to the line and subtract it from the initial reactive power. The difference defines the new reactive power for the circuit.

1. Delta connection

For 50 Hz each capacitor reactance is

$$X_C = \frac{1}{(2)(\pi)(50)(0.000100)} = 31.8\ \Omega$$

Find power of the capacitor bank:

$$I_{Ph} = \frac{V_{Ph}}{X_C} = \frac{400}{31.8} = 12.5\ \text{A}$$

$$I_L = \left(\sqrt{3}\right)(12.5) = 21.65\ \text{A}$$

$$Q_{Cap} = \left(\sqrt{3}\right)(400)(21.65) = 15{,}000\ \text{VAR}$$

Find the apparent power of the circuit before capacitors are entered:

$$\text{Apparent power} = S = \frac{\text{Actual power}}{pf} = \frac{12{,}000}{0.75} = 16{,}000\ \text{VA}$$

(from Equation 9.10)

$$\text{Reactive power} = Q = S\sqrt{1 - pf^2} = (16{,}000)\sqrt{1 - 0.75^2} = 10{,}583\ \text{VAR}$$

New reactive power (after capacitors are added) = 10,583 − 15,000 = −4417 VAR

$$\text{New apparent power} = \sqrt{12{,}000^2 + (-4417)^2} = 12{,}783\ \text{VA}$$

New power factor (for delta connection) = $\frac{12{,}000}{12{,}783} = 0.93$ (leading power factor)

2. Wye connection

The same procedure must be repeated for wye connection. Certain variables do not change, and the calculation does not need repetition. Thus,

$X_C = 31.8\ \Omega$, Initial reactive power = 10,583 VAR.

Find power of the capacitor bank (for wye connection):

$$I_{Ph} = \frac{V_{Ph}}{X_C} = \frac{400/\sqrt{3}}{31.8} = 7.3\ \text{A}$$

$$I_L = I_{Ph} = 7.3\ \text{A}$$

$$Q_{Cap} = \left(\sqrt{3}\right)(400)(7.3) = 5000\ \text{VAR}$$

(Note the important fact that the power this time is 3 times smaller than when the capacitors were delta connected.)

Find the new reactive power of the circuit after capacitors are added:

$$\text{New reactive power} = 10{,}583 - 5000 = 5583 \text{ VAR}$$

$$\text{New apparent power} = \sqrt{12{,}000^2 + (5583)^2} = 13{,}235 \text{ VA}$$

$$\text{New power factor (for wye connection)} = \frac{12{,}000}{13{,}235} = 0.91$$

Pay attention to the fact that, though the values of the power factor in the two cases are very near, there is a big difference between the two cases. When the capacitors are delta connected, the resulting circuit is capacitive and the current is leading the voltage, but when they are wye connected, the circuit is still inductive and the current is lagging the voltage. In the first case if a larger capacitor is selected, it has an inverse effect of reducing the power factor, whereas in the second case a larger capacitor improves the power factor further.

9.5.4 Star-Delta Switch

The above example shows a number of facts that in practical applications use is made of them. These are

1. For the same load, the current in the line and the power consumption are smaller (one third, to be more specific) if the load is connected in wye rather than delta.
2. By the same token, a generator can deliver more power if the windings are delta connected.

For the above reasons, in order to reduce the initial current of certain applications, particularly three-phase electric motors that have a high **inrush current**, a special type of switch is used that connects a three-phase load to electricity in two stages. In the first stage a wye connection is made by a first turn of the switch, and with a further turn the connection is changed to delta. This arrangement is schematically illustrated in Figure 9.21. First, points R, S, and T, which are the free ends of the load (winding) are connected to A′, B′, and C′. This makes a wye connection, but if R, S, and T are connected to A, B, and C (i.e., what is done after a further turning of the switch), then the load is delta connected to the three-phase line.

Inrush current: Relatively high current that a motor initially experiences when connected to electricity (at zero speed). Current decreases as the motor speeds up.

A star-delta switch, however, is a mechanical system, and it belongs to the technology of the past. Nowadays, this technology is replaced with electronic switching; instead, for starting motors a variable frequency drive (VFD) is employed, which also offers other advantages. A variable frequency drive is based on AC-to-DC and DC-to-AC conversion, discussed in Chapter 20, which can also be regarded as a frequency converter. With such a device the frequency of AC electricity can be varied. Hence, as we discuss in Chapter 11, the rotational speed of a motor, which

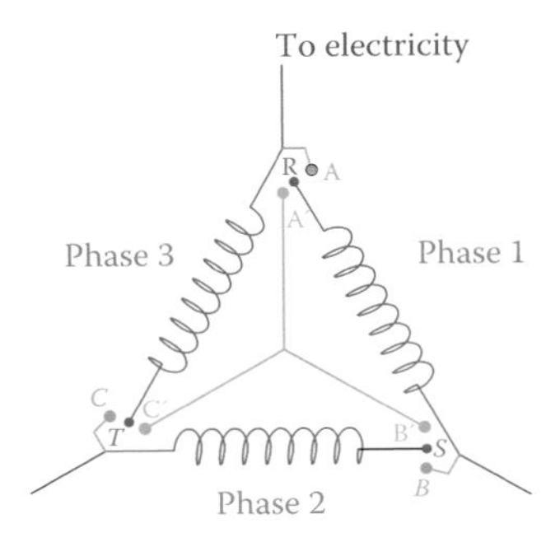

Figure 9.21 Schematic of a star-delta switch for starting three-phase motors.

depends on the line frequency, can be altered. Furthermore, because this is an electronic device, it can be used for soft starting of a motor by limiting the current. Thus, it provides the same advantage as the star-delta switch. VFDs come in various capacities (e.g., for driving a 0.5 hp motor or a 500 hp motor). They must be capable of handling the required power by a motor; hence, their physical size depends on the power that they can provide to a motor. A very small example of such a device is shown in Figure 9.22.

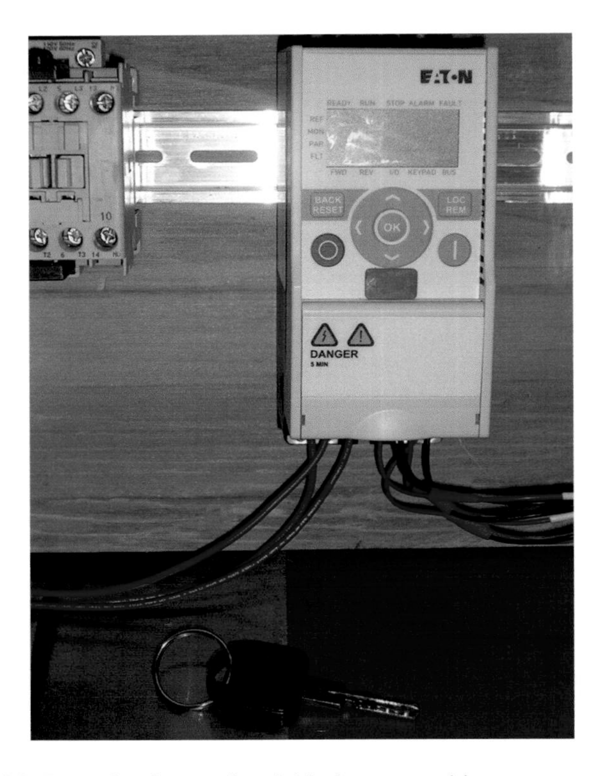

Figure 9.22 Example of a small variable frequency drive.

9.6 Use of Vectors in Three-Phase Electricity

We have already seen that using vectors could help tackle AC problems, displaying phase relationships, and make it easier. In the same fashion, in three-phase systems, using vectors greatly facilitates finding line currents and the phase angles for different loads. Moreover, the relationship between the line voltage (current) and phase voltage (current) and why one is $\sqrt{3}$ times the other are more clearly demonstrated by the use of vectors.

Figure 9.23 depicts three vectors denoting the three voltages in a three-phase system, which are 120° out of phase with each other. These could be the line voltages or the phase voltages. Note that the vectors shown clearly satisfy the principal relationship in Equation 9.1; that is, if they are added together their vector sum is zero. All other voltage or current values can be shown on the same graph. In this way the phase difference between variables will be graphically shown.

Figure 9.24a shows the windings of a three-phase generator with wye connection, which allows defining a reference point for measuring voltages. The three-phase voltages are called V_A, V_B, and V_C, respectively. Also, the three line voltages are identified as V_1, V_2, and V_3, as shown. The line voltages are obtained from the phase voltages as follows:

$$\begin{aligned} V_1 &= V_A - V_B \\ V_2 &= V_B - V_C \\ V_3 &= V_C - V_A \end{aligned} \tag{9.14}$$

V_A, V_B, and V_C are shown in Figure 9.24b, with arbitrarily and intentionally showing V_A along a horizontal line; V_B and V_C, are then at 120° and 240°, respectively, lagging V_A.

Figure 9.24b also shows the formation of V_1 from V_A and V_B, It shows clearly the value and phase relationship between V_A and V_1. In the same

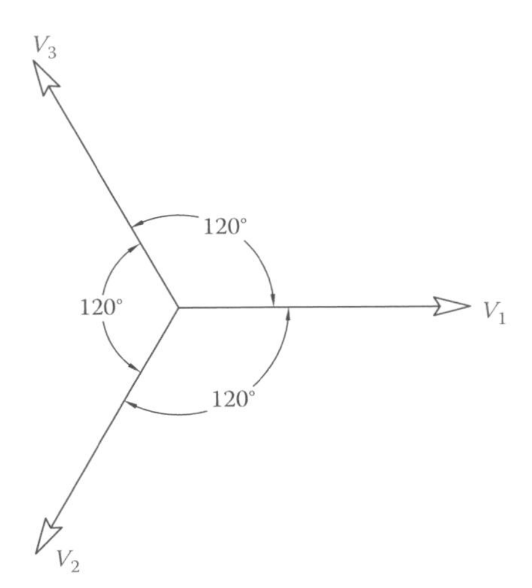

Figure 9.23 Representation of the three voltages in a three-phase system.

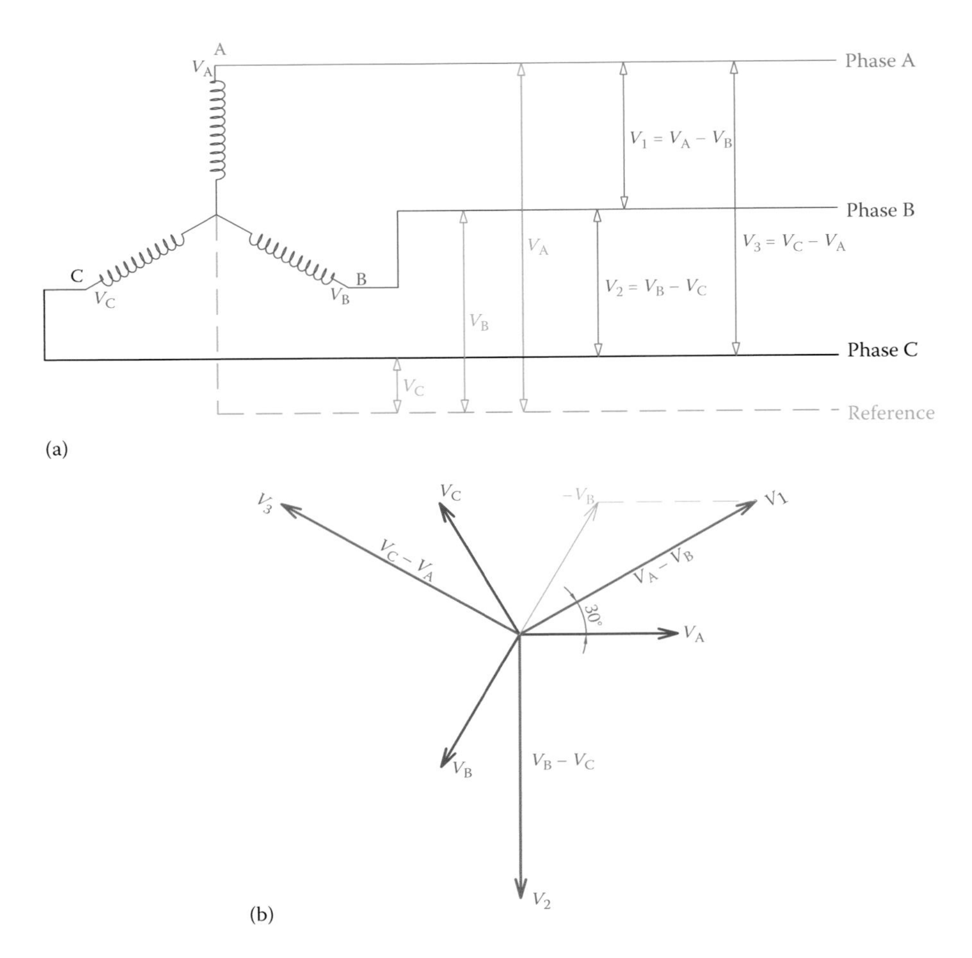

Figure 9.24 Vector representation of voltages in three-phase system. (a) A three-phase system. (b) Its phasor diagram for voltages and currents.

way, V_2 and V_3 can be obtained (the details are repeats of the same process as for V_1 and are not shown). Figure 9.24b clearly shows that each line voltage is 30° leading its corresponding phase voltage, and it is larger by a ratio of $\sqrt{3}$. (The length ratio can be depicted by trigonometry relations, but it is not shown here.)

The same graph as in Figure 9.24b is repeated with a different notation in Figure 9.25a and b. The purpose is to indicate that as far as the length relationships and the angle relationships are clearly shown, the vector representation can be as in Figure 9.25a. The fact that the corresponding vectors for line voltages appear in the form of a triangle has nothing to do with delta and wye connection methods. The line voltages in Figure 9.24b can equally be shown as in Figure 9.25a, with no fault. The vectors for line voltages must indicate that

1. They are 120° out of phase with each other.
2. They have 30° phase difference with phase voltages.
3. They are $\sqrt{3} = 1.73$ times in length compared to phase voltages.

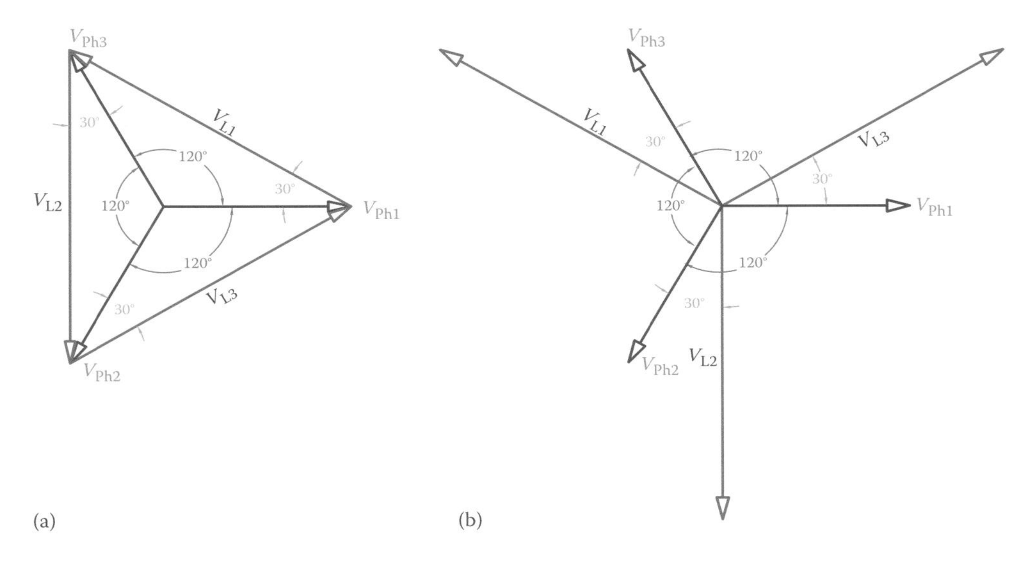

Figure 9.25 Phase and line voltages relationships. Either (a) or (b) can be used. Both show the 30° phase difference between line voltage and phase voltage.

On the basis of the above mentioned vector representation, three vectors for currents can also be shown on the same graph. Depending on the type of load (if resistive, inductive or so on) and the way they are connected, they can be in phase or out of phase with line voltages and phase voltages. It all depends on the loads, their values, and how they are connected. This is more clearly exhibited by an illustrative example. For a balanced load all the current vectors are equal and all the phase angles for the three phases are the same.

Example 9.13

A three-phase load consists of three identical elements. Each element (for each phase) consists of an inductor and a resistor in series. The load is wye connected and V_L = 440 V. If R = 3 Ω and X_L = 4 Ω, find the current in the elements, the voltages across each resistor and across each inductor.

Solution

For V_L = 440 V the phase voltage (across the element) is 440/1.73 or 254 V. Also, the impedance for each element is

$$\sqrt{3^2 + 4^2} = 5\ \Omega$$

and the (line or phase) current is

$$254 \div 5 = 50.8\ \text{A}$$

The voltages across the resistor and the inductor are, respectively,

$$V_{Res} = 3 \times 50.8 = 152.4\ \text{V}$$

$$V_{Ind} = 4 \times 50.8 = 203.2\ \text{V}$$

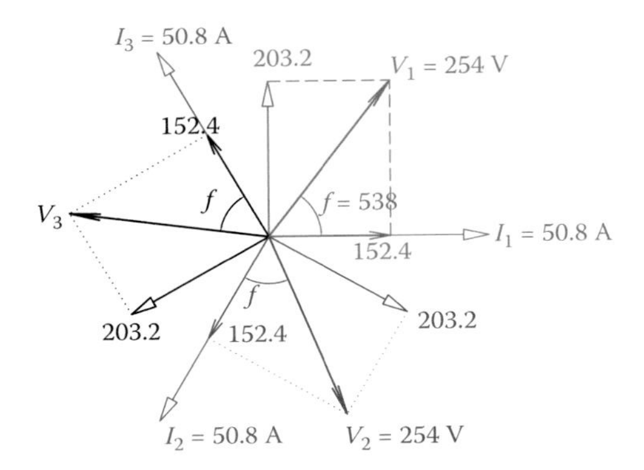

Figure 9.26 Phasor diagram showing the vector relationships between voltages and currents.

The phase angle for this load is 53° (e.g., from *R/Z*), and as usual for inductive load the current is lagging the voltage. Figure 9.26 shows the various vectors corresponding to the currents and voltages in this example. The graph shown in Figure 9.26 is called the phasor diagram for the circuit and graphically shows the size and phase relationships for various voltages and currents. Because the load is balanced in this case, everything is repeated three times but also rotated 120°. The process is similar to what we have already seen for single-phase AC electricity. Note that the voltage across the resistor is in phase with the current and the voltage across the inductor leads the current by 90°, resulting in a 53° phase difference between the current and the voltage. Also, note that the vector sum of the three currents I_1, I_2, and I_3 is zero, thus conforming to Equation 9.8, as it is for the voltages V_1, V_2, and V_3.

If there were capacitors in the circuit, the treatment would be exactly the same, bearing in mind that their effect is the opposite of that of the inductors and that they influence the phase angles. Also, resonance can occur (see Section 8.13).

9.6.1 Current Relationships

Similar to the way various voltages (three-line voltages and three-phase voltages) could be represented by vectors, various currents can also be shown as vectors. In this way their relationships can be graphically illustrated. For balanced loads all the line currents are similar and all phase currents are also similar. The magnitude and phase relationships are also the same for all three phases, and, therefore, calculations can be carried out for one phase and generalized for other phases. This, nevertheless, is not true for unbalanced loads. For unbalanced loads, finding the magnitudes and phase relationships for currents cannot be done without the use of vectors.

Figure 9.27 shows the line and phase currents for a delta-connected load. Whereas formerly we considered only two currents, namely, the line current and phase current, here we need to distinguish these two currents for

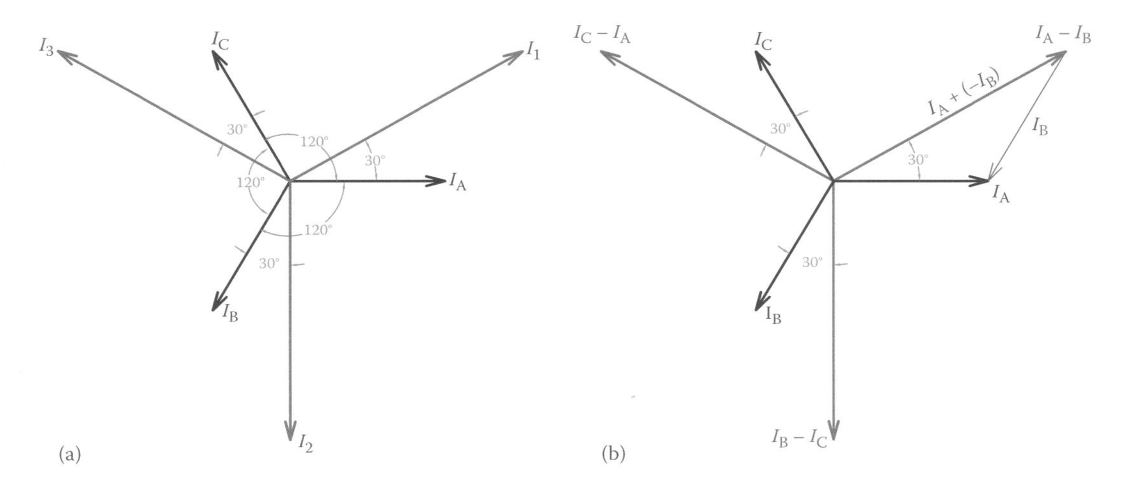

Figure 9.27 General relationships between phase and line currents in delta connection. (a) Phase and line currents. (b) Line currents in terms of phase currents.

the three phases. For this reason, as shown in Figure 9.27, the line currents are called I_1, I_2, and I_3, while the phase currents are named I_A, I_B, and I_C (similar to what was done for the associated voltages). The graphs shown correspond to a balanced load and, as can be seen, in the same way that there is a phase difference of 30° between line voltages and phase voltages when they are not the same (i.e., for wye connection), there is a 30° phase difference between line currents and phase currents in delta connection (i.e., when they are not the same. In wye connection these currents are the same).

In Figure 9.27, three-phase currents are shown equal to each other, corresponding to a balanced load. In general, the three currents of each phase can be different. The relationships between three-line currents and three-phase currents, as shown in Figure 9.27, are as follows

$$\begin{aligned} I_1 &= I_A - I_B \\ I_2 &= I_B - I_C \\ I_3 &= I_C - I_A \end{aligned} \tag{9.15}$$

Figure 9.27b shows the details of vectors for I_1. The relationships in Equations 9.15 can be used to find the line currents if the phase currents are known and vice versa. It provides also the general method for tackling problems with unbalanced loads.

9.7 Unbalanced Loads

Many loads in a three-phase system can consist of individual single-phase loads connected between the neutral line and one of the phase lines. In general, a single-phase load can be connected to two phase lines, but this is less likely to happen in practice. The best example is the distribution of electricity in cities. Almost all the houses receive single-phase electricity

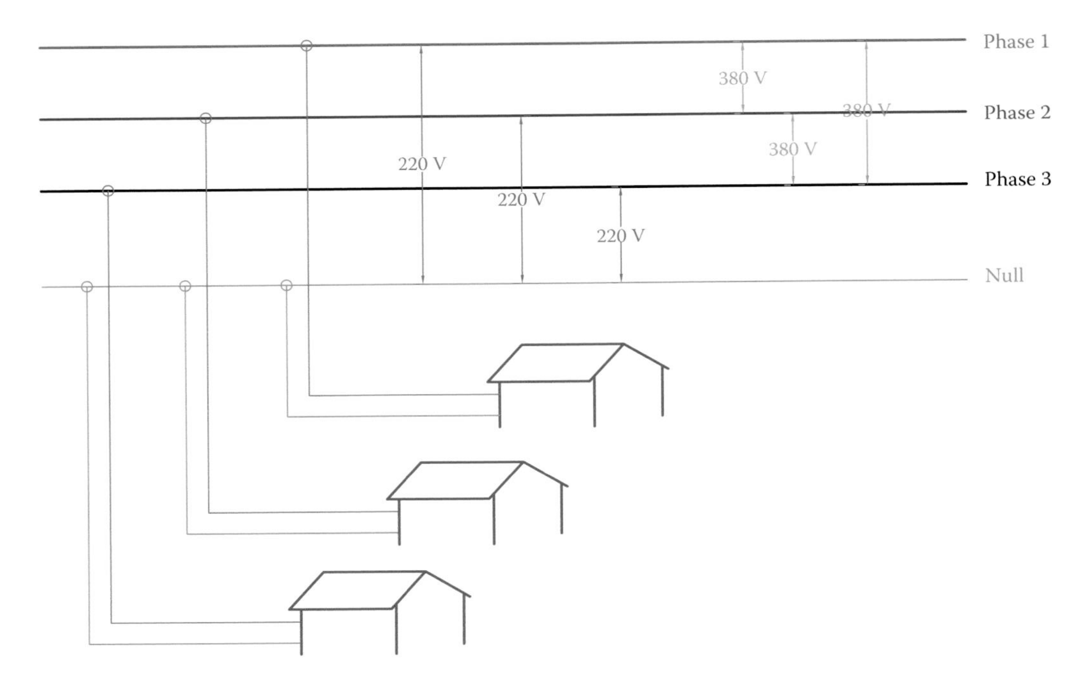

Figure 9.28 Electricity distribution in cities from a three-phase system.

because there is no need for three phases and all the domestic appliances are single-phase loads. This is schematically shown in Figure 9.28 for 220/380 V systems. (For distribution in North America, see Section 12.2.)

For an unbalanced load the individual currents for each phase must be found as for a single phase. This gives the values for phase currents. If the loads are wye connected (in the majority of cases), these individual values define the currents in the three-phase wires. The neutral wire in such a case may carry a nonzero current that can be determined by the vector sum of the calculated currents.

In the rare case of having unbalanced loads connected in delta the relationships in Equations 9.15 can be employed.

The following examples help to better understand the cases for unbalanced loads.

Example 9.14

In a three-phase, four-wire system the line voltage is 415 V and the resistive loads of 10, 8, and 5 kW are connected in Y. Calculate the line currents.

Solution

In this example the loads are all resistive. So, there is no phase difference, but the load values are different. First, we find the phase voltage:

$$V_{Ph} = 415 \div 1.73 = 240 \text{ V}$$

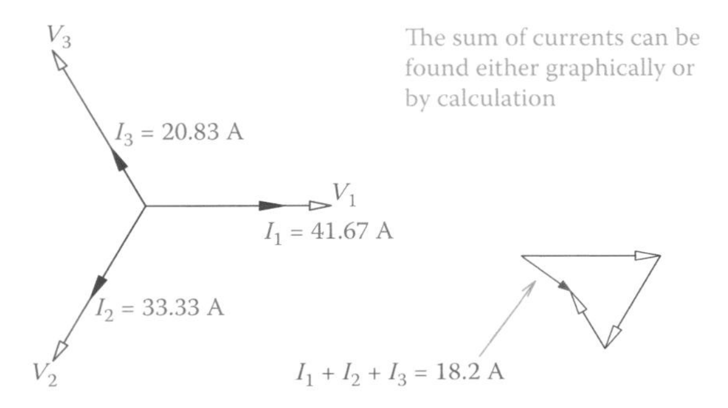

Figure 9.29 Solution of Example 9.14.

and the three-phase currents are

$$I_A = 10 \times 1000 \div 240 = 41.67 \text{ A}$$

$$I_B = 8 \times 1000 \div 240 = 33.33 \text{ A}$$

$$I_C = 5 \times 1000 \div 240 = 20.83 \text{ A}$$

The above values are already for line currents because of wye connection. What is important here to notice is the current in the neutral line, which is not zero because the loads are not balanced. This current can be found from the (vector) sum of the three line currents. Figure 9.29 shows the sum of the calculated currents and the nonzero current of the neutral line.

From the above values of the line currents the neutral line current can be found (by using a graphics method or by calculation) to be 18.2 A.

Example 9.15

Three different single-phase loads are connected in delta (each one between two phase lines) across a 480 V, 60 Hz three-phase line. Determine the line current and show the current and voltage in the vector form (phasor diagram). The loads are as follows:

Load 1: 24 kW, with $pf = 1$ (phase angle = 0, just active power)
Load 2: 24 kVA, with $pf = 0$, lagging (current lags voltage, inductive load)
Load 3: 24 kVA, with $pf = 0.5$, lagging (current lags voltage, more inductive than capacitive load, as well as resistive load).

Solution

Power magnitude for all the three loads is 24, but the loads are of different nature. Thus, the currents are equal in magnitude, but each one has a phase difference with its respective voltage. If, as in Figure 9.26, the phase currents are denoted by I_1, I_2, and I_3 and the line currents are represented by I_A, I_B, and I_C then

$$I_A = 24 \times 1000 \div 480 = 50 \text{ A}$$

$$I_B = 24 \times 1000 \div 480 = 50 \text{ A}$$

$$I_C = 24 \times 1000 \div 480 = 50 \text{ A}$$

These are the currents in the loads. A power factor of 0 for the second load implies a 90° lagging phase difference and a power factor of 0.5 for the third load implies a 60° ($\cos^{-1} 0.5 = 60°$), also lagging, phase difference. The first load has no phase difference.

Figure 9.30b illustrates the three phase currents with respect to the three voltages (implying their respective phase differences). Figure 9.30c shows how the line currents are then graphically determined from the relationships in Equations 9.15. This could be done numerically, instead. The results are depicted in Figure 9.31. Values for line current are

$$I_1 = 96.6 \text{ A}$$

$$I_2 = 70.7 \text{ A}$$

$$I_3 = 50 \text{ A}$$

Also, as can be observed from Figure 9.30, I_1 lags the voltage V_1 by 15°, I_2 lags V_2 by 46°, and I_3 is in phase with its voltage V_3.

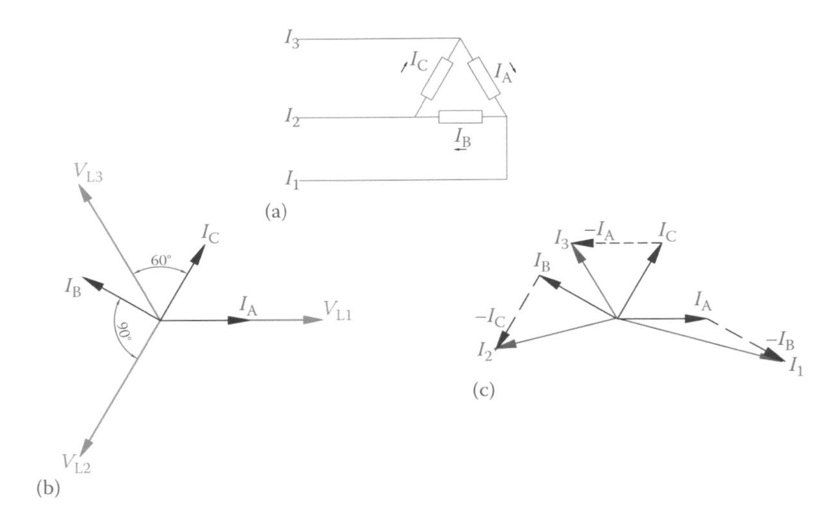

Figure 9.30 Current vectors in Example 9.15. (a) Definition of line and phase currents. (b) Phase currents as calculated. (c) Resulting line currents.

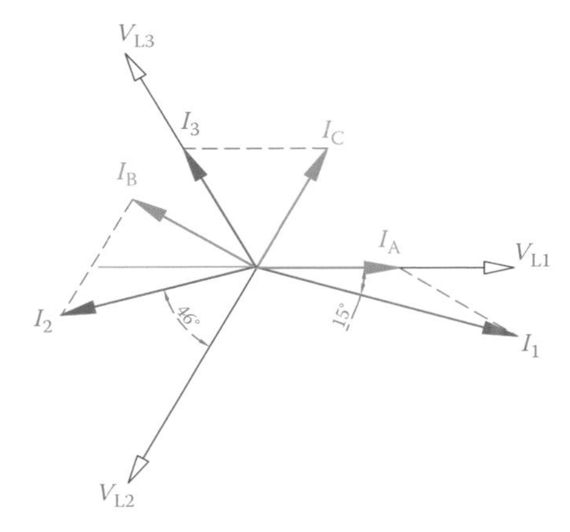

Figure 9.31 Line currents and phase differences with their voltages for Example 9.15.

9.8 Chapter Summary

- All commercial electricity production and industrial usage are in three-phase AC.
- Three-phase electricity is more efficient and needs fewer wires than equivalent single-phase AC.
- Three-phase electricity transmission requires three lines, but a fourth line, called the neutral line, can also be used.
- Three-phase load has three similar elements.
- In a three-phase system the voltages between each pair of wires are not in phase with each other. They are out of phase by 120°.
- Connection of a generator or a load to the three lines can be performed in two different ways: delta connection and wye connection.
- Current in the transmission lines is called line current. The current in the load elements is called phase current.
- The voltage between each pair of wires is called line voltage. The voltage across the load elements is called phase voltage.
- In wye connection the phase current and the line currents are equal.
- In wye connection the voltage across each load element is less than the line voltage.
- In delta connection the voltage across each load element is equal to the line voltage.
- In delta connection the phase current is less than the line current.
- A balanced load has exactly the same load types and values for each of the three load elements.
- All three-phase loads are balanced loads.
- For a balanced load the current in the three wires are the same in magnitude but are out of phase by 120°.
- If a load is connected in delta, it takes more current and more power from a circuit compared with if wye connected.

- A wye-delta switch is used to connect a load first with wye connection before changing to delta connection to reduce the inrush current.
- Vector representation is a convenient way of showing the phase relationships in a three-phase system.
- A vector diagram showing all the voltages and currents in a three-phase system is called a phasor.

Review Questions

1. What are the main differences between single-phase and three-phase systems?
2. What are the advantages of three-phase systems over single-phase electricity?
3. Can a three-phase load be connected to single-phase electricity? Why or why not?
4. How many wires are required for transmitting three-phase electricity?
5. Why are three-phase systems more common in industry?
6. What is meant by phase current and phase voltage?
7. How is the load connected to electricity in delta connection?
8. What is the difference between delta connection and wye connection?
9. What is the difference between wye connection and star connection?
10. What is the voltage between each pair of wires in a three-wire system called?
11. Can a three-phase system have four wires?
12. Is phase current larger or line current?
13. Is phase voltage larger or line voltage?
14. Can three different single-phase loads be connected to three-phase electricity?

15. Does the term apparent power apply to three-phase electricity?

16. Is the electricity at home single phase or three phase?

17. What is the difference between a three-phase load and a single-phase load?

18. How can you make a three-phase capacitive load out of three capacitors?

19. Can three single-phase sources be put together to make a three-phase source? Why or why not?

20. Does power factor correction apply to three-phase electricity?

21. In which of the four cases shown in Figures 9.12 and 9.13 is the current in the wires lowest for the same load elements, if the phase voltage generated by the source is 220 V?

Problems

1. A three-phase heater is 10 kW and the power line is 220/380 V. The heater is delta connected. Find the current in the line.

2. If after a repair by mistake the heater in Problem 1 is wye connected, what is the power it delivers?

3. If 1 W = 0.483 cal/sec, how much is the difference between the calories expected and the calories that the heater provides when it is not correctly connected, in one hour?

4. The fuse (three of them) used in the heater of Problem 1 must allow 20 percent more than the current taken by the heater, what is the size of the fuse to be used?

5. If one of the fuses in the heater of Problem 1 blows out and disconnects its line, find out how much is the current in the lines and how much is the power delivered by the heater?

6. Each element of a three-phase load is a 10 Ω resistor. For each of the four scenarios shown in Figures 9.12 and 9.13, find the current in the wires connecting the load to the generator if the phase voltage in the generator is 220 V.

7. Each element of a three-phase load can be assumed to consist of an inductor and a resistor in parallel. The load is

delta connected. If the line current is 35 A, the line voltage is 400 V, and the power factor is 0.55, find the resistance of the resistive part of each load element for both wye and delta connection.

8. If for Problem 7 a bank of delta-connected capacitors is used to increase the power factor to 100 percent, what is the size of each capacitor if the operating frequency is 50 Hz?

9. If a set of delta-connected 100 μF capacitors are used for the load of Problem 7, what is the power factor?

10. For Problem 9, draw the phasor diagram before and after exerting the capacitors.

11. A 12 kW wye-connected resistive load and a 15 kVA delta-connected load are fed by the same circuit. What is the apparent power of this circuit, and what is the power factor if the power factor of the second load is 70 percent? The line voltage is 480 V.

12. Three different loads are connected in delta across a 400 V, 50 Hz three-phase line. Determine and show the current and voltage in the vector form (phasor diagram). The loads are as follows:
Load 1: 28 kW, with $pf = 1$
Load 2: 22 kVA, with $pf = 0.75$, leading
Load 3: 18 kVA, with $pf = 0.85$, lagging

Advanced Learning: Mathematical Representation of Three-Phase Systems

Introduction

Alternating current electricity is generated by electric generators, which are rotary machines. This is why the voltage and current in AC electricity have a sinusoidal form. Therefore, mathematically an AC voltage can be defined by a sine or cosine function. Often, a cosine function is used, but because in this book all figures representing an AC voltage start at zero, we use a sine waveform for this purpose in order to accord.

Voltage is the principal variable in an AC circuit, which results in the current in a load to have also the same waveform, either in phase with the voltage or out of phase with it, depending on the load.

In this section we include the main formulation for the mathematical representation of an AC system, with the emphasis on three-phase electricity.

Expression for AC Voltage

Time variation of a single-phase voltage can be represented by

$$V = V_{\text{peak}} \sin(\omega t) \tag{A9.1}$$

where V is the amplitude at each instant, V_{peak} is the peak value of the voltage, which is a constant, ω is the frequency expressed in radians per second (and not in Hz), and t denotes the time. With the change of t the voltage V traces a sinusoidal waveform. The relationship between ω and f (the frequency in Hz) is

$$\omega = 2\pi f \tag{A9.2}$$

When another waveform, such as current, is in phase with the voltage in Equation A9.1 it can be expressed as

$$I = I_{\text{peak}} \sin(\omega t) \tag{A9.3}$$

and if it is lagging the voltage by an angle θ, it can be written as

$$I = I_{\text{peak}} \sin(\omega t - \theta) \tag{A9.4}$$

Similarly, if the current leads the voltage by θ, the expression is

$$I = I_{\text{peak}} \sin(\omega t + \theta) \tag{A9.5}$$

Note that since ω is in rad/sec, ωt is in radians and, thus, θ must be defined in radians (and not in degrees).

Three-Phase System

In a three-phase AC system, there are three voltage values that are out of phase with each other by 120°. Denoting the three voltages by V_A, V_B, and

V_C, and assuming a zero value for V_A at time $t = 0$, the expressions for the three phase voltages are (note that 120° = 2π/3 rad.):

$$V_A = V_p \sin(\omega t) \quad \text{(A9.6)}$$

$$V_B = V_p \sin(\omega t - 2\pi/3) \quad \text{(A9.7)}$$

$$V_C = V_p \sin(\omega t - 4\pi/3) = V_p \sin(\omega t + 2\pi/3) \quad \text{(A9.8)}$$

where V_p denotes the peak voltage and ω is as defined in Equation A9.1.

(Equally, we could use cos function, but then at time $t = 0$ V_A is at its peak.)

The above three voltages are the outputs of a generator in a three-phase system. It can be seen that at each instant of time the sum of all the voltages is equal to zero. This property, in fact, allows us to put the three lines together to make a wye connection; otherwise, we cannot (and should not) join three points with a voltage difference between them. Also, from these equations one can find the expressions for the voltage difference between each two phases, as it is used in delta connection.

For both tasks, showing that the sum of the three voltages is equal to zero and finding the voltage difference between the phases, we need to use the trigonometric relationships for the sum of angles. These relationships are repeated here for convenience.

$$\sin(x + y) = \sin x \cos y + \cos x \sin y \quad \text{(A9.9)}$$

$$\cos(x + y) = \cos x \cos y - \sin x \sin y \text{ (cos terms are + and sin terms are –)} \quad \text{(A9.10)}$$

Hence,

$$\sin(\omega t - 2\pi/3) = \sin(\omega t)\cos(2\pi/3) - \cos(\omega t)\sin(2\pi/3) \quad \text{(A9.11)}$$

$$\sin(\omega t + 2\pi/3) = \sin(\omega t)\cos(2\pi/3) + \cos(\omega t)\sin(2\pi/3) \quad \text{(A9.12)}$$

The sum of the three voltages accordingly is

$$V_A + V_B + V_C = V_p [\sin(\omega t) + \sin(\omega t)\cos(2\pi/3) - \cos(\omega t)\sin(2\pi/3) + \sin(\omega t)\cos(2\pi/3) + \cos(\omega t)\sin(2\pi/3)] = V_p [\sin(\omega t) + 2\sin(\omega t)\cos(2\pi/3)]$$

But cos (2π/3) = −0.5 and 2 cos (2π/3) = −1.
Therefore,

$$V_A + V_B + V_C = V_p [\sin(\omega t) - \sin(\omega t)] = 0 \quad \text{(A9.13)}$$

The following lines derive the values for phase-to-phase voltages, denoted by V_1, V_2, and V_3. We are going to use Equations A9.10 to A9.12. Also, the constant values can be substituted for. That is, $\sin(2\pi/3) = \frac{\sqrt{3}}{2}$ and $\cos(2\pi/3) = -0.5$, $\cos(-2\pi/3) = -0.5$, etc. Some mathematical manipulation is necessary to express the results in a useful form.

$$V_1 = V_A - V_B = V_p[\sin(\omega t) - \sin(\omega t)\cos(2\pi/3) + \cos(\omega t)\sin(2\pi/3)]$$

$$= V_p\left[\frac{3}{2}\sin(\omega t) + \frac{\sqrt{3}}{2}\cos(\omega t)]\right]$$

The expression in the bracket can be written as

$$\sqrt{3}\left[\frac{\sqrt{3}}{2}\sin(\omega t)-\left(-\frac{1}{2}\right)\cos(\omega t)\right]=\sqrt{3}[\sin 120°\sin(\omega t)-\cos 120°\cos(\omega t)]$$

which can be simplified to $-\sqrt{3}\cos(\omega t + 120°)$.

As a result,

$$V_1=-\sqrt{3}\,V_p\cos(\omega t+120°) \quad \text{(A9.14)}$$

$$\begin{aligned} V_2 &= V_B - V_C \\ &= V_p[\sin(\omega t)\cos(2\pi/3)-\cos(\omega t)\sin(2\pi/3) \\ &\quad -\sin(\omega t)\cos(2\pi/3)-\cos(\omega t)\sin(2\pi/3)] \\ &= V_p[-2\cos(\omega t)\sin(2\pi/3)] = -\sqrt{3}\,V_p\cos(\omega t) \end{aligned}$$

Thus,

$$V_2=-\sqrt{3}V_p\cos(\omega t) \quad \text{(A9.15)}$$

$$V_3 = V_C - V_A = V_p[\sin(\omega t)\cos(2\pi/3)+\cos(\omega t)\sin(2\pi/3)]-\sin(\omega t)]$$

$$= V_p\left[-\frac{3}{2}\sin(\omega t)+\frac{\sqrt{3}}{2}\cos(\omega t)\right]$$

A treatment similar to what was done for V_1 leads to

$$\begin{aligned} V_3 &= V_p\sqrt{3}\left[-\frac{\sqrt{3}}{2}\sin(\omega t)-\left(-\frac{1}{2}\right)\cos(\omega t)\right] \\ &= \sqrt{3}V_p[\sin(-120°)\sin(\omega t)-\cos(-120°)\cos(\omega t)] \\ &= -\sqrt{3}V_p\{\cos[\omega t+(-120°)]\} \end{aligned}$$

or

$$V_3=-\sqrt{3}V_p\cos(\omega t-120°) \quad \text{(A9.16)}$$

The expressions given in Equations A9.14, A9.15, and A9.16 clearly show the important facts for V_1, V_2, and V_3: (1) that their variation with time is sinusoidal, (2) that these values are larger than V_A, V_B, and V_C by a factor of $\sqrt{3}$ and (3) that they have a phase difference of 120° from each other.

We may want to express V_1, V_2, and V_3 as sine functions rather than cosine. In this way we can directly compare them with V_A, V_B, and V_C. The following lines indicate this modification, in which we use the two basic relationships:

$$\cos x = \sin(90° - x)$$

$$-\sin x = \sin(-x)$$

In this regard,

$$V_1 = -\sqrt{3}\,V_p \cos(\omega t + 120°) = -\sqrt{3}\,V_p \sin(90° - \omega t - 120°)$$
$$= \sqrt{3}\,V_p \sin(\omega t + 30°) \qquad \text{(A9.17)}$$

$$V_2 = -\sqrt{3}V_p \cos(\omega t) = -\sqrt{3}V_p \sin(90° - \omega t)\sqrt{3}V_p \sin(\omega t - 90°) \qquad \text{(A9.18)}$$

$$V_3 = -\sqrt{3}V_p \cos(\omega t - 120°) = -\sqrt{3}V_p \cos(\omega t + 240°)$$
$$= -\sqrt{3}V_p \sin(90° - \omega t - 240°) = \sqrt{3}V_p \sin(\omega t + 150°) \qquad \text{(A9.19)}$$

Figure A9.1 illustrates the vectors associated with V_A, V_B, and V_C, and their relationship with V_1, V_2, and V_3. Figure A9.1 is a modified version of Figure 9.24b in order to accord with using sine function.

When a balanced three-phase load is powered by a voltage source expressed by Equations A9.6 to A9.8 or A9.14 to A9.16, the resulting current has a sinusoidal waveform. Employing the same approach as used for Equation A9.13, it can be proved that the sum of the currents at each instant equals zero.

$$I_A + I_B + I_C = 0$$

or

$$I_1 + I_2 + I_3 = 0 \qquad \text{(A9.20)}$$

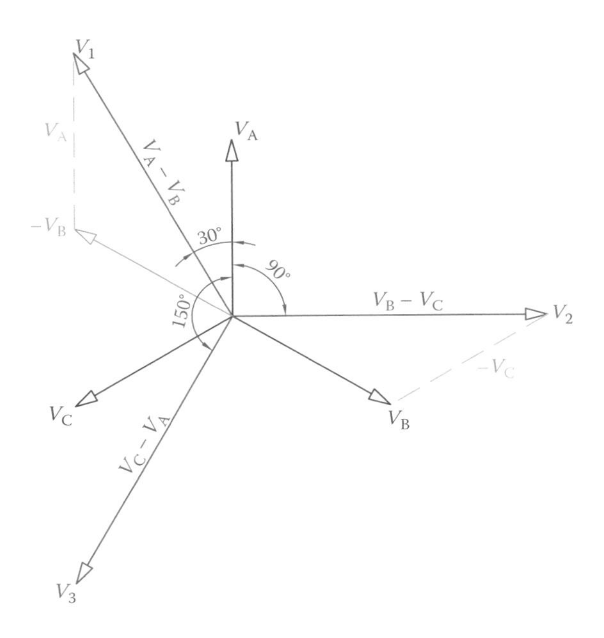

Figure A9.1 Vector representation of phase and line voltages based on mathematical relationships.

10 Transformers

OBJECTIVES: After studying this chapter, you will be able to

- Describe the principles of operation of all transformers
- Define the terms step-up and step-down
- Select a transformer power rating based on the loads to be powered
- Understand practical concerns when using transformers
- Explain the differences and similarities between single-phase and three-phase transformers
- Understand and use terms employed for transformers
- Understand the four different configurations for connecting three-phase transformers
- Calculate efficiency of a transformer based on its daily usage data
- Explain familiar tests for measuring transformer data
- Use the data from tests to determine transformer losses

New terms:
All-day efficiency, autotransformer, copper loss, core loss, core loss current, core-type transformer, energy efficiency, hysteresis, ideal transformer, isolation transformer, leakage (in transformer), magnetizing current, open-circuit test, pad mount (transformer), pole mount, shell-type transformer, short-circuit test, step-down transformer, step-up transformer, tertiary, tertiary winding, transformer percent impedance, turns ratio, voltage regulation, volts per turn, zigzag transformer

10.1 Introduction

One of the salient features of AC electricity is that it can be transformed easily and in a relatively cheap way compared to DC electricity. Without this feature, electricity usage would be very limited.

As we have discussed so far, there are four parameters associated with AC electricity, voltage, current, power, and frequency. The main function of a transformer is changing the voltage of AC electricity, either increasing or decreasing it. Nevertheless, as we will see in this chapter, because of power balance, as a result of voltage change current also undergoes a change. The only parameter that is not affected is the frequency.

The applications of transformer are plentiful, and transformers of all sizes are made. Here *size* can refer to the physical size, but it can also reflect the power because the larger the power capacity of a transformer, the larger it is in size. For example, a transformer for powering a halogen lightbulb (halogen lightbulbs use 12 V electricity) is small and can handle about 20 W of power (their rating is about 20 W). The transformer of a 2 MW wind turbine must be capable of handling 2 MW of power, and it is considerably larger.

10.2 An Illustrative Example

We start this chapter with an example, which illustrates the importance and the effect of changing voltage, particularly for power transmission.

The importance of a transformer is more pronounced for high powers, although at low-power level, there are millions of small transformers in all household and industrial products. This is a simple example but conveys the importance of using transformers.

Example 10.1

A 100 hp load (e.g., a three-phase motor) is fed by a 400 V line. The load is delta connected to the line, and its power factor is 0.8. If the resistance of each of the three wires connecting the load to the mains is 0.5 Ω, (1) find the voltage drop and the power loss of the wires and (2) compare these values with the case if the voltage was 4000 V.

Solution

On the basis of the relationships for power in a three-phase system (see Section 9.5.2), the current in the line can be found.

1. For $V = 400$ V,

$$I = \frac{P}{\sqrt{3}V.pf} = \frac{(100)(746)}{(1.73)(400)(0.8)} = 134.6 \text{ A}$$

For this current the voltage drop in the line is (by Ohm's law)

$$V_{\text{drop}} = \Delta V = (0.5)(134.6) = 67.3 \text{ V}$$

and the power used in the three wires is

$$P_{\text{loss}} = (3)(0.5)(134.6^2) = 27{,}173.6 \text{ W}$$

2. For $V = 4000$ V

$$I = \frac{P}{\sqrt{3}V.pf} = \frac{(100)(746)}{(1.73)(4000)(0.8)} = 13.46 \text{ A}$$

For this current the voltage drop in the line is (by Ohm's law)

$$V_{\text{drop}} = \Delta V = (0.5)(13.46) = 6.73 \text{ V}$$

and the power used in the three wires is

$$P_{\text{loss}} = (3)(0.5)(13.46^2) = 271.74 \text{ W}$$

It can be observed that the current in the 400 V line is so high that it leads to a voltage drop of 67.3 V, which implies that the voltage in the line must be 467.3 V so that the load gets 400 V; otherwise, the voltage at the load would be smaller than 400 V. Moreover, a large amount of energy is lost in the form of heat, which is undesirable.

The comparison of the results for 4000 V illustrates that in this case the voltage drop is only 6.73 V, and the power loss in the line is under 272 W.

It is very important that the voltage drop and the consumed energy by the feeding lines be minimized by using the proper voltage for transmission of electricity, and this is done by transformers.

10.3 Single-Phase Transformer

In the same way that single-phase AC and three-phase AC are different, although both are alternating current electricity, single-phase and three-phase transformers cannot be interchanged. Nevertheless, most of what can be said for single-phase transformers is later on extended to three-phase transformers because they work based on the same principle.

Transformers work based on the mutual induction between two windings. As we have learned, when electricity is applied to a coil a magnetic field is developed (see Section 7.2). Because in AC the voltage (and thus the current) continuously alters, the resulting magnetic field is variable. However, if a coil is placed in a varying magnetic field, it is similar to a wire that moves in a magnetic field; then a voltage is induced in it.

In what follows in this section, everything said mainly refers to a single-phase transformer, though for the sake of simplicity the term "single phase" is not repeated.

10.3.1 Transformer Fundamental Construction

In the simplest form a transformer is made up of two windings which share a common core. The core must be closed and its material be a ferromagnetic metal; it can be of various shapes, but the shape does not play a significant role. Some shapes are preferable to the others, either because of manufacturing or because they can result in a better performance. The most common shapes for a core are rectangular and circular. From the manufacturing viewpoint, a rectangular shape is easier to work with. Figure 10.1 shows a transformer with a rectangular core. In this configuration, called a **core-type transformer**, the two windings are on the

Core-type transformer: Category for the construction of a transformer in which the core has the shape of a rectangular frame, and the windings are normally on the opposite sides of the rectangle.

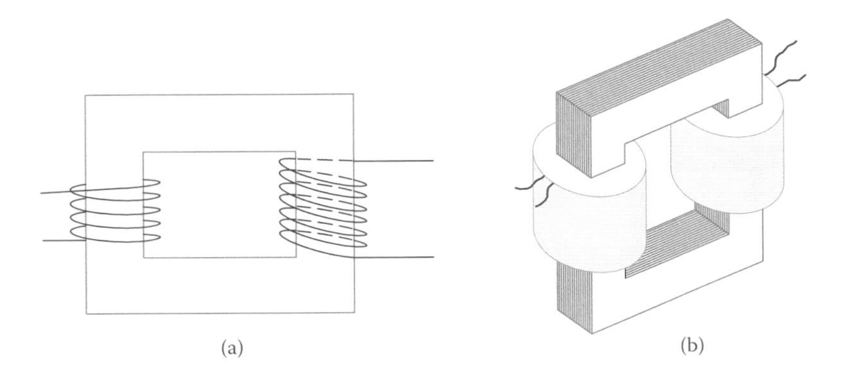

Figure 10.1 Basic structure of a transformer (core type). (a) Schematics. (b) Pictorial representation.

Shell-type transformer: Category for the construction of a transformer in which the core has the shape of a figure 8 (i.e., with a bridge in the middle of a rectangular frame).

Step-down transformer: Transformer with secondary winding having fewer turns than the primary winding, thus decreasing voltage.

Step-up transformer: Transformer with secondary winding having more turns than the primary winding, thus increasing voltage.

opposite sides (legs) of the core. This is not necessary, and both can be on the same leg. In fact, most of the small transformers have a core as shown in Figure 10.2, and both windings are placed—usually side by side—on the middle leg. This is a called a **shell-type transformer**.

The two windings are called "primary" and "secondary," depending on their role. The primary winding is connected to the voltage that we want to change. The secondary winding provides the required voltage. In other words, the primary winding is connected to the input voltage and the secondary winding is the output and is connected to the output circuitry. To show a transformer in a circuit, the transformer symbol is used, as depicted in Figure 10.3.

If the primary winding has more turns than the secondary winding, the transformer decreases the voltage and it is a **step-down transformer**. If the primary winding has fewer turns than the secondary winding, the output voltage is higher than the input voltage and the transformer is called a **step-up transformer**. The numbers of wire turns in the primary and secondary windings are represented by N_1 and N_2, respectively.

For a step-down transformer, $N_1 > N_2$ and $V_1 > V_2$.
For a step-up transformer, $N_1 < N_2$ and $V_1 < V_2$.

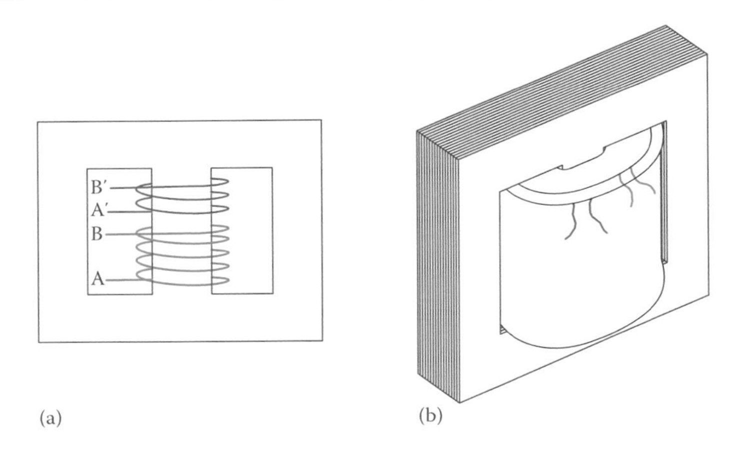

Figure 10.2 Most of transformers have a shell-type structure. (a) Schematics of a shell-type transformer. (b) Pictorial view.

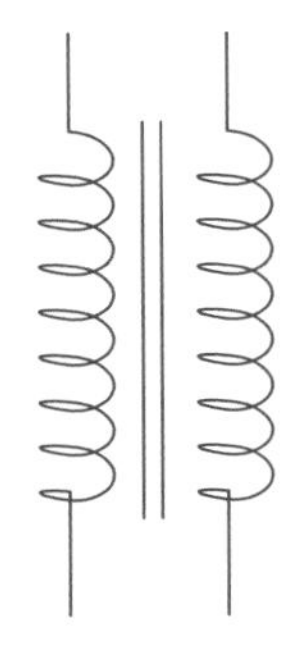

Figure 10.3 Symbol for a transformer.

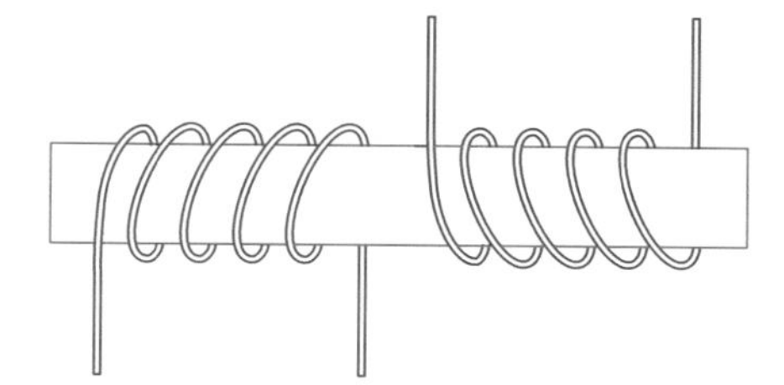

Figure 10.4 Two possible ways of winding a wire around a core.

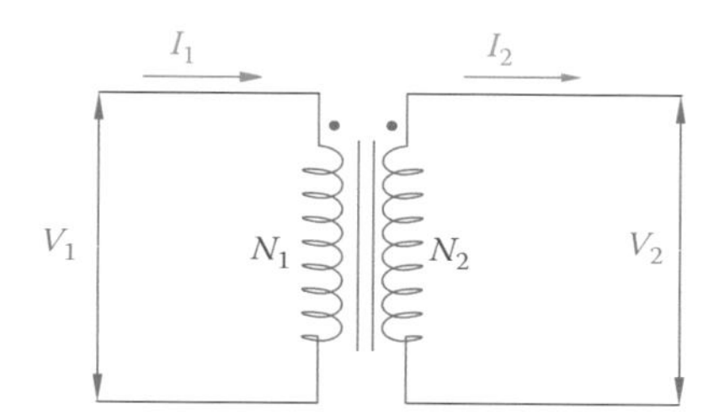

Figure 10.5 Number of turns *N*, voltage *V* and current *I* in the primary and secondary circuits of a transformer.

Note that a wire can be wound around a core in two directions, as shown in Figure 10.4. As a result, depending on the directions of both windings in a transformer, at a given instant and compared to the primary voltage, the current in the secondary winding can be in one or the opposite direction. In other words, the secondary voltage can be in phase with the primary voltage, or it can be 180° out of phase with it. It is, therefore, important to pay attention to the direction of turns in the winding of a transformer. In Figure 10.2, for instance, both windings have the same direction. To show the polarity in the drawings of transformer circuits, a dot is put at one side of each winding indication the phase relationship, illustrating also the relative directions of the primary and secondary currents. This is indicated in Figure 10.5. Thus, for example, considering a sinusoidal waveform, when the side with a dot in the primary winding is at its maximum value the side with a dot in the secondary winding is at its maximum value. In practice, the terminals are marked by letters such as H and X on the primary and secondary referring to the dot side to indicate this polarity. Figure 10.5 also indicates the number of turns *N*, the voltage *V*, and the current *I* and its direction for a given instant, identified by subscripts 1 and 2 for the primary and the secondary circuits, respectively.

10.3.2 Multi-Output Transformers

It is possible to have multiple secondary windings on the same transformer. This provides various voltages from one source voltage, thus reducing the cost and space. This is very common in electronic devices such as a television that need various voltages for their operation. This is done in very small transformers. In larger transformers and three-phase transformers it is not normally necessary to have multiple voltages. Figure 10.6 shows the schematics of a multi-output transformer.

Figure 10.7 shows a very small, 20 W, transformer. It converts 120 V into 6 and 12 V. In this sense, it is a two-output transformer. A 4 kVA industrial transformer is shown in Figure 10.8.

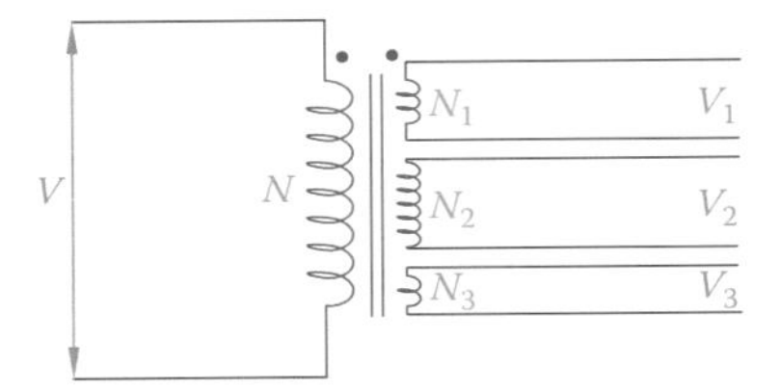

Figure 10.6 Multi-output transformer.

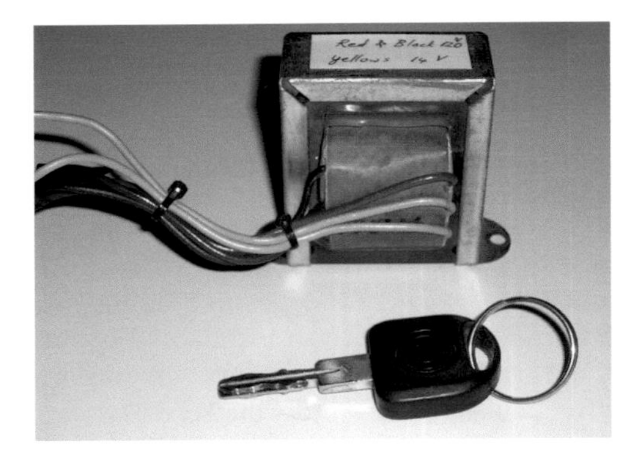

Figure 10.7 A 20 W single-phase transformer.

Figure 10.8 A 4 kVA single-phase transformer.

10.4 Ideal Transformer

In all transformers, one phase or three phase, the primary winding gets energy from the mains and the secondary winding is connected to the load(s). If the secondary circuit is turned off by a switch, there is no load on the secondary winding. Nevertheless, the primary winding is still connected to electricity and forms a closed circuit. In this case the primary winding behaves as a coil with a core; a current flows through it that (1) warms the winding and (2) warms the core as a result of eddy currents and hysteresis.

Eddy current is a local movement of electrons in a ferromagnetic material when placed in a magnetic field. In a solid piece of iron or other ferromagnetic metal, there can be many local circuits of eddy current in various directions, depending on the variation of the magnetic field. This unwanted current not only warms up a core, it consumes energy and also affects the magnetic property of the core metal.

To reduce the eddy currents and its effects, all the core materials of transformers and motors are made up of laminated metal. Each lamination has a layer of paint or nonconductive wax on it so that it is electrically isolated from its neighbors. In this way, eddy currents are limited to those only inside of one laminated metal. Figure 10.9 shows what a laminated metal looks like. This greatly reduces the amount of energy loss by eddy current and the resulting heat. Nevertheless, eddy currents cannot be entirely stopped and the associated energy loss cannot be reduced to zero. There are other types of loss in transformers, such as hysteresis and mechanical loss (energy changed to noise). But these are relatively smaller than eddy current loss in a well-designed transformer.

The loss due to the current in the winding(s) is called **copper loss**, and the losses due to eddy currents, hysteresis, and so on are referred to as **core loss** because they occur in the transformer core, and they are independent of how much the current is (but they increase as frequency goes up).

To develop the electrical relationships between the primary and secondary transformers, all losses are assumed to be zero. This simplifies the relationships and facilitates developing the pertinent equations. Such an assumed transformer in which all the losses are neglected is called **ideal transformer**.

Copper loss: Amount of power loss in a transformer that corresponds to the resistance of the wire winding and it depends on the load current (the percentage of loading).

Core loss: Amount of power loss in a transformer that corresponds to the quality of design and core material and is independent of the load current.

Ideal transformer: When in a transformer all the losses are assumed to be zero and, as a result, input power equals output power.

Figure 10.9 Laminated metal for transformers and motors.

In an ideal transformer the output power and the input power are equal. That is, all power received by the primary winding is delivered to the secondary winding.

In dealing with any device, including transformers, one needs to bear in mind that it is always the load that determines how much power is required.

In an ideal transformer all the losses are assumed to be zero. As a result, the power in the primary side is equal to the secondary side power.

10.4.1 Voltage Relationship

Referring to Figure 10.5, the primary winding is connected to a supply voltage V_1 and the secondary voltage V_2 is applied to a load. In general, a load can be resistive, inductive, capacitive, or a combination of these. A current I_2 will flow in the secondary winding. On the basis of this current the primary winding curries a current I_1.

The relationship between the primary and secondary voltages is based on the ratio of the number of turns in the primary and secondary windings. The **turns ratio** in a transformer is the ratio of turns in the primary winding to that of the secondary winding and is denoted by a.

Turns ratio: Ratio of number of turns in the secondary and primary windings of a transformer.

The following equation is the fundamental relationship for an ideal transformer:

$$\frac{V_1}{V_2} = \frac{N_1}{N_2} = a \tag{10.1}$$

where V_1 and V_2 are the primary and secondary voltages, respectively; N_1 and N_2 are the number of turns in the primary and secondary windings, and a is the turns ratio. For most transformers the turns ratio is fixed and cannot change. When $a > 1$, the secondary voltage is smaller than the primary voltage, thus a step-down transformer. When $a < 1$, the secondary voltage is larger than the primary voltage and the transformer is a step-up transformer.

Equation 10.1 shows that a transformer with a given turns ratio, for instance 10, can divide its input voltage by 10 at the secondary winding. This transformer can theoretically also multiply its input voltage by 10. For instance, if the side with the lower number of turns is connected to 120 V, the secondary winding has 1200 V at its terminals. This implies an important and serious issue: in working with a transformer, special care must be taken for its correct connection to the source and loads. Wrong connection can easily lead to damage and injuries.

In a step-up transformer the secondary voltage is higher than the primary voltage. In a step-down transformer it is the reverse.

Take note that although theoretically one can connect a transformer for stepping up or stepping down a given voltage, and there are transformers designed for working both ways, in practice, design considerations such as wire thickness and transformer power rating determine the limitation for the use of a transformer. A transformer cannot necessarily be connected to any arbitrary voltage or in an arbitrary fashion.

In working with a transformer, special care must be taken for correct connection of its primary and secondary to the outside circuits.

Example 10.2

The primary winding of a transformer has 1000 turns. If this transformer is to be used for changing 120 V input to 24 V output, how many turns are needed for the secondary winding?

Solution

It follows from Equation 10.1 that the number of secondary winding turns is

$$N_2 = \frac{V_2}{V_1} N_1 = \left(\frac{24}{120}\right)(1000) = 200$$

In using transformers, often the turns ratio is mentioned in the form of *a:1* (a to 1) or *1:a* (1 to a). For example, in a 2:1 transformer the primary voltage is twice the secondary voltage.

In conjunction with Equation 10.1 one can understand that in a transformer the ratio

$$\frac{V}{N} = \frac{V_1}{N_1} = \frac{V_2}{N_2}$$

is a constant. This constant is called **volts per turn** and determines how many volts there are per each turn of either the primary or the secondary winding. The use of this constant is in the design stage of a transformer.

Volts per turn: Number indicating the value of volts for each turn of winding in a transformer. This value can be obtained from either the primary side values or the secondary values by dividing the voltage by the number of turns.

10.4.2 Current Relationship

The relationship between the primary and secondary currents for an ideal transformer is based on the power relationship; that is, the power in the primary side is equal to the power consumed in the secondary side. Always, in AC electricity the consumed power refers to the apparent power, the product of voltage and current. Writing

$$S = V_1 I_1 = V_2 I_2$$

leads to

$$\frac{V_1}{V_2} = \frac{I_2}{I_1} \quad \text{or} \quad \frac{I_2}{I_1} = \frac{N_1}{N_2} = a \tag{10.2}$$

This equation also implies that in a transformer the side with the higher voltage has a smaller current and vice versa. The secondary current I_2 can always be found from the secondary voltage and the load impedance connected to the secondary. Having the current in the secondary, then the current I_1 in the primary can be found. This is for an ideal transformer, and I_1 represents the current in the primary due to the load only. If there is no load connected to the secondary side, this current is zero, too. In other words, in an ideal transformer the current I_1 as obtained from Equation 10.2 does not include the no-load current and the core loss currents (due to eddy currents in the core, mentioned earlier) that exist in a real transformer.

Because the current in the secondary and primary windings are not the same, the wire sizes for the two windings are not the same. The current in the winding with the lower number of turns is always higher than the current in the other winding. Thus, the lower voltage (higher current) side always has a thicker wire than the higher voltage (lower current) side. This can be a good way to judge the connections if in doubt.

For a step-down transformer, secondary current is higher and the winding wire is thicker than that of the primary winding.

Example 10.3

A step-down transformer is used to change 220 to 110 V. If a resistive load consuming 500 W is connected to the 110 V side, what is the current due to this load in the secondary and primary windings?

Solution

Because this is a 2:1 transformer, current in the primary winding is half of the current in the secondary winding. From the power of the load and the voltage in the secondary side the current I_2 can be found

$$I_2 = \frac{P}{V_2} = \left(\frac{500}{110}\right) = 4.54 \text{ A}$$

from which the current in the primary side follows

$$I_1 = (0.5)(4.54) = 2.27 \text{ A}$$

10.5 Autotransformer and Isolation Transformer

Autotransformers and isolation transformers are made for both single-phase and three-phase applications. They are not different in principle from other transformers but are meant for a special purpose.

Autotransformer: Transformer with only one winding per phase (instead of two). Primary and secondary windings are part of the same wire winding.

10.5.1 Autotransformer

An **autotransformer** has only one winding with one or more taps that serves for both primary and secondary windings. In this way a part of the

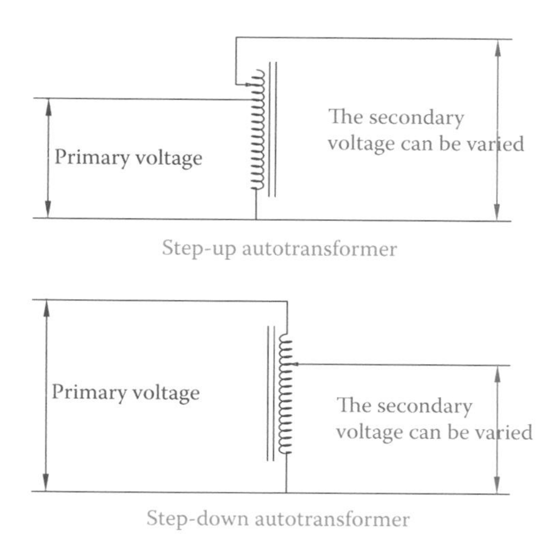

Figure 10.10 Principle of operation of an autotransformer.

winding is shared between the primary and secondary of the transformer. Secondary winding can be selectable by taps or a slider, which provides an adjustable voltage. Application of such a transformer is numerous when a voltage must be only slightly changed. For example, normally toward the end of a distribution line there is a noticeable voltage drop. To compensate for this voltage drop, the line voltage must be raised to its nominal value. Consider the houses at the end of a street where the line voltage must be 120 V. If the voltage has dropped to 105 V, then it is necessary to increase the voltage by 15 V to bring it to the nominal level. In such a case an autotransformer is used.

The winding is tapped at the proper point for the desired voltage ratio. An autotransformer can be used for both stepping up and stepping down voltage. Figure 10.10 shows the concept for a single-phase autotransformer.

Small autotransformers can often be used when a variable voltage is required. This is obtained by a slider that for the secondary voltage connects to different points on the winding. In the larger three-phase transformers, there is a tap changer that moves from one connection to the other for regulating the voltage, as required. In these transformers this operation is in the form of make and break (meaning that a new connection is made before the previous connection is removed). Otherwise, sparking occurs under high current, which is undesirable and unsafe. A small single-phase autotransformer is depicted in Figure 10.11.

Note that in an autotransformer the primary and secondary currents flow through the same winding. But the direction of the currents at each instant is such that the primary and secondary currents do not add together for the common part of the winding.

10.5.2 Isolation Transformer

Sometimes in electrical applications it is necessary that a device be isolated from the rest of the circuit feeding it. That is, the device receives a

Figure 10.11 A small autotransformer.

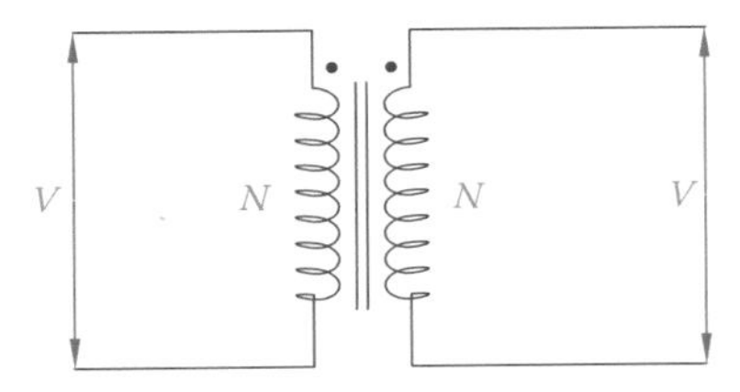

Figure 10.12 Insulation transformer schematics.

Isolation transformer: Transformer with the same number of turns in the primary and secondary windings, thus not changing voltage. Its purpose is only to protect a device from possible high currents in the primary side by isolating it from being directly connected to the mains.

voltage while not directly in parallel with other devices. In such a case an **isolation transformer** is employed. The schematic of an isolation transformer is shown in Figure 10.12. In an isolation transformer the number of turns of the primary and secondary windings is the same. As a result, the secondary voltage is the same as the primary voltage. Nevertheless, if by a fault in the primary circuit the current suddenly jumps up, the device connected to the secondary side is not affected. The isolation transformer, thus, is used when a device must be protected from damage.

Pad mount (transformer): Transformer that is mounted on a flat surface (a flat slab) as opposed to those which are designed to be mounted on an electric pole (pole mount).

Pole mount: Type of transformer which is suitable for mounting on an electric pole.

10.6 Transformer Mounting

While small transformers used with a device (such as a television) are mounted inside the device, large size transformers are stand-alone and must be somehow supported where they stand. In this sense, transformers are divided into **pad mount** and **pole mount**. As the name implies, a pad mount transformer is mounted on a platform on a horizontal surface, whereas a pole mount transformer is more suitable for mounting on the top of distribution line poles.

Figure 10.13 depicts a pad mount transformer of a wind turbine. The figure shows the transformer radiators for cooling the oil inside which the body of the transformer resides. Figure 10.14 shows three pole mount transformers.

Figure 10.13 A wind turbine pad mount transformer.

Figure 10.14 Pole mount transformers.

10.7 Three-Phase Transformers

For three-phase electricity transformation it is possible to use three separate but similar single-phase transformers, or a three-phase transformer (see both cases in Figures 10.13 and 10.14). A three-phase transformer is more compact and more efficient. The single core in a three-phase transformer can have various forms by combining the cores of single-phase transformers. Figure 10.15 shows the more common core forms. In practice, however, in addition to the basic structure, there are other concerns to be taken into account in order to increase the efficiency (see Section 10.9) and protection of a transformer. A cutaway of a medium-sized, three-phase transformer is shown in Figure 10.16.

In a three-phase transformer the three primary windings are connected to the supply voltage as a three-phase load for the power supply circuit; so, they can have star or delta (Y or Δ) connection. Independent of how the primary side is connected, the secondary windings can be connected together forming either a star or a delta. This implies that there are four

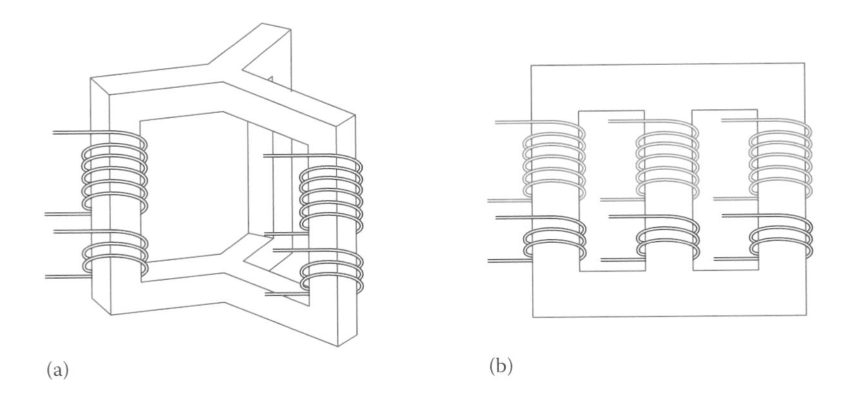

Figure 10.15 Basic structure of three-phase transformer. Transformer core structure can be as in (a) or in (b).

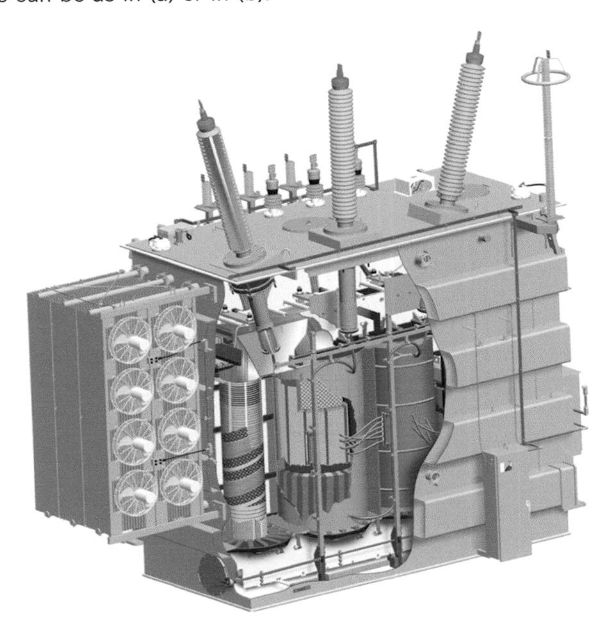

Figure 10.16 Construction of a three-phase transformer. (Courtesy of SPX Transformer Solutions, Inc. All rights reserved.)

ways that a three-phase transformer can be used in a circuit. The four ways are referred to as delta-delta, delta-star, star-delta, and star-star; these are shown in Figure 10.17. Thus, the voltage ratio not only depends on the turn ratio of the windings but also on what way the connections are made.

The choice of configuration depends on the application. A delta-delta configuration is used in many industrial (factories and manufacturing) applications. In electricity transmission that might not be the best choice, depending on various other factors. In practice, there are other concerns that are taken into account. Grounding for faults and the existence or excitation of harmonics are among these issues.

If the higher voltage has delta connection and the lower voltage side is connected in wye and has a ground, this is referred to *delta high-grounded*

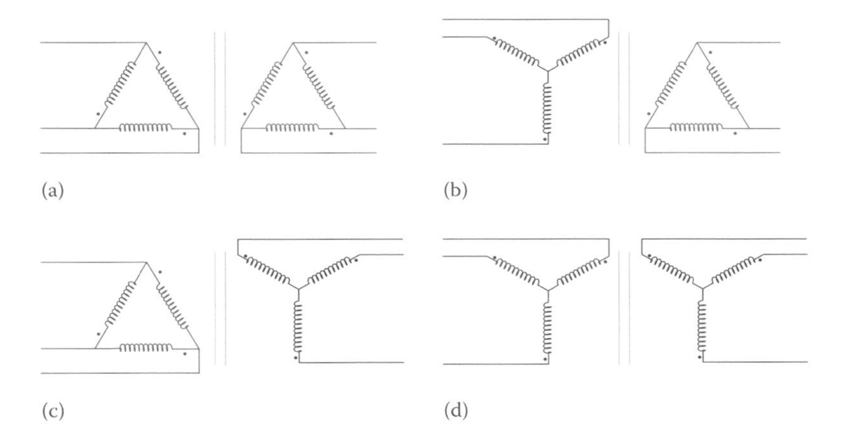

Figure 10.17 Four different configurations for primary and secondary windings connection in a three-phase transformer. (a) Delta – delta. (b) Wye – delta. (c) Delta – wye. (d) Wye – wye.

wye low. This is the standard configuration used by utilities for transmission and distribution. Some (large) transformers in power generation and distribution have auxiliary windings in addition to the main windings. This smaller winding is referred to as **tertiary**. In the case that the two main windings are wye connected because both provide a reference point for grounding, the **tertiary winding** is delta connected, and its purpose is to provide a path for the third harmonics. The purpose of the tertiary winding can be for providing a neutral point for grounding; for example, when both main windings are delta connected. A Y-connected tertiary then provides a reference for grounding. Another way of generating a neutral point for grounding is using a **zigzag transformer** (see Appendix C).

Tertiary winding: Auxiliary winding in a three-phase transformer used for providing a grounding point or for making a path for flow of certain currents, depending on the case.

Zigzag transformer: Special three-phase transformer in which the primary and secondary windings have the same number of turns and are combined in a particular way such that they cancel the magnetic effect of each other. It can be used for various purposes including fault detection.

In the two cases where both windings are connected the same way, the voltage ratio can be used, similar to the case of a single-phase transformer. In the other two cases the situation is different. Each winding in the primary is matched with a winding in the secondary. Figure 10.18 helps to understand this fact. In addition to the change in voltage ratio for these cases, there is a phase difference between the primary and secondary voltages. There is a 30° phase shift for star-delta and delta-star combination.

The current can always be found from the power relationships. The key issue is that for an ideal transformer the load on the secondary of a transformer determines the power that the transformer must deliver. In real transformers the power loss also must be considered. This is discussed in Section 10.9.

In three-phase transformers, voltage ratio depends not only on the turns ratio but also on the way (wye or delta) the primary and secondary windings are connected to their circuits.

For star-delta and delta-star combination in three-phase transformers, there is a 30° phase shift between the two sides of a transformer.

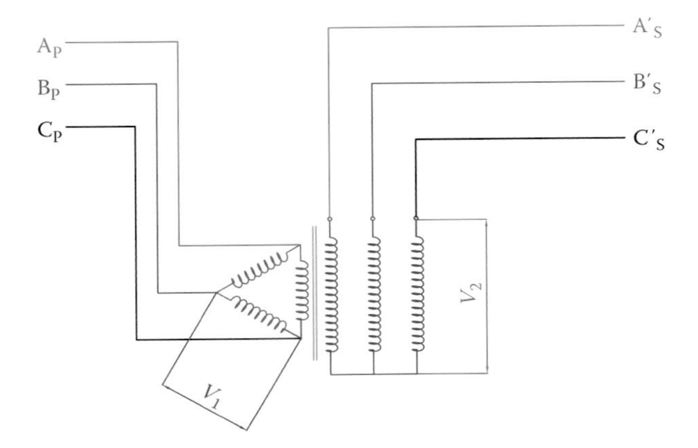

Figure 10.18 The voltage relationship between windings in a three-phase transformer.

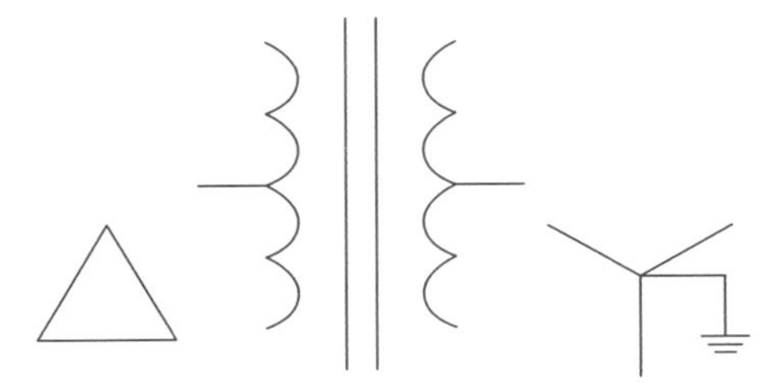

Figure 10.19 Symbol for showing three-phase transformer and connection configuration.

In practice, to show the connection configuration of three-phase transformers, a symbol as shown in Figure 10.19 is used. This figure shows that the primary is delta connected and the secondary is wye connected and is grounded at neutral. For three-phase transformers, high voltage terminals are marked by H_1, H_2, and H_3, and the other ones are marked by X_1, X_2, and X_3. H′ and X′ may be used for indicating the other side of each winding.

When connecting three-phase transformers or when connecting three single-phase transformers to form a three-phase transformer, special care must be taken in the correct order of connecting the 12 available terminals. Any connection made in the wrong order can lead to damaging the transformer and other equipment. An identification of various terminals and a sketch of how they must be connected, based on the required configuration is of paramount importance.

Example 10.4

The turns ratio in a step-up transformer is 1:2.65. The primary line voltage is 4160 V and the primary windings are connected in Δ. Determine the secondary line voltage (1) if the secondary windings are also connected in Δ and (2) if the secondary windings are connected in Y.

Solution

1. Because the primary and secondary windings have the same type of connection, the secondary line voltage is multiplied by the turns ratio:

   Secondary line voltage = (4160)(2.65) = 11,024 V

2. Voltage between the two ends of each secondary winding is 2.65 times of that of the primary, i.e., 11,024 V. Line voltage is the voltage between each two of phase lines, which is $\sqrt{3}$ times larger (see Chapter 9). Thus, the line voltage for this configuration is

   $$(11{,}024)(1.73) = 19{,}071 \text{ V}$$

Example 10.5

The primary winding of a transformer is Y connected to 11,000 V line. If the secondary winding is Δ connected, and the turns ratio in the transformer is 15:1, what is the secondary line voltage?

Solution

Because the primary is Y connected, the voltage affecting each winding is $\sqrt{3}$ times smaller. This smaller voltage is transformed and reduced by the turns ratio.

$$(11{,}000) \div \sqrt{3} = 6350.85 \text{ V}$$

$$6350.85 \div 15 = 423.4 \text{ V}$$

Thus, the secondary line voltage is 423.4 V.

Example 10.6

A three-phase load takes 25 A from the secondary of a step-down transformer supplying 418 V line voltage. If the transformer secondary windings are delta connected and its primary windings are Y connected, find the current in the primary of this transformer. The transformer turns ratio is 20:1.

Solution

The nominal apparent power for the load is $\sqrt{3}(418)(25) = (1.73)(10{,}450) = 18{,}100$ VA.

Because the secondary is delta connected and the primary is Y connected, the line voltage in the primary is

$$\sqrt{3}(418)(20) = 14{,}480 \text{ V}$$

Because for three-phase circuits apparent power = $S = \sqrt{3}\, V_L\, I_L$, the current in the primary circuit due to this load is

$$18,100 \div \left(\sqrt{3} \times 14,480\right) = 0.72 \text{ A}$$

Example 10.7

A three-phase delta-delta transformer is used to convert 14,400 to 380 V. Find the turns ratio.

Solution

Because nothing is mentioned about the two given voltages, they are the line voltages of the primary and secondary sides. Because both windings are delta connected, the ratio between the line voltages can directly determine the required turns ratio.

$$\text{Turns ratio} = a = 14,400 \div 380 = 37.9.$$

10.8 Transformer Applications and Power Rating

A transformer is usually designed for a specific purpose with a specific power capacity and a frequency range. This determines many mechanical and electrical parameters involved, such as core dimensions, number of turns, current, and wire size, as well as the type and amount of cooling, if necessary.

For all electrical transmission and for stepping down the voltage for all domestic and industrial devices the frequency is either 50 or 60 Hz, but transformers are also used for audio and video signals and for radio frequency (RF) applications. Audio transformers work at audio frequencies (50–20,000 Hz) and transformers for RF work at much higher frequencies. As the frequency increases, the size decreases. Also, the number of turns in each winding decreases. For this reason, the transformers for high-frequency signals are much smaller than those for 50 and 60 Hz. Most of transformers of this type are single-phase transformers and work at very small power, even at a fraction of 1 W (any necessary amplification of signals is performed after all the processing). Figure 10.20 illustrates two of these transformers in an electronic circuit.

On the other side of the spectrum, transformers for transmission and distribution of electricity may carry millions of watts of power. These transformers, normally three phase, need cooling for taking away the heat that is generated in their windings and core. An example of such a transformer is depicted in Figure 10.21. A large transformer that need cooling can be air cooled or oil cooled. Oil-cooled transformers are inside an oil-filled container. The oil circulates and removes the heat generated in the transformer. They might need a radiator and other components such as a pump, a fan, and an automatic system to start and stop these devices when necessary, based on the temperature.

Figure 10.20 Very small and low-power transformers in electronic circuits.

Figure 10.21 A wind farm transformer. (Courtesy of SPX Transformer Solutions, Inc. All rights reserved.)

While small transformers are rated in watts, like 500 W, the large industrial transformers are rated in volt-amperes (VA) (e.g., see Figures 10.8 and 10.21). Rating represents the maximum capacity of a transformer.

Like most machinery a transformer has the highest efficiency (see Section 10.9) if it is used at the rating it is designed for. Although it is possible to use a higher rated transformer at lower powers, the core losses makes it less efficient at lower powers.

10.9 Transformer Efficiency

Efficiency of any device or a system is the ratio of the output power to the input power. That is, the ratio of the power taken from the system to the power given to the system (or consumed by the system). On the basis of the definition, for an ideal transformer, the secondary power is the same as the primary power. That is, the output and input powers are equal. This implies an efficiency of 100 percent. The ideal transformer is for the purpose of determining the minimum voltage and current.

For a real transformer the efficiency cannot be 100 percent. Transformers, in general, have good efficiency. For well-designed large transformers it can reach up to 98 percent. For such a transformer in service for 24 hours per day even that 2 percent shortage from 100 percent can translate to a sizable amount of money in the long run. For this reason, whenever necessary, the efficiency of a transformer must be determined.

Hysteresis: Property of keeping an effect after its cause has ceased. This is particularly noticeable in electromagnets and electric machines that maintain the effect of magnetization.

Leakage (in transformer): Part of the (magnetic) flux produced by the primary winding that is lost and does not reach (link) the secondary of a transformer, due to structural imperfection.

Two main categories of power loss in transformers are core loss and copper loss (see Section 10.4). These two represent the energy that turns into heat, for which a transformer must be cooled. The main reason for core loss is eddy current as described in Section 10.4. In addition to eddy current the losses are due to **hysteresis** and **leakage** of magnetic energy. These all depend on the design, material, and construction quality of a transformer. Hysteresis loss is due to the nature of alternating current magnetizing effect on ferromagnetic metals. Each time the magnetic field is reversed, there is a chance for losing a small amount of energy. Leakage implies a tiny amount of magnetism lost or leaked out while flowing through the core. Core loss is almost independent of load, but copper loss depends on the current, thus it is load dependent. Input power to a transformer is equal to the output taken from it plus these two losses. In this sense, the efficiency of a transformer can be defined as

$$\text{Power efficiency} = \frac{\text{Output power}}{\text{Input power}} = \frac{\text{Output power}}{\text{Output power} + I^2R\ \text{loss} + \text{core loss}} \tag{10.3}$$

where the second term in the denominator represents the copper loss. It is written this way to indicate that it is load dependent. When the load on the secondary increases, the current goes up, and this loss goes up, too.

If the information about the core loss and the copper loss is known, the efficiency of a transformer can be found for each load. Note that both the primary and secondary windings contribute to the copper loss. In practice, the efficiency of a transformer for a given period can be found from a more practical equation as follows. The value found this way represents the **energy efficiency** of a transformer.

Energy efficiency: Overall efficiency of a transformer in a period of time, for example, in one year; That is, the ratio of the energy provided by a transformer in a length of time divided by the energy consumed by the transformer in the same period.

$$\text{Energy efficiency} = \frac{\text{Output energy for a given period}}{\text{Input energy for the same period}} \tag{10.4}$$

This equation deals with the measurement of the consumed energy by the transformer and the energy supplied to the transformer loads, instead of dealing with power components that might be difficult to measure. Usually, a period of 24 hours is used, and the term **all-day efficiency** is employed:

All-day efficiency: Average efficiency of a transformer during a 24-hour period.

$$\text{All day efficiency} = \frac{\text{Output energy for 24 hours}}{\text{Input energy for 24 hours}} \tag{10.5}$$

Note that even if the efficiency of a transformer is very good at full load, if it works at a fraction of its load, efficiency drops to a lower level. This is a problem that can be observed particularly in wind turbines because a wind turbine does not have the same rate of generation at all times. Its power generation depends on the available wind and its speed. In selecting a transformer for a wind turbine this fact must be taken into consideration.

Example 10.8

When a 8.5 kW load with a power factor of 0.82 is connected to a 10:1, 14.1 kVA single-phase transformer, the secondary current is 250 A. and the transformer efficiency is 94 percent.

(1) What is the current in the primary? (2) What are the losses in the transformer? and (3) What is the primary voltage?

Solution

1. Because the current in the secondary is known, the current in the primary can be found from the turns ratio and the efficiency.

$$I_1 = \frac{I_1}{a} \cdot \frac{1}{\text{eff}} = \left(\frac{250}{10}\right)\left(\frac{1}{0.94}\right) = 26.6 \text{ A}$$

2. Assuming the transformer has the same power factor as the load, the active power supplied by the transformer is

$$\frac{8500}{0.94} = 9043 \text{ W}$$

The transformer loss, therefore, is

$$9043 - 8500 = 543 \text{ W}$$

3. The apparent power taken by the transformer is

$$\frac{9043}{0.82} = 11{,}028 \text{ VA}$$

and based on the primary current the applied voltage is

$$\frac{11{,}028}{26.6} = 414.6 \text{ V}$$

Example 10.9

In the transformer of Example 10.8, the core loss is 300 W. What is the copper loss at full load under the same conditions, if the transformer efficiency at full load is also 94 percent?

Solution

The deliverable active power at full load is

$$\text{Power} \times \text{Efficiency} \times pf = (14{,}100)(0.94)(0.82) = 10{,}868 \text{ W}$$

and the power loss is

$$(14{,}100)(1 - 0.94)(0.82) = 694 \text{ W}$$

Because the core loss is independent of load, it remains at the same value as before. The copper loss, therefore, is

$$694 - 300 = 394 \text{ W}$$

Another way to look at this problem is to observe that in the previous problem the output power of the transformer was 11,027 VA, which translates to

$$\frac{8500}{10{,}868} = 0.782 \rightarrow 78.2\%$$

That is, the transformer was at 78.2 percent load, having a primary current of 25.6 A. At full load, thus, the primary current is

$$\frac{25.6}{0.782} = (25.6)(1.28) = 32.79 \text{ A}$$

and because the copper loss is proportional to the square of current, thus,

$$\text{Copper loss} = (543 - 300)(1.28)^2 = 398 \text{ W}$$

Example 10.10

Average power taken from a transformer is 25 kW, 10 hours per day. The energy consumption in 24 hr according to a meter is 280 kWhr. What is the all-day efficiency of this transformer?

Solution

On the basis of Equation 10.5, all-day efficiency can be directly determined as

$$\frac{(25{,}000)(10)}{280{,}000} = 0.89 \rightarrow 89\%$$

Example 10.11

The efficiency of a 750 kVA transformer is 98 percent. If the average duty cycle of this transformer when in service for 24 hrs, 7 days per week is 85 percent, find the amount of energy turned to heat each year, in term of BTU and calories. If each kilowatt-hour of energy costs \$0.08, determine how much is the cost of the energy lost this way (1 joule = 0.2388 cal; 1 joule = 1 watt-sec; 1 watt = 3.412 BTU/hr).

Solution

We first calculate the amount of electrical energy lost per year in kilowatt-hour.

Energy lost per year = (750)(1 − 0.98)(0.85)(365)(24) = 11,690 kW-hr
Cost of lost energy = (11,690)(0.08) = \$935.2
1 watt = 3.412 BTU/hr → 1 watt-hr = 3.412 BTU
11,690 kW-hr = (11,690,000)(3.412) = 39,886,280 BTU
11,690 kW-hr = (11,690,000)(3600) = 42,084,000,000 J = 42,084,000 kJ
42,084,000 kJ = (42,084,000)(0.2388) = 10,049,659 kcal

Another measure for performance of a transformer is its **voltage regulation**. Voltage regulation reflects how large or small is the change of voltage (due to voltage drop inside a transformer) when the load on a transformer changes, say from full load to 80 percent. Voltage regulation is expressed in percent and is determined from

Voltage regulation: Percent value representing how good a transformer is, obtained from the ratio of voltage loss to the transformer nominal voltage. The smaller this number is, the better the transformer.

$$\text{Voltage regulation} = \frac{\text{No load voltage} - \text{Full load voltage}}{\text{No load voltage}} \times 100 \qquad (10.6)$$

The smaller the voltage regulation, the better the transformer because it implies that there is smaller copper loss in the transformer. For an ideal transformer the voltage regulation is 0 percent because it has no loss and no voltage drop.

Example 10.12

A transformer is used to change 4160 to 120 V. If the voltage across a load is 116 V at full load, what is the voltage regulation of this transformer?

Solution

The turns ratio of the transformer is

$$a = \frac{4160}{120} = 34.67$$

Thus, the primary voltage at full load is

$$(116)(34.67) = 4022 \text{ V}$$

and the voltage regulation is

$$\frac{4160 - 4022}{4160} \times 100 = 3.3\%$$

10.10 Transformer Circuits

Loads of any type (resistive, inductive, and capacitive) can be connected to the secondary of a transformer. Thus, in general, the secondary of a transformer has a current that, depending on the values of R, L, and C, has a phase difference with its voltage. Depending on this current, primary current and its phase difference with the primary voltage can be determined. In what follows we discuss the way this is done, which is much more complicated than the relationship for the ideal transformer. The analysis is always done for a single-phase branch but can be extended with the rules in Chapter 9 for three-phase loads, even if not balanced.

10.10.1 Transformer Model

A simplified electrical model of a single-phase transformer or one phase of a three-phase transformer is shown in Figure 10.22. Primary winding has a resistance R_1. Also, in addition to the main winding inductance, an inductance L_1, which at the working frequency exhibits a reactance X_1 is added for the magnetic leakage of primary winding. Similarly, secondary winding has a resistance R_2 and an inductance L_2, the reactance of which at the working frequency is X_2. Core losses can be shown in the form of a resistance in a parallel branch with the winding. So, load change does not affect the current through this branch.

If a voltage V_1 is applied at the terminals of the primary winding, a current I_1 flows in the primary winding. This current consists of I_c, a current assumed for the core loss, called the **core loss current**, and I_m, the **magnetizing current**. Owing to the voltage drop in the winding the voltage magnetizing the core is E_1, which is smaller than V_1. Because of the mutual inductance between the windings, an AC voltage is available across the terminals of the secondary winding. With the secondary being open, this voltage is equal to E_2, which is found from the voltage relationships (Equation 10.1) for the ideal transformer. But, if a load is connected to the secondary winding this voltage reduces to V_2 ($V_2 < E_2$) because of voltage drop in the secondary winding (R_2 and X_2).

Core loss current: Current in the primary winding of a transformer that corresponds to the core losses; that is, it is equal to the core loss power divided by the primary voltage. It exists even if there is no load connected to a transformer.

Magnetizing current: Part of the current in a wire winding that corresponds to creating a magnetic field.

The current I_2 in the secondary is $V_2 \div Z_2$, where Z_2 is the load impedance. Before we can continue with the corresponding relationships for this

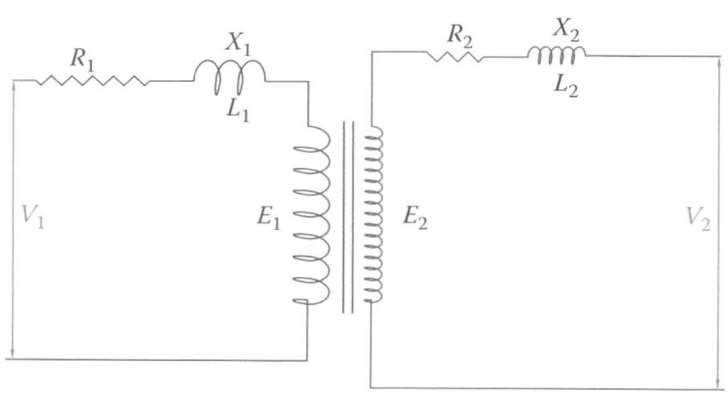

R_1 X_1 L_1 V_1 E_1 R_2 X_2 L_2 E_2 V_2

Figure 10.22 Transformer electrical model.

model and see how to find the current in the transformer primary due to a load, we need to associate the secondary parameters to the primary, as discussed in Section 10.10.2.

10.10.2 Primary-Secondary Load Interchange

It is customary for transformers to refer the loads in the secondary, to the primary. This implies replacing the secondary loads by their equivalent values, as if they were connected in series with the primary winding instead of the transformer. Writing

$$Z_2 = \frac{V_2}{I_2}$$

and substituting for V_2 and I_2 from Equations 10.1 and 10.2 in terms of V_1 and I_1 leads to

$$Z_2 = \left(\frac{N_2}{N_1}\right)^2 \cdot \frac{V_1}{I_1}$$

This relationship implies that the equivalent of Z_2 if substituted in the primary circuit is a value proportional to the square of the turns ratio. If the value is denoted by Z_2', then

$$Z_2' = \left(\frac{N_1}{N_2}\right)^2 Z_2 = a^2 Z_2 \tag{10.7}$$

Z_2' is said to be the *secondary load as seen from the primary*. This equation is general and is also valid for resistance, capacitance, and inductance, as well as the winding parameters (R_2 and X_2 in Figure 10.22). Moreover, the inverse of the same thing is valid if a value must be referred to the secondary from the primary circuit.

Example 10.13

If the turns ratio in a transformer is 10, what is the resistance as seen by the primary of a 5 Ω load that is connected to the secondary, assuming an ideal transformer?

Solution

It follows directly from Equation 10.7 that

$$R_2' = (10^2)(5) = 500\ \Omega$$

Example 10.14

A three-phase transformer changes 14,270 to 240 V. A 24 kW heating load is connected to each secondary. What is the resistance of this load as seen by the primary? Assume an ideal transformer.

Solution

The turns ratio for this transformer is

$$a = \frac{N_1}{N_2} = \frac{V_1}{V_2} = \frac{14{,}270}{240} = 59.45 \cong 60$$

Resistance of the load is

$$R_2 = \frac{V^2}{P} = \frac{240^2}{24{,}000} = 2.4\ \Omega$$

$$R_2' = (60^2)(2.4) = 8640\ \Omega$$

Note that the secondary current is $\frac{V_2}{R_2} = \frac{240}{2.4} = 100$ A.

On the basis of the current relationships the primary current for this load is $\frac{100}{60} = 1.67$ A.

And if the primary current is calculated from the primary voltage and its assumed load of 8600 Ω resistor, it is $\frac{14{,}270}{8640} = 1.65$ A. The slight difference observed here is because of the round-off errors (a is not exactly 60).

On the basis of the aforementioned discussion, Figure 10.23 depicts the electrical model of a transformer when the effects of the secondary winding resistance and leakage reactance are referred to the primary. We see that when the secondary is open there is no current in the secondary (I_2 is zero), but I_c and I_m still exist. Also, there is no voltage drop due to $\left(\frac{N_1}{N_2}\right)^2 R_2$ and $\left(\frac{N_1}{N_2}\right)^2 X_2$. The only voltage drop in V_1 is due to R_1 and X_1 and the currents I_c and I_m.

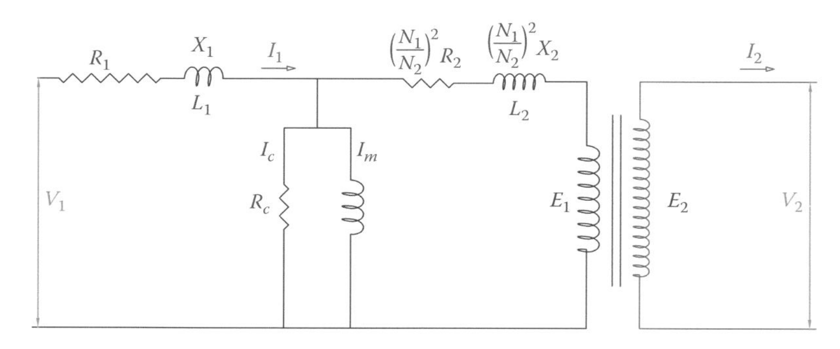

Figure 10.23 Transformer equivalent circuit when the secondary parameters are referred to the primary.

When there is no load, the current in the primary winding is denoted by

$$I_{1,\text{No load}} = I_c + I_m \tag{10.8}$$

This is the current that is ignored in an ideal transformer. When the secondary of a transformer is open, the transformer behaves like an inductor.

Open-circuit test: Test for measuring current in the primary of a transformer when the secondary winding is not connected to any load. This measurement determines the core loss (the losses for magnetization, leaks, and eddy current, which are load independent) of a transformer.

10.10.3 Measurement of Copper Loss and Core Loss

Core loss and copper loss at rated current are experimentally found for a transformer by two well-accepted standard tests. These are called **open-circuit test** and **short-circuit test**.

In the open-circuit test the secondary of a transformer is open, and there is little current in the primary (therefore, the copper loss is very little). To a good approximation, the reading of a wattmeter that measures the consumed power represents the core loss (the copper loss in this test is negligible).

Short-circuit test: One of the two important tests for transformers. Short-circuit test measures the copper losses of a transformer. The secondary winding is shorted for this test.

The short-circuit test measures the copper loss. For this test the secondary winding is short circuited and a low voltage is applied to the primary terminals. This voltage is much smaller than the rated voltage; it is experimentally adjusted such that the primary winding current reaches its maximum rated value. (If this test is done with the rated voltage, the transformer immediately overheats and gets damaged.) With a good approximation the wattmeter reading in this test represents the copper loss because this current is much more than that in the open-circuit test. Hence, a small part of this current corresponding to core loss (which still exists) can be ignored.

For a transformer, the open-circuit test determines the core loss and the short-circuit test determines the copper loss.

In the short-circuit test, the voltage that causes the rated current in the transformer primary has a specific significance. The percent ratio of this voltage to the rated voltage is called **transformer percent impedance** (denote by *Z%*). This number defines two things:

Transformer percent impedance: Percent ratio of the voltage in the short-circuit test of a transformer that causes the rated current to the rated voltage.

1. How much voltage drop in the primary exists under full rated current
2. How much the maximum secondary current is that a transformer can handle if a short circuit occurs in the secondary.

Item 2 is based on what secondary current causes 100 percent voltage drop in the primary winding. For industrial transformers smaller than 1 MVA the value for percentage impedance is around 5 percent, but as the power and the voltage rating increase this value increases up to 22 percent. The smaller this value is the better the transformer is. The following example shows a practical application of the percentage impedance.

Example 10.15

Percentage impedance of a single-phase 15 kV to 415 V transformer is 5 percent. If its rated power is 30 kVA, find (1) what the voltage drop in the primary is and (2) what the maximum short circuit current for the secondary is.

Solution

1. Maximum voltage drop in the primary at the rated voltage is

$$(15{,}000)(0.05) = 450 \text{ V}$$

2. Full load current in the secondary is

$$\frac{30{,}000}{415} = 72.29 \text{ A}$$

and the maximum short circuit current is

$$\frac{72.29}{0.05} = (20)(72.29) = 1446 \text{ A}$$

The information from the latter part can be used for selection of breakers and fault protection devices.

10.11 Chapter Summary

- All transformers work based on mutual induction between two windings.
- Windings must have a common core; core is made out of laminated ferromagnetic metal.
- A transformer can increase or decrease voltage.
- A step-up transformer increases voltage, and a step-down transformer decreases voltage.
- The input side of a transformer is called primary, and output side is called secondary.
- The key issue in a transformer is that the primary side power is equal to the secondary side power and the power losses.
- Power losses are categorized to core loss (due to eddy current, hysteresis, etc.) and copper loss (due to current flow in the wire windings).
- In an ideal transformer the losses are ignored. Primary power is equal to the secondary power.
- Rated values of a transformer are for the operating voltage and frequency, and maximum power that a transformer is designed for.
- A three-phase transformer can have a single core or it can be made up of a bank of three identical transformers.

- In a three-phase transformer the primary and secondary can be independently connected with star or delta configuration.
- There are four different configurations for connecting three-phase transformers, delta-delta, delta star, star-delta, and star-star.
- In delta-star and star-delta connections, there is a 30° phase difference between the primary circuit and the secondary circuit.
- Attention to polarity and primary/secondary sides is very important in making connections in transformers.
- There are two important tests for measuring transformer parameters, open-circuit test and short-circuit test. The former measures the core loss, and the latter measures the copper loss.
- The efficiency of a transformer is the ratio of output power from the secondary to the input power the primary takes from mains.
- All-day efficiency of a transformer is based on its output and input in a period of 24 hours.
- Transformer percent impedance can be used to see how much voltage drop occurs in the primary at full load. It also serves for a measure of the maximum current in a shorted secondary.

Review Questions

1. What is the main purpose of a transformer?
2. What is the basic construction of a transformer?
3. What are primary and secondary windings?
4. Is the primary voltage larger or smaller than the secondary voltage?
5. What does increase in a step-up transformer?
6. Define an ideal transformer.
7. Define an isolation transformer.
8. Is it possible in a step-down transformer to have a lower current in the secondary? Why or why not?
9. What is turns ratio?
10. Define volt per turn.
11. Is the volt per turn larger in the primary or in the secondary?
12. What are the two categories of main losses in a transformer?
13. If frequency changes, does it affect the performance of a transformer? Explain.

14. In a real transformer, where does the difference between the primary and secondary powers go?

15. Is the current larger in a transformer when it has a load or when it does not have a load?

16. What is the efficiency of a transformer?

17. What is a typical acceptable figure for the efficiency of a transformer?

18. What is the energy efficiency?

19. What is all-day efficiency?

20. What is the voltage regulation of a transformer?

21. Why is the core in a transformer laminated?

22. What is copper loss?

23. What are the two tests for a transformer? What are determined by these tests?

24. Can one use three single-phase transformers in place of a three-phase transformer?

25. What are the four configurations that a three-phase transformer can be connected?

26. Is there any phase difference between the primary and secondary voltages in a single-phase transformer?

27. Is there any phase difference between the primary and secondary voltages in a three-phase transformer?

Problems

1. The primary winding of a 10:1 single-phase transformer has 500 turns. Assuming an ideal transformer, what is the number of turns of the secondary?

2. If the transformer in Problem 1 provides 415 V, what is the primary voltage?

3. If the rated power of the transformer in Problem 1 is 15 kVA, what is the maximum secondary current?

4. A 500 W resistive load is connected to the secondary of a step-down transformer. Primary and secondary voltages are 220 and 110 V. What are the currents due to this load in the secondary and primary windings, assuming an ideal transformer?

5. The turns ratio in a step-up transformer is 8.4. The primary line voltage is 4160 V, and the primary windings are star connected. Determine the secondary line voltage (1) if the secondary windings are also star connected and (2) if the secondary windings are delta connected.

6. The primary winding of a transformer is Y connected to a 4100 V generator. If the secondary winding is Δ connected and the turns ratio in the transformer is 14.6:1, how much is the secondary line voltage?

7. A three-phase load is connected to a 208/120 V line by a delta configuration. The line is fed from a 600 V Y/Δ transformer. Find the current in the primary of this transformer if the load current is 75 A.

8. A single-phase transformer is part of a bank to transform 125 to 21 kV. If a Δ/Y configuration is employed, determine the line to neutral voltage in the secondary and the turns ratio of the transformer.

9. A three-phase delta-star transformer is used to convert 11,400 to 480 V. Find the turns ratio.

10. A load with a power factor of 0.85 is connected to the secondary of a 12,000/480 V, 48 kVA single-phase transformer. Secondary current is 62.50 A, and the transformer efficiency is 95 percent. (1) What is the current in the primary? and (2) What are the total losses in the transformer?

11. In the transformer of Problem 10, the core loss is 500 W. What is the copper loss at full load under the same conditions if the transformer efficiency at full load is also 95 percent?

12. During a 24-hour period the average power taken from a transformer is 30 kW. The corresponding energy consumption in 24 hr is 840 kW-hr. What is the all-day efficiency of this transformer?

13. A three-phase transformer rated at 75 kVA is used to feed a plant. The plant contains a number of motors, lighting, and auxiliary devices that altogether consume 70 kW energy. The bill paid at the end of a 30 day month for this plant is $5760. If the cost of electricity is $0.08 per kW-hr, what is the energy efficiency of this transformer?

14. A three-phase transformer is used to step down 4160 to 208 V. If the line to line voltage measured is 200 V at full load, what is the voltage regulation of this transformer?

15. Turns ratio in a three-phase transformer is 26:1. A 25 Ω load is connected to the secondary. What is the resistance of this load as seen by the primary?

16. A transformer is used for stepping up 7200 to 34,400 V. What is the impedance as seen by the primary for a 240 mH inductor connected to the secondary?

17. A transformer changes 660 to 240 V. If a 3 kW load is connected to the secondary, assuming an ideal transformer, what is the resistance of this load as seen by the primary?

18. The percentage impedance of each phase of a three-phase 480 to 120 V transformer is 5 percent. The transformer rated power is 4 kVA. (1) What is the voltage drop in the primary? and (2) What is the maximum short circuit current that the secondary protection circuit should handle?

19. Core loss and copper loss for a transformer have been determined to be 200 and 1700 W at full load. The full load secondary current is120 A. What is the total loss of this transformer at 80 percent load?

20. A 26:1 three-phase transformer is rated 45 kVA. If the primary voltage is 12,680 V, what is the voltage drop in the primary if the secondary voltage is 473 V? What is the rated primary current for this transformer?

21. A three-phase transformer with the turns ratio 5:1 is used to power a plant. The supply line voltage is 2400 V. If the secondary circuit allows a maximum current of 20 A, find what the maximum power that can be drawn is for each of the following cases of the transformer connection: (1) delta-delta, (2) delta-wye, (3) wye-delta, and (4) wye-wye.

Advanced Learning: Real Transformer Circuit Analysis

The phase difference between the voltage and current in the secondary and primary circuits of a transformer makes it rather difficult to analyze loads and determine power loss and voltage drop. In this section a simplified version for tackling transformer circuits is discussed. A few examples are included for illustrating the matter.

We refer to Figure 10.22 and the discussion in Section 10.10.1. We learned that if an AC voltage V_1 is applied to the primary winding, an AC voltage is available across the terminals of the secondary winding. If there is no load on the transformer, secondary current I_2 is equal to zero and this voltage is equal to $E_2 = E_1 \div a = \frac{N_2}{N_1} E_1$.

The following two equations constitute the relationships for a transformer:

$$V_1 = E_1 + ZI_1 \tag{A10.1}$$

$$I_1 = I_c + I_m + \frac{N_2}{N_1} I_2 \tag{A10.2}$$

ZI_1 represents the total voltage drop due to the load and the core and magnetizing losses. The variables in Equations A10.1 and A10.2 are not in phase with each other; thus, the values must be added in the vector form (see Figure A10.1a).

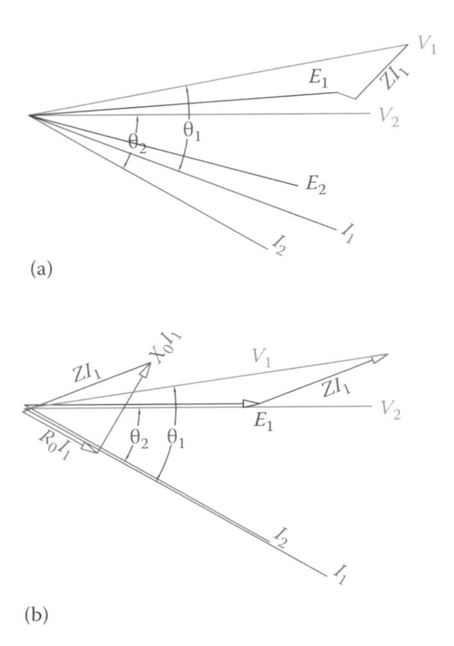

Figure A10.1 Phase relationships between I_1, I_2, V_1, and V_2 in a simplified real transformer. (a) Real situation: I_1 and I_2 are not in phase; V_2 and E_1 are not in phase. (b) Simplified form: I_1 and I_2 are assumed in phase; V_2 and E_1 are assumed in phase.

When a load is added to the transformer (secondary), secondary current I_2 is not zero anymore, and the secondary terminal voltage is no longer E_2; I_2 and V_2 depend on the type of load (inductive, resistive, capacitive) and the losses, which depend on the parameters of the transformer.

When the secondary has a current in it, the effects of R_2 and L_2 must also be brought into consideration. (They contribute to the voltage drop in the primary, which results in a decrease in V_1, E_2, and V_2.) The effect of the transformer parameter can be represented by Z_0 and the load impedance is denoted by Z_2. Z_0 is the total impedance due to the resistance and inductance of both primary and secondary windings (see Figure 10.23 and Equation 10.7). (Note that the R and X terms are added in the vector form. A bar on the letter denotes this.)

$$Z_0 = \overline{R_0} + \overline{X_0} = \overline{(R_1 + R_2')} + \overline{(X_1 + X_2')} \quad \text{(A10.3)}$$

The total load Z_1 on the primary is (when the transformer load is referred to the primary)

$$Z_1 = Z_0 + Z_2' = Z_0 + a^2 Z_2 \quad \text{(A10.4)}$$

and the secondary and primary currents can be found, respectively, from

$$I_2 = \frac{V_2}{Z_2} \quad I_1 = \frac{V_1}{Z_1}$$

when V_l and V_2 are known.

Example A10.1

Data for a 3:1 step-down three-phase transformer are R_1 = 2.3 Ω, R_2 = 0.07 Ω, X_1 = 8.47 Ω, and X_2 = 0.304 Ω per phase. If the three-phase balanced load on the secondary can be represented by a 42 Ω resistor (Z_2 = 42) per phase, find the total impedance of the primary.

Solution

For this transformer, all calculation is done for one phase because the load is balanced.

$$R_0 = R_1 + R_2' = 2.3 + (3)^2(0.07) = 2.93\ \Omega$$

$$X_0 = X_1 + X_2' = 8.47 + (3)^2(0.304) = 10.206\ \Omega$$

$$Z_2' = \left(\frac{N_1}{N_2}\right)^2 Z_2 = (3)^2(42) = 378\ \Omega$$

In this example the load is purely resistive because it is represented by a resistor. Together with R_0 the total resistive value of the load is 2.93 + 378 = 380.93 Ω. The reactance values are at 90° lagging. Thus, the total impedance of the primary is

$$Z_1 = \sqrt{380.93^2 + 10.206^2} = 381.067\ \Omega$$

Example A10.2

In Example A10.1 if the full load current in the primary is 89 A,

1. What is the applied primary voltage?
2. What is the voltage drop in the primary at full load?
3. What is the no load primary current?
4. What is the power rating of this transformer?

Solution

1. $V_1 = (381.067)(89) = 33{,}915$ V
2. Voltage drop $= \Delta V = Z_0 I_1 = \left(\sqrt{2.93^2 + 10.206^2}\right)(89)$

   $$= (10.618)(89) = 945\text{ V}$$
3. For this question with good approximation we may say that the ratio between the no load current to the full load current is the same ratio for the impedance values in 1 and 2. Thus,

   $$I_{1,\text{No Load}} = (89)\left(\frac{10.618}{381.067}\right) = 2.48\text{ A}$$
4. Primary voltage of this transformer is 33,915 V (nominally 33,000 V), and its full rate current is 89 A. Its power rating, therefore, is

   $$\sqrt{3}(33{,}915)(89) = 5{,}222{,}000\text{ VA} = 5222\text{ kVA}$$

In a real transformer a thorough analysis of the relationships between V_2 and E_1 is cumbersome, because of the phase differences as shown in Figure A10.1a and the core losses that are current independent. Here we use a simplified version based on certain assumptions that, to a great degree, are acceptable. The assumptions are as follows:

1. Primary and secondary currents are in phase with each other.
2. Voltages E_1 and V_2 are in phase with each other.
3. Core losses are ignored, because they are relatively small.

On the basis of these assumptions, the phase relationships between various transformer variables become simpler, as can be seen comparing Figure A10.1a and b. In this figure, ZI_1 is the effect of copper losses that is subtracted from V_1 to find E_1. The third assumption allows replacing Z by Z_0 in the figure.

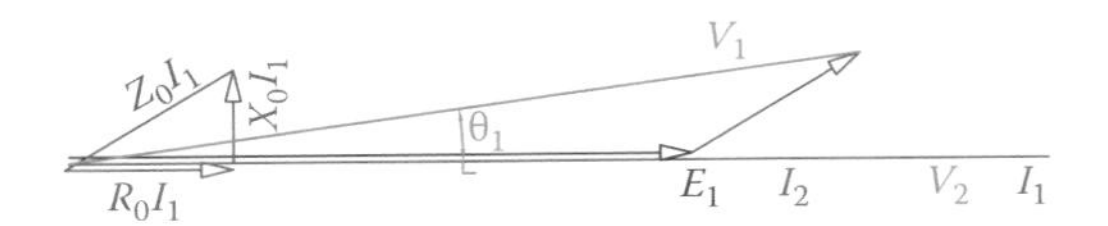

Figure A10.2 Phase relationships between I_1, I_2, V_1, and V_2 when $\theta_2 = 0$.

Note that even if $\theta_2 = 0$, θ_1 is not zero. When the load is purely resistive, or in the case of no load, $\theta_2 = 0$ and I_1 coincides with E_1 (as well as I_2 coinciding with V_2). Figure A10.2 illustrates a case when $\theta_2 = 0$.

In Figure A10.1b the terms R_0I_1, X_0I_1, and Z_0I_1 are exaggerated. In reality, as can be observed from Example A10.2, the value for Z_0I_1 is much smaller than V_1. In this respect, the difference between angles θ_1 and θ_2 is very small, particularly when θ_2 is nonzero. This allows us to approximate the relationship in Equation A10.1 to arithmetic sum instead of vector sum (see Figure A10.2).

Example A10.3

In a three-phase 33,000 V/11,000 V transformer a load takes 150 A from the secondary. If, the primary voltage is 33,200 V, determine (1) What the voltage V_2 is, (2) How much power the load takes, and (3) What the efficiency of the transformer is. Assume the same values of resistance and reactance as in Example A10.1 for the transformer.

Solution

1. Preliminary steps for calculation of R_0 and X_0 are the same as in Example A10.1 and are repeated here for convenience.

$$R_0 = R_1 + R_2' = 2.3 + (3)^2(0.07) = 2.93\ \Omega$$

$$X_0 = X_1 + X_2' = 8.47 + (3)^2(0.304) = 10.206\ \Omega$$

$$Z_0 = \sqrt{2.93^2 + 10.206^2} = 10.618\ \Omega$$

Because this is a step-down transformer, current in the primary winding is smaller than I_2:

$$I_1 = \left(\frac{N_2}{N_1}\right) I_2 = \left(\frac{1}{3}\right)(150) = 50\ \text{A}$$

2. Voltage drop in the primary winding is

$$\Delta V = (10.618)(50) = 531\ \text{V}$$

The secondary voltage, therefore, is

$$(33,200-531)\left(\frac{1}{3}\right)=10,890\ \mathrm{V}$$

and the power taken by the load is

$$S=\sqrt{3}(10,890)(150)=2,829,305\ \mathrm{VA}=2829\ \mathrm{kVA}$$

3. Efficiency is the ratio of the output power over the input power

$$\mathrm{Eff}=\frac{\sqrt{3}(10,890)(150)}{\sqrt{3}(33,200)(50)}\times 100=98.4\%$$

11

AC Motors and Generators

OBJECTIVES: After studying this chapter, you will be able to

- Describe the way most AC motors work
- Explain the difference between various AC motors
- Define what a rotating magnetic field is
- Recognize the difference between an AC motor and a DC motor
- Understand that AC motors and AC generators have the same structure
- Explain synchronous speed of an AC machine and understand various standard synchronous speeds
- Associate synchronous speeds with frequency
- Formulate power and other relationships for AC motors
- Understand the difference between single- and three-phase motors
- Calculate current for motors in order to select fuses
- Understand and apply relationships for power correction to motors
- Judge if a motor and a generator are compatible

New terms:
Alternator, asynchronous machine, breakdown torque, capacitor-run motor, copper loss, core loss, dynamic braking, full load torque, locked rotor current (LRC), locked rotor torque (LRT), pull-out torque, pull-up torque, reluctance, reluctance force, rotating magnetic field, rotor, shaded pole, slip, slip speed, split-phase motor, squirrel cage machine, stator, stray loss, synchronous motor, synchronous speed, universal motor, windage, wound rotor induction machine (WRIM)

11.1 Introduction

In Chapter 7 we discussed DC motors and DC generators. We discussed how a DC machine works and that DC motors and generators have the same structure, and we can simply call them DC machines. Moreover, we discussed how motors and generators work based on Lorentz force and Faraday's law, which could be interpreted as the interaction between an electric field and a magnetic field that could be created by a winding or a permanent magnet.

The way an AC machine works, in general, is not exactly the same way DC machines work, although, again, the structure of an AC motor and an AC generator is the same. The latter statement implies that if an AC machine is rotated by a prime mover (mechanical energy is given to it), it generates electricity (generator), and if it is provided with electrical power, it delivers mechanical power (motor). Despite this similarity, a machine can be made for one purpose only, either as a motor or as a generator. One could say, in general, that AC machines work based on the interaction between two magnetic fields. One of them can be a permanent magnet in some machines.

In this chapter we learn the principal aspects of AC motors and generators (AC machines). Because understanding of the subject is easier with three-phase motors, we start with three-phase machines.

11.2 Main Categories of AC Machines

AC machines (generators and motors) initially fall into single phase (denoted also by 1 ϕ or one phase) and three-phase motor (3 ϕ). In each of these categories, then, we have synchronous and asynchronous (also called induction) machines. Induction machines are of two types, wound rotor and squirrel cage. There is, moreover, another type of motor that is called a **universal motor**. A universal motor is a single-phase motor (no three-phase counterpart) that works with both AC and DC; thus, it's called universal. A universal motor is, in fact, a series wound DC motor. We will see how, in accord with the principle based on which it functions, it can use AC electricity as well as DC.

Universal motor: Type of motor that can work with both AC and DC. A series-wound DC motor has this capability.

There are other types of motors that are either very specific in the way they work or are used for very specific applications. Just to name a few, there are linear motors, which have a linear motion, rather than rotational motion, ultrasonic motors that are very small in size and capacity, hysteresis motors, step motors, and brushless DC motors (which is, indeed, an AC motor with a DC-powered driver). Except for the linear motors, which can be used for traction of railed vehicles, the others are normally small (fraction of a horsepower). Also, the term *machine* cannot be applied to these motors because they cannot work as generators or even if they do, nobody uses them as generators.

Before one can learn about how an AC motor works, it is necessary to understand the concept of **rotating magnetic field**. All three-phase motors work based on a rotating magnetic field; that is, a magnetic field that, though not visible, has a rotational motion.

Rotating magnetic field: Magnetic field with constant strength but rotating about an axis, formed in the stator of three-phase machines when connected to electricity. Such a rotating filed is developed because the three windings are physically apart by 120°.

11.3 Rotating Magnetic Field

Referring to electromagnets (see Section 7.3.2) if a coil is connected to a DC source a magnetic field is generated in the coil, the direction of which is found according to Figures 7.6 and 7.7. Consider now if the source is AC with a frequency of 50 Hz. Again, a magnetic field is generated, but this time its north and south pole swap places rapidly with the change in the current direction (100 times per second).

Now, consider three pairs of similar but separate windings that are connected to the three phases of a three-phase supply, each pair connected to one phase. Because the voltages in the three phases are not in phase with each other, the magnetic fields, accordingly, are not reaching their maximum and minimum strength at the same instants. Figure 11.1 shows these windings that are physically placed 120° from each other. Because three-phase voltage has a sinusoidal pattern, the strength of each magnetic field traces a sinusoidal pattern, which continuously varies from zero to a maximum and back to zero, and then changing direction, repeating the

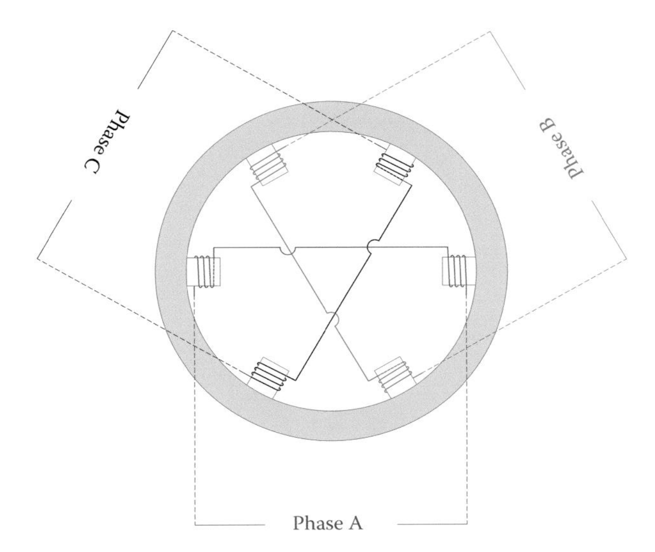

Figure 11.1 Three windings connected to three voltages of a three-phase system.

same pattern in the opposite direction. We may schematically show the resulting magnetic field by two opposite vectors that depict the direction and only the maximum and minimum lengths, corresponding to the peak values. As said before, because voltages in a three-phase system are 120° out of phase, the variations of the three fields are not coincident, and their maximum and minimum values do not occur at the same time.

Figure 11.2 illustrates the direction of the magnetic field for the maximum positive and the maximum negative values of the current in phase A, the maximum currents for phased B and C, and for some negative current in phase B.

At each instant of time the resulting magnetic field is obtained by adding together the three individual magnetic fields developed in the three windings. For this addition, however, as well as the magnitude, direction of the magnetic fields must be brought into account because the windings are at 120° from each other. In other words, addition implies the vector sum of the three magnetic fields. Because the magnetic fields are not constant, at each instant of time the resultant magnetic field has a different direction within 360°. Some of these positions are shown in Figure 11.3. It can be graphically and mathematically shown that the resultant has the same magnitude, but its direction continuously changes. In fact, it can be seen that the resulting magnetic field rotates with time. Speed of rotation is constant and for the arrangement in Figure 11.1 (one pair of windings for each phase) is the frequency of the AC line. That is, if the line frequency is 50 Hz, the speed of the rotating magnetic field is 50 revolutions per second.

For three winding pairs, as was the case in Figure 11.1, for each one cycle of current, the resulting magnetic field rotates by one revolution. The same arrangement can be practiced with two sets of windings (six pairs) connected to three phases in a correct order and each one being physically at 60° from the neighboring windings (instead of 120°). In such a case, it can be seen that for each full cycle of current variation the magnetic field revolves half a revolution.

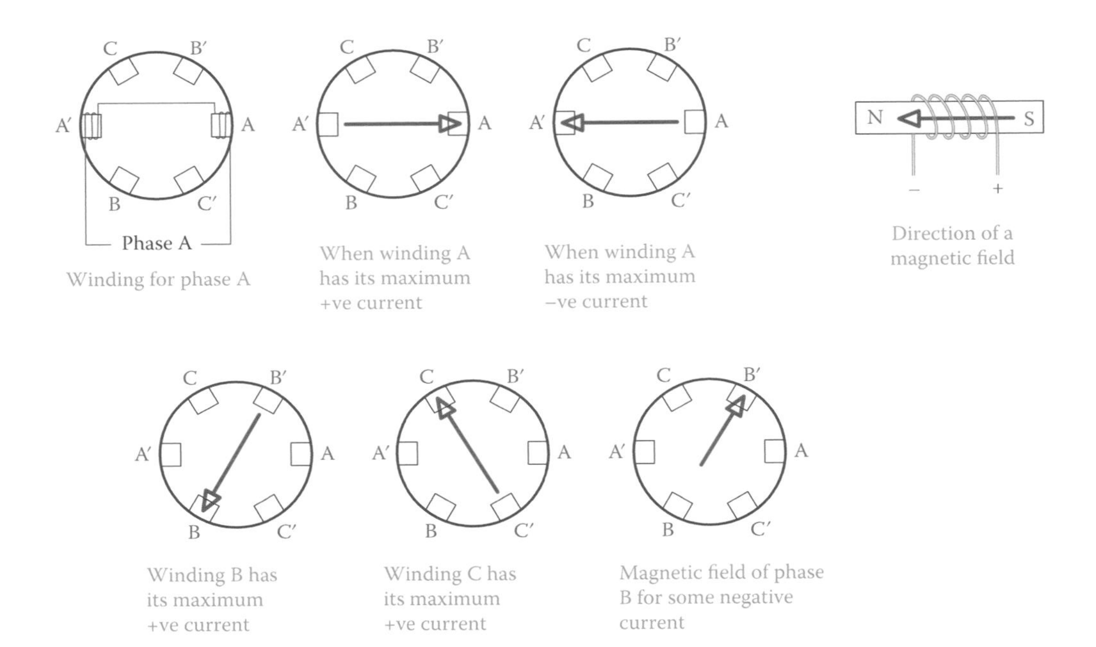

Figure 11.2 Direction and magnitudes of magnetic fields of windings in Figure 11.1 at a number of times.

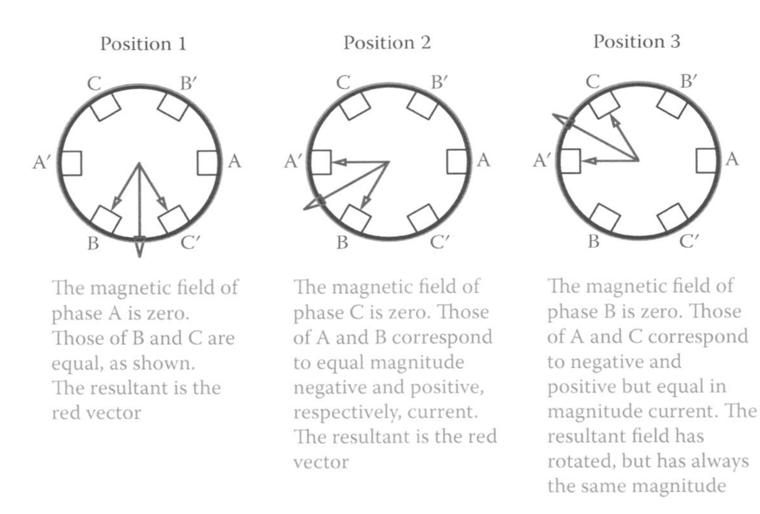

Figure 11.3 Illustration of three positions of the rotating magnetic field.

By increasing the number of windings and connecting them in a proper manner, the number of revolutions per second of the magnetic field can be modified. This is what is done in AC machines to change the operating rpm. The relationship between the number of poles (each pair of windings counts for one set of north and south poles) and the rotating speed is

$$N_S = \frac{60f}{p} \tag{11.1}$$

In this equation, N_S is the rotational speed in revolutions per minute (rpm), f is the line frequency in Hertz, and p is the number of pairs of poles per phase. That is, for six windings as shown in Figure 11.1, for instance, p is equal to 1. The subscript S in N_S represents the **synchronous speed**. For any AC machine, synchronous speed is determined by the frequency of the line it is connected to because the number of poles is a physical constant when the machine is built. In some machines the arrangement of poles can be modified; in such a case a machine can have two speeds (one twice the other). This is done outside the machine by a special switch.

Synchronous speed: Special speed for an AC machine depending on its number of poles and the frequency of the AC line.

> In three-phase machines a rotating magnetic field is developed by the stator windings.

For lower-speed machines (higher number of poles) poles are distinct, but for a lower number of poles windings are distributed around the periphery of a cylindrical support (Figure 11.4). The set of windings forming the magnetic poles is called the **stator**. These windings do not rotate but create the rotating magnetic field. Figure 11.4 shows the stator winding of a three-phase generator. The generated magnetic field has its strongest effect in the cylindrical space inside the winding, where the rotor is inserted, and its axis of rotation coincides with the stator axis.

Stator: Stationary part of AC electric machines, as opposed to the rotor that rotates.

Direction of rotation of the magnetic field depends on the order the windings are connected to a three-phase circuit. If (any) two connections are interchanged, the direction of rotation of the magnetic field reverses.

The rotating part of an AC machine is called a **rotor**. It has a cylindrical form and is supported by bearings; it resides inside the stator and has a very small clearance from the inner diameter of the stator windings. Remember that in DC machines two different terms are normally used instead of stator and rotor. Figure 11.5 illustrates a cutaway of a three-phase machine; notably, it shows the rotor that sits inside the rotating magnetic field. This machine is much smaller than that in Figure 11.4, and all the inside components are on the same frame, whereas for large machines

Rotor: The rotating parts of any motor, generator, turbine, and so on. In a propeller-type turbine, the motor consists of the hub and the blades.

Figure 11.4 Stator windings of a three-phase generator. (From Hemami, A., *Wind Turbine Technology*, 1E ©2012 Delmar Learning, a part of Cengage Learning Inc. Reproduced with permission from http://www.cengage.com/permissions.)

Figure 11.5 Cutaway of a three-phase motor. (Courtesy of ABB Inc.)

each component is separately supported. Most small machines have fins at one end of the rotor structure, which function as a blower when rotating, forcing air across the hot components (mainly the stator windings) in order to carry the heat away. This is also shown in Figure 11.5.

> If (any) two connections of a three-phase system to the stator of an AC machine are interchanged, direction of rotation of the rotating magnetic field reverses.

In order for AC machines to work, the stator must be connected to electricity. Unlike DC machines, it is not necessary to also connect the rotor to electricity. When this connection is made, a rotating magnetic field is created inside the body of the stator. It is like a rotating magnet that can pull and move ferrous metals with it.

> The stationary part of an AC machine is called the *stator.*

Example 11.1

The stator of a three-phase AC machine has a total of 24 poles. What is the rotating speed of the magnetic field developed in this machine when it is connected to 60 Hz line?

Solution

Because the machine is three-phase, the number of poles per phase is 8. Because poles have to be in pairs of north and south, then there are four pairs of poles; thus, p = 4. Using $p = 4$ and $f = 60$ in Equation 11.1 gives

$$N_S = \frac{(60)(60)}{4} = 900 \text{ rpm}$$

TABLE 11.1 Synchronous Speed for Various Number of Poles

Number of pole pairs per phase	1	2	3	4	5	6	8	9	10	16	20	40
Total number of poles	6	12	18	24	30	36	48	54	60	96	120	240
Synchronous rpm for 60 Hz	3600	1800	1200	900	720	600	450	400	360	225	180	90
Synchronous rpm for 50 Hz	3000	1500	1000	750	600	500	375	333	300	187	150	75

11.4 Synchronous Machines

A synchronous machine has a rotating magnetic field, as described in Section 11.3, in its stator when connected to a line of the compatible voltage and a magnet in the rotor. If the magnet in the rotor is rotated by mechanical means, we have a generator, otherwise, a motor. A synchronous machine is characterized by its synchronous speed, as defined by Equation 11.1. Because the common operating frequencies in the world are 50 and 60 Hz, the synchronous speeds can be uniquely defined for the number of poles of a machine.

Note that in Equation 11.1, p is the number of *pairs of pole per phase*. Alternatively, this formula can be written as

$$N_S = \frac{60f}{P/6} = \frac{360f}{P} \tag{11.2}$$

where P is the *total number of poles* in the stator (P = 6p). Table 11.1 shows the synchronous speeds for 50 and 60 Hz for common synchronous machines.

11.4.1 Synchronous Motors

Synchronous motors always have a constant speed that is determined by the relationship in Equations 11.1 or 11.2. This is a desired requirement for certain loads that no matter how much the load varies (the torque and power requirements), speed must be constant. Moreover, the synchronous motor has another property that is desirable. It can be run at any power factor. Thus, if alone, it can be run at pf = 1, or if run in a circuit that has many other types of motors it can run at a leading power factor to compensate for the lagging power factor of the other motors (other AC motors have a lagging power factor). In this sense, while running, it can induce a capacitive effect in the circuit to which it is connected.

Synchronous motor: Type of AC motor (usually very large size) that is insensitive to load change and always runs at a fixed (synchronous) speed, unless the load becomes larger than the motor capacity (when it stops working).

A synchronous motor is not self-starting.

Despite the desirable properties of a synchronous motor, it is only used when the requirement justifies for the extra cost of a synchronous motor. A synchronous motor is much more expensive than other type of motor of equal power, and for this reason it is made only in large sizes. The main reason for the high cost of a synchronous motor is its inability to self-start. That is, a synchronous motor is not self-starting. This means that it needs to be started by some additional means. The reason a synchronous motor is not self-starting is because of how it functions.

Figure 11.6 shows how a synchronous motor works. Suppose that there is a rotating magnetic field inside the stator of a three-phase machine. This is like having a large magnet that is rotating. Furthermore, assume that you could manage to put a magnet inside the stator, its axis coincided with that of the magnetic field. Because the magnetic field has a rather high rotational speed, the magnet stays stationary. If the magnetic field was not rotating, the magnet would orient itself aligned with it.

Now, if the above magnet is initially helped to rotate so that it reaches a speed around that of the magnetic field, then it will snatch the magnetic field and latch to it, hence, continuing to rotate with the magnetic field. If the external auxiliary mover is disconnected now, the magnet continues to rotate with and at the same speed as the magnetic field, getting its power from the magnetic field.

The above description clearly indicates that for a synchronous motor to work it is necessary first to rotate the rotor and bring its speed up to near the synchronous speed. Then, after this is done, the motor continues to run without the need of the external starter. For larger machines the starter can be a pneumatic or another electric motor that can move the unloaded synchronous motor and bring its speed up to within a few percent of the synchronous speed. In the smaller motors, this can be another type of AC motor, sometimes embedded in the same frame. A special switch will disengage this auxiliary motor after latching to the synchronous speed has taken place.

It also follows from the preceding discussion that a synchronous machine works based on the interaction between two magnetic fields (the rotating magnetic field developed by the current through the windings and

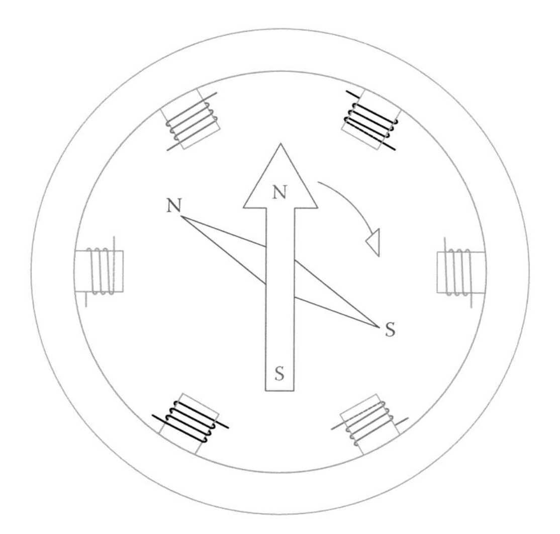

Figure 11.6 Principle of operation of synchronous motor.

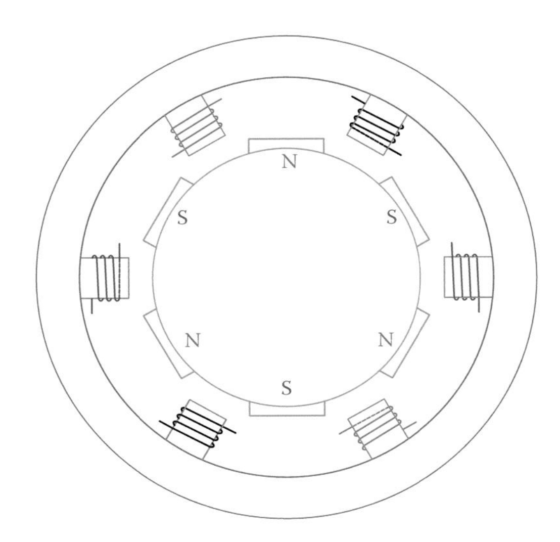

Figure 11.7 The rotor magnet is uniformly distributed around the rotor.

that of the magnet). Although the description of how a synchronous motor works suggests that the rotor of a synchronous motor must be a magnet, it is possible to interchange the rotor and stator roles and have a stationary magnet at the stator. In the larger machines, normally, the rotor is a magnet; otherwise, large currents have to be passed through the brushes and slip rings. It can be a permanent magnet or it can be an electromagnet that is powered by a DC source. For very large motors (and generators) the DC source can be a DC generator embedded within the structure of the synchronous machine (sharing the same shaft), which is solely used for magnetizing the rotor. Furthermore, instead of a commutator for the DC generator, an electronic converter can be used, as discussed later on in Chapter 20. For those machines with the magnet rotor, only the stator is connected to the power line. In smaller motors the stator can be a permanent magnet and the rotor is connected to the outside line through three slip rings.

The magnet in the rotor of a synchronous machine is not a single magnet (one pair of north and south poles). To have more uniformity of motion and torque, the rotor consists of a number of magnets uniformly placed around the rotor, as schematically shown in Figure 11.7, based on the stator number of poles.

The rotor in synchronous AC machines is not connected to electricity.

11.4.2 Synchronous Generator

A three-phase synchronous generator is the standard type for conventional power generation at a commercial level. Except in wind turbines, or in very isolated cases, power plants employ a synchronous generator, generally called an **alternator**. A synchronous generator has exactly the same structure as a synchronous motor. Because generators are usually very large (up to 1 GW for some nuclear power plants), the rotor is the

Alternator: Generator of alternating current (a machine that produces alternating current electricity when turned).

magnet that rotates. In this way, slip rings are avoided (slip rings need regular maintenance because of the wear of the brushes and sparking under high currents). As the magnetic field of the rotor cuts the wires of the three sets of windings in the stator, three-phase electricity is developed in the stator windings. The rotor, again consisting of a set of magnets as shown in Figure 11.7, must be rotated by a prime mover at precisely the machine synchronous speed (depending on the machine's number of poles) to generate electricity at a desired frequency.

The formulae for synchronous speed still apply, but this time the frequency is a function of the rotating speed. Thus, if a machine with a total of P poles is rotated at N rpm, the frequency of the generated electricity is

$$f = \frac{N \cdot P}{360} \tag{11.3}$$

So, for example, if a machine with a total number of 18 poles is rotated at 1800 rpm, the frequency of the generated electricity is 90 Hz, and if rotated at 900 rpm, the frequency is 45 Hz. To generate electricity at standard frequencies of 50 and 60 Hz, thus, the generator must be rotated only at synchronous speeds listed in Table 11.1. For $P = 18$ this speed is 1000 rpm for 50 Hz and 1200 rpm for 60 Hz.

Table 11.1 illustrates that for a given frequency the lower the number of poles, the faster the generator must be turned. In this respect, generators that are energized by steam and gas turbines have a smaller number of poles compared to those powered by hydraulic turbines because steam and gas turbines run at a higher speed than hydraulic turbines. For instance, in a water turbine driven generator the total number of poles can be 96 ($P = 96$); that is, 16 pairs per phase ($p = 16$). In order for the electricity to have a frequency of 60 Hz, the turbine must run at 225 rpm.

The number of poles in a generator determines its structure and diameter size owing to the fact that all the poles must be accommodated around the periphery of the generator stator. Accommodating more poles requires a larger diameter. For the same power capacity, generators with fewer poles are slimmer and longer, and those with more poles are larger in diameter but shorter in length.

When frequency is not important (e.g., for heating elements or when the generated power is to be converted to DC), it is thus possible to run a generator at any speed. For instance, in a certain mode of operation of wind turbines a synchronous generator produces electricity at a frequency determined by the wind speed, but this is converted to DC before it is converted back to AC with the desired frequency.

11.4.3 Characteristic Curves of Synchronous Machines

In many synchronous machines, in particular the larger ones, the rotor magnetic field is not supplied by a permanent magnet; instead, it is created by an electromagnet powered by a DC generator. In this way the strength of the rotor magnetic field can be controlled by varying the DC

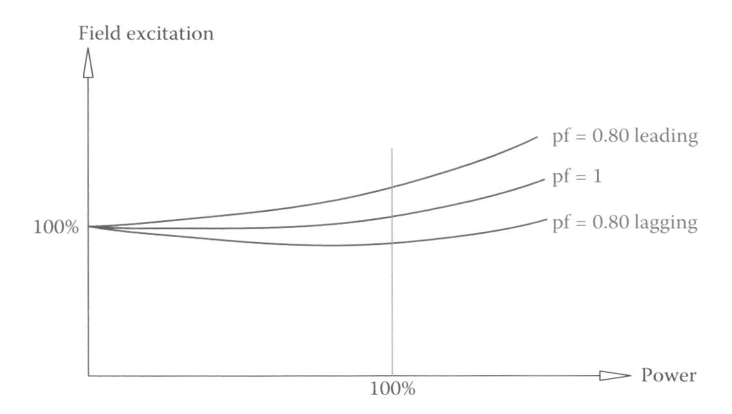

Figure 11.8 Over- and under-excitation of a synchronous machine.

voltage. This voltage is applied to the rotor winding. Increasing this voltage increases the excitation current through the rotor winding.

In the majority of applications (no matter a generator or a motor) the stator voltage is constant, but the power demand from a generator or the mechanical power of a motor can vary. For a constant stator voltage as power demand increases, the rotor excitation current must be increased accordingly. This is shown by the middle curve in Figure 11.8. Nevertheless, it is possible to overexcite the rotor by increasing the DC voltage fed to the rotor winding. It is also possible to provide less voltage to (under-excite) the rotor. The effect of field over-excitation in a synchronous machine is to run it (both generators and motors) with a leading power factor and the effect of field under-excitation is to run the machine with a lagging power factor. Typical curves for these conditions are also depicted in Figure 11.8 for pf = 0.8.

As a motor, this property can be used to correct the power factor of a circuit. Specifically, in the cases that there are many motors connected to a circuit because the motors behave like inductors and cause a circuit to lag, if one or more of the larger motors is a synchronous motor it can improve the power factor of the circuit when running. For this purpose the motor(s) must run with a leading power factor (see Section 11.8.5).

11.4.4 Single-Phase Synchronous Machines

Whereas single-phase synchronous generators are very common for small electricity production in fixed or mobile units, they do not have a place in commercial electricity production. The structure of a single-phase alternator can be the same as its three-phase counterpart except that it has only one set of windings instead of three sets. In this sense, Equation 11.1 can also be used for a single-phase generator (but not Equation 11.2).

Alternatively, instead of having the magnet in the rotor and rotating, it can be in the stator and stationary, like in DC machines. In such a case, the rotor has a winding in which electricity is generated and is transferred to outside through slip rings. This is shown in Figure 11.9. Note that the structure of such a machine and how electricity is generated is exactly the

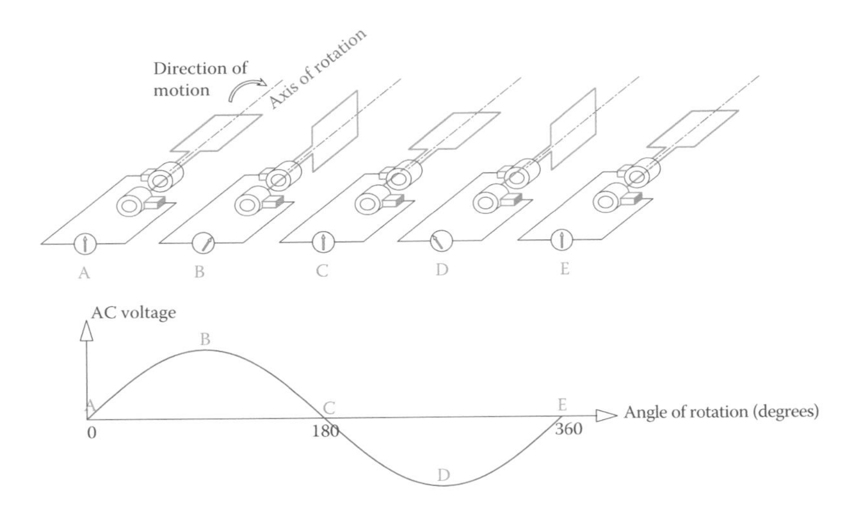

Figure 11.9 Single-phase alternator similar to a DC machine without commutator.

same as happens in a DC generator, with the exception of the commutator, which converts the otherwise AC electricity to DC electricity. In Figure 11.9, variation of the generated voltage during one turn of the rotor (for a single pole machine) is illustrated and identified by letters A to E. In this sense, a DC generator and a single-phase alternator can have identical construction with the exception of the commutator versus a pair of slip rings, which helps to distinguish them from each other.

As for single-phase synchronous motors with the structure as explained earlier, they are not built. There are, nevertheless, some single-phase motors (such as reluctance motor) that have a constant speed. Because of this constant speed property, some people categorize them as synchronous motors, whereas they work based on a different principle than was described in this section (see Section 11.7).

11.5 Asynchronous or Induction Machines

Asynchronous machine: Another name for induction motor (see induction machines).

Asynchronous machines are more known as induction machines because their principle of operation is based on induction. Recalling Faraday's law from Chapter 7 (see Section 7.5), if a wire is moved in a magnetic field, then a voltage difference is generated between its two ends. We may also say a voltage or an electromotive force is induced in the wire. Remember, also, that it is the relative motion between a wire and a magnetic field that counts; that is to say, the wire can be stationary while the magnetic field moves.

This principle, together with the rotating magnetic field, constitutes the basis for operation of induction motors and generators. Hence, one can say the difference between synchronous and induction machines lies in their rotors. In other words, it lies in how the magnetic field in the rotor is made. In this sense, the structure of the stator for a synchronous machine and an induction machine is the same, and only the rotors differ from each other. That is to say, in principle, one can interchange the stators of two similar (in size and power) synchronous and induction machines.

There are two types of induction machines. The difference again comes from the structure of the rotor, and the two types are named based on the rotor structure. One type is called **squirrel cage machine** and the other is called **wound rotor induction machine (WRIM)**. The rotor of the squirrel cage machine has no winding, and there is no need for the rotor to be electrically connected to any electricity. The rotor of the latter, however, as the name implies, has windings. The windings must be connected to circuits outside of the rotor, though not to the three voltage lines. The arrangement for windings on the rotor, which rotate, to be connected to outside circuitry that is stationary is done through slip rings. Slip rings, as briefly described in Section 7.4, are metallic circular rings mounted and rotating with the rotor shaft and connected to the rotor windings but insulated from the rotor body. Brushes are spring loaded to make good contact with the surface of the rings. They do not rotate and are connected to the outside circuitry. Remember that in DC machines we had brushes and commutators. The difference between slip rings and commutators is that the former is made of one piece of metal, whereas the latter consists of many isolated pieces of metals around a ring. In both cases they are isolated from their shafts. Also, for DC machines, there is only one commutator, but for AC machines there are two (for single phase) or three (for three-phase) rings.

Squirrel cage machine: Type of alternative current induction machine in which the rotor winding has little resistance and thus carries very high current. To withstand high currents, the rotor structure is modified and more resembles a cage than a winding.

Wound rotor induction machine (WRIM): Type of induction machine for AC in which the rotor has wire winding. The windings are accessible through slip rings. Another type is (squirrel) cage machine that has no wire winding and has no slip rings.

11.5.1 Three-Phase Wound Rotor Induction Machine

Consider Figure 11.10 in which a one loop wire (for simplicity) is placed inside a rotating magnetic field.

Notice that the single wire loop is connected to a resistor and together they form a closed loop. Also, notice that the resistor is external to the

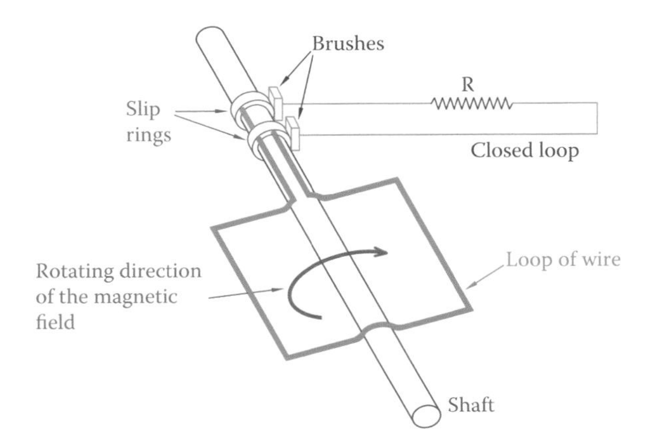

Figure 11.10 One loop of rotor winding connected to outside circuit through slip rings. (From Hemami, A., *Wind Turbine Technology*, 1E ©2012 Delmar Learning, a part of Cengage Learning Inc. Reproduced with permission from http://www.cengage.com/permissions.)

wire loop and its connection to the wire loop is through slip rings that are mounted on the shaft holding the loop. The loop ends are fixed to the slip rings, and two brushes make the connection between the slip rings and the external circuit.

Rotation of the stator magnetic field is equivalent to moving the wire in a stationary field. The following lines describe what happens as a result:

1. Because the magnetic field is moving, it induces a voltage in the wire.
2. Because the wire ends are connected to the resistor and form a closed circuit, a current is developed in the loop (including the resistor), proportional to the induced voltage.
3. Because of the current in the loop, a force is generated that pushes each side of the loop wire in opposite directions, thus creating a torque. This torque makes the wire loop and its shaft rotate.
4. This is what happens in a wound rotor AC induction motor. There is no connection to the electricity for the rotor winding (the loop wire), but the rotor windings make a closed circuit through the external resistor.
5. We see that the torque develops if there is a current in the rotor windings. If there is no current, the torque diminishes. In other words, as long as there is a current in the rotor winding, motion exists. This current exists when there is a relative motion between the magnetic field and the rotor winding. If the rotor runs at the same speed as the magnetic field, then there is no relative motion. For this reason, to maintain current, in a motor the rotor always runs slower than the magnetic field.
6. For simplicity and clarity, in Figure 11.10 and the above description, there was only one loop. Other loops at different angles can be connected in parallel with the loop shown, using the same slip rings.
7. One loop, or a few loops in parallel with each other (two connections), corresponds to a single-phase machine. But, the discussion is equally valid for three-phase machines, having three separate loops (or three sets of parallel loops). The winding in the rotor of a single-phase machine has two terminals that must be connected to the outside of the rotor through two slip-ring-brush sets. Three-phase rotor windings have three terminals and need three slip rings.

The current flowing in the rotor winding cannot be direct current; thus, it has an AC nature (see Section 7.6 for DC machines) at a certain frequency.

The outside resistor can be used for various purposes (e.g., control of the current in the rotor winding when necessary). It can also modify the performance characteristic of a machine, as described in Section 11.5.2.

Normally, winding rotors is a costly job and the slip-rings-brush sets add to the cost and need repairs. For these reasons, the wound rotor induction machines are relatively costly. Their counterpart squirrel cage machines are much cheaper and economically preferred.

11.5.2 Squirrel Cage Induction Machine

Referring to Figure 11.10, suppose that the outside resistor is replaced by a piece of wire with no resistance. This immediately causes a significant increase in the current in the rotor windings. To compensate for the higher current, thickness of the wires in the rotor windings must accordingly be increased. In addition to the wire thickness increase, we observe that in such a case, then, there is no need for a set of slip rings because the external resistors do not exist anymore and, thus, the wire loops can be closed inside the rotor rather than outside the rotor. This latter reality has been the basis of the **squirrel cage machines**, in which the extra cost of the slip rings and brushes have been eliminated.

The windings of a squirrel cage machine are a number of thick bars of copper, brass, or aluminum (to carry very high currents) that are shorted together (forming parallel components) at both ends. Because the metal bars are connected to two circular rings at their ends, connecting them together, they form a cage shape, as shown in Figure 11.11. The name "squirrel cage," thus, stems from the shape of the conducting bars.

This is the simplest form of a cage, where the bars have circular cross section; they are parallel to each other and parallel to the rotor shaft. The bars can be slanted (like a twisted cage) or they can have a different cross section rather than being circular. These are variations of models of squirrel cage machines for various reasons or purposes. The aluminum or copper cage is embedded in a ferrous material to increase the permeability of the rotor winding. The ferrous medium is laminated (like the metals in a transformer) to prevent or reduce eddy currents, thus reducing heat and losses. The cage, therefore, is not hollow. A hollow rotor would have a much less efficient performance. The picture of a tiny squirrel cage motor, illustrating the squirrel cage rotor, is shown in Figure 11.12a. (This motor is one phase, since it is so small.) A three-phase 1.5 hp squirrel motor is illustrated in Figure 11.12b.

Most of the motors used in industry are induction motors. Out of those the majority are of the squirrel cage type. This is because of the obvious economic reason; as was mentioned earlier, beyond the initial higher

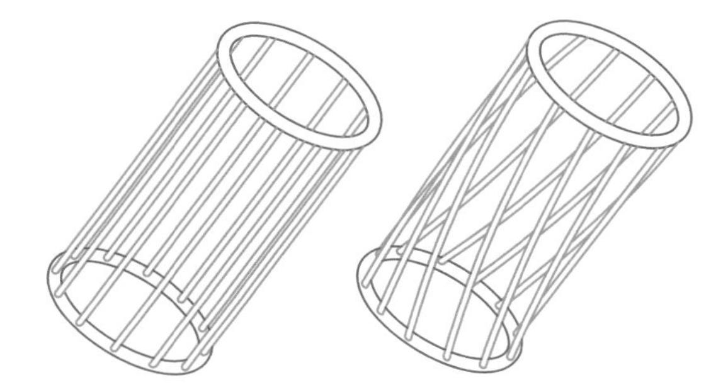

Figure 11.11 Cage structure of squirrel cage induction machine. (From Hemami, A., *Wind Turbine Technology*, 1E ©2012 Delmar Learning, a part of Cengage Learning Inc. Reproduced with permission from http://www.cengage.com/permissions.)

(a) (b)

Figure 11.12 Squirrel cage induction machine: (a) very small single-phase motor and (b) 1.5 hp three-phase motor.

cost, the wound rotor induction machines need maintenance work on their brush and slip rings, whereas the squirrel cage machine is relatively maintenance free. Hence, unless necessary for any technical reason, after a squirrel cage motor has acceptable performance, it will be the candidate for a job.

As a generator, one can say that almost the only application of induction generators is in wind turbines. The intrinsic performance of this machine matches well with the nature of wind, having a variable speed and not being in our control. Formerly, almost all of the wind turbine generators were of squirrel cage type, but, gradually more and more wound rotor induction generators are being used because of their advantages in grasping more power from the wind, despite their extra cost. Their advantages outweigh the higher cost (learn more about this in Chapter 21).

11.5.3 Characteristics of Three-Phase Induction Machine

Similar to other types of electric machines, a three-phase induction machine can work as a generator and as a motor. For this machine, however, because the stator must be connected to the three-phase circuit, the difference between being a motor or functioning as a generator lies in the speed of the rotor. In general, if the rotor speed is higher than the synchronous speed, then it behaves as a generator, and if the rotor speed is less than the synchronous speed, it becomes a motor. The synchronous speed is determined by the line frequency and the number of poles of the stator winding. The developed rotating magnetic field, after the stator is electrically connected, revolves at the synchronous speed. This causes the rotor to follow the rotating magnetic field and rotate (thus, a motor), but, if the rotor shaft is given mechanical energy to rotate faster than the speed of the magnetic field, then the machine behaves as a generator.

Figure 11.13 shows a typical characteristic curve of an induction machine. It involves the torque and speed relationships. The curve consists of two almost symmetrical curves, one representing the motor operation and one associated with the generator operation. This curve can be shown

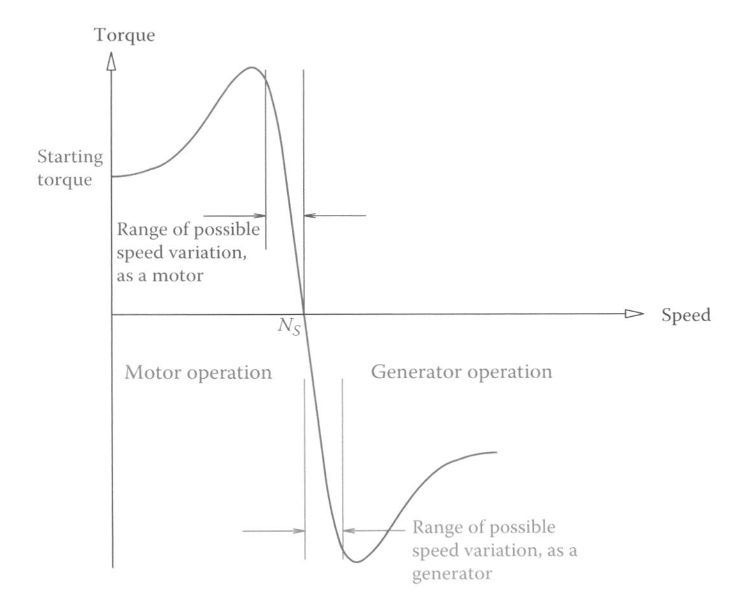

Figure 11.13 Torque-speed characteristic curve of a three-phase induction machine.

differently (swapping the coordinates) or can be continued from motor side for other characteristics (such as dynamic braking, where the direction of the developed torque is to the opposite of rotation, thus, a braking action), but here we are only interested in the following information that can be observed from the curve:

- If the rotor speed is less than the synchronous speed (N_S), the machine acts as a motor; it delivers torque to a load. The amount of torque changes based on a given speed of the load.
- Only part of the curve, which is almost linear, is good for operation. For both functions, as a motor and as a generator, outside of this linear segment is not suitable for operating the machine.
- As a motor, the machine has a starting torque (at zero speed); thus, the machine is a self-starter.
- In the linear (operating) region as a motor, as torque increases, the speed decreases, and vice versa. At the synchronous speed N_S, no torque can be delivered by the motor.
- Within the linear region as a generator, if speed is increased, torque demand also increases, and vice versa. This is a self-regulatory behavior; it prevents a machine from going into a runaway state.
- If the machine runs at a speed equal to N_S, it does not produce any power (because the torque is zero at this point, and power = torque × speed).

The above observations imply that as a motor the speed is always less than the synchronous speed, and as a generator, the machine must always be run above the synchronous speed.

Except in a particular mode, as discussed later in Chapter 21, until recently, many wind turbines used an induction generator. This is an intelligent choice because the variable wind speed can move the operating point of an induction generator along the linear portion of the generator characteristic curve. In this way, with higher wind speeds a generator turns faster and produces more power without affecting the synchronous speed.

Slip speed: Difference between the speeds of the rotor and the rotating magnetic field in an AC induction machine.

The curve depicted in Figure 11.13 has a typical shape. The exact form of the curve depends on the physical properties of each individual machine, including the resistance in the windings and the external circuit for a WRIM.

The difference between the speed of the rotating magnetic field (the synchronous speed) and the rotor speed is called the **slip speed**, and the ratio of this difference to the synchronous speed is called **slip**. The magnitude of slip, expressed in percentage, is usually small, say below 3 percent. It can be determined from

Slip: The fact that the rotor of an induction machine does not rotate with the same speed as the rotating magnetic field (turning faster in a generator and slower in a motor).

$$S = \frac{N_S - N}{N_S} = 1 - \frac{N}{N_S} \tag{11.4}$$

where N_S is the synchronous speed and N is the rotor speed. For a generator $N_S < N$ and, thus, S becomes negative. Note, that as a generator, no matter how much N is larger than N_S the frequency of the electricity injected into the AC line still corresponds to N_S.

The rotational speed N of a motor, and thus the slip, depends on the operating point, which depends on a load torque requirement (so, for a motor, if torque is higher, speed is lower). The same is true for a generator; an equilibrium point is reached when the torque demand on the shaft is fulfilled by the generator prime mover.

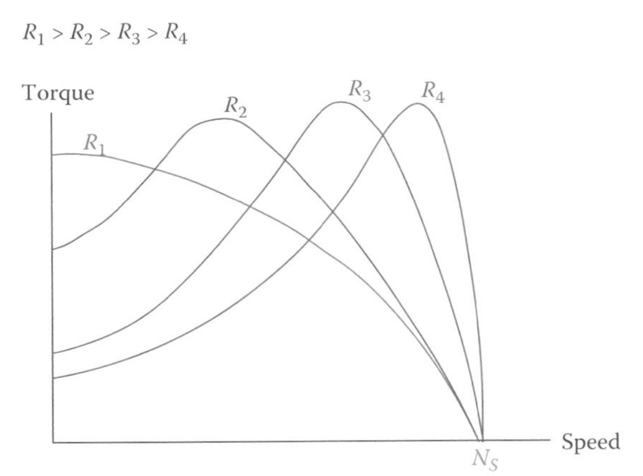

Figure 11.14 Change in the form of the torque-speed curve due to external resistance.

In a wound rotor induction machine the external resistor R (see Figure 11.10) can be modified. This alters the characteristic curve of a machine (both as a generator and as a motor). Figure 11.14 illustrates the effect of this change (only the motor part is shown). The effect is in slip and torque at zero speed. More about this is discussed in Section 11.6.

In a squirrel cage machine, there is no external resistor. However, the same effect (change in characteristic curve) can be obtained by altering the shape (not being round) and position (nearer or farther from center) of the bars of the cage. Sometimes two sets of bars are used; and it is called double-cage machine.

In a three-phase motor, there are three connections to be made. If (any) two connections are interchanged, direction of rotation reverses.

11.5.4 Single-Phase Induction Machine

Most of the smaller AC motors, including those used in some domestic appliances, are single-phase induction motors. They are more likely squirrel cage induction motors rather than wound rotor induction motor (WRIM) because the latter is more expensive and is not made smaller than a certain size.

A single-phase induction machine works based on the same principle as its three-phase counterpart. However, there is a clear difference. Because a single-phase motor has only one set of windings in the stator (and not three), there is no rotating magnetic field developed in the stator. It is rather a back and forth magnetic field. In other words, a magnetic field along one direction, whose strength varies between a maximum positive and a maximum negative value, in the same way that a sine wave amplitude varies.

Not having a rotating magnetic field, one does not expect that such a machine work at all. Nevertheless, the reality is that it works. In fact, a single-phase motor was first detected when one of the three lines of a three-phase machine was accidentally disconnected. The motor continued working, but with a lower power.

For the same physical size, a three-phase induction motor is more powerful and also it works more smoothly than a single-phase motor. The torque-speed characteristic curve of a one-phase induction motor is shown in Figure 11.15. Its behavior in the operating region (the linear segment of the curve) is the same as the curve for the three-phase machine, but this curve passes through the origin of the torque-speed coordinates. This implies that the one-phase induction motor has no starting torque. In other words, it is not a self-starter, and, thus, it needs to be started. Once started, however, it continues to rotate. Its rotational speed is, similar to the three-phase induction machine, lower than the synchronous speed by the slip magnitude following the relationship in Equation 11.4. The slip magnitude depends on the resistive torque from a load.

Because the single-phase induction motor has no starting torque, it needs an auxiliary device to start its motion (see Section 11.6.2). After

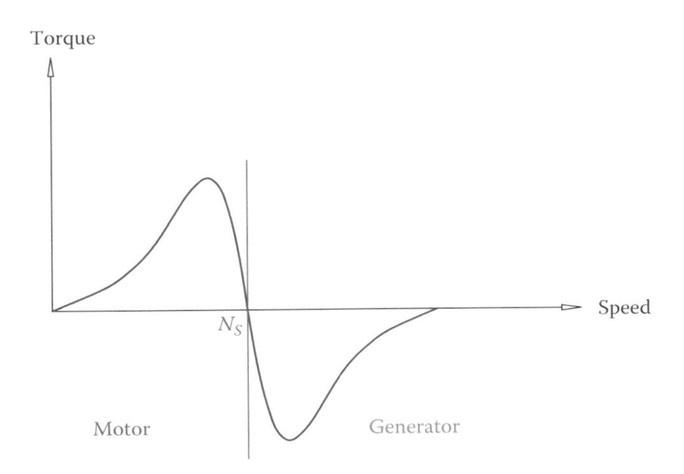

Figure 11.15 Torque-speed characteristic curve of a single-phase induction machine.

the rotation in one direction has started, the motor continues to speed up to near the synchronous speed. The final speed is determined by the load torque.

As a generator, it is not usual to have a single-phase induction machine used as a generator, except in experimental works and in very limited and isolated cases.

11.6 Induction Motors

Locked rotor torque (LRT): Starting torque in a motor (corresponding to when the rotor speed is zero) or the torque if the rotor is prevented from turning or cannot turn the load.

In many industries, motors are used for powering pumps, compressors, fans, machine tools, presses, conveyor belts, and so on. In most of these applications induction motors are employed. For this reason it is worth covering induction machines used as motors.

11.6.1 Classes of Electric Motors

Pull-up torque: The minimum torque, after starting, in an induction motor that can be provided by the motor. If this torque is not sufficiently higher than that of a load the motor cannot accelerate the load fast enough (see the characteristic curve of a three-phase induction motor).

Characteristics of torque and power requirement of loads connected to motors are not the same. Some require higher starting torque, and some can be started with lower torque. When a motor starts spinning its load, their speed starts from zero and gradually increases to the operating rpm. The operating rpm is when the torque demand of the load and the torque provided by the motor become equal, and a state of equilibrium is reached. In other words, the operating speed/torque is where the characteristic curve of a load and the characteristic curve of its drive meet. This is shown in Figure 11.16, which also depicts a few terms for the characteristic curve of an induction motor.

The starting torque is called **locked rotor torque** because the speed is zero. If a motor cannot overcome the initial resistive torque on its shaft, it remains at that zero speed (as if the shaft is locked). Associated current in the motor winding is high. **Pull-up torque** refers to the torque at the

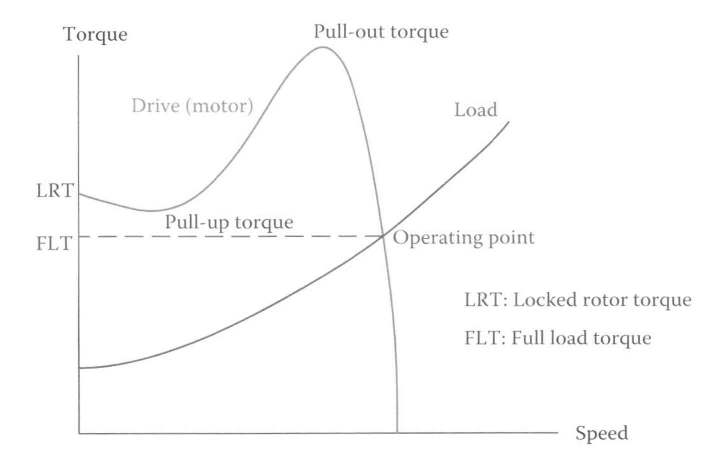

Figure 11.16 Operating point of a load.

point that a motor accelerates and the speed goes up. **Pull-out torque**, also called **breakdown torque**, is the maximum torque that a motor can deliver. If load torque is more than breakdown torque at the corresponding speed, motor speed drops and the motor fails to bring the load to its normal operating speed, at which **full load torque** must be provided.

For the load shown in Figure 11.16, starting torque is lower than operating torque. But this is not always the case and certain loads, like a crane, need a higher torque at the start. A motor must be able to handle starting torque of a load without overheating or giving up. In order not to overdesign motors for various loads and be able to select the appropriate motor for each job, the National Electrical Manufacturers Association (NEMA) has defined a number of standards for size (frame dimensions), characteristics, and electrical insulation properties of electric motors. (Other equivalent organizations are Underwriters Laboratories [UL] in North America, and International Electrotechnical Commission [IEC] for European countries.) For performance, these are called classes of motors. The most common classes of motors are class A, B, C, and D, which define the expected performance with respect to starting torque, slip percentage, and the like. For a full description of motor classes and other standards, a reader need to refer to NEMA publications.

The most common motor class for many applications is class B, which is characterized by a slip of not exceeding 5 percent and a locked rotor torque of 150 percent the full load torque. The motor current for starting is 600 percent of the full load current. This motor is good for applications such as fans, blowers, and compressors that do not need a high starting torque. Class A has a higher starting torque than class B, and its slip is about the same as that of B, with a maximum of 5 percent. Class C motor has a higher locked rotor torque of about 225 percent of the full load torque but has a higher slip. Class D design is for loads demanding a high starting torque, such as cranes and hoists. The starting torque is 280 percent of the full load torque, and the maximum slip is between 5 and 13 percent. Figure 11.17 illustrates the typical curves for NEMA class motors.

Pull-out torque: The same as the breakdown torque. The maximum torque that an induction motor can provide before reaching its operating speed. The operating speed is within a region where the motor continues to work at full-load torque and the speed variation is proportional to the motor torque (see the characteristic curve of an induction motor.

Breakdown torque: The same as the pull-out torque. The maximum torque on an induction motor speed–torque characteristic curve that can be provided by the motor before reaching the operating speed, where the motor continues to work at full-load torque (see the characteristic curve of an induction motor).

Full load torque: The torque provided by an induction motor at its operating speed. This is the point of intersection of the load line and the motor characteristic curve, and must be sufficiently smaller than the pull-out torque to avoid motor stall.

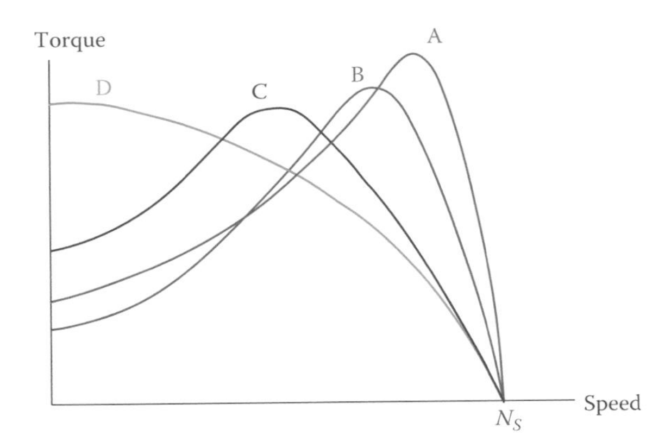

Figure 11.17 Typical characteristic curves of the four most common NEMA motor classes.

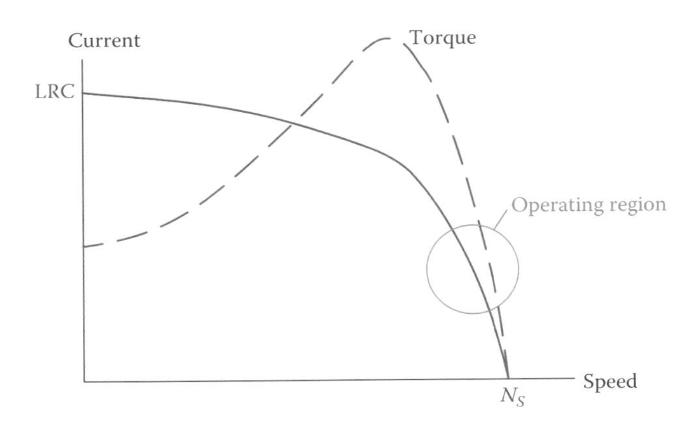

Figure 11.18 Typical current curve in an induction motor.

In an induction motor the stator is connected to electricity. For three-phase motors it can be wye connected or delta connected, as we have discussed in Chapter 8. When the motor is turned on, current starts from the highest value of **locked rotor current** and decreases to the rated current. Figure 11.18 shows a typical curve for this variation.

Locked rotor current (LRC): Current in a motor corresponding to when the rotor speed is zero (the current at starting a motor or if it is prevented from turning); this is the highest current that the winding in a motor experiences.

11.6.2 Starting of Single-Phase Induction Motors

As said before (see Figure 11.15), a one-phase motor has no starting torque and cannot start rotation on its own when powered. Therefore, it requires some boosting for the instant it is connected to electricity to start turning in the direction it is designed for. After it starts spinning in one direction, it continues running in the same direction. There are a number of ways to start a single-phase induction motor, and these motors are categorized and named by their starting method. The auxiliary device can be connected through a centrifugal switch, which turns it off after the motion starts or it can stay on as long as the motor runs. Motor performance and cost of the auxiliary device are not equivalent for the various starting methods.

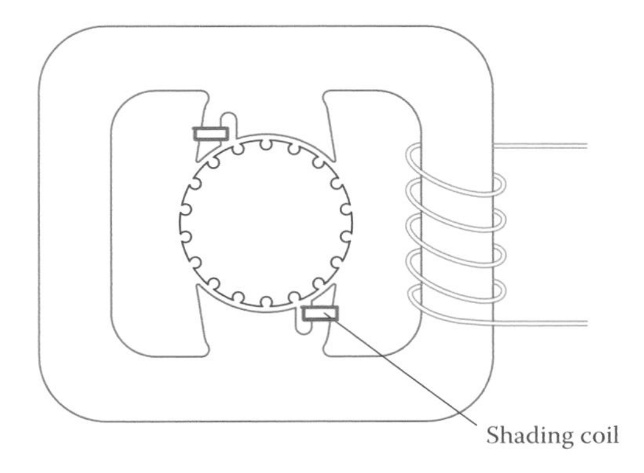

Figure 11.19 Shaded pole motor.

In a **split-phase motor** a smaller winding is physically put at 90° from the main winding. It is later disconnected after the rotor reaches about 25 percent of its normal speed. Mutual effect of the two windings creates the necessary torque for the motor to start. In very small motors a cheaper method is used, called a **shaded pole**. In a shaded pole motor an asymmetry is introduced in the magnetic field of the stator by adding two copper straps to each pole, as shown schematically in Figure 11.19. This asymmetry creates the necessary torque to start the motor.

Split-phase motor: Type of single-phase AC motor in which a starting torque is generated by an additional winding in the rotor.

Shaded pole: Type of single-phase AC motor (usually small motors of a fraction of horsepower) based on the creation of an initial (starting) torque by magnetic shading of its poles.

It is very common to use a capacitor in parallel with an induction motor to start it. The phase difference created in the current caused by the capacitor generates a torque that spins the motor. In a **capacitor-start motor** the capacitor is disconnected by a centrifugal switch after the motor starts. The torque generated this way is larger than that of a split phase motor.

In a **permanent split capacitor motor**, there is no centrifugal switch and the capacitor stays in parallel with the motor for the entire time the motor runs. It has the extra advantage of improving the power factor of the motor circuit. A **capacitor-start capacitor-run motor** has two capacitors, one to start the motor and one aimed for power factor correction. In this sense, this motor is a combination of the capacitor-start motor and the permanent split capacitor motor.

Capacitor-run motor: Alternative current type of motor that needs a capacitor in parallel with it in order to work.

11.6.3 Dynamic Braking

The complete speed-torque characteristic curve for an induction machine is shown in Figure 11.20. The additional part in Figure 11.20, compared with that shown in Figure 11.13, depicts the behavior of the machine when the rotor spins in the opposite direction of the rotating magnetic field. If during the operation of a motor the direction of the magnetic field is reversed, the torque on the shaft tries to rotate the rotor in the opposite way, thus adding to the resistive torque on the shaft of a motor. This property is called **dynamic braking** and can be used in motors to stop a motor faster when it must be stopped.

Dynamic braking: Using the back EMF in a motor for the purpose of braking.

As an example, for a large motor turning a fan, it can take say 10 min before the motor stops after it is switched off. This is because of the large inertia of the motor and its load. Making use of dynamic braking helps

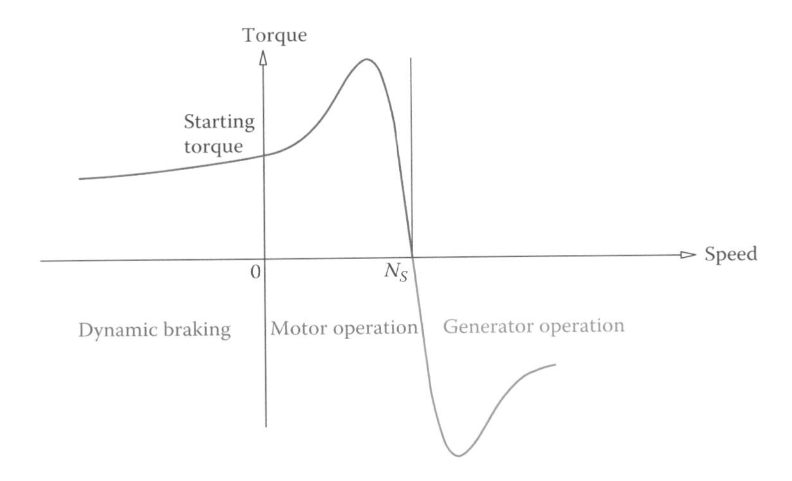

Figure 11.20 Complete characteristic curve of an induction machine.

bring a motor to a stop much faster. In applying dynamic braking, nevertheless, electricity must be completely disconnected at the moment the motor comes to a stop; otherwise, it starts turning (and speeding up) in the other direction, which is not desirable. Also, a large current flows in the stator winding while dynamic braking is activated.

11.7 Other Motors

Although at the industrial level motors are mostly of the types already discussed, at the consumer level, particularly in the single-phase and smaller category, there exist other types of motors. For example, the motors in a vacuum cleaner or a refrigerator, or inside a CD player, are more likely none of the ones we have studied so far. It is worth briefly describing some of these motors here. We consider only those that can be categorized as AC motors. These motors are universal motor, reluctance motor, and hysteresis motor. Other motors, such as stepper motor (or stepping motor) and brushless DC motors, cannot be directly considered as AC motors, and, hence, they are not discussed.

A universal motor is more common than the other two and is used in many household appliances that need a high starting torque. The reluctance motor and hysteresis motor work based on other principles than have been discussed so far. These motors are sometimes categorized under single-phase synchronous motors because they operate at a constant synchronous speed.

11.7.1 Universal Motor

A universal motor is by no means in the same category as the other motors described in this chapter. However, because many of the small (below 1 hp) motors, in particular those in the common tools and appliances such as a drill or a vacuum-cleaner, are of this type, it is worth mentioning them here. As said before, a universal motor is a series wound DC motor and, therefore, has the same characteristic curve as described in Chapter 7. A universal motor can be distinguished from a single-phase squirrel cage motor by the presence of brushes and a commutator. The rotor may be hidden inside and not be recognizable even in a squirrel cage (where there are no brushes and no commutator).

A universal motor has a smoother operation when fed by DC than when operated on AC electricity. Efficiency is higher and also the brushes last longer.

The following helps to explain how a series wound DC motor can work on both DC and AC electricity. Figure 11.21 can be helpful to clarify the matter. This is not possible in a shunt wound DC machine. You may need to review the Lorentz force and series wound machines in Chapter 7 (see Section 7.4) before continuing.

When connected to DC electricity,

1. Direction of the field is constant because the polarity of the field winding does not change.
2. Polarity of each loop of the armature winding must change when the loop passes the position indicated as XX′ in Figure 11.21. If the polarity does not change, then the torque generated after a loop has turned 180° will be in the reverse direction. The arrows in the figure show the direction of the forces at various positions of a loop denoted by AB.
3. This polarity change is done through the commutator.

When connected to AC electricity,

1. Polarity of field winding is switched after each half cycle.
2. Polarity of the current in a loop of the armature winding also switches after each half cycle.

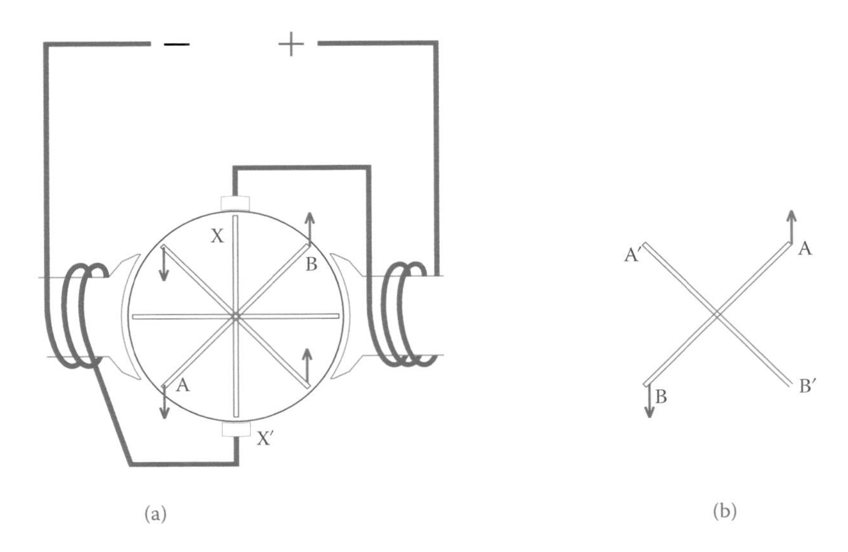

Figure 11.21 Series wound motor connected to DC electricity. (a) Series wound motor connected to DC electricity. (b) Forces exerted to one loop of the armature winding.

The mutual effect of the simultaneous changes in the field direction and the current direction cancel each other. The net effect is like having no change in either; that is, what a DC supply provides.

A further polarity change through the commutator is still needed and is carried out as normal for maintaining the direction of the developed torque.

11.7.2 Reluctance Motor

Reluctance force: Force on a ferromagnetic material caused by a magnetic field to minimize the path of magnetic flux.

Reluctance: Equivalence of resistance in an electric circuit in a magnetic circuit; resistance to magnetic flux.

A reluctance motor works on the basis of the **reluctance force**. A reluctance force, as depicted in Figure 11.22, tries to pull or push a ferromagnetic metal into the magnetic path, so that there is minimum reluctance (or magnetic resistance) in the path. This motor has the same structure as a squirrel cage motor, with the exception that the rotor is deformed in order not to be round or uniformly symmetric. As a result of the shape change of the rotor, when near the synchronous speed, it can snatch into synchronism and continue its motion at the synchronous speed. Thus, the motor starts as an induction motor until it reaches the synchronous speed, at which the principle of operation is **reluctance** and not induction. Figure 11.23 shows schematics of some typical examples of the rotors for a reluctance motor.

The torque-speed characteristic curve of a reluctance motor and a comparison with that of an induction motor are shown in Figure 11.24. If in operation the magnitude of load torque increases beyond a certain value, then the reluctance motor continues to run as an induction motor.

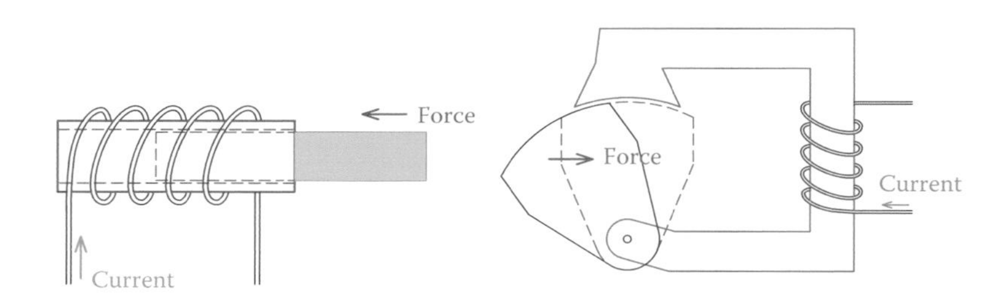

Figure 11.22 Exhibition of reluctance force. (From Hemami, A., *Wind Turbine Technology*, 1E ©2012 Delmar Learning, a part of Cengage Learning Inc. Reproduced with permission from http://www.cengage.com/permissions.)

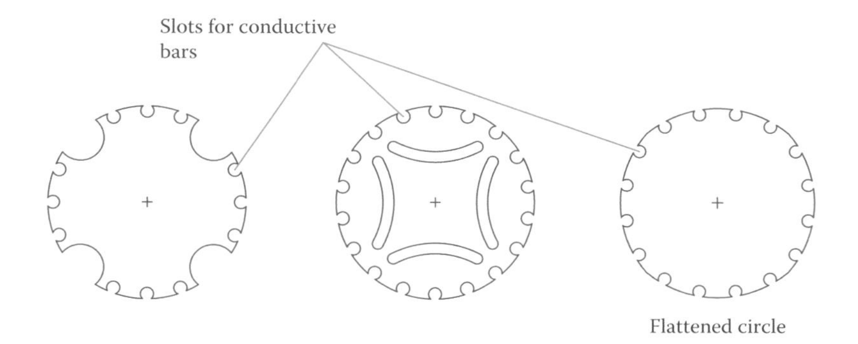

Figure 11.23 Forms of the rotors of reluctance motor.

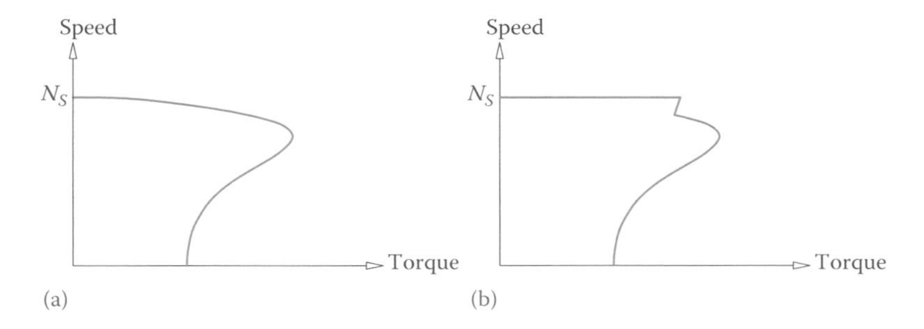

Figure 11.24 Comparison of the characteristic curve of (a) an induction motor with (b) reluctance motor.

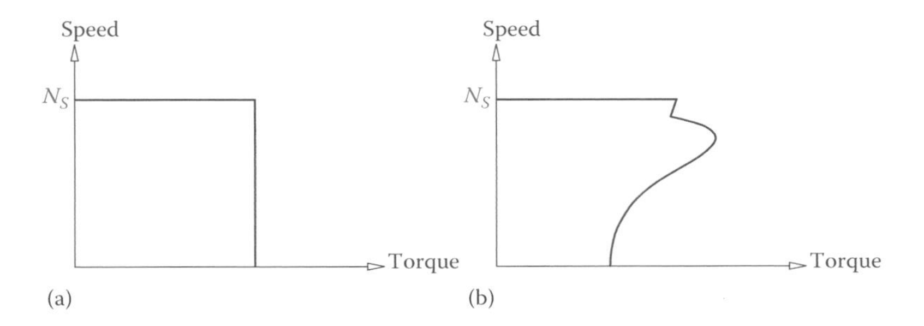

Figure 11.25 Characteristic curve of a (a) hysteresis motor and a (b) reluctance motor.

11.7.3 Hysteresis Motor

The stator of this motor is similar to that of an induction motor (only single phase). The rotor is a uniform cylinder made out of magnetically hard steel (a type of steel that keeps magnetism; it is good for permanent magnet, but not for an electromagnet). Torque developed in the rotor is due to hysteresis effects on the rotor (hysteresis effect is keeping the magnetism). Thus, the torque is constant. This motor has a much better starting torque than the same size reluctance motor. It smoothly increases the speed of a load to the synchronous speed. Pricewise, it is more expensive than a reluctance motor.

Hysteresis motor is characterized by the following features: (1) constant torque, (2) smooth operation, and (3) quiet operation.

Figure 11.25 depicts the typical torque-speed characteristic curve for a hysteresis motor.

11.8 Notes on Application of Electric Machines

An electric machine either converts electric power to mechanical power (motor) or mechanical power to electrical power (generator). It is always power balance between load and drive that determines the equilibrium and operating conditions. If there are controllers for any parameter, such as speed control, they will include their effect with respect to the controlled parameter, but still there must be equilibrium between power of the mechanical system and the electrical system.

For a generator the prime mover (e.g., steam turbine, gas turbine, and wind turbine) provides the mechanical power and the generator is a load for the prime mover. The load on the generator is electrical in this case. For a motor the device connected to it constitutes its load. Thus, for a motor, the load is mechanical.

It is always load that determines power demand, unless drive has reached its maximum capacity and cannot provide more power.

11.8.1 Mechanical Power

We have already seen expressions for electric power in terms of current and voltage for DC circuits and single-phase and three-phase AC circuits (see Chapters 6, 8, and 9). Electrical power demand is always the product of current and voltage; in DC electricity this determines power in watts, and in AC circuits it is apparent power (in volt-amps) that must be provided for a load.

In a mechanical system, power is defined by the parameters in a motion, that is, speed and effort causing that speed. In a linear motion, velocity (speed) is along a straight line and the effort is the component of the force in the velocity direction. This is shown in Figure 11.26. In a rotational system, effort is a torque and speed is the angular velocity. Thus, for a mechanical lifting operation with constant velocity (see Figure 11.26b)

$$\text{Power} = \text{Velocity} \times \text{Weight}$$

and for any rotating shaft with constant angular velocity

$$\text{Power} = \text{Angular velocity} \times \text{Torque}$$

Therefore

$$P = vF \quad \text{(for linear motion)} \tag{11.5}$$

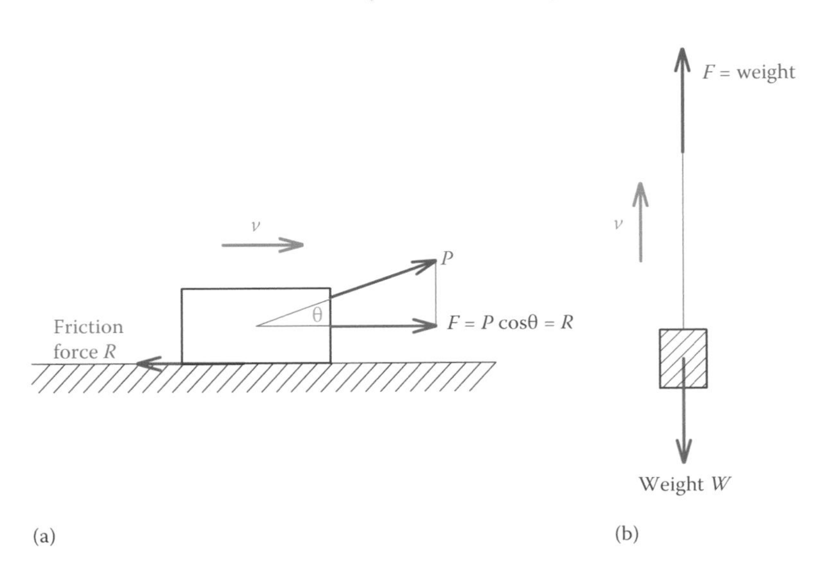

Figure 11.26 Power in linear motion. (a) Horizontal motion with constant velocity. (b) Vertical lifting motion with constant velocity.

$$P = \omega T \quad \text{(for rotating loads)} \tag{11.6}$$

where P is power, v is the vertical motion velocity, F is the weight of the objected lifted with a constant velocity, ω (omega) is the angular velocity, and T is the torque.

Note that Equation 11.5 is not only for vertical lifting. F can be any resistive force that driving force must overcome. In most cases, where no lifting is involved, this can be the force necessary to overcome friction forces in a system.

Equations 11.5 and 11.6 are complete as shown if P is expressed in watt, v in meters per second (m/s), F in Newton (N), ω in radians per second (rad/s), and T is in Newton-meter (N.m). If the unit for any of the parameters is changed, then a multiplier accordingly appears in the equation. Common customary units for power are watt, kw, and horsepower (hp); velocity can be expressed in m/s and ft/s; force can be in Newton or in pound; torque can be measured in Newton-meter or foot-pound; rotational speed may be expressed in rad/s or revolutions per minute (rpm). For convenience, Equations 11.5 and 11.6 with some other units are given below. The numbers are rounded; for more precision, refer to conversion tables with more decimal digits.

For lifting and linear motion loads,

$$P\ (\text{kW}) = 0.001\ v\ (\text{m/s}) \times F\ (\text{N})$$

$$P\ (\text{hp}) = 0.001341\ v\ (\text{m/s}) \times F\ (\text{N})$$

$$P\ (\text{hp}) = 0.0018164\ v\ (\text{ft/s}) \times F\ (\text{lb})$$

$$P\ (\text{kW}) = 0.00135451\ v\ (\text{ft/s}) \times F\ (\text{lb})$$

and for rotational loads,

$$P\ (\text{kW}) = 0.001\ \omega\ (\text{rad/s}) \times T\ (\text{N.m})$$

$$P\ (\text{kW}) = 0.0001047198\ \omega\ (\text{rpm}) \times T\ (\text{N.m})$$

$$P\ (\text{hp}) = 0.000140375\ \omega\ (\text{rpm}) \times T\ (\text{N.m})$$

$$P\ (\text{hp}) = 0.00019014232\ \omega\ (\text{rpm}) \times T\ (\text{ft.lb})$$

$$P\ (\text{kW}) = 0.000141846232\ \omega\ (\text{rpm}) \times T\ (\text{ft.lb})$$

In the above relationships the calculated value is the power essential to move or power a load as desired. The power to be provided is more than the required power because of efficiency (see Section 11.8.2). A motor powering its load then must have a higher power to cover for the losses. For a generator, power demand from electrical loads in a circuit determines the power requirement from the generator. Then the prime mover must be more powerful to deliver the power the generator requires as well as the losses. Thus,

$$\text{Power to be provided} = \frac{\text{Power demand by a load}}{\text{Machine efficiency}} \tag{11.7}$$

Example 11.2

Torque required to yaw a wind turbine nacelle is 500,000 Nm (369,131 ft.lb). There are eight similar motors to carry out the yaw motion. If the nacelle has to be turned at a speed of 2 rpm, what is the power requirement from each of the eight yaw motors?

Solution

Angular speed is given in rpm, and the torque is in Nm. Hence, power for each motor can be found as

$$\frac{(0.000140375)(2)(500,000)}{8} = 17.5\ \text{hp}$$

Example 11.3

A motor is used to drive a pump. If at the operating conditions the torque is 150 Nm, the pump rotates at 1150 rpm and its efficiency is 80 percent, what is the pump power requirement? If the pump is to be driven by an AC motor with 90 percent efficiency, what should be the power rating of the motor?

Solution

We may choose horsepower for the unit of power. The net power required by the load (water flow) is

$$P_W = (0.000140375)(150)(1150) = 24.22\ \text{hp}$$

Then the power requirement to run the pump is

$$P_P = \frac{24.22}{0.8} = 30.26\ \text{hp}$$

And the power necessary for the motor is

$$P_M = \frac{30.28}{0.9} = 33.6\ \text{hp}$$

The rating of the motor should be more than 33.6 hp.

Example 11.4

A conveyor belt is driven by two identical electric motors, each one rated at 12.5 kW. At full load of the belt the two motors work at their rated power. If efficiency of the motors is 90 percent and the belt moves with a speed of 1.4 m/s, what is the resistance force F in the conveyor belt?

Solution

Power provided by the motors to the conveyor belt is

$$P_M = (2)(12.5)(0.90) = 22.5 \text{ kW}$$

Because this power is used to move the belt, then

$$P_B = (0.001)(1.4)(F) = 22.5$$

and the resistive force in the conveyor system is

$$F = \frac{22.5}{(0.001)(1.4)} = 16{,}071.4\ N$$

which is 3616.5 lb.

11.8.2 Efficiency

No system can be found that is 100 percent efficient, meaning that output power from it is equal to input power to it. Any device, particularly delivering energy of some sort, has an input and an output. Input is power taken by the device to operate; output is what it delivers in the same form or other form of energy or work. Motors and generators fit well into this general definition.

In all electrical, mechanical, and electromechanical systems and devices, there is always some loss of energy in the form of heat due to friction in the moving parts or current in the electrical circuitry. In motors the losses are of two types: those that are fixed, no matter what the load on the motor is, and those that are a function of the load and increase as the current through the machine increases. Among fixed losses are

- Friction in the bearings.
- **Windage**. This is the aerodynamic losses associated with ventilation fans and rotating parts.
- **Core loss**. In both stator and rotor cores, some energy is lost owing to eddy current. The iron core gets warm, meaning that some energy is lost in the form of unwanted heat.

Among the variable losses are

- **Copper loss**, the energy loss in the form of heat in the windings of both the stator and the rotor. This loss is proportional to the square of the current flowing in the winding (or cage bars in a squirrel cage machine).
- **Stray loss**, the losses of various types that cannot be included in the other categories, including that in the air gap of a machine. It is proportional to the square of rotor current and counts for 1 to 2 percent of the total input power.

Always, a system input must make up for these losses before an output can be delivered. In this sense, input power P_i to a system is always greater than output power P_o from the system. The ratio of output to input is the efficiency, which is always less than 1. Normally, efficiency is expressed in

Windage: Aerodynamic losses associated with ventilation fans and rotating parts in an electric motor.

Core loss: Amount of power loss in a transformer or an electric machine that corresponds to the quality of design and core material and is independent of the load current.

Copper loss: Amount of power loss in a transformer or an electric machine that corresponds to the resistance of the wire winding and it depends on the load current (the percentage of loading).

Stray loss: Losses of various origins, including that in the air gap, in an electric machine. These losses cannot be included in the other categories for a machine.

percentage; thus, output to input ratio is multiplied by 100. It is common to use the Greek letter η (eta) to denote efficiency.

$$\text{Eff} = \eta = \frac{P_o}{P_i} \times 100 \tag{11.8}$$

where P_o is the output power and P_i is the input power. In an electric machine, because some of the losses are load dependent, efficiency is not constant. In the design stage of any device or machine, effort is made to have the maximum efficiency. This is usually at the rated operating conditions and for the full load of a machine. The efficiency drops from its maximum value when a machine is not working at its full load capacity.

For motors and generators, normally, the larger a machine, the higher is its efficiency. One obvious reason is that for these machines even 1 percent change in efficiency counts for a large amount of energy involved; thus, considerable time is devoted to its proper design.

Example 11.5

A 10 kW motor is used to turn a shaft at 720 rpm. The motor uses its full power for this load. If the load applies a 120 Nm torque on the shaft, what is the motor efficiency at this load?

Solution

The power required to turn the shaft is

$$P_{\text{Load}} = (0.0001047198)(720)(120) = 9.05 \text{ kW}$$

and the motor efficiency is

$$\eta = \frac{9.05}{10} \times 100 \cong 90\%$$

Example 11.6

An overhead crane uses a 20 kW DC motor for its lifting operation. Efficiency of the motor is 95 percent and efficiency of the gear system driven by the motor is 85 percent. If a 5000 lb load is lifted by this crane, what is its speed in ft/s?

Solution

According to Equation 7.8, total efficiency is

$$\eta_t = (0.85)(0.95) = 0.81$$

based on which, it follows from Equation 11.3 that

$$v = \frac{(0.81)(20)}{(0.00135451)(5000)} = 2.385 \text{ ft/s}$$

11.8.3 Duty Cycle

Not all motors work continuously during their operation. A pump in a process plant may work 24 hours in a day, but a motor in a passenger lift works only when people use the lift. Similarly, motors in the yaw system of a wind turbine are in operation only when the turbine has to be yawed into the wind. The above pump and its motor are said to have a 100 percent duty cycle, whereas the motor of the turbine yaw system can have a duty cycle of only 20 percent or even less. Some other motors can have more regular work than that of a passenger lift or a wind turbine yaw system. For instance, the fan for the cooling of the gearbox oil in a wind turbine can turn on and off on a continuous basis. Then, by study of the working conditions, one can figure out that on average the fan works 60 percent of the time during summer, for example. The duty cycle of the motor turning the fan, therefore, is 60 percent.

Knowing the duty cycle is important for the selection and operation of a motor. If a motor does not have 100 percent duty cycle, then it has the opportunity to cool down after each cycle of operation, whereas a motor with 100 percent duty cycle does not have such an opportunity. Also, energy consumption of a motor depends on the time that it is in operation.

Other matters also come into consideration for a motor that turns on and off intermittently. Among these are the power factor and the means for power factor correction of the circuit from which a motor is fed, and the spike in the voltage that the motor creates in the circuit when turned on, and its effect on other loads. Each of these must be considered, particularly for large motors, during the design and selection of motors for a particular job.

11.8.4 Back Electromotive Force

Any electric machine has windings, and any winding has both electric resistance and inductive reactance, in series with each other. In this sense the equivalent electrical circuit for each phase winding of the stator of a three-phase electric machine is a resistor and an inductor in series. Resistance of the wire winding is very low. Also, because frequency is rather low (50 and 60 Hz), the value for inductive reactance is also quite low. For example, for a 30 hp, four pole motor running on 460 V, 60 Hz electricity the values for stator and rotor resistance and reactance (see Figure 11.27a) are $R_S = 0.641\ \Omega$, $X_S = 0.750\ \Omega$, $R_{RS} = 0.300\ \Omega$, $X_{RS} = 0.500\ \Omega$, and $X_M = 26.3\ \Omega$. These are the values as measured or calculated based on the winding inductance.

Connecting a winding, with such a low ohmic resistance and inductive reactance values, to a voltage source of, say, 415 V results in a very high current. In reality, the effect of the turning rotor in a motor significantly reduces such an otherwise high current. The scenario is true if the stator of a motor without its rotor in place is connected to electricity or if the rotor is prevented from turning (braked rotor). In this regard a motor is similar to a transformer; the stator wining is equivalent to the primary winding of the transformer and the rotor winding equivalent to the transformer secondary winding. If the rotor is prevented from turning, the machine behaves as a transformer with shorted secondary circuit; that is, a high current is expected. This is indeed the inrush current of a motor

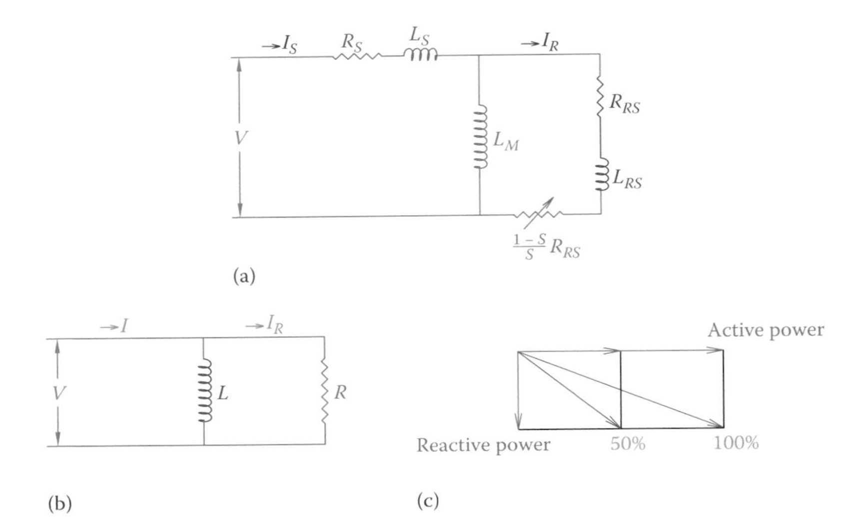

Figure 11.27 (a) Model of an induction machine. (b) Model based on active and reactive elements and their currents. (c) Power relationships at full and half load.

when the rotor starts to turn, which can be several times larger than the normal current.

Nevertheless, as we have learned (see also Chapter 7), an electric machine when turned generates electricity. That is, for a motor, at the same time it is rotating as a motor, it also behaves as a generator and generates a voltage in its windings. This voltage is in the opposite direction of the voltage turning it and opposes the applied voltage.

The electromotive force (voltage) generated in a motor winding is called back electromotive force (BEMF) or counter electromotive force (CEMF). If this BEMF in a single-phase motor (or each phase of a three-phase motor) is represented by E, and the voltage applied to a motor is denoted by V, then using Ohm's law, we have

$$V - E = ZI \tag{11.9}$$

where Z is the impedance of the winding. The BEMF greatly reduces the net voltage across the winding and the current through the winding. Value of E depends on motor speed (as well as other parameters) and becomes larger as speed increases. If the rotor is prevented from turning, then E drops to 0.

(Note that for a generator the same thing happens, but this time the electromotive force E is larger than the terminal voltage V by the value ZI.)

11.8.5 Power Factor Correction

As we have seen in this chapter, the synchronous motor can be run at any desired power factor by changing the excitation current. The induction motor runs with a lagging power factor. Also, we discussed in Chapter 8 that a low power factor is not desirable because it implies loss of energy and a voltage drop in the transmission and distribution lines.

In many industries a large number of motors are used for various processes. Most of these motors are induction motors; thus, it is essential to improve the low power factor of the circuit feeding these motors. The simplified model of a running induction motor is shown in Figure 11.27a. This is for each phase in a three-phase motor. R_S is the stator winding resistance, L_S is the stator inductance, R_{RS} and L_{RS} are the rotor winding resistance and inductance as seen by the stator (in the same way that the secondary winding values of a transformer are referred to the primary winding), L_M is the magnetizing inductance of the stator, and I_S and I_R are stator and rotor currents, respectively. The last term is a function of R_{RS}. It is variable and depends on the rotor rpm, which can be defined by the slip S, as shown in Figure 11.27a. This model is good for studying the behavior of an induction motor.

For the purpose of power factor correction a simpler model that resembles a resistor in parallel with an inductor is employed, as illustrated in Figure 11.27b. The inductor corresponds to the magnetization effect and the reactive power, and R corresponds to the active power. The power factor of this circuit is improved by inserting a capacitor of proper size in parallel with the other two components (see Chapter 8). To improve the power factor of a motor, the capacitor is put in parallel with the motor. For three-phase motors a bank of three similar capacitors is used. The capacitors can be connected in delta or they can be wye connected, independent of the way the motor is connected.

Reactive power of a motor corresponds to the power required to create the rotating magnetic field. This power is almost constant and independent of whether or not a motor is loaded. Because of this, because a motor can be run at full load or at a percentage of its capacity, its power factor varies and depends on its load. This is shown in Figure 11.27c for full load and 50 percent load of a motor. It is impossible to have a fixed capacitor for power factor correction at all different loading condition of a motor. It is also possible to use a set of capacitors to take care of the power factor correction for a few motors at the same time. It is, generally, very expensive to have a system of capacitors that connect and disconnect according to the load changes in a system, unless cost is justified. Therefore, it is important to have the most suitable capacitor that can compensate for different conditions of a circuit, as a whole. This needs a detailed study of any particular system and its operating conditions.

In the following examples a few scenarios exemplify some general cases of power factor correction for motors.

> The three capacitors for power factor correction of a three-phase motor can be delta connected or wye connected, independent of the way the motor is connected.

Example 11.7

Three 470 μF capacitors are connected to a 415 V, 60 Hz line. What is the reactive power of the set if (1) they are wye connected, (2) they are delta connected?

Solution

The capacitive reactance of each capacitor is

$$X_C = \frac{1}{(2)(3.14)(60)(470)(10^{-6})} = 56.44\ \Omega$$

1. For wye connection the voltage across each capacitor is $\frac{415}{\sqrt{3}}$

   and thus the corresponding current is

$$I_\phi = I_L = \frac{415}{56.44\sqrt{3}} = 4.245\ \text{A}$$

   and the reactive power can be found in one of the usual ways (see Chapter 9) to be

$$Q = 3I_\phi V_\phi = (3)(4.245)\left(\frac{415}{\sqrt{3}}\right) = 3051.49\ \text{VAR}$$

2. For delta connection the corresponding current (phase current) is

$$I_\phi = \frac{415}{56.44} = 7.352\ \text{A}$$

   and the (reactive) power is

$$Q = 3I_\phi V_\phi = (3)(7.353)(415) = 9154.49\ \text{VAR}$$

   which is 3 times the previous value for wye connection.

As can be seen, in wye connection the capacitors are subject to a lower voltage (1.73 times smaller), but their reactive power is 3 times smaller. This is important in selecting capacitors for power factor correction.

Example 11.8

A 20 hp motor is connected to 415 V line. The motor power factor is 0.75 at full load. If the resistance of each of the three wires connecting the motor to the main circuit is 0.5 Ω, determine voltage drop and power loss in the wires for full load operation and compare them with when the power factor is improved to 0.95.

Solution

Current in the wires can be found from the power relationships

$$P = \sqrt{3}\,VI\cos\phi = (1.73)(415)(0.75)I = (20)(746)$$

From the above relationship, I is calculated to be

$$I_{75} = 27.68 \text{ A}$$

Voltage drop and the power loss in the wires are, thus,

$$\text{Voltage drop} = (0.5)(27.68) = 13.84 \text{ V}$$

$$\text{Power loss} = (3)(0.5)(27.68)^2 = 1149 \text{ W}$$

If the power factor is corrected to 0.95, then the new current is

$$I_{95} = \frac{(20)(746)}{(1.73)(415)(0.95)} = 21.85 \text{ A}$$

And the voltage drop and power loss are, respectively,

$$\text{Voltage drop} = (0.5)(21.85) = 10.92 \text{ V}$$

$$\text{Power loss} = (3)(0.5)(21.85)^2 = 716 \text{ W}$$

(Note: For this problem the nominal voltage of 415 V is assumed to be available at the consumption point. Otherwise, the problem becomes more complicated if the supply line is 415 V before the voltage drop.)

Example 11.9

The power factor of a single-phase 20 hp motor at full load is improved to 0.95 by a set of auxiliary capacitors. If the motor works at 50 percent load, what is the power factor?

Solution

Referring to Figure 11.27c the power to magnetize the stator and the power used by the accompanying capacitor(s) is reflected in the reactive power of the motor. Power that converts to mechanical work is reflected in the active power. The active power, therefore, is 20 hp at full load. We use two methods for solving this problem. One can always use the method that is of preference to him/her.
First method:

Full load

$$P = \text{Active power} = (20)(746) = 14{,}920 \text{ W}$$

$$S = \text{Apparent power} = \frac{P}{\text{pf}} = \frac{14{,}920}{0.95} = 15{,}705 \text{ VA}$$

$$Q = \text{Reactive power} = \sqrt{S^2 - P^2} = \sqrt{15{,}705^2 - 14{,}920^2} = 4904 \text{ VAR}$$

50% load

$$P = (10)(746) = 7460 \text{ W}$$

$$Q = 4904 \text{ VAR (unchanged)}$$

$$S = \sqrt{P^2 + Q^2} = \sqrt{7460^2 + 4904^2} = 8928 \text{ VA}$$

$$\text{pf} = \frac{P}{S} = \frac{7460}{8928} = 0.84$$

Alternative method:
We keep horsepower for the unit of power measurement. From the value of the power factor at this load the reactive power can be determined using the trigonometric relationships (see Figure 11.27c).

$$\text{Reactive power} = Q = \text{Active power} \times \tan\theta = P \tan\theta$$

$$\theta = \cos^{-1}(0.95) = 18.2°$$

$Q = (20)(\tan 18.2°) = (20)(0.3287) = 6.5737$ (note that the unit for Q is not VAR now, because P was not in watts)

At 50 percent load, active power is reduced to 10 hp.

$$\theta_{New} = \tan^{-1}(6.573 \div 10) = \tan^{-1}(0.6573) = 33.3°$$

and the new power factor is

$$\text{pf} = \cos 33.3° = 0.84$$

Example 11.10

Three three-phase motors are connected to a 415 V, 60 Hz circuit. The power and power factor of these motors are 8, 10, and 20 kW, and 0.65, 0.70, and 0.75, respectively. These motors work together, and a set of three capacitors can be used to improve the circuit power factor to 0.95. If the capacitors are connected in delta, find the capacitance of each capacitor in the set.

Solution

For convenience and better understanding, the solution to this problem is divided into three steps.

Step 1:

From the given values for the (active) powers and the power factors of the three motors we are able to find the apparent power and reactive power of each motor. We indicate the values for each motor by subscripts 1, 2, and 3 for 8, 10, and 20 kW motor, respectively.

$$S_1 = \frac{8}{0.65} = 12.31 \text{ kVA}$$

$$S_2 = \frac{10}{0.70} = 14.29 \text{ kVA}$$

$$S_3 = \frac{20}{0.75} = 26.67\ \text{kVA}$$

$$Q_1 = \sqrt{12.31^2 - 8^2} = 9.35\ \text{kVAR}$$

$$Q_2 = \sqrt{14.29^2 - 10^2} = 10.20\ \text{kVAR}$$

$$Q_3 = \sqrt{26.67^2 - 20^2} = 17.64\ \text{kVAR}$$

Total apparent power = 53.26 kVA
Total active power = 38 kW
Total reactive power = 37.2 kVAR

It is not necessary to find the power factor for the three motors together, but out of interest, one may want to see how much it is. It is

$$\text{pf}_T = \frac{38}{53.26} = 0.71$$

which corresponds to a phase angle of about 45°.

Step 2:

On the basis of the desired power factor, determine the total apparent power for the same active power of the motors.

$$S_{\text{New}} = \frac{38}{0.95} = 40\ \text{kVA}$$

and the new total reactive power is

$$Q_{\text{New}} = \sqrt{40^2 - 38^2} = 12.5\ \text{kVAR}$$

The difference between the previous value and the new value of the reactive power is the power of the capacitors to be added to the circuit.

Capacitor power = 37.2 – 12.5 = 24.7 kVAR

Step 3:

Based on the power for the capacitors, find their capacitance, taking into account the way they are going to be connected, delta or wye.

For delta connection, the line current is (based on Equations 9.12 and 9.13)

$$I_L = \frac{(24.7)(1000)}{\left(\sqrt{3}\right)(415)} = 34.36\ \text{A}$$

But the current in each branch containing a capacitor is $\sqrt{3}$ times smaller (phase current). Thus, the capacitor current is

$$I_C = \frac{34.36}{\left(\sqrt{3}\right)} = 19.84 \text{ A}$$

Based on this current, the capacitive reactance for each capacitor is thus

$$X_C = \frac{415}{19.84} = 20.785\ \Omega$$

and the capacitance of each capacitor is

$$C = \frac{1}{2\pi f X_C} = \frac{1}{(2)(3.14)(60)(20.785)} = 1.276 \times 10^{-4} \text{ F} = 128\ \mu\text{F}$$

If we face difficulty finding the right capacitor, we may want to see what size capacitor is needed if wye connection is employed.

For wye connection, the voltage across each capacitor is smaller $\left(V_\phi = \frac{415}{\sqrt{3}} = 239.6 \text{ V}\right)$, but we still need to have the same reactive power for the capacitor bank. In such a case, the line current (and the current in each capacitor branch) is still the same as before:

$$I_C = 34.36 \text{ A}$$

leading to

$$X_C = \frac{239.6}{34.36} = 6.97\ \Omega$$

The capacitor size is thus

$$C = \frac{1}{(2)(3.14)(60)(6.97)} = 3.8 \times 10^{-4} \text{ F} = 384\ \mu\text{F}$$

The capacitors to add to the circuit for wye connection are three times larger in size, which result in a capacitive inductance three times smaller. The voltage rating of the capacitors is nevertheless $\sqrt{3}$ times smaller. The current in the lines connecting to the capacitors is still the same as in delta connection.

(Note that if the same capacitor as for delta connection is connected in Y configuration, the power would be three times smaller. But since the capacitors are three times larger, the current is three times larger and the power is the same as when 128 μF capacitors are delta connected.)

Example 11.11

For the expansion of the plant in the previous example a 30 kW (40 hp) motor is needed. A decision has been made to use a synchronous motor and retire the capacitor banks. All the motors are delta connected. Determine at what power factor the synchronous motor must work so that the power factor of the entire system at full load is 0.95.

Solution

For ease of reference, all the pertinent values of the three induction motors are shown in the following table:

	Active Power, kW	Reactive Power, kVAR	Apparent Power, kVA	pf	Phase Angle, degrees
Motor 1	8	9.35	12.31	0.65	49.5
Motor 2	10	10.20	14.29	0.70	45.6
Motor 3	20	17.64	26.67	0.75	41.4
All motors	38	37.20	53.26	0.71	45

With the addition of the new (synchronous) motor the total active power is (38 + 30 =) 68 kW, and on the basis of the desired pf of 0.95, apparent and reactive powers can be determined.

$$S = \frac{68}{0.95} = 71.58\ \text{kVA}$$

$$Q = \sqrt{71.58^2 - 68^2} = 22.35\ \text{kVAR}$$

To bring 37.20 lagging reactive power to 22.35 lagging reactive power, the amount of leading reactive power to be added is

$$37.20 - 22.25 = 14.95 \cong 15\ \text{kVAR}$$

This reactive power must be provided by the synchronous motor. The synchronous motor, therefore, must work with a leading power factor of

$$\text{pf} = \frac{30}{\sqrt{30^2 + 15^2}} = \frac{30}{33.54} = 0.89$$

11.8.6 Nameplate Information

Any motor has a nameplate attached to it that gives the rated values and other pertinent information about the motor and its operation. The following list gives the most common information found on the nameplate of a motor. Most of the items are self-explanatory.

Type	High Efficiency	Frame	286T
hp	38	Service factor	1.1
Amp	44	Volts	415
rpm	1750	Hz	60
Duty	Continuous	Cos ϕ	0.89
Class Ins.	F	NEMA design B	
3-ph Y 4-pole			

The type is a categorization of various motors based on the form and size of the squirrel cage bars. The frame number and class of insulation are other NEMA classifications similar to design B. These deal with design of a motor for various ambient conditions, like humidity and existence of combustible gases. There are also many other NEMA specifications such as TEFC (Totally-Enclosed, Fan-Cooled), but they are out of the scope of this book. For more details, refer to NEMA publications or other standards for motor. Service factor is the percent of the maximum allowed power that a motor can withstand for a short period of time without overheating and getting damaged.

Among the information for a motor is the direction of rotation. It is shown on the motor frame (not on the nameplate) somewhere near the shaft.

11.9 Chapter Summary

- There are two principal categories for many AC motors, synchronous motors and asynchronous motors.
- Asynchronous motors are better known as induction motors.
- A synchronous motor always has a constant speed.
- A synchronous motor always turns at the synchronous speed, which depends on the line frequency.
- An induction motor always runs at a lower speed than the synchronous speed.
- In induction machines the rotor is either a wound rotor or a squirrel cage rotor.
- Synchronous speed for a machine depends on the number of poles (in addition to the line frequency).
- Any AC machine is either single or three-phase.
- AC motors do not have a commutator. This is the way to distinguish between AC and DC motors. However, a universal motor works with both AC and DC electricity.
- A universal motor is a series wound DC motor, but because of its electrical structure, it can work with AC, too.
- AC motors and AC generators have the same structure.
- All three-phase machines work based on *rotating magnetic field*.
- In most AC machines, particularly the large ones, the stator is connected to the electricity circuit not the rotor.

- A three-phase machine is smaller and works smoother than its same power single-phase counterpart. In any electric machine, power is proportional to the product of voltage and current.
- In a mechanical system, power is proportional to the product of speed and either force (for linear systems) or torque (for rotational systems).
- In a three-phase motor, swapping two connections changes the direction of rotation. A motor must have been designed for operating in both directions.
- In a single-phase motor one cannot change the direction of motion.
- Single-phase induction motors need an auxiliary device for their start. They are categorized and named based on their starting method.
- Power for a generator is in volt-amp, but for a motor it is in kW or horsepower.

Review Questions

1. What are the two main categories of AC machines?
2. Is there a motor that can work with both AC and DC electricity?
3. What are the units for measuring power in motors?
4. What is the difference in construction between a motor and a generator of the same category?
5. How can you judge if a motor and a generator are compatible?
6. What is a synchronous speed?
7. Is there only one synchronous speed? If yes, what is it?
8. Does a synchronous motor run at a synchronous speed?
9. What are the parameters defining a synchronous speed?
10. What is the part of an AC machine that rotates called?
11. What is the rotating magnetic field?
12. What is the other name for an induction machine?

13. If a motor has six terminals (six wires to be connected to electricity), does it have six poles?

14. How many terminals does a single-phase machine have?

15. What is the main difference between a synchronous machine and an induction machine in operation?

16. How do you know if a motor is AC or DC?

17. In a motor, what does the term operating conditions mean?

18. What do the terms rated voltage, rated power, and rated frequency mean?

19. How do you know if an AC motor is three or single phase?

20. Why is the starting current in a motor higher than its rated current?

21. If a motor is made for 60 Hz, can you use it with 50 Hz electricity? Why or why not?

22. Why must a synchronous generator always be run at a synchronous speed?

23. Mention some of the synchronous speeds for 60 Hz.

24. Describe some of the synchronous speeds for 50 Hz.

25. Can a three-phase induction motor start by itself?

26. If a motor cannot start by itself, what can be done to start the motor?

27. For a single-phase motor, what are the usual starting methods?

28. If the rated speed of an AC motor is 1100 rpm, what type of a motor is it?

29. What happens if you connect two of the three wires of a three-phase motor to a three-phase line?

30. What happens if you connect two of the three wires of a three-phase motor to a single-phase line of compatible voltage?

Problems

1. Find the synchronous speed of an induction motor with 24 poles, when connected to a 415 V, 60 Hz line.

2. What is the rotating speed of the motor in Problem 1 if it has a 4 percent slip?

3. What is the rotating speed of the motor in Problem 2 if it is connected to a 50 Hz line?

4. If an induction motor rotates at 960 rpm, can you determine the line frequency?

5. Determine the total number of poles in the motor in Problem 4.

6. Find the reactive and real (active) power in a single-phase motor that when connected to 120 VAC, the current is 1 A, and the power factor is 0.8.

7. A 2 kW, 220 V, 50 Hz, single-phase AC motor has a power factor of 0.77 at full load. What is the current for the line connecting the motor?

8. What is the current for the motor in Problem 7 at 80 percent load if the power factor for this load is 0.72?

9. In Problem 8 if the power factor is improved to 0.95 (for 80 percent load), what is the power factor when the motor is at full load?

10. Three of the motors in Problem 7 are used and work together on a 380 V, 50 Hz circuit. A single capacitor is used to correct the power factors of the three at full load to 95 percent. Determine the capacitance of the capacitor.

11. If after the insertion of the capacitor one of the motors in question 10 is turned off, what is the line power factor?

12. If each line connecting the motors in Problem 10 has 0.5 Ω resistance, what is the voltage drop in the line before and after the power factor is corrected to 0.95?

13. A three-phase motor is delta connected to a 415 V, 60 Hz line. If the line current is 25 A and the motor power factor is 0.75, how much power is consumed by the motor?

14. Determine the size of the bank of three capacitors to improve the power factor of the motor in Problem 13 to 0.95. The capacitors are delta connected.

15. If two of the motors in Problem 13 work together, what is the size of capacitors to correct the power factor to 0.95 if delta connected?

16. For the motors in Problem 15 compare the voltage drop in the line between the two cases that the pf is or is not corrected. The total resistance of each wire is 1 Ω.

17. If the efficiency of the motor in Problem 13 is 93 percent, how much mechanical power is available on its shaft?

18. At a particular time, a wind turbine absorbs 1750 kW of power from wind. Efficiency of the gearbox is 90 percent and efficiency of the generator is 95 percent. What is the output of the generator?

19. If the generator in Problem 18 works at a power factor of 0.94, what is the current in the line between the generator and its transformer if the generator is fully loaded (all its output is used)? The generator voltage is 600 V, and it is delta connected.

20. What is the maximum current the generator in Problem 19 can deliver when it absorbs only 1 MW power from wind? All the other numbers are unchanged.

12

Electric Power Transmission and Distribution

OBJECTIVES: After studying this chapter, you will be able to

- Define and use terms used in the electric power industry
- Explain what a substation is
- Describe why it is necessary to transmit electricity at an elevated voltage
- Describe the difference between transmission and distribution
- Calculate the resistance of a length of wire
- Explain why transmission line conductors have inductance and capacitance
- Determine inductance and capacitance of transmission lines
- Describe what transposition in transmission line is and how it is done
- Select conductors carrying a given electric current under given conditions
- Explain why it is necessary to ground electric equipment
- Describe different ways that equipment can be grounded
- Use Thevenin's and Norton's theorems for solving network problems
- Use the principle of superposition for network problem solving

New terms:
ampacity, bolt short circuit, circuit breaker, circular mil, corona, generator step up (GSU) transformer, geometrical mean distance (GMD), geometrical mean radius (GMR), grounding transformer, negative sequence, neutral grounding resistance (NGR), node, positive sequence, reactor, resistivity, specific resistance, substation, superposition, temperature coefficient, transposition, zero sequence

12.1 Introduction

Nowadays, in many large cities, electricity is provided by a grid that is powered by power plants that can be at a significant distance from the city. In modern countries, there are one or more grids for all electric power that are fed at various points where power plants are situated. For example, North America has three major grids: the Western Interconnection, the Eastern Interconnection, and the Electric Reliability Council of Texas (or ERCOT) grid. Larger power plants can be hydro, nuclear, or gas, oil, and coal fired. Wind and solar farms can be regarded as relatively small power plants. Also, there can be a lot of small hydro and diesel power plants that contribute to a grid power only during the peak hours.

Thus, a grid is an electricity power transfer carrier. Depending on its power transfer capacity and geographical extension, a grid can operate at various voltages, and as such, it must receive power from all generators feeding it and deliver to all consumers powered by it. This can imply transforming electric power from one level to another, sometimes several times, to be able to make connections. Even with only one power supply at one end of a grid and only one consumption location at the other end, it is necessary to raise the voltage from the generator site to a higher level and lower it after transmission at the consumer site. This is essential for the purpose of reducing energy waste and voltage drop, as demonstrated by Example 10.1 in Chapter 10.

In this chapter the pertinent matters associated with transmission and distribution of electricity and the necessary equipment are discussed.

12.2 Power Transmission and Distribution

Example 10.1 clearly demonstrated for a small load how the energy loss and voltage drop can be reduced by raising the voltage for transmission. This is, however, only possible to a certain limit, particularly at the level of consumer products. Operating voltage at domestic level in the world ranges from 100 V (in Japan) to 240 V (in the United Kingdom). This is the range accepted and operating in all countries (the operating voltages are 100, 110, 115, 120, 125, 127, 220, 230, and 240 V). From the production and distribution viewpoint, however, this is not an acceptable range at all. If the production, transmission and distribution are carried out at this low voltage, probably more than 98 percent of the production is lost in the transmission lines.

The reason is obvious. For the same power, if voltage goes up, current goes down. In this respect, both the voltage drop (IR) and the power used in the lines (RI^2) decrease, as current decreases. Thus, the remedy is to increase the voltage for transmission as much as possible. Evidently, nevertheless, there is a cost involved and technical issues arise and how much the voltage can be increased is not an arbitrary choice. Different parts of a grid, moreover, can have different voltages.

Voltage chosen for a grid is based on decisions at the time of its construction but later on a grid can expand and integrate with other grids. The more common values for grid voltages are 34, 69, 138, 161, 230, 345, 500, and 765 kV. (Recently, a 1000 kV line was installed in China.) Table 12.1 shows the classification of the standard voltages in the United States.

Output of generators at large power plants is between 4000 and 20,000 V. This is further raised to a higher voltage, that of the grid, in one or more stages. Then, at the consumer site, the high voltage of the grid is lowered, again in one or more stages, before finally reaching houses and streets with a final stage when the voltage is lowered to one of the aforementioned values.

Substation: Point in the electric distribution network where transformers and switchgears are installed.

12.2.1 Substation

Raising or lowering line voltage is carried out in a **substation**. In addition, all switching, regulations, monitoring, and protection of the line take place

TABLE 12.1 Standard Nominal Three-Phase System Voltages per ANSI C84.1

Voltage Class	Three Wire, V	Four Wire, V
Low voltage	240	208 Y/120
	480	240 Y/120
	600	480 Y/277
Medium voltage	2400	4160 Y/2400
		8320 Y/4800
	4160	12,000 Y/6930
	4800	12,470 Y/7200
	6900	13,200 Y/7620
	13,800	13,800 Y/7970
	23,000	20,780 Y/12,000
	34,500	22,860 Y/13,200
	46,000	24,940 Y/14,400
	69,000	34,500 Y/19,920
High voltage	115,000	
	138,000	
	161,000	
	230,000	
Extra high voltage	345,000	
	500,000	
	765,000	
Ultra high voltage	1,100,000	
	1,200,000	

in a substation. The main device for changing voltage is a transformer. The first transformer after a generator is the **generator step up (GSU) transformer**. Figure 12.1 shows a schematic of a small grid. This grid is fed only by one power plant. Larger grids have numerous feeding power stations. Any later installment to either feed this grid or take energy out of it must do so through a substation. In Figure 12.1, G represents a generator (power plant), GSU stands for generator step up, used for a stepping up transformer just after the generator to raise the generators voltage to that of the line it connects to, and CB denotes **circuit breaker**. A circuit breaker is an interrupter device as part of the protective means for part or all of a circuit. In case of malfunction or abnormal situations in a circuit, it cuts off electricity.

Generator step up (GSU) transformer: Transformer just after a three-phase generator in power generation plants that increases the voltage from the generator to match the voltage of the power line to the plant.

Circuit breaker: Same as a breaker, to disconnect a device or a circuit from power source to prevent damage.

In addition to transformers a substation is equipped with other devices for various functions. Among the devices in a substation are lightning arresters and air break switches. An air break switch is a switch for high voltages in transmission lines. Because opening and closing of these switches is accompanied by electric arc flash, it is necessary to suppress the arc. In an air break switch, compressed air is used for this purpose. Points A, B, C, and D in Figure 12.1 signify switchgear substations, where no voltage change takes place, but different branches meet.

As can be seen, a grid can have many circuit breakers at various points. Circuit breakers are of various types and they act based on some criteria, the existence of which are sensed by the appropriate sensors somewhere in the circuit. The action of sensors, normally called relaying, is set to

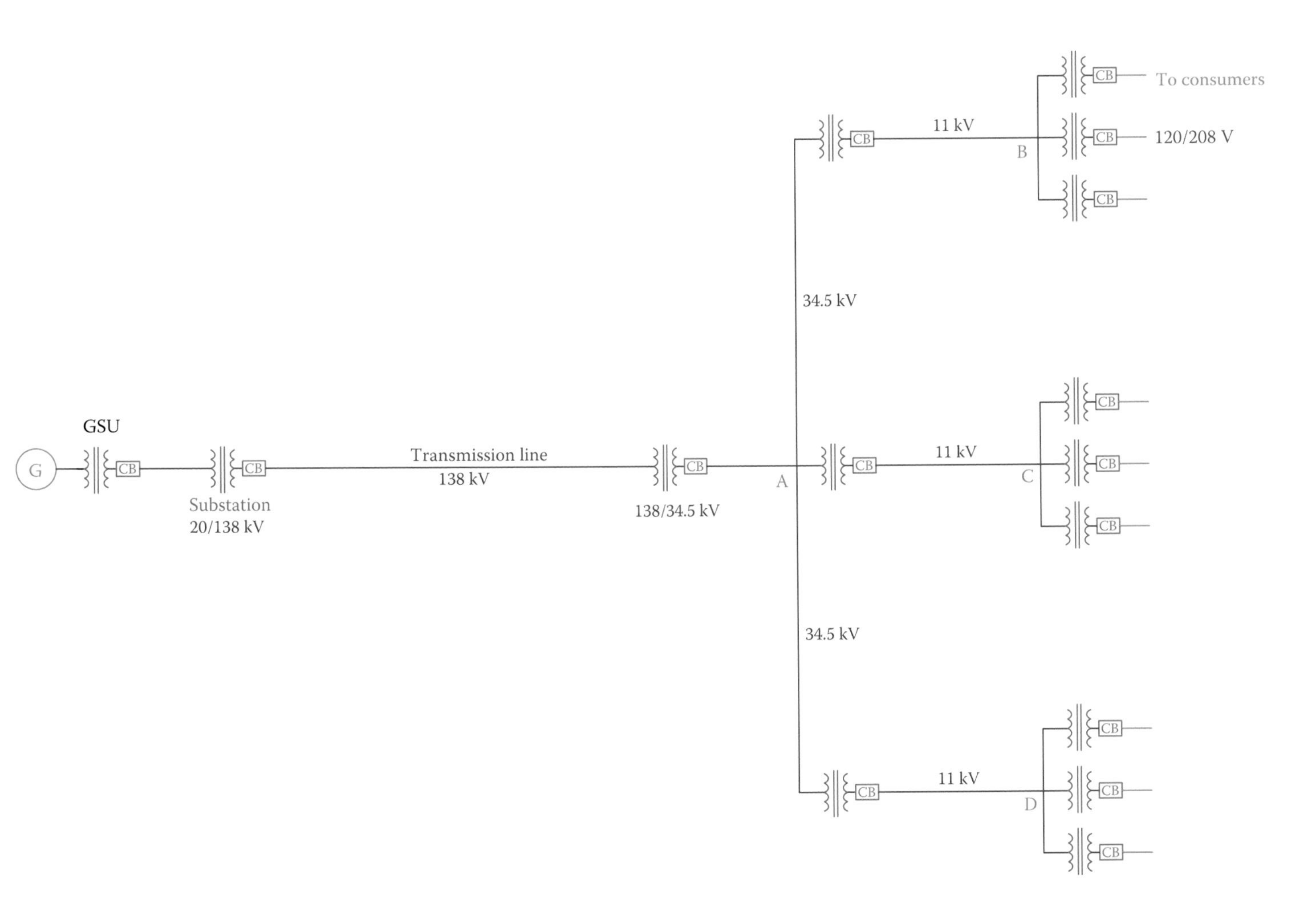

Figure 12.1 General representation of a simple electric grid.

Figure 12.2 Substation. (From Hemami, A., *Wind Turbine Technology*, 1E ©2012 Delmar Learning, a part of Cengage Learning Inc. Reproduced with permission from www.cengage.com/permissions.)

measure and detect overvoltage, undervoltage, overcurrent, overfrequency, underfrequency, overtemperature, open phase, reverse power flow in motors, and so on.

Figure 12.2 shows a small substation.

Electricity to houses and small buildings is single phase. In North America the nominal voltage for electricity in the buildings is 110 V (but depending on region and municipalities it can be between 110 and 125 V). Nevertheless, for certain appliances that need more power and take more current, such as the kitchen stove (range) and dryer, to lower current, the required voltage is 220 V (nominal). Common practice in North America is to deliver to each house a center-tapped 220 V line. The center-tapped line is grounded and acts as the neutral for building wiring. In this way, while 220 V is available for those devices that need it, the voltage difference between the center line and each other line is 110 V. Wiring for various parts of a house is almost evenly distributed between the two half voltage sources, and the stove and dryer are connected to the full voltage. Figure 12.3 shows the schematics of this wiring. This method of wiring is not practiced in other countries with 220 V electricity.

At this point it is worth noting that the special wiring of the lights for staircases and corridors that can be turned on-off by two switches at the two ends of the pathway. For this purpose, two DPST (double-pole single-throw; see Chapter 5) switches are used. Figure 12.4 shows this wiring scheme. It is also possible to turn a light on and off from more than two points. In this case the middle switch(s) must have four connections; each time, two alternative terminals pairs are connected together.

12.3 Electric Cables

Electric cables or conductors are basically of two types, but each type has numerous categories. Conductors are either for overhead transmission lines or for underground installation. Overhead conductors are bare wire

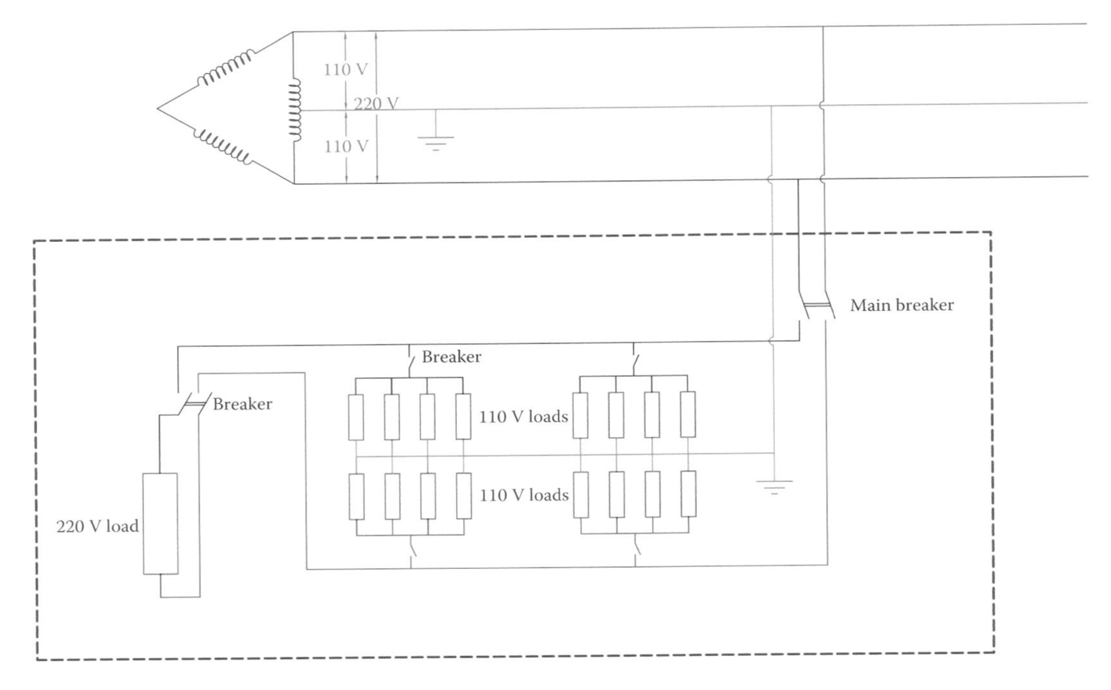

Figure 12.3 North American wiring scheme for residential buildings.

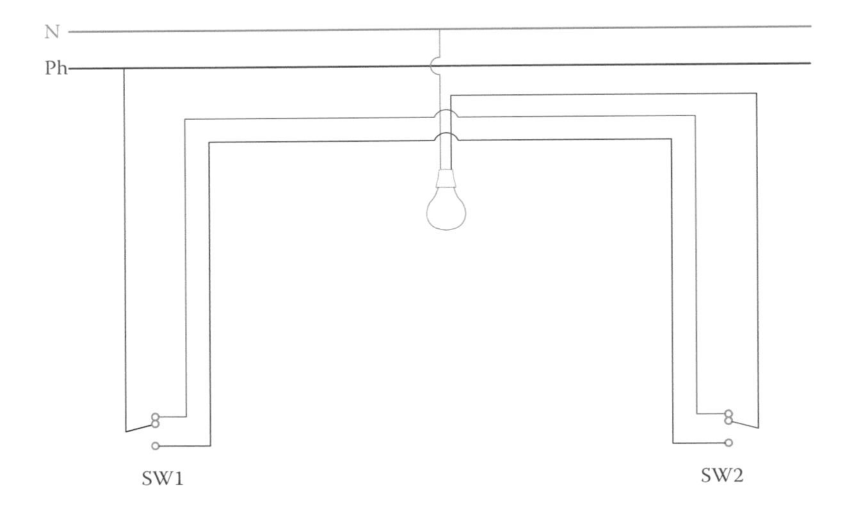

Figure 12.4 Two-switch staircase light wiring arrangement.

and do not have insulation except at residential areas where contact with trees and other objects is possible, whereas underground conductors cannot be without insulation.

Overhead cables are cheaper because they do not have the problems with inclusion of insulation material and the required properties in their manufacturing process. But, they must be mechanically stronger because they are suspended from the poles/towers, and they must support their own weight, for which they must withstand a great tension. All conductors expand owing to heat generated in them when carrying current. This causes overhang conductors to sag more, which sometimes can result in contacts with lower lines or trees.

As we see in this chapter, we distinguish between transmission and distribution of electric power. Transmission is from the power plant to the consumer site where distribution takes place. Distribution takes place at different stages and at different voltages. Normally, the distribution lines conductors, for the low voltage that electricity comes to buildings, are made out of copper. Copper is heavier and more expensive than aluminum, which is an alternative for electrical wire material.

12.3.1 Overhead Conductors

For transmission lines nowadays the conductors are made of aluminum. Compared with copper, aluminum has less conductivity and less strength. For this reason, for the same power transmission, aluminum conductors must be thicker. Also, for overhead lines their strength can be reinforced by steel.

Theoretically, conductors can be made out of rigid bars. However, for transportation this is impossible and they should be flexible. See Figure 12.5. To increase flexibility of thicker conductors, they are made of several strands. As the current carrying capacity requirement of cables increases, more strands are added, and accordingly more reinforcement is necessary. The customary number of strands in overhead cables is as follows:

$$1 - 7 - 19 - 37 - 61 - 91, \text{ and } 127$$

Note that each layer has six strands more than the layer inside it. For example, $19 = 1 + 6 + (6 + 6)$ and $37 = 1 + 6 + 12 + (12 + 6)$, etc. See Figure 12.6.

There are several categories of aluminum conductors. These are some of the more common aluminum conductors: all aluminum conductor (AAC), all aluminum alloy conductor (AAAC), aluminum conductor alloy reinforced (ACAR), aluminum conductor steel reinforced (ACSR), aluminum conductor steel supported (ACSS), aluminum conductor carbon fiber reinforced (ACFR), and gap-type aluminum conductor steel reinforced (GTACSR).

Figure 12.5 For various reasons including transportation, conductors must be flexible. (From ©2014 Southwire Company, LLC. All rights reserved. PPI reproduction is pursuant to Southwire Company, LLC's express permission.)

Figure 12.6 Typical bare wire conductor. (Courtesy of General Cable.

There are other categories as well. The names are self-explanatory, to some extent, but can change from company to company. Not all conductors have strands with circular cross section. Figure 12.7 shows a cable with oval shape strands. Also, to increase the conductivity of cables for the same cross section, some cables have trapezoid shape strands that form circular layers, which resemble tubes of different diameters inside each other (see Figure 12.8). In this way, more use of space (thus, more conductivity) is made out of the same conductor diameter.

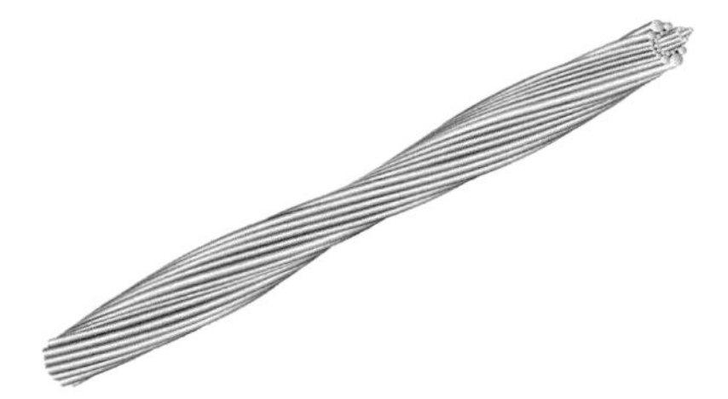

Figure 12.7 Conductor with oval cross section. (From

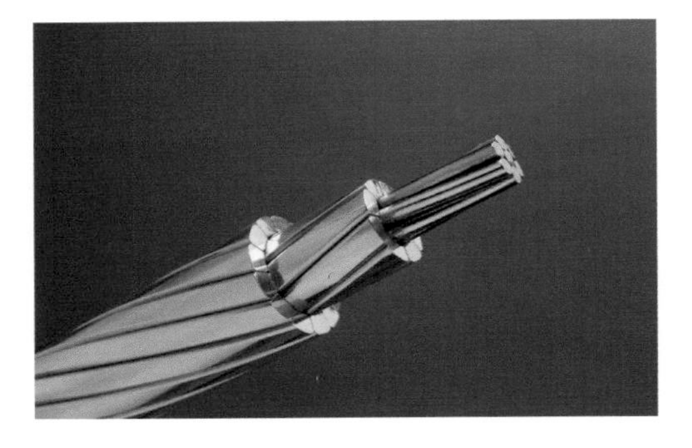

Figure 12.8 Conductor with trapezoid strands. (Courtesy of General Cable.

Figure 12.9 Aluminum conductor with composite core. (From ©2014 Southwire Company, LLC. All rights reserved. PPI reproduction is pursuant to Southwire Company, LLC's express permission.)

More recently, carbon fiber reinforcement cables have been introduced; instead of steel, these cables have strands of carbon fiber composite material in the middle. Carbon fiber composite cable (CFCC) offers desired properties such as less weight and smaller thermal expansion compared with steel. It has 1/5 of the weight and 1/12 of thermal expansion of those of steel, for a price. Figure 12.9 shows a picture of such a cable.

A gap-type conductor consists of a core of high strength steel surrounded by a small gap filled with temperature resistant grease. Stacked around this gap are the trapezoid shape stands of aluminum.

In the seven-strand conductor there are six aluminum strands around one steel cable. The number of steel strands depends on the specification of a particular conductor. For a 37-strand cable, there are 30 aluminum and 7 steel strands, but for the 61-strand cable the number of steel strands can be 7 or 19 and the rest are aluminum. There are always exceptions, and these depend on the manufacturer.

Electric poles and supporting structures come in different forms and sizes, mainly based on the voltage of the power they transmit. Some typical ones are shown in Figure 12.10. For the final distribution to consumers' poles of approximately 12 m (40 ft) are used, and the height of larger structures varies between 18 and 42 m (60 and 140 ft).

12.3.2 Electrical Properties of Electric Wires and Cables

The principal electrical property of a piece of metal is the resistance R that it shows to the flow of electrical current. It is very important to know the resistance of electric conductors when they are used in electric circuits. Knowing R allows one to determine voltage drop and the energy transformed into heat in parts of an electric circuit, in motor windings, and so forth. In addition to this property, for wires and cables, there is another property that determines how much current is allowed to pass through a conductor. This property, called **ampacity** (made from the two words "ampere" and "capacity"), defines the *current capacity* of an inductor based on the heat that is generated owing to electrical current, the structure and material of the conductor, and ambient temperature. It is easy to understand, therefore, that whereas the resistance of a wire can be almost constant, its ampacity depends on the temperature and some other working conditions, and it cannot be a constant.

Ampacity: A word (name) made of *Amp*ere and Cap*acity*. It is used to define the current capacity of standard conductors (wires) in different working conditions for safe operation. Ampacity is determined based on the heat generated in a conductor due to the current through it.

Resistance of a piece of any material (even an insulator) to the flow of electricity is proportional to its length and inversely proportional to its cross-section area. So, for example, if the length of a wire doubles,

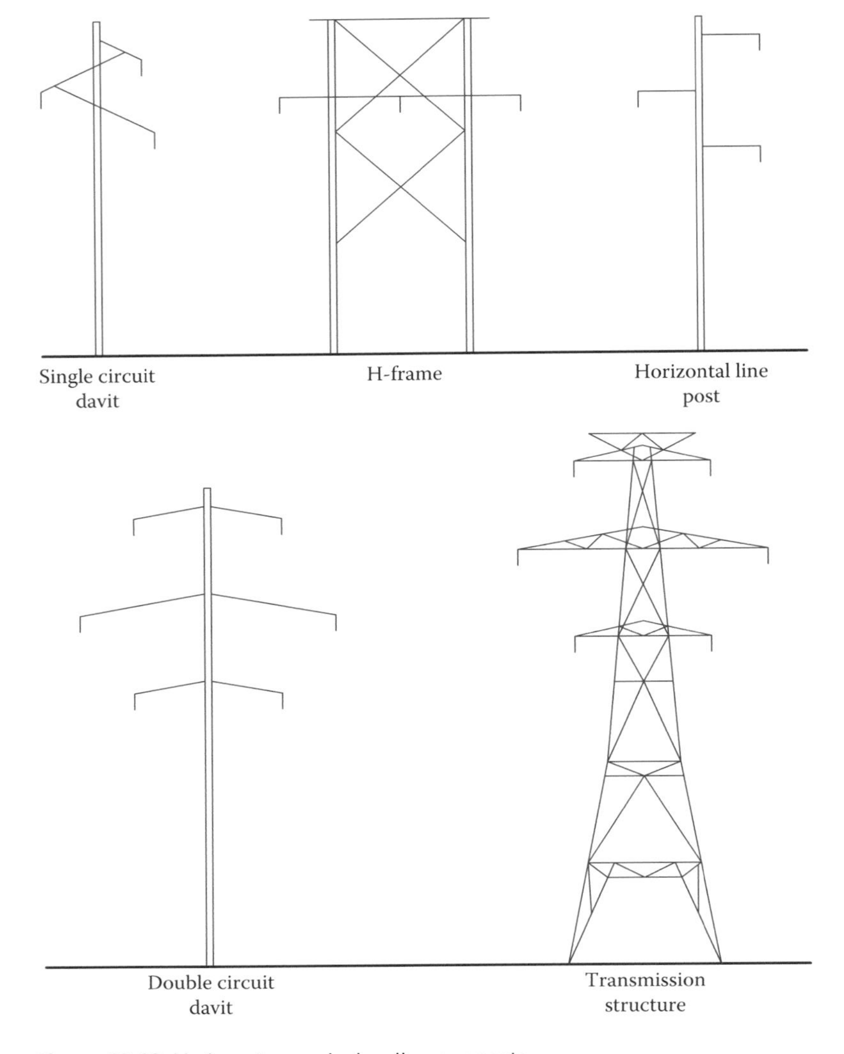

Figure 12.10 Various transmission line supports.

Specific resistance: Same as resistivity: the electric resistance of a specific size (based on the measurement system) of a metal or material.

Resistivity: Same as specific resistance: the electric resistance of a specific size (based on the measurement system) of a metal or material.

its resistance doubles, but if cross section doubles, resistance halves. R depends also on the material; for example, copper is a much better conductor than iron. This is why wires are made up of copper and not iron. The effect of the material is designated by the Greek letter ρ (rho, pronounced *ro*), which represents the resistance of a piece of the material with specific dimensions. The value ρ is called the **specific resistance** or **resistivity** of a substance.

On the basis of the aforementioned discussion, resistance R of a piece of wire can be found from

$$R = \rho \frac{l}{A} \tag{12.1}$$

where ρ is the specific resistance, l is the length, and A is the cross-section area of the wire. Table 12.2 shows the specific resistance of some metals and nonmetals in metric system. The numbers shown are for the measurements made at 20°C. For other temperatures, values must be corrected by using the temperature coefficients in column 3 of Table 12.2.

Table 12.2 also shows the conductivity of materials. Conductivity is the inverse of resistivity. It is an indicator of how conductive a material is. Its unit in metric system is, thus, 1/ohm-meter. Also shown in the table is **temperature coefficient**, which represents how much the specific resistance of a metal changes with temperature. This implies that ρ does not have a constant value. Values given in tables are usually for the resistivity at room temperature (20°C). The relationship to calculate specific resistance of a metal at a different temperature than that known is

Temperature coefficient: Numerical value (positive for metals) representing how much the specific resistance of a material changes with temperature.

$$\rho_2 = \rho_1[1 + \alpha_1(t_2 - t_1)] \tag{12.2}$$

where t_2 is the temperature at which we need to know the specific resistance ρ_2, t_1 is the temperature at which the value is known (ρ_1), α_1 is the temperature coefficient at t_1.

Equation 12.2 can be readily substituted by

$$R_2 = R_1[1 + \alpha_1(t_2 - t_1)] \tag{12.3}$$

where R_1 is the known value of the resistance at t_1 and R_2 is the resistance at t_2.

The variation of the temperature coefficient α can be found from

$$\alpha_1 = \frac{\alpha_0}{1 + \alpha_0 t_1} \tag{12.4}$$

α_0 is the temperature coefficient at zero degrees.

TABLE 12.2 Specific Resistance (Resistivity) of Some Metals and Insulators

Material	Resistivity ρ, Ωm	Temperature Coefficient per Degree Celsius	Conductivity $\sigma \times 10^7/\Omega m$
Silver	1.59×10^{-8}	0.0061	6.29
Copper	1.68×10^{-8}	0.0068	5.95
Aluminum	2.65×10^{-8}	0.00429	3.77
Tungsten	5.6×10^{-8}	0.0045	1.79
Iron	9.71×10^{-8}	0.00651	1.03
Platinum	10.6×10^{-8}	0.003927	0.943
Manganin	48.2×10^{-8}	0.000002	0.207
Lead	22×10^{-8}	...	0.45
Mercury	98×10^{-8}	0.0009	0.10
Nichrome (Ni, Fe, Cr alloy)	100×10^{-8}	0.0004	0.10
Constantan	49×10^{-8}	...	0.20
Carbon[a] (graphite)	$3–60 \times 10^{-5}$	−0.0005	...
Germanium[a]	$1–500 \times 10^{-3}$	−0.05	...
Silicon[a]	0.1–60	−0.07	...
Glass	$1–10{,}000 \times 10^{9}$	...	...
Quartz (fused)	7.5×10^{17}	...	...
Hard rubber	$1–100 \times 10^{13}$	...	...

[a] The specific resistance of these nonmetallic materials highly depends on the impurities in them.

Specific Resistance

Specific resistance is electrical resistance of a specific size of a material, and it is shown by the Greek letter ρ. In the metric system, ρ is the resistance of a piece of metal (10 mm × 10 mm × 10 mm), that is 1 cm long and has a cross section of 1 cm^2 (see figure below). Because in the metric system the units for the length and area are meter and square meter, respectively, ρ is defined in ohm-meter (Ω.m) so that when inserted in the relationship for resistance $R = \rho \frac{l}{A}$ the resultant is Ω.

$$\Omega.\text{m}\frac{\text{m}}{\text{m}^2} \rightarrow \Omega$$

In the US customary system, ρ is defined by the resistance of 1 foot of metal with a cross section of 1 **circular mil** (CM). See Appendix D for the definition of circular mil. The unit of measurement for ρ, therefore, is ohm-circular mil per foot (Ω.CM/ft).

Circular mil: Unit for measuring the thickness (cross section) of wires. A CM is the area of a circle whose diameter is one mil (1/1000 of an inch).

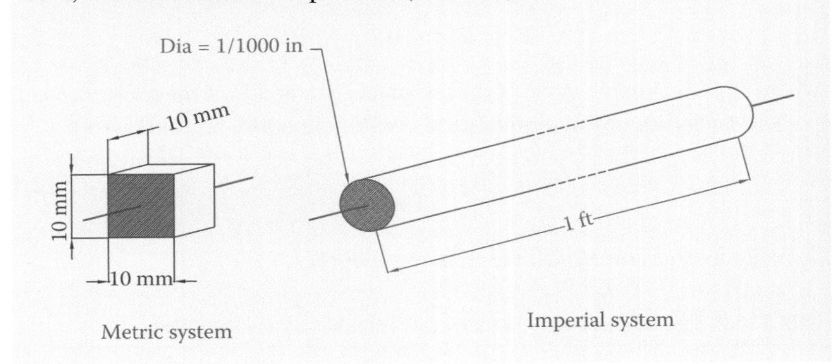

Example 12.1

What is the resistance of a copper wire 50 m long and having a cross-section area of 16.78 mm^2 (AWG 5 wire)?

Solution

The resistivity of copper is 1.6779 (10^{-8}) ohm-m. Thus,

$$R = \rho \frac{l}{A} = (1.6779)(10^{-8})\left(\frac{50}{16.78 \times 10^{-6}}\right) = 0.05\ \Omega$$

Example 12.2

If the specific resistance of copper in metric system is 1.6779 × 10^{-8} (Ω.m), find what is it in the imperial system (Ω.CM/ft).

Solution

We should calculate the resistance of a foot long (0.3048 m) of the cable with a diameter of 1 mil (1/1000 inch).

$$\text{Length} = 1\text{ ft} = 0.3048\text{ m}$$

$$\text{Diameter} = 1\text{ mil} = 0.001\text{ in} = 0.0000254\text{ m} = 25.4 \times 10^{-6}\text{ m}$$

$$\text{Area} = \pi\frac{d^2}{4} = (3.14)\frac{(25.4)^2(10^{-6})^2}{4} = (3.14)(25.4 \div 2)(10^{-6})^2$$

$$= 5.067 \times 10^{-10}\text{m}^2$$

$$R = \rho\frac{l}{A} = (1.68)(10^{-8})\left(\frac{0.3048}{5.067 \times 10^{-10}}\right) = 10.11\ \Omega$$

Thus, the specific resistance of copper in the imperial system is 10.11 Ω.CM/ft.

In the above example the number in the second bracket can be used for conversion between values of specific resistance in the metric system and in the imperial system. Note that in the above example and in Table 12.2, specific resistance is (Ω.m) but sometimes it can be given in (Ω.cm).

Noting that

$$(10^{-8})\left(\frac{0.3048}{5.067 \times 10^{-10}}\right) = 6.0115 \cong 6$$

shows that when the specific resistance of a material in the metric system is given in Ω.m and in the form of a number multiplied by 10^{-8}, we may easily find its value in Ω.CM/ft by multiplying the first number by 6.

Table 12.3 shows the specific resistance of a few metals in the imperial system of measurements.

Specific resistance of a material in Ω.m in the form of (xxx)(10^{-8}) can be converted to Ω.CM/ft by multiplying the portion xxx by 6. The inverse conversion is done by dividing by 6.

Example 12.3

The diameter of a lightbulb filament is 0.05 mm, and it is made out of tungsten with the specific resistance of 5.6×10^{-8} Ω.m at the room temperature. If the filament length is 5 cm, what is the resistance of the filament at the room temperature?

Solution

$$\text{Length} = 0.05\text{ m}$$

$$\text{Radius} = 0.025\text{ mm} = 0.025 \times 10^{-3}\text{ m}$$

TABLE 12.3 Resistivity of Some Metals in Ω.CM/ft

Material	Specific resistance Ω.CM/ft
Silver	9.6
Copper	10.3
Aluminum	17
Gold	14
Tungsten	33.8
Nickel	52
Platinum	66
Constantan	295
Mercury	590
Nichrome	675
Carbon	22,000

$$R = (5.6)(10^{-8})\frac{0.05}{(3.14)(0.025)^2(10^{-3})^2} = 1.426\ \Omega$$

Example 12.4

The resistance of a piece of wire 10 ft long is 100 mΩ. If the wire is made out of copper, find its thickness in CM.

Solution

The specific resistance of copper in imperial system is 10.11 Ω.CM/ft. From Equation 12.1 it follows that

$$A = \frac{\rho l}{R} = \frac{(10.11)(10)}{(0.1)} = 1011\ \text{CM} \quad \text{(wire thickness)}$$

The diameter of the wire is

$$\sqrt{1011} = 31.8\ \text{mil}$$

Example 12.5

A piece of tungsten wire has a circular cross section with a diameter of 22 mils. What length of this metal makes a resistance of 10 Ω?

Solution

Diameter = 22 mil = 0.022 in = 0.0005588 m = 5.588×10^{-4} m

Cross-section area = $(3.14)(5.588/2)^2(10^{-4})^2 = 2.4525 \times 10^{-7}$ m^2

It follows from the resistance relationship that

$$l = \frac{RA}{\rho} = \frac{(10)(2.4525 \times 10^{-7})}{(5.5 \times 10^{-8})} = 44.6 \text{ m}$$

Length in English units = $44.6 \times 3.28 = 146.3$ ft

Example 12.6

If the resistance of a piece of tungsten wire at 20°C is 5 Ω, what is its resistance at 1200°C?

Solution

From Table 12.3 we see that the specific resistance and the temperature coefficient α for tungsten at 20°C are respectively 5.6×10^{-8} and 0.0045. Applying those in Equation 12.2 leads to

$$\rho_{1200} = 5.6 \times 10^{-8} [1 + 0.0045 \times (1200 - 20)]$$
$$= 5.6 \times 6.31 \times 10^{-8} = 35.336 \times 10^{-8}$$

The specific resistance is 6.31 times larger. The resistance is, thus, larger by the same ratio. Therefore,

$$R_{1200} = (5)(6.31) = 31.55 \ \Omega$$

You realize that when a lightbulb is on, the temperature of the filament is much more than room temperature, and, therefore, the resistance of its filament is several times larger than when at room temperature.

The standard wire sizes are given in Table 12.4 for both aluminum and copper wires. Table 12.4 also indicates the electric resistance per 1000 m and 1000 ft of these wires. AWG stands for American wire gauge.

12.3.3 Insulated Conductors and Ampacity

In the previous section we discussed the resistance of materials to electric current. We learned that almost all conductors are from copper or aluminum, and those from aluminum are reinforced by steel to give them mechanical strength when hanging from electric poles and towers. We also noticed that the sizes of wires are standardized and one must select a size among those that are commercially available. A different question that needs to be answered is how much current is allowed for a particular conductor. Ampacity (ampere capacity) is the answer to this question. Ampacity determines how many amperes of electric current a conductor can handle safely.

TABLE 12.4 American Wire Gauge Table

AWG Gauge	Conductor Diameter, Inches	Conductor Diameter, mm	Ohms per 1000 ft	Ohms per km
0000	0.46	11.684	0.049	0.16072
000	0.4096	10.40384	0.0618	0.202704
00	0.3648	9.26592	0.0779	0.255512
0	0.3249	8.25246	0.0983	0.322424
1	0.2893	7.34822	0.1239	0.406392
2	0.2576	6.54304	0.1563	0.512664
3	0.2294	5.82676	0.197	0.64616
4	0.2043	5.18922	0.2485	0.81508
5	0.1819	4.62026	0.3133	1.027624
6	0.162	4.1148	0.3951	1.295928
7	0.1443	3.66522	0.4982	1.634096
8	0.1285	3.2639	0.6282	2.060496
9	0.1144	2.90576	0.7921	2.598088
10	0.1019	2.58826	0.9989	3.276392
11	0.0907	2.30378	1.26	4.1328
12	0.0808	2.05232	1.588	5.20864
13	0.072	1.8288	2.003	6.56984
14	0.0641	1.62814	2.525	8.282
15	0.0571	1.45034	3.184	10.44352
16	0.0508	1.29032	4.016	13.17248
17	0.0453	1.15062	5.064	16.60992
18	0.0403	1.02362	6.385	20.9428
19	0.0359	0.91186	8.051	26.40728
20	0.032	0.8128	10.15	33.292
21	0.0285	0.7239	12.8	41.984
22	0.0254	0.64516	16.14	52.9392
23	0.0226	0.57404	20.36	66.7808
24	0.0201	0.51054	25.67	84.1976
25	0.0179	0.45466	32.37	106.1736
26	0.0159	0.40386	40.81	133.8568
27	0.0142	0.36068	51.47	168.8216
28	0.0126	0.32004	64.9	212.872
29	0.0113	0.28702	81.83	268.4024
30	0.01	0.254	103.2	338.496
31	0.0089	0.22606	130.1	426.728
32	0.008	0.2032	164.1	538.248
Metric 2.0	0.00787	0.200	169.39	555.61
33	0.0071	0.18034	206.9	678.632
Metric 1.8	0.00709	0.180	207.5	680.55
34	0.0063	0.16002	260.9	855.752
Metric 1.6	0.0063	0.16002	260.9	855.752
35	0.0056	0.14224	329	1079.12
Metric 1.4	0.00551	0.140	339	1114

(*Continued*)

TABLE 12.4 (CONTINUED) American Wire Gauge Table

AWG Gauge	Conductor Diameter, Inches	Conductor Diameter, mm	Ohms per 1000 ft	Ohms per km
36	0.005	0.127	414.8	1360
Metric 1.25	0.00492	0.125	428.2	1404
37	0.0045	0.1143	523.1	1715
Metric 1.12	0.00441	0.112	533.8	1750
38	0.004	0.1016	659.6	2163
Metric 1	0.00394	0.1000	670.2	2198
39	0.0035	0.0889	831.8	2728
40	0.0031	0.07874	1049	3440

Ampacity is determined based on the amount of heat generated in a conductor owing to the current and the fact that this heat must be taken away, so that the conductor temperature does not increase anymore beyond a certain safe level. Otherwise, the buildup of heat can cause a problem. In this sense, the ambient temperature and the nature of conductor and its surroundings are the parameters that affect the ampacity rating of a conductor. In this respect, the same cable has more ampacity when in the air than when in a conduit.

Figure 12.11 illustrates a three-conductor underground cable. These cables come in different sizes with one, two, and three conductors and various insulation materials. Good electrical insulation is absolutely necessary for underground cables.

Overhead lines are normally in the air and cooled by streams of free air, whereas the underground cables are either in a conduit or buried underground. Also, their insulation decreases the rate of cooling compared to bare wires. The ambient temperature can be –30°C or +30°C, for instance, and it can vary significantly from winter to summer. Selection of cables must be based on the worst case scenario and the highest ambient temperature.

On the basis of the material and other conditions the ampacity of conductors are standardized. Tables 12.5 and 12.6 are examples of the ampacity of conductors according to National Electric Code (NEC) standard. These

Figure 12.11 A three-conductor underground cable. (Courtesy of CME Wire and Cable, Inc.)

tables are only samples. Refer to NEC publications for a complete set of standards. Table 12.5 corresponds to ampacities of up to three conductors in a conduit, and Table 12.6 is for the same types of conductors when hanging in free air. In both cases the line voltage cannot exceed 2000 V.

As can be seen from these tables, ampacities are restricted for temperatures shown (30°C in this case) and must be adjusted for other temperatures based on the correction factors given in Table 12.7. The following examples show how these tables can be used. Letter combinations identified in the table are for the insulation types. Table 12.8 exemplifies the pertinent information. For full list of these types one must consult National Electric Code publications.

Example 12.7

Find the allowable current in a copper wire gauge AWG-8 for free air installation in an ambient reaching 45°C. The wiring is for 416 V (below 2000 V). Assume wire type TW.

Solution

From Table 12.6 we see three columns for copper wire. In the first column indicating TW and UF types, we find number 60 for gauge 8 wire. This implies that the maximum allowable current for this wire when the ambient temperature is 30°C is 60 A. For an ambient temperature of 45°C, however, this number must be adjusted by the coefficient given under "correction factors" in the same table. The corresponding factor for 45°C can be found to be 0.71. The maximum allowable current is, thus, $60 \times 0.71 = 42$ A.

Note that the ambient temperature may be less than 45°C most of the time, but one needs to consider the worst case when calculating the ratings.

Current capacity calculations for cables must be performed based on the worst case scenario.

Example 12.8

Determine the required wire gauge for 40 A current in a conduit where two of such copper wires are to be installed. The ambient temperature does not go over 38°C. The operating voltage is below 2000 V. Assume wire type TW.

Solution

From Table 12.7 we find that for temperature range 36°C–40°C a correction factor of 0.82 must be considered. If the desired current of 40 A is divided by this number, we get $40 \div 0.82 \cong$ 50 A. Now we must find the standard wire corresponding to the

TABLE 12.5 Allowable Ampacities of Insulated Conductors in Conduits

Size AWG or kcmil	Temperature Rating of Conductor (See Table 310.104[A])						Size AWG or kcmil
	60°C (140°F)	75°C (167°F)	90°C (194°F)	60°C (140°F)	75°C (167°F)	90°C (194°F)	
	Types TW, UF	Types RHW, THHW, THW, THWN, XHHW, USE, ZW	Types TBS, SA, SIS, FEP, FEPB, MI, RHH, RHW-2, THHN, THHW, THW-2, THWN-2, USE-2, XHH, XHHW, XHHW-2, ZW-2	Types TW, UF	Types RHW, THHW, THW, THWN, XHHW, USE	Types TBS, SA, SIS, THHN, THHW, THW-2, THWN-2, RHH, RHW-2, USE-2, XHH, XHHW, XHHW-2, ZW-2	
	Copper			Aluminum or Copper-Clad Aluminum			
18[a]	–	–	14	–	–	–	–
16[a]	–	–	18	–	–	–	–
14[a]	15	20	25	–	–	–	–
12[a]	20	25	30	15	20	25	12[a]
10[a]	30	35	40	25	30	35	10[a]
8	40	50	55	35	40	45	8
6	55	65	75	40	50	55	6
4	70	85	95	55	65	75	4
3	85	100	115	65	75	85	3
2	95	115	130	75	90	100	2
1	110	130	145	85	100	115	1
1/0	125	150	170	100	120	135	1/0
2/0	145	175	195	115	135	150	2/0
3/0	165	200	225	130	155	175	3/0
4/0	195	230	260	150	180	205	4/0
250	215	255	290	170	205	230	250

(*Continued*)

TABLE 12.5 (CONTINUED) Allowable Ampacities of Insulated Conductors in Conduits

	Temperature Rating of Conductor (See Table 310.104[A])						
	60°C (140°F)	75°C (167°F)	90°C (194°F)	60°C (140°F)	75°C (167°F)	90°C (194°F)	
	Types TW, UF	Types RHW, THHW, THW, THWN, XHHW, USE, ZW	Types TBS, SA, SIS, FEP, FEPB, MI, RHH, RHW-2, THHN, THHW, THW-2, THWN-2, USE-2, XHH, XHHW, XHHW-2, ZW-2	Types TW, UF	Types RHW, THHW, THW, THWN, XHHW, USE	Types TBS, SA, SIS, THHN, THHW, THW-2, THWN-2, RHH, RHW-2, USE-2, XHH, XHHW, XHHW-2, ZW-2	
Size AWG or kcmil	Copper			Aluminum or Copper-Clad Aluminum			Size AWG or kcmil
300	240	285	320	195	230	260	300
350	260	310	350	210	250	280	350
400	280	335	380	225	270	305	400
500	320	380	430	260	310	350	500
600	350	420	475	285	340	385	600
700	385	460	520	315	375	425	700
750	400	475	535	320	385	435	750
800	410	490	555	330	395	445	800
900	435	520	585	355	425	480	900
1000	455	545	615	375	445	500	1000
1250	495	590	665	405	485	545	1250
1500	525	625	705	435	520	585	1500
1750	545	650	735	455	545	615	1750
2000	555	665	750	470	560	630	2000

Source: Reprinted with permission from NFPA 70®-2014, *National Electrical Code*®, Copyright © 2013, National Fire Protection Association, Quincy, MA. This reprinted material is not the complete and official position of the NFPA on the referenced subject, which is represented only by the standard in its entirety.

[a] Refer to 240.4(D) for conductor overcurrent protection limitations.

TABLE 12.6 Allowable Ampacities of Insulated Conductors in Free Air

	Temperature Rating of Conductor (See Table 310.104[A])						
	60°C (140°F)	75°C (167°F)	90°C (194°F)	60°C (140°F)	75°C (167°F)	90°C (194°F)	
	Types TW, UF	Types RHW, THHW, THW, THWN, XHHW, ZW	Types TBS, SA, SIS, FEP, FEPB, MI, RHH, RHW-2, THHN, THHW, THW-2, THWN-2, USE-2, XHH, XHHW, XHHW-2, ZW-2	Types TW, UF	Types RHW, THHW, THW, THWN, XHHW	Types TBS, SA, SIS, THHN, THHW, THW-2, THWN-2, RHH, RHW-2, USE-2, XHH, XHHW, XHHW-2, ZW-2	
Size AWG or kcmil	Copper			Aluminum or Copper-Clad Aluminum			Size AWG or kcmil
18	–	–	18	–	–	–	–
16	–	–	24	–	–	–	–
14[a]	25	30	35	–	–	–	–
12[a]	30	35	40	25	30	35	12[a]
10[a]	40	50	55	35	40	45	10[a]
8	60	70	80	45	55	60	8
6	80	95	105	60	75	85	6
4	105	125	140	80	100	115	4
3	120	145	165	95	115	130	3
2	140	170	190	110	135	150	2
1	165	195	220	130	155	175	1
1/0	195	230	260	150	180	205	1/0
2/0	225	265	300	175	210	235	2/0
3/0	260	310	350	200	240	270	3/0
4/0	300	360	405	235	280	315	4/0
250	340	405	455	265	315	355	250

(Continued)

TABLE 12.6 (CONTINUED) Allowable Ampacities of Insulated Conductors in Free Air

	Temperature Rating of Conductor (See Table 310.104[A])						
	60°C (140°F)	75°C (167°F)	90°C (194°F)	60°C (140°F)	75°C (167°F)	90°C (194°F)	
	Types TW, UF	Types RHW, THHW, THW, THWN, XHHW, ZW	Types TBS, SA, SIS, FEP, FEPB, MI, RHH, RHW-2, THHN, THHW, THW-2, THWN-2, USE-2, XHH, XHHW, XHHW-2, ZW-2	Types TW, UF	Types RHW, THHW, THW, THWN, XHHW	Types TBS, SA, SIS, THHN, THHW, THW-2, THWN-2, RHH, RHW-2, USE-2, XHH, XHHW, XHHW-2, ZW-2	
Size AWG or kcmil	Copper			Aluminum or Copper-Clad Aluminum			Size AWG or kcmil
300	375	445	500	290	350	395	300
350	420	505	570	330	395	445	350
400	455	545	615	355	425	480	400
500	515	620	700	405	485	545	500
600	575	690	780	455	545	615	600
700	630	755	850	500	595	670	700
750	655	785	885	515	620	700	750
800	680	815	920	535	645	725	800
900	730	870	980	580	700	790	900
1000	780	935	1055	625	750	845	1000
1250	890	1065	1200	710	855	965	1250
1500	980	1175	1325	795	950	1070	1500
1750	1070	1280	1445	875	1050	1185	1750
2000	1155	1385	1560	960	1150	1295	2000

Source: Reprinted with permission from NFPA 70®-2014, *National Electrical Code*®, Copyright © 2013, National Fire Protection Association, Quincy, MA. This reprinted material is not the complete and official position of the NFPA on the referenced subject, which is represented only by the standard in its entirety.

Note: Refer to 310.15(B)(2) for the ampacity correction factors where the ambient temperature is other than 30°C (86°F).

[a] Refer to 240.4(D) for conductor overcurrent protection limitations.

TABLE 12.7 Temperature Correction Factors for Tables 12.5 and 12.6

Ambient Temperature (°C)	Temperature Rating of Conductor			Ambient Temperature (°F)
	60°C	75°C	90°C	
10 or less	1.29	1.20	1.15	50 or less
11–15	1.22	1.15	1.12	51–59
16–20	1.15	1.11	1.08	60–68
21–25	1.08	1.05	1.04	69–77
26–30	1.00	1.00	1.00	78–86
31–35	0.91	0.94	0.96	87–95
36–40	0.82	0.88	0.91	96–104
41–45	0.71	0.82	0.87	105–113
46–50	0.58	0.75	0.82	114–122
51–55	0.41	0.67	0.76	123–131
56–60	–	0.58	0.71	132–140
61–65	–	0.47	0.65	141–149
66–70	–	0.33	0.58	150–158
71–75	–	–	0.50	159–167
76–80	–	–	0.41	168–176
81–85	–	–	0.29	177–185

Source: NFPA 70®-2014, *National Electrical Code®*, Copyright © 2013, National Fire Protection Association, Quincy, MA. Reprinted with permission. This reprinted material is not the complete and official position of the NFPA on the referenced subject, which is represented only by the standard in its entirety.

Note: For ambient temperatures other than 30°C (86°F), multiply the allowable ampacities specified in the ampacity tables by the appropriate correction factor shown above.

given conditions and with 50 A ampacity. We need to look in Table 12.5 under the column for the type of wire TW, but this ampacity of 50 A cannot be found in the table. Hence, we need to choose the gauge corresponding to the higher current of 55 A. The required wire is, thus, gauge 6.

12.4 Fault Detection and Protective Measures

Faults in an electric system can happen because of various reasons. Some faults in electricity distribution are due to natural causes such as lightning and are inevitable. Others can be due to deterioration of components as a result of age or abuse, and so on. What is important in operation is to detect a fault when it happens and have a proper action to prevent the harmful consequences, such as fire, damage, and destruction of equipment. In this section, as far as the scope of the book allows, electric faults in a power system are discussed.

12.4.1 Electric System Faults

In a three-phase power system, a fault, in general, can be a result of a short circuit or an inadvertent open circuit in one or two of the phases. A short

TABLE 12.8 Example of Definitions for Various Insulated Cables

Trade Name	Type Letter	Maximum Operating Temperature	Application Provisions	Insulation	Thickness of Insulation: AWG or kcmil	mm	mils	Outer Covering[2]
Thermoplastic and fibrous outer braid	TBS	90°C 194°F	Switchboard and switchgear wiring only	Thermoplastic	14–10	0.76	30	Flame-retardant, nonmetallic covering
					8	1.14	45	
					6–2	1.52	60	
					1–4/0	2.03	80	
Extended polytetrafluoroethylene	TFE	250°C 482°F	Dry locations only; only for leads within apparatus or within raceways connected to apparatus, or as open wiring (nickel or nickel-coated copper only)	Extruded polytetrafluoroethylene	14–10	0.51	20	None
					8–2	0.76	30	
					1–4/0	1.14	45	
Heat-resistant thermoplastic	THHN	90°C 194°F	Dry and damp locations	Flame-retardant, heat-resistant thermoplastic	14–12	0.38	15	Nylon jacket or equivalent
					10	0.51	20	
					8–6	0.76	30	
					4–2	1.02	40	
					1–4/0	1.27	50	
					250–500	1.52	60	
					501–1000	1.78	70	
Moisture- and heat-resistant thermoplastic	THHW	75°C 167°F	Wet location	Flame-retardant, moisture- and heat-resistant thermoplastic	14–10	0.76	30	None
					8	1.14	45	
					6–2	1.52	60	
		90°C 194°F	Dry location		1–4/0	2.03	80	
					213–500	2.41	95	
					501–1000	2.79	110	
					1001–2000	3.18	125	

(*Continued*)

TABLE 12.8 (CONTINUED) Example of Definitions for Various Insulated Cables

Trade Name	Type Letter	Maximum Operating Temperature	Application Provisions	Insulation	Thickness of Insulation AWG or kcmil	mm	mils	Outer Covering[2]
Moisture- and heat-resistant thermoplastic	THW	75°C 167°F 90°C 194°F	Dry and wet locations Special applications within electric discharge lighting equipment; limited to 1000 open-circuit volts or less (size 14–8 only as permitted in 410.68)	Flame-retardant, moisture- and heat-resistant thermoplastic	14–10	0.76	30	None
					8	1.14	45	
					6–2	1.52	60	
					1–4/0	2.03	80	
					213–500	2.41	95	
					501–1000	2.79	110	
					1001–2000	3.18	125	
	THW–2	90°C 194°F	Dry and wet locations					
Moisture- and heat-resistant thermoplastic	THWN	75°C 167°F	Dry and wet locations	Flame-retardant, moisture- and heat-resistant thermoplastic	14–12	0.38	15	Nylon jacket or equivalent
					10	0.51	20	
					8–6	0.76	30	
					4–2	1.02	40	
	THWN–2	90°C 194°F			1–4/0	1.27	50	
					250–500	1.52	60	
					501–1000	1.78	70	

(*Continued*)

TABLE 12.8 (CONTINUED) Example of Definitions for Various Insulated Cables

Trade Name	Type Letter	Maximum Operating Temperature	Application Provisions	Insulation	Thickness of Insulation: AWG or kcmil	Thickness of Insulation: mm	Thickness of Insulation: mils	Outer Covering[2]
Moisture-resistant thermoplastic	TW	60°C 140°F	Dry and wet locations	Flame-retardant, moisture-resistant thermoplastic	14–10	0.76	30	None
					8	1.14	45	
					6–2	1.52	60	
					1–4/0	2.03	80	
					213–500	2.41	95	
					501–1000	2.79	110	
					1001–2000	3.18	125	
Underground feeder and branch-circuit cable—single conductor (for type UF cable employing more than one conductor, see Article 340)	UF	60°C	See Article 340	Moisture-resistant	14–10	1.52	60[6]	Integral with insulation
		140°F			8–2	2.03	80[6]	
					1–4/0	2.41	95[6]	
		75°C 167°F[5]		Moisture- and heat-resistant				

circuit can happen owing to physical wear of insulation as a result of age, thermal stress and fatigue, high voltage, high current, harsh environment, abuse, and the like. An open circuit can occur owing to lightning, overload, load surge, malfunction of a device, loss of synchronization, and so on. In either case, there is a significant imbalance between the currents in the three phases.

Depending on the severity of a fault, currents of several times magnitude can flow through wires, transformers, and a generator on the power side. If the fault persists, it can cause other failures in a system by overheating and damaging parts, wires insulations, and insulators. In a power system it can cause fire or in the best situation if the protective devices act right and cut off the electricity (clearing the fault), a lot of consumers will be left without electricity until the cause of the fault is found and removed. In a **bolt short circuit**, that is, when a current carrying line is solidly shorted without any impedance in between, currents of several hundred times are not uncommon.

Bolt short circuit: Short circuit by direct contact between two lines of different voltages.

A fault in a power system is in one of the following forms: one phase shorted to ground, two phase lines shorted together, two phase lines shorted to ground, all phases shorted to ground, one line open, and two lines open.

In the case of any of these faults, if a system is not provided with the proper protective means, the consequences are costly. The importance of a protective measure is realized when a fault occurs. Thus, to reduce the risk of losing a power system and its equipment, or the electrical equipment in a consumer site, each side (generation and consumption) needs to be protected by the proper means. At the very lowest level and very fundamental to all protective means is grounding of an electric circuit at the source and at the consumption sites.

12.4.2 Grounding

Grounding is the major feature of the protective means for a power system, an electric device, and personnel. The other feature is clearing a fault, which is done by devices that break a circuit and clear the fault by removing the faulty device or portion of the circuit containing a fault.

For equipment, grounding causes the fault currents find their way through ground, and not through the equipment parts and windings. For personnel, grounding causes the voltage of a faulty equipment frame/case to stay at the ground level, so that if a person touches the frame, there is no potential difference between the body part in contact with the equipment and the ground.

Grounding implies physically connecting a point of a device or circuit to the earth, which provides a common reference point for zero voltage and a path for return of a current that otherwise has no path to flow through, to the source. For a power system, grounding is connecting the neutral point of the system to the earth, and for equipment, grounding implies connecting the noncarrying conductive material(s) such as motor frames and enclosures to earth.

In normal operation of electric systems the current through ground is negligible. This current may be due to unbalanced loads in a three-phase system, but it is still relatively small, especially that the soil resistance comes into effect. It is only when a fault happens that the connections to ground and the earth play their role in the currents that seek a path to follow. In such a case a high current is flowing in the ground wire of a grounded system. Table 12.9 shows the specific resistance of various earth materials.

For most electrical equipment, grounding is performed by a solid or bolt connection to ground, whereas for a system, such as the electric network, generators (like in all wind turbines), and in larger and expensive equipment, grounding can be either solidly or through a current limiting device (see Figure 12.12).

The purpose of a current limiting device is to reduce the current that will flow through a device, and a system or its parts in case of a fault, which can cause damage. A common current limiting device for this purpose is a **neutral grounding resistance (NGR)**. Such a resistor, nonetheless, must be carefully selected based on a number of criteria including the involved voltage and the allowed delay for the protecting device (e.g., a circuit breaker). A low resistance neutral grounding or a high resistance

Neutral grounding resistance (NGR): An added resistance to the grounding strap to limit the (otherwise high) current flow through a device or a circuit in case of a fault.

TABLE 12.9 Specific Resistance (Resistivity) of Soil

Type of Soil	Specific Resistance ρ, Ωm
Loam	1–50
Clay	20–100
Sand and gravel	50–1000
Surface limestone	100–10,000
Shale	5–100
Sandstone	20–2000
Granites, basalt	10,000
Slates	10–100

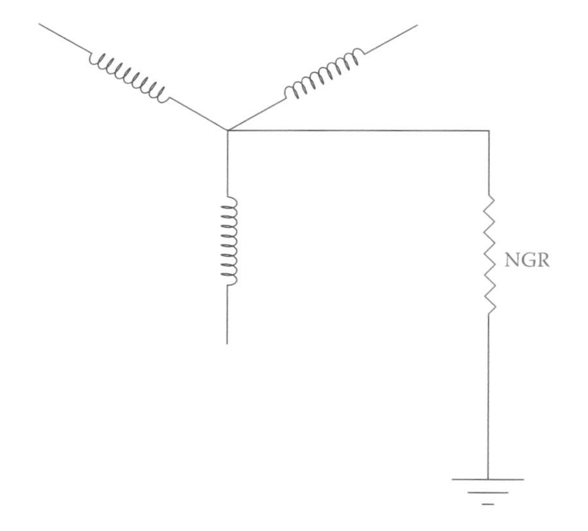

Figure 12.12 Neutral Grounding Resistor.

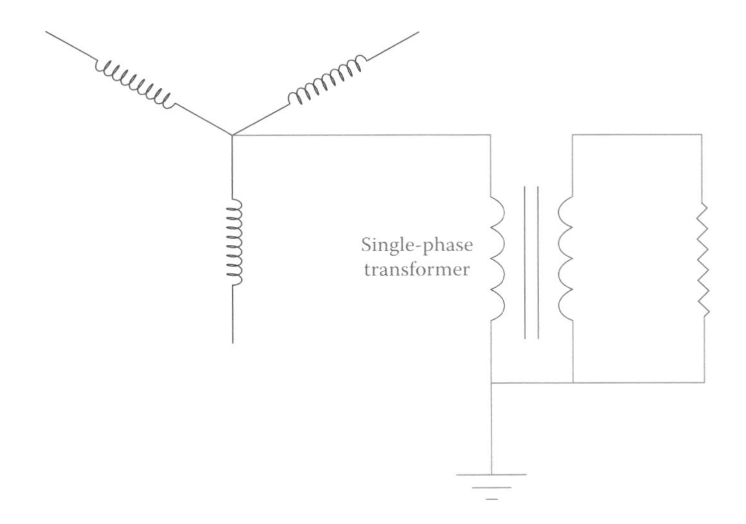

Figure 12.13 Using a single phase transformer in conjunction with a grounding resistor.

neutral grounding can be employed. In the former, the allowed current is higher. A current of 50 A or more (up to 400 A) is allowed for 10 sec. In the latter the maximum allowed current is 25 A for potentials of 600 V or less. A current of 5 A is more common. For higher voltages (13.8 kV and up), solid grounding is normally employed.

Instead of directly connecting a grounding resistor to the ground bus, a single-phase transformer can be employed, as shown in Figure 12.13. In this case a smaller resistor is used based on the transformer turns ratio.

It is important to have the grounding point on the secondary of a grounding transformer and as near to it as possible.

Recall from Chapter 8 that, as its specific property, an inductor resists a change in the current through it. This property makes it a candidate for current limiting, so it is desirable for grounding. As another means for grounding, a **reactor** can be used. A reactor is a winding (inductor) with or without a core, dry or oil immersed, which is specifically designed for a number of purposes, including grounding. Among other reactor applications are inrush-current limiting (for capacitors and motors), harmonic filtering, reactive power compensation, reduction of ripple currents, damping of switching transients, flicker reduction, load balancing, and power conditioning.

Reactor: Wire winding with or without a core, dry or oil immersed, which is specifically designed for a number of purposes, including grounding in a three-phase system.

12.4.2.1 What Point Must Be Grounded?

So far, we have talked about grounding. But, what point must be grounded? As we discussed in Chapter 9, the center point in a star connection of three-phase systems has a zero voltage with respect to all the phase lines. This point is, thus, a good reference for grounding. This is a point that, when it exists, it is grounded in all power facilities. When this point does not exist, in a delta connection, for example, such a point is created by an auxiliary transformer.

When this point is grounded in the source, and the loads have a similar point grounded, these points are connected to each other through the earth and any current coming from the load to ground finds its return way to the source through the earth. This can be a small current in the case of unbalanced three-phase loads, or a high current in the case of a fault. Remember that a substation is a load to a generator or to an upstream substation. In the same fashion a substation is a source for the loads or for a downstream substation.

In a grid, each line must be individually protected from faults.

12.4.2.2 Grounding in Power Transmission

We have seen in Chapter 10 that three-phase transformers can be used in four different configurations (see Section 10.6). These are delta-delta, delta-wye, wye-delta, and wye-wye. Each of these combinations has its advantages and disadvantages. A wye connection provides a suitable point for grounding. A delta connection provides a path for capturing third-order harmonics, which is quite important in power transmission and distribution.

It is more common to have one side delta connected and the other side wye connected to benefit from both advantages. However, it might be necessary to use a **grounding transformer** for the side with delta connection. The standard configuration by utilities for power transmission is "delta high – grounded wye low." This implies that the higher voltage side is delta connected and the lower voltage side is star connected and it is grounded. This has the advantage that the low voltage bus has always a ground reference. Moreover, the common practice in traditional power plants (thermal and hydraulic plants) is to have the generator-step-up (GSU) transformer delta connected at the generator side (lower voltage) and the transformer secondary star-connected and grounded. A grounding transformer is then used for the generator to provide a solid ground for fault protection. This is depicted in Figure 12.14. A grounding transformer must have the correct

Grounding transformer: Transformer used only for providing a reference point for connection to ground, in order for detection of faults in power generation and distribution.

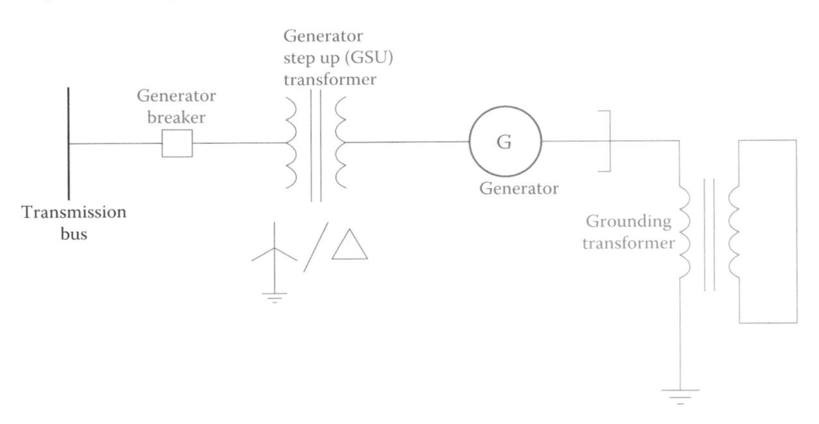

Figure 12.14 Common arrangement for generators in traditional power plants.

size and rating for its purpose. Thus, its selection must be based on a careful study, and it is an additional cost to any facility.

In wind farms, however, because there are many wind turbines in a wind farm, the transformer at the generator side is star connected and grounded and the collector side (high voltage) is delta connected.

> The standard configuration by utilities for power transmission is "Delta high – grounded wye low."

A grounding transformer is a special wye-delta transformer that is designed for this purpose only. The secondary is not connected to any load and the primary is connected to the three-phase line at the location the grounding is desired. In the case of significant unbalance primary currents (a fault), very high currents circulate in the secondary windings. The transformer, therefore, must have been designed and rated for handling the high currents.

Also, a zigzag transformer can be used for grounding. The description of a zigzag transformer is given in Appendix C.

> A grounding transformer is a special wye-delta transformer with its secondary not connected to any load. It is designed for the purpose of grounding only.

12.5 Transmission Lines

This section briefly describes matters concerned with transmission lines in electrical networks. Knowledge of the material described here is very useful for the technical personnel in electricity and its generation and distribution. Nevertheless, a thorough discussion of the subjects is out of the scope of this book, and the enthusiast reader must refer to other sources.

12.5.1 Transmission Lines Effects

In addition to the ohmic resistance in the wires and electric conductors, which we have discussed in this and Chapters 4–6, conductors carrying AC electricity have inductive and capacitive effects. Consider, for instance, a pair of transmission lines that extend for several miles and compare them to two plates forming a capacitor. You will see that there is a similarity in terms of the *two metallic surfaces*, being *at a distance from each other*, with a *common cross-section area* separated by a *non-conductive material* (air). Thus, the two conductors form a capacitor. The same thing is true for the magnetic effect of the current in each conductor on itself and on the other conductor. Thus, there exists inductance between the two conductors.

The amount of capacitive and inductive effects might be small and negligible for short wires, for example, in the household wiring having a low voltage, but they are not negligible for the very long transmission lines and the high voltages they carry.

Capacitive and inductive effects of the transmission lines, thus, need to be included in the calculation of their current and voltage drop. If these effects are added, the schematic representation of an electric network between load(s) and the source(s) is similar to Figure 12.15b, instead of Figure 12.15a. Resistances and inductances are in series with each other along the length of a conductor, but the capacitive effects are shown as parallel. These effects are indicated for one line only, but they correspond to both conductors. In practice, there are usually more than two conductors, particularly for three-phase lines. But, for three-phase systems the effects of two conductors do not add up as in the case of single-phase wires because there are no return lines carrying the exact same currents at the same moment.

Values for ohmic resistance can be readily defined for each conductor in terms of its length, cross-section area, the conductor material, and its specific resistance. These data are usually available for any cable in terms of the resistance for each meter of the conductor, or 1000 ft of it, as we have seen in the tables in Section 12.3.3.

Capacitor value and inductor value for a set of conductors, however, depend on the factors particular to their setup. Depending on conductors spacing, their radii, the way they are laid out, the number of conductors, and how many are in each bundle sharing the same current, a value can be determined for the inductance of a given length (e.g., 1 km or 1 mile) of a transmission line. Similarly, a value can be obtained for the capacitance. In this way, more precise values for the current, voltage drop, lost energy, and so on can be found for a transmission line of a given length.

Corona: Corona is a local electric discharge around the conductors of high-voltage transmission lines, when the (especially humid) air is ionized. It is manifested by a bluish glow and a small humming sound. This phenomenona is accompanied with a slight loss of power used for making the sound and the glow.

Another phenomenon that causes loss in transmission lines is **corona**. Corona is the effect of potential difference (voltage) on the air surrounding the transmission lines, ionizing the air. This effect reflects in the form of a glow coming out of lines. Corona depends on the voltage and spacing between the wires. The amount of energy lost this way, however, is small and is often neglected.

Details of calculation of the equivalent inductor and capacitor for various line arrangements and spacing conditions is lengthy and involves mathematics that are out of the scope of this book. We may only use the results accompanied by some examples. To comprehend the subject, nevertheless, a reader is expected to have acquaintance with logarithm of

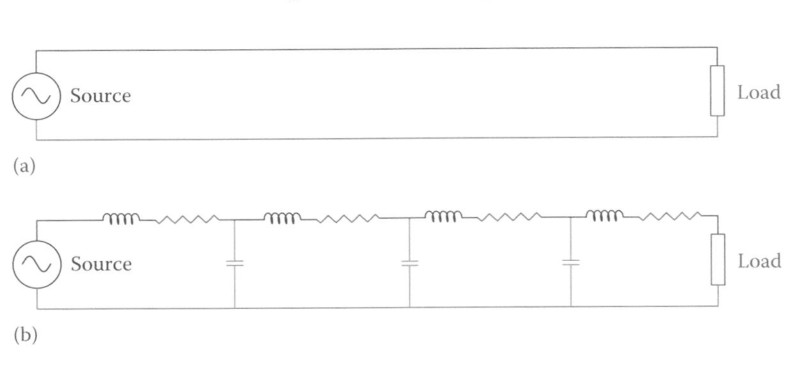

Figure 12.15 Typical representation of a transmission line. (a) Two long transmission wires for AC. (b) Assumed capacitors and inductors to represent the capacitance and inductance effects of the nearby wires in AC transmission.

a number because all calculations involve natural (or Naperian) logarithm (see the boxed text in this chapter for logarithm).

Figure 12.16 illustrates the relative spacing of possible transmission lines arrangements (compare with Figure 12.10). Figure 12.16a shows a general case where the three lines of a three-phase system are at locations 1, 2, and 3. These lines, furthermore, are at general distances D_{12}, D_{13}, and D_{23} from each other. In the discussions that follows it is assumed that the loads on the three phases are the same and we are dealing with balanced loads. Except in the case of Figure 12.16c where all lines are equidistance from each other, values for the inductance and capacitance of a length of the lines are not the same. As a result, voltage drops in the three lines are different. Consequently, at the load site the voltage at the three lines carrying three phases are not the same. Also, the phase relationships can be disturbed and the three phases will no longer be 120° from each other.

It is very undesirable if the voltages arriving at a site from the same source are different and a proper three-phase system is not delivered at the line end. In many arrangements for transmission lines, it is not common, and sometimes impossible, to have the lines at the corners of an equilateral triangle (e.g., Figure 12.16b and d). To overcome this problem, it is common practice to rotate the lines, so that each of the phases occupies the same physical position equally along the length of the transmission

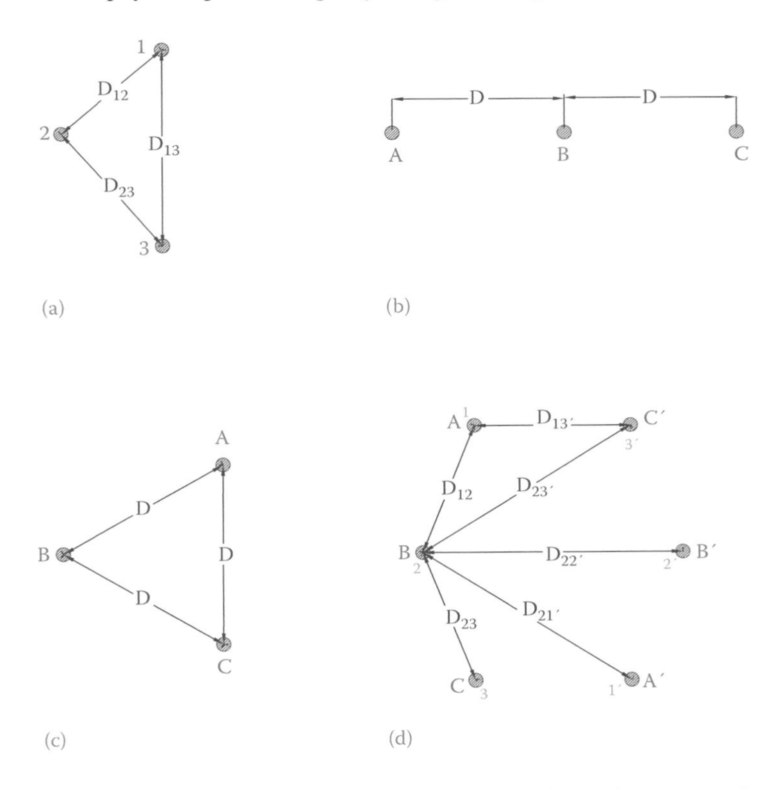

Figure 12.16 Conductor positions in some common three-phase transmission lines. (a) In three corners of a triangle. (b) Along one horizontal line. (c) At the three corners of an equilateral triangle. (d) Double-circuit three-phase, as on transmission structures.

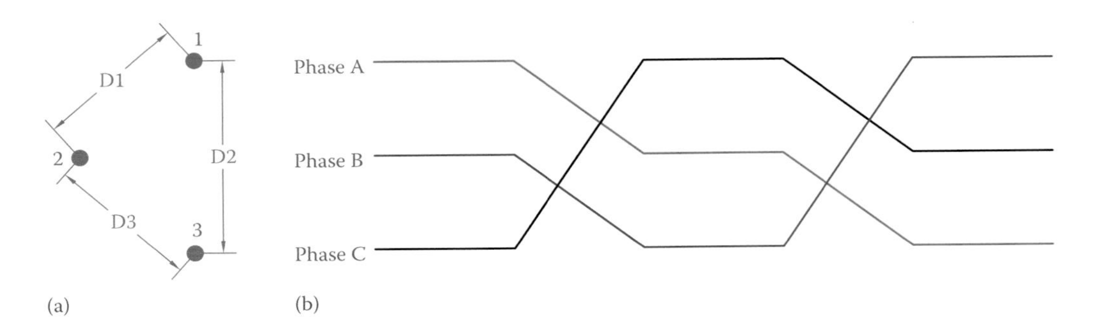

Figure 12.17 Transposition in transmission lines. (a) Physical position of the three-phase lines. (b) Interchanging the wire positions at some point along the line.

Transposition: Special arrangement in the transmission of three-phase power, in which the wires are rotated such that the location of wires for each phase is physically interchanged so that each wire equally occupies the same location as the other wires within the length of a line.

Geometrical mean distance (GMD): A term in power transmission lines corresponding to the spacing between the two (in single phase) and three or six (in three phase) power lines used to determine the inductance and capacitance associated with lines.

Geometrical mean radius (GMR): Term in power transmission lines corresponding to the size (radius) of power lines used to determine their associated inductance and capacitance.

line. In this way, the total inductance and capacitance of all three lines will be equalized. This rotation is called **transposition**, and a line treated this way is said to be transposed. This is illustrated in Figure 12.17, which illustrates that each of the three-phase lines A, B, and C are placed equally in the three physical positions denoted by 1, 2, and 3 along the length of the transmission line. When all the three phases of a line equally occupy the three positions, so that all the voltage differences cancel each other, the line is said to be completely transposed.

In all that follows we assume that the lines are solid, circular, and all have the same radius r. The distances are denoted by D with one or two indexes when necessary (like in D_{12}). The numbers calculated for inductance and capacitance are for a given length, and, thus, for the whole length of a transmission line they must be multiplied by the line length. Results of the calculations are per phase for three-phase lines. In single-phase systems the numbers must be doubled for inductance and halved for capacitance because there are two lines and one is the return of the other line.

Before going ahead we need to define the two following terms. In fact, these are the parameters that change, from one line arrangement to another. **Geometrical mean distance (GMD)** is a number corresponding to the equivalent distance between two or more lines (cables). **Geometrical mean radius (GMR)** is a number corresponding to the effective radius of one or more lines.

12.5.2 Line Inductance

The following general formula can be used for finding the inductance of each line in a transmission system. If three phase, the line is assumed to be transposed and the load(s) balanced.

$$L = (2 \times 10^{-7}) \ln\left(\frac{\text{GMD}}{\text{GMR}}\right) \quad (12.5)$$

GMD and GMR are as defined above, and they vary on a case-by-case basis. The coefficient (2×10^{-7}) is for metric system, and the answer is in henries per meter. For the imperial system a conversion is necessary, as

Logarithm of a Number

Using logarithms is a useful tool in many engineering and other subjects. It makes calculations and solution to problems easier. Here we briefly explain what a logarithm is and show some examples. There are two logarithms, and we may say any positive number has two logarithm values. The difference between them is the base number, and it is possible to change from one logarithm to the other.

The two logarithms are decimal logarithm, which has a base 10, and the natural logarithm whose base is another number. We first start with the decimal logarithm, and when you understand it well, it is easy to understand the other. Indeed, logarithm is the opposite of exponential power.

DECIMAL LOGARITHM

To show the decimal logarithm of a number, the symbol "log" is used, such as log 10, log 20, etc. Consider the following relationships. You may want to check them on a calculator:

$$10^{0.1} = 1.2589$$

$$10^{0.2} = 1.5849$$

$$10^{0.3} = 1.9953$$

$$10^{0.5} = 3.1623$$

$$10^{1} = 10$$

$$10^{2} = 100$$

$$10^{2.5} = 316.23$$

$$10^{3} = 1000$$

In conjunction with the above relationships and in the same order you can write

$$\text{Log } 1.2589 = 0.1$$

$$\text{Log } 1.5849 = 0.2$$

$$\text{Log } 1.9953 = 0.3$$

$$\text{Log } 3.1623 = 0.5$$

$$\text{Log } 10 = 1$$

$$\text{Log } 100 = 2$$

$$\text{Log } 316.23 = 2.5$$

$$\text{Log } 1000 = 3$$

From the above examples it is clear that if $y = 10^x$, then $x = \log y$.

Now that the notion of logarithm function is clear for you, try some negative numbers for the x and see the result on y. All values found this way are smaller than 1.0. Also, remember that negative numbers do not have logarithm. You will get an error message if try to find the log of a negative number on the calculator.

NATURAL OR NAPERIAN LOGARITHM

Instead of 10, the base for the natural (or Naperian) logarithm is the following number, which is denoted by e (always lowercase). You can see it on a calculator keyboard.

$$e = 1 + \frac{1}{1!} + \frac{1}{2!} + \frac{1}{3!} + \frac{1}{4!} + \frac{1}{5!} + \frac{1}{6!} + \frac{1}{7!} + \frac{1}{8!} + \frac{1}{9!} + \ldots = 2.718282$$

The symbol for natural logarithm is ln, like ln 10, ln 20, etc. The following show some examples for natural logarithm.

$e^{0.2} = 1.221$	$\ln 1.221 = 0.2$
$e^{0.5} = 1.649$	$\ln 1.649 = 0.5$
$e^{0.7} = 2.01$	$\ln 2.01 = 0.7$
$e^{1} = 2.718$	$\ln 2.718 = 1$
$e^{2.3026} = 10$	$\ln 10 = 2.3026$
$e^{4.605} = 100$	$\ln 100 = 4.605$
$e^{-0.2} = 0.8187$	$\ln 0.8187 = -0.2$
$e^{-1} = 0.3679$	$\ln 0.3679 = -1$

There are a number of useful rules for using logarithm, but we restrict this discussion and leave them out. You can find all those in a mathematics book. You notice that for negative powers the answer is smaller than the base. Also, notice the relationship between the two logarithm values

$$\ln x = 2.3026 \log x \quad (x \text{ can be any positive number})$$

$$\log x = 0.4343 \ln x$$

will be seen later in some examples. We consider GMD and GMR for the following three cases:

1. Case of single phase (two lines):

$$\text{GMD} = D = \text{the distance between the lines}$$

$$\text{GMR} = r' = \frac{r}{e^{1/4}} = 0.7788r$$

where r is the wire radius. Because in this case the return line has the same characteristic and current as the main line, the value obtained for L must be doubled to represent the closed loop.

For single-phase lines the values calculated for inductance from Equation 12.5 must be doubled.

2. Case of three-phase single circuit line (Figure 12.16a, b, and c):
   For all of the above cases

$$\text{GMD} = \sqrt[3]{D_{12} \times D_{13} \times D_{23}} \tag{12.6}$$

$$\text{GMR} = \frac{r}{e^{1/4}} = 0.7788r \tag{12.7}$$

(D_{12}, D_{13}, and D_{23} are the distances between lines as shown in Figure 12.16, r is the radius of the cables.)

3. Case of double circuit three-phase lines (Figure 12.16d):
   Note the way numbers and letters are assigned to the three-phase lines, and the definition of various distances.

$$\text{GMD} = \sqrt[3]{D_{\text{AB}_{\text{eq}}} \times D_{\text{AC}_{\text{eq}}} \times D_{\text{BC}_{\text{eq}}}} \tag{12.8}$$

where

$$D_{\text{AB}_{\text{eq}}} = \sqrt[4]{D_{12} \times D_{12'} \times D_{1'2} \times D_{1'2'}}$$

$$D_{\text{AC}_{\text{eq}}} = \sqrt[4]{D_{13} \times D_{13'} \times D_{1'3} \times D_{1'3'}}$$

$$D_{\text{BC}_{\text{eq}}} = \sqrt[4]{D_{23} \times D_{23'} \times D_{2'3} \times D_{2'3'}}$$

$$\text{GMR} = \sqrt[3]{\text{GMR}_{\text{A}} \times \text{GMR}_{\text{B}} \times \text{GMR}_{\text{C}}} \tag{12.9}$$

where

$$\text{GMR}_{\text{A}} = \sqrt{\frac{r}{e^{\frac{1}{4}}}(D_{11'})}$$

$$\text{GMR}_{\text{B}} = \sqrt{\frac{r}{e^{\frac{1}{4}}}(D_{22'})} \tag{12.10}$$

$$\text{GMR}_{\text{C}} = \sqrt{\frac{r}{e^{\frac{1}{4}}}(D_{33'})}$$

The values obtained from the preceding relationships determine the inductance of each line of a three-phase system. The values are in henries per meter.

Note that $\frac{1}{e^{\frac{1}{4}}} = \frac{1}{e^{0.25}} = \frac{1}{1.2840} = 0.7788$. Thus, in all the above equations, this value can be directly used ($e = 2.7183$).

After L is determined, the inductive reactance of the line is determined from the familiar relationship $X_L = 2\pi fL$.

(For all distance definition between lines, see Figure 12.16d; all lines are assumed to be the same size with a radius r.)

> Values obtained for L and X_L are per meter of each line. For the inductance and inductive reactance of the entire line, they must be multiplied by the line length.

Example 12.9

Find the inductance for a pair of single-phase lines, where their distance is 2.5 ft and their diameter is 0.2576 in (gauge 2 wire). What is the inductive reactance per mile of this line for 60 Hz electricity?

Solution

$D = 2.5$ ft $= 0.7622$ m, $r = 0.2576 \div 2 = 0.1288$ in $= 0.00327152$ m

$$\ln \frac{0.7622}{(0.7788)(0.003221545)} = 5.7$$

$L = (2)(2 \times 10^{-7} \times 5.7) = 2.28 \times 10^{-6}$ H/m (For single phase the value given in Equation 12.5 must be doubled.)

$X_L = 2\pi fL = (2)(3.14)(60)(2.28 \times 10^{-6}) = 0.86 \times 10^{-3}\ \Omega$/m

$(0.86 \times 10^{-3})(1609) = 1.38\ \Omega$/mile (1 mile = 1609 m)

(Note that because we calculate the logarithm of a ratio, as far as the two values have the same units, no further unit conversion is necessary.)

Example 12.10

Three wires of 25 mm diameter are used in a transmission line arranged as shown in Figure 12.16b. If the distance *D*, as shown in the figure, is 2.5 m, find the inductance of the wire per phase and the inductive reactance for a completely transposed line 4 km long carrying 50 Hz electricity.

Solution

On the basis of the definition of the distances between the lines, we observe that

$D_{12} = D_{23} = 2.5$ m, $D_{13} = 5$ m, $d = 25$ mm, $r = 12.5$ mm $= 0.0125$ m

$$\text{GMD} = \sqrt[3]{(2.5)(2.5)(5)} = 3.1498 \text{ m}$$

$$\text{GMR} = (0.7788)(0.0125) = 0.009735 \text{ m}$$

$$\ln\frac{\text{GMD}}{\text{GMR}} = \ln\frac{3.1498}{0.009735} = \ln 323.554 = 5.7794$$

$$L = 2 \times 10^{-7} \times 5.7794 = 1.1559 \times 10^{-6} \text{ H/m}$$

$$X_L = 2\pi fL = (2)(3.14)(50)(1.1559 \times 10^{-6}) = 3.63 \times 10^{-4}\ \Omega/\text{m}$$

For 4 km,

$$X_L = (4000)(3.63 \times 10^{-4}) = 1.45\ \Omega$$

Example 12.11

A transmission line uses six similar conductors of 4 cm diameter, as shown in Figure 12.16d. If the vertical distance between each pair of adjacent wires is 3 m, horizontal distance for the adjacent pairs is 5 m and for middle conductors is 8 m (as shown in Figure 12.18), find the inductance of the line. The line is completely transposed.

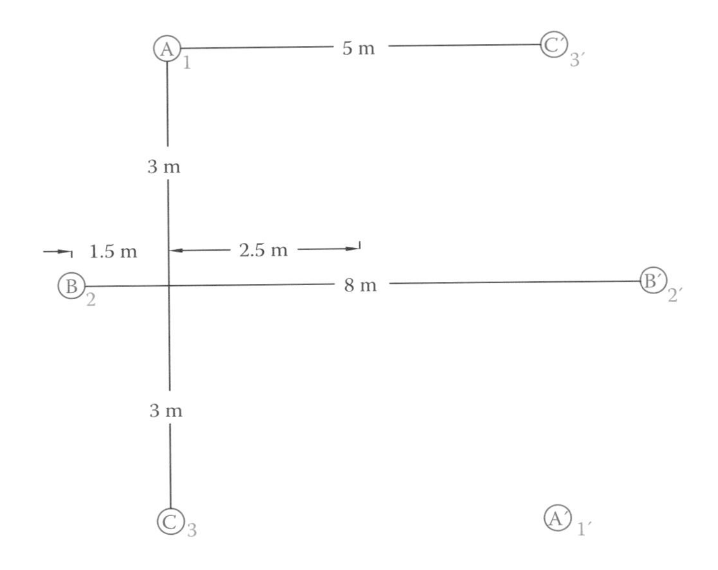

Figure 12.18 Wire position arrangement for Example 12.11.

Solution

We can do this calculation in four steps:

Step 1: First, we need to find the distances between various lines, from geometry.

$$D_{12} = \sqrt{3^2 + 1.5^2} = 3.354\,\text{m}$$
$$D_{12'} = \sqrt{3^2 + 6.5^2} = 7.159\,\text{m}$$
$$D_{1'2} = D_{12'} = 7.159\,\text{m}$$
$$D_{1'2'} = D_{12} = 3.354\,\text{m}$$
$$D_{13} = 3 + 3 = 6\,\text{m}$$
$$D_{13'} = 5\,\text{m}$$
$$D_{1'3} = 5\,\text{m}$$
$$D_{1'3'} = D_{13} = 6\,\text{m}$$
$$D_{23} = D_{12} = 3.354\,\text{m}$$
$$D_{23'} = D_{12'} = 7.159\,\text{m}$$
$$D_{2'3} = D_{12'} = 7.159\,\text{m}$$
$$D_{2'3'} = D_{12} = 3.354\,\text{m}$$
$$D_{11'} = \sqrt{6^2 + 5^2} = 7.810\,\text{m}$$
$$D_{22'} = 8\,\text{m}$$
$$D_{33'} = D_{11'} = 7.810\,\text{m}$$

Step 2: Find the equivalent values for D_{AB}, D_{AC}, and D_{BC} and calculate GMD.

$$D_{AB_{eq}} = \sqrt[4]{(3.354)(7.159)(7.159)(3.354)} = 4.900\,\text{m}$$

$$D_{AC_{eq}} = \sqrt[4]{(6)(5)(5)(6)} = 5.477\ \text{m}$$

$$D_{BC_{eq}} = \sqrt[4]{(3.354)(7.159)(7.159)(3.354)} = 4.900\,\text{m}$$

$$\text{GMD} = \sqrt[3]{4.900 \times 5.477 \times 4.900} = 5.085\,\text{m}$$

Step 3: Find GMR from GMR_A, GMR_B, and GMR_C.

$$\text{GMR}_A = \sqrt{\frac{0.04}{1.284}(7.810)} = 0.493\,\text{m}$$

$$\text{GMR}_B = \sqrt{\frac{0.04}{1.284}(8)} = 0.499\,\text{m}$$

$$\text{GMR}_C = \text{GMR}_A = 0.493\,\text{m}$$

$$\text{GMR} = \sqrt[3]{0.493 \times 0.499 \times 0.493} = 0.495\,\text{m}$$

Step 4: Find the inductance.

$$L = (2 \times 10^{-7})\ln\left(\frac{5.085}{0.495}\right) = 4.66 \times 10^{-7}\ \text{H/m}$$

Example 12.12

If the wire in Example 12.9 has 0.1563 Ω resistance per 1000 ft, what is its impedance per mile?

Solution

$$\text{Resistance per mile} = R = (0.1563)(5280)(0.001) = 0.825\ \Omega$$

$$\text{Inductance per mile} = X_L = 1.38\ \Omega \quad \text{(found in Example 12.9)}$$

$$\text{Impedance per mile} = \sqrt{0.825^2 + 1.38^2} = 1.61\ \Omega$$

12.5.3 Line Capacitance

Line capacitance of electric conductors can be found in the same way that their inductances were formulated for various line arrangements. The only difference is in the value of geometric mean radius, as far as the formulas are concerned. In addition, the effect of the ground comes into picture for capacitance. However, this effect is only a small fraction of the values of capacitance due to the wires themselves and can be ignored unless the cables are very near to the ground.

The value of GMR in Equation 12.7 for the case of capacitance is the actual radius of the conductor. In this respect, for all the calculations of capacitance the term $\frac{1}{e^{\frac{1}{4}}} = \frac{1}{e^{0.25}} = \frac{1}{1.2840} = 0.7788$ (when it appears in the equations for inductance) must be replaced by 1. The general formula for capacitance of conductors in the metric system is

$$C = \frac{(2\pi \times 8.854 \times 10^{-12})}{\ln\left(\frac{\text{GMD}}{\text{GMR}}\right)} \tag{12.11}$$

and the answer found this way is in F/m (farads per meter). For any length of conductors this value must be multiplied by the length.

For the simple case of two wires, GMR is simply the radius of the wires and GMD is their distance. The following examples determine the capacitance for some of the same conductors as previously seen in this section.

The value obtained from Equation 12.11 is for the capacitance between each phase line and the neutral in a three-phase balanced system. For single phase the calculated value must be divided by 2 to find the capacitance between the hot line and neutral. In the above equation the effect of earth is assumed small and is not included.

Example 12.13

Find the capacitance for a pair of single-phase lines, where their distance is 2.5 ft and their diameter is 0.2576 in.

Solution

For the simple case of two wires GMD = distance between the two wires, and GMR = radius of each wire (assuming the same radius). Because this is for a single phase, the value obtained from Equation 12.11 must be halved or, instead, the multiplier 2 in the numerator should be dropped. Moreover, because the logarithm of a *ratio* is found, it is only necessary that the units for the numerator and the denominator be the same.

$D = 2.5$ ft $\quad r = 0.2576 \div 2 = 0.1268$ in $= 0.0105667$ ft

$$\ln\frac{2.5}{(0.0105667)} = \ln(236.593) = 5.466$$

$C = (\pi \times 8.854 \times 10^{-12}) \div (5.466) = 5.088 \times 10^{-12}$ F/m $= 5.1 \times 10^{-6}$ μF/m

For any 1 km of this line, capacitance is 0.0051 μF, and for each mile it is 0.0082 μF.

Example 12.14

A transmission line uses six similar conductors of 4 cm diameter, as shown in Figure 12.18, with the shown spacing between them. Find the capacitance of the line. The line is completely transposed.

Solution

The line configuration and size here are the same as in problem of Example 12.11. It is recommended that you refresh your memory about what was done there to determine the line inductance. Steps 1 and 2 of Example 12.11 can be repeated to find GMD. These are not duplicated here. The final answer is the same as obtained before:

$$\text{GMD} = \sqrt[3]{4.900 \times 5.477 \times 4.900} = 5.085\,\text{m}$$

A new GMR, however, must be found. GMR values for the three phases depend on the spacing $D_{11'}$, $D_{22'}$, $D_{33'}$, based on Equations 12.10. These values also have been already determined in Example 12.11 and are

$$D_{11'} = D_{33'} = 7.810\,\text{m}$$

$$D_{22'} = 8\,\text{m}$$

Using these values in Equation 12.10, but without the factor 0.7788, leads to

$$\text{GMR}_A = \sqrt{(0.04)(7.810)} = 0.559\,\text{m}$$

$$\text{GMR}_B = \sqrt{(0.04)(8)} = 0.566\,\text{m}$$

$$\text{GMR}_C = \text{GMR}_A = 0.559\,\text{m}$$

$$\text{GMR} = \sqrt[3]{0.559 \times 0.566 \times 0.559} = 0.561\,\text{m}$$

and the capacitance can be found to be

$$C = \frac{(2\pi \times 8.854 \times 10^{-12})}{\ln\left(\dfrac{5.085}{0.561}\right)} = 2.52 \times 10^{-11}\,\text{F/m} = 2.52 \times 10^{-5}\,\mu\text{F/m}$$

In practice, for medium length lines (typically 80–250 km, 50–156 mi) after the values of the resistance, inductive reactance, and capacitive

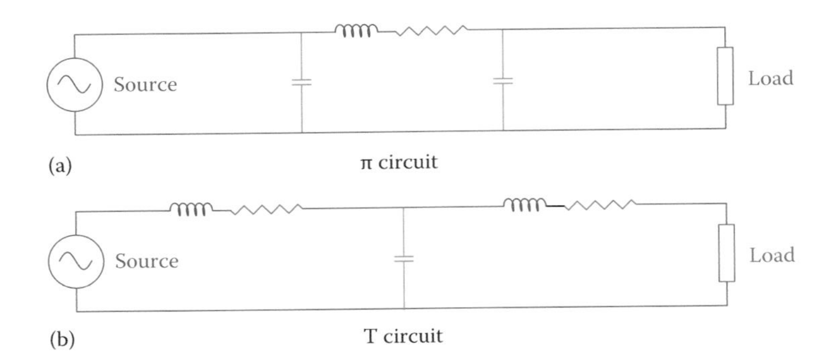

Figure 12.19 Modeling of transmission lines. (a) π model. (b) T model.

reactance are found, the line is modeled as a π circuit or a T circuit, as shown in Figure 12.19. In the π circuit model, the capacitance of each capacitor is half of the total calculated value. Likewise, in T circuit the values of impedances are half of the calculated value.

12.6 Rules for Network Problems

Analysis of circuits described in Chapters 6 and 8 were for simple electric networks containing a combination of components that could be broken into series and parallel resistors (in DC) or RLC's (in AC) and reduced to a final single loop. Not all the circuits are so simple and can be analyzed with only those rules. For example, most of the electronic circuits, at the small-scale voltage, and the power transmission and distribution in electrical grid, at the large scale, have more complicated circuits. For instance, consider the circuits depicted in Figure 12.20. None of these circuits can be simplified by the rules of series and parallel components. These circuits are characterized by having more than one loop. (Note that here we consider an RLC parallel circuit as one loop containing the three elements because we already know how to handle it).

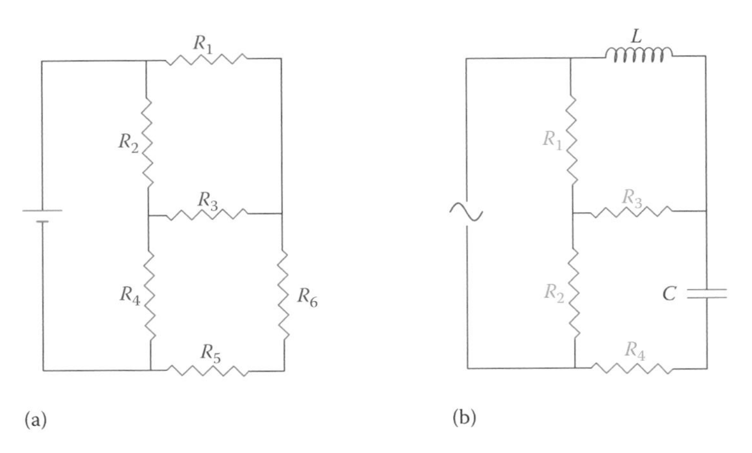

Figure 12.20 Examples of electric networks for the analysis of which we need new rules. (a) DC circuit. (b) AC circuit.

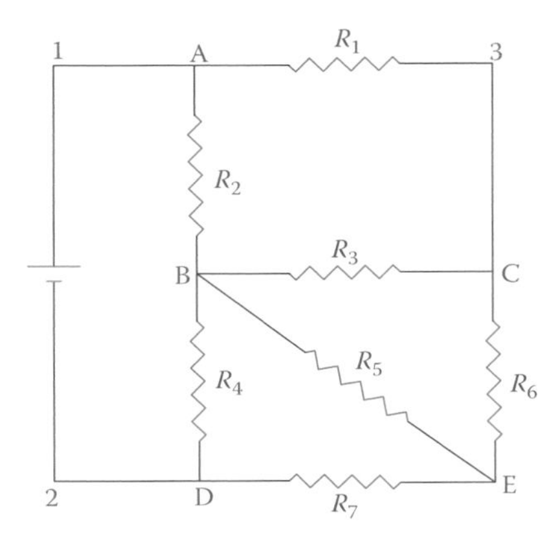

Figure 12.21 A circuit with multiple loops and nodes.

In this section we consider some additional laws and theorems that can be employed for more involved circuits and networks. The first step in all problems, however, is to reduce a circuit to its simplest form by the rules that we have already studied for series and parallel components. If the reduced circuit has only one loop, then the new rules will not be necessary. Otherwise, the new rules will be implemented to solve for the unknowns of the circuits.

In a circuit with more than one loop there nodes also exist. A **node** is a point in a circuit where three or more branches meet. Points A, B, C, D, and E in the circuit in Figure 12.21 are nodes. Quite a number of loops can be identified in this circuit. These are 1AD2, 13E2, A3CB, A3ED, BCED, BCEB, BEDB, A3CEB, 1A3CBD2, 1A3CED2, 1ABCED2, and 1ABED2. Not all of the loops contain the power supply (or a power supply in case there is more than one).

Node: Any point in an electric circuit where more than two lines (branches) join each other.

12.6.1 Kirchhoff's Voltage Law

Problems similar to those in Figure 12.20a and b can be solved by employing Kirchhoff's laws. We need to find some or all the currents in each of the seven resistances and/or finding the potential difference between various points, such as V_{AB}, V_{AC}, V_{BC} and so on.

Kirchhoff's voltage law states that the algebraic sum of all voltages across the components in a loop is equal to the algebraic sum of the source voltages (there might be more than once source) in the loop. Attention must be paid to the polarity and the sign of a voltage. This is the reason for stating algebraic sum because some of the values can be negative. The voltage sign in a component depends on the direction of the current. For clarity, this is shown in Figure 12.22a. For a source the voltage sign is negative for the direction toward the positive terminal.

For those loops that contain no source the sum of the voltages across the components is zero because there is no source, and, consequently, their sum is zero. In Figure 12.23, three loops out of six possible loops are shown. Application of Kirchhoff's voltage law to these loops reveals that

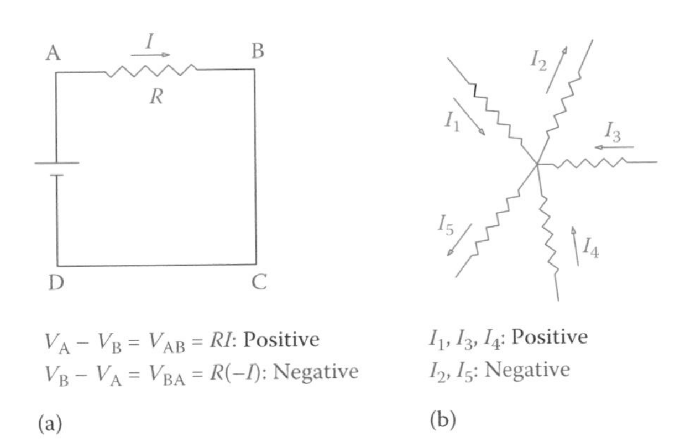

Figure 12.22 Voltage and current sign designation. (a) Voltage difference signs for a loop. (b) Current signs for a node.

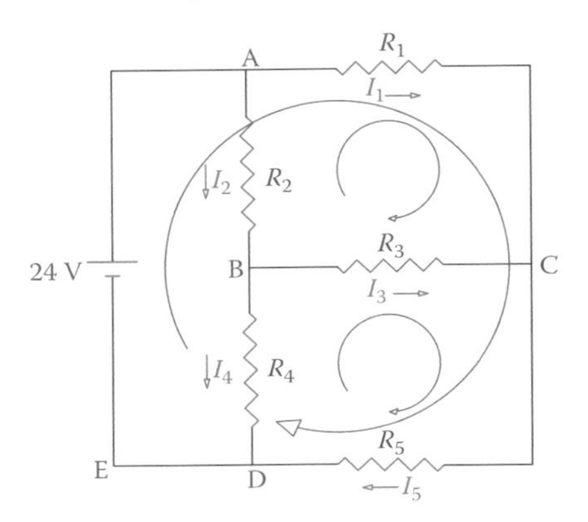

Figure 12.23 Loops and currents in a network.

$$V_{AC} + V_{CE} = 24$$

$$V_{AC} + V_{CB} + V_{BA} = 0$$

$$V_{BC} + V_{CD} + V_{DB} = 0$$

An alternative way of expressing Kirchhoff's law is "The algebraic sum of all the voltages in a loop is equal to zero." This statement involves the sign assignment for all components including the sources.

> The sum of all the voltages in a loop is equal to zero considering a positive sign for the voltage in a forward direction of current and negative in the reverse direction of current for all components but opposite to this for the sources.

12.6.2 Kirchhoff's Current Law

Kirchhoff's current law states that the algebraic sum of all the currents in a node is zero. In other words, the sum of all currents coming to a node is equal to the sum of all currents leaving it. The currents coming toward a

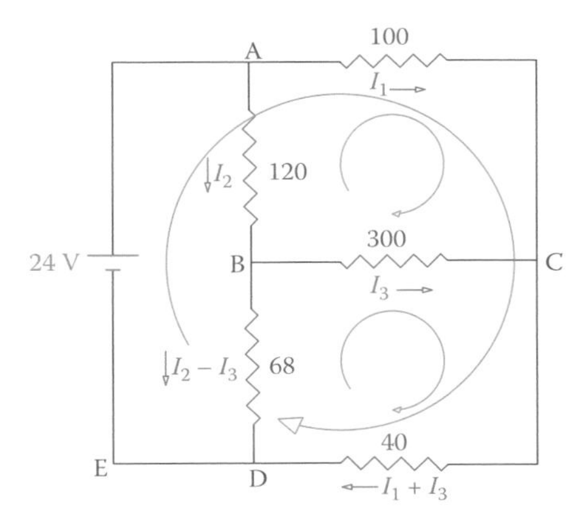

Figure 12.24 Circuit, for example, of Kirchhoff's laws application.

point are considered positive and the currents leaving the node are considered negative, as depicted in Figure 12.22b.

In the circuit of Figure 12.23, there are five resistors and each one has a different current. Applying the Kirchhoff's current law to nodes A, B, C, and D reveals that

$$\text{At A,} \quad I_1 + I_2 = I \text{ (circuit current)}$$

$$\text{At B,} \quad I_3 + I_4 = I_2$$

$$\text{At C,} \quad I_1 + I_3 = I_5$$

$$\text{At D,} \quad I_4 + I_5 = I$$

The Kirchhoff's current law defines the relationships between various currents in a circuit because these currents are not independent of each other. In this sense, in the circuit of Figure 12.23 one can assume only three unknown currents I_1, I_2, and I_3 because I_4, I_5, and I may be defined in terms of I_1, I_2, and I_3, as shown in Figure 12.24. The following example shows the application of Kirchhoff's voltage and current laws.

> The algebraic sum of all the currents in a node is equal to zero, considering the currents toward a node as positive and currents leaving a node as negative.

Example 12.15

In Figure 12.24, find the current in the circuit and the voltage at point B (assuming voltage at D is zero). All the resistor values are in Ω.

Solution

Various loops and currents can be defined for any such problem but all lead to the same solutions. A set of currents are defined and shown in Figure 12.24. We need to write the Kirchhoff's law for the selected loops. As there are only three independent currents, only three equations are necessary and sufficient.

$$140I_1 + 40I_3 = 24 \qquad \text{(From loop ACDEA)}$$

$$100I_1 - 120I_2 + 40I_3 = 0 \qquad \text{(From loop ACBA)}$$

$$40I_1 - 68I_2 + 40I_3 = 0 \qquad \text{(From loop BCDB)}$$

Any known method can be used for the simultaneous solution of the above three equations. The results are

$$I_1 = 170.0 \text{ mA}$$

$$I_2 = 128.8 \text{ mA}$$

$$I_3 = 4.90 \text{ mA}$$

The total current in the circuit is

$$I_1 + I_2 = 298.8 \text{ mA}$$

and the voltage at point B can be found from the voltage drop in the 120 Ω resistor to be

$$V_B = 24 - (120)(0.1288) = 8.544 \text{ V}$$

12.6.3 Thevenin's Theorem

Thevenin's theorem is particularly helpful when multiple power supplies exist in a circuit. In Figure 12.25, for example, there are two power supplies, and we need to find the current through the resistor R. This could be part of a larger circuit, which provides 12 and 6 V across the points shown in the figure. Refer to Figure 12.25 for understanding the statement of Thevenin's theorem.

Thevenin's theorem states that the current through a resistor R connected across any two points A and B in an active network is obtained by dividing the potential difference between A and B (with R disconnected) by $R + r$, where r is the resistance of the circuit between A and B (with R disconnected) and the sources of the circuit replaced by their internal resistances.

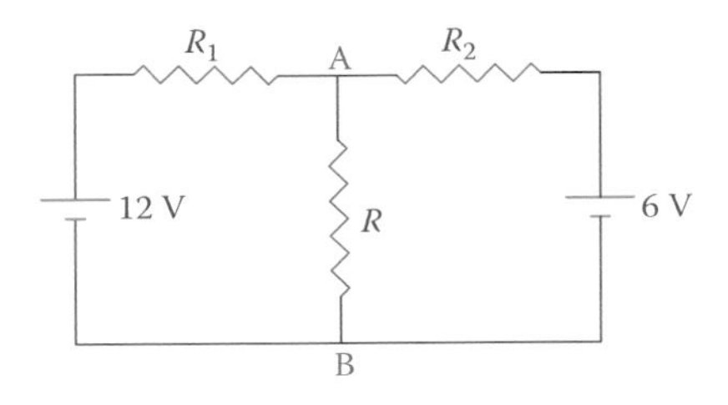

Figure 12.25 Application of Thevenin's theorem.

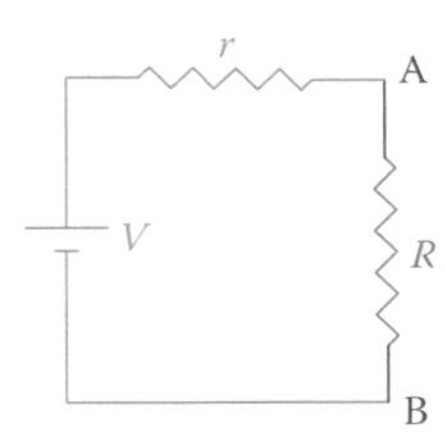

Figure 12.26 Equivalent circuit by Thevenin's theorem.

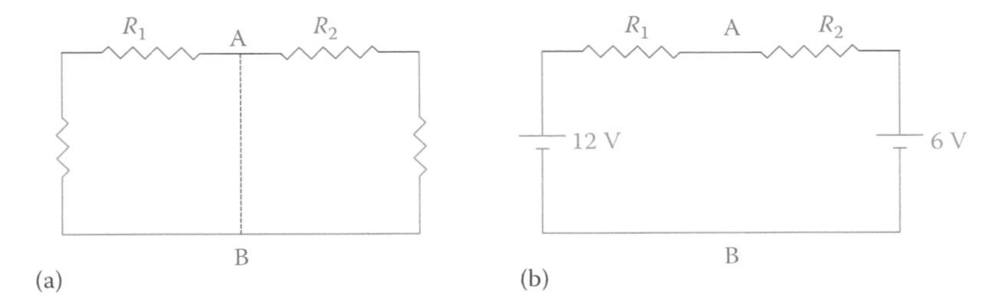

Figure 12.27 Steps in implementing Thevenin's theorem. (a) Finding equivalent r_{Th} between points A and B. (b) Finding the voltage difference V_{Th}, between points A and B.

Active network implies that the circuit considered is not dead and has EMF sources within it (EMF implies a voltage source). The internal resistance of sources corresponds to resistances that may exist in the power sources but can be zero (see Appendix A).

In other words, Thevenin's theorem implies that to find the current through R (or the voltage difference between A and B), replace this circuit by an equivalent circuit as shown in Figure 12.26. In this circuit, there are two unknown values of which need to be determined: the resistance r and the voltage V.

To find r,

1. Disconnect R (see Figure 12.27a).
2. Replace each voltage source by its internal resistance.
3. Remove any current source.
4. Determine the resistance between A and B.

To find V,

1. Disconnect R (see Figure 12.27b).
2. Find the voltage between A and B.

Example 12.16

Part of a circuit can be represented by three resistors and two power sources as shown in Figure 12.28. If R_1, R_2, and R (corresponding to Figure 12.25) are, respectively, 80, 120 and 32 Ω, find the current in the 32 Ω resistor and the voltage across it. The internal resistances of the sources are zero.

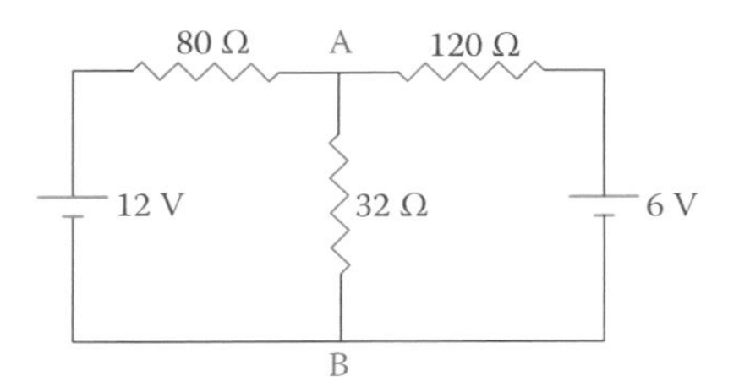

Figure 12.28 Circuit of Example 12.16.

Solution

By Thevenin's theorem this circuit can be replaced by a voltage V and a resistor r, as shown in Figure 12.29a.

The resistor r can be found by following the three above mentioned steps. Because the source internal resistances are zero, the two branches in parallel with the 32 Ω resistor contain 80 and 120 Ω. Thus the resistor r is calculated as

$$\frac{1}{r} = \frac{1}{80} + \frac{1}{120} = \frac{2.5}{120} = \frac{1}{48}$$

or

$$r = 48\ \Omega.$$

The voltage V may be calculated as the voltage at A (assuming $V_B = 0$) based on a voltage drop in R_1 (the 80 Ω resistor) in the circuit of Figure 12.30 (32 Ω resistor removed). The current in this circuit is

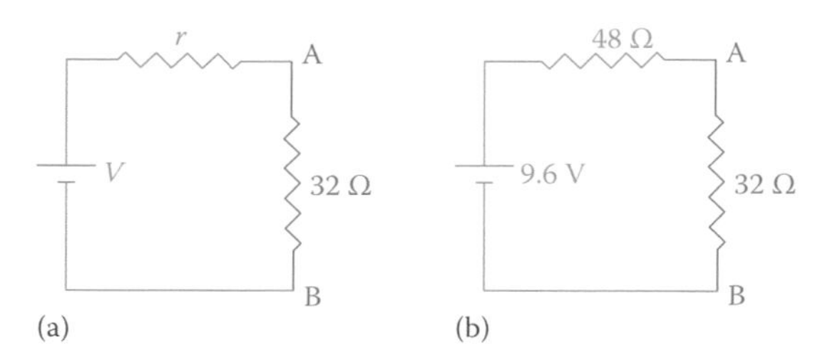

Figure 12.29 Thevenin's circuit for Example 12.16. (a) Thevenin's equivalent circuit. (b) Values for Thevenin's model.

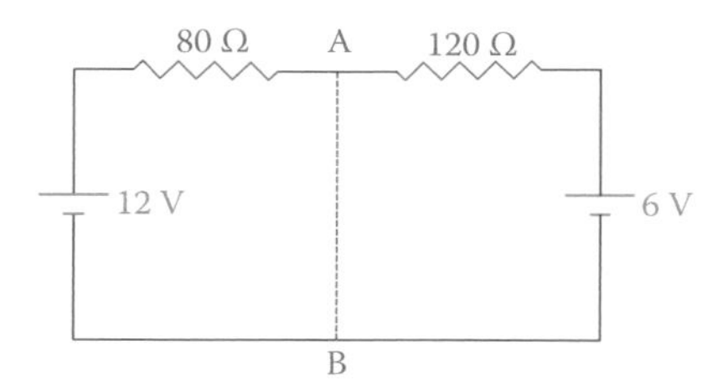

Figure 12.30 Finding voltage at A Thevenin's equivalent circuit.

$$\frac{12-6}{80+120} = \frac{6}{200} = 0.03 \text{ A}$$

and the voltage drop in R_1 is, accordingly, 80 × 0.03 = 2.4 V, which implies that the voltage at A is 12 − 2.4 = 9.6 V. The resulting circuit is indicated in Figure 12.29b. On the basis of this voltage and the resistor r (Thevenin's equivalent circuit) the current through the 32 Ω resistor is 9.6 ÷ (48 + 32) = 0.12 A.

Thevenin's theorem is not only for circuits with multiple sources. It can be employed for a circuit similar to that shown in Figure 12.23 (and solved by using Kirchhoff's laws) when we are more interested to see the current in a resistor's value is subject to change. If for various values of a resistor we assume we have to repeat all the calculation, it is tedious. By using Thevenin's theorem we can readily find the current for each value of the resistor.

Example 12.17

For the circuit shown in Figure 12.24 and repeated in Figure 12.31a we want to find the equivalent Thevenin's circuit as seen from the 40 Ω resistor.

Solution

The two points across the 40 Ω resistor are called M and N for referencing.

1. Find the resistor r: Shorting the two sides of the battery leads to the circuit shown in Figure 12.31b. We need to find the single resistor that is between points M and N, neglecting the 40 Ω resistor. This is shown in Figure 12.31b. We need to rearrange this circuit to a clearer form, so that it can be simplified and reduced to a single resistor. We note that points A and C are the same for this circuit; between M and A, there is a 100 Ω resistor; and there exists a 300 Ω resistor between M and B. The circuit in Figure 12.31b can be rearranged as shown in Figure 12.31c, which can be easily managed. Point M is shown twice for better clarity.

   For these four resistors we can write

$$\frac{1}{r} = \frac{1}{100} + \frac{1}{300 + \left(\dfrac{1}{\dfrac{1}{68} + \dfrac{1}{120}}\right)} = 0.01291$$

   from which the value of r can be found as $r = 77.45$ Ω.

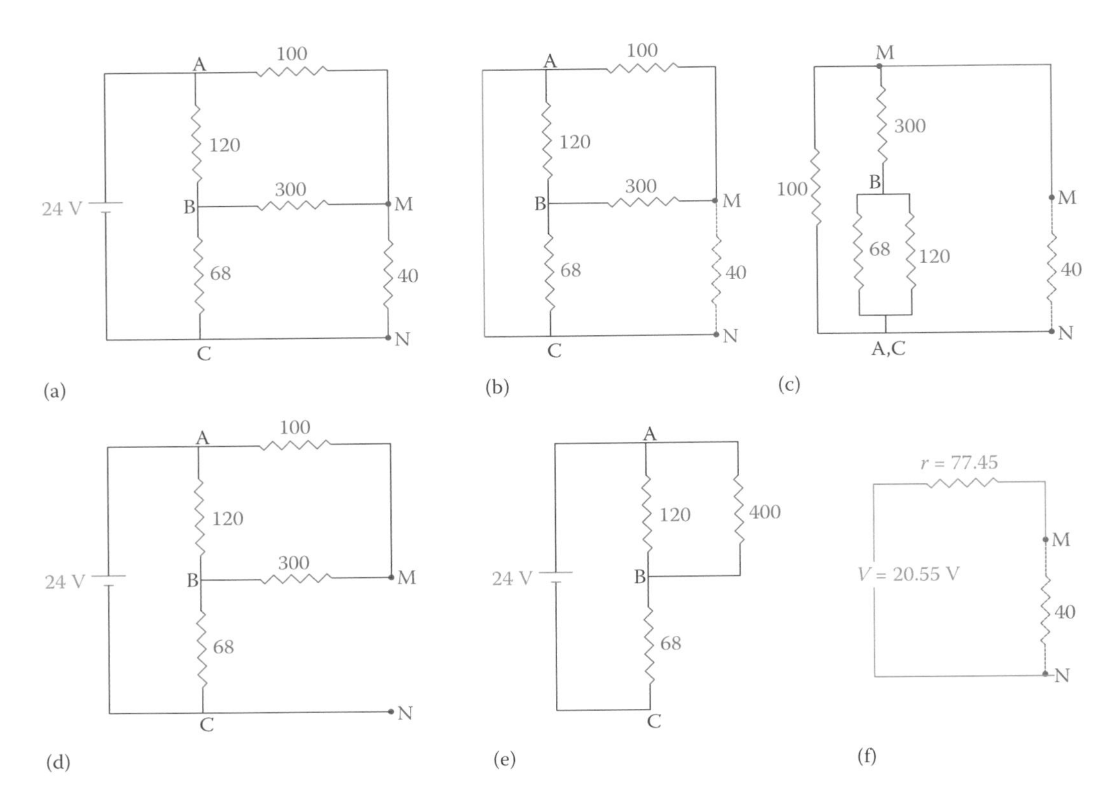

Figure 12.31 Steps for finding the Thevenin's circuit in Example 12.17. (a) Original circuit. (b) Replacing voltage sources by their internal resistance. (c) Finding the total resistance across the load points. (d) Removing the load to find the voltage at load points. (e) Finding voltage difference across the load points (load removed). (f) Final result.

2. Find the voltage V: We need to find the voltage between points M and N when the 40 Ω resistor is removed. Removing this resistor from the circuit leads to that shown in Figure 12.31d, which can be further reduce to the circuit depicted in Figure 12.31e to find the voltage at point B assuming the voltage at C is zero (connected to the negative terminal of the battery). The voltages V_{AB} and V_{BC} can be found from the voltage divider relationships (see Chapter 6).

$$V_{AB} = (24)\left(\frac{93.3}{93.3+68}\right) = 13.8 \text{ V}$$

(93.3 Ω is equivalent resistance of 120 and 400 Ω resistors in parallel)

$$V_{BC} = 24 - 13.8 = 10.2 \text{ V}$$

We need further to find the voltage at point M. This voltage can be obtained by observing that in the branch containing the 100 Ω resistor and 300 Ω resistor the 13.8 V voltage is divided between these two resistors. The voltage at M, thus can be determined to be

$$V_M = 24 - \frac{1}{4}(13.8) = 20.55\ \text{V}$$

which implies that $V_{MN} = 20.55$ V.

The final result, the Thevenin's circuit, is shown in Figure 12.31f.

As a test, let see the current in the 40 Ω resistor based on the above analysis. This current is equivalent to $I_1 + I_3$ in Example 12.15.

From Thevenin's theorem:

$$I_{40} = 20.55 \div (77.45 + 40) = 0.175\ \text{A} = 175\ \text{mA}$$

From Example 12.15:

$$I_1 + I_3 = 170 + 4.90 = 174.9\ \text{mA}$$

Consider, for instance, if the 40 Ω resistor has to be replaced by a 50 Ω resistor. The new current in this resistor is

$$I_{50} = 20.55 \div (77.45 + 50) = 0.161\ \text{A} = 161\ \text{mA}$$

12.6.4 Principle of Superposition

The principle of **superposition** is another useful method that can be used in many problems. It has a broad application in many topics. In the context of this discussion it can be employed for solving problems similar to those tackled by Thevenin's theorem when multiple sources exist. The principle is that a problem is broken into a number of overlapping but simpler cases. At the end, the solutions of the various cases are combined together. In our case of two sources in the same circuit, we break down into two overlapping problems, each one dealing with the same circuit, but only one source. We solve the simpler problems, and at the end, we add up the separately found currents for each component. As an example, we consider the same problem in Example 12.16.

Superposition: A method of analysis of problems containing multiple parameters or conditions (inputs for example) by supposing that each parameter exists without the presence of the others, and then the different outcomes (of applying each parameter) are added together.

Example 12.18

Referring to Figure 12.28, repeated in Figure 12.32a, part of a larger circuit can be represented by three resistors and two power sources, as shown. Find the current in the 32 Ω resistor.

Solution

We first consider the circuit without the 6 V source. The resulting circuit is shown in Figure 12.32b. The corresponding current I_1 in the circuit can be found as

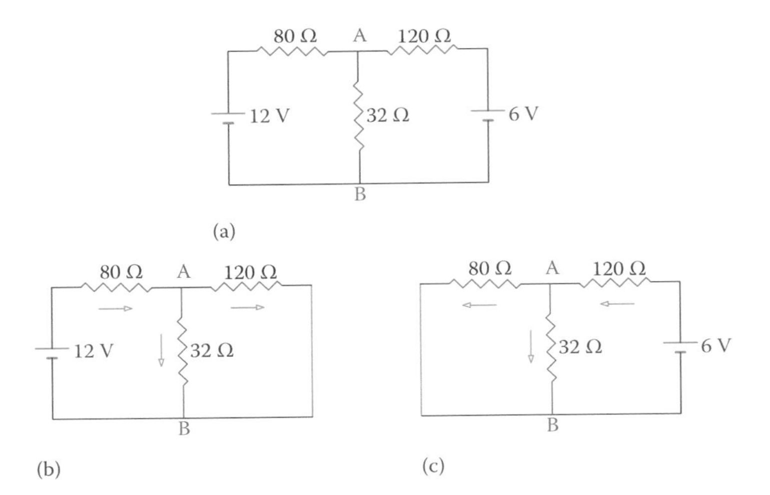

Figure 12.32 Separating the sources in the circuit of Figure 12.27. (a) Original circuit. (b) The 6 V battery is removed. (c) The 12 V battery is removed.

$$I_1 = \frac{12}{80 + \dfrac{1}{\dfrac{1}{32} + \dfrac{1}{120}}} = 0.114 \text{ A}$$

This is the current in the 80 Ω resistor due to the 12 V battery. Only part of this current ($I_{1,32}$) flows through the 32 Ω resistor, which can be determined from the resistors' ratio:

$$I_{1,32} = (0.114)\left(\frac{120}{120 + 32}\right) = 0.090 \text{ A}$$

Next, we consider the circuit without the 12 V source and follow the same process (Figure 12.32c).

$$I_2 = \frac{6}{120 + \dfrac{1}{\dfrac{1}{32} + \dfrac{1}{80}}} = 0.042 \text{ A}$$

This is the current through the 120 Ω resistor due to the 6 V battery. The part $I_{2,32}$ passing through the 32 Ω resistor is determined as

$$I_{2,32} = (0.042)\left(\frac{80}{80 + 32}\right) = 0.030 \text{ A}$$

The total current through the 32 Ω resistor is obtained by adding the two partial currents. Therefore,

$$I_{32} = 0.090 + 0.030 = 0.120 \text{ A}$$

This is the same answer as obtained by Thevenin's theorem.

Thevenin's theorem has the advantage that in case this resistor is subject to variation, we do not need to repeat all the calculation every time. We just simply enter the changed value of the resistor in the equivalent circuit.

> The advantage of using Thevenin's or Norton's theorems is more pronounced when the value of a resistor is subject to change. It is not necessary to repeat all calculations every time for a new value of a variable resistor.

Note that the directions of the two currents to be superimposed for the 32 Ω resistor are the same. For the other resistors this is not the case, and if we were to find the current in one of those (e.g., in the 80 Ω resistor) the two currents had to be subtracted from each other, according to the arrows shown in Figure 12.32b and c.

12.6.5 Norton's Theorem

Norton's theorem is a method similar to the Thevenin's theorem for network analysis by reducing a circuit as seen by a load (i.e., all other components in the circuit around that load) to only one source and one resistor. The difference is that in Thevenin's theorem a circuit is reduced to a voltage source and a resistor that is in series with the load. In Norton's theorem the circuit is reduced to a current source and a resistor in parallel with the load. This is shown in Figure 12.33, together with the Thevenin's equivalent circuit. The subscript Th has been added to the Thevenin's parameters V and r, for clarity. The Norton's parameters carry a subscript N.

The procedure to find the Norton's equivalent circuit is the same as described for finding the Thevenin's equivalent circuit. The difference is for the current source, which provides a (constant) current instead of a voltage to the circuit. This is described below. Suppose that R_L is the resistance of the load for which we want to find the Norton's equivalent circuit as seen by the load. We need to find the resistance R_N and the current I_N.

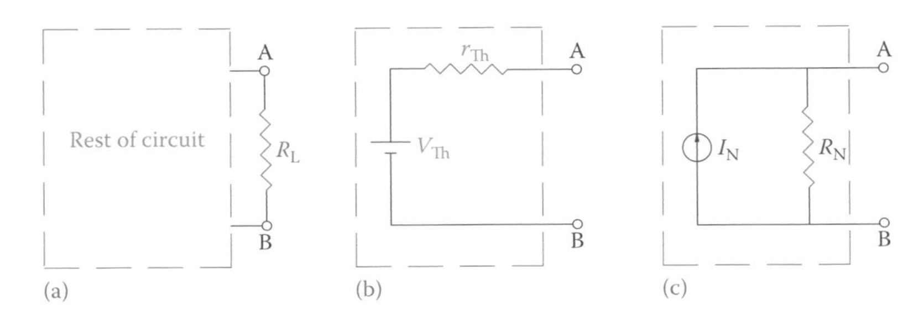

Figure 12.33 Comparison of Thevenin's and Norton's equivalent circuits. (a) Original circuit. (b) Thevenin's circuit. (c) Norton's circuit.

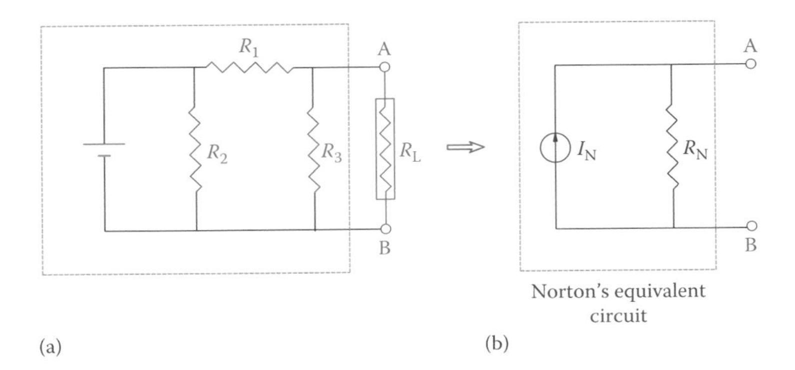

Figure 12.34 Norton equivalent of a circuit as seen from a load between points A and B. (a) Original circuit. (b) Norton's equivalent circuit.

To find R_N,

1. Disconnect R_L (see Figure 12.34a).
2. Replace each voltage source by its internal resistance.
3. Remove any current source if it exists.
4. Determine the resistance between A and B.

To find I_N,

1. Short R_L (short circuit between A and B).
2. Find the current flowing from A to B.

In fact, there is a correspondence between the values obtained for r_{Th}, R_N, and between V_{TH} and I_N. These are

$$R_N = r_{Th} \tag{12.12}$$

$$I_N = \frac{V_{Th}}{r_{Th}} \tag{12.13}$$

Example 12.19

Find the Norton's equivalent circuit for the circuit shown in Figure 12.28, as seen from the 32 Ω resistor. The internal resistances of the voltage sources are zero. The figure is repeated here for convenience.

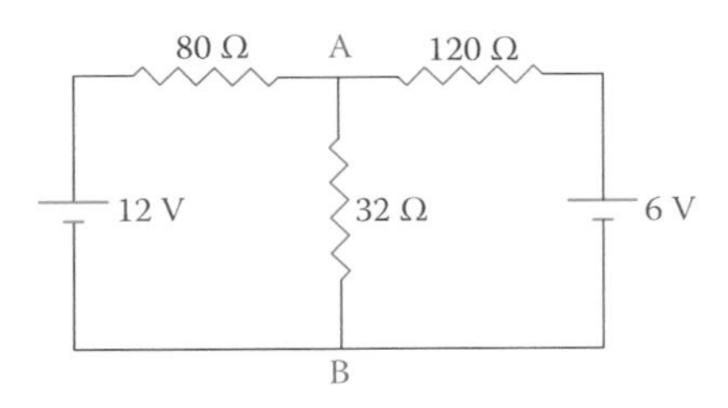

Figure 12.28 Circuit of Example 12.19.

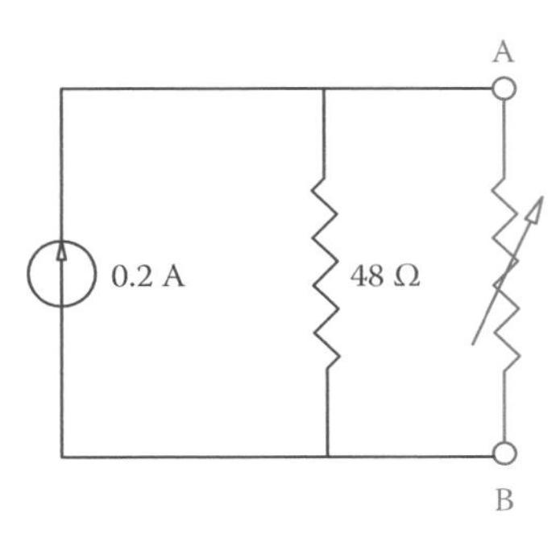

Figure 12.35 Norton's Equivalent for the circuit in Figure 12.28.

Solution

The equivalent resistance R_N is exactly found in the same way as in Example 12.16 and is not repeated here. Its value is 48 Ω.

The current from A to B when A and B are shorted together can be determined easier if the principle of superposition is used. First, the 6 V battery is removed. Because A is shorted to B the 120 Ω resistor does not receive any current and the current between A and B is 12 ÷ 80 = 0.15 A. Next, the 12 V battery is removed and the current between A and B is 6 ÷ 120 = 0.05 A. Thus, the total current is 0.15 + 0.05 = 0.20 A. The resultant Norton's circuit is shown in Figure 12.35.

The current in the 32 Ω resistor is determined from considering that the total current shared between the 32 Ω resistor and R_N (=48 Ω) is 0.2 A. This current is divided between them and based on their values ratio the smaller resistor take the higher current. Thus,

$$I_{32} = (0.2)\left(\frac{48}{48+32}\right) = 0.12 \text{ A}$$

If instead of 32 Ω the value of this resistor changes to 64 Ω, for instance, the new value of the current is

$$I_{32} = (0.2)\left(\frac{48}{48+64}\right) = 0.086 \text{ A}$$

Note that in finding I_N, it would be easier for this and some problems to use the relationship in Equation 12.13, if we already have the value of V_{Th}.

$$I_N = \frac{V_{Th}}{R_{Th}} = \frac{9.6}{48} = 0.2 \text{ A}$$

12.7 High Voltage Direct Current Transmission

For very long distance transmission of electricity, as we discussed in this chapter, AC electricity carrying conductors introduce inductive and capacitive reactance. These two can become a burden and cause limitations on transmission lines, particularly for underground and underwater lines, where the distance between cables is much smaller than the overhead lines. For this reason and in some cases for strategic reasons (e.g., maintaining an undisturbed supply of electricity to cities), transmission in the form of direct current has been considered and is practiced. Obviously, line voltage must be high to avoid large voltage drop as well as too much loss of energy in the line.

For high voltage direct current (HVDC) transmission, at the power side (sending end) the voltage of DC electricity must be raised to some high level (e.g., a few hundred thousand volts). After transmission to the destination, this voltage must be reduced to the distribution (and finally consumption) level. Because no direct use of transformers is possible, these operations are much more complicated and costly than the counterparts in AC electricity. There are advantages and disadvantages for HVDC transmission, but a major factor is the cost that makes it prohibitive in many cases, unless the cost can be justified. The total cost for HVDC can well reach 10× of the cost for the same power AC transmission. As for example, the cost of HVDC is in the order of \$10 million per mile, whereas for AC line it is in the range of \$1 million per mile.

Among the advantages is the lack of length limitations as mentioned above (no line induction and capacitance) and that the voltage at the receiving end is not affected by phase distortion, which is possible with AC lines. Also, networks with different frequencies can be interconnected.

Among the disadvantages is the fact that no branching is possible in between the two ends, unless another costly station for voltage conversion is provided. For AC transmission it is relatively very inexpensive to add a new substation and branch into or from an existing grid.

Presently, there are a number of HVDC lines in the United States and elsewhere. These are in cases where advantages overcome the disadvantages, or there is no alternative for HVDC.

12.8 Chapter Summary

- Energy is transmitted from the generation site to the consumption site before being distributed among the consumers.
- To reduce loss of energy and voltage drop in transmission lines, voltage is raised, so that for the same power the line current is smaller.
- Transmission and distribution voltages are standardized and are categorized to low, medium, high, extra high, and ultrahigh voltage.
- Most conductors for transmission lines today are made of stranded aluminum and reinforced by steel cables.

- A substation is where voltage is changed or where various lines meet. It is equipped with switches and breakers and protection devices.
- The important electrical parameters of transmission lines are their electrical resistance, inductance, and capacitance. By itself this is like a small load between the two ends of transmission lines.
- Resistance in a conductor depends on its thickness, length, and material. Inductance and capacitance depend on the thickness, wire spacing, and line length for all conductors in a transmission line.
- Ampacity is the current carrying capacity of a line. This depends on the resistance and the generated heat in the conductor as well as how this heat can be disposed of, thus on the ambient temperature.
- The same conductor has different ampacities under different conditions.
- Conductor sizes are standardized. Electrical wires are made of copper or aluminum. The electrical resistance and weight of standard wires are tabulated and can be directly embedded in calculations.
- An electrical grid or network consists of many interconnections of lines between various electric sources and electric loads.
- For analysis of currents and voltages in different sections of more complex networks, use can be made of a number of methods that allow reducing the networks to simpler circuits. These are Kirchhoff's voltage and current laws, Thevenin's theorem, Norton's theorem, and the principle of superposition.
- HVDC transmission is an alternative method for long distance when AC transmission faces limitation, but the cost is considerably higher.

Review Questions

1. What are the main reasons to raise voltage for electricity transmission?
2. How does raising voltage affect the voltage drop in an electric line?
3. What are the three parameters that influence the resistance of electric wires?
4. How does raising voltage affect power consumed in transmission lines?
5. What is a substation? Name some of the major equipment in a substation.
6. Is 220 V considered a high voltage? What about 11,000 V?

7. Why can't the output of a power station be directly connected to a transmission line?

8. Is there any difference between transmission and distribution lines? Describe.

9. Is there any difference between transmission and power lines? Describe.

10. What happens if one ignores the inductance and capacitance of transmission lines?

11. Is the voltage the same at the two ends of a transmission line? Why or why not?

12. Does the phase difference between current and voltage stay the same along a transmission line? Why or why not?

13. If the length of a line doubles, does its resistance double, too? Why or why not?

14. What is ampacity? What factors affect ampacity?

15. If the ambient temperature changes from –10°C to +10°C, does ampacity go up or down?

16. What is transposition in transmission lines? Why is it necessary?

17. What are the two main parameters affecting the inductance of power lines?

18. Why is grounding important? What is the main objective of grounding?

19. Where does the electricity go when a ground current exist? Does it dissipate in the ground?

20. What are the advantages of Thevenin's theorem and Norton's theorem?

21. What does the Kirchhoff's voltage law imply?

22. What does the Kirchhoff's current law imply?

23. What does the principle of superposition imply?

24. What does HVDC mean, and what is it?

Problems

1. The diameter of a round copper wire is 0.7620 mm. What is the resistance of a piece of this wire 150 m long? The specific resistance of copper in metric system is 1.6779×10^{-8} Ω.m.

2. If the area of a copper wire is 0.0285 mm^2, what is the resistance of 1 km of this wire?

3. If the specific resistance of a metal in imperial system is 265 Ω.CM/ft, find its specific resistance in the metric system?

4. Tungsten has a specific resistance of 5.6×10^{-8} Ω.m at room temperature (20°C). Find its specific resistance at 1800°C. Find the corresponding information from Table 12.2.

5. What is the resistance of a tungsten wire with a cross-section area of 1 mm^2 and 1 m long at 1000°C?

6. Using numbers in Table 12.2 determine if the resistance of a copper wire 20 m long is more or an aluminum wire 15 m long if both have the same cross-section area.

7. What length of a tungsten filament of a lightbulb with a 0.05 mm diameter is required for 10 Ω resistance at 20°C temperature?

8. The resistance of a piece of wire 1000 ft long is 2 Ω. If the wire is made out of copper, find its diameter in inches.

9. For the wire in Problem 8 find the thickness in CM (circular mil).

10. The diameter of a piece of round tungsten wire is 32 mils. What length of this metal makes a resistance of 20 Ω?

11. If the resistance of a length of aluminum wire at 20°C is 2 Ω, what is its resistance at 100°C? Use the information in Table 12.2.

12. From Table 12.4 find the allowable current in a TW type copper wire gauge AWG-12 for free air installation in an ambient reaching 100°F. The voltage is below 2000 V.

13. For the wire in Problem 12 find the resistance and the weight for 5 km of this wire.

14. If the wire in Problem 12 must be used in a conduit, what is the ampacity under the same ambient conditions?

15. Determine the wire gauge for 60 A current in a conduit where two of such copper wires are to be installed. The ambient temperature does not go over 25°C. The operating voltage is below 2000 V and the wire type is TW.

16. For a pair of single-phase lines, 3 ft apart gauge 2 wire is used. Find the inductance of 500 m of this line? Use Table 12.4 for the required information.

17. If the line in Problem 16 carries 60 Hz electricity, what is its inductive reactance per mile?

18. If the wire in Problem 16 is used for transmission of three-phase 60 Hz electricity and the conductors are completely transposed, find the inductance per mile if the three lines are on the corners of a triangle and 3 ft apart from each other.

19. Three cables of 40 mm diameter are used in a transmission line arranged as shown in Figure 12.16b. The distances between the lines, are 3, 3, and 6 m, respectively. Find the inductance of the line per phase and the inductive reactance for a completely transposed line 100 km long for 50 Hz.

20. A six conductor transmission line uses 2 in diameter aluminum cables. The lines are arranged in two rows each consisting of three cables as shown in Figure 12.16b. If the vertical distance between each pair of adjacent wires is 10 ft and the horizontal distance for the adjacent pairs is 15 ft, find the unit inductance of the line per phase. The line is completely transposed.

21. If the line in Problem 20 is 20 mi (32.18 km) long, what is its inductance per phase?

22. If the wire in Problem 20 has 0.1563 Ω resistance per mile, what is its impedance per mile per phase, ignoring the capacitance? The line frequency is 60 Hz.

23. Find the capacitance per mile for the line in Problem 18.

24. Find the capacitive reactance per mile for the line in Problem 18.

25. Find the impedance per mile for line in Problem 18 if the ohmic resistance per mile is 0.3 Ω.

26. Determine the capacitance for the transmission line in Problem 20.

27. Find the current change in resistor R_L in the circuit of Figure P12.1 when R_L changes between 51 Ω ± 20%.

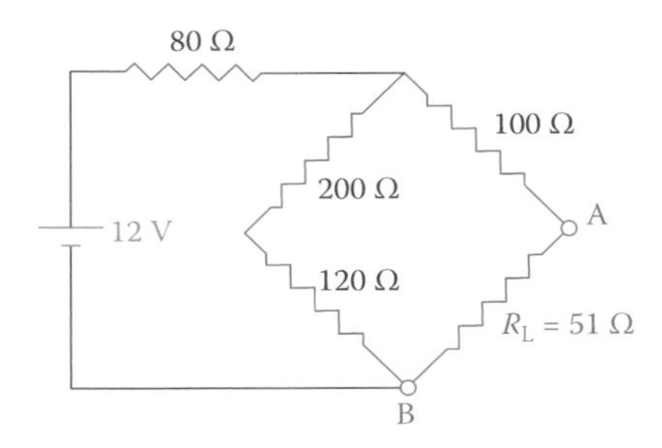

Figure P12.1 Circuit for Problem 27.

28. Repeat Problem 27 using Thevenin's theorem (find the Thevenin's equivalent circuit as seen from the 51 Ω resistor). Assume the internal resistances of the power supplies to be zero.

29. Modify the circuit in Figure P12.2 to its simplest form that you can.

30. For the circuit shown in Figure P12.3, find the equivalent Thevenin's circuit as seen from the 500 Ω resistor.

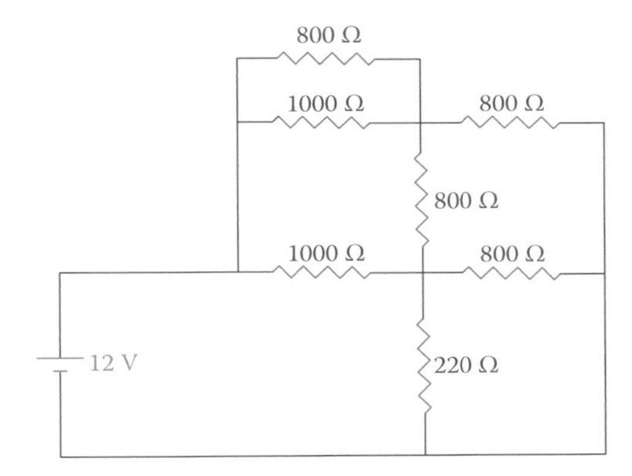

Figure P12.2 Circuit for Problem 29.

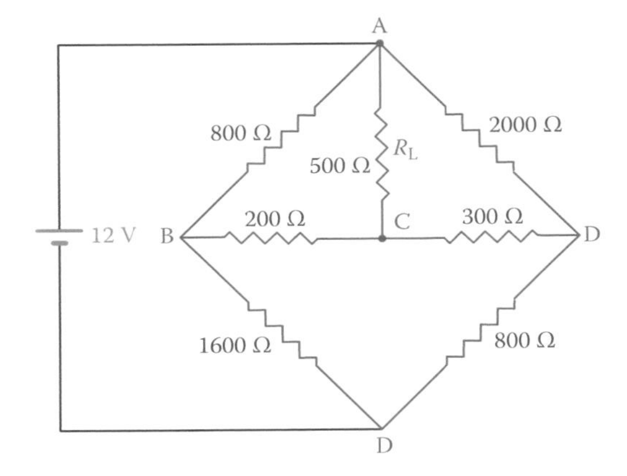

Figure P12.3 Circuit for Problem 30.

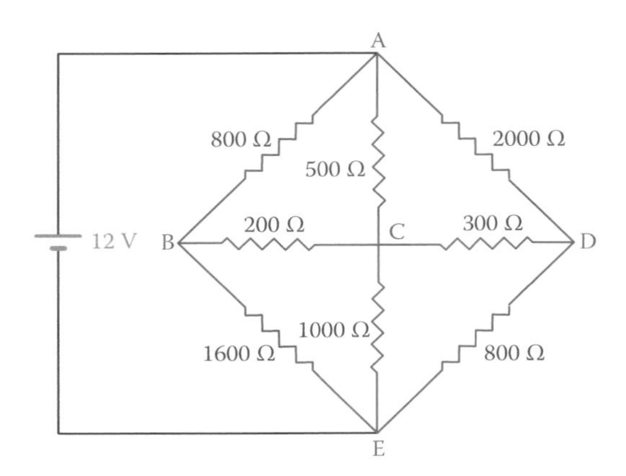

Figure P12.4 Circuit for Problem 32.

31. Find the Norton's equivalent circuit for the circuit of Problem 30.

32. Find the current in each resistance in the circuit as shown in Figure P12.4 by eploying Kirchoff's laws.

Advanced Learning: Positive, Negative, and Zero Sequence for Fault Analysis

When a fault occurs in part of a line, it must be detected and acted upon. Detection is performed by some sort of sensor, measuring current, voltage, etc. Selection of sensors with correct and proper rating and able to act correctly without getting damaged is important. This section is added so that an advanced reader becomes familiar with three important terms that are used for the analysis of fault situations.

Because various faults can occur in an electric network and at different places (e.g., at a generator site, a load site, a transformer, and any point of a transmission line), each fault must be detected and the proper action be taken. A fault introduces a large disturbance in a network and causes a large imbalance in the system. Even if a system has an unbalanced load prior to a fault, the imbalance situation caused by a fault dominates the normal operating condition.

Thus, detection of a fault starts with an analysis of the unbalanced loads. The terms positive sequence, negative sequence, and zero sequence are used to define the imbalance condition in an electric network. The full discussion is out of the scope of this book. So we confine this section to only the definitions.

So far, we have considered any three-phase system to consist of three voltages and the corresponding currents with 120° phase difference between them. In a more general scope, any such system can be considered to consist of those three components (which constitute the **positive sequence components**) and six other components. The above three components rotate in a positive direction. Three similar components rotating in the opposite direction are called **negative sequence components**, and finally another set of similar components that do not rotate are called **zero sequence components**. The advantage of this expansion is that an unbalanced system can be expressed in terms of three sets of three balanced systems, which are its positive, negative, and zero sequence components.

Figure A12.1a shows three vectors A, B, and C. They can represent three unbalanced voltages or three unbalanced currents in a three-phase electric circuit. Associated with their values, three sets of vectors that are shown in Figure A12.1b can be found. These vectors are specific to the vector conditions in Figure A12.1a. They are the zero (a_0, b_0, and c_0), negative (a_2, b_2, and c_2), and positive (a_1, b_1, and c_1) components, respectively, for A, B, and C. Figure A12.1c shows the composition of each vector (A, B, and C) of the components from the three sequence sets depicted in Figure A12.1b. That is,

$$A = a_0 + a_1 + a_2 \qquad B = b_0 + b_1 + b_2 \qquad C = c_0 + c_1 + c_2$$

Note that the vectors making the positive sequence components and the negative sequence components are at 120° from each other, whereas the zero sequence components are at 0° from each other.

Not necessarily all the components must have nonzero values. For instance, in a generator delivering electricity to a balanced load, only

Positive sequence: One of three balanced components of any unbalanced condition in a three-phase system. The positive sequence comprises three vectors with a positive sense of direction of rotation. Balanced components are used for detection of faults, particularly ground faults, in three-phase generators, loads and transformers (see also negative sequence).

Negative sequence: One of three sets of balanced components (represented by equal size vectors, 120° apart from each other) of any unbalanced condition in a three-phase system. The imbalance condition is more noticeable when a fault happens. Thus, decomposition to balanced components is used for detection of faults in three-phase generators, loads, and transformers. The sense of direction for negative sequence is negative, as opposed to the direction for positive sequence.

Zero sequence: One of the three balanced components when any unbalanced condition (or load) in a three-phase system can be decomposed into balanced components. The zero sequence comprises three equal vectors that together with positive sequence and negative sequence vectors can represent the unbalanced condition. See also the positive sequence and negative sequence.

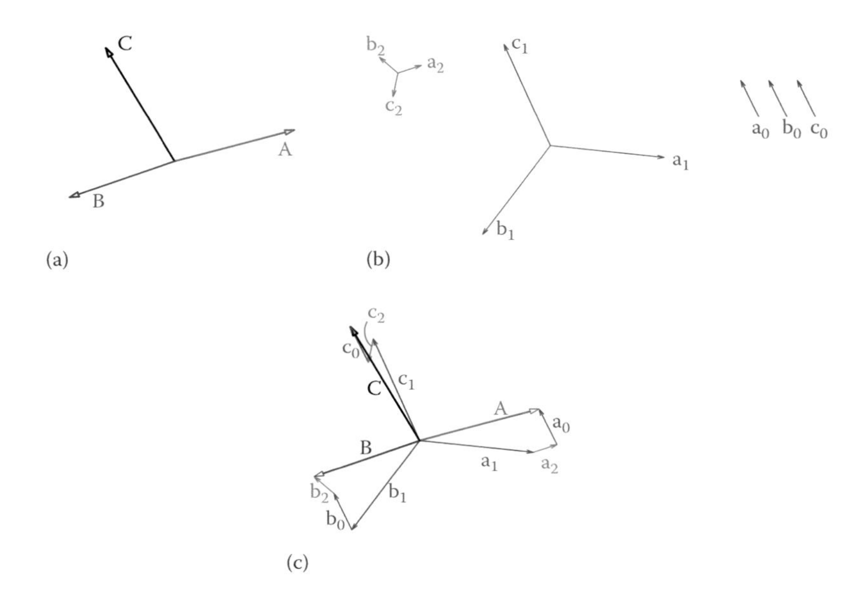

Figure A12.1 Decomposition of an unbalanced condition to three sets of balanced components. (a) Three vectors of currents in a three-phase system. (b) Three sets of vectors representing positive (a_1,b_1,c_1), negative (a_2,b_2,c_2) and zero (a_0,b_0,c_0) sequences. (c) Result of decomposition to sequence components.

the positive sequence values exist and the others are zero. The nonzero components obtain values if the load is unbalanced or a fault occurs.

The zero sequence components are very helpful in dealing with ground faults. The values of zero sequence components for various faults for any device, such as a generator or a setup like a transformer with delta-delta connection, are analyzed and determined. The results can be used for the deployment of the appropriate relaying (sensing) and protective devices.

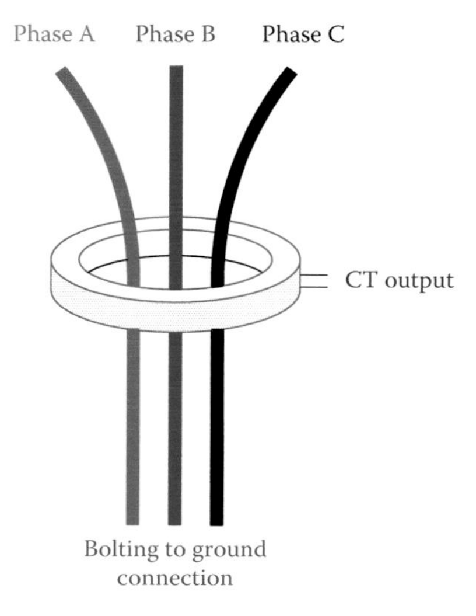

Figure A12.2 Current transformer for ground fault detection.

For detection of ground faults a current transformer (CT) is used. The three lines in a Y connection pass through the winding of a current transformer before being connected together and to the ground, as shown in Figure A12.2. The currents in the three lines add up together in this way and are measured by the current transformer. This is, in fact, the neutral current. In the normal conditions, particularly for balanced loads the current measured this way is zero, because the three currents in the three lines are 120° out of phase. When a ground fault occurs, the neutral current is not zero anymore. A high current in the neutral causes a voltage in the CT winding. This current can be high and a CT must be rated for it.

13

Electronic Components, Functions, and Devices

OBJECTIVES: After studying this chapter, you will be able to

- Describe what a semiconductor material is
- Understand the atomic structure of a semiconductor material
- Explain the difference between an electric and an electronic device
- Explain N-type and P-type materials
- Describe what the term hole implies in the context of electronics fundamentals
- Describe the terms doping, dopant, and doped material
- Name and explain common functions performed by electronic circuits
- Explain the term solid-state
- Understand the difference between analog and digital electronics
- Become familiar with some terms that you will learn later in the book
- Explain how in electronic circuits AC and DC can be combined together

New terms:
amplification, amplifier, amplitude modulation, analog, cathode ray tube, demodulation, digital, dopant, doped material, doping, filter, filtering, frequency modulation, holes, inverter, N-type material, P-type material, rectification, rectifier, semiconductor, semiconductor material, solid-state

13.1 Introduction

The previous 12 chapters of the book dealt with electricity, electrical devices, and the relationships between the electrical entities associated with generation and consumption of electricity. The rest of the book is concerned with electronics and electronic devices. Whereas all the relationships and rules studied thus far stay valid for all the circuits and devices that work with electricity, they are not sufficient for the study of electronic devices and their behavior.

To understand, design, and repair electronic devices, study of electronics is essential. For example, consider a TV set. While it is an electric device and needs electricity to work, it is an electronic device. This implies that it is much more complicated than a merely electric device, such as a motor. It has many complex circuits, and it is much more difficult and time-consuming to diagnose a problem and repair compared with a motor.

This chapter is devoted to the introduction of electronic components and simple devices (more precisely, only semiconductor electronic devices) as compared with electrical devices. An electronic circuit is the interconnection of electronic components. Electronic devices are plenty, but they are made up of basic components. Understanding the basic components is the key to the diagnosis and repair of more complex devices.

The discussion of electronics in this book is only at the introductory level. Nevertheless, it paves the way for those enthusiasts who would like a more profound study of the subject.

13.2 Electric and Electronic Devices

In general, we can categorize electric and electronic devices on the basis of the components used in them. Although nowadays many electric appliances are enhanced by electronic devices to better control their behavior or to run them more efficiently, there is still a distinct line between what is considered an electronic device and what is considered an electric device. For instance, many motors nowadays are run by a driver rather than being directly connected to electricity. The speed of an induction motor can be varied or controlled by a driver, for example. The driver in this case is an electronic device made up of electronic (and electric) components, whereas the motor is an electric device. As a second example, in the past a kettle to boil water was only a resistive element that could have a safety device to disconnect from electricity if something went wrong, as well as a switch. A new kettle may have additional features such as a timer, different temperature setups and various lights and a buzzer to indicate its status. All these extras normally involve electronic circuits.

As a distinction between the two categories, we may say that in electronic devices the flow of electrons is controlled in a desired manner, whereas in electric devices the electrons just flow when there is a path to follow and a potential difference to cause the flow.

Some of the most fundamental electronic components are described in Chapters 14–18 of this book. Any electronic device consists of a combination of these basic components; however, any such device, by itself, can constitute a component in a larger device. In this way, there is no limit to how an electronic device can grow in the number of components in it. This is particularly noticeable in an integrated circuit (IC), which has a miniature structure. Nowadays, an IC with one billion components is not uncommon. For example, the ICs in the computers and cellular phones can contain ten million components, while the size can be only 5 cm × 5 cm (2 in × 2 in), or even smaller.

While not all electronic devices necessarily consist of a large number of components, in general, they contain many components. For this reason, one can observe that many consumer and industrial electronic devices nowadays consist of a number of modules. A module, often containing a large number of subcomponents in itself, is designed to perform a particular function (e.g., amplification). In this way, if a module is diagnosed to malfunction, it is replaced by a new one, without spending time on pinpointing the bad component(s) in the module. Thus, the time spent fixing a fault, which is more crucial from an operational viewpoint, is reduced. The faulty module can be sent for repair later on.

The larger the number of components in a module, the more complex it is, and, accordingly, its design and/or repair is more complicated and requires more knowledge and takes more time. For this reason, repair of any specific module is left to specialized personnel, who are equipped with the correct equipment and often work for or are trained by the manufacturer of the piece.

Examples of basic electric components, as we have studied in Chapters 4–6, 8, and 9, are resistors, capacitors, and inductors. Examples of basic electronic components, on the other hand, are diodes, transistors, and vacuum tubes. Various types of tubes in the past were employed for the functions in radio, television, and many other apparatuses. Almost all of these are replaced by various transistors and integrated circuits because of obvious advantages for performing the same functions. The main advantages are voltage requirement, size, and energy consumption. For instance, a transistor compared with a vacuum tube of the same category works with only a few volts (instead of a few hundred volts), it size and weight can be 1/50 of that of the tube, and its energy consumption, accordingly, can be less by the same ratio. Transistors and diodes are made of **semiconductor material** and are referred to as **semiconductors**. Because of the aforementioned advantages and the fact that their packaging for any device (or as a module) is much smaller than those using vacuum tubes, they are called **solid-state**; and any device employing them (not using any vacuum tube) is called a solid-state device. Semiconductors are discussed later in this chapter.

Another basic distinction can be made between electric components and electronic components. Whereas, in general terms, electric components can work with AC and DC, electronic components can only work with DC. As a result, an electronic device (which is built of electronic components) works only with DC. Note that often, in practice, AC is modified to DC for many electronic devices. This not only does not nullify the aforementioned statement, it reinforces it.

All electronic devices work only with DC electricity.

13.3 Categories of Electronic Devices

All the electronic devices primarily can be categorized into **analog** and **digital**. Although the firmware at the fundamental level is always analog and follows the rules of analog devices, the way each category is designed and the relationships governing their behavior at the application level cannot be mixed up. In this regard, analog electronics and digital electronics are separated from each other and require two entirely different classes of discussions and knowledge. By the same token, an analog signal is different from a digital signal, though the two can be converted from one form into the other, as will be discussed in Chapter 22.

Semiconductor material: Man-made material by adding elements with five covalent electrons or three covalent electrons to silicon or germanium (with four covalent electrons). The electric properties of such a material can be controlled. It is the base material for all semiconductor devices.

Semiconductor: Term referring to a semiconductor device or a semiconductor material.

Solid-state: Electronic circuits and devices being made of semiconductors, as opposed to those made out of vacuum tubes.

Analog: Any natural phenomenon, which has a continuous nature and is not discrete, like the output of a sensor and a measurement device (e.g., temperature, current, voltage).

Digital (1): Variable that has discrete (not continuous) values, as opposed to analog. It could be the result of digitization of an analog value.

Digital (2): Representation by readable numbers.

There has been so much progress in digital electronics in the recent years that an analog version of a digital device can seem to be obsolete or on the verge of obsolescence and sometimes cannot even be perceived by the younger generation. As an example of the two, one can refer to the older TV sets using the **cathode ray tube** as an analog device versus the flat screen TVs using liquid crystal display (LCD) or light emitting diode (LED), which is a digital device. As another example, recall the telephones with a rotary dialing device, rather than push buttons.

Cathode ray tube: Vacuum tube for showing images in devices such as TV and oscilloscope, nowadays replaced by LCD and other types of screens.

It is very common to have a hybrid device, parts of which are digital and parts are analog. For instance, consider the final stage for sound in any radio, TV set, or audio device is the signal entering the loud speaker or the output of a microphone, which must be amplified. Both these are analog signals, and the immediate circuit to work on them must be an analog circuit, whereas the rest of the device can be digital.

Figure 13.1 illustrates the difference between an analog and a digital signal. The signal can be the voltage across a resistor or the output of a sensor. The figure clearly illustrates that for conversion of an analog signal to digital, initially time is broken into short intervals, and each interval is treated separately. This implies that any variable (signal) is considered a constant for a short period of time, separate from its preceding and succeeding values. The remarkable merit is that the information for each interval can be expressed by a number. This is the number that can be stored, processed, or manipulated in a digital computer.

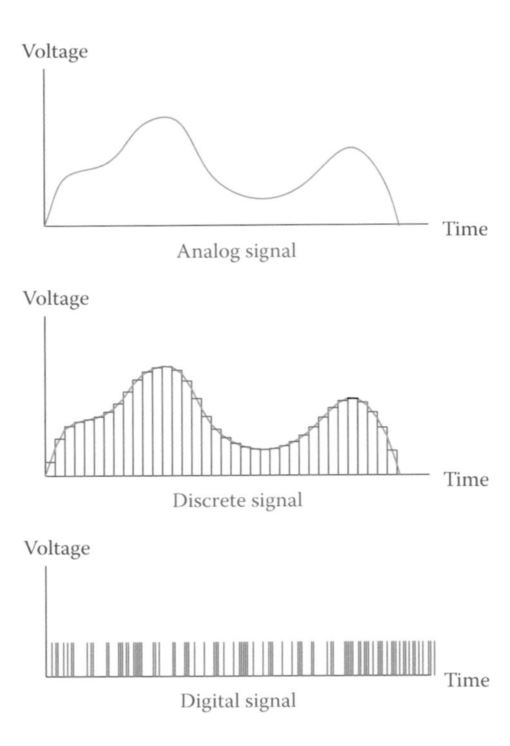

Figure 13.1 Analog and digital signals.

Digital information can be represented by numbers like computer data. These numbers can be easily processed and stored.

13.4 Semiconductor Materials

We remember from Chapter 3 that for almost all metals the number of valence electrons in their atom is less than 4. Also, metal is a conductor of electricity. When this number is more than 4, the material is nonmetallic and is an insulator to electric current. Certain atoms have 4 valence electrons, such as carbon (C), germanium (Ge), silicon (Si), lead (Pb), and tin (Sn).

Whereas lead and tin are metals and they conduct electricity, carbon is not and cannot be called a metal. It is not hard and shiny as the metals. However, carbon is a conductor of electricity and is used in many applications. The most common electrical application of carbon is in the manufacturing of brushes for DC machines and some AC machines, as we discussed in Chapters 7 and 11. Also, earlier microphones were made of carbon.

In the case of germanium and silicon the story is different and they are not conductors. Nevertheless, having 4 valence electrons gives them a specific property that no other element has, and, as a result, they are the basis of almost all the solid-state electronic devices such as transistors. Germanium and silicon are called **semiconductor**. These semiconductor materials are the main ingredients for all the semiconductor devices and components. Note that the term semiconductor refers to both a semiconductor element and a **semiconductor device**, as we learn later in this chapter.

The specific property of germanium and silicon is shown when they are mixed with some other material such as boron (B) with 3 valance electrons or antimony (Sb), arsenic (As), and phosphorus (P) with 5 valence electrons. Note that when substances are mixed with each other, even in a very tiny amount, there are millions of atoms that interact with each other. As a result of mixing the two dissimilar substances, they are bound to end up with extra electrons or missing electrons.

When very small amounts of an element with 5 valence electrons are forced to mix up with pure germanium or silicon, at the atomic level the resulting crystal has some extra electrons. This is schematically illustrated in Figure 13.2a. Similarly, when very small amounts of an element with 3 valence electrons is added to pure germanium or silicon, the resulting crystal made up of the mixture of the two elements can have electron deficiencies or missing electrons in its construction, as shown in Figure 13.2b.

The action of mixing up pure germanium or silicon with a little of another substance is called **doping**, which implies adding some impurity to a pure semiconductor material. A pure semiconductor that has received some (always relatively very small) amount of impurity in this way is called a **doped material**. The material that is used for doping a semiconductor is called **dopant**.

The interesting result of doping a semiconductor is the formation of a substance with excess electrons, which has negative charge, or a substance

Doping: Act of adding a very small amount of dopant to pure silicon or germanium.

Doped material: Pure silicon or germanium to which a very small impurity has been added.

Dopant: Very small impurity (from elements with three or five valance electrons) added to pure silicon or germanium to form a semiconductor material.

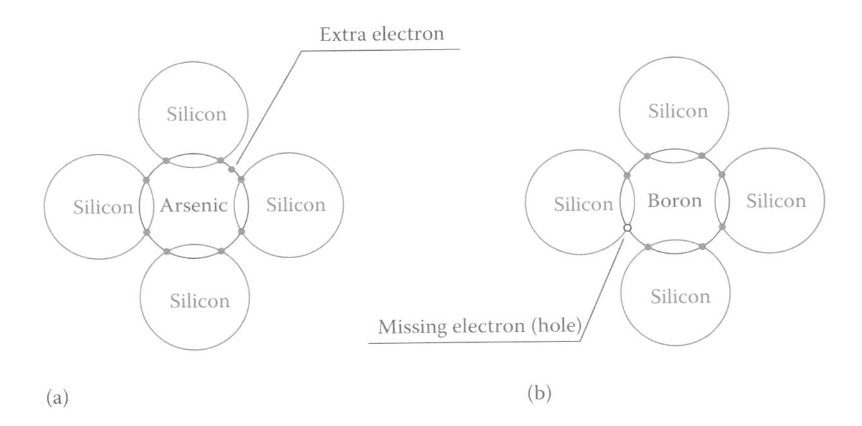

Figure 13.2 (a) N- and (b) P-type materials.

Hole: State of missing electrons in an atom in a semiconductor material, equivalent to having a positive charge. In the same way that electric current can be attributed to flow of electrons, in a semiconductor device one may use flow of holes, when appropriate.

N-type material: Semiconductor material carrying extra electrons, thus having negative charge an attracting positive charge, as opposed to P-type semiconductor material.

P-type material: Semiconductor material missing some electrons, thus carrying *holes* and having positive charge, as opposed to N-type semiconductor material having negative charge.

with electron deficiency, which has positive charge. Instead of using the term electron deficiency a semiconductor is said to have **holes**. A doped semiconductor with excess electrons (carrying negative charge) is called **N-type material**, and a doped semiconductor with holes (having positive charge) is called **P-type material**.

In other words, a semiconductor material with holes has positive charge and can attract negative charges, and a semiconductor with excess electrons has negative charge and can attract positive charges.

> A P-type semiconductor material has positive charge. An N-type semiconductor material has negative charge.

13.5 Functions of Electronic Devices

If we try to summarize the functions of electric devices, we can categorize them to electromechanical devices and heating elements. As the name implies, electromechanical devices convert the mechanical energy to electrical energy (generators) and vice versa (motors). The heating devices convert the electrical energy to heat and light. Transformers can be considered as auxiliary devices to change voltage. They can be considered electrical devices that alter the electrical property of electrical energy.

When it comes to electronic devices, a lot of other functions are performed, which cannot be done by electric devices. These are the many functions based on which today's technology is built, on the top of the list of which modern communication stands. Among these functions we may name some such as amplification, rectification, signal generation, timing, modulation and demodulation, filtering, packing and storage.

All these functions are performed by devices ultimately made from simple components when they are logically put together. All the relationships for voltage, current, and power, as we have already discussed, may be used for the analysis of the electronic circuits at the basic level. Also, because electronic circuits usually comprise a large number of elements in series and parallel combinations, the Norton and Thevenin theorems, Kirchhoff's law, and the principle of superposition may be employed for their analysis.

One must notice, moreover, that in addition to the complexity of electronic circuits many electronic components have more than two terminals, unlike electric components that mainly have two terminals. Having more than two terminals categorizes them as input-output device, and, therefore, the relationships between their inputs and outputs must also be brought into consideration.

While a study of the electronic circuits for the aforementioned functions and their analysis and diagnosis is outside the scope of this book, a definition of these functions is given in the following below. This helps to have a general idea about each function, which can be very helpful in practice in dealing with the diagnosis and repair of faulty equipment.

13.5.1 Amplification

Amplification is the multiplication of the strength of a signal by a (normally greater than 1) number. Strength here ultimately translates into the power of a signal. This is because an electrical signal at any instant has some voltage and current, and power is the product of voltage and current. Amplification can imply voltage amplification (where the current does not change) or current amplification (where voltage does not change). The latter case is commonly called power amplification, though it is the current that is amplified. Voltage amplification can be better understood by observing Figure 13.3 in which the voltage of signal A is amplified by 2 times. At each instant the voltage of the signal B is 2 times that of the corresponding point on A. Note that any signal can be amplified. Figure 13.4 illustrates an example of an AC signal (a sinusoidal waveform) that has been amplified.

Amplification: Enlarging the intensity of a signal. For an electric signal, intensity can be the power, through the voltage, the current or both.

In practice, it is not possible to have simultaneous voltage and current amplification, though this can be done in steps. Also, in order not to lose data or deviate from the original signal, amplification has limitations and high amplification must be achieved in more than one step.

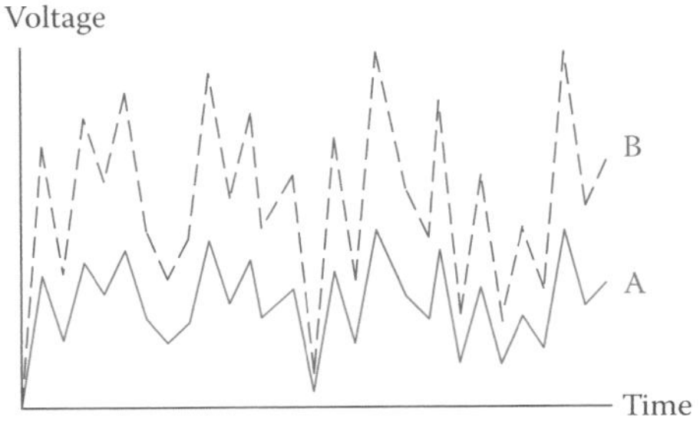

Figure 13.3 Voltage amplification of a signal.

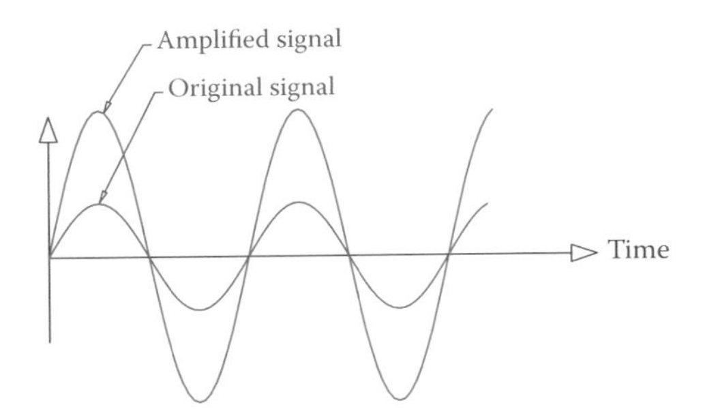

Figure 13.4 Amplification of a sinusoidal signal.

In many applications such as in music, voice, and TV broadcasting and reception there is more than one stage of amplification, and both voltage amplification and power amplification are performed but separately and in different stages. The objective is to boost the level of the weak signal received by the antenna, or generated by a microphone, to a powerful signal to match and run through the loudspeaker. Amplification is performed by an **amplifier**.

Amplifier: A (normally electronic) device that amplifies (enlarges the strength of) an electric signal.

It is very important not to confuse the function of a transformer with that of an amplifier. Whereas a step-up transformer multiplies the input voltage by a constant, the output current is decreased by the same ratio. Thus, the power transfer by a transformer remains theoretically unchanged (practically, the power is reduced because of efficiency). In a voltage amplifier, while the output voltage is increased, its current is not decreased; thus, the power of an amplified signal is increased. This increase in power is provided by the amplifier, which adds power to its input signal. This does not happen in a transformer, and no power increase is involved. More details for amplification are given in Chapters 17 and 18.

13.5.2 Rectification

Rectification: Converting AC to DC.

Rectification is obtaining DC electricity from AC sources. For all electronic and electrical devices that need DC to operate, direct current can be supplied by rectified AC. There is more than one way to do this. The quality of the DC signal, the power requirement, and other factors determine the way this conversion can be performed. A device that converts AC to DC is called a **rectifier**. The opposite action, that is, obtaining AC from a DC source, is achieved by an **inverter**. Rectification and inversion are widely used in certain renewable energy technologies. This is discussed in Chapters 20 and 21.

Rectifier: Electrical device that converts AC to DC.

Inverter: Electronic device that generates AC electricity from DC. In the simplest form the generated AC is square wave.

13.5.3 Signal Generation

In many industrial and domestic applications it becomes necessary to generate a signal with a desired profile. For example, in a TV receiver, different sorts of signals are required for various functions of the TV. Another example can be addressed for the operation of inverters used in some wind

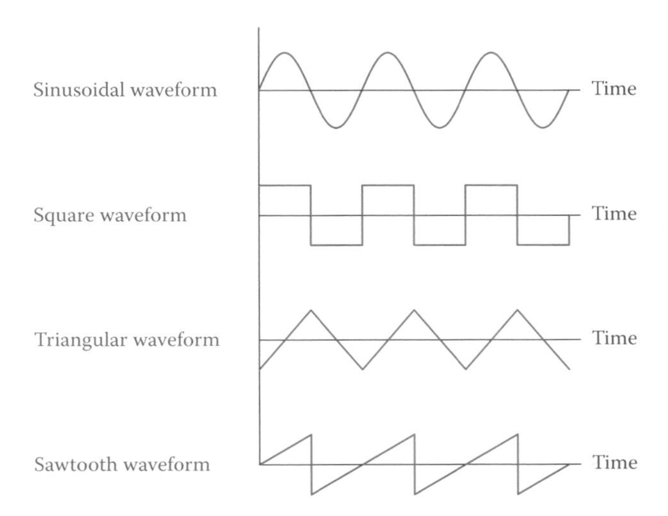

Figure 13.5 Some common waveforms.

turbines and in solar energy where alternating current with a sinusoidal waveform must be generated from a DC source. For this purpose a model sine wave must first be generated. Examples are plentiful. Figure 13.5 depicts some of the more common signals.

13.5.4 Timing

In various applications where sequential operations have to be managed or proceed in a timely manner, or with a required time delay between them, time keeping and measure of the elapsed time becomes vital. A digital watch is the most common example; it works based on counting the cycles of a waveform. But, the more fundamental level of that operation is the generation of a waveform with a desired period. Unless a higher precision is required, in its simplest form a simple RC circuit is employed for generating a delay. This can be enhanced by other circuits for any desired time keeping function.

Examples of time keeping can be seen in digital systems such as computers, digital watches, and cellular phones, and in analog devices such as in signal generators and analog TV sets.

13.5.5 Modulation and Demodulation

Modulation and demodulation are among the backbones of telecommunication and broadcasting. But, they have other applications, too. To send low-frequency signals such as music and speech over the air (radio and TV broadcasting) or through intercontinental communication lines, they must be carried by electromagnetic waves with elevated frequencies. There is no other way than loading the information to be sent to a carrier signal (the one with high frequency). The action of loading or transferring the variation of the desired signal to the carrier signal is called

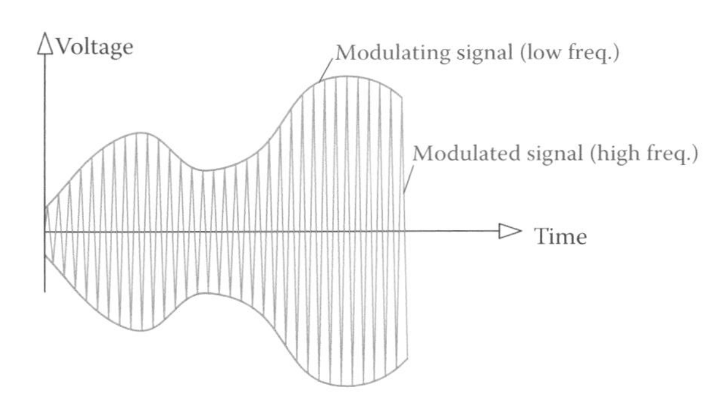

Figure 13.6 Amplitude modulation.

Modulation: Act of combining the information contained in a waveform (like voice) to a carrier waveform with much higher frequency (like radio frequency) for transmission.

Demodulation: Opposite action to modulation. Action of separating the original modulating signal from a modulated signal by filters.

Amplitude modulation: A way of embedding a signal to be transmitted (called the modulating signal; for example, music for radio broadcast) to a signal called carrier (for example, radio frequency electromagnetic waves) that can be transmitted through antennas across the air (or empty space) to a distant destination. In amplitude modulation, the amplitude of the carrier signal is modified by the effect of the modulating signal. AM radio uses this type of modulation (see also FM: frequency modulation).

Frequency modulation: A way of loading the information contained in a signal (like electric waveforms for music and voice) to another type of signal, with a much higher frequency, which is capable of being transmitted through antennas (like electromagnetic signal). In frequency modulation, the frequency of the latter is changed according to the variations in the former signal. FM radio uses this type of modulation (see also amplitude modulation).

modulation. In general, we may say modulation is changing the properties of a waveform with the contents of a signal for some desired purpose. This is shown below for analog signals. In Chapter 20 we see other applications, too (we can see *pulse width modulation* in the operation of rectifiers and inverters). **Demodulation** is the opposite action. It is the action of separating the modulating signal from the modulated signal. In this way the original signal is extracted and reconstructed from the carrier signal.

Figure 13.6 shows an example of modulating a signal. In this case the modulated signal amplitude varies according to the variation of the modulating signal, whereas the frequency stays constant. This is called **amplitude modulation**. AM radio stations use this kind of modulation. AM stands for amplitude modulation. In Figure 13.6, for simplicity, a triangular wave is shown for the modulated signal. Normally, a sinusoidal waveform is employed.

It is also possible to inject the information contained in the modulating signal in the frequency of the carrier. This is shown in Figure 13.7. The amplitude of the carrier signal is constant, but its frequency changes. The variation introduced in the frequency of the carrier waveform is proportional to the variation of the modulating signal. This is called **frequency modulation** (FM). FM radio stations and TV broadcasters use this type of modulation. Again here, for the sake of simplicity, a triangular waveform is used.

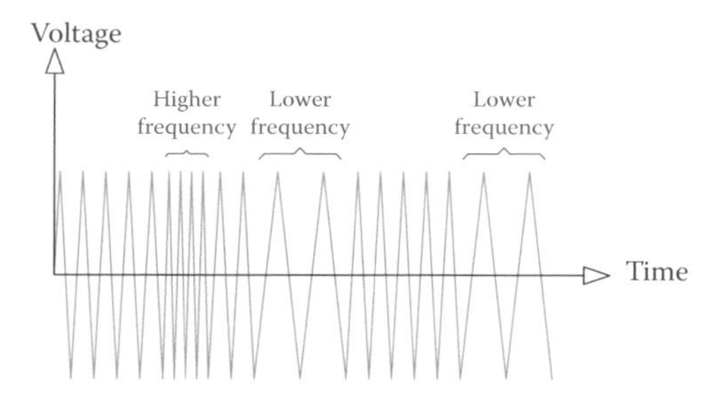

Figure 13.7 Frequency modulation.

13.5.6 Filtering

Similar to other functions, **filters** play a big role in various electronic circuits, without which it would be impossible to perform many tasks. **Filtering** is the action of removing the unwanted signals or separating various components of a signal. The best example is the separation of various waves with different frequencies received by the antenna in radio and TV receivers. The antenna does not distinguish between the electromagnetic waves available in the area. It is a series of filters that pick up a desired channel and separate it from the rest. The picked-up signal is then demodulated, amplified, and sent to the loudspeaker or directed to other processes for display on the screen.

Filter: Device to separate, for selection or rejection, electric signals of certain property. Selection criteria can be frequency or sudden deviation in voltage and current.

Filtering: Act of separating unwanted electric signals (normally with various frequencies) or reducing the effect of sudden and abrupt changes in voltage or current (smoothing) when there is only one frequency.

In the majority of cases, filtering is based on the frequency of a signal. That is, the mixed signals are separated according to their frequency. Similar to amplification, filtering may be performed in a number of stages to narrow down the signal with the desired frequency.

Often, a signal can contain noise. Noise is the unwanted signals that mix up with the signal of interest. This often happens with measurements; for example, measurement of wind speed or gearbox oil pressure and temperature in a wind turbine. The signals coming out of the measuring devices (sensors) are usually contaminated by other signals of various sources, which are considered as noise in measurement. This noise must be filtered out and removed from the main signal. A noisy signal is not reliable and can lead to faulty and improper results.

13.5.7 Packaging and Storage

Most of you are more likely familiar with CDs, DVDs, and memory sticks. These are storage media for digital signals. Prior to the digital era, almost all signals were analog, and they were stored on magnetic tapes such in the cassettes for voice and music and VCR video cassettes for picture, in addition to other formats.

While discussion of any of these devices is out of the scope of this book, thinking about and obtaining knowledge about the way things are done, or how the technology has developed, helps us to appreciate the works of those who have made contribution to the progress of knowledge and technology.

Packaging is only associated with digital communication, broadcasting, and storage. Data gathered by a microphone after a fraction of a second in a cellular phone, for instance, are put together in the form of packages before being delivered to the antenna. Similarly, the receiver in a digital TV, radio, and cellular phone receive the data in packages. Remember that, as mentioned earlier, the information is converted to a series of numbers. A package has more of these numbers. Packaging allows the data to be compressed, thus, taking less space and time for its process and storage.

For digital storage, no matter if it is for a file in your computer or a movie on a DVD, at the very fundamental level the problem comes to storing a one bit of data. Information to be stored is 1 or 0. There are billions of these 1 and 0 data to be stored. And, as you may guess, the order that these are stored and the speed of processing the information are the capabilities of the technology that bring us to the status that we currently are.

13.6 Blending DC and AC Signals

Although in working with electrical appliances, AC and DC are completely separate from each other and cannot be mixed, in electronics and electronic devices it is quite possible, and sometimes essential, to blend AC and DC together. This is especially true for radio and TV communications. Often, only the variations in a signal are of importance, because they represent the embedded information. For instance, in the signals picked up by a microphone the variation of electric signal represents the input sound. But, the picked-up signals are a mixture of AC and DC. Here, AC signal does not necessarily mean a sinusoidal waveform with a fixed frequency; it implies a compound signal with various frequencies and at variable amplitudes whose variation is above and below zero value, such that its mean value becomes zero. Such a compound signal consists, in fact, of many different sinusoidal waveforms with different amplitude and frequency. For simplicity of understanding and analysis, however, a single sinusoidal signal is considered.

Figure 13.8 illustrates when an AC signal B and a DC signal A are added together. The resultant in this case is a DC signal (because it is always positive) whose valued fluctuates between a minimum and a maximum. In practice, it is sometimes essential to form such a signal for processing data. But, later the DC portion must be separated (filtered),

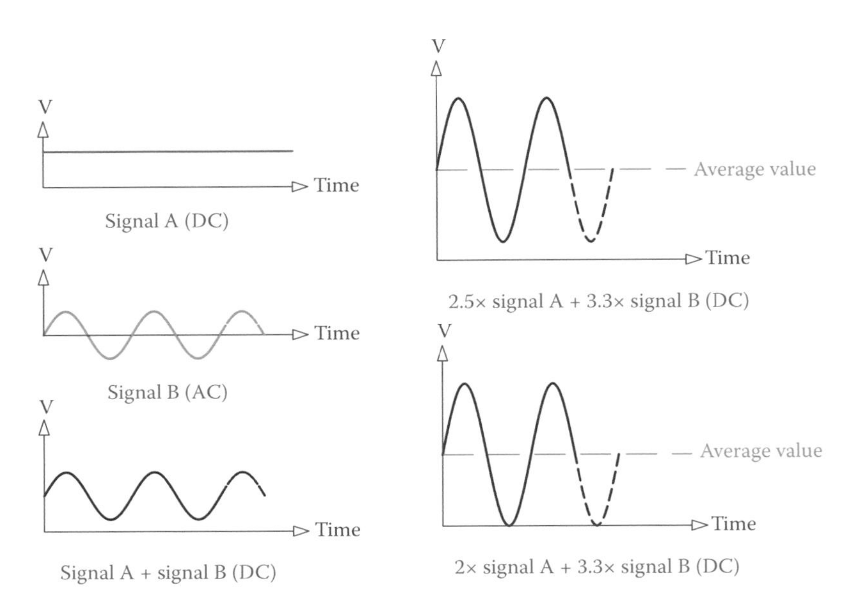

Figure 13.8 Blending AC and DC signals.

because it represents useless or even harmful current in a device. For instance, if a signal with too much DC content is sent to a speaker, the DC content only appears as a hum and unnecessarily heats up the coils of the speaker. A capacitor usually serves as a filter to remove the DC content.

13.7 Chapter Summary

- An element that is neither a metal nor an insulator of electricity and has four valence electrons is called a semiconductor.
- A component or device that is made out of semiconductor material is a semiconductor device.
- An electronic device is usually more complicated than an electric device, and its function is based on controlled flow of electrons.
- Electronic devices can perform functions that electric devices cannot.
- Electronic circuits use both electronic components and electric components (such as resistor, capacitor, and inductor).
- Doping means adding a small amount of impurity to a pure semiconductor element.
- When a semiconductor material (with 4 valence electrons) is doped with a material with 5 valence electrons, an N-type semiconductor material is formed. An N-type material has excessive electrons.
- When a semiconductor material is doped with a material with 3 valence electrons, a P-type material is formed. A P-type material has *holes*. Hole means electron deficiency.
- An N-type material has negative charge and attracts positive charges (rejects negative charges).
- A P-type material has positive charge and attracts negative charges (rejects positive charges).
- Amplification of a signal implies generating the same signal with a higher voltage or current. An amplifier amplifies a signal.
- Filtering implies removing unwanted components from an electric signal or separating various components of a signal.
- Modulation is the act of combining a low-frequency signal with a high-frequency signal to be able to transmit the information within the low-frequency signal. Radio waves have high frequency and they transfer sound and speech (low-frequency signals). Some properties of the high-frequency signal change, usually the frequency or amplitude. In this way the high-frequency signal (carrier) carries the information contained in the low-frequency signal. High-frequency signal is called carrier and is suitable for broadcasting.
- AM and FM in radio broadcasting stand for amplitude modulation and frequency modulation, respectively.

- There are two types of electronic circuits and devices: analog and digital. Their corresponding details and laws can be studied under analog electronics and digital electronics.
- In electronic circuits it is common and sometimes necessary to add AC and DC signals together (blending AC and DC signals).

Review Questions

1. Describe what specifies a semiconductor material in terms of their valence electrons.
2. Give an example of an element with 4 valence electrons, which is not a metal, but it is conductor of electricity.
3. What is the process of adding some impurity to a semiconductor material called?
4. What happens when some impurity is added to a semiconductor material?
5. Is an electronic circuit entirely made of electronic components? Why or why not?
6. Explain N-type and P-type materials.
7. What is a "hole" and how does it related to electrons?
8. What does doping mean? What is a dopant?
9. Can carbon be doped? Why?
10. When an electric signal is amplified what property of it is changed?
11. Can we say a step-up transformer amplifies the voltage of a signal? Why or why not?
12. Can we say a signal with negative voltage cannot be amplified? Why or why not?
13. What is the main feature of a digital signal? Why is it different from an analog signal?
14. What is meant by solid-state?
15. What does the term modulation mean?

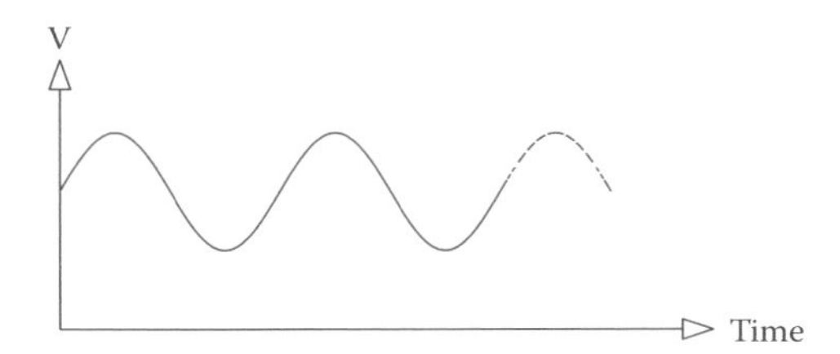

Figure P13.1 Signal addressed in Question 20.

16. What is the main function of a filter?

17. What is the unwanted electrical signal usually called?

18. What is the difference between amplitude modulation and frequency modulation?

19. Name a few electronic storage systems?

20. Is the signal shown in Figure P13.1 AC or DC?

14

Diode

OBJECTIVES: After studying this chapter, you will be able to

- Explain a PN junction
- Understand forward and reverse bias
- Define what a diode is
- Understand the structure of a diode
- Define the main function of a diode
- Recognize a diode among other electrical and electronic components
- Recognize a diode in schematic diagrams of electronic devices
- Describe the characteristic curve of a diode
- Use a diode in a circuit
- Describe how to check a diode
- Determine if a diode is good or if it is damaged
- Understand that any diode has a certain rating that must be respected when used
- Combine diodes when necessary
- Explain what a light-emitting diode is

New terms:
Anode, avalanche current, breakdown point, breakdown voltage, carrier, cathode, depletion region, forward bias, light-emitting diode (LED), liquid crystal display (LCD), peak inverse voltage (PIV), peak reverse voltage (PRV), reverse bias

14.1 Introduction

A diode is the simplest semiconductor component. It is made of only two pieces from different semiconductor materials. Nonetheless, you may find diodes in almost all electronic circuits. Its function is to let electric current flow in one direction but to prevent flow in the opposite direction. This is a very important and useful property. This function is similar to a check valve in hydraulics. The check valve depicted in Figure 14.1a opens when the pressure in the left side of the pipe is greater than that in the right-hand side, but it prevents any fluid going from the right side to the left. The applications of a check valve are plenty; for instance, at the entrance of the pipe in a well connected to a pump, a check valve is used in order to prevent water from flowing back to the well (Figure 14.1b). Also, to protect a pump from being under high pressure when not working, a check valve can be installed at its outlet.

Similarly, in electronic circuits, diodes can be used for protection of a device against a reverse current in case that there is such a possibility (examples can be seen in Chapter 20). A diode makes a short circuit across a device an easier path for the current. A more common application of a diode is in rectifiers, as will be discussed in Chapter 16.

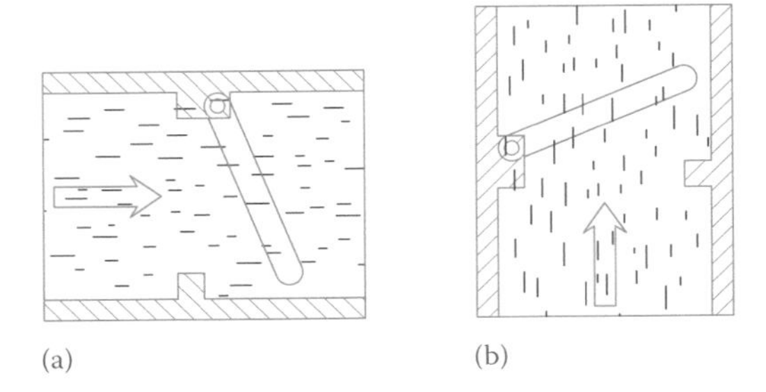

Figure 14.1 Check valve allowing flow only in one direction. (a) If the pressure on the right exceeds the pressure on the left, the valve closes. (b) If a pump sending water up stops, water is prevented from going back.

A diode is the simplest semiconductor component.

Before the structure of a diode, its applications and circuits are described we need to know what a PN junction is and what properties it has.

14.2 PN Junction

In Chapter 13 we learned about N-type and P-type semiconductor materials. These are the basis of all solid-state (made up of semiconductors) electronic components. Here we see what happens when a P-type and an N-type material are combined.

A PN junction is formed when an N-type material and a P-type material are joined together. This is a through process in manufacturing of semiconductor materials, which must be carried out in a very clean environment, called the *clean room*.

Figure 14.2 shows an N-type and a P-type material put together. A PN junction is not a physical object that one can buy; it is merely the internal structure of diodes, transistors, and other semiconductor components where a P-type material and an N-type material are joined.

Where the two types of materials contact each other, there is a small boundary where the holes (with positive charge) attract the electrons (with negative charge). As the attracted electrons gather near this boundary,

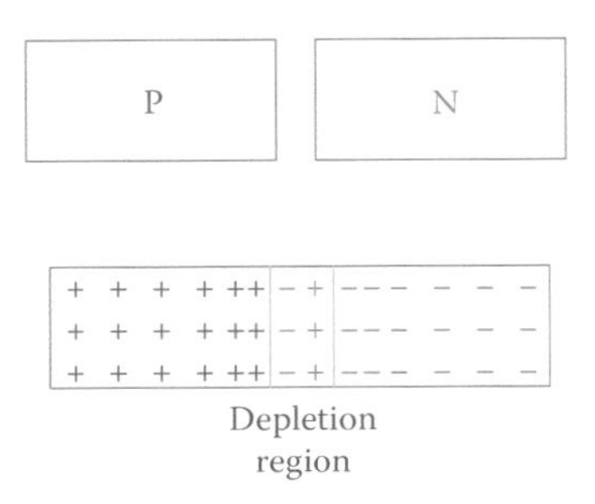

Figure 14.2 PN junction.

their negative charges accumulate and then they repel any additional electron to become attracted by the holes. Similarly, the holes also gathered on the other side of the boundary, repel any other holes from coming near. The area formed this way is called **depletion region**, also depicted in Figure 14.2.

The flow of electrons and holes to the depletion region continues until they balance each other and a state of equilibrium is reached. In this sense, the depletion region would be thinner or thicker, depending on the amounts of impurities when manufacturing the PN junction.

Notice that in the study of electronics, in the same way that electric current is considered to be due to the flow of electrons, sometimes the flow of holes is considered, instead. For simplicity, the term **carrier** is used as the cause of current. A carrier, thus, can be electrons or holes.

Depletion region: Narrow width area in a semiconductor junction just between the N-type and P-type materials.

Carrier: The entity (either electrons or holes) that the majority of an electric current in a semiconductor device can be attributed to. For example, if 94% of current is due to movement of electrons and 6% is due to hole, then electrons are the major carrier. (See another definition for carrier in Section 20.6.3.)

14.3 Forward and Reverse Bias

When a PN junction is connected to DC electricity, there are two possibilities: either the P-type side is connected to the positive and N-type side connected to the negative terminal, or vice versa. In the first case, shown in Figure 14.3a, the P side of the PN junction is positive with respect to the N side. This configuration is called **forward bias**, and the junction is said to be forward biased. The case that the P side is connected to the negative terminal and the N side connected to the positive terminal is called **reverse bias** (and the junction is said to be reverse biased). In general, one side can be more positive (or less negative) with respect to the other side. On the basis of this definition, inside a circuit, any PN junction is connected in either forward bias or reverse bias configuration. This is always the case, and one can determine from the voltage relationships whether any PN junction at any time is forward biased, or reverse biased.

When a PN junction is forward biased, it conducts because the polarity of the circuit helps a current to flow through the junction (see Figure 14.4a). But, in general, when a PN junction is in reverse bias, electrons are repelled when approaching the depletion region and the flow of electricity is, thus, blocked, as shown in Figure 14.4b.

We recall from Chapter 13 that the N- and P-type materials can be based on silicon or germanium. Moreover, various dopants can be used for doping. As a result, the characteristics of PN junctions vary based on the materials from which a junction is made. The proportion of the impurity also has a direct effect on the thickness of the depletion region

Forward bias: Applying a more positive voltage to the P-type side of a PN-junction than that to the N-type.

Reverse bias: Opposite of forward bias: connecting the P-type side of a PN-junction to a less positive or more negative voltage than the voltage connected to the N-type side.

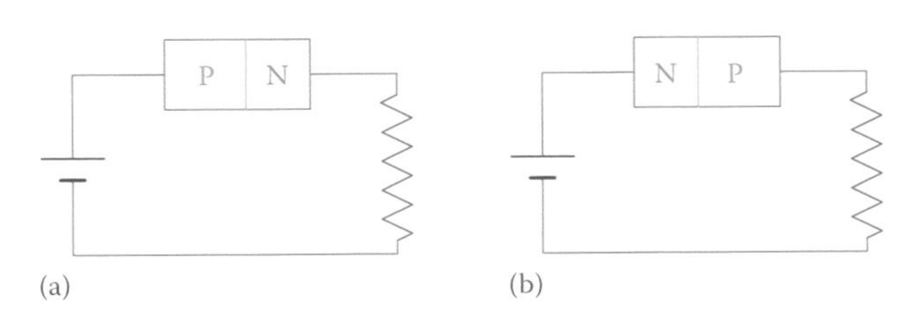

Figure 14.3 Forward (a) and reverse (b) bias of a PN junction.

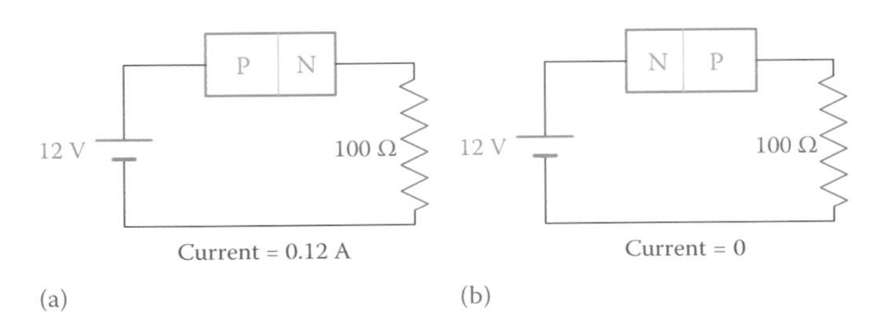

Figure 14.4 Conduction in a PN junction: (a) forward bias and (b) reverse bias.

and thus on the properties of a PN junction and the device using the PN junction.

> Properties of a PN junction depend on the materials and their proportions used in its manufacturing.

14.4 Diode

A diode is the simplest semiconductor device and consists of only one PN junction. It has two wire terminals for connection to the external circuits. A diode can be connected in a circuit in the forward bias or reverse bias configuration. Figure 14.5 illustrates the symbol as well as the physical shape of the most common diodes. The arrow in the symbol shows the direction of the current for the positive toward negative convention. That is, the current direction from P side to the N side. The P side of a diode is called the **anode**, and the N side is called the **cathode**. When inside a circuit, for the majority of applications the anode must be connected to the positive terminal (or to a more positive voltage), whereas the cathode is connected to the negative terminal (or to a less positive voltage). In the physical shape shown in Figure 14.5, the side denoted by a bar is the cathode. There are other physical shapes for diodes, too, especially for more specific diodes. In practice, the cathode can also be identified by the length of the connection wires. The cathode side connector is always shorter.

Anode: Positive side terminal in a diode and a thyristor.

Cathode: Negative side terminal of a diode or similar devices, as opposed to the anode.

Figure 14.6 depicts other physical shapes for diodes. The anode and cathode sides are also identified.

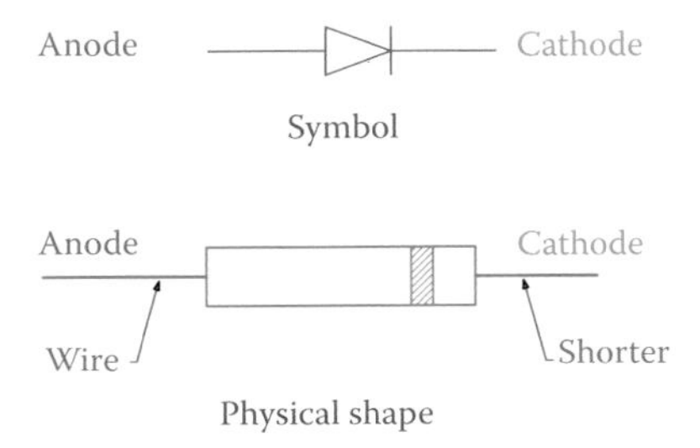

Figure 14.5 Diode symbol and the most common physical shape.

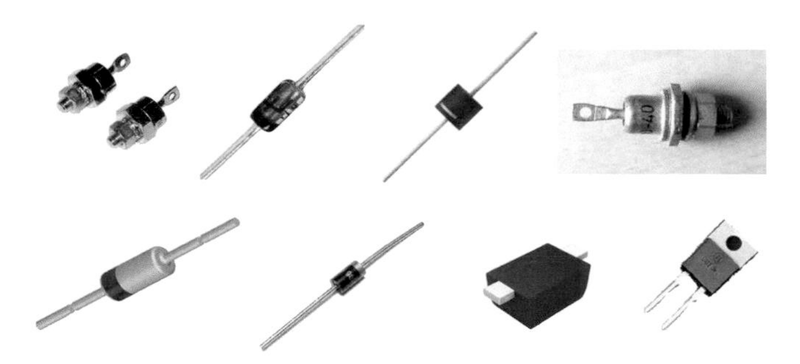

Figure 14.6 Other shapes of diodes.

For forward bias, the anode must be connected to the positive side (positive terminal or the more positive voltage) in a circuit.

When a diode is inserted in a circuit and is forward biased depending on the material from which the diode is made (silicon or germanium), there is a small voltage drop across the diode. This is due to the resistance to electron flow as a result of the formation of the depletion region. This voltage drop must be brought into consideration in any calculation of current through a diode and its circuit. For a silicon-based diode this voltage drop is around 0.7 V, and for a germanium-based diode it is around 0.3 V. In all examples in this chapter, unless otherwise stated, we always assume a silicon-based diode.

14.5 Diode Characteristic Curve

A typical characteristic curve for a diode is shown in Figure 14.7. This curve shows the variation of the diode current versus the voltage across the diode. On the right-hand side of the vertical axis (the current axis) a diode is forward biased because the voltage applied across it is positive. The left-hand side corresponds to reverse bias. When a diode is forward biased, the current across it has an abrupt increase as the voltage increases. This represents a small resistance to the current flow. On the contrary, for reverse bias the diode exhibits a large resistance and the current is, thus, very small (negligible). When reverse biased, the diode can withstand the voltage across it up to a certain limit before breaking down. When the voltage surpasses the **breakdown voltage**, an ordinary diode gets damaged and there is a sudden increase in current. When a diode is damaged, it becomes useless and cannot withstand any voltage. The point at which breakdown happens is called the **breakdown point**, and the high current after this point is called the **avalanche current**.

In general, one can say that a diode characteristic curve consists of three lines. This is true except around the origin, which is different. Remember that a voltage of 0.7 V is required for a silicon-based PN junction to conduct (0.3 V for germanium). This implies that the line corresponding to the forward bias region does not start from the origin, but from 0.7 V point on the horizontal axis. Note that in Figure 14.7 the slopes of the curves are

Breakdown voltage: Voltage at which a semiconductor device changes behavior or gets damaged when reverse biased.

Breakdown point: Point in the characteristic curve of a semiconductor device where the applied voltage is beyond the limit that the device can stand without a sudden change in behavior or without getting damaged.

Avalanche current: Relatively higher current in a semiconductor device after breakdown occurs.

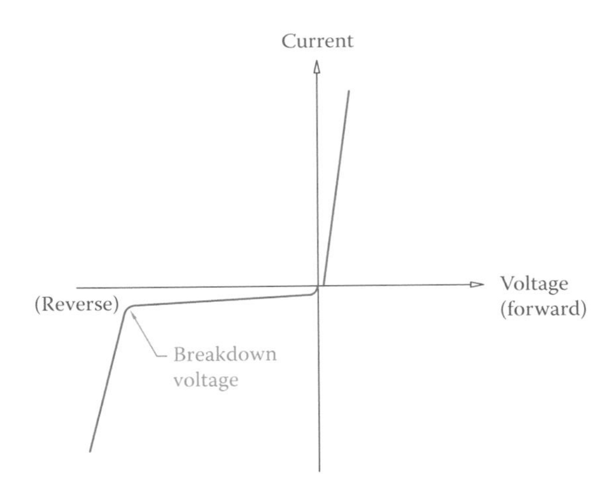

Figure 14.7 Diode characteristic curve.

moderated for clarity. In reality, the slopes of these lines are more vertical or more horizontal than those shown.

Peak inverse voltage (PIV): It is the maximum reverse bias voltage that a diode can withstand without getting damaged (breaking down). This is important in selection of diodes in the design of rectifiers. It is also called Peak Reverse Voltage (PRV).

Peak reverse voltage (PRV): The same as Peak Inverse Voltage (PIV).

Similar to any other device or electric component, a diode has ratings for maximum voltage and maximum current. For example, a small diode may have a rating of 1 A; thus, it should not be used in a circuit with more than 1 A current (unless it is used in parallel with some other diodes). See Section 14.8. A higher rated diode may be suitable for a circuit with, say, 200 A current. Figure 14.8 shows a special diode for power electronics, with a high current capacity.

In addition to the above ratings a diode has a rating for the reverse voltage. This is the maximum voltage that a diode can withstand in the reverse bias before exhibiting an avalanche current and getting damaged. For any application the **peak inverse voltage (PIV)**, also called **peak reverse voltage (PRV)**, rating of a diode must be considered for the maximum voltage across the diode in the reverse bias condition.

When in a circuit, a diode needs a small voltage across it in order to conduct.

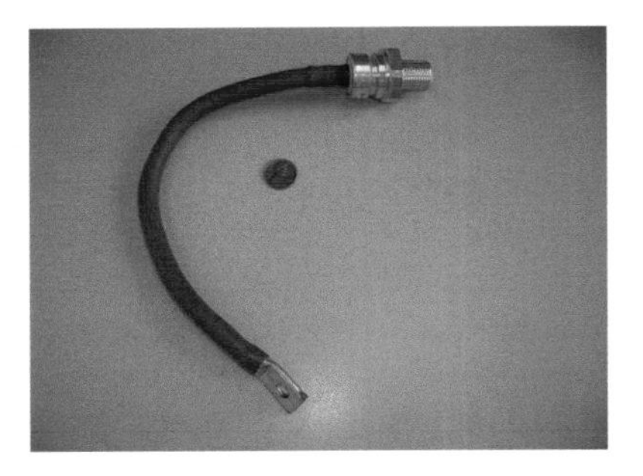

Figure 14.8 A special high current diode.

When using a diode in a circuit, its voltage and current ratings and also its peak reverse voltage rating must be brought into consideration.

14.6 Testing a Diode

A diode may be damaged by being subjected to currents or voltages beyond its rated values. In such a case, similar to a resistor or a capacitor, a damaged diode either becomes short, or it becomes totally nonconductive (open). If there are diodes in a circuit that does not function correctly, the diodes should be checked for damage.

The simplest way to test if a diode functions or it is damaged is to measure its resistance with an ohmmeter. A diode should exhibit little resistance to the electric current in one direction (when the diode is connected to the ohmmeter terminals) and a large resistance when the connections are switched. If the diode conducts in both directions or if it shows an infinite (or very large) resistance in either way it is connected, then it is damaged. In the former case, the diode is shorted, and in the latter case it is open.

Alternatively, if a digital multimeter (DMM) is employed, it has a special setting for testing diodes. In most digital multimeters this function can be selected by turning the selector knob to the position where a diode sign shows (see Figure 14.9). If the meter red lead is connected to the diode anode side and the black lead to the cathode side, then the meter must show a reading around 0.555. If the connections are reversed, the meter must show a very large reading, denoted by OL (for overload). Any reading other than these implies a damaged diode.

Note: Never connect a diode directly to a voltage source. This can cause a current through the diode more than its rating. Even if the voltage is low, say 6 V, because the resistance of a diode in forward bias configuration is very small, a high current can flow through the diode, which can damage it. Always put a resistor in series with the diode.

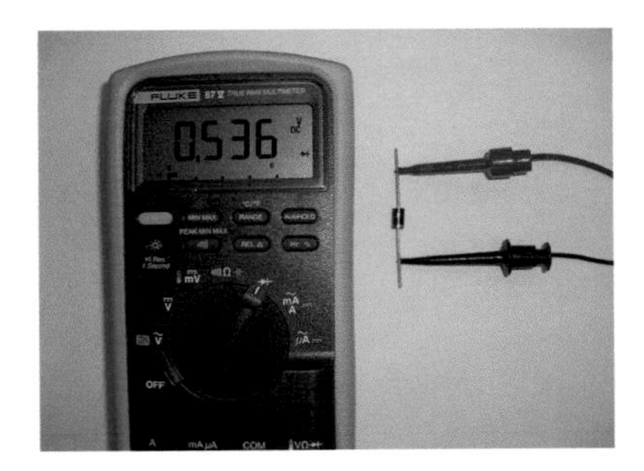

Figure 14.9 Testing a diode with a digital multimeter.

14.7 Diodes in a Circuit

The main application of a diode is for rectification. It can also be used for protection of part of a circuit or a device. Both of these applications are discussed in Chapters 16 and 20. Here we want to see the effect of inserting diodes in a circuit. The following examples show how a diode can affect a circuit.

Example 14.1

In the circuit shown in Figure 14.10, find the current. Assume a silicon-based diode.

Solution

Because of the voltage drop in the diode, as mentioned earlier, the voltage across the resistor is only 11.3 V (and not 12 V). Thus, the current is

$$11.3 \div 100 = 0.113 \text{ A} = 113 \text{ mA}$$

Example 14.2

Find the current in each of the following circuits in Figure 14.11, assuming a silicon-based diode.

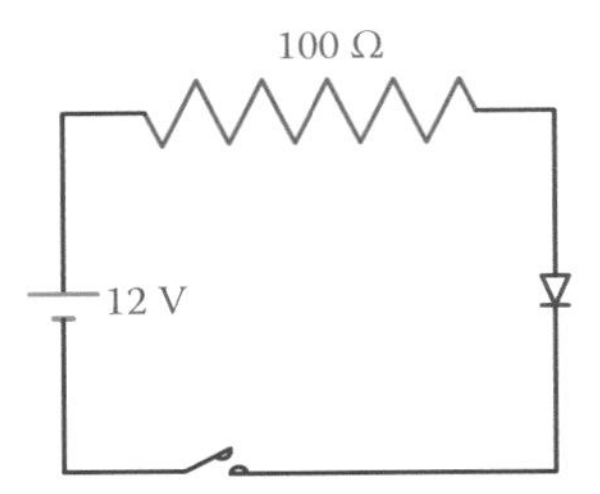

Figure 14.10 Diode in a simple circuit.

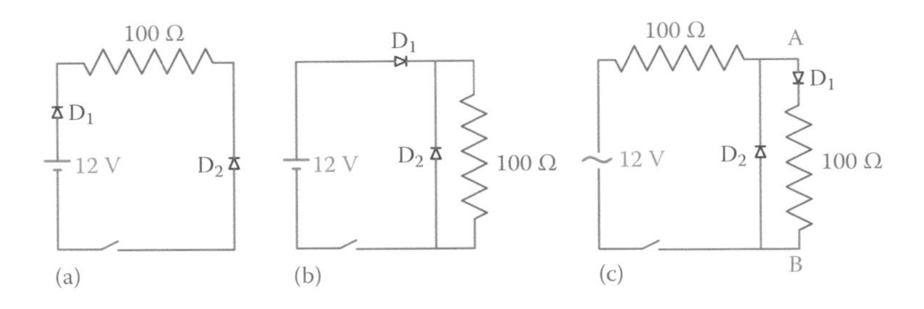

Figure 14.11 Circuits for Example 14.2. (a) Both diodes in series with the load. (b) One of the diodes is in parallel with the load. (c) AC power supply.

Solution

1. There is no current flowing in the resistor because diode 2 is reverse biased.
2. The answer for this problem is the same as that for Example 14.1. The role of the diode 2 is nothing because it has a reverse bias and does not conduct.
3. Note that the source in this case is AC. The current is different for each half cycle. When point A is more positive with respect to B, the current is

$$11.3 \div 200 = 0.0565 \text{ A} = 56.5 \text{ mA}$$

But for the next half cycle when point B is positive and point A is negative, the current is

$$11.3 \div 100 = 0.113 \text{ A} = 113 \text{ mA}$$

This is because one of the resistors has no current in this case (current is blocked by diode D_1). Note that the current as calculated is the effective value of the current and for each half cycle the instantaneous current varies between zero and a maximum value.

14.8 Diodes in Series and Parallel

As mentioned earlier, any diode has a certain voltage and current rating. For example, if a diode has a rating of 1 A and 40 V, it cannot be connected to 120 V or be used in a circuit whose current might exceed 1 A. Nevertheless, it is possible to put diodes in series with each other. Thus, if we need to use a diode with 40 V rated voltage in a circuit where the voltage is 120 V, we can put three of these diodes in series together, as shown in Figure 14.12. Obviously, in such a case the voltage drop in the circuit due to the diodes is 3 times that of each diode; that is, if three silicon-based diodes are in series a voltage drop of 2.1 (3 × 0.7) must be considered across the diodes.

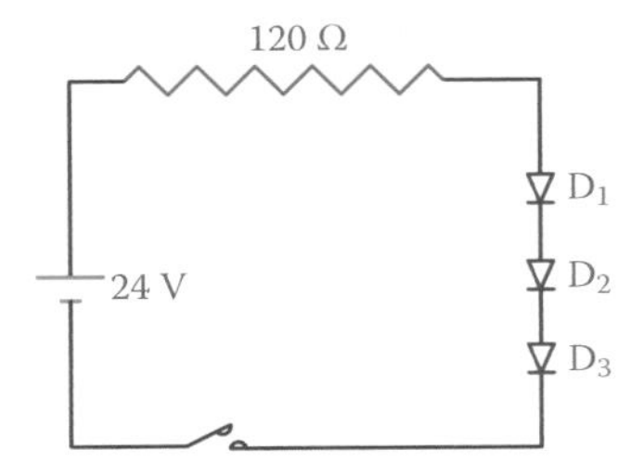

Figure 14.12 Diodes in series.

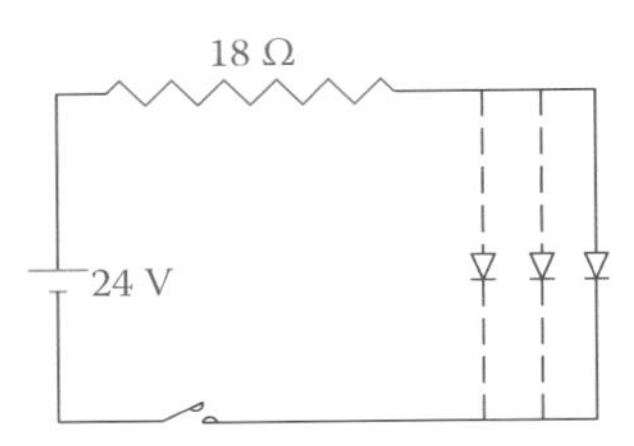

Figure 14.13 Diodes in parallel.

Example 14.3

Find the current in the resistor in the circuit of Figure 14.12.

Solution

The total voltage reaching the resistor is 24 – 2.1 = 21.9 V. The current in the resistor is, thus,

$$21.9 \div 120 = 0.1825 \text{ A} = 182.5 \text{ mA}$$

Similar to putting the diodes in series to add their voltage handling capability, they can be put together in parallel to increase their total current capacity. If the rating of each diode is 1 A, the rating of two diodes in parallel is twice as much. Normally, in putting diodes in parallel it is better that they are all similar; that is, they have similar ratings. In this way any current will be equally divided between them.

Example 14.4

If the current rating of each diode in Figure 14.13 is 0.5 A, find how many of these diodes are required for the circuit.

Solution

The total voltage reaching the resistor is 24 – 0.7 = 23.3 V. The total current in the resistor is, thus,

$$I = 23.3 \div 18 = 1.29 \text{ A}$$

The number of required diodes is 3 (3×0.5 A = 1.5 A > 1.29 A).

Light-emitting diode (LED): Diode that when subject to a DC voltage with correct polarity (positive to anode) and magnitude (around 5 V) emits light. The color of the light depends on the material used; it can be red, orange, green yellow, blue, and so on.

14.9 Light-Emitting Diode

A **light-emitting diode (LED)** is a special type of diode that when forward biased emits light, like a tiny lightbulb. Its current and the voltage drop across it are, however, much less than an incandescent lightbulb of equal luminance. Figure 14.14 depicts an example and the symbol for an

Figure 14.14 Light-emitting diode and its symbol.

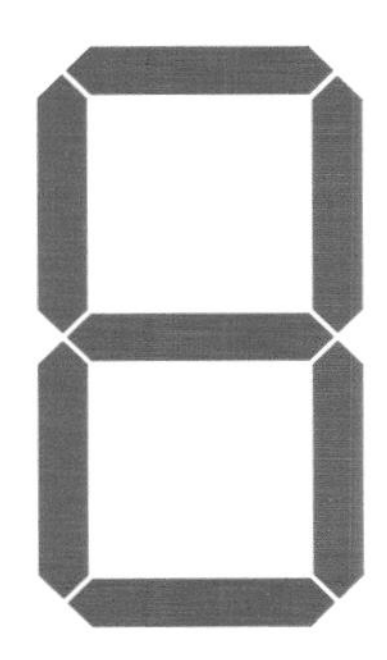

Figure 14.15 Seven-segment LED.

LED in a circuit. The applications of LEDs are plenty, including readout of measured values and signal display. Depending on the material used, different color LEDs are made. Earlier LEDs were not so bright and could not be used in certain applications such as airplane cockpit instruments because they could not be seen in the presence of bright light or sunlight. But today's LEDs are much more advanced, and they are employed even in flat screen TVs. If an LED is reverse biased, it does not light up. Note that an LED is a diode; it requires a forward voltage of slightly more than 0.7 V to light up, but if you connect it to a higher voltage, a large current flows through it, which can damage the LED. You need to drop the voltage by putting it in series with a resistor. In all LEDs, also, the cathode connection is usually smaller in length, as shown in Figure 14.14.

An LED should not be confused with a **liquid crystal display (LCD)**, which works on a completely different principle and is not a diode. An LCD works based on the obstruction of a background light. For this reason, it needs an illumination at its back. In some flat panel TV sets this illumination is provided by LEDs (instead of a more common fluorescent backlight).

Liquid crystal display (LCD): A kind of technology for display in calculators, TV, monitors, and so on, based on blocking a backlight from passing through to a viewer according to the electric voltage applied to each unit in the display.

Figure 14.15 shows the diodes in a seven segment LED display. A seven-segment light-emitting diode can be used to create all digits from 0 to 9. It has seven LEDs each one in the form of a bar. By turning the correct combination of a set of them a single digit number is displayed. They can be combined together for more digits.

14.10 Chapter Summary

- A diode is the simplest semiconductor device.
- A diode contains one PN junction.
- The main function of a diode is allowing electrical current to flow in one direction but to prevent its flow in the opposite direction.

- A term used for the cause of current in a semiconductor device is "carrier," which can refer to both electrons and holes.
- Diodes are made in various shapes.
- There is a limit for the voltage across a diode in the reverse bias configuration, for which a current flow can be blocked. If the limit is surpassed, the so-called breakdown voltage, a diode gets damaged.
- When forward biased, in order to conduct a silicon-based diode needs a minimum of 0.7 V across it. A germanium-based diode needs 0.3 V.
- When damaged, a diode becomes short or open.
- A diode is tested by an ohmmeter.
- Most digital multimeters have a special setting for checking diodes.
- Diodes can be put together in series or in parallel.
- LED stands for light-emitting diode; it is a diode that emits light when forward biased.
- LEDs come in different colors, depending what material is used to construct them.
- An LED should not be confused with an LCD, which stands for liquid crystal display. It does not emit light and needs an auxiliary source for light.

Review Questions

1. What is a diode used for?
2. What is an anode in a diode?
3. When is a diode forward biased?
4. What is a diode made of?
5. What instrument can one use to test a diode?
6. What happens if a diode is connected to a relatively high voltage with forward bias?
7. What happens if a diode is connected to a relatively high voltage with reverse bias?
8. Does a diode conduct if a voltage of 0.2 V is applied across it? Why or why not?
9. What is the maximum current that a diode can handle?
10. Can diodes be connected together in series? Why or why not?

11. For what purpose would you put diodes in parallel with each other?

12. What does the characteristic curve of a diode represent?

13. What is the breakdown voltage in a diode?

14. What is an LED?

15. What is an LED used for?

16. How can you test an LED?

17. If an LED in a circuit does not light up, what could be the problem?

18. Can you put a number of LEDs in parallel?

19. What does the rating of a diode imply?

20. Can you mix up germanium- and silicon-based diodes in a circuit? Why or why not?

Problems

In all problems, when necessary, assume a silicon-based diode.

1. What is the current in Figure P14.1?

2. What is the minimum voltage to have a current of 20 mA in the circuit of Problem 1?

3. If you have 10 silicon-based LEDs in series with each other, what is the minimum voltage required to light them up?

4. What is the difference between Figure P14.2a and b, in terms of the circuit current?

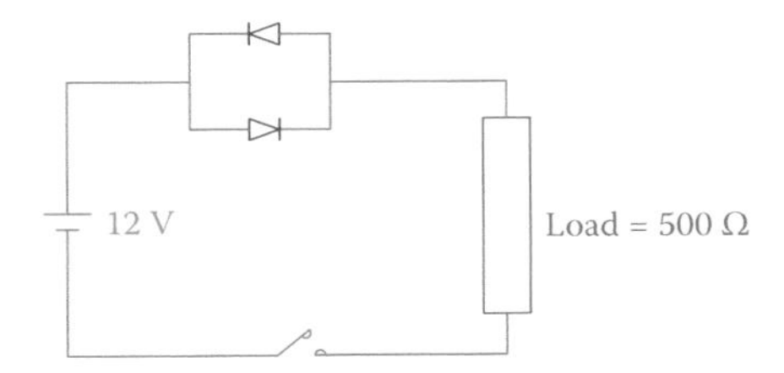

Figure P14.1 Circuit of Problem 1.

5. Find the current in the 50 Ω resistor in Figure P14.3.

6. Draw the current variation for a full cycle of the AC current in Figure P14.4.

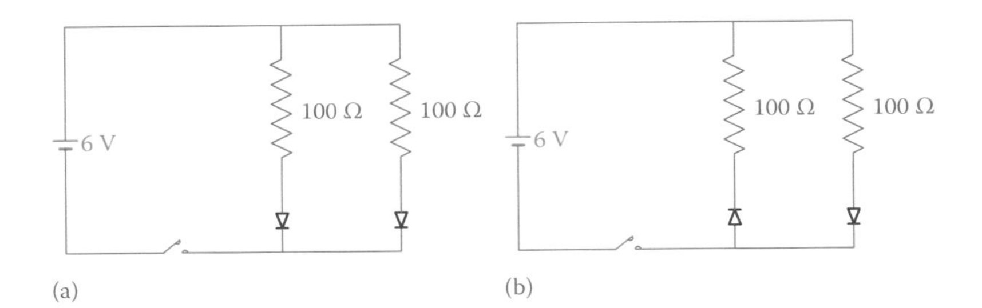

Figure P14.2 Circuit of Problem 4. (a) Diodes are parallel. (b) Diodes are antiparallel.

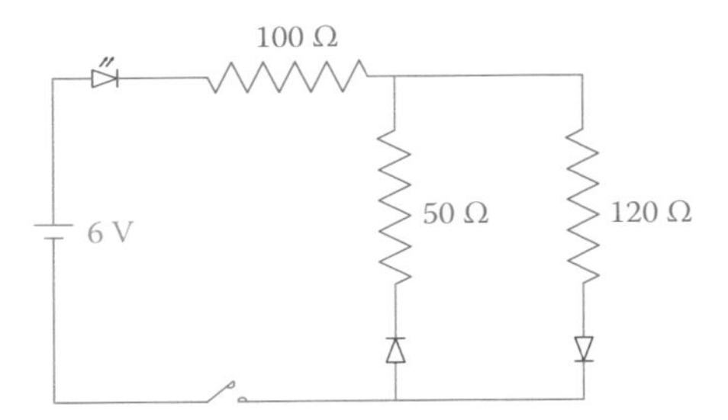

Figure P14.3 Circuit of Problem 5.

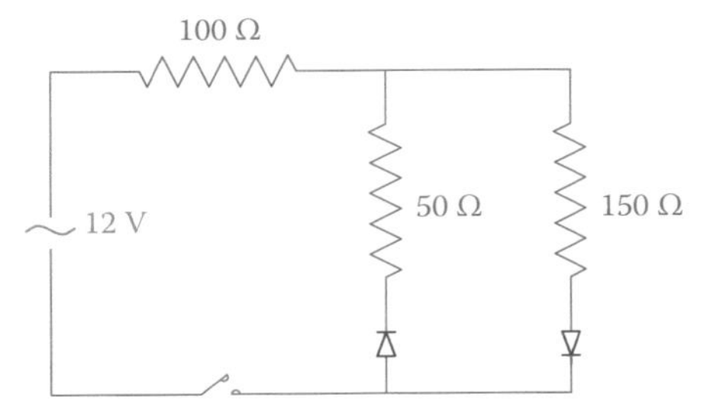

Figure P14.4 Circuit of Problem 6.

15

Regulating Diodes and Applications

OBJECTIVES: After studying this chapter, you will be able to

- Describe what a zener diode is
- Describe the differences between a regular diode and a zener diode
- Use a zener diode in a circuit
- Define how to combine zener diodes and for what purpose
- Use a zener diode for voltage regulation
- Recognize a zener diode in a circuit
- Describe what a diac is and what it is used for
- Explain the characteristics of a diac
- Describe the application of a silicon-controlled rectifier (SCR)
- Define the difference between a diode and a SCR
- Describe the relationship between diac and triac
- Recognize the application of a triac
- Draw circuit diagrams the professional way

New terms:
Breakover voltage, diac, silicon-controlled rectifier (SCR), triac, voltage regulation, zener current, zener knee, zener region

15.1 Introduction

In Chapters 16 to 21, some electronic devices, which are more pertinent to the objective of this book, will be explained. It is not the intention, nor within the scope of the book, to introduce all electronic devices. Nevertheless, because of their importance, two regulating diodes are discussed in this chapter. These two devices are used for voltage regulation in DC circuits and current regulation in AC circuits.

The zener diode is widely used for voltage regulation in DC circuitry. It is a diode with two layers of semiconductor material. The **triac** is another electronic device use in AC circuitry; it controls current for the soft start of motors and reducing inrush current, and operation of a dimmer switch, for instance. These two devices are not of the same category as far as their structure is concerned. But, because of their function, control and regulation, it is reasonable to include them in one chapter.

Triac: Type of gated diode for alternative current. Its main terminals do not connect until the gate allows connection or the voltage across it has reached certain value.

The triac is a switching device, meaning that it has a gate through which it can be turned on and off. Gated or switching devices are discussed in Chapter 19. They are used for various applications covered in Chapters 20 and 21.

15.2 Zener Diode

Similar to the ordinary diode that we discussed in Chapter 14, there are other types of diodes that are commonly and frequently used in electronic circuits for industrial and domestic applications. One of these is the zener diode.

Zener current: Same as the avalanche current (the higher current after breakdown occurs in a semiconductor device) in a zener diode.

Zener knee: Sharp curve in the characteristic curve of a zener diode where the breakdown voltage is reached and breakdown occurs.

Zener region: Region in the characteristic curve of a zener diode where the breakdown has occurred because of the reverse voltage surpassing a certain value.

The zener diode is a special type of diode that is designed to work in reverse bias and in the so-called zener region of the diode characteristic curve. This region is after the reverse-biased voltage has exceeded the breakdown voltage (breakdown point). If the reverse bias voltage is less than the breakdown voltage, or if the zener diode is forward biased, it acts as an ordinary diode. That is, in forward bias it allows current, and in reverse bias it blocks current. After this voltage has surpassed the breakdown point (in reverse bias), the diode falls in the zener region, where it conducts without getting damaged. Current in this region was called avalanche current in Chapter 14. For a zener diode it is also called **zener current**. As soon as the voltage decreases, the diode retains its nonconducting condition and gets back to its normal properties. This specific property of being operational in the reverse bias and with avalanche current is given to the zener diode by a rich doping of the semiconductor material. Moreover, by controlling the amount of doping, the thickness of the depletion region in the PN junction and the breakdown voltage can be set to any value.

Figure 15.1 shows the characteristic curve of a zener diode, which is more or less similar to that of an ordinary diode. The range after the **zener knee** (breakdown voltage) until the rated voltage of a zener diode is reached is called the **zener region**. In this region a small change in voltage

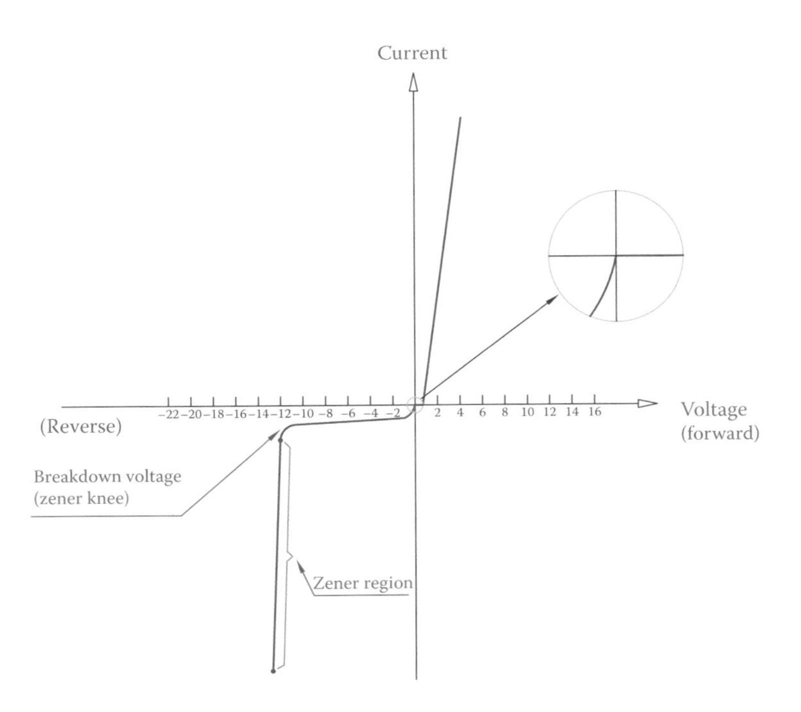

Figure 15.1 Zener diode characteristic curve.

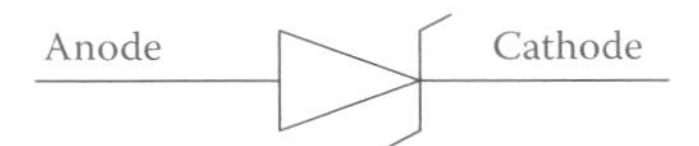

Figure 15.2 Symbol for a zener diode.

leads to a large change in current. For instance, a change of 1 V in voltage can cause the current to increase by 10 times.

A zener diode works when reverse biased. Otherwise, it behaves as an ordinary diode.

As can be understood from the characteristic curve, in this region, when the reverse voltage is smaller in absolute value than the zener breakdown voltage, the current is very small and confined to the so-called leakage current, which is negligible and can be ignored. But for higher reverse voltages, there is an abrupt and large increase in current. In other words, whereas the voltage remains almost constant, the current has a considerable variation. This is the significant property of a zener diode, which can be employed.

A zener diode has a different symbol to differentiate it from other diodes. This symbol is shown in Figure 15.2. The physical shape of a zener diode can be the same as that of a regular diode.

15.3 Zener Diode Application

The main application of a zener diode is in **voltage regulation**. Voltage regulation implies keeping a voltage at a desired value irrespective of current and other factors that tend to change it. It is very important for many devices to protect them from overvoltage (and excessive current, as a result). The specific property of a zener diode is that in its operating range the voltage drop across it is constant (it has an internal self-regulatory characteristic). This could be understood from the characteristic curve in Figure 15.1, which can be expressed differently: large current change in a zener diode causes only a small variation in voltage. That is, the voltage drop across it is almost constant. In this respect, if a zener diode is in parallel with another component, or device, the voltage across that device is kept constant, because of the zener diode.

Voltage regulation: Controlling electric voltage in a circuit or a device to remain within a desired range (small variation) under different load conditions.

The main application of a zener diode is in voltage regulation.

Figure 15.3 depicts a load in parallel with a zener diode, both in series with the 200 Ω resistor R. Notice the polarity of the zener diode. The role of the resistor R is to provide the leverage for current adjustment. The zener diode is a current governed device. If we remove R, the whole voltage is applied to the load.

Zener diodes come in different voltages, for example, 5.6, 7.5, 10, and 16 V, just to name a few. For each purpose the right size diode (voltage and rated current) must be employed (see Section 15.5 for a combination of zener diodes in series and parallel).

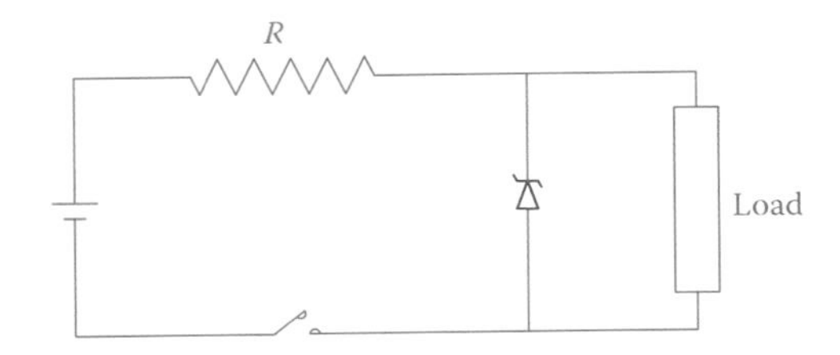

Figure 15.3 Zener diode in a circuit, regulating load voltage.

15.4 Application Examples

The following examples show the application of a zener diode and the concerns for the proper rating. Note the polarity of the circuits in the reverse biasing of the zener diode. If a zener diode is used in forward bias, it performs as a regular diode.

In all the examples, there is a resistor in series with the zener diode and the load. The zener diode is in parallel with the load. When necessary, we refer to this resistor as the series resistor.

> If a zener diode is forward biased, it behaves as an ordinary diode.

Example 15.1

In Figure 15.4 the 100 Ω resistor is the load and the diode used is a 4.7 V zener diode. In each of the three cases, find the current in the load (note polarity difference of the zener diode in cases 2 and 3).

Solution

1. Following Ohm's law the current is

$$I = 12 \div 150 = 0.08 \text{ A}$$

The voltage drop in the 50 Ω resistor is (50)(0.08) = 4 V.

2. The zener diode is forward biased, and it behaves as a regular diode. The voltage drop across it is only 0.7 V.

Current in the load = 0.7 ÷ 100 = 0.007 A = 7 mA
Current in the 50 Ω resistor = (12 – 0.7) ÷ 50 = 0.226 A = 226 mA
Current in the diode = 226 – 7 = 219 mA

3. The zener diode is reverse biased and performs as a voltage regulator

Current in the load = 4.7 ÷ 100 = 0.047 A = 47 mA
Total circuit current = Current in the 50 Ω resistor = (12 – 4.7) ÷ 50 = 0.146 A = 146 mA
Current in the diode = 146 – 47 = 99 mA

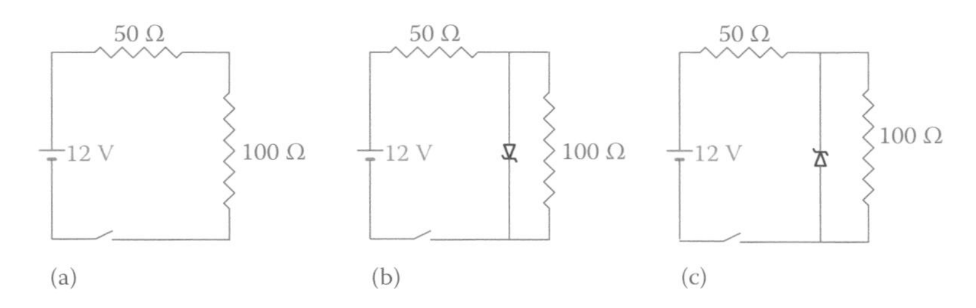

Figure 15.4 Example 15.1. (a) No zener diode in the circuit. (b) Zener diode is forward biased. (c) Zener diode is reverse biased.

Example 15.2

If the load in Example 15.1 is changed to 200 Ω, what is the change in the circuit current?

Solution

Circuit current is always the current in the 50 Ω resistor. In case 1, $I = 12 \div 250 = 0.048$ A. But, we get the same values for the circuit current in cases 2 and 3 as in Example 15.1, since in case 2 the load is shorted by the diode and in case 3 the zener diode regulates the voltage across the load to be 4.7 V, as before. The current in the load is, however, half of the previous values (for cases 2 and 3).

Example 15.3

If the value of the 50 Ω resistor in Example 15.1 is changed to 200 Ω, what is the change in the circuit current, in case 3 only?

Solution

The current in the 50 Ω resistor must always be greater than the load current because it is the sum of the current is the zener diode and the load. If we calculate the currents in the load and in the 50 Ω resistor, we have

Load current = $4.7 \div 100 = 0.047$ A
Current in the 50 Ω resistor = $(12 - 4.7) \div 200 = 0.036$ A

But, we observe that this current is smaller than the load current; thus, *this is not a possible case*. A value of 200 Ω resistor is not acceptable.

Example 15.3 shows that in selecting a series resistor for inclusion in the circuit when a voltage regulator is used, care must be taken not to have an improper size resistor. In such a case the circuit does not function as desired. The action of a zener diode is that it absorbs the extra current of the circuit that would flow through the load. But, in the case of Example 15.3, there is no extra current.

For the circuit just discussed, the current is $12 \div 300 = 0.04$ A and the zener diode does not play any role.

When using a zener diode, it is important to remember that

1. Voltage across the load (in case the zener diode is removed) must be more than the zener diode value for voltage regulation to happen.
2. Current in the series resistor (the total circuit current) must be higher than the load current.
3. We cannot entirely eliminate the series resistor.

Example 15.4

For case 3 in Example 15.1, how much is the maximum value that the series resistance R can have?

Solution

Maximum value of the resistor R must be such that the current in R does not go below the load current. In this sense, minimum current in R is the load current. The load current, thus, can define the maximum value of this resistor.

$$V_R = 12 - V_Z = 12 - 4.7 = 7.3 \text{ V}$$

$$I_R = I_{\text{Load}} = 4.7 \div 100 = 0.047 \text{ A}$$

$$R = 7.3 \div 0.047 = 155\ \Omega$$

For regulating the voltage across a load we cannot entirely eliminate the series resistor.

15.5 Zener Diodes in Series and Parallel

In the same way that diodes could be put in series, if a larger regulated voltage is required a number of (compatible) zener diodes can be put together in series. For example, in Figure 15.5, three zener diodes are put together to provide a larger regulated voltage for the load.

Also, it is possible to put zener diodes in parallel to increase the current capacity. Figure 15.6 depicts an example.

15.6 Diac and Triac

Diac and triac are two devices, usually used together for current control in AC circuits. The names are made based on "diode for AC" and "triode for AC." Although you now know about solid-state diode, the counterpart of that in vacuum tubes was also called diode; a tube with two elements

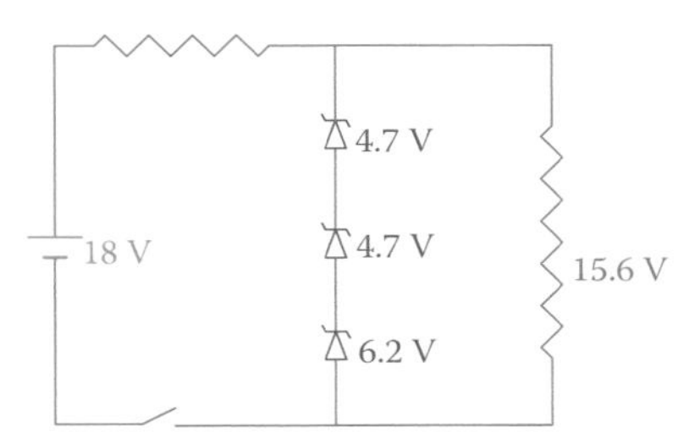

Figure 15.5 Zener diodes in series.

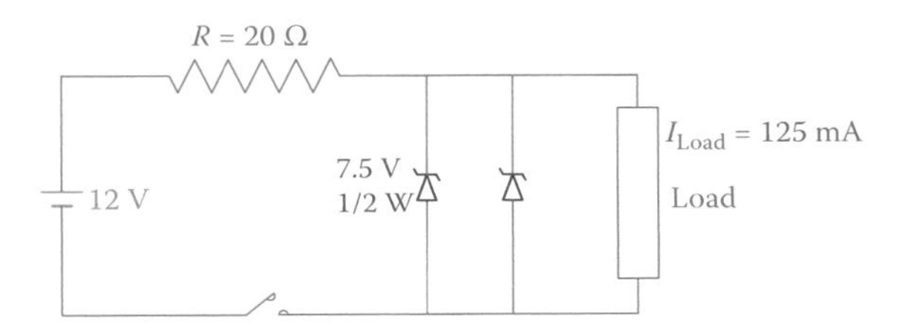

Figure 15.6 Zener diodes in parallel.

(electrodes) inside it, plus a heating element. That was a vacuum tube used for rectification. In conjunction with it, there were triode, tetrode, and pentode with 3, 4, and 5 electrodes inside the tube. These were used for amplification and other functions, which are performed by various transistors and other semiconductor devices nowadays.

15.6.1 Diac

Diac is made up of three layers of N-type and P-type semiconductors, like a transistor (which are discussed in Chapter 17), but because of its doping it does not behave as a transistor. It is more like two antiparallel diodes (two diodes in reverse direction and parallel to each other). Thus, it is a bidirectional device and conducts in both directions, suitable for AC electricity. Nevertheless, it is not like two ordinary diodes, and conduction takes place when the applied voltage to it reaches a minimum value called **breakover voltage** (V_{BR}).

Diac: Type of semiconductor device. It acts as a diode for alternating current, restricting the current until certain voltage is reached and then it conducts normally.

The symbol and schematic structure of a diac are shown in Figure 15.7. As can be observed, a diac has no cathode and the two terminals are called anode 1 and anode 2, or alternatively MT_1 and MT_2. MT stands for main terminal.

Breakover voltage: Voltage below which a diac or a triac is not conducting, but when the voltage surpasses that conduction starts.

Figure 15.8 shows the characteristic curve of a diac. It is a symmetrical curve, meaning that it behaves the same way for current in either direction. The breakover voltage of diacs is around 30 V (typically 20–40 V). When the applied voltage surpasses V_{BR}, a diac sharply conducts and the current is abruptly increased. Later, if the current falls, conduction continues until the circuit current drops below the holding current (I_H).

A diac is used for switching because it can deliver a sharp and clear signal when conduction starts. It serves for turning on the gate of a triac.

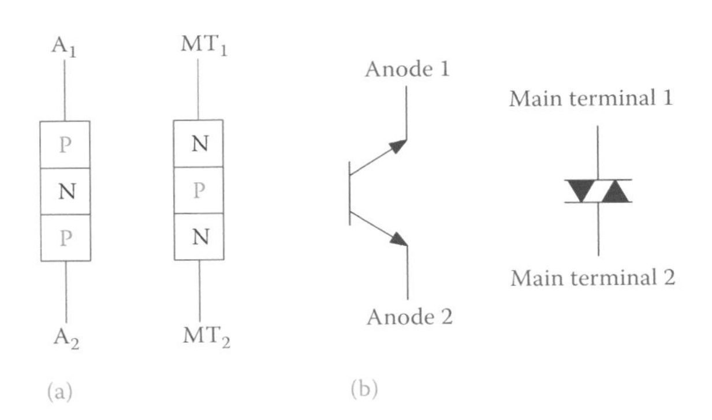

Figure 15.7 (a) Basic structure and (b) symbols for diac.

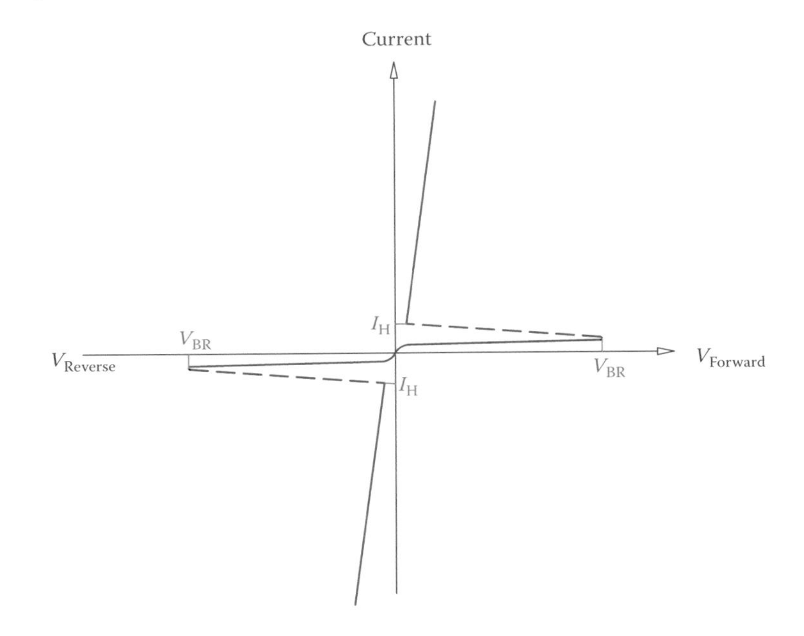

Figure 15.8 Diac characteristic curve.

15.6.2 Triac

Silicon-controlled rectifier (SCR): A family of gated diodes or transistors that can be controlled (turned on and off) through their gates.

Before learning about a triac, let's describe a notation about a concept employed in many controlled semiconductor devices. A triac can be considered a member of a family of devices called **silicon-controlled rectifier**, abbreviated to **SCR**. In a later chapter we are going to learn more about SCRs and their applications, but here we want to introduce the concept of a gate. A silicon-controlled rectifier is like a diode with a gate. The gate is a control element through which the performance of a diode can be controlled. That is to say, unlike a normal diode, a silicon-controlled rectifier allows a current from its anode to its cathode only when the gate allows that. A silicon-controlled rectifier is a four-layer PNPN semiconductor, and it has three terminals, anode, cathode and gate.

Figure 15.9 illustrates the characteristic curve of a member of the SCR family. Considering only the forward bias half of the diac characteristic

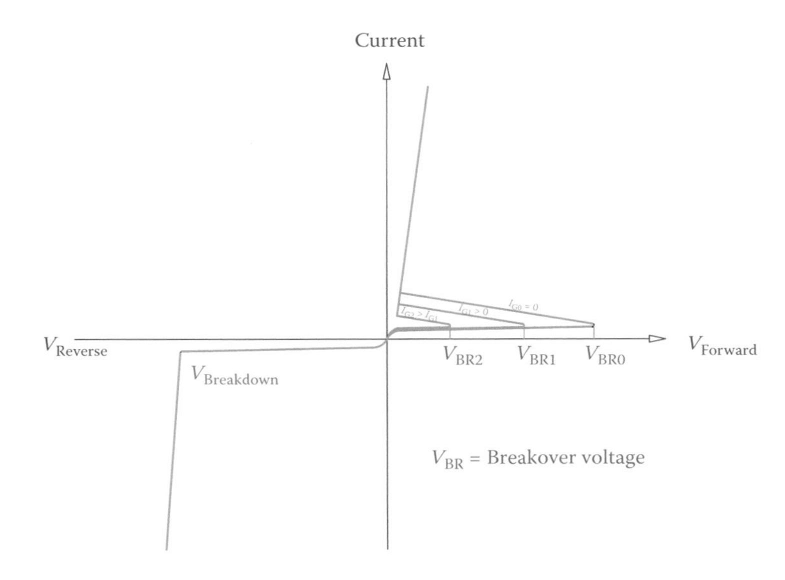

Figure 15.9 Characteristic curve of a silicon-controlled rectifier.

curve, this SCR has a similar performance except that with the current through the gate the behavior of the SCR changes. For this device a current flow from its anode to its cathode takes place only after the anode-cathode voltage has surpassed a minimum breakover voltage, but the breakover voltage is not fixed and can be controlled by the gate current I_G.

In Figure 15.9 the breakover voltage V_{BR0} is needed to start conduction in the SCR if the gate current I_G is 0. For a positive I_{G1} the necessary voltage to start the SCR is V_{BR1}, which is smaller than V_{BR0}. Similarly, if the gate current increases to I_{G2}, the device can start conducting at a lower breakover voltage V_{BR2}. In this way the gate provides a means for control of the current through the SCR.

A triac is equivalent to two SCRs as shown in Figure 15.9 put antiparallel to each other and with their gates connected together. In this sense, it has three terminals, and it can perform in both directions, like a diac. We may say that a triac is a diac with a gate. Its performance in each direction is similar to a controllable diac, and it is used for AC electricity. Its terminals are called MT_1 and MT_2, or anode 1 and anode 2, similar to a diac. The schematics of its structure and its symbol are shown in Figure 15.10. As can be seen, its symbol is the same as that of a diac, with an added part for the gate.

Figure 15.11 shows one situation of the characteristic curve of a triac in comparison with that of a diac, which is almost symmetric for current in both directions. In order for a triac to be turned on, either a rather high voltage must be applied to it or it must be turned on by applying a voltage to its gate (see Figure 5.9). By changing the gate voltage, the minimum voltage between its two anodes (breakover voltage) to turn it on can be controlled.

Because a triac can work in either direction for current, its switching action falls into one of the following four modes: note that when A_1 is connected to the mains positive terminal A_2 is connected to the negative terminal and vice versa.

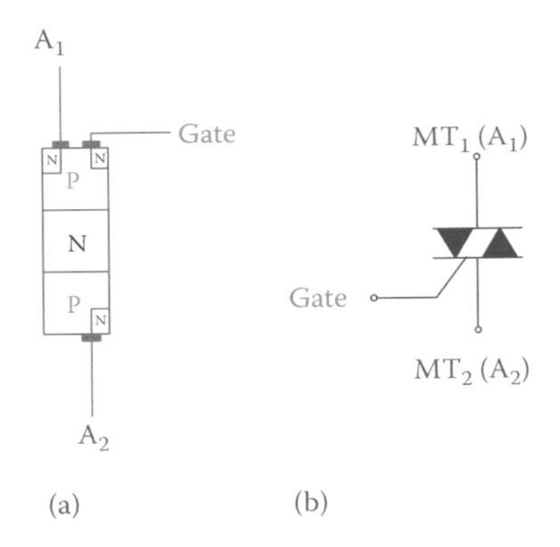

Figure 15.10 (a) Basic structure and (b) symbol of triac.

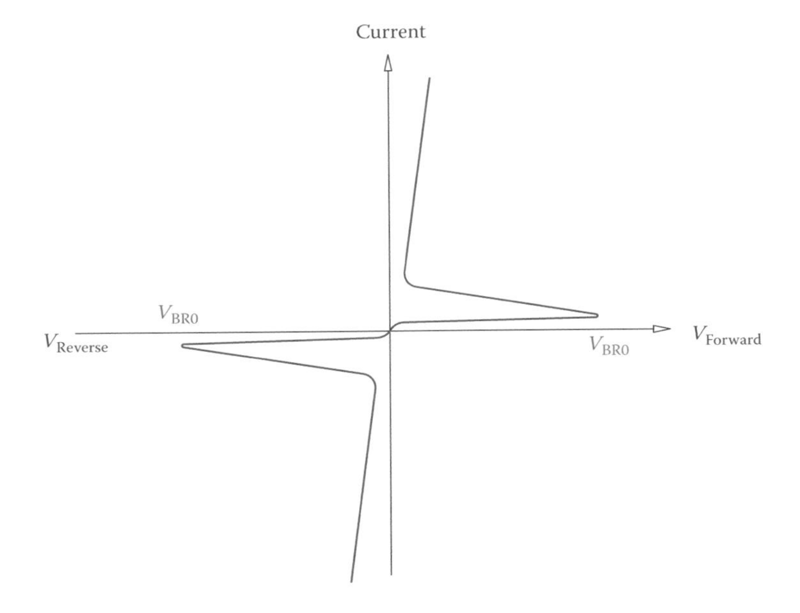

Figure 15.11 Characteristic curve of a triac.

1. A_1 is +, and gate signal is +
2. A_1 is +, but gate signal is –
3. A_1 is –, but gate signal is +
4. A_1 is –, and gate signal is –

A triac can work in all cases. Only the necessary gate signal is smaller in modes 1 and 4 because the voltage difference between anode 1 and anode 2 is similar to the voltage difference between the gate and anode 2.

Note that because the point of zero crossing of an AC waveform implies a zero current, a triac can turn off until the voltage across the anodes and the gate current allow conduction. In this sense, the general behavior of a triac can look as shown in Figure 15.12, for which the average voltage is smaller than if the current had a full waveform. This is the way that a triac can regulate the current in its circuit.

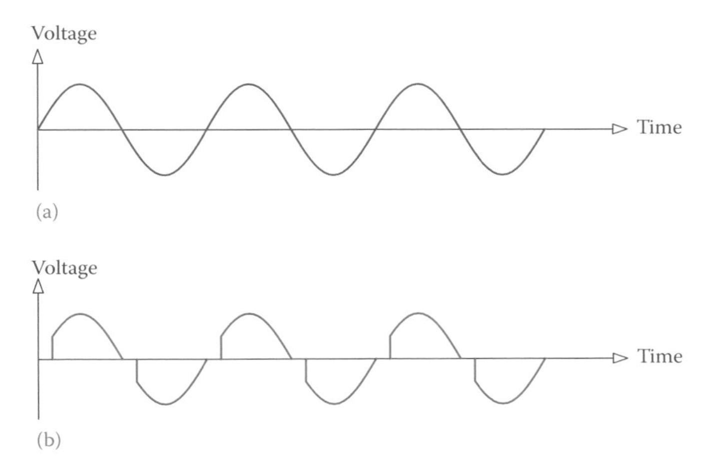

Figure 15.12 Effect of inserting a triac in an AC circuit. (a) Waveform without triac. (b) Waveform after triac is added.

Note that the waveform shown in Figure 15.12 is for a resistive load in the triac circuit. Phase shift is involved with inductive and capacitive loads.

Triacs are used for clean switching in place of electromechanical switches. They can be employed for soft starting of motors of small and medium size, particularly universal motors, lowering the inrush current in loads and many AC applications that need soft switching. High inrush current is possible for motors because their rotor initial rpm is zero, even for heating resistors and incandescent lightbulbs, because their resistance before they get hot is much lower. Among home appliances, the vacuum cleaner, refrigerator, washing machine, and so forth have universal motors that nowadays are controlled by triacs. A light dimmer also works with a triac, and the automatic voltage switch in many devices that must work with 120/220 V uses a triac.

There is a disadvantage with triac that they can turn on without being wanted by voltage spikes in the mains. Thus, this must be taken into account, and safety measures must be included for their usage in any electrical circuitry.

15.6.3 Testing a Triac

A triac can be tested by a meter. Either analog or digital meters may be used. Put the selector switch for resistance (ohm). Because a triac is an open circuit until it is turned on, a good triac shows a high resistance between its two terminals until it is triggered, when the high resistance falls to a small value or zero.

To test a triac, connect anode 2 of the triac to the positive (red) lead and anode 1 to the negative (black) lead. In fact, it does not matter how you connect, as long as the gate is free. The meter should read a large resistance or OL if a digital meter is used. For an analog meter it is recommended that first the meter is put on a large range ohm value, such as X100.

Now you connect the gate to anode 2 (if you know which one is anode 2) or to any of the terminals. The ohmmeter reading should go down to a small value (you may want to change to a smaller range on an analog meter for more precision). The reading should stay low even if you clear the gate connection.

Next, start over with swapping the red and black lead connections to the two anodes. The reading must show a high magnitude again. Make a connection between the gate and anode 2. Then the reading should fall and stay low, again, even if the gate connection is broken.

If the test results conform to the above description, the triac is good; otherwise, it is damaged. This test is good for typical triacs that do not have very sensitive gates. A small range (like X1) of an analog ohmmeter can damage a triac with very sensitive gate.

15.7 Presentation of Electronic Circuits Diagrams

Thus far in this book we have been following the common method for presenting an electrical or electronic circuit. This method is used for teaching because it facilitates the process of understanding the way components are in a circuit and the relationship between them (parallel, in series, having the same voltage, etc.). In industry, it is more customary to reduce the number of lines in a drawing by omitting the return line to the battery. This is particularly practiced for electronic circuits that usually contain more components. Instead, use is made of a common *ground*, where all the return lines to the power supply are shown as a connection to *ground*. In this way, the circuit of Figure 15.3 is depicted as in Figure 15.13. There is no difference between the two circuits, except the location of the switch, which is indeed better in Figure 15.13.

In Figure 15.13 the end of the 200 Ω resistor is connected to the positive side of a 24 V DC power supply, through the switch. The negative port of the power supply is connected to the circuit *ground*, which can be a chassis or a common point. Normally, the switch is not shown because it is a part of the power supply.

Hereafter, this way of presentation of circuits will be followed in the rest of the book.

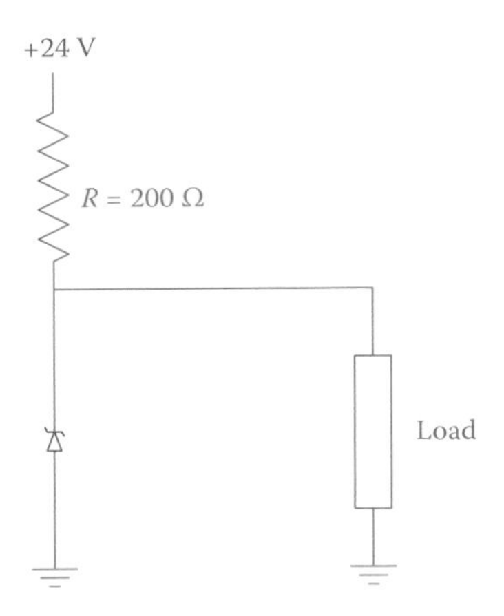

Figure 15.13 A more common way of exhibiting an electronic circuit.

15.8 Chapter Summary

- The main application of a zener diode is voltage regulation.
- A zener diode functions in reverse bias.
- If a zener diode is connected in forward bias configuration, it acts as an ordinary diode.
- The characteristic curve of a zener diode looks the same as that of an ordinary diode. The breakdown voltage is also called the zener knee.
- Voltage across a zener diode remains constant if a zener diode is subject to a higher (reverse bias) voltage across it. This constant depends on any particular diode.
- A load whose voltage is desired to be regulated is connected in parallel with a zener diode of the appropriate voltage.
- A zener diode cannot increase a voltage to regulate (the regulated voltage cannot be larger than the power supply voltage).
- Zener diodes come in various voltages and with different current ratings. A zener diode can be damaged if the current through it is over its current rating.
- Zener diodes can be put together in series to provide regulation for higher voltages (than their rating).
- Zener diodes can be put together in parallel to provide more current capacity for a circuit.
- A diac is like a diode for AC, but it conducts only if the voltage across its terminals reaches a certain level called the *breakover voltage*.
- When a diac conducts, it continues conduction as long as its current does not fall below the *holding current*.
- SCR stands for silicon-controlled rectifier. It behaves like a diode, but with a control capability through its gate.
- A triac behaves like a gated diac. It can be turned on by applying a voltage to its gate.
- A triac is used for current control of AC loads.
- Many domestic and small and medium size industrial appliances use triacs for soft starting.

Review Questions

1. What is the importance of a zener diode?

2. What is the main difference between a diode and a zener diode?

3. What is the main difference between characteristic curves of a diode and a zener diode?

4. How do you connect a zener diode in a circuit?

5. Can you use a zener diode instead of a diode? Why or why not?

6. What can stand against using a zener diode instead of a regular diode?

7. What is meant by voltage regulation?

8. Why is voltage regulation important?

9. Can you use a 7.5 V zener diode when the power supply is 6 V? Explain why or why not.

10. In the case raised in Question 9 what is the effect of the zener diode when inserted in a circuit?

11. What do diac and triac stand for?

12. What is the difference between a diode and a diac?

13. What is a SCR?

14. What are the three elements of a SCR?

15. What is the difference between a diac and a triac?

16. What is the main application of a diac?

17. What are the main applications of a triac?

18. Name three appliances at home that may use a triac.

19. Draw the symbols of a diac and a triac.

20. In what way are practical wiring diagrams different than what we have used so far?

Problems

1. How much is the current in R_2 in the circuit of Figure P15.1 if $R_1 = 450\ \Omega$ and $R_2 = 1000\ \Omega$?

2. How much is the current in R_2 in the circuit of Problem 1 if a resistor of 510 Ω is put in parallel with R_2?

3. Find the maximum value that R_1 can have in Problem 1.

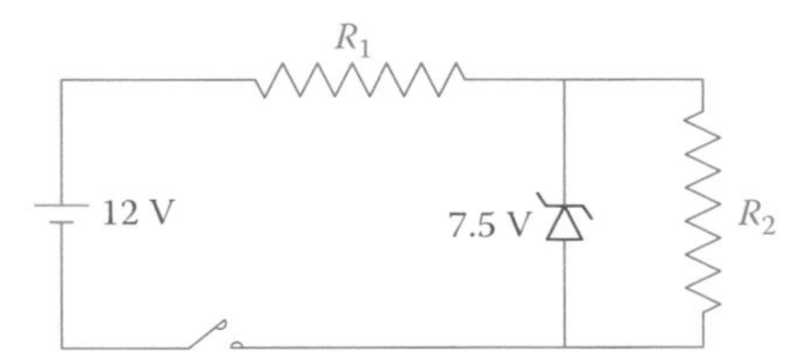

Figure P15.1 Circuit of Problem 1.

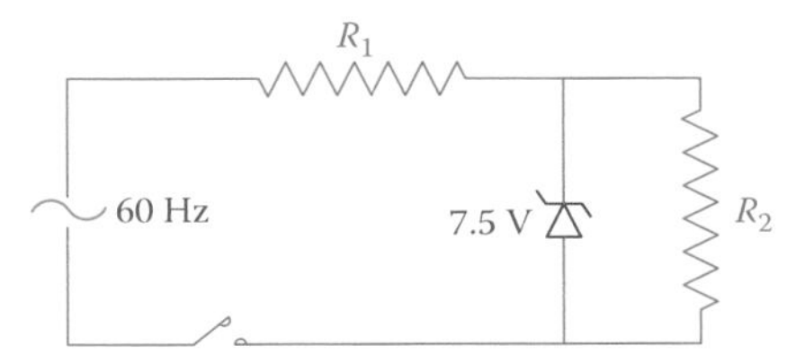

Figure P15.2 Circuit of Problem 5.

4. In Problem 1, find the total current if the battery voltage drops to 11 V?

5. Figure P15.2 shows a circuit whose source is AC. Plot the variation of the current in R_2 for one cycle of the alternating current.

16

Diode Rectifiers and Filters

OBJECTIVES: After studying this chapter, you will be able to

- Describe how AC is converted to DC
- Explain the quality of direct current as obtained from a rectifier
- Define ripple and how one can reduce ripples
- Understand the significance of filtering in rectified electricity
- Describe how one can add filters to a rectified waveform
- Explain what a half-wave rectifier and a full-wave rectifier are
- Define bridge rectifier
- Make a diode rectifier for single-phase AC
- Make a diode rectifier for three-phase AC
- Construct a power supply with filtered and regulated output
- Define various types of filters
- Define and use low pass filters, band pass filters, high pass filters, and band reject filters

New terms:
Attenuation, band pass filter, band reject filter, band stop filter, bandwidth, bridge rectifier, full-wave rectifier, half-wave rectifier, high pass filter, low pass filter, ripple, ripple factor, surge, surge suppressor, tank circuit

16.1 Introduction

A rectifier is a device to convert AC electricity to DC. Except batteries that are small sources of DC electricity, or the rechargeable ones that store electricity, most of the domestic devices that work with DC electricity use rectifiers. At the industrial level, there are industries that need DC electricity to run DC motors or processes that can function only with DC; they either must have their own generators or must obtain their DC requirement from AC sources by rectifiers.

Both single-phase and three-phase AC can be converted to DC. For domestic products and small applications, single-phase rectifiers suffice, but for large loads at the industry level, such as electroplating, electrolytic metal refining and high voltage direct current (HVDC) transmission and smaller ones like DC motor drives, three-phase converters are employed. Converter is a term that is used for both rectifier and inverter (inverter does the opposite job of providing AC from DC; discussed in Chapter 20).

In the simplest form a rectifier consists of a number of diodes, and therefore, we can call it a diode rectifier. And, as you can guess, there are other type(s), as will be discussed in Chapter 20. Diode rectifiers are simpler than the other types that use switching devices. Hereafter in this chapter the word diode is omitted for simplicity because the discussion implies that only diode rectifiers are concerned. The most common and widely used single-phase rectifier is the **bridge rectifier**, but **full-wave rectifiers** and **half-wave rectifiers** can also be used.

Bridge rectifier: Full-wave rectifier of AC using four diodes (for single phase) or six diodes (for three phase) to obtain direct current from alternating current.

Full-wave rectifier: Rectifier in which both half cycles of an alternating current waveform are rectified and delivered to the output as DC, as opposed to a half-wave rectifier in which only one half of each cycle reaches the output.

Half-wave rectifier: Simplest type of rectifier for alternating current, consisting of only one diode (for single phase) and three diodes (for three phase) that block the negative half cycle of AC, so that only the positive half cycles are passing to the output.

Ripple: Fluctuations in a rectified AC waveform. Fast fluctuations in an electrical value, such as voltage, that is supposed to be constant.

16.2 Half-Wave Rectifier

Figure 16.1 shows the schematic of a half-wave rectifier, which is the simplest and lowest quality type of rectifier. But, it demonstrates the principle of how rectifiers work. It consists of only one diode inserted in an AC circuit. As a result, for each full cycle of the alternating current the diode conducts only for half of the cycle but blocks the current for the other half. The outcome is shown in Figure 16.1c, which is a DC voltage as seen by a load.

The performance of a half-wave rectifier is very poor, and the DC voltage has a lot of fluctuation. The DC voltage here is, in fact, a series of half sinusoidal pulses (a pulse is a short duration DC signal). This variation of the voltage level, called a **ripple**, can be smoothed out to some extent by employing a filter. Note that the variation of voltage reflects in a load, depending on what the load consists of. Unless otherwise said, the load for this rectifier is everything that is connected in the circuit and is represented by R in Figure 16.1. All that we have learned about series and parallel components comes into view. An average value for the DC voltage, which lies somewhere between zero and maximum (peak values of the AC voltage), can be determined.

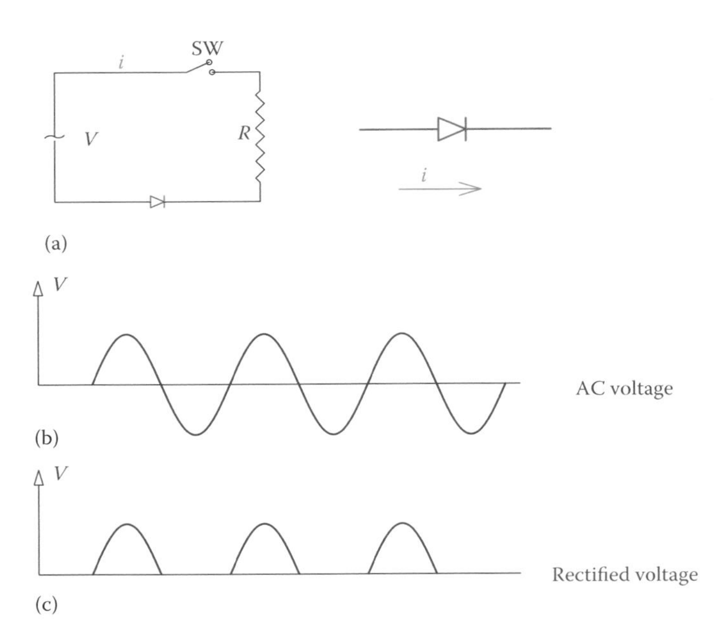

Figure 16.1 Half-wave rectifier. (a) AC circuit. (b) AC voltage across resistor without the diode. (c) Voltage across resistor when diode is added to the circuit.

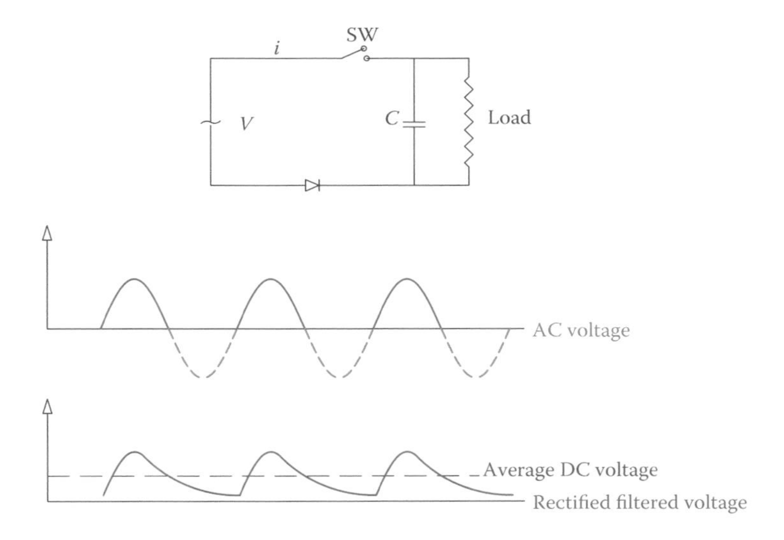

Figure 16.2 Schematic of a half-wave rectifier and the effect of a filter on the load voltage.

The most common filter is one or more capacitors connected across the positive and negative poles of the DC voltage, that is, in parallel with the load. Half-wave rectifiers are used only in applications for which a crude DC voltage is acceptable, like battery charging. A half-wave rectifier with filter and its output to the load (filtered output) are shown in Figure 16.2. As can be seen, as a result of the capacitor (filter), output voltage is not as before, meaning that the voltage does not vary between zero and the peak value of the alternating current. It varies between a minimum and a maximum. Variation of the voltage in the filtered output is between the peak value and a nonzero positive magnitude. The average DC value in this case is larger than in the unfiltered case. The larger the capacitor, the larger is the minimum value and the difference between the minimum and the maximum (the ripple) is smaller. The average DC value, as a result, is higher.

In a half-wave rectifier the ripple amplitude is rather high. The ripple frequency is the same as the frequency of the input AC signal.

> To convert DC to AC a half-wave rectifier eliminates the negative half in each cycle of AC.

16.3 Full-Wave Rectifier

The half-wave rectifier uses only a half cycle of an AC waveform. A full-wave rectifier has two diodes, and its output uses both halves of the AC signal. During the period that one diode blocks the current flow the other diode conducts and allows the current. The schematic of a full-wave rectifier is shown in Figure 16.3, where the unfiltered output voltage is also illustrated. The AC source is shown as a transformer. This is the reality

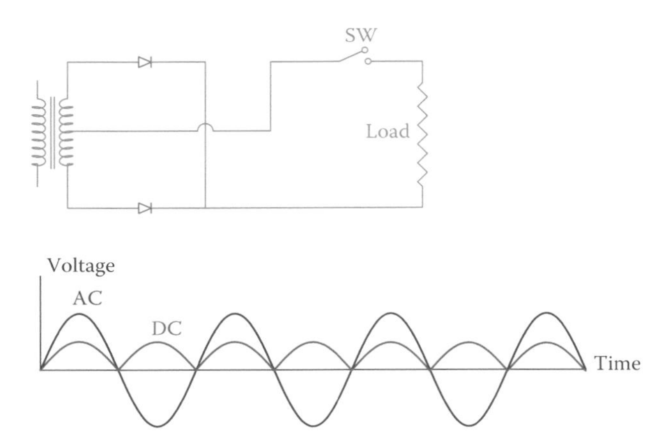

Figure 16.3 Schematics of a full-wave rectifier.

in many rectifiers. First, the voltage is lowered (or increased) to a desired value, and then it is sent to the rectifier.

As can also be observed from Figure 16.3, the two similar sides of the diodes are connected together and are connected to one side of the load. The other side (of the load) is connected to the center point of the transformer secondary winding. This implies that the transformer must have a center tap and it is required that access to this point is available. Furthermore, in a full-wave rectifier the DC voltage obtained corresponds to only half of the supplied voltage. Thus, for a direct conversion of the mains supply of 120 V to DC one needs a 1:2 transformer with a center tap. This is one of the drawbacks of a full-wave rectifier.

The average value of the unfiltered DC voltage obtained this way is only 45 percent of the effective voltage of the transformer secondary. In this sense, if the effective (rms) voltage in the transformer secondary of Figure 16.3 is 240 V, for instance, the average value of the rectified (DC) voltage is

$$DC_{AV} = (0.45)(V_{Eff}) = (0.45)(240) = 108 \text{ V}$$

Practically, this average value is not so useful except in simple and cheap battery chargers. This is because in most cases, in practice, a capacitor (or other filter; see Section 16.8) is used to reduce ripples. The effect of a filter added to a circuit (the simplest is a capacitor in parallel with the rectifier load) is shown later.

The average DC value of the output of a full-wave rectifier is twice as much as a comparable (having the same rectified pulse peak value) half-wave rectifier because it has twice as many pulses. The frequency of its ripples is also twice. The filtered output has much less ripple content than that of the half-wave rectifier. The frequency of ripples is 2 times the frequency of the mains.

> Ripple is the rapid fluctuations in the voltage of DC electricity obtained from rectified AC.

16.4 Bridge Rectifier

A bridge rectifier is similar to having two full-wave rectifiers together to obtain the full voltage of the source in the output instead of half. In addition to the voltage relationship, another advantage is, thus, no need for a center tap point. It uses four diodes as shown in Figure 16.4. Pay attention to the way the four diodes are connected together and to the circuit. At each half cycle, two of the diodes conduct and two of them block the current. The resulting rectified waveform, as seen by a load, is similar to those shown for full-wave rectifier, with the exception that the voltage this time is double of that with a full-wave rectifier, all conditions being the same. Figure 16.4 shows the direction of current for one half cycle. Note that we have used the common way for showing electronic circuits; thus, the current path is completed through the ground.

> The current through a load connected to a full-wave rectifier or a bridge rectifier flows in one direction only, as if all negative half cycles of the alternative current are converted to positive.

If you trace the flow of current, you will observe that no matter which side of the transformer is at a higher voltage the current through the load is always in one direction. That is, it is direct current. Normally, a capacitor is used to filter the ripples. Always for all rectifiers the higher the capacitance of this capacitor, the better is the filtering effect. The average voltage of the unfiltered DC voltage can be determined from

$$V_{AV} = \frac{2V_{peak}}{\pi} = \frac{2\sqrt{2}V_{Eff}}{\pi} = 0.90\,V_{rms} \tag{16.1}$$

Thus, for a 120 V effective AC voltage the average of the unfiltered output DC voltage is 90 percent of 120 V, that is, 108 V. Nevertheless, after inserting a capacitor this voltage can increase. For a pure resistive load the filtered DC voltage after putting a capacitor in parallel with the load is

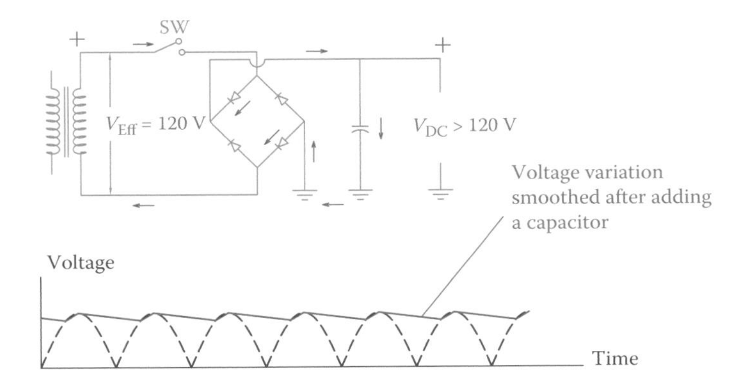

Figure 16.4 Schematics of a bridge rectifier.

$$V_{DC} = V_{peak}\left(1 - \frac{1}{2fRC}\right) \tag{16.2}$$

where R is the load resistance, C is the capacitance of the filter, and f is the frequency of the ripples. This equation shows that a larger capacitor, or a larger load, gives a much smoother DC volt; it also indicates that with the same capacitor a better result is obtained if the frequency of the ripples is higher (e.g., comparing single-phase bridge rectifier with three-phase bridge in Section 16.6).

The effect of filters can be better observed from Figures 16.5 and 16.6, which show the comparison of the filtered and the unfiltered voltage for a particular case. These figures correspond to 14 V AC effective voltage (40 V peak-to-peak) shown on an oscilloscope. Reading on the left is the effective value of the ripples and reading on the right is the average DC value. (Note that always some percentage of the voltages is dropped in the diodes.) Figure 16.6 illustrates the same rectified waveform. The ripples voltage has dropped to 0.4 V, whereas the DC voltage is 16.7 V for the same loading condition.

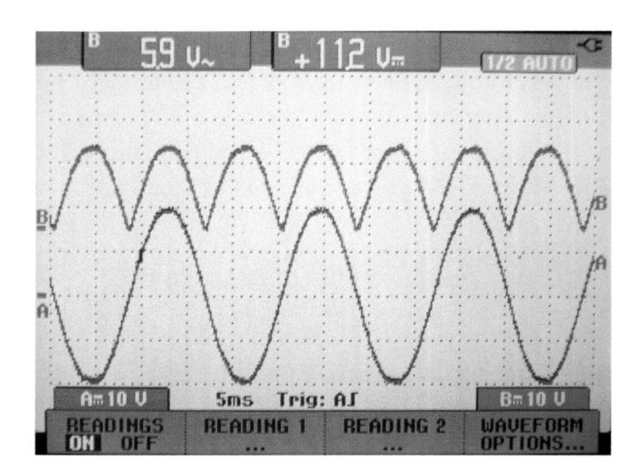

Figure 16.5 Unfiltered output of a bridge rectifier.

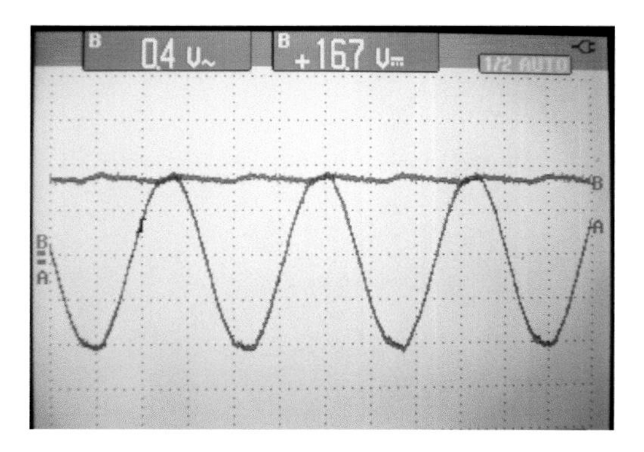

Figure 16.6 Filtered output of the DC output shown in Figure 16.5.

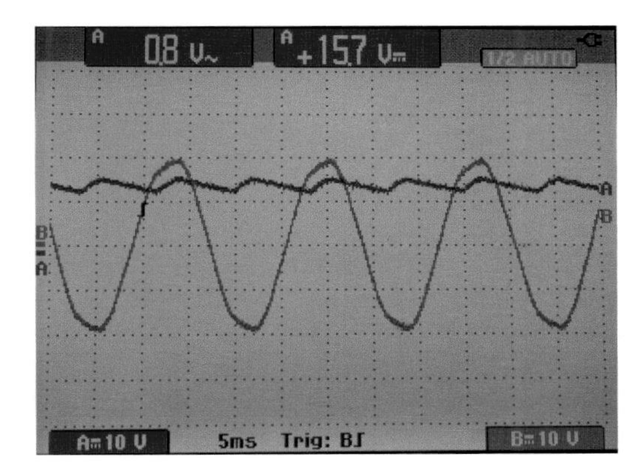

Figure 16.7 Effect of increased load on the DC voltage shown in Figure 16.6.

How much the average DC voltage is and how much ripple remains in the rectified DC depend on the nature of the load, its power consumption (circuit current), and the filter (capacitance of the capacitor), as can be determined from Equation 16.2. Figure 16.7 depicts the effect of increasing the all resistive load (increase in current) for the same case in Figure 16.6.

A bridge rectifier is practically the most common and most frequently used rectifier for single-phase AC. Nowadays it is possible to buy the four diodes integrated together in one package, as shown in Figure 16.8. They come in different shapes. It can be larger than a single diode, but the size depends also on the voltage and current rating (power). It has two input terminals to connect to AC and two output terminals, which provide the DC electricity. Any capacitor for filtering and the load are connected to the DC side.

In practice, the power rating of a rectifier and the maximum voltage are the main consideration for selection of a proper rectifier. As in DC, power is the product of voltage and current. Thus, for a particular application the rectifier diodes must stand the applied voltage and the circuit current.

When diodes are used in a rectifier in each half cycle of the AC signal, they are subject to a negative voltage across them when they are reverse biased. A diode must withstand the peak inverse voltage (see Chapter 14).

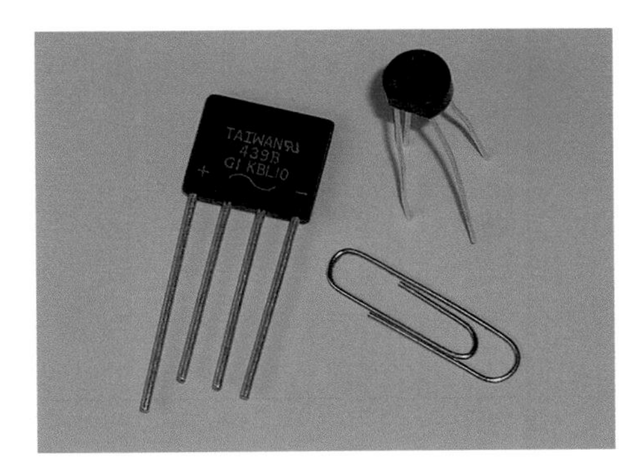

Figure 16.8 Bridge rectifier integrated circuits.

For a single-phase bridge rectifier this voltage is around 1.57 of the DC voltage.

16.5 Ripple and Filters

Ideally, the output of a rectifier must be a perfect DC and look as smooth as a straight line on an oscilloscope, similar to the output from a battery. This implies no fluctuation in the value of the voltage. Because of the nature of rectification, nevertheless, the unevenness of the rectified sinusoidal waveform is present. The percentage of this ripple with respect to the voltage value, however, can be small or large. And the smaller this percentage is, the better it is. For example, the filtered output of a bridge rectifier in Figure 16.5 has a much smaller ripple than the unfiltered output, but it is not as smooth as DC from a battery. The measure of quality of the DC output of a rectifier can be defined in the terms of the ratio of the variation in the voltage (ripple) to the voltage value. For example, if for 24 V DC the ripple is 1 V, (i.e., the voltage varies between 23.5 and 24.5 V), then we may calculate this ratio to be 1/24, or about 4 percent. Alternatively, one may express this ratio in terms of the peak value of the AC voltage, in case it is more appropriate, like 2 V ripple on 40 V peak.

Recall from Chapter 13 (see Section 13.5.6) that one of the functions of filters is to remove unwanted signals. Here, the ripple is an unwanted part of a DC signal that we want to remove. As said, a capacitor is the most common type of filter for this case. Also, capacitors are relatively cheap. There are, nevertheless, other types of filters, a few of which are discussed here.

As you remember, when a capacitor is inserted in a circuit (see Section 8.6), it charges and discharges, based on the polarity of the voltage at its terminals and the previous charge in the capacitor. In doing so, a capacitor introduces a delay in a voltage to develop or change. It all depends on the capacitance of a capacitor and the frequency of a signal (how fast it wants to change).

In the same way that a capacitor acts as a blocking agent for a change in voltage an inductor was a blocking means for a change in current. In this sense, an inductor can also be used as a filter. In fact, a combination of capacitor plus an inductor and sometimes a resistor construct the few types of filters, as we discuss here.

Figure 16.9 illustrates the schematic of a bridge rectifier with an inductor as filter. Note that contrary to the capacitor filter that was in parallel

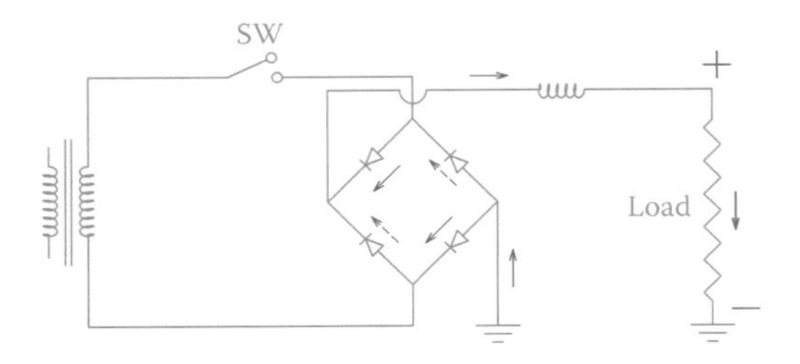

Figure 16.9 Inductor as a filter for the rectified AC.

with load, an inductor filter is in series with the load. Normally, a capacitor is cheaper than an inductor, and therefore the majority of single-component filters use a capacitor.

Other types of filters consisting of capacitors and inductors are L-type filters, T-type filters, and π-type filters. The reason for these names is the resembling of the branches containing the filter elements to an L, T, or the Greek letter π. Figures 16.10 through 16.12 depict different forms of each category. In these figures the diodes (the rectifier part) are not shown. The two terminals for each filter are connected to the output of the rectifier. Note that in T-type and π-type filters a resistance can be employed instead of an inductor. We see more of this later in Section 16.8.

A filter is sensitive to frequency. It can remove all signals of some ranges of frequencies. It attenuates the input signals having those frequencies.

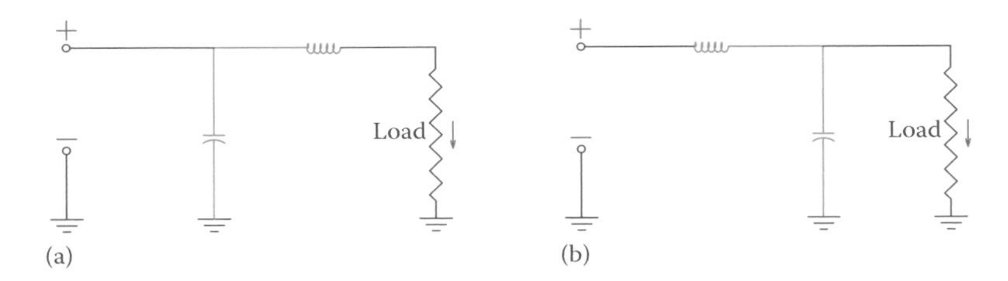

Figure 16.10 L-type filters. The inductor and capacitor form a letter L. (a) Inductor is put after capacitor. (b) Inductor is put before capacitor.

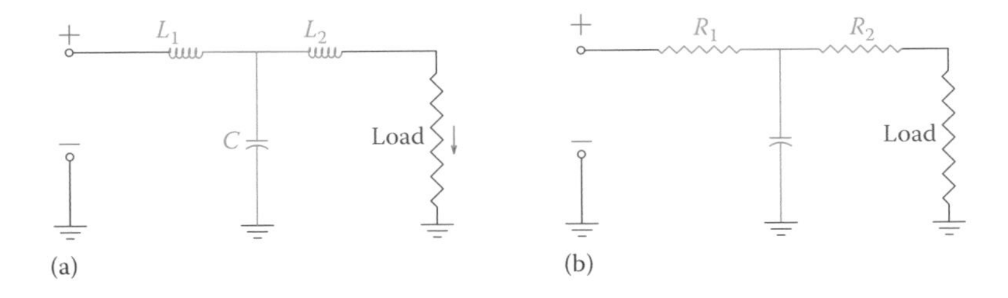

Figure 16.11 T-type filters. The three filter components form a letter T. (a) Two similar inductors are used. (b) Two similar resistors are used.

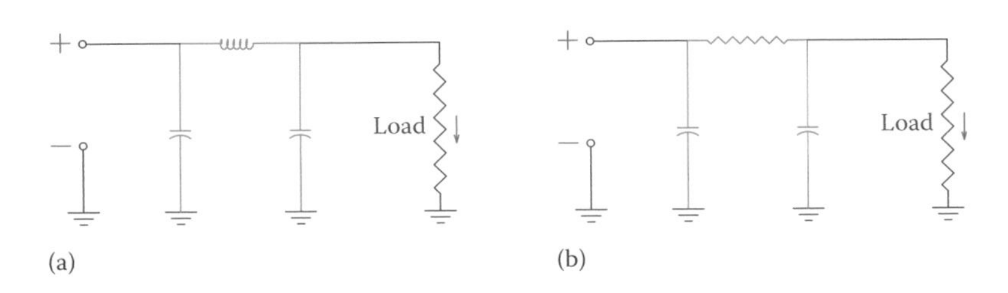

Figure 16.12 The three filter components form the π-type filters. (a) LC filter. (b) RC filter.

16.6 Three-Phase Rectifiers

Full-wave bridge rectifier: Same as bridge rectifier.

At the industrial level, where more power is required, a three-phase rectifier is employed. In this way, power is taken from all the three phases, but a single DC supply is obtained. Figure 16.13 depicts the diode arrangement and connections of the most common three-phase rectifier. There are other arrangements, too. To make a distinction among other designs, the design in Figure 16.13 is referred to as a bridge rectifier, full-wave rectifier or **full-wave bridge rectifier** by different people. For this configuration, connection to the mains is provided by three-phase lines and no provision is available for a neutral line, similar to a delta connection. This arrangement uses six diodes. Other arrangements are also possible, which use only three diodes or 12 diodes, but the three-phase rectifier in Figure 16.13 is more common and more efficient. We do not consider other designs. Similar to all other rectifiers, the output voltage can be filtered for ripple reduction/elimination. Also, as discussed in Section 16.7, the output voltage can be regulated for better quality.

The average DC voltage from a three-phase rectifier under discussion is approximately 0.95 times the peak value of the line voltage. That is, $0.95 \times \sqrt{2} = 1.35\times$ the nominal (effective) voltage of the supply line voltage. The DC voltage thus is larger than the AC supply peak voltage, as shown in Figure 16.13b. At each instance the voltage difference across the load is determined by the most positive and most negative voltages at the three-phase lines. For example, during the period between 0 and 30° of each cycle, according to the figure shown, the voltage in phase C is the most positive and that in B is the most negative. The load voltage, therefore, is the total voltage difference between the positive value of C and the negative value of B. All of these values are represented by the vertical lines shown at 10° intervals, for 360° (one cycle) in the figure. These line segments are repeated at the bottom of the figure, showing the DC voltage.

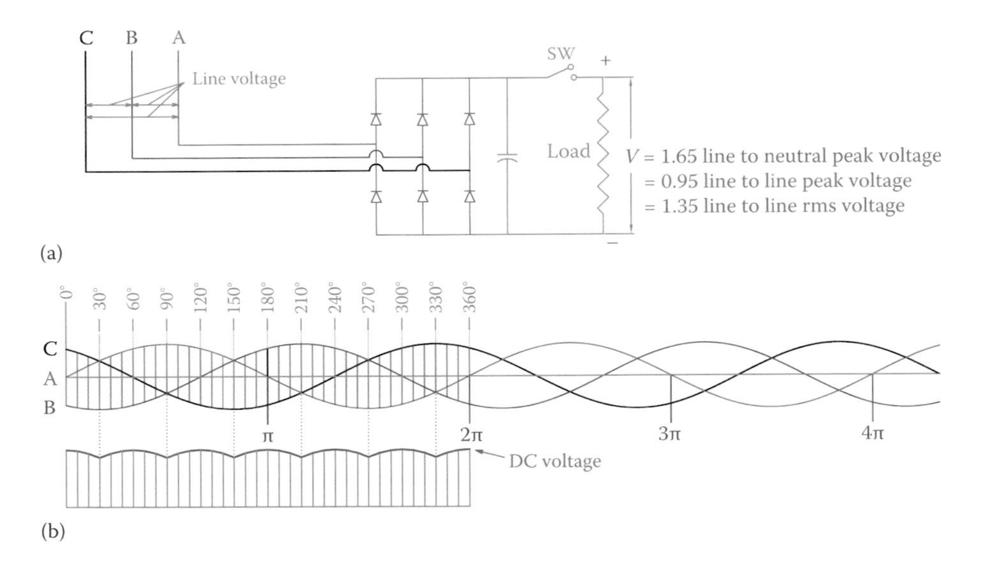

Figure 16.13 Schematics of a three-phase rectifier. (a) Arrangement and direction of diodes. (b) Resulting DC electricity at the rectifier output.

It is easy to see from Figure 16.13 that the frequency of ripples in a three-phase rectifier is 6 times the frequency of the mains supply. Also, compared to the single-phase bridge rectifier, the **ripple factor** is much smaller. Three-phase rectifiers are revisited in Chapter 20, where more powerful and versatile rectifiers and power converters are studied.

Ripple factor: The ratio of the AC content (peak-to-peak value in volts) of a rectified voltage to its DC content (which is the average value of the rectified voltage). The smaller this ratio is, the better is the rectified value.

In a DC voltage provided by a three-phase rectifier the ripple frequency is 6 times the supply frequency.

In a three-phase rectifier the peak inverse voltage that appears across each diode is equal to the peak of the line (indicated as line-to-line in Figure 16.13, for consistency) voltage. In practice, a safety factor of 2.5 is also considered and for a line voltage v diodes that can withstand $2.5v$ are used.

16.7 Small DC Power Supplies

Because all electronic devices can only work with DC electricity, almost all of them have a built-in power supply or they have an external power supply that provides DC electricity at the required voltage and power. Examples are plenty, at home and in industry. For instance, radio, television, cellular phone, computer, electronic guitar, and so on, all are powered by AC line, whereas AC is converted to DC.

The basic component of a power supply is a rectifier (single phase for all the above applications) of the types discussed earlier, which is connected to the secondary of a transformer. The transformer supplies the necessary AC electricity at the correct voltage.

In addition to the rectifier, a power supply is most likely equipped with a filter of some type, and it may have a voltage regulator, too. Because the line voltage may have some fluctuation, and normally the functions of electronic devices can vary with temperature, the output of a basic rectifier may not be sufficiently good and stable. To keep the output voltage at a desired level, a power supply may be equipped with a voltage regulator. In this way, the operating condition (e.g., louder music or quieter music) and the ambient temperature have less effect on the functioning of a device.

Voltage regulation was discussed in Chapter 15, and now you know how a zener diode is employed for this purpose. All power supplies do not necessarily have a voltage regulator. Usually, the more sensitive devices and those which are more expensive and have better performance need to have a voltage regulator.

Many devices do require protection from overvoltage and overcurrent. Sudden changes in voltage level in the form of spikes are very common in the electricity line due to various reasons, including the on-off transition of larger loads. A power supply may have provisions for protection of itself as well as filtering these unwanted spikes from being passed to its load. These protection provisions become an integrated part of a power supply. The unwanted spikes are considered as noise. They are of short

Surge: Sudden and short-term increase of voltage in a line, for instance, during a few seconds after a large load is taken off the line.

Surge suppressor: Device or arrangement to stop or reduce the effect of a surge.

duration, which is, as you know, associated with high frequency like most other noises.

Also, when power is initially applied to a device, the inrush current through the device can be several times higher than the normal current. This initial higher current is called **surge**. A voltage surge in a line can also originate from other loads on the line. A power supply may also be equipped with a **surge suppressor**, which is a circuit to absorb the higher current and not pass it through to the load. Figure 16.14 illustrates a simple adaptor for charging a battery (in a particular device). Compare it with the power adaptor in Figure 16.15 that has a number of extra features such as overvoltage and overcurrent protection as well as fuses.

Remember that a capacitor reactance becomes smaller as the frequency increases. Consequently, at relative higher frequencies a capacitor can act as a conductor. For protecting an electronic device, or a component in a device, from overvoltage at relative higher frequencies a capacitor is often used in parallel with the item. Figure 16.16 shows an example. In this circuit all protective capacitors are in parallel with various components and

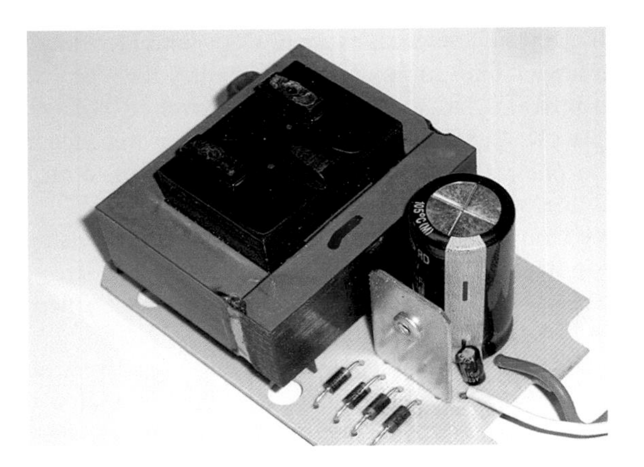

Figure 16.14 Example of a simple DC power adaptor.

Figure 16.15 Example of a power adaptor with voltage regulation and protection circuits.

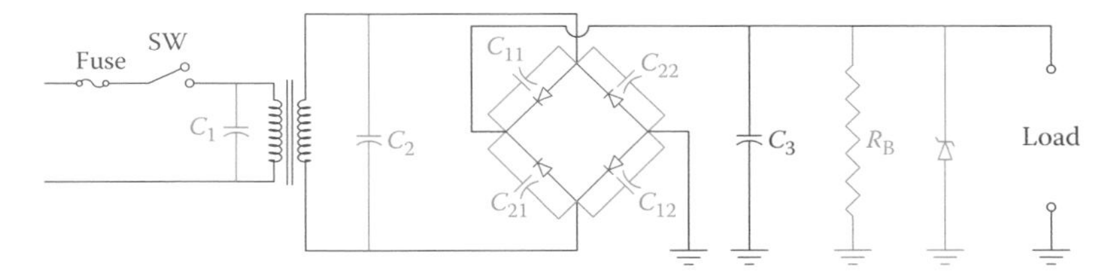

Figure 16.16 Protection and voltage regulation arrangements in a power adaptor.

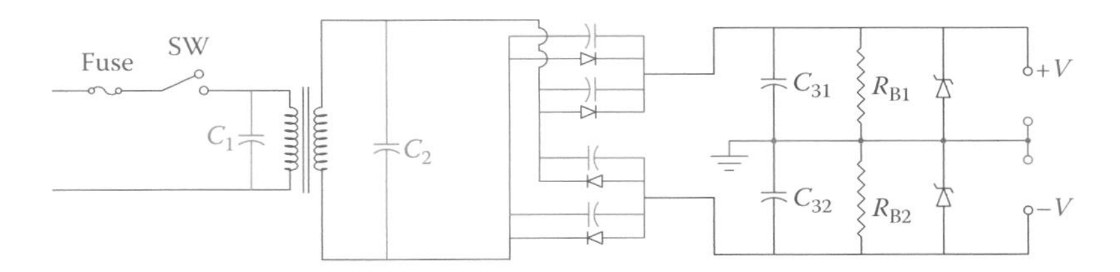

Figure 16.17 DC power supply with positive and negative voltage.

bypass any higher-frequency voltage that might arrive. C_1 is to protect the primary winding, and C_2 does the same for the secondary winding. C_{11}, C_{12}, C_{21}, and C_{22} are to protect the diodes. Notice that R_B is added to this circuit, which is in parallel with the load. This resistor serves as a permanent load to the power supply. It is customary to have such a permanent load because it

1. Provides a discharge path for the filter capacitor.
2. Reduces the surge current and protects the diodes.

This resistor has a magnitude much larger than that of the load.

In practice, the physical arrangement of the diodes does not need to be as shown in Figure 16.16. Here, for the sake of simplicity and similarity with the previous figures for a bridge rectifier, we have shown it this way. In Figure 16.17 a bridge rectifier is shown differently. But notice that the way the four diodes are connected still represents a bridge rectifier. The difference between Figures 16.16 and 16.17 is that in the latter the capacitor filters, the zener diodes and resistor R_B are divided into two, and the middle point is grounded. In this way, a negative voltage (with respect to ground) is created and a power supply arranged this way can provide both positive and negative voltage. This is necessary for certain circuits.

16.8 More on Filters

The filters discussed so far have been used to take care of removing the ripple and smoothing the output voltage from a rectifier. There are other aspects of filters that we did not consider but were pointed out in Chapter 13. We are going to elaborate on these features.

The frequency of ripples in each of the preceding cases is constant. What if a general signal is made out of components at different

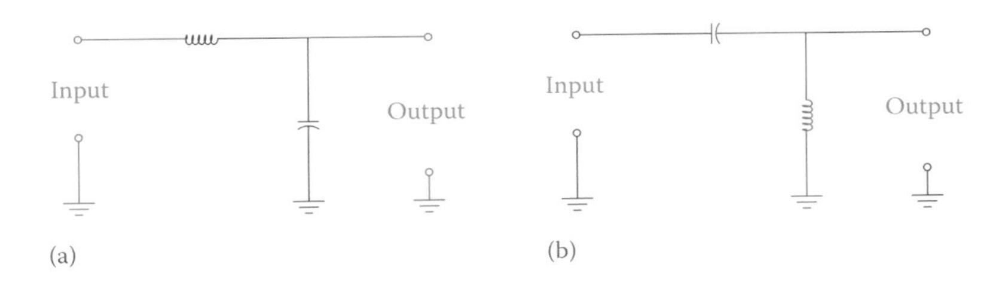

Figure 16.18 Interchanging L and C in an L-type filter. (a) Capacitor is parallel with load. (b) Inductor is parallel with load (load is connected to the output).

frequencies? Examples of such a scenario are plenty. Speech and sound are good examples. The audible range of sound for a human being can be between 100 Hz and 20 kHz (though for most people the more realistic range is between 500 and 5000 Hz). If sound is converted to an electric signal through a microphone, the same frequencies in the sound appear in the electric signal. But, reproducing a recorded signal is not immune to noise from various sources.

Recalling the behavior of a capacitor or an inductor, their apparent resistance (reactance) changes with frequency. An inductor exhibits more reactance as the frequency increases, whereas for a capacitor, the opposite is true (a review of Chapter 8 before proceeding further on this section is recommended). We now consider the filter part of the previous circuits in this chapter and treat a filter as an input-output device. We further consider that we may have different ways of combining components to construct filters, as exemplified by the circuits in Figures 16.10 and 16.11.

Let start with the filter in Figure 16.10b, which is repeated in Figure 16.18a, and compare its behavior with that in Figure 16.18b. We are going to use the terms high- and low-frequency signals but bear in mind that high and low are relative to each other and do not refer to any particular value. You may assume that a high-frequency signal has a frequency 10 times larger than that in question.

16.8.1 High Pass and Low Pass Filters

In the circuit of Figure 16.18a for low-frequency signals the resistance exhibited by the reactance of the inductor is low, whereas for high-frequency signals this resistance is high. For the capacitor it is the reverse; for high frequencies the capacitor acts as a shorted connection (low resistance), but for low frequencies it has a relatively sizeable reactance value. The result of this behavior of the capacitor and the inductor is that if a combined signal consisting of high- and low-frequency components appears at the input terminal of this filter, the low-frequency components are passed to the output, whereas the high-frequency components are shorted to the ground and do not pass.

On the contrary, for the circuit shown in Figure 16.18b the opposite happens. The low-frequency signals are encountered by high resistance on the way because of the reactance of the capacitor. Additionally, the inductive reactance of the inductor is low, which leads to a low voltage across the output terminals. Therefore, the low-frequency signals are blocked from reaching the output, but the high-frequency signals pass. On the basis of their performance and the action on an input signal the filter in Figure 16.18a is called a **low pass filter**, and the one in Figure 16.18b is called a **high pass filter**.

Low pass filter: Type of electric filter that blocks the higher frequencies in a signal consisting of various frequencies and allows the lower frequencies to pass to the output.

High pass filter: Type of electric filter that blocks the lower frequencies in a signal consisting of various frequencies and allows the higher frequencies to pass to the output.

> A filter selects or removes signals based on their frequency and based on the filter specification.

Another way of looking at what happens in the filters in Figure 16.18 is to consider them as voltage dividers. The input voltage is divided into two, one part across the component on the horizontal segment and one part across the component shown on the vertical segment and grounded. The output has the same voltage as the component on the vertical segment, which is a capacitor in the first filter and an inductor in the second filter. For high frequencies the output of the first filter is shorted; thus, this is a low pass filter. In a circuit as shown in Figure 16.18b the capacitor is shorted for high frequencies and the whole input voltage appears across the inductor; thus, it is a high pass filter. The inductor exhibits a large reactance to the high-frequency signal and does not short it to the ground.

Filters of other types exhibit similar behavior. Other high pass filters are shown in Figure 16.19, and other low pass filters are shown in Figure 16.20.

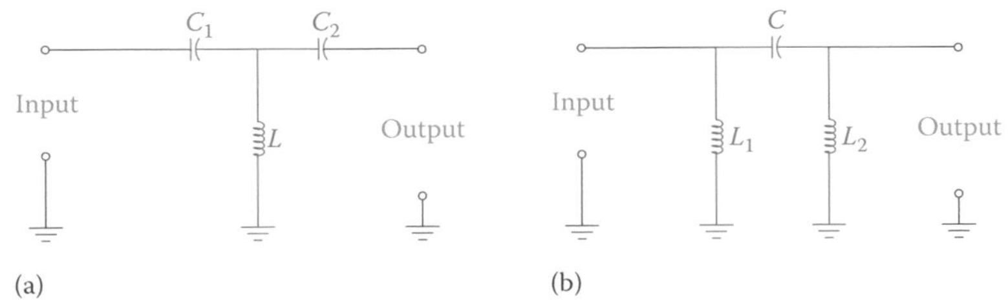

C_1 C_2 Input L Output (a)

C Input L_1 L_2 Output (b)

Figure 16.19 Other high pass filters. (a) T-type high pass filter. (b) π-type high pass filter.

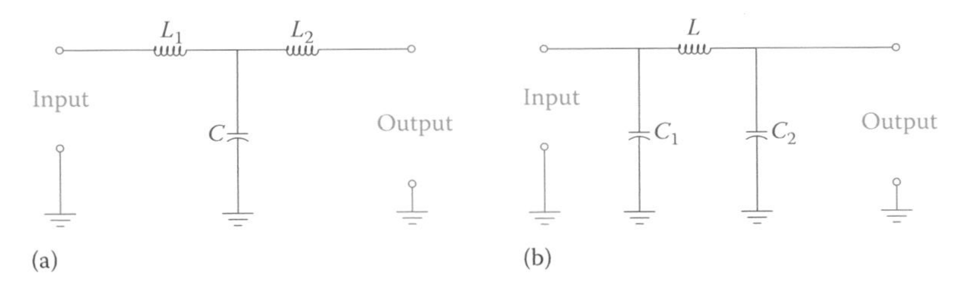

Figure 16.20 Other low pass filters. (a) T-type low pass filter. (b) π-type low pass filter.

The performance of any filter in passing high- or low-frequency signals and rejecting (or suppressing) the other ones can be shown by a curve. There is never a sharp line separating which frequencies will pass and which ones will be blocked because the distinction between high and low is relative and the transition is gradual. Figure 16.21 illustrates the transition. It shows the output voltage with respect to the change in frequency for a low pass filter. Similarly, Figure 16.22 depicts the behavior curve of a high pass filter. These two figures exhibit the typical forms of the associated curves. There are no numbers given either for the frequency or voltages. A particular filter (with values of the capacitance and inductance given) has a specific curve that reflects the frequencies that are on the *pass* side, those on the *suppress* side, and the ones in the transient region.

As we can see, a filter may never eliminate a signal by 100 percent. It reduces the strength or power of a signal by reducing its voltage or current (and thus the power associated with it). In the figures the voltage was used, but depending on a circuit, it could have been the current. The act of reducing the effect of a signal (i.e., suppressing a signal) is called **attenuation**. A high pass filter, for instance, has no or little attenuation of the high-frequency signals, but the low-frequency signals are attenuated to a great degree.

Attenuation: Reducing the intensity of an electric signal to match measurement and other devices.

You notice that in all the aforementioned filters there are two elements that stand between the input and the output. Those are shown as "1" and "2" in Figure 16.23, which represents a general form of filters. For passing

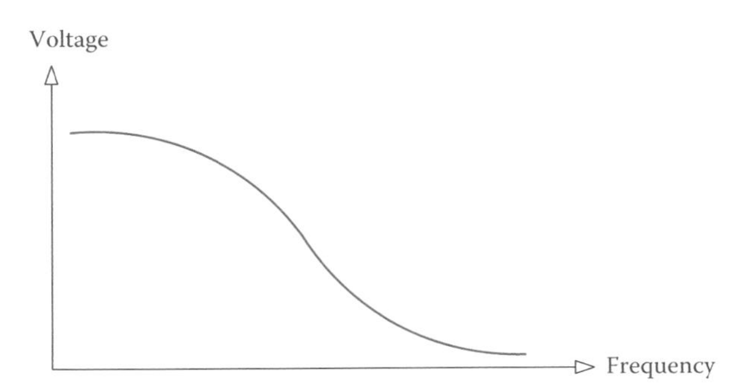

Figure 16.21 Typical characteristic curve of a low pass filter.

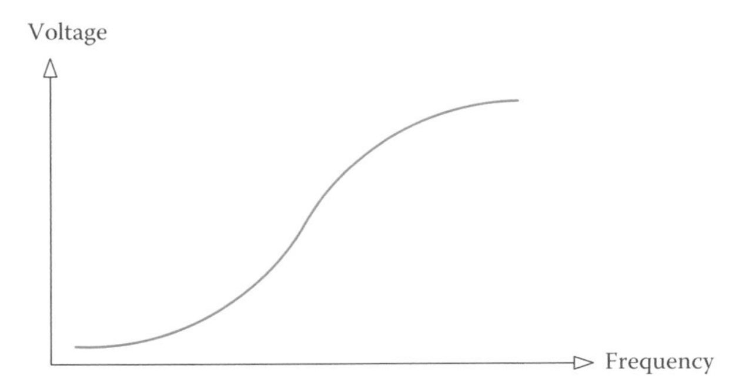

Figure 16.22 Typical characteristic curve of a high pass filter.

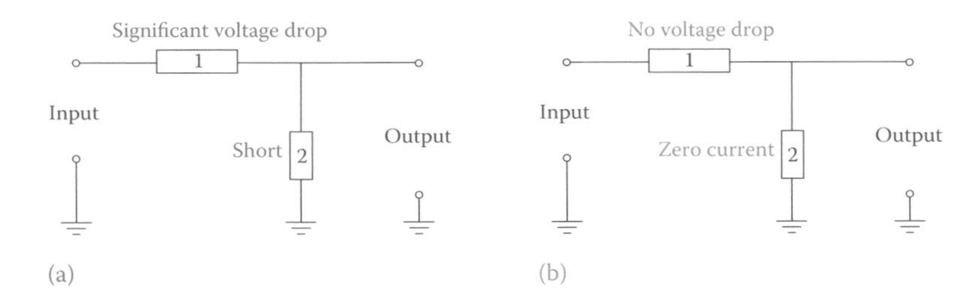

Figure 16.23 General characteristic for (a) block a signal or (b) pass a signal behavior of filters.

a signal from input to output, ideally, element 1 must have no resistance and element 2 must have zero current. On the contrary, for blocking a signal, element 1 must have significant voltage drop and element 2 must be short circuit (thus, no voltage across it).

16.8.2 Band Pass and Band Reject Filters

In addition to high pass and low pass filters, there are two more classes of filters: band pass and band reject. When inside a circuit, a **band pass filter** exhibits small or no resistance against current for signals within a finite range of frequencies and high resistance for other frequencies. As a result, all signals with frequencies in that range pass to the output. All other signals with frequencies outside of the range are affected, and their voltage level is reduced.

The opposite of band pass filter is **band reject filter**, which lets all signals with frequencies outside a particular range pass to the output, but those signals having frequencies within that range be blocked from reaching the output.

Figure 16.24 depicts a characteristic curve of a typical band pass filter, and Figure 16.25 shows that of a band reject filter. The range of frequencies involved is called the **bandwidth** (BW) of a filter. The smaller

Band pass filter: Type of electric filter that lets signals in a range of frequencies pass from the input side to the output. All other signals with frequencies outside this range are blocked.

Band reject filter: Opposite of band pass filter. All the signals with frequencies in a defined range are blocked from reaching the output.

Bandwidth: Range of frequencies in a band pass or band reject filter.

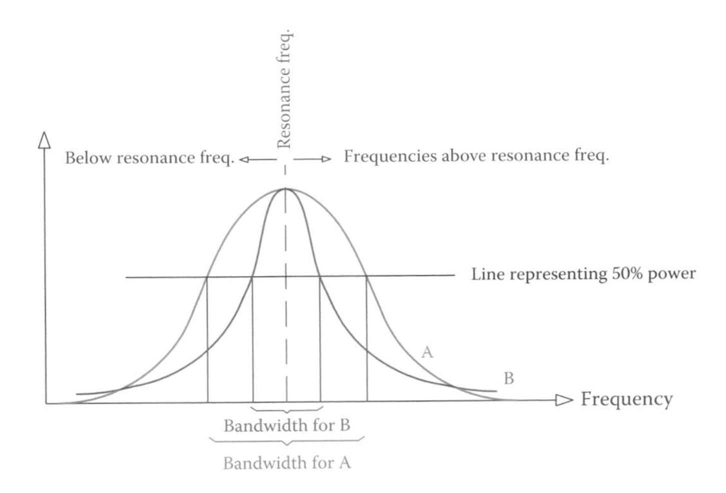

Figure 16.24 Typical characteristic curve of a band pass filter.

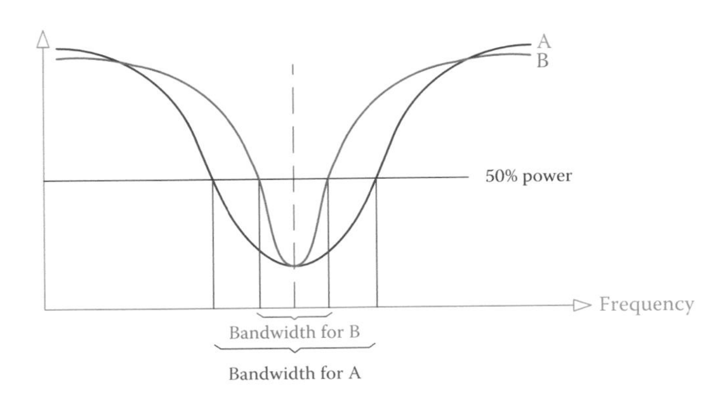

Figure 16.25 Characteristic curve of a band reject filter.

a bandwidth is, the higher the selectivity of a filter in passing the desired frequencies or rejecting the unwanted signals.

Band pass and band reject filters work based on the resonance frequency of a capacitor-inductor pair. We recall from Chapter 8 (see Section 8.13) that for an inductor and a capacitor, either in parallel or in series, there is a resonant frequency that can be determined from the values of inductance L and capacitance C. For each circuit we have observed a number of properties. A parallel LC circuit at resonance, especially when used in filters, is known as a **tank circuit**. At resonance, for a capacitor and an inductor forming a tank circuit the total current in their main circuit (not in the tank circuit that has a large local current) becomes minimum (see Section 8.13.2), meaning that the impedance of the branch containing the tank circuit is maximum. As a result, the voltage drop across the tank circuit is high.

Tank circuit: Part of an electric circuit consisting of an inductor and a capacitor in parallel with each other, used in filters.

On the contrary, at resonance the voltage drop across the series capacitor-inductor is minimum (see Section 8.13.1), implying that they (the capacitor and inductor together) have negligible impedance for the current flow. If a set of series capacitor-inductor and parallel capacitor-inductor combination having the same resonance frequency is used in a filter structure (L-type, π-type, and T-type), as shown in Figure 16.26, then the filter affects a band (range) of frequencies. All the filters in Figure 16.26 have the same behavior of a passing filter (based on Figure 16.23b) for a band of frequencies.

Figure 16.26 represents three designs for a band pass filter. If the parallel LC and the series LC blocks are swapped the result is a band reject filter. Figure 16.27 shows three different designs of band reject filters. In all circuits shown in Figures 16.26 and 16.27 the resonance frequency of the capacitor and inductor in parallel must be the same as the resonance frequency of those in series, so that the arrangement behaves as a band pass or band reject filter. A band reject filter is also called a **band stop filter**.

Band stop filter: Another name for band reject filter.

Note that the filters shown in Figures 16.26 and 16.27 are not the only designs for band pass and band reject filters. If the components in the horizontal segments (in L, T, and π) are removed, the remaining part still acts as a band pass or band stop filter.

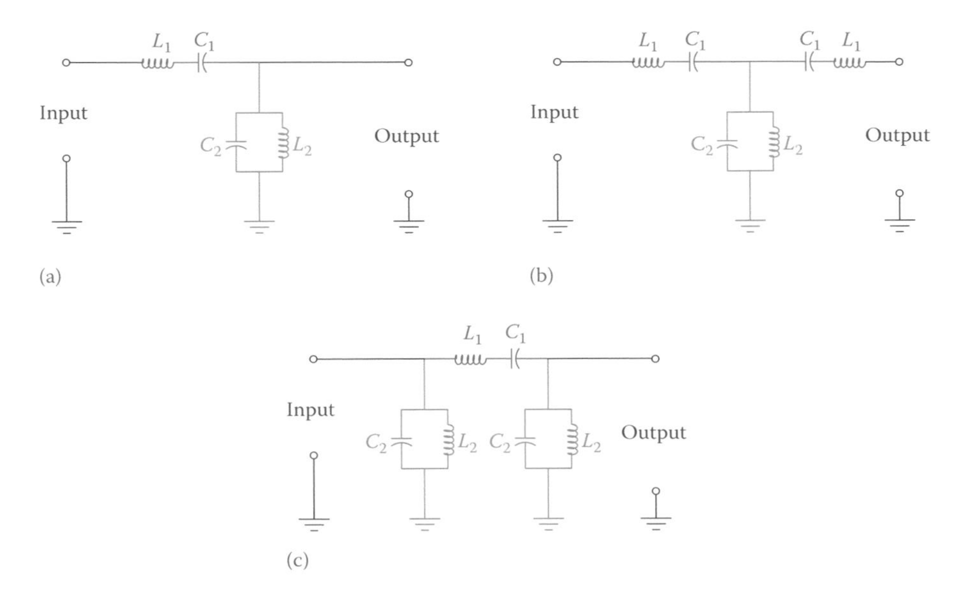

Figure 16.26 Structures of band pass filters. (a) L-type. (b) T-type. (c) π-type.

> For band pass and band reject filters the resonance frequency of the series inductor-capacitor and that of the parallel inductor-capacitor must be the same.

If the resonant frequency of the L and C components in a filter is denoted by f_{Res}, the bandwidth for that filter is such that f_{Res} is in the middle. The bandwidth is defined by two frequencies, one higher and one lower than f_{Res}. The boundaries are determined by those frequencies for which the output voltage drops by about 30 percent. That is, the frequencies at which the output voltage will be 70.7 percent of the output voltage for resonance frequency. In Figures 16.24 and 16.25 the characteristics of two band pass filters A and B are shown. A line corresponding to 50 percent power denotes the 70 percent voltage output at resonance (since power is proportional to square of voltage, and $0.707^2 = 0.5$). Accordingly, the bandwidth for each filter is depicted on the basis of the intersections of the 50 percent power line and each curve.

The bandwidth of a filter depends on the so-called quality (or quality factor) Q of the inductor-capacitor used in its structure. The higher the value of Q, the narrower the bandwidth. Q is a number and has no unit. Higher value of Q implies better quality.

16.8.3 Quality Q

Bandwidth and quality (factor) are related to each other by

$$\text{BW} = \frac{f_{\text{Res}}}{Q} \tag{16.3}$$

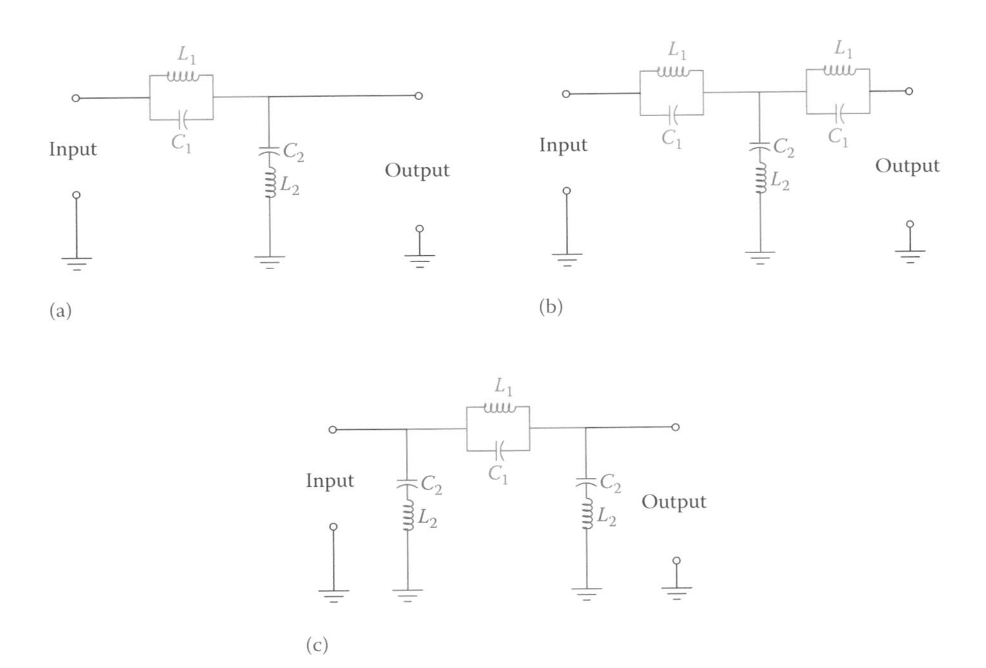

Figure 16.27 Structures of band reject filters. (a) L-type. (b) T-type. (c) π-type.

Thus, if the resonance frequency and the quality of a filter are known its bandwidth can be determined.

On the basis of the preceding discussions, a band pass or band reject filter is made of a set of capacitors and inductors, either in parallel or in series with each other. The quality of any filter, therefore, is defined by the quality of these two main elements.

Quality of a capacitor is normally high (based on leakage), and, consequently, the quality of a filter is determined by that of the inductor. Because in reality an inductor is a coil, it has some resistance in its wire winding, though it can be small. This resistance is shown as a resistor r in series with the inductor (note that we have shown this resistor with a lowercase to make it distinct from any other resistor R). Quality of an inductor is defined by the ratio of its reactance to its resistance, at any given frequency.

$$Q = \frac{X_L}{r} \tag{16.4}$$

For a filter X_L is the reactance at the resonance frequency.

Example 16.1

Quality of a 3.3 μF capacitor is 300. This capacitor is used in a tank circuit at the output of a band pass filter. If the coil inductance and resistance are 1 mH and 0.4 Ω, respectively, find the bandwidth of the filter.

Solution

First, we find the resonance frequency for the capacitor and the inductor.

$$f_{\text{Res}} = \frac{1}{2\pi\sqrt{LC}} = \frac{1}{(2)(3.14)\sqrt{(0.001)(0.0000033)}} = 2770 \text{ Hz}$$

Then for the resonant frequency we find the inductor reactance

$$X_L = (2)(3.14)(2770)(0.001) = 17.4\ \Omega$$

The value for inductor Q now can be determined

$$Q_L = \frac{17.4}{0.4} = 43.5$$

Because $Q_L < Q_C = 300$, the Q for the tank circuit is taken as that of the coil; thus, $Q = 43.5$, and the filter bandwidth is

$$\text{BW} = \frac{2770}{43.5} = 63.7 \cong 64 \text{ Hz}$$

Thus, this filter affects the frequencies between 2738 and 2802 Hz.

Note that for the cases that the quality of the inductor is higher than the quality of the capacitor (like in the above example), Equations 16.3 and 16.4 can be combined together and the bandwidth can be defined in terms of the specifications of the inductor. In such a case,

$$\text{BW} = \frac{r}{2\pi L} \tag{16.5}$$

16.9 Chapter Summary

- DC electricity can be obtained from AC using a rectifier.
- A single-phase half-wave rectifier eliminates the negative half of a full cycle of AC to obtain DC. It uses one diode.
- A single-phase full-wave rectifier uses two diodes, each one conducting for half of each cycle. Current through a load is always in one direction, as if all negative half cycles are converted to positive by the rectifier.
- A single-phase bridge rectifier has an output like a full-wave rectifier. It uses four diodes, each pair conducting for half of a cycle. It is the most commonly used single-phase rectifier.
- A three-phase bridge rectifier uses six diodes.
- Fluctuations in the voltage of DC electricity from a rectifier is called ripple.

- Ripples are undesirable, and for many applications they must be removed or smoothed out.
- Frequency of ripples in a rectifier output depends on the type of rectifier.
- Ripples are removed or their intensity is reduced by filters.
- Capacitors and inductors are the main components for filtering.
- L filters, T filters, and π filters are the most common filters.
- Filters are also used for separating signals of different frequencies.
- A low pass filter delivers low-frequency signals and blocks high-frequency signals from reaching its output.
- A high pass filter delivers higher-frequency signals to its output and blocks lower-frequency signals.
- A band pass filter delivers the signals within a range of frequencies and blocks all signals with other frequencies from reaching its output.
- The function of a band reject filter is opposite to the band pass filter and allows all signals except those within a particular frequency range to pass to the output.
- The bandwidth of a filter is all those frequencies between two bounds, which a filter is able to separate from the rest.
- The bandwidth of a filter is inversely proportional to the quality of a filter. The higher the quality is, the narrower the bandwidth.
- The quality of a tank circuit is dominated by quality of the inductor.
- Filters are used also to remove noise from measured signals.

Review Questions

1. Describe how AC is converted to DC.
2. What is the principal function of a rectifier?
3. Why do we need rectifiers?
4. What is the main component of a rectifier?
5. What is ripple and how is it coming into DC?
6. How can one reduce ripples?
7. What is the difference between a half-wave rectifier and a full-wave rectifier?
8. What is the difference between a full-wave rectifier and a bridge rectifier?
9. What is meant by filtering?

10. What are the main components of filters?

11. How many diodes are used in a single-phase bridge rectifier?

12. How many diodes are used in a three-phase bridge rectifier?

13. Draw the schematic of a single-phase bridge rectifier.

14. Draw the schematic of a three-phase bridge rectifier.

15. What is the difference between a filter in a rectifier as shown in Figure 16.16 and one shown in Figure 16.20?

16. What is the difference between a low pass filter and a high pass filter?

17. What is meant by *band* in a band pass or band reject filter?

18. What components do you need to make a 12 V DC power supply?

19. If you construct a power supply, can you use it to start your car with? Why or why not?

20. Can you charge the battery of your car with the power supply in the previous question?

21. Can you say that a band pass filter is made up of a low pass filter and a high pass filter? Why or why not?

22. What is the difference between a band pass filter and a band reject filter?

23. What is the bandwidth of a filter?

Problems

1. For the circuit shown in Figure 16.1, draw the output voltage if the direction of the diode is reversed.

2. In a bridge rectifier, what happens if the direction of all diodes is reversed?

3. What happens in a three-phase rectifier if three of the diodes from one side are removed?

4. What is the bandwidth of a 0.1 mH coil if used in a tank circuit with a 4.7 μF capacitor? The resistance of the coil is 0.5 Ω. The capacitor quality is sufficiently high.

5. Referring to Figures 16.6 and 16.7, determine the percentage ripple ratio in each case.

17

Transistor

OBJECTIVES: After studying this chapter, you will be able to

- Explain what a transistor is and how it works
- Describe the structure of a transistor
- Define PNP and NPN transistors
- Recognize various transistors of different physical shapes
- Name the three elements of a transistor
- Explain the terms, common base, common collector, and common emitter
- Explain the two primary functions of a transistor
- Define what biasing a transistor is
- Describe bias conditions for a transistor to work
- Define cutoff state of a transistor
- Explain how a transistor can be used for switching
- Explain how a transistor can be used for amplification
- Understand the importance of correct biasing
- Describe heat sink

New terms:
Base, biasing a transistor, bipolar junction transistor, collector, common base, common collector, common emitter, cutoff, emitter, heat sink, junction transistor

17.1 Introduction

A transistor is a semiconductor device with many applications. It is, in fact, the mother of all the other semiconductor electronic devices that we use or see around us today. Before transistors were invented, all electronic devices were made with vacuum tubes. Still one can find devices such as stereo amplifiers and radio receivers that use vacuum tubes, but they are rare and expensive, in addition to being bulky and heavy in comparison to solid-state electronics that use transistors and semiconductor technology. There are various types of transistors, and we will study their fundamentals in this chapter. The simplest form of transistor is a **junction transistor**, also called **bipolar junction transistor**. The reason for calling it junction transistor is because it contains two PN junctions. In other words, from the structure viewpoint it is similar to two diodes put together during the manufacturing process. Nonetheless, note that a transistor is a completely different device than a diode, meaning that one cannot put two diodes together and expect to have a transistor.

Junction transistor: Same as bipolar junction transistor, which includes any ordinary transistor.

Bipolar junction transistor: Another name for junction transistor.

17.2 Junction Transistor

Recalling that a PN junction has a P side and an N side, imagine you want to put two of them together. It is easy to understand that depending on how they are put together, two basic types of transistors can result, as shown in Figure 17.1.

Accordingly, the two types of a junction transistor are PNP and NPN. This classification is based on the polarity in the structure. There are other types of classification, based on transistor functioning and performance in a circuit.

Base: One of the three principal terminals of a junction transistor (ordinary transistor).

Emitter: One of the three main terminals of a common (not FET) transistor.

Collector: One of the three main terminals of an ordinary transistor.

As can be easily understood from Figure 17.1, a transistor has three terminals. This is different from the DC components that we have studied so far. In a transistor, irrespective of the type (PNP or NPN), the middle terminal is called the **base**. The other two terminals are called the **emitter** and the **collector**. The internal structure of a transistor and the amount of doping in each P and N material (which determine the thickness of the depletion region in a PN junction) determine which side is the emitter and which is the collector. The emitter junction is much higher doped than the collector junction. Figure 17.2 shows the symbols for junction transistors, where B, C, and E identify the base, emitter, and collector, respectively.

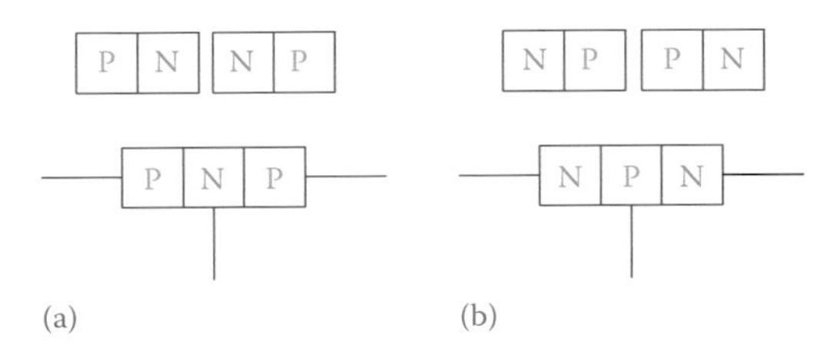

Figure 17.1 Two basic types of transistors: (a) PNP and (b) NPN.

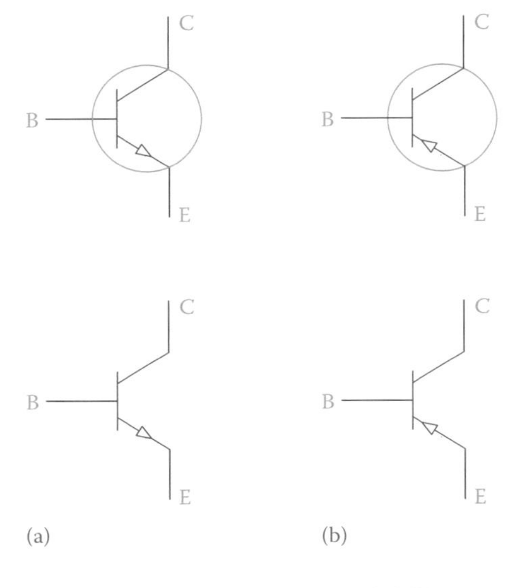

Figure 17.2 Symbols for (a) NPN transistor and (b) PNP transistor.

The emitter is identified by an arrow, which shows the direction of current. Note the difference between the two symbols. The direction of the arrow in the two transistor types are from positive to negative, or better say from P to N, where P and N refer to the semiconductor material type.

Note that for the sake of simplicity in the rest of this chapter we drop the suffix "junction" and only use the word "transistor."

17.3 Transistor Size and Packaging

Transistors come in various packages (their physical form), the most common of which are shown in Figure 17.3. Their size is comparable with paper clips, as shown in Figure 17.4. Like any other device, a transistor has ratings for current, voltage, and power. These are defined in the data sheet of each transistor and must be respected when used in electrical and electronic circuits. Transistors may get hot when operating. In such a case a **heat sink** may be attached (screwed or glued) to its body. A heat sink is not part of a transistor; it is used for any component that can get hot in operation. It is a piece of metal (normally aluminum) with a large area (so that it can easily exchange heat with its surrounding environment), sometimes corrugated, which is attached to the body of a component that can get hot. This is a common practice, particularly in electronic devices. A heat sink can be air cooled or in very high power devices can even be water cooled. Figure 17.5 shows a resistor with a large heat sink. Part of the reason for different packaging of transistors is the heat that they produce in operation. Only three of the transistors in Figure 17.3 can accept a heat sink.

Heat sink: Piece of aluminum attached to an electronic component to provide a larger area of contact with air, so that the heat produced in the component is removed from it better and faster and overheating and damage is prevented.

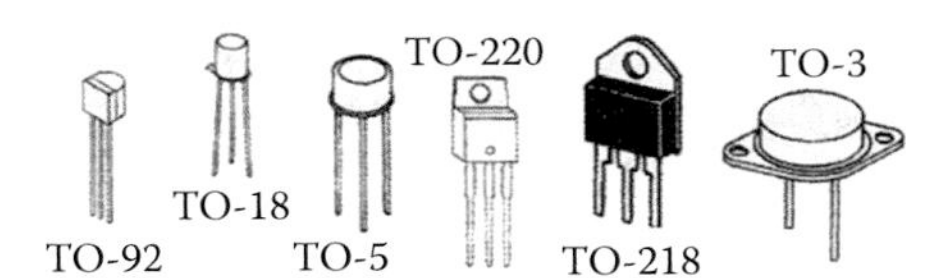

Figure 17.3 Examples of physical shape (packaging) of transistors.

Figure 17.4 Indication of the size of a typical transistor.

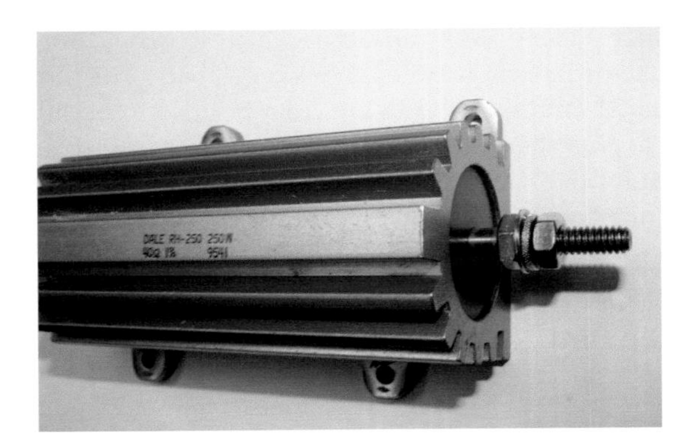

Figure 17.5 Example of a heat sink for a resistor.

For transistors with three in-line terminals (all three in a row; see Figure 17.3), usually, the connection in the middle is the base, and the other two could be marked or unmarked. But, this is not always the case. One must determine the base, emitter, and collector terminals before use.

17.4 Transistors Connection and Biasing Configuration

Because a transistor has three terminals, three different voltages can appear at each terminal when inside a circuit. In other words, on the basis of the currents in the circuit the voltage at each terminal can be different than the other two. The normal operating condition of a transistor requires that the junction between the collector and the base be reverse biased and the junction between the emitter and the base be forward biased. In this sense, the correct connections for a PNP and an NPN transistor are different. Two possible cases are shown in Figure 17.6a and b, for NPN and PNP transistors, respectively.

In the NPN transistor (see Figure 17.6a) the base is a P-type material and the collector is an N-type material. When the voltage at the collector terminal is larger than that at the base terminal, the collector-base junction is reverse biased. Also, in this transistor the voltage at the B terminal is larger than the voltage at the E terminal. Thus, the base-emitter junction is forward biased. Note that a 0 voltage means that the emitter is connected to the reference voltage, which can be the negative terminal of the DC power supply, or the ground voltage.

Similarly, in Figure 17.6b the voltages at the B, E, and C terminals are +15, +20, and +2 V, respectively. This implies that the voltage at B (N side of the base-collector junction) is greater than the voltage at C terminal (thus, reverse bias) and the voltage at B is smaller than the voltage at E terminal (thus, forward bias).

> For a transistor to function the condition is that (1) the emitter-base junction be forward biased and (2) the collector-base junction be reverse biased.

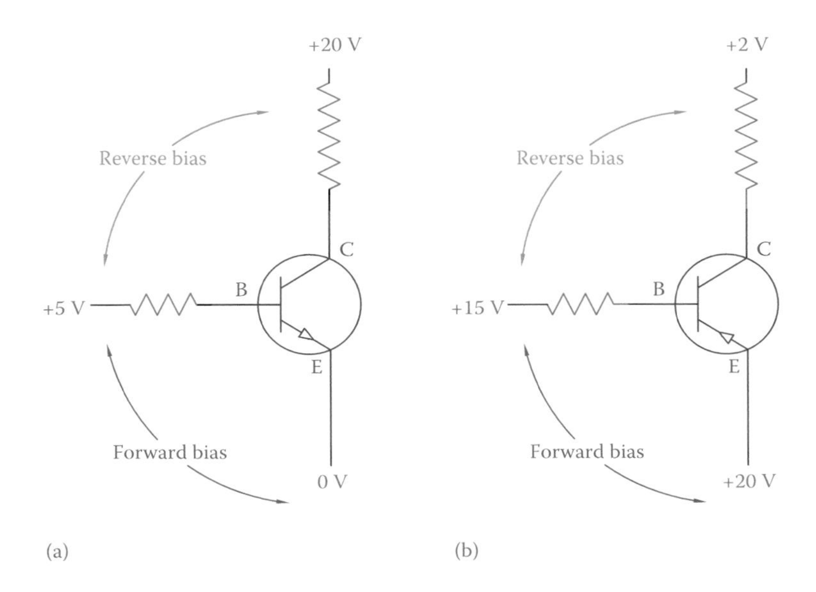

Figure 17.6 Correct connection of (a) NPN and (b) PNP transistors.

17.5 Modes of Operation

As we will learn gradually, most applications of a transistor are for switching and for amplification. In switching, a transistor is used as a switch, meaning that it can turn on and off a device or part of a circuit. As for amplification, a transistor can be used as a voltage amplifier, a current amplifier, or a power amplifier. In both switching and amplification, there is an input signal to the transistor and an output signal from the transistor. For each signal (input and output), two connections are necessary, as shown in Figure 17.7. This figure symbolizes a transistor as an input-output device, each one with an internal resistor between the two terminals. When a signal is introduced to the input terminals, the corresponding output signal is generated and is available at the output terminals. Because a transistor has only three physical terminals to the outside world, in order to have two input lines and two output lines, two of the terminals can be assumed to be internally connected, through an internal resistor. In other words, one of the terminals must be shared between the input and the output.

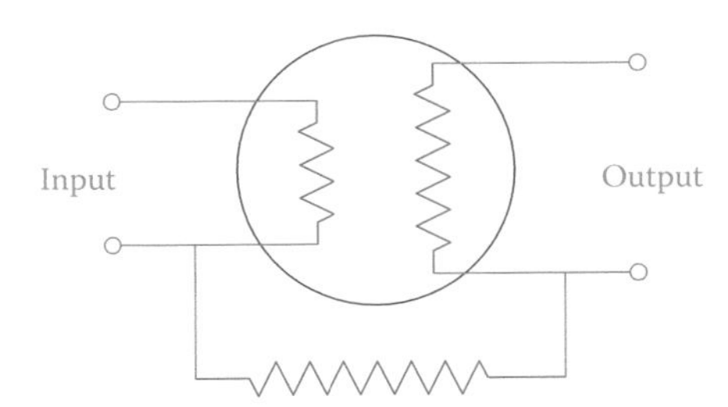

Figure 17.7 Input and output concept in a transistor.

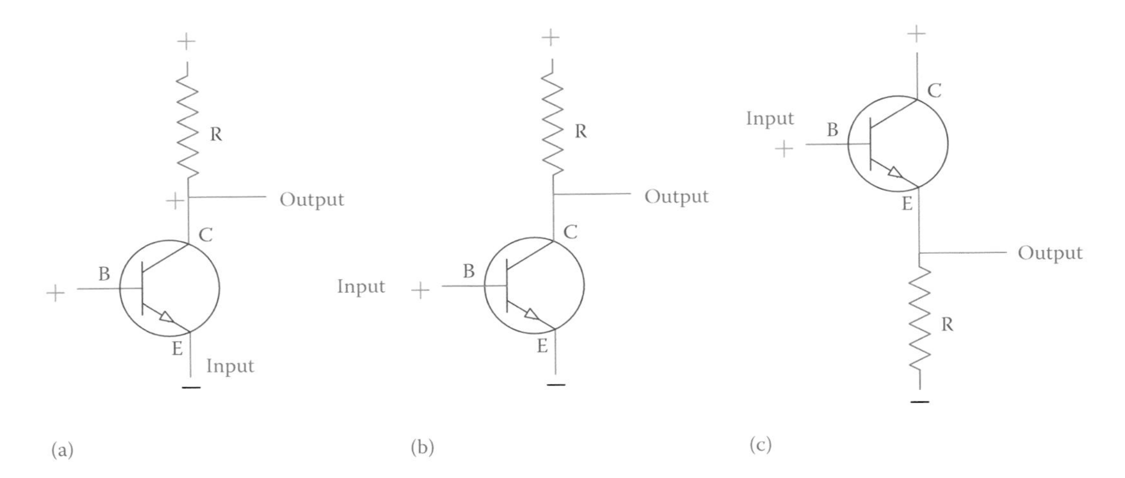

Figure 17.8 Three modes of operation of transistors: (a) common base, (b) common emitter, and (c) common collector.

Common base: One of the three configurations for using transistors in a circuit, where the base is a part of both the input and the output.

Common emitter: One of the three configurations for using transistors in a circuit, where the emitter is a part of both the transistor input and its output.

Common collector: One of the three configurations for using transistors in a circuit, where the collector is a part of both the transistor input and the output.

In a circuit containing a transistor, in order that the correct biasing as mentioned above is respected, the transistor can be used in three different ways. These different ways are based on which terminal is shared between the input and output and define three modes of operation of transistors. These three modes are **common base**, **common emitter**, and **common collector**. As the name implies, in each mode one of the terminals is common for both input and output.

Figure 17.8 shows the three modes of operation for an NPN transistor. The same three modes are possible if a PNP transistor is used, but they are not shown here. In Figure 17.8a the input signal is introduced between the emitter and the base and the output is taken between the base and the collector. The + and – signs shown are for the voltage, which must be such that the corresponding junctions, have the correct bias. It is obvious that the voltages and currents must be compatible so that the base can be connected to both the input circuit and the output circuit. The same principle applies for the common-emitter and common-collector configurations. Later, we see that both correct biasing and voltage compatibility are achieved by having the appropriate resistors or other components in the circuits of each of the three terminals.

17.6 How a Transistor Works

The function of a transistor is like a gated device to the electric current. If the gate is open, there is a flow, and if the gate is closed, there is no flow. The role of the transistor input signal is to open and close the gate, as desired. The output is, thus, controlled by the input. When the gate is open, energy is delivered to the output (e.g., to a load). From the power viewpoint the important fact in the operation of a transistor is that the input power is much smaller than the output power.

Figure 17.9 shows a transistor circuit. The collector is connected to the positive side of a 12 V battery through a resistor R_2, and the emitter to

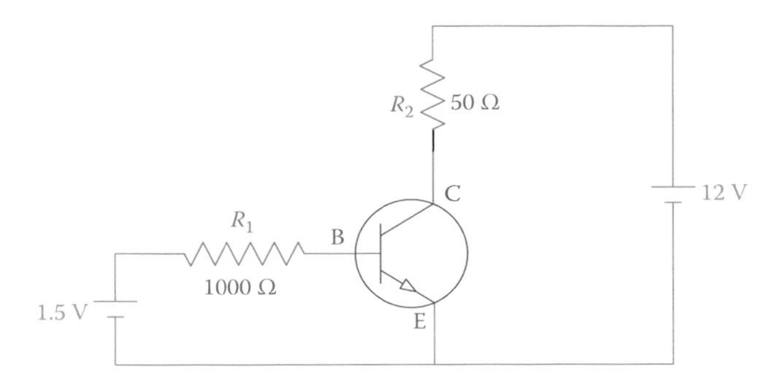

Figure 17.9 Connecting a transistor to electric supply.

the negative side. At the same time a smaller (1.5 V) battery is connected between the base and the emitter through the resistor R_1 and a switch. The polarity of this battery is such that the base-emitter junction is forward biased, that is, positive side to the base. In this sense, there are two loops formed for the electric flow, one between the collector and the emitter, and one between the emitter and the base (emitter is the common terminal).

While the collector-emitter loop is closed (and there is no switch), no current flows from the collector (positive side) to the emitter. Also, in the base-emitter circuit, there is no current when the switch is open. Now, if the switch is closed, a small current of 0.8 mA (that is, [1.5 V – 0.7 V]/1000 Ω, assuming a silicon-based transistor) can flow through R_1 from base to the emitter. This current from the base to emitter can cause conduction in the transistor between the collector and the emitter (the second loop). Its function is, in fact, similar to opening a gate, as a result of which a current of around 240 mA (i.e., 12 V/50 Ω) flows from the collector to the emitter. In other words, closing the switch in the first loop causes a current in the second loop.

Consider now the power in the two loops, by multiplying the loop current and its voltage. It is evident that the power in the second loop (12 V × 0.240 A = 2.88 W) is much greater than that of the first loop (1.5 V × 0.0008 A = 0.00120 W).

The above explanation shows the way a transistor works. That is to say, the functioning of a transistor is based on currents in the two loops and not the voltages. This implies that *a transistor is a current driven device.* Note that in this simple analysis we have ignored the internal resistances in the junctions between the three elements (base, collector, and emitter) of a transistor. In reality, there is always some resistance between these elements. It is also imperative to note that the transistor, itself, does not generate power. All the power necessary for amplification comes from the battery (the power supply).

> When there is no current between the base and the emitter in a transistor, the transistor is off. When a current flows between the base and the emitter, a current also flows between the collector and the emitter.

> A transistor is a current-driven device.

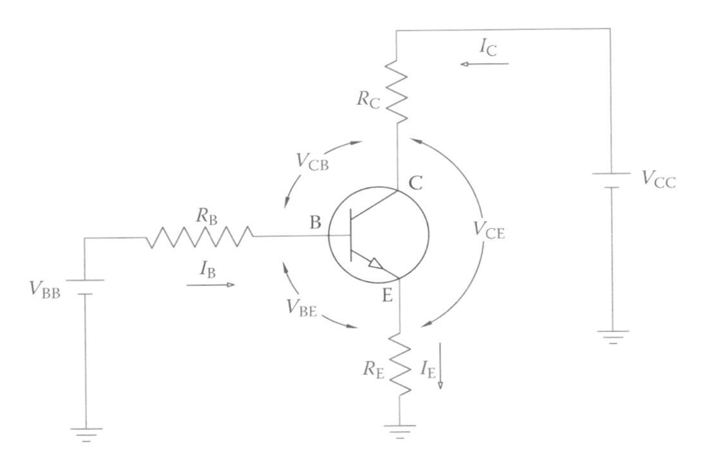

Figure 17.10 Definition of voltages and currents in a transistor.

17.7 Transistor Properties

Figure 17.10 depicts the definitions for various voltages and currents in a transistor. The voltage between the base and the emitter is denoted by V_{BE}. The voltage between the collector and the emitter is denoted by V_{CE}, and similarly the voltage between the collector and the base is denoted by V_{CB}. Likewise, the currents flowing through the base, the collector and the emitter are, respectively, represented by I_B, I_C, and I_E. Moreover, as indicated in Figure 17.10, the total voltage supplying power to the collector is called V_{CC} and that corresponding to the base is called V_{BB}. By the same token in other transistor configurations a supply voltage may be used for the emitter. In such a case this voltage is represented by V_{EE}.

On the basis of the discussion in Section 17.6 and the voltage drop in a PN junction (0.7 for silicon and 0.3 V for germanium), if V_{BE}, is less than 0.7 V (assuming a silicon transistor), no matter how much the voltage between the collector and the emitter is (provided that it is within the rating of the transistor and does not cause a breakdown), no current flows from collector to the emitter. In such a case the transistor is said to be in the **cutoff** state.

Cutoff: Status of a transistor when it is not operating because of very low (negligible) current.

As soon as V_{BE} becomes larger than 0.7 V, a current (I_B) flows from B to E. This is the current through the base. This current turns the transistor on and causes conduction between the collector and the emitter. Thus, the current I_C flows through the collector toward the emitter. The current in the emitter, I_E, is the sum of I_B and I_C.

The current in the emitter is the sum of the current in the base and the current in the collector.

It is important to understand the operational characteristics of a transistor. In addition to the fact that it behaves as a gated device, we must also know that *a transistor behaves like a variable resistor*. This is concerning the voltage V_{CE} and the current from the collector to the emitter (I_C). When a transistor is conducting, the current flow from the collector to the emitter is the major component of interest. This is the output current. There is a voltage drop between the collector and the emitter when this current exists. Nevertheless, this voltage drop cannot be determined based on the assumption that there is a constant resistance between the collector and emitter terminals inside the transistor. The value of such a resistance is not constant. This matter will become clearer later.

A transistor behaves like a variable resistor as far as the relationship between the collector-emitter voltage and collector current is concerned.

As has been shown in Section 17.5 and Figure 17.9, a noticeable characteristic of a transistor is that I_C is much larger than I_B (it can be 100 times or 200 times larger). Moreover, the ratio I_C/I_B, which is almost constant for a transistor in a given circuit, is denoted by the Greek letter β (beta) and can be used for calculation of I_C when I_B is known, or vice versa. β depends on the particulars of a circuit, transistor properties, and temperature (by particulars of a circuit we mean the components, their values and the circuit arrangement). The minimum and maximum values for parameters of a transistor are normally given in its data sheet.

$$I_C = \beta I_B \tag{17.1}$$

Moreover, because β is normally large (typically, $50 < \beta < 300$), then

$$I_E = I_B + I_C = \beta I_B + I_B = (\beta + 1)\, I_B \approx \beta I_B = I_C \tag{17.2}$$

Equation 17.2 implies that, for simplification, in many calculations I_E is set equal to I_C. Note that the electrical values must match with reality; in practice, there are always maximum values for current and power that a device can handle, based on its internal structure and rating. In other words, one cannot rely only on mathematical relationships for a device that is bound by its physical properties, thus exhibiting limit values for current, voltage, and power that it can handle.

Example 17.1

Consider the transistor in the circuit of Figure 17.9. If the value of β for this transistor is 100 and $R_C = 50\ \Omega$, what is the voltage V_{CE}?

Solution

The base current is

$$I_B = (1.5 - 0.7) \div 1000 = 0.0008\ \text{A}$$

The collector current is β times (100×) the base current

$$I_C = 100 \times 0.0008 = 0.08\ \text{A} = 80\ \text{mA}$$

and the voltage drop in R_C is

$$V_{Rc} = 50 \times 0.080 = 4\ \text{V}$$

V_{CE} is the difference between the power supply voltage and the voltage drop in R_C:

$$V_{CE} = 12 - 4 = 8 \text{ V}$$

Example 17.2

For the transistor in Example 17.1, what is the internal resistance between the collector and the emitter for the given condition?

Solution

On the basis of the voltage $V_{CE} = 8$ V and the current $I_C = 0.08$ A, the internal resistance between the collector and the emitter is

$$8 \div 0.08 = 100 \ \Omega$$

Note that this resistance value is not constant, and changes for other operating conditions.

Example 17.3

If in the circuit of Example 17.1, the resistance R_C is substituted by a 200 Ω resistor, what will be the voltage V_{CE}?

Solution

For this case it can be seen that the voltage drop in R_C is 16 V (200 × 0.08). This is greater than the total V_{CC} voltage (12 V), which is impossible. This reflects a bad design, in which the transistor cannot function correctly, or as expected. The voltage drop in R_C cannot exceed 12 V.

17.8 Transistor Main Functions

Two principal functions of a transistor are switching and amplification. Both of these have been demonstrated in the discussion of Section 17.5. The rest of this chapter is devoted to an explanation of how a transistor can be employed for these functions. First, in order to better understand what is meant by switching and amplification, we examine the following two examples.

17.8.1 Switching Function

Suppose that the current in a motor when working is 30 A. Dissimilar to a household device that you turn on and off by a simple switch, for such a relatively large load a **relay** is used to turn the motor on and off. A relay consists of an electromagnet that can work with a low voltage (for better safety), for instance, 12 V. The electromagnet takes little current and can easily be activated even from a distance away without a concern about voltage drop in the wire. When activated, two (or more) sets of reeds make contacts. These reeds are connected to the motor and the electric supply (higher voltage). In this way, electricity is applied to the motor through an indirect switch. A person's hand doesn't have to even be near to the higher voltage line. A relay is shown in Figure 17.11 together with a dime (10 cent American or Canadian coin), for size comparison.

Figure 17.12 illustrates a schematic for a relay with only two connections. This relay can be used in two ways: when activated a connection is made, or an existing connection is broken open. In the relay schematic "NO" stands for **normally open** and "NC" denotes **normally closed**. For making a connection, for instance in the case of a motor, the lines are connected between NO and the common connector (COM). Similarly, if we need to stop a connection when the switch is activated, it is connected between the NC and the COM terminals. Figure 17.13 illustrates the use of a relay for switching a motor on and off.

The advantages of using a relay are twofold:

1. A person is not directly in contact with a switch carrying a relatively high voltage and/or a high current, which is accompanied by a spark when contact is made or broken.
2. The switch can be installed in a remote position without the necessity for thicker wires of the motor circuit to be long. Thus, more expensive wiring and larger voltage drop in the circuit (and consequently waste of energy) are avoided.

Whereas a relay is a mechanical device, a transistor can perform the same job electronically and without any moving parts. It is smaller, too.

Figure 17.11 Picture of a small relay with eight sets of connections.

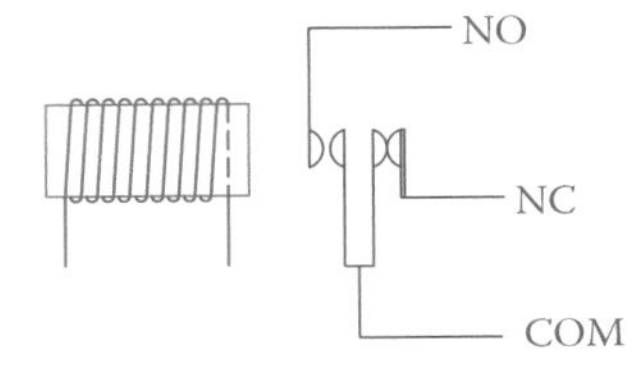

Figure 17.12 Schematic of the operation of a relay.

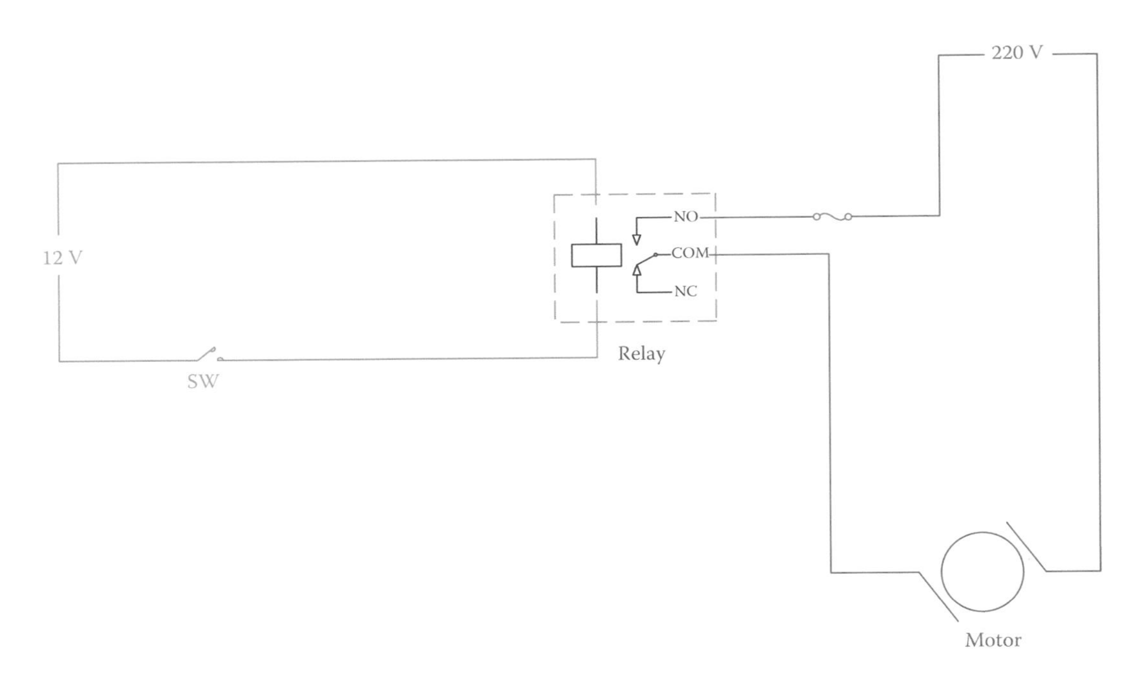

Figure 17.13 A relay used for activating a motor.

17.8.2 Amplification Function

Figure 17.14 shows a scenario in which by opening and closing a small valve and allowing a small flow of water a larger amount of water flow is controlled. This example shows another important function of a transistor, which is amplification of a signal. Amplification is a major part of the function of almost all electronic devices, such as radio, TV, phone, voice and data recorders.

As illustrated in Figure 17.14, the round gate G can move up and down, thus adjust the flow from the channel C that connects to a water source. The gate movement is provided by a rotational motion of the gate assembly about axis xx′. Because of the counterweight on the other side of the axis the gate can move up by a slight downward force on the right. The weight of the gate keeps it normally in the closed position and the counterweight W is such that little effort is required to move the gate about its axis. The bottom hole in the container B, attached to the lever bar between the gate and its counterweight, lets any water remaining in it be emptied and directed downstream through the funnel D. In this sense, the gate will close unless extra downward force exists on the right-hand side. By opening the valve S a small flow of water runs into B. This causes water to accumulate in B and make the counterweight side heavier, thus, pulling the gate upward and opening it. Consequently, a large flow of water is established from upstream to the E part of the channel. When the valve S is closed again, the water in B drains out and the gate goes back to its closed position.

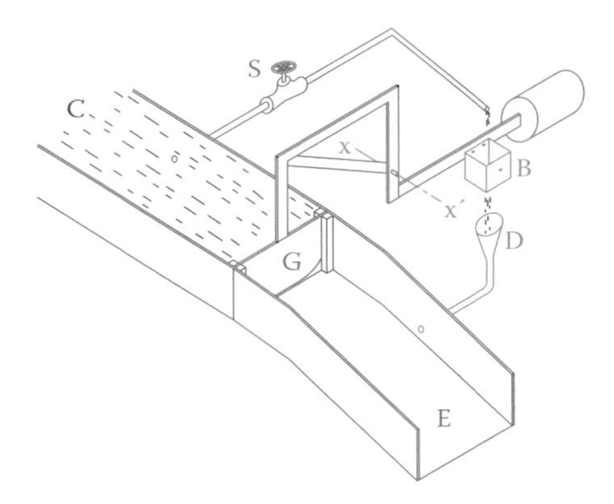

Figure 17.14 A hydraulic analogy of amplification function of a transistor.

By controlling the flow of water through the valve S we may control a larger flow of water through channel C. In other words, the variation of the output in channel C is proportional and similar to the smaller flow controlled by the valve S.

In an amplifier a low power signal is duplicated with much higher power. In other words, the input signal drives an amplifier to generate a more powerful signal with the same pattern at its output. More power implies higher voltage, higher current, or higher both voltage and current. The power that is included in the output must be provided by the power supply powering a circuit.

In an amplifier the power added to a signal is provided by the power supply.

17.9 Transistor as a Switch

In the following circuit the transistor acts as a switch, in the same way that a relay works. The advantage of a transistor to a relay is that a transistor is lighter, has less noise (if any at all), has no moving components, it is faster, consumes less energy and it is cheaper. Note that in a complicated system, such as the operation of an elevator or a train, there can be hundreds or sometimes more than a thousand relays. In the past, these relay systems were common. In addition to taking a much larger space and making noise by each relay when making contacts, after a while bugs could also make nests in the warm space between them and gradually disrupt the operation (the term *bug* and *debug* in computer terminology stems from here).

In the circuit shown in Figure 17.15 an NPN transistor is shown. The collector-base junction is reverse biased, as required, and a current never flows between the collector and the base. The base-emitter junction is forward biased, but when the switch is open there is no current between the base and the emitter. Consequently, there is no current between the collector and the emitter because the transistor acts like an insulator.

If the voltage at point A (where the load is connected) is measured, it shows 25 V because there is no voltage drop in the 200 Ω resistor (due to carrying no current). The voltage difference across the load is zero and the load is not energized.

Closing the switch lets a current flow between the base and emitter. As mentioned earlier, this action turns on the internal collector-emitter connection and causes an electric current to flow from collector to the emitter. The magnitude of this current depends on the resistance R, which is 200 Ω here. Ignoring the small voltage drops in the transistor, the C-E current is 125 mA (25 ÷ 200, based on Ohm's law). The current flow of 125 mA causes a voltage drop of 25 V in the 200 Ω resistor. In such a case, if the voltage at point A is measured it must be 0 V.

Comparing with when the switch was open, now the voltage difference across the load is 25 V, and, therefore, the load is energized (note that when the load is connected to point A, the current through the emitter is the sum of the current through the 200 Ω resistor and the load current).

The scenario shown in Figure 17.15 is a practical example of the switching action of a transistor. Although it is a very simple example, it shows the principle of the way a transistor can be used for this type of application. A similar circuit can be used for other applications such as counting people entering a library or the cars passing through a pathway by replacing the manual switch by a photocell (a resistor whose resistance changes with light and can be used as a light sensor), or by an automatic switch of a different type. The switching action of a transistor can be at very high frequency, like the radio, TV, and communication signals with megahertz frequencies.

In practice, we may want to avoid using more than one voltage source. This is very reasonable. Figure 17.16 illustrates the same setup as in Figure 17.15, but the voltage at the base is derived from the same power source supplying the load and the collector. In this arrangement the resistors R_2 and R_3 are so selected to provide the necessary voltage at the transistor base when the switch is closed. R_2 and R_3 resistors on the left establish a voltage divider. When the switch is closed a voltage equal to

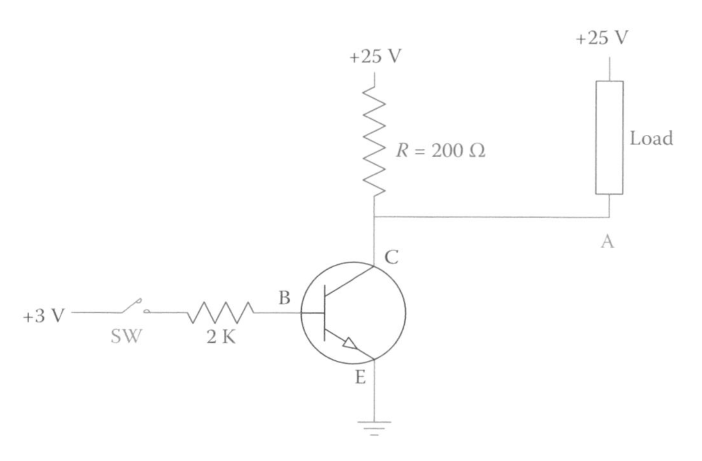

Figure 17.15 Transistor used as a switch.

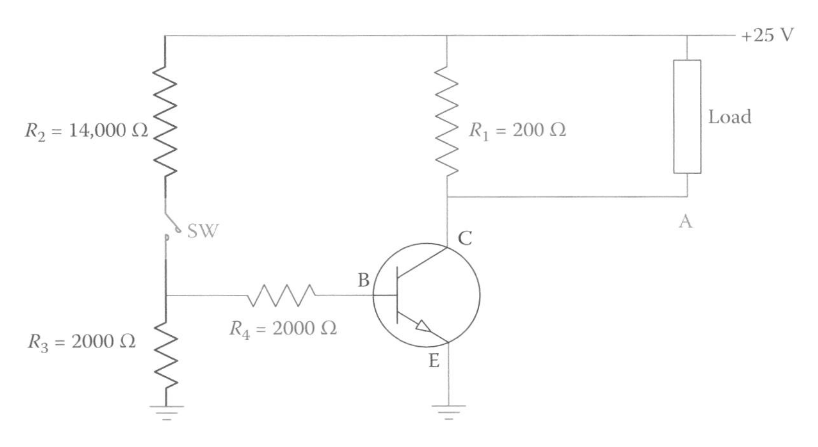

Figure 17.16 Providing base voltage by a voltage divider.

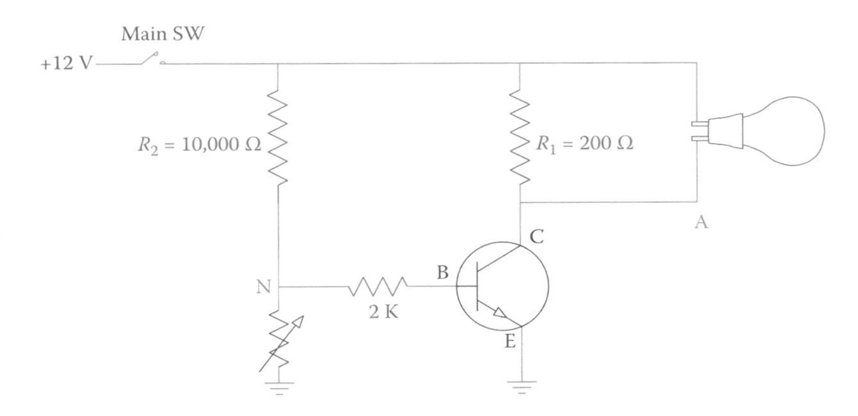

Figure 17.17 An automatic switch for turning a light on at night.

1/8th (2000/[14,000 + 2000]) of the total voltage provided to the load is applied to the base.

The above circuit can be used for automatic on-off switching of a lightbulb, if a photocell is inserted as a switch in the circuit. A photocell is like a variable resistor. When there is no light, it has a high resistance and in the presence of light its resistance drops. Consider the circuit shown in Figure 17.17. During the day the photocell (shown as a variable resistor) acts as a conductor, and, thus, the voltage at point N is zero (N is shorted to the ground). During the night, however, the resistance is considerable and the voltage at point N is equal to the voltage drop across the photocell resistance. As a result, current flows between the transistor base and emitter; this causes the transistor to turn on, as described.

For other applications, alternatively, the voltage at the base terminal can be provided by a sensor that generates a signal with sufficient voltage. (Remember that the junction between B and E requires 0.7 V, if made out of silicon, to conduct. The voltage at B, thus, must be more than that for a current to flow.)

17.10 Transistor Amplifier

A transistor can amplify the current or voltage of a signal. Amplifying current implies that the transistor output current is higher than its input current. Amplifying voltage implies that the voltage variation of the output is higher than the voltage variation of the input. When put together, a number of transistors can serve as a power amplifier (amplifying both voltage and current). For instance, a very weak electromagnetic signal received by the antenna of a radio has little voltage and cannot cause any significant current in a circuit. It must be amplified by hundreds of times in order to be audible through a loudspeaker. Similarly, a weak signal picked up by a sensor in many industrial applications, is amplified before it can be used for further processing. The circuit in Figure 17.18 illustrates a simple example of how the voltage of a simple sinusoidal signal can be amplified. In other words, if a sinusoidal signal is introduced as the input of a transistor, the transistor output is a similar signal (having the same form and frequency) with a higher voltage. We study the variation of the voltage across the load and compare it with the input voltage.

The input is a sinusoidal voltage with a peak-to-peak value of 2 V introduced at point K, which is connected to the transistor base through a 47 kΩ resistor. The other wire for this signal is connected to the ground. The role of the capacitor before point K is very important. It filters (prevents from passing) any DC content that may be blended with the input signal (see Section 13.6). The corresponding values for the input (the sinusoidal waveform) and the output (voltage across the load) for each 30° of the one cycle are entered in Table 17.1, together with some other values, as discussed.

Assuming that this transistor is made out of silicon, because in a silicon base transistor there is always a voltage drop of about 0.7 V between the base and the emitter, the voltage at K must be such that it never goes below

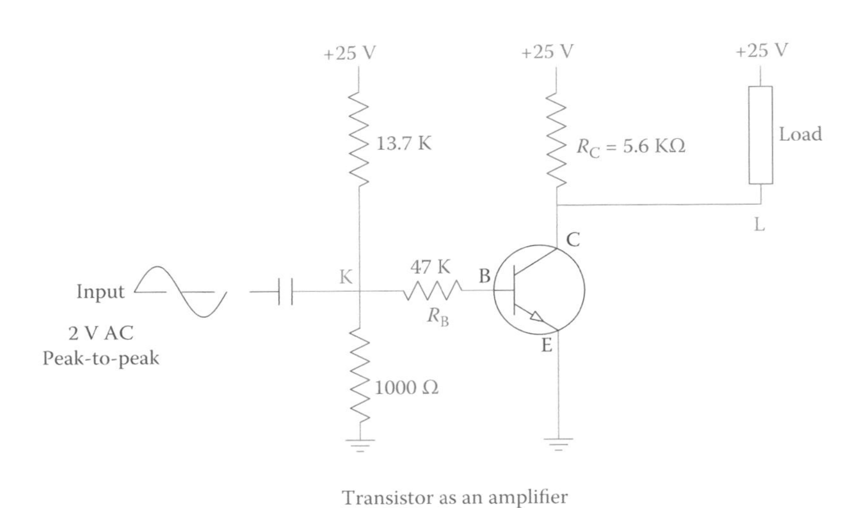

Figure 17.18 Example of voltage amplification by a transistor.

TABLE 17.1 Variation of Input versus Output Voltage for the Circuit of Figure 17.18

Input signal angle (degree)	0	30	60	90	120	150	180	210	240	270	300	330	360	One cycle
Input signal value, V	0	0.50	0.87	1.00	0.87	0.50	0	−0.50	−0.87	−1.00	−0.87	−0.50	0	Sinusoidal values × 1
Voltage at K (V_{BB})	1.7	2.20	2.57	2.70	2.57	2.20	1.70	1.20	0.83	0.70	0.83	1.20	1.7	$V_{BB} = V_{inp} + 1.7$
Voltage across R_B	1.00	1.50	1.87	2.00	1.87	1.50	1.00	0.50	0.13	0	0.13	0.5	1.0	$V_{BB} - 0.7$
Base current, I_B, μA	21.3	31.9	39.7	42.6	39.7	31.9	21.3	10.6	2.85	0	2.85	10.6	21.3	$I_B = (V_{BB} - 0.7) \div 47\ k$
I_C, mA	2.13	3.19	3.97	4.26	3.97	3.19	2.13	1.06	0.29	0	0.29	1.06	2.13	$I_C = 100\ I_B$
Voltage across R_C, V	1 1.93	17.87	22.23	23.83	22.23	17.87	11.91	5.96	1.60	0	1.60	5.96	11.91	V_{RC}
Output voltage V_{CE}	13.09	7.13	2.77	1.17	2.77	7.13	13.09	19.04	23.40	25	23.40	19.04	13.09	$25 - V_{RC}$

0.7 V. Otherwise, the transistor goes to cutoff state and stops conducting. In this example the 13.7 kΩ and 1000 Ω resistors form a voltage divider to provide 1.7 V DC at point K. This value of 1.7 V has been specifically selected such that for the most negative value of the sinusoidal wave the base-emitter junction is still forward biased and its V_{BE} is not less than 0.7 V $(1.7 - 1/2 \times 2) = 0.7$ V; the input is 2 V peak to peak.

The 2 V peak-to-peak value AC signal and the 1.7 V DC signal add together and, as a result, the voltage at point K has a variation between 2.7 and 0.7 V.

Table 17.1 shows the following values for 12 different points (360° at 30° intervals):

1. Input signal angle at various instants during one cycle.
2. Input signal voltage.
3. Voltage at point K.
4. V_{BE} (Voltage at K – 0.7 V).
5. Base current, in micro ampere (μA), based on Ohm's law V_{BE}/470 kΩ.
6. Collector current in mA, assuming 100× the base current ($I_C = \beta I_B = 100 I_B$).
7. Voltage drop in resistor R_C (see Figure 17.18).
8. Voltage at point L. This is V_{CE}, because the emitter is grounded, V_{CE} = 25 V – voltage drop in the 5.6 kΩ resistor.

Notice that here we have intentionally connected the load across the 5.6 kΩ resistor. The variation of the voltage across the load is sinusoidal, that is, similar to the input. Although this way of connecting the load is possible, it is not the only way to do so. The load could be connected between the ground and point L, which indeed represents the output for this amplifier, based on Figure 17.8. This matter is further discussed in Chapter 18 (see Section 18.3). For this configuration both the output voltage and the voltage across the load have sinusoidal forms, but the two voltages have 180° phase difference.

Figure 17.19 shows the relationship between the voltages of the input and output signals (rows 2 and 8 of Table 17.1). There are a number of important points that can be observed from Figures 17.18 and 17.19:

1. The transistor is used in a common-emitter configuration.
2. The input signal has a sinusoidal waveform.
3. The input signal was blended with a positive DC value to create a varying DC (varying in value, but otherwise always positive) signal.
4. The output is also always positive, but it has a variation in the form of a sinusoidal wave.
5. The output variation (sinusoidal pattern) has the same frequency as the input.
6. The peak to peak of the output voltage variation has a larger value than that of the input signal.
7. When the input signal is at its maximum, the output signal is at its minimum, and vice versa. That is, the output is 180° out of phase with the input.

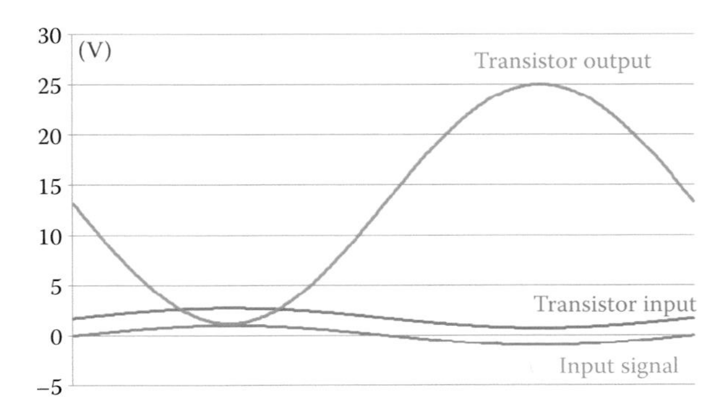

Figure 17.19 Input and output relationship in a transistor when used in common emitter design.

In the same way that the input signal was blended with a constant DC voltage, the output signal has a DC content, as seen in Figure 17.19. To remove the DC content of the output signal, a capacitor is added to the circuit at a point where the output signal is taken (point L). The revised circuit is as shown in Figure 17.20. This capacitor of appropriate value can filter out the DC content (block the DC current). The outcome of this filtration is a signal that varies between −11.9 and +11.9 V, $\left(\frac{25-1.2}{2}\right)$, that is, its peak-to-peak value is about 23.8 V. Figure 17.21 depicts the voltages at points H, K, L, and M of the circuit in Figure 17.20; the amplification, however, is shown smaller in order fit the page. The input signal is introduced between point H and the ground, and the output signal is taken between point M and the ground. The sinusoidal input signal was amplified by 11.9 times (the ratio between peak value of the output and the peak value of the input).

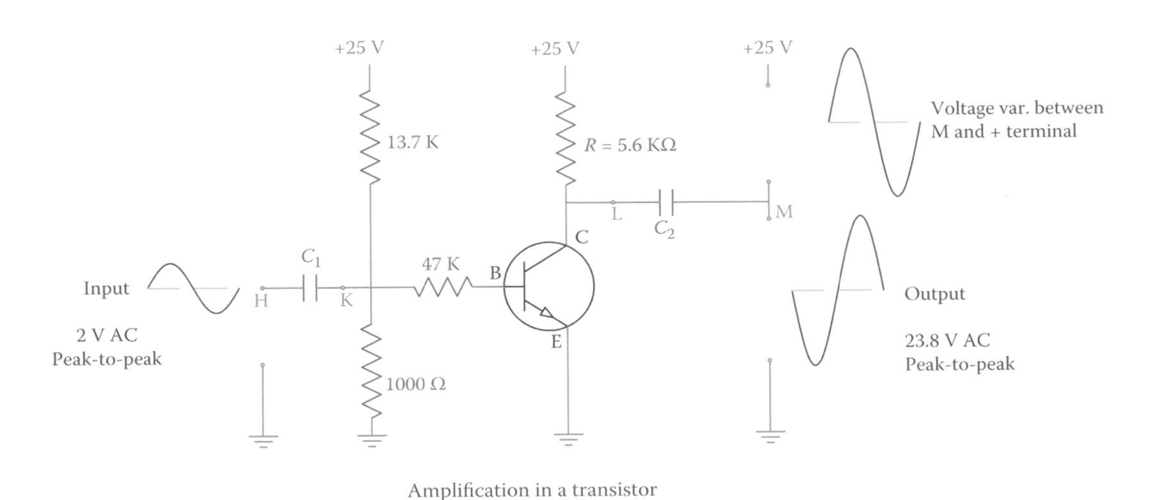

Figure 17.20 Capacitor C_2 is added to the circuit in Figure 17.19.

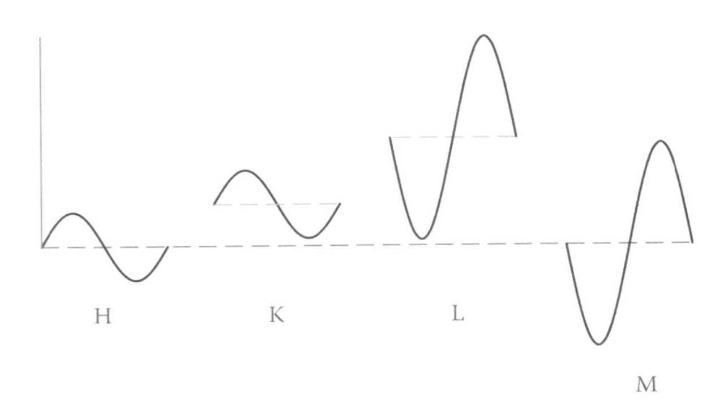

Figure 17.21 Voltage forms at various points of the circuit in Figure 17.20.

17.11 Biasing a Transistor

The action of blending the input signal with a constant DC value (observation 3 in the previous section) is extremely important and needs more elaboration. One must understand the motivation for it. Notice that as a result of this DC value, which was added to the input signal, V_{BE} (the base-emitter voltage) always has a positive value greater than 0.7 V (we assumed a silicon transistor). To better understand the subject, first, we consider a similar scenario, but instead of blending the input with 1.7 V DC, we add only 0.7 V, using a smaller resistor instead of 1000 Ω in Figure 17.18. The resultant of the voltage values at different points (similar to those in Table 17.1) associated with various instants of one cycle of the sinusoidal input signal are shown in Table 17.2 for this case, and the corresponding graph is depicted in Figure 17.22.

As can be observed, two main differences exist between the results of this case and the previous case: (1) only half of the signal is amplified and the other half is clipped (cut), and (2) the variation of the output signal is only between 25 and 13 V. That is, the power added to the input signal is only half of the previous case. The behavior of the transistor in this case is not necessarily unwanted. There are cases when such a performance is desired. This will be further studied in the classification of amplifiers in Chapter 18.

Biasing a transistor: Connecting the three terminals of a transistor to the appropriate voltage to adjust/select the operating region of a transistor.

The action of adjusting the V_{BB} in a transistor circuit so that the amplification be as desired is called **biasing a transistor**. The discussions about transistor amplification in Sections 17.10 and 17.11 are mainly to learn how a transistor works. In practice, there are many more issues to be concerned about, and it is more customary to include an emitter resistor that can also adjust the base current I_B. This is shown in Figure 17.23. Note that current through R_E corresponds to both I_B and I_C. The emitter current is obtained by dividing the emitter voltage by the value of the resistance R_E, and only a fraction of this current (1/β) passes through the base-emitter junction (see Equation 17.1). The typical value for R_E is 1/10th of the value of R_C. Observe other differences between this and the previous figures. The voltage provided at point K is partly dropped across the BE junction and partly dropped in R_E. We will discuss more about amplifier circuits in the next chapter.

TABLE 17.2 Variation of Input versus Output Voltage for the Transistor in Figure 17.20, with 0.7 V DC at Point K

Input signal angle (degree)	0	30	60	90	120	150	180	210	240	270	300	330	360	One cycle
Input signal value, V	0	0.50	0.87	1.00	0.87	0.50	0	−0.50	−0.87	−1.00	−0.87	−0.50	0	Sinusoidal values × 1
Voltage at K (V_{BB})	0.7	1.20	1.57	1.70	1.57	1.20	0.70	0.20	−0.17	−0.30	−0.17	0.20	0.7	$V_{BB} = V_{inp} + 1.7$
Voltage drop across R_B	0.00	0.50	0.87	1.00	0.87	0.50	0.00	−0.50	−0.87	−1.0	−0.87	−0.5	0.0	$V_{BB} - 0.7$
Base current, I_B, μA	0	10.6	18.4	21.3	18.4	10.6	0	0	0	0	0	0	0	$I_B = (V_{BB} - 0.7) \div 47$ k
I_C, mA	0	1.064	1.843	2.128	1.843	1.064	0	0	0	0	0	0	0	$I_C = 100\ I_B$
Voltage across R_C, V	0	5.96	10.32	11.91	10.32	5.96	0	0	0	0	0	0	0	V_{RC}
Output voltage V_{CE}	25	19.4	14.68	13.09	14.68	19.04	25	25	25	25	25	25	25	$25 - V_{RC}$

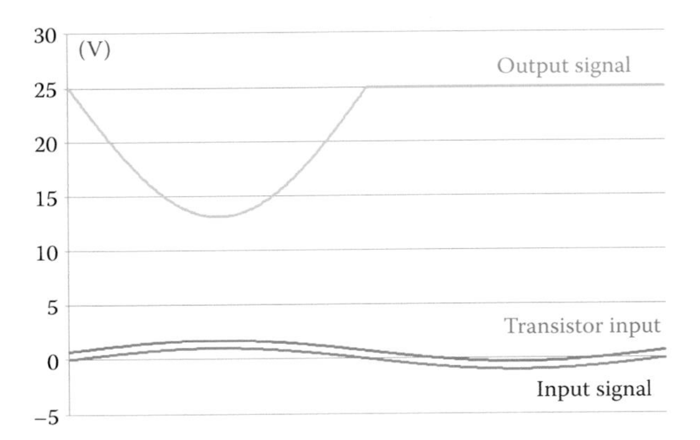

Figure 17.22 Clipping the output signal by a transistor due to different biasing.

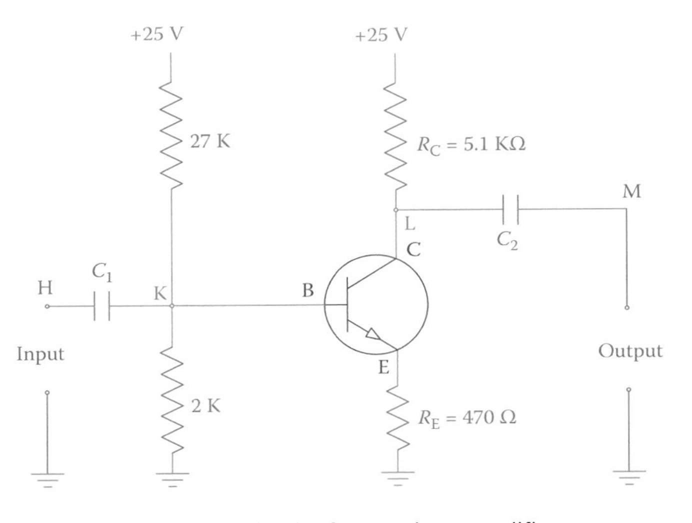

Figure 17.23 More common circuit of a transistor amplifier.

17.12 Chapter Summary

- A junction transistor is made of two PN junctions.
- Depending on how the two PN junctions are joined together, the PNP and NPN transistors are made.
- Transistors come with different packaging, and they are small.
- A transistor has three elements: base, collector, and emitter.
- A transistor has three terminals connected to its three elements, and it is an input-output device.
- As an input-output device, one of the terminals must be shared between the input and output.

- The three operating configurations, common base, common collector, and common emitter stem from which terminal is shared between the input and output.
- The primary functions of a transistor are switching and amplification.
- For a transistor to function the collector-base junction must be reverse biased, and the emitter-base junction must be forward biased.
- Without a minimum voltage between the base and the emitter, a transistor is cutoff. It does not conduct. This minimum voltage depends on if a transistor is germanium based or silicon based.
- When there is a current between the base and the emitter, a transistor conducts, meaning that the collector and emitter are internally connected and can make a closed loop with the external circuit.
- Current in the emitter is the sum of the base current and the collector current.
- In a transistor the current in the collector is much higher than the current in the base. The ratio between these two currents is almost constant for a transistor circuit. This value is denoted by the Greek letter β (beta) and depends on transistor characteristics and the circuit it is used in, as well as operating temperature.
- A transistor behaves like a variable resister between its collector and emitter. This affects the relationship between the collector-emitter voltage and the collector current (between collector and emitter).
- A heat sink is a flat or corrugated piece of metal used for components that become hot during operation, like a transistor. It can be attached to the body of a component to dissipate the heat faster and cool the item.

Review Questions

1. What are the two primary classifications of transistors?

2. What are the three elements in a transistor?

3. What are the principal functions of a transistor?

4. How many terminals does a transistor have?

5. Name the three configurations under which a transistor can function.

6. Describe the correct bias condition for a transistor to function.

7. What is the difference between a PNP and an NPN transistor?

8. What is the difference between the symbols for a PNP and an NPN transistor?

9. Is the current in the collector of a transistor greater or the current in the emitter?

10. Is the current in the base of a transistor greater or the current in the emitter?

11. When a signal is amplified by a transistor what property of the signal becomes larger?

12. If a signal is amplified by a transistor, where does the power added to signal come from?

13. What is meant by cutoff?

14. If two lines are needed for input and two lines for output, how does a transistor work as an input-output device with only three terminals?

15. Can you guess from what the name "transistor" is made of?

16. Is the transistor in Figure P17.1 a PNP transistor or an NPN transistor?

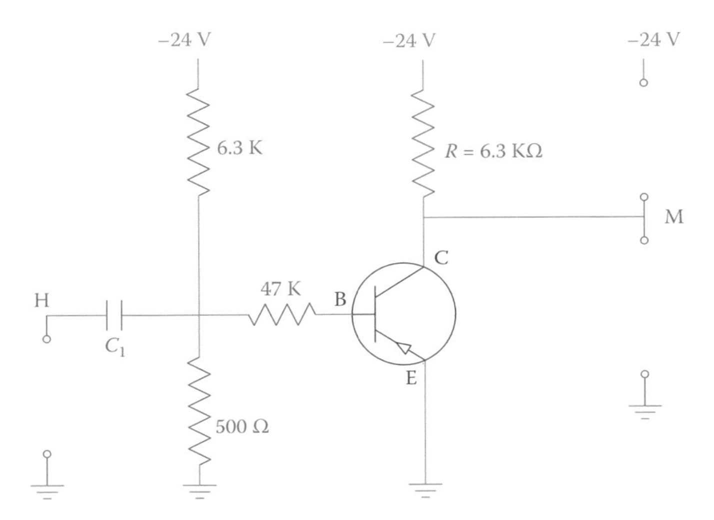

Figure P17.1 Circuit of Problem 1.

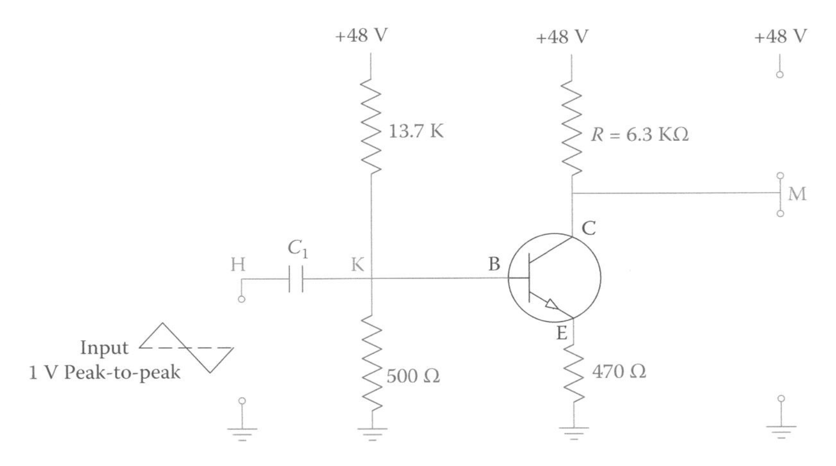

Figure P17.2 Circuit of Problems 3 and 4.

Problems

1. Find the internal resistance between the collector and the emitter for the transistor in Figure P17.1 if $I_C = 3$ mA.

2. If $\beta = 120$ for the transistor in Problem 1, find the current I_B.

3. Without the input considered, how much is the voltage V_{BB} (voltage at K) in the transistor shown in Figure P17.2?

4. For the transistor in the circuit shown in Figure P17.2 make a table similar to Table 17.1.

18

Transistor Circuits

OBJECTIVES: After studying this chapter, you will be able to

- Describe the characteristic curves of transistors
- Explain other ways that a transistor amplifier can be made
- Define various classes of amplifiers and their differences
- Use transistors in circuits
- Use a transistor for voltage or current amplification
- Define emitter follower
- Explain the input and output impedances
- Explain impedance matching
- Define voltage and current gains
- Describe coupling in amplifiers and the ways this is carried out

New terms:
Active region, breakdown voltage, cascaded, conduction angle, coupling (in amplifier), current gain, emitter follower, fixed bias, impedance matching, input impedance, load line (in transistor), output impedance, power gain, quiescent point, saturation, self-bias, swing, voltage gain

18.1 Introduction

In Chapter 17 we discussed how a transistor can be used as a switch and how it can amplify a signal. We also discussed the importance of correct biasing for transistors, in addition to many facts about transistors and their operation. There is still more to learn about transistors and their applications. One must have a good understanding of the contents of Chapter 17 before continuing with this chapter, where we discuss more about transistors and other ways that a transistor can be biased, as well as the three transistor configurations that are used for amplification. We also discuss the characteristic curves of transistors and various classes of amplifiers.

18.2 Transistor Characteristic Curves

A fundamental property of a transistor and how it works was pointed out in Figure 17.9, where a transistor is correctly biased by two power supplies. Although now we know how to use a single power supply to provide the necessary voltages for base current and collector current, here, for the sake of simplicity, we use the same figure with some modification for this discussion. Figure 18.1 shows a simple circuit of a transistor, in which the 1.5 V battery and the resistance R_B determine the base current I_B, and the 24 V battery together with R_C define the collector current I_C. We are interested in determining the variation of the collector current I_C. This current can be varied either by changing the base current I_B or the

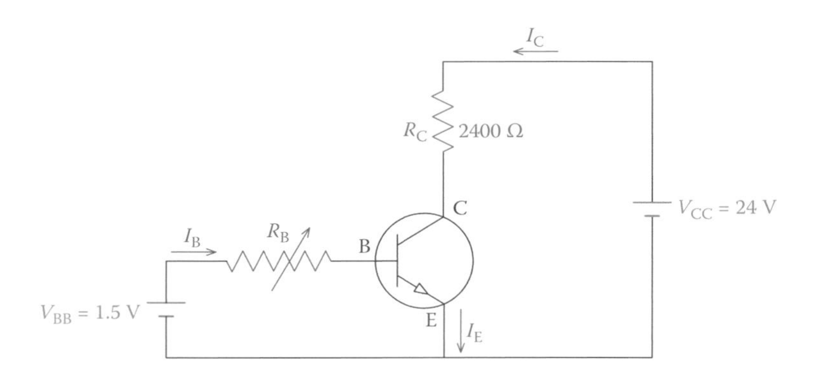

Figure 18.1 Simple circuit for a transistor operation.

collector-emitter voltage V_{CE} (the voltage between the collector C and the emitter E). See Figure 17.10. The base current can be varied by the variable resistor R_B.

The characteristic curves of a transistor provide the relationship between collector-emitter voltage and collector current for different values of the base current. Because there are two parameters that affect I_C, a set of individual curves shown together denote various operating conditions. A typical curve is shown in Figure 18.2a, and a set of these curves are depicted in Figure 18.2b. Each individual curve depicts the variation of I_C versus the value of collector-emitter voltage (V_{CE}) for a fixed value of base current I_B. When I_B is zero, a transistor is cutoff, and it does not conduct no matter how much voltage is applied to the collector; any collector current is due to leaks, is very small, and is negligible. In both Figure 18.2a and b the curve corresponding to $I_B = 0$ is exaggerated for clarity. The area under the curve corresponding to $I_B = 0$, shaded in Figure 18.2a, represents the region where a transistor is cutoff and is not conducting.

Saturation (in a transistor): The state of a transistor at which the collector current has reached its maximum value for the present collector-emitter voltage, and cannot increase further by only increasing the base current I_B.

For each nonzero value of I_B the collector current starts from zero when the collector-emitter voltage is zero. A transistor starts conducting and the collector current increases rapidly when $V_{CE} > 0$. The area around this abrupt change in I_C, also shaded in Figure 18.2a, corresponds to when a transistor is in **saturation**. Saturation implies that the collector current has reached its maximum value for that collector-emitter voltage and cannot increase further by increasing the base current I_B. For example, consider point M corresponding to $V_{CE} = V_M$ in Figure 18.2b. For this point, I_C has reached its maximum and cannot be increased by increasing I_B. In contrast, an increase in I_B can move point N to N′, both corresponding to a collector-emitter voltage V_N.

The meaning of transistor saturation is better demonstrated in Figure 18.3, in which the scale of the horizontal axis is augmented, so that the line segments with sharp slopes can be better displayed. Two characteristic curves, corresponding to two base currents I_{B1} and I_{B2} are shown. Suppose that the collector-emitter voltage is 2 V. On both curves the corresponding point is A. This implies that if the base current is increased to I_{B2}, but the V_{CE} is still 2 V, the collector current does not change. The collector current increases only if the V_{CE} increases, for instance, to 4 V, for which the operating point moves from A to B.

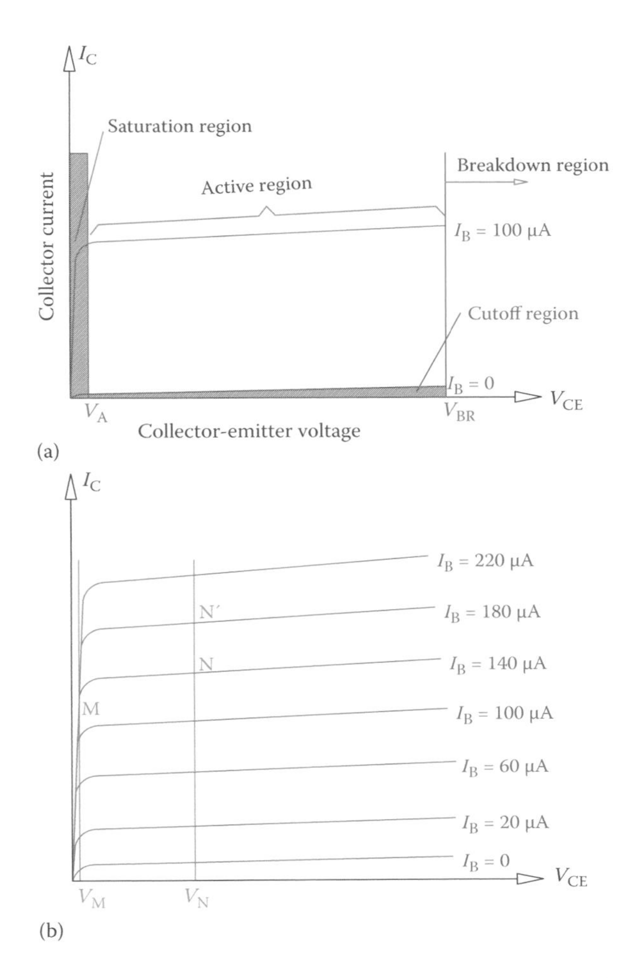

Figure 18.2 Collector current versus collector voltage characteristic curve of a transistor. (a) For one value of base current. (b) For multiple values of base current.

Active region: An area in the characteristic curve of a transistor, in terms of collector–emitter voltage and collector current values that the transistor can function. If any of these values falls outside of its range a transistor falls in the saturation region or cutoff region and cannot function (see Figure 18.2a).

Breakdown voltage: Voltage at which a semiconductor device changes behavior or gets damaged.

When saturated, a transistor cannot operate as expected. In normal operation, transistors function in the **active region**, the area that the characteristic curve is a segment of an almost horizontal straight line. In this region increasing collector-emitter voltage has little effect on the collector current. In other words, the transistor exhibits a large resistance in this region, so that increasing voltage has little effect on the current through it. This resistance is variable because it depends on the value of I_B (for each value of I_B the ratio V_{CE}/I_C is different).

The active region is between the two voltages denoted by V_A and V_{BR} in Figure 18.2a. If V_{CE} surpasses the **breakdown voltage** V_{BR}, the transistor gets damaged, and if $V_{CE} < V_A$, the transistor is in saturation state.

The operating point of a transistor is a point on these curves, corresponding to a given I_B and a given value for V_{CE}. A transistor, nevertheless, cannot work at all the possible points that can be found on the characteristic curves. This is because of the physical limitations of a transistor in handling a collector current without getting overheated and damaged. The

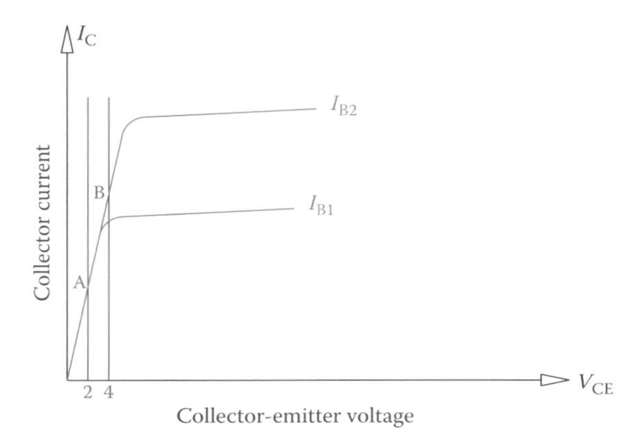

Figure 18.3 Typical set of characteristic curves for a transistor.

limiting power boundary is shown in Figure 18.4 by the dashed curve for a typical transistor. It is possible to run the transistor in all the points to the left of the dashed curve, but not on the region on the right side of this curve. In this region, either I_C is high or V_{CE} is high, resulting in relative high power consumption in a transistor, which converts to heat.

The operating point for a transistor (represented by Q in Figure 18.4) under the operating conditions governed by the supply voltage V_{CC} and the base voltage V_{BB} (see Figure 18.1) and the resistances R_B and R_C is at the intersection of lines corresponding to V_{CE} and the base current. For a constant set of values for V_{CC}, V_{BB}, and R_C if R_B is varied the value of I_B, and consequently I_C and V_{CE}, change. Each pair of I_C and V_{CE} defines an operating point denoted by Q. When as a result of varying R_B, while the parameters V_{CC}, V_{BB}, and R_C are kept constant, the values of the collector-emitter voltage V_{CE} and collector current I_C vary, this point Q moves on a straight line AB, as shown in Figure 18.4.

V_{CE} is obtained from the supply voltage V_{CC} minus the voltage drop in R_C (and any other resistor in the collector-emitter loop connected to V_{CC},

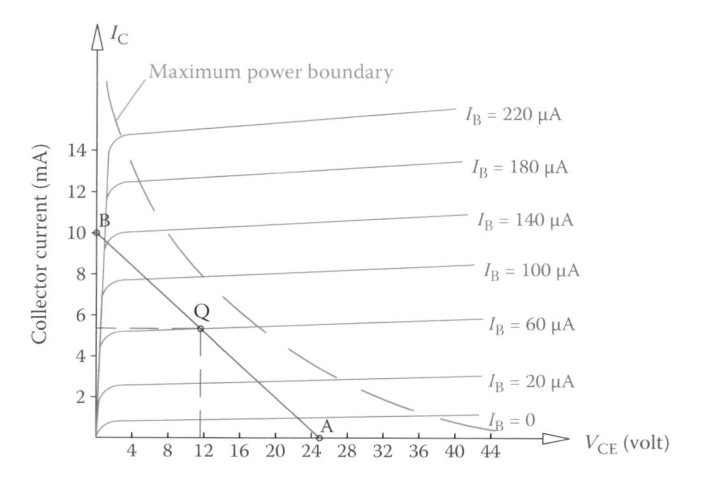

Figure 18.4 Operating point and boundary curve for the maximum power capacity of a transistor.

as we see later). In the cutoff state the current through R_C is zero and no voltage is dropped in R_C (and any other resistor in series with R_C in the same loop). Consequently, all the applied voltage (V_{CC}) appears at the collector. This defines point A of the line, where V_{CE} is maximum. Also, if a transistor is conducting, but there is no internal resistance between C and E, this defines the maximum current that the collector can have (point B of the line). This maximum current can be found by dividing V_{CC} by all the resistors in the loop (only R_C in Figure 18.1). Thus, for a given supply voltage V_{CC} the collector current can vary between zero (at point A) and a maximum value (at point B), as the line AB shows. The intersection of this line with one of the transistor characteristic curves defines the operating point Q. The line AB is called the **load line**. The load line, thus, represents all the possible locations of point Q for a transistor in a given circuit with constant V_{CC} and resistance(s) in the collector-emitter circuit.

Load line (in transistor): Line showing the operating points of a transistor on a transistor characteristic curve, based on the supply voltage and the resistors in the transistor circuit. This line is between the points corresponding to the maximum collector voltage and maximum collector current.

Example 18.1

For the transistor shown in Figure 18.1, whose characteristic curves are shown in Figure 18.4, if the base current is 60 μA, find the collector current and the value of ß.

Solution

The maximum current through the collector is defined by dividing the supply voltage (24 V) by the resistor R_C. Thus, point B end of the load line on the current (vertical) axis is at

$$24 \div 2400 = 0.01 \text{ A} = 10 \text{ mA}$$

The other end of the load line (point A) is on the horizontal axis at 24 V. This line is the same as shown in Figure 18.4 and intersects the curve corresponding to 60 μA base current at point Q (as shown). Dropping a perpendicular from Q to the vertical axis gives the current in the collector. On the basis of the figure this current is 5.5 mA.

The value for ß can be found from dividing collector current by the base current.

$$5.5 \text{ mA} \div 60 \text{ μA} = 5500 \text{ μA} \div 60 \text{ μA} = 91.67 \approx 92$$

Example 18.2

If in Example 18.1 the supply voltage changes to 30 V and R_C is 2200 Ω (other parameters remaining unchanged), how much are I_C and V_{CE}?

Solution

A new load line needs to be drawn on the same graph of Figure 18.4. The intersection of this line with the line corresponding to I_B = 60 μA defines the values of I_C and V_{CE}.

Point A for this new line is at V_{CE} = 30 V, thus between 28 and 32 (not shown), and point B is at I_B = 30 ÷ 2200 = > 13.6 mA. If you draw this line you will find its intersection with 60 μA base current, which corresponds to I_C = 5.6 mA and V_{CE} = 17.4 V.

18.3 Common-Emitter Amplifier

As explained in Section 17.5 and depicted in Figure 17.8, there are three modes of operation of transistors: common emitter, common base, and common collector. Consequently, there are three ways a transistor can be employed as an amplifier, and, therefore, we have three categories of amplifiers: common-emitter amplifier, common-base amplifier, and common-collector amplifier.

In each category the common element of a transistor is shared between the input and the output. The example we studied in Chapter 17 had the structure of a common-emitter amplifier, because the emitter was grounded, the input was introduced between the base and ground, and the output was taken from the collector. Nevertheless, the load was not connected to the points where normally is considered for the output. This implies that in electronics designs there are always different ways to perform the same function.

A common-emitter amplifier is the most widely used category among the three. Figure 18.5 illustrates a common-emitter amplifier. In this figure, as seen before, for biasing the transistor and supplying the base with a suitable voltage, use is made of a voltage divider between the supply voltage and the ground. Moreover, as pointed out in Chapter 17 (see Section 17.11 and Figure 17.23), R_E is added to adjust the base current, as opposed to R_B in Figure 18.1. The input is introduced between the base and ground, and the output is taken between the collector and ground. The output voltage is equal to V_{CE} plus the voltage drop in R_E. Notice the phase relationship between the input and the output. They are 180° out of phase.

In practice, instead of the circuit in Figure 18.5a, the circuit in Figure 18.5b is used, in which C_E is added. In this circuit R_E adjusts the voltage at E as necessary. The role of the bypass capacitor C_E is to provide a ground at the emitter for AC signals. In this way the voltage at the emitter is pure DC (without variation) particularly at higher frequencies for which the reactance of C_E is small and acts as a short to ground (for AC signals).

In ordinary applications the resistor R_E is normally chosen such that the voltage across it is around one volt. Note that because R_E is in the emitter circuit the voltage across it determines the emitter current I_E. Also, because I_B is normally very small, the biasing voltage can be selected by noting that it must compensate for the voltage drop in R_E and the drop across the BE junction. In practice, a good value for R_E is around 10 percent of R_C.

> In a common-emitter amplifier the input and the output signals are 180° out of phase.

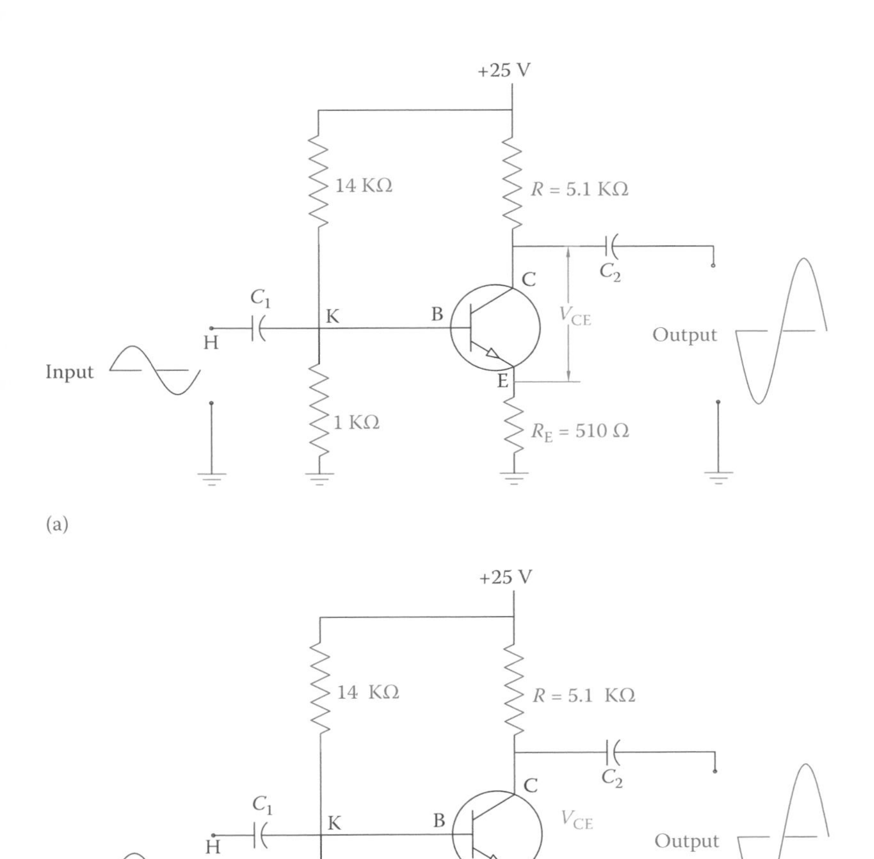

Figure 18.5 Schematic of a common-emitter amplifier. (a) Typical resistance values for a general application transistor. (b) Capacitor C_E is added to emitter to provide a zero ground for AC signals.

Example 18.3

Draw the load line for the transistor amplifier circuit shown in Figure 18.6a. If the operating point Q is in the middle of the load line, find I_C, V_{CE}, and the voltage at the emitter.

Solution

The load line is shown in Figure 18.6b. The point corresponding to the maximum V_{CE} is at 25 V, and the value of maximum I_C, defining the other end of the load line is

$$I_{\text{C-max}} = \frac{25}{2200 + 200} = 0.0104 \text{ A} = 10.4 \text{ mA}$$

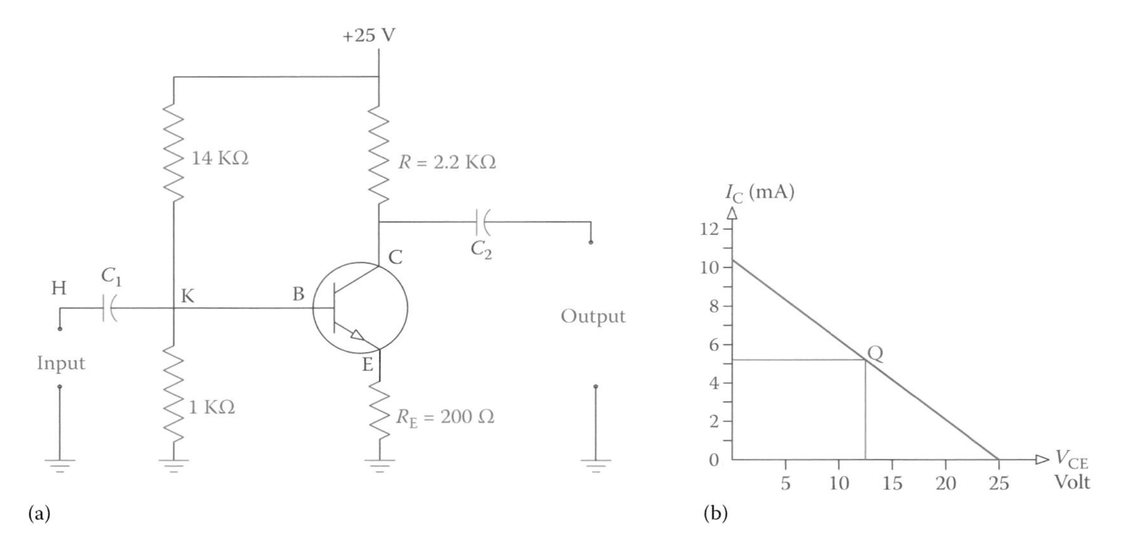

Figure 18.6 (a) Transistor circuit and (b) the load line, for Example 18.3.

At the midpoint of the load line values of both I_C, V_{CE} are half of the maximum values. Thus, $I_C = 5.2$ mA and $V_{CE} = 12.5$ V. Because $I_E \approx I_C$ we can say with good approximation that the emitter voltage is equal to the voltage drop in the 200 Ω resistor. Thus,

$$V_E \approx (200)(5.2) \div 1000 = 1.04 \text{ V}$$

Fixed bias: Type of biasing in a transistor in which a fixed resistor is connected between the base terminal and the power supply, thus providing a fixed voltage at the base.

Self-bias: Type of biasing in a transistor in which a fixed resistor is connected between the base terminal and the connector, thus providing a voltage at the base that varies with the current through the collector.

We have already seen how a transistor can be biased. At this point we introduce two more ways that the transistor in Figure 18.5 can be biased. These are shown in Figure 18.7a and b. Notice that the emphasis is on the biasing methods. Observe the difference between the three ways of biasing, by comparing Figures 18.5b, 18.7a, and b. All of these alternatives are used in practice. The terms used for distinguishing between the various ways of biasing as shown are **fixed bias** for Figure 18.7a and **self-bias** for Figure 18.7b. The amplifier in Figure 18.5 is voltage divider biased. In a fixed bias configuration the voltage at point K is a fixed value, determined by the supply voltage and the resistors involved. In a self-bias configuration the voltage at K is not fixed, but it is of the correct polarity to forward bias the base-emitter junction.

18.4 Common-Base Amplifier

In the same way that we have seen for a common-emitter amplifier so far, it is possible to use a transistor with common-base configuration for amplification purposes. The conditions for correct biasing do not change. Only

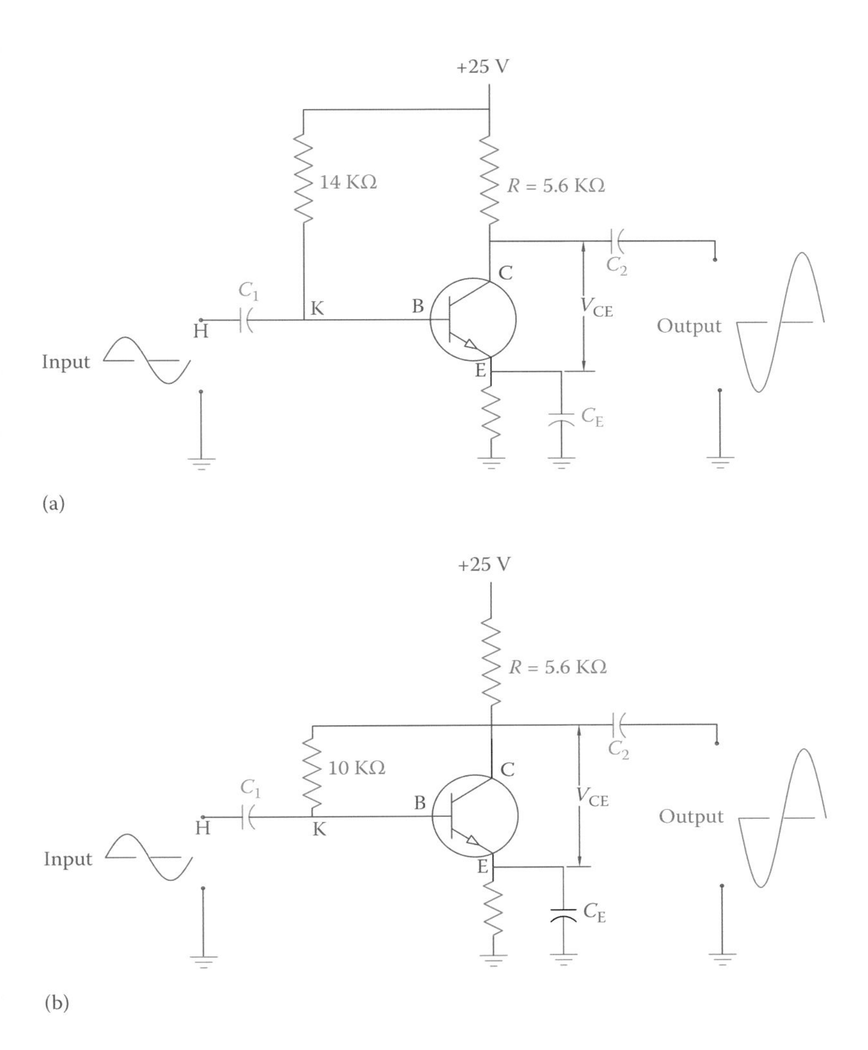

Figure 18.7 Common-emitter amplifiers with (a) fixed bias and (b) self-bias.

the base must be common between the input and the output. Figure 18.8 illustrates a typical case for a common-base amplifier. As you recall, the base-emitter junction must be forward biased; thus, the base voltage for an NPN transistor must be higher than the emitter voltage. In this respect, if the base is at zero voltage, the emitter voltage needs to be negative. In such a case, a power supply with both positive and negative polarity, similar to that depicted in Figure 16.17 (see Section 16.7) may be utilized to provide the necessary biasing voltages.

To avoid the necessity for two power supplies, the base can be biased by one of the alternative methods so that it is still forward biased while the emitter voltage is raised to zero. This is shown in Figure 18.8b.

As also shown in Figure 18.8, for a sinusoidal input the output signal is in phase with the input signal. This is a characteristic of the common-base

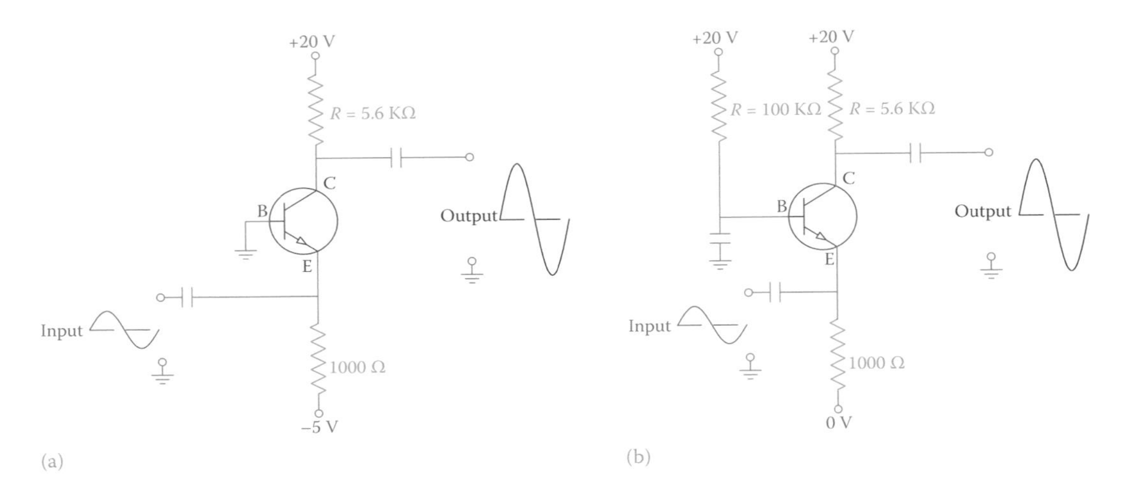

Figure 18.8 Schematic of a common-base amplifier. (a) Base is directly grounded (zero voltage), or (b) it can have a DC voltage (a capacitor is required), then emitter can be at zero voltage.

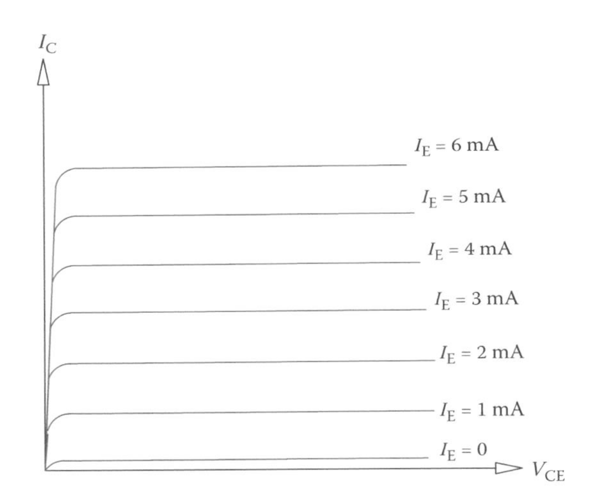

Figure 18.9 Characteristic curves for a common-base amplifier.

amplifier. For a common-base amplifier the characteristic curves define the variation of collector-emitter voltage versus collector current for various emitter currents. In the active region the collector current stays constant no matter how much V_{CE} changes (this is obvious, because $I_C \approx I_E$). This fact is reflected in the curves shown in Figure 18.9, each of which is almost a horizontal line segment.

> In a common-base amplifier the input and output signals are in phase.

18.5 Common-Collector Amplifier

In a common-collector amplifier, as the name implies, the collector is common between the input and the output; thus, the input signal is introduced between the base and the collector, and the output signal is taken between the emitter and the collector.

Figure 18.10 illustrates the schematics of a common-collector amplifier. The collector has no resistance attached to it and is grounded through a capacitor C_2. The role of the capacitor C_2 is to block the DC voltage of the power supply from ground. Otherwise, the supply is directly grounded. In the configuration shown, thus, the input is applied between the base and ground, and the output is taken from the emitter and ground. The roles of the capacitors C_1 and C_3 are, again, to isolate DC voltages and signals from being transferred from the input to the base and reaching to the output (see Section 17.10).

A characteristic of the common-collector amplifier is that the output signal is in phase with the input signal, as shown in Figure 18.10. Also, for this configuration the output current is I_E, which is much larger than the input current I_B. In fact, a common-collector amplifier is used for amplifying current because the output current is a direct amplified value of the input current. That is, the output obtained from the emitter is following the pattern of the input to the base, but with a higher current capability. For this reason the common-collector amplifier is called an **emitter follower** (emitter follows the input).

Emitter follower: Name given to common collector amplifier.

Another characteristic of the common-collector amplifier is that the output voltage is (slightly) smaller than the input voltage. This can be seen from the voltage relationships between the points H and G (the ground),

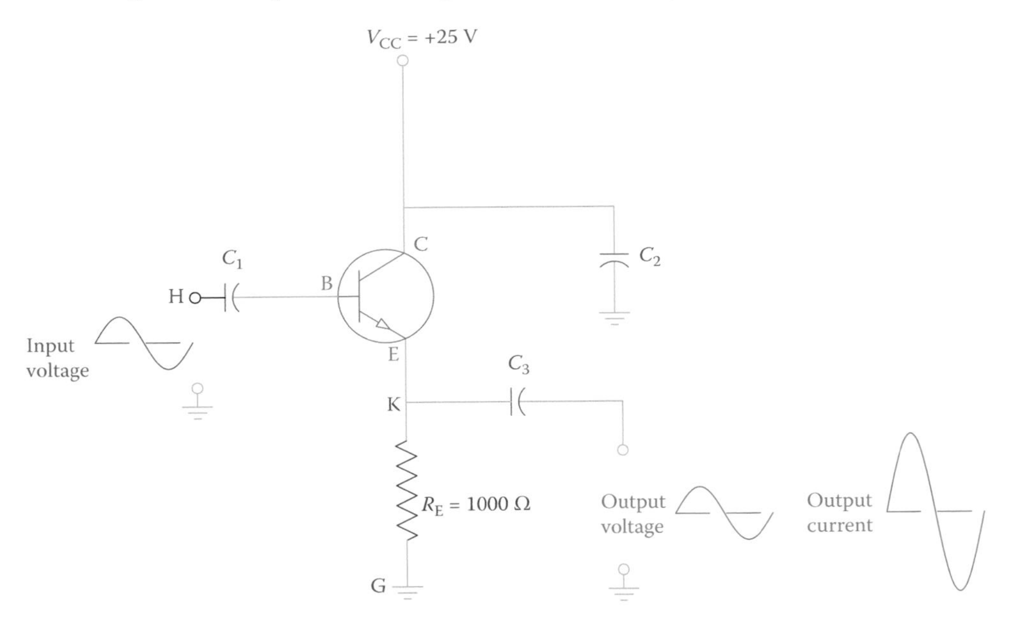

Figure 18.10 Schematics of a common-collector amplifier (emitter follower).

Emitter follower or common-collector amplifier does not amplify voltage. It is a current amplifier.

and K and G in Figure 18.10 ($V_{HG} > V_{KG}$, considering the voltage drop between B and E junction in the transistor).

18.6 Current Gain, Voltage Gain, and Power Gain in Amplifiers

From the study of amplifiers it can be seen that in an amplifier a transistor is governed by the input signal and generates an output signal that (1) has the same frequency as the input and (2) its voltage or its current capacity (or both) are several times larger than those of the input signal.

In practice, on the basis of the requirement, the voltage of a signal, the current of a signal or both must be amplified. If both the current and the voltage need to be amplified, two or more amplifiers must be cascaded together when at least one is for voltage amplification and one is for current amplification (see Section 18.9).

Current gain: Ratio of the output current variation to the input current variation in an amplifier.

Voltage gain: Ratio of output voltage variation (output swing) to the input voltage variation (input swing) in an amplifier.

Swing: Total change (the difference between the maximum and minimum values) of the input or the output in an amplifier.

Power gain: Ratio of output power to the input power in an amplifier.

The quantitative amplification of current and voltage are called **current gain** and **voltage gain**, respectively. They represent the ratio of output current (voltage) variation to the value of the input current (voltage) variation. The term **swing** is employed to refer to the total variation of voltage or current. Because in most applications the input signal is not DC, it is standard practice to consider a sinusoidal waveform to study the characteristics and gains of an amplifier. In this respect, the current gain and voltage gain can be measured by the ratio of the peak-to-peak value of the sinusoidal output signal (output swing) to the peak-to-peak value of the sinusoidal input signal (input swing).

In addition to the current gain and voltage gain, we also have **power gain**. Power gain determines how much the output power is larger than the input power. Because power is the product of voltage and current, power gain is determined from multiplication of current gain and voltage gain.

An amplifier current gain, voltage gain and power gain are denoted by A_C, A_V, and A_P, respectively. The following relationships summarize the definitions for the three gain values in amplifiers. The Greek capital letter Δ (delta) is employed to represent the total change or swing. For an AC signal the total change is the peak-to-peak value.

$$\text{Current gain } (A_C) = \frac{\Delta I_{out}}{\Delta I_{in}} \tag{18.1}$$

$$\text{Voltage gain } (A_V) = \frac{\Delta V_{out}}{\Delta V_{in}} \tag{18.2}$$

$$\text{Power gain } (A_P) = A_C \times A_V \tag{18.3}$$

TABLE 18.1 Characteristics Summary of the Three Types of Amplifiers

	Common Emitter	Common Base	Common Collector
Current gain	Normal	Less than 1	Normal
Voltage gain	Normal	Normal	Less than 1
Phase relationship (input-output)	180° out of phase	In phase	In phase

As we have seen so far, the three types of amplifiers have different characteristics. Table 18.1 summarizes the properties of these single amplifiers. The word *normal* is used to refer to a practically acceptable value for amplification.

Example 18.4

Referring to the amplifier whose characteristic curves are shown in Figure 18.11, the output is taken from the collector. If the input voltage variation is between 40 and 160 mV and the output voltage changes between points M and N, as shown, how much is the amplifier voltage gain.

Solution

The output has a swing of 14.6 V – 6.8 V = 7.8 V, and the input has a swing of 0.160 V – 0.040 V = 0.120 V. The voltage gain, thus, is

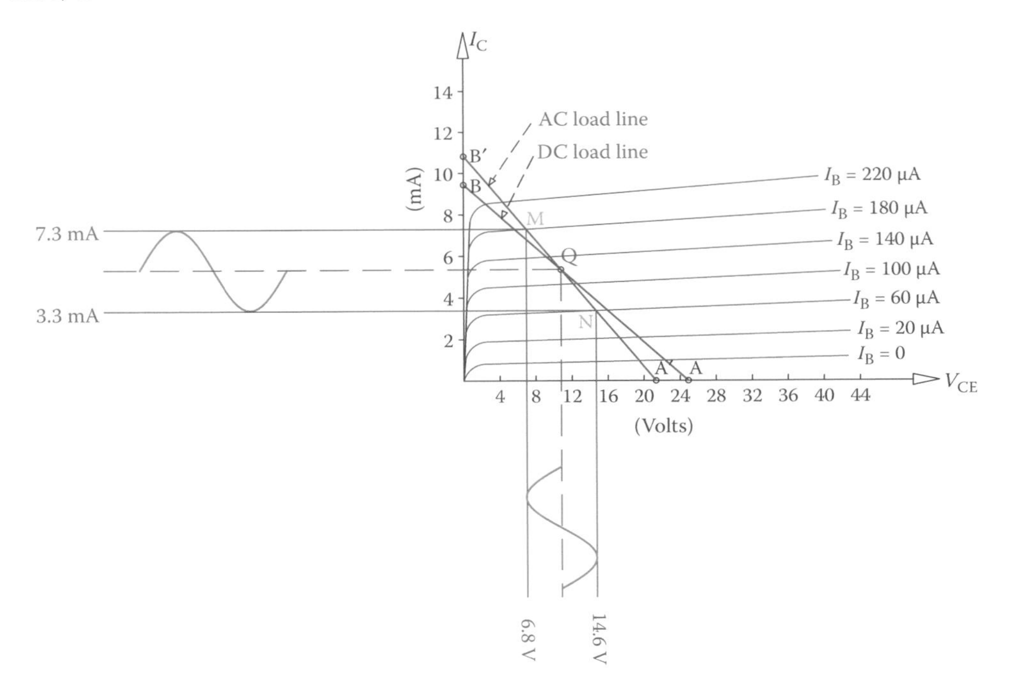

Figure 18.11 Relative positions of DC and AC load lines.

$$\text{Voltage gain} = \frac{7.8}{0.12} = 65$$

18.7 Amplifier Classes

Transistors are widely used in radio, TV, sound recording and play back, communications, and so on. In practice, often a few single transistor amplifiers are cascaded together in order to obtain a desired level of amplification of voltage and power. The discussion about specific application of amplifiers is outside the scope of this book. Thus, the discussion below is for single transistor amplifiers.

There are a number of classes of amplifiers, but four of them are more common. These are classes A, B, C, and AB. Before we discuss the difference between them and the reason why they are needed, we study the effect of biasing a transistor when its input is an AC signal. As usual, we consider a sinusoidal waveform as the signal to be amplified. This is a reinforcing of what we have already seen along the discussions in Sections 17.10 and 17.11.

Consider a common-emitter (CE) amplifier, as shown in Figure 18.5. We have already defined the load line and what it implies. First, without going into details, there are two load lines for a transistor, one corresponding to DC operation and one corresponding to AC operation, called **AC load line**. The AC load line is for when the input signal is AC and varies; the DC load line is for the case that the input signal has not much rapid change and can be considered to be DC. For a given Q point of a transistor in a circuit the position of the AC load line has a slight clockwise rotation about point Q with respect to the DC load line, as shown in Figure 18.11. Line AB in this figure represents the DC load line and line A′B′ is the AC load line.

Quiescent point: Point on the load line of a transistor in a circuit. This point corresponds to a condition that the circuit input does not change (stays quiescent).

Point Q, thus far referred to as operating point, is called the **quiescent point**, because it determines the "no signal" condition of a transistor. This point is defined by the bias condition of the base (assumed CE amplifier) governed by the resistances used in the circuit, the internal resistances of a transistor and the supply voltage. In other words, if there is no input to the transistor the collector current (I_C) and the collector-emitter voltage (V_{CE}) are those corresponding to point Q. In Figure 18.11 this point corresponds to $I_C = 5.3$ mA. Now, suppose that the input signal is sinusoidal and causes the collector current change between 3.3 and 7.3 mA. This causes the operating point of the transistor to move on the AC load line between points M and N (see Figure 18.11) and generates a sinusoidal output signal whose peak-to-peak value is between 6.8 and 14.6 V, as depicted in Figure 18.11. As far as M and N are on the line A′B′ and do not pass over the two ends of this line segment, the transistor can amplify the input signal and reproduce a complete sinusoidal signal at its output.

Figure 18.12 shows a similar representation to that depicted in Figure 18.11. The characteristic curves and the DC load line have been removed for clarity, and only the AC load line is shown. The biasing condition leads to a quiescent point Q, almost in the middle of the load line.

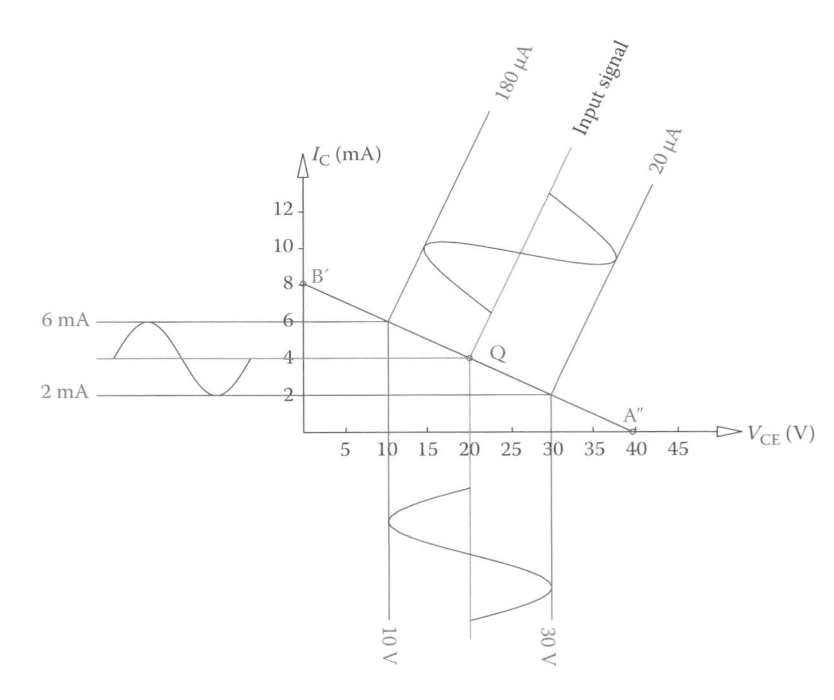

Figure 18.12 Effect of an AC input on the transistor collector values.

Suppose now that a sinusoidal input signal is introduce at the amplifier base. When blended with DC bias voltage, it causes the base current to swing between 20 and 180 μA; that is, a swing of 160 μA. This causes the operating point of the transistor to travel along the load line. The variation of the sinusoidal input signal causes the collector current to assume values between 2 and 6 mA. In other words, the collector current has a swing of 4 mA. Consequently, the collector-emitter voltage has a swing of 20 V (between 10 and 30 V), as depicted in Figure 18.12. For the operating condition shown, an input signal with larger peak-to-peak values leads to a larger swing in the base current, and consequently, a larger variation in the collector current. Nevertheless, the maximum allowable collector current swing is between 0 and 8 mA. Any larger signal that forces the collector current beyond its extremities causes the transistor either to cutoff or to saturate. In both these conditions a transistor cannot perform as expected.

Figure 18.13 depicts the same transistor with three different bias conditions. Figure 18.13a is the same as Figure 18.12 except that the input signal representation is removed. It is repeated for ease of comparison with Figure 18.13b and c. We want to draw attention to the important fact that if the quiescent point Q is not in the middle, even the signal that causes a 4 mA swing in the collector current can run the transistor into cutoff or an unfeasible condition. As a result, for those values that the transistor can function it produces an output with the proper voltage, and for those values outside the operational conditions, it does not function. The outcome, consequently, is an incomplete or truncated output, as shown in Figure 18.13b and c. The same conclusion was shown in a different way in the discussion in of Chapter 17 (Section 17.11).

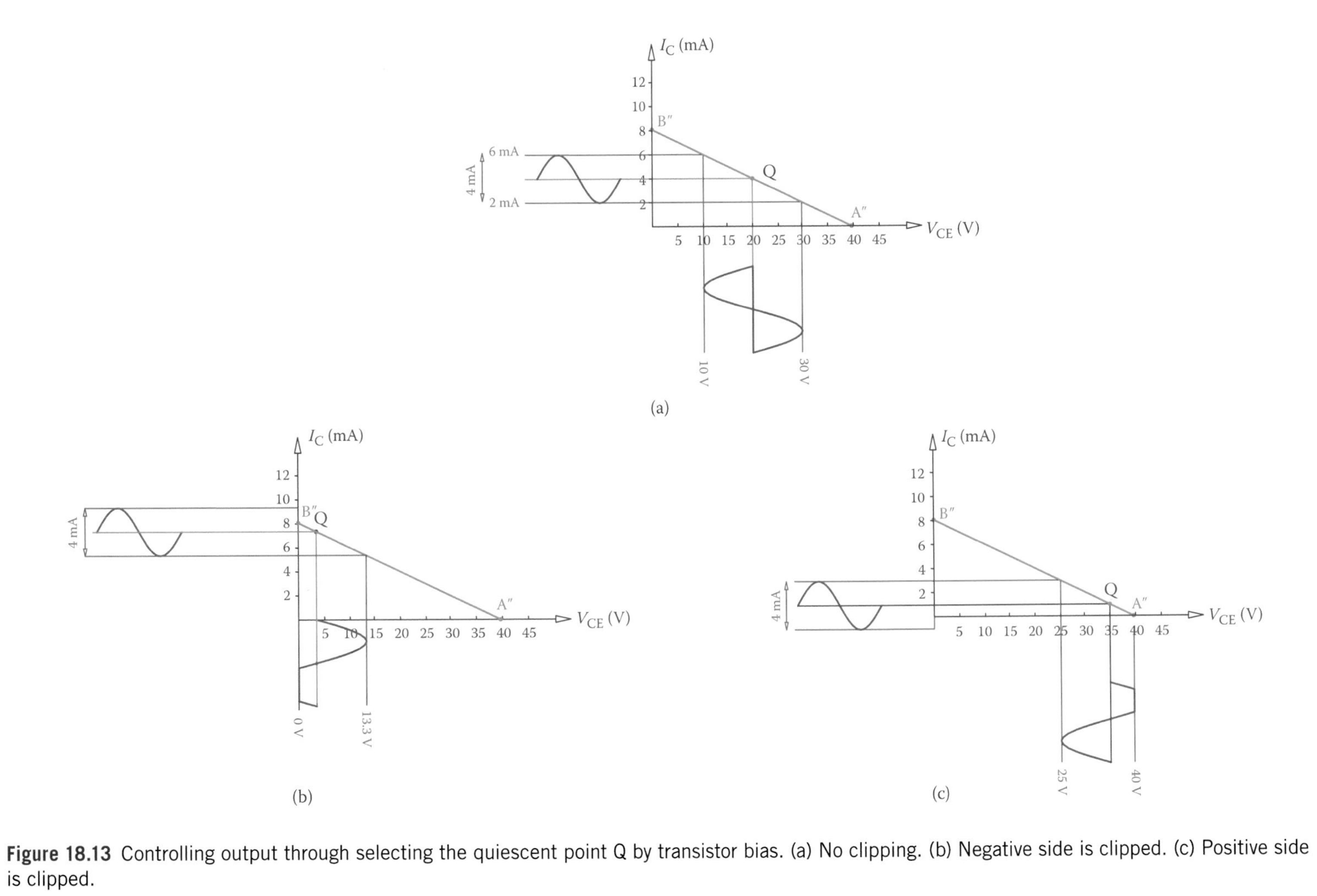

Figure 18.13 Controlling output through selecting the quiescent point Q by transistor bias. (a) No clipping. (b) Negative side is clipped. (c) Positive side is clipped.

The operating conditions depicted in Figure 18.13b and c are not necessarily useless or unwanted. In fact, from certain viewpoints they are desirable. For instance, a transistor is not conducting 100 percent of the time, and it finds time to cool off in each cycle when it is in the cutoff state. The amplifier classes stem from the operating bias condition (initial positioning of point Q). There are more than four classes of amplifiers, but four of them are more common, as described below.

Class A Amplifier

A class A amplifier amplifies the entire input signal in the same manner as we have already seen in Figures 17.19 and 18.12. This is through biasing a transistor in such a way that the range of voltages in the signal never run the transistor into cutoff. This implies that the transistor is always conducting, and, therefore, a good portion of the power consumed is converted to heat in the circuit components and the transistor, itself.

Class B Amplifier

A class B amplifier amplifies only 50 percent of the input signal; thus, operating only 50 percent of the time, just in the same way that we have already seen in Table 17.2 and Figure 17.22. This is done, as seen before, by proper biasing of the transistor to operate only for one half, either the positive parts or the negative parts, of the input. When amplification of the whole input signal is desired, two amplifiers are used back to back; each one acts only on half of the input signal. Then the two halves are put together (the two transistors are normally of two different types for this application; one is PNP and the other is NPN). The advantage of such an arrangement is that each transistor works only for half of the time and is cutoff for the other half. In this way the heat generated is less than in a class A amplifier, and higher power can be handled. A class B amplifier, thus, has a better efficiency than class A.

Class AB Amplifier

Although from an efficiency viewpoint a class B amplifier is preferred to class A, it suffers from introducing some degree of distortion to a signal, which is not acceptable in many cases, such as for sound and video signals. A third class, which overcomes this drawback of class B is class AB amplifier. Class AB has the same structure as class B, with slight modification, so that it operates at more than 50 percent of time (clips the output by less than 50 percent). This improves the performance when used in pairs.

Class C Amplifier

A class C amplifier has a bias such that more than 50 percent of the input signal is clipped, and the rest appears in the output. There are many applications for which class C amplifier is sufficiently acceptable or even more appropriate. The energy loss of class C amplifier is, therefore, less than that of class B.

To define the performance of an amplifer with respect to its class of operation, a term **conduction angle** (or **angle of flow**) is employed. The

Conduction angle: The same as "angle of flow."

Angle of flow: Measure of percentage of time (based on 100 percent for 360°) for one cycle of a sinusoidal waveform during which an amplifier is in an operating state (transistor is conducting in a transistor amplifier.)

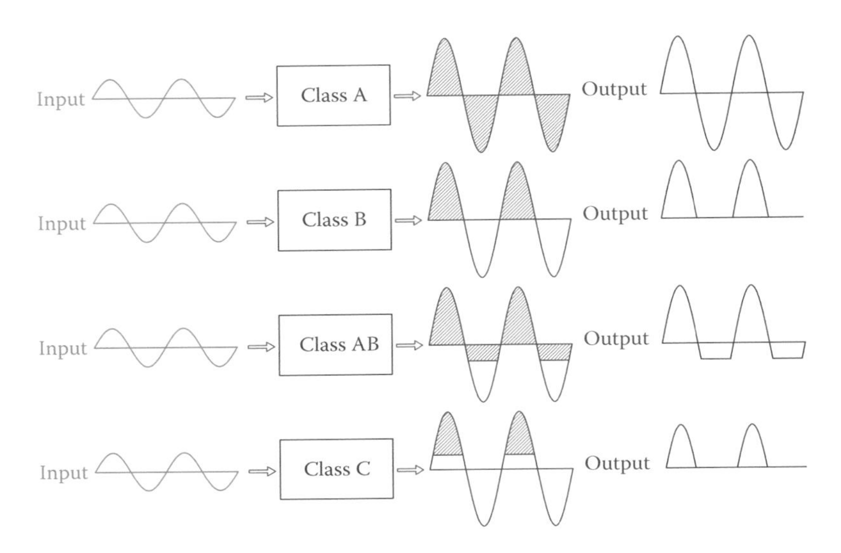

Figure 18.14 Input-output relationship of four classes of amplifiers.

coduction angle defines what percentage of a signal is amplified. Using a sinusoidal signal, 100 percent corresponds to 360° conduction angle, and 50 percent corresponds to a conduction angle of 180°. In this sense, class A and class B amplfiers have conduction angles of 360° and 180°, respectively. Accordingly, the condution angles of class AB and class C amplifers are more than 180° and less than 180°, respectively. The output of the four classes of amplifiers to a sinusoidal input is shown in Figure 18.14.

Comparing the above four classes of amplifiers, it can be seen that the initial position of the quiescent point Q on the load line determines the class. The amplitude of the input signal (input swing) comes into consideration for class A, AB, and C. For class B, point Q must be at either of the two ends of the AC load line.

18.8 Input and Output Impedance

In all amplifiers, there is an input signal from an outside source that is connected to the amplifier input, which is amplified and delivered to an output device. For instance, in a wind turbine the wind speed is measured by a type of sensor. The measurement value will be in the form of a voltage or current that varies with change in the wind speed. If this signal is small, it needs to be amplified before being processed for integration in the turbine control software. Examples of this nature are plenty in industry. Normally, when there is a measurement of any varying parameter, the measured signal needs amplification.

The discussion here is not specific to an amplifier. It is true for many devices with an input or/and an output. We refer to an amplifier as a good example because so far we have gained some knowledge about them.

Figure 18.15a illustrates the schematics for the input and output connection to an amplifier. The amplifier is connected to the input device through point terminals A and B and to the output device through points

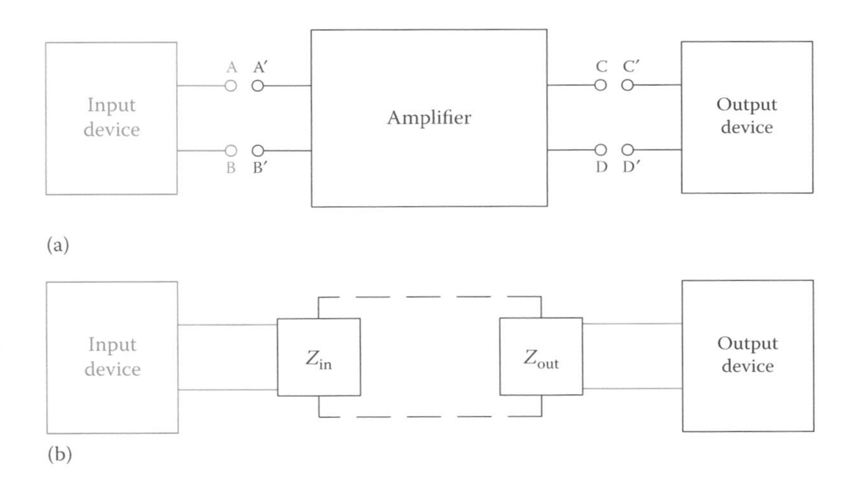

Figure 18.15 Definition of input and output impedances. (a) Amplifier with its input and output connections. (b) Amplifier circuit as seen from the outside device.

> For maximum power transfer the impedances of two devices that are to be connected must be equal to each other.

C and D. Because an amplifier has various components, its effect on any external device makes it look as if that device is connected to a circuit consisting of resistors, inductors, and capacitors, having a measureable impedance value. That is, the amplifier appears as an impedance to the external device and exhibits some impedance value toward the circuit of that device, as shown in Figure 18.15b.

The **input impedance** is the value of impedance an amplifier exhibits to its input device, and the **output impedance** is the value of impedance an amplifier exhibits to its output device. These values are specific to an amplifier and are independent of the external devices. When connecting to an external device, the input and output impedances of an amplifier are in series with the impedance of that external device. In this respect, the voltage that arrives to the amplifier input, or is delivered to the output device, is determined from the voltage divider that is formed this way. This is schematically shown in Figure 18.16.

Input impedance: Value of impedance an amplifier (or a circuit, in general) exhibits to its input device, so that the device sees the amplifier as an impedance of that value.

From a practical viewpoint the values of input impedance (Z_{in}) and output impedance (Z_{out}) are very important. The importance is in matching the impedance values between each two devices that must be connected together. In the same way that an amplifier exhibits impedance to the input and output devices, the input and output devices exhibit impedance to the amplifier. This is true, indeed, for any two devices that must be connected together. For maximum power transfer between two devices that are to be connected their impedances must be equal. If they are not exactly the same, their values must not be too much different. If instead of maximum power transfer, some other criterion is more important, for instance, for maximum voltage gain, then this rule does not need to be enforced.

Output impedance: Equivalent impedance an electronic device (such as an amplifier) exhibits to the device connected to its output.

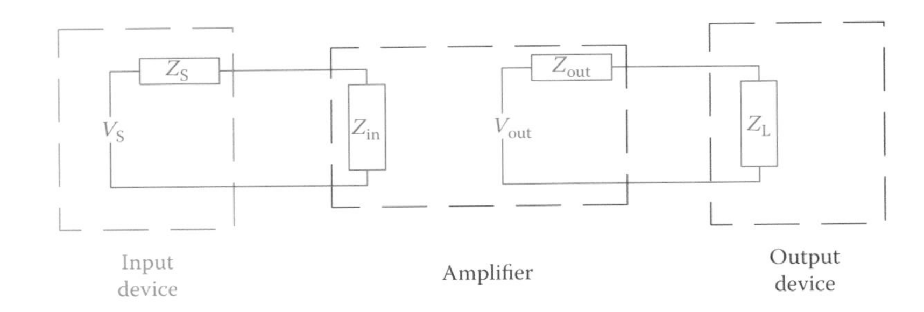

Figure 18.16 Input (output) impedance is in series with the external device impedance.

18.9 Coupling and Impedance Matching

Cascaded: Way of putting devices of the same functionality together in succession to each other for enhancing their effect, such that the output from one is the input to the next, like two or more amplifiers.

Coupling (in amplifier): Act of properly connecting an amplifier to its input or output or connecting two amplifiers in cascade.

Impedance matching: Selecting suitable values for resistors, capacitors and/or other components so that the output from one device matches the input of the following device for a desired purpose such as maximum power transfer or other criteria.

In many electronic applications in which a signal must be amplified, it becomes necessary to carry out this process in a number of stages and through more than one amplifier. In such applications a number of single amplifiers, as we have seen so far, are put in series with each other. This means that the output from the first amplifier serves as the input for the second, and the output from the second amplifier is the input for the third, and so forth. The amplifiers are said to be **cascaded** together, and each amplifier constitutes a stage of amplification.

The last stage is normally a power amplifier (amplifying current). For example, if there are only two amplifiers, the first one is a voltage amplifier and the second one is a current amplifier (and not the reverse). When two or more single amplifiers are cascaded together, the final gain of the system is obtained by multiplying the gains of the individual stages. For instance, for a three stage amplifier the final power gain is

$$A_P = (A_P)_1(A_P)_2(A_P)_3 \tag{18.4}$$

where A_P stands for power gain, and the subscripts 1, 2, and 3 refer to the power gain for each stage.

Connecting the output of each amplifier to the next stage is called **coupling**. For efficient working of the set of amplifiers in cascade it is important that their impedances are appropriate for connecting together. In other words, they must match together. For example, if maximum power transfer is concerned, the mating impedances must be equal. Other objectives may require other conditions. This is called **impedance matching**. It is also true for the first and the last stages that connect to external devices. The output impedance of an input device must match the input impedance of an amplifier. Similarly, the input impedance of an output device must match the output impedance of an amplifier.

There are a number of coupling methods. These are direct coupling, RC coupling, LC coupling (or impedance coupling), and transformer coupling.

In a direct coupling the output of one stage is directly connected to the input of the next stage. This is shown in Figure 18.17, where two NPN common-emitter amplifiers are directly coupled together. In Figure 18.17 the only connection between the two transistor amplifiers is by the line MN.

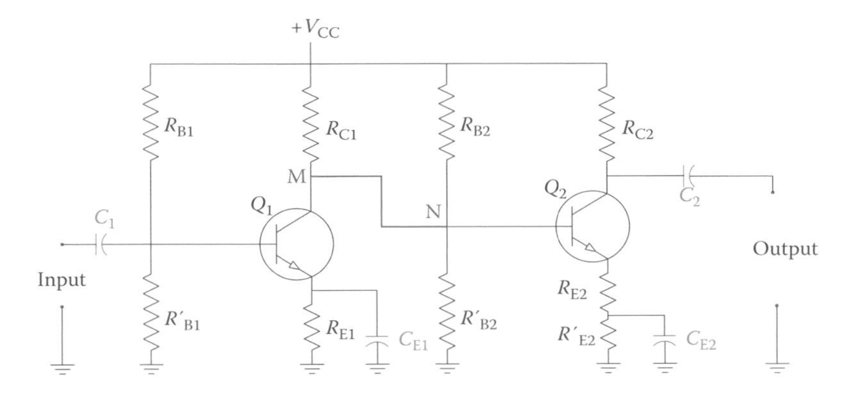

Figure 18.17 Direct coupling between transistors Q_1 and Q_2.

Figure 18.18 illustrates an RC coupling or (resistive-capacitive coupling). This circuit is exactly the same as in Figure 18.17 with the exception that a capacitor has been replaced for the line MN. In this configuration the only link between the two transistors is the capacitor C.

A third type of coupling is obtained if in Figure 18.18 the resistor R_{C1} is replaced by an appropriate inductor. Such an arrangement is called impedance coupling. This circuit is not shown because it is similar to that in Figure 18.18 except the inductor replacing the collector resistance of transistor Q_1. In operation, this circuit is sensitive to frequency. For higher-frequency applications this circuit is preferred to the previous ones.

Finally, one version of a transformer-coupled two-stage amplifier is shown in Figure 18.19 (there are other alternative ways for this). Using a transformer has the advantage of impedance matching as well as increasing the efficiency of a system by replacing resistors that consume electricity and convert it to heat. It also allows the inversion of polarity of the output signal, if required. Impedance matching is very important, as mentioned before, and it is very common to achieve this by using transformer coupling. The relationships between the current, voltage, and power of the primary and secondary circuits work here for appropriate selection of a desirable transformer. The frequency range of signals must also be

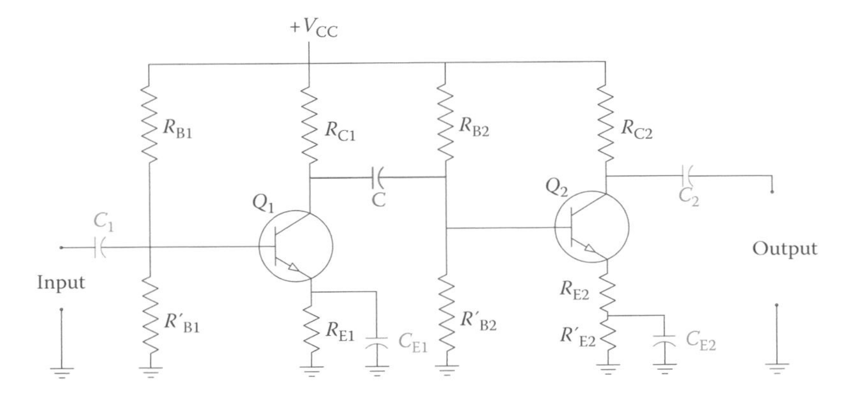

Figure 18.18 RC coupling between two common-emitter amplifiers.

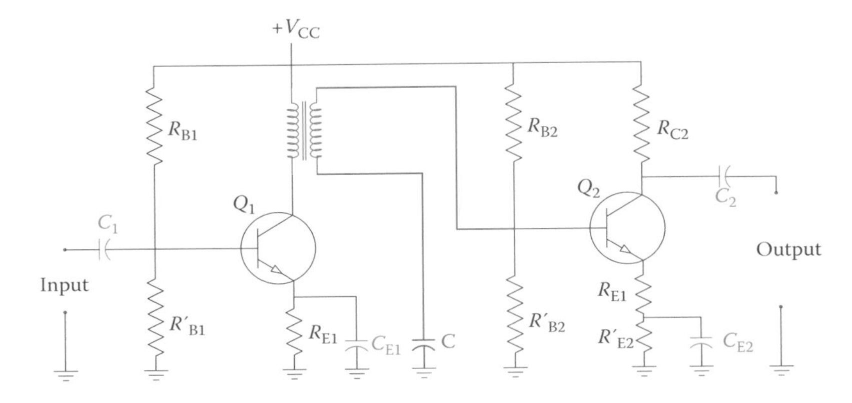

Figure 18.19 Transformer coupling.

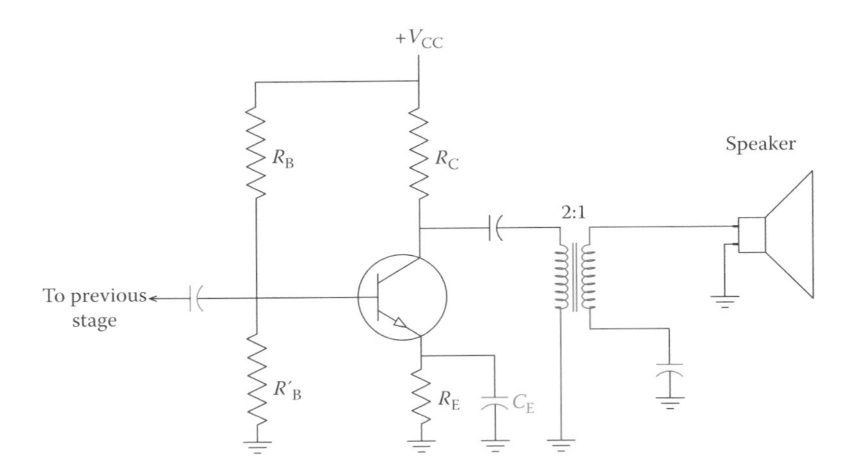

Figure 18.20 Impedance matching for the speaker of an audio system.

brought into consideration because this type of coupling is also sensitive to frequency.

A common example of impedance matching by a transformer takes place in many audio devices. The input impedance of many loudspeakers is 8 Ω, but some come with 4 or 16 Ω. If the output impedance of the last stage of the audio device, usually, a power amplifier (amplifying current), is 16 Ω then there is a mismatch if a 4 Ω speaker or an 8 Ω speaker is connected to it. By transformer coupling this mismatch can be corrected, as seen in Figure 18.20. Example 18.5 is a simple case showing the effect of a mismatch.

Example 18.5

Consider a 50 W (nominal) amplifier that requires a 4 Ω speaker to deliver about 50 W at its output. What happens if you connect an 8 Ω speaker to this amplifier?

Solution

The impedance of a speaker normally reflects its impedance as a result of the coil resistance and its inductive reactance at the standard frequency of 1000 Hz. From Ohm's laws and simple calculations (just using the relationships between voltage, current, and power, and considering the speaker as a resistance) we may conclude that the voltage provided by the amplifier at its terminal to connect to a speaker is

$$V = \sqrt{PR} = \sqrt{(50)(4)} \approx 14\ \text{V}$$

If a 4 Ω speaker is connected to this amplifier, the current in the speaker coil is

$$I = \frac{14}{4} = 3.5\ \text{A}$$

and the product of voltage and current is about 50 W.

Now, if instead of a 4 Ω speaker an 8 Ω speaker is connected the current is half of the previous value and the power delivered by the speaker (product of voltage and current) is only about 25 W. Thus, the amplifier is underutilized and is not delivering its full power.

Suppose now that the speaker is connected to the amplifier through a small audio transformer of 1:1.5 ratio. The transformer increases the voltage by the ratio of 1.5 to 21 V.

The new current in the speaker is

$$I_{\text{New}} = \frac{21}{8} = 2.625\ \text{A}$$

and the power drawn by the speaker is

$$\text{Power}_{\text{New}} = (21)(2.6) = 55\ \text{W}$$

This new power is more like what one expects from this amplifier. The more precise ratio for 50 W power delivery is 1:1.4 (or more precisely $\sqrt{2}$), but purposely we selected 1:1.5 for this discussion.

18.10 Testing Transistors

A transistor can be tested to see if it is good or faulty. An ohmmeter, analog or digital, can be used for this purpose. If a digital multimeter is employed, the check diode may be used. The resistance between each two terminals (base, collector, and emitter) should be checked. Because a transistor is made of two junctions, the same thing that was mentioned about diodes is valid here, too. The forward bias resistance must be low, and the reverse bias resistance must be very high (OL in a digital meter).

Remember that in an NPN transistor the base-emitter junction is forward biased if the base is connected to + and emitter is connected to –; similarly,

TABLE 18.2 Summary of Ohmmeter Reading for an NPN Transistor

Terminals	Connections		Meter Reading
Emitter–base	Base to + (red)	Emitter to – (black)	0.4–0.7
	Base to – (black)	Emitter to + (red)	Very high (OL)
Collector–base	Base to + (red)	Collector to – (black)	0.4–0.7
	Base to – (black)	Collector to + (red)	Very high (OL)
Collector–emitter	Collector to + (red)	Emitter to – (black)	Very high (OL)
	Collector to – (black)	Emitter to + (red)	Very high (OL)

TABLE 18.3 Summary of Ohmmeter Reading for a PNP Transistor

Terminals	Connections		Meter Reading
Emitter–base	Base to + (red)	Emitter to – (black)	Very high (OL)
	Base to – (black)	Emitter to + (red)	0.4–0.7
Collector–base	Base to + (red)	Collector to – (black)	Very high (OL)
	Base to – (black)	Collector to + (red)	0.4–0.7
Collector–emitter	Collector to + (red)	Emitter to – (black)	Very high (OL)
	Collector to – (black)	Emitter to + (red)	Very high (OL)

the base-collector junction is forward biased if the base is connected to positive and the collector is connected to –. The positive lead of a meter is always red, and the negative lead is always black. The expected meter reading for an NPN transistor is, thus, as shown in Table 18.2. The numerical values shown in the table correspond to when the check diode option of a meter is used.

Because a PNP transistor has an opposite structure, the directions of forward and reverse bias invert. Table 18.3 shows a summary of the meter reading for such a transistor. In each case if the readings are different, then the tested transistor is faulty.

18.11 Chapter Summary

- Each transistor has a set of characteristic curves.
- Characteristic curves represent the variation of the collector current versus the collector-emitter voltage.
- Characteristic curves also show the saturation and cutoff regions for a transistor.
- Saturation is a state that a transistor's collector current has reached its maximum value for a given base current and cannot go higher unless the collector-emitter voltage is increased.
- Cutoff is a state that a transistor does not conduct, and there is no collector current (the current between collector and emitter is zero).
- A load line of a transistor is a straight line connecting the point for maximum collector-emitter voltage to the point with maximum collector current, for a given V_{CC}, R_C, and R_E.
- If the voltage applied to the collector of a transistor is so high that V_{CE} exceeds the breakdown voltage, the transistor gets damaged.

- There are a number of ways to bias a transistor. Among these are self-bias, fixed bias, and voltage divider bias.
- The quiescent point is the operating point of a transistor (on its load line) corresponding to no input.
- In a common-emitter amplifier the input is applied between the base and ground and the output is taken between the collector and ground.
- In a common-emitter amplifier the output is 180° out of phase with the input.
- A common-emitter amplifier is the most commonly used amplifier.
- In a common-base amplifier the input is applied between the emitter and ground and the output is taken between the collector and ground (base is grounded).
- In a common-base amplifier the output is in phase with the input.
- In a common-collector amplifier the input is applied between the base and ground and the output is taken between the emitter and ground. The collector is grounded through a capacitor or it may not be grounded.
- In a common-collector amplifier the output is in phase with the input.
- A common-collector amplifier is called an emitter follower because the emitter follows the input signal variation.
- In an amplifier, voltage gain is the ratio of the output voltage variation range (peak to peak if sinusoidal) to the input voltage variation range.
- In an amplifier, the variation range for a voltage or current is called swing.
- In an amplifier, current gain is the ratio of the output current swing to the input current swing.
- In an amplifier, the power gain is obtained by multiplying voltage gain by current gain.
- In an amplifier, conduction angle or angle of flow defines what percentage of each cycle of a sinusoidal input waveform is amplified and appears at the output.
- In a class A amplifier the whole cycle of an AC input signal is amplified. The angle of flow is 360°.
- In a class B amplifier, only one half cycle of an AC input signal is amplified. The angle of flow is 180°.
- In a class AB amplifier, more than one half cycle of an AC input signal is amplified. The angle of flow is between 180° and 360°.
- In a class C amplifier, less than half cycle of an AC input signal is amplified. The angle of flow is less than 180°.
- When two devices are connected together, each one sees the other as an impedance.
- The input impedance of an amplifier is the impedance value that an input device is subjected to when connected to the amplifier.
- The output impedance of an amplifier is the impedance value that an output device is subjected to when connected to the amplifier.
- Cascaded amplifiers are a number of amplifiers in series with each other, so that the output of each is the input for the next one.

- Impedance matching is necessary for efficient functioning of two devices that are connected together.
- Coupling implies how the electronic devices are connected together. Coupling can be direct, through a capacitor and resistor (RC coupling), through an inductor and capacitor (impedance coupling) or through a transformer.

Review Questions

1. What are the characteristic curves of transistors?
2. What is meant by saturation? Where is the saturation region on the characteristic curve graph?
3. What is meant by cutoff?
4. How many configurations of transistor amplifier do exist?
5. What is the main difference when a common-emitter amplifier is compared to other amplifier configurations?
6. In a common-emitter amplifier where are the input and output connections?
7. What is the difference between class A and class B amplifiers?
8. What is the difference between class B and class AB amplifiers?
9. What is conduction angle?
10. What is meant by swing?
11. What is amplified in an amplifier?
12. What is amplified in a power amplifier?
13. What is the difference between voltage gain and current gain?
14. What type of amplifier has a voltage gain less than 1?
15. What is the load line? Is there any difference between load lines for DC and for AC?
16. Which amplifier type is the most commonly used?

17. What is an emitter follower?

18. What is the difference between input and output impedances in an amplifier?

19. Is impedance matching important? Why or why not?

20. Name two types of coupling in amplifiers and other electronic devices.

Problems

1. Refer to Figure 18.12. If this graph corresponds to a common-emitter amplifier, from the given data in the figure, find the current gain of this amplifier.

2. In an amplifier the sinusoidal input voltage has a swing of 5 mV, while the peak-to-peak current variation is 48 μA. The output voltage has a swing of 40 mV, and the output current variation is between a minimum of 0.1 mA to a maximum of 0.7 mA. Find the power gain for this amplifier.

3. Referring to Figure 18.11, consider the AC load line. In the current position of point Q the amplifier works as class A. Where is/are the possible point(s) for point Q so that the amplifier becomes class B?

4. In Figure P18.1, if the collector current has a minimum of 1 mA and a maximum of 4 mA, what are the minimum and maximum values for collector-emitter voltage?

5. In the amplifier of Problem 4, if the input is sinusoidal, what is the peak value of the output signal?

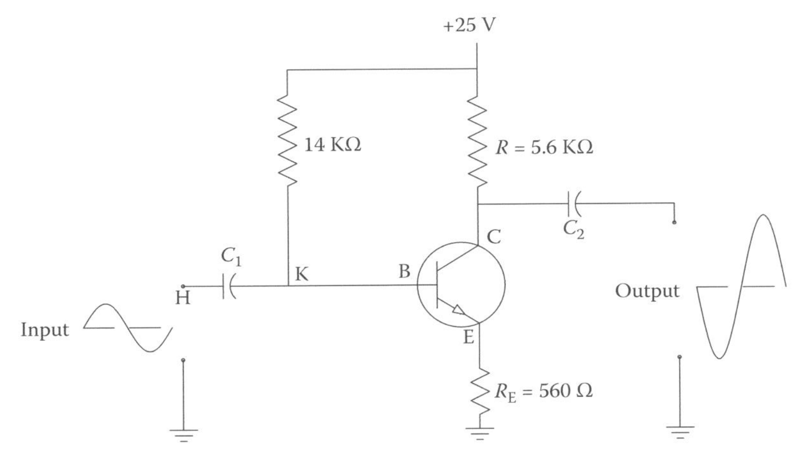

Figure P18.1 Circuit for Problem 4.

19

Switchable Diodes and Gated Transistors

OBJECTIVES: After studying this chapter, you will be able to

- Define the principal structure of a field effect transistor
- Understand the elements in the family of the field effect transistor
- Describe the advantages and shortcomings of each class of transistor
- Explain the correspondence between a bipolar junction transistor and a field effect transistor terminals
- Recognize and use the terms FET, JFET, MOSFET, and the difference between D-MOSFET and E-MOSFET
- Recognize and use the terms IGBT, thyristor, GTO, SCR, and IGCT
- Understand and use the symbols for all the new devices: JFET, MOSFET, IGBT, and SCR
- Describe the correct bias conditions for the aforementioned devices

New terms:
Drain, enhancement, field effect transistor (FET), forced commutation, gate, gate turn off (GTO) thyristor, insulated gate bipolar transistor (IGBT), insulated gate field effect transistor, integrated gate commutated thyristor (IGCT), junction field effect transistor, metal-oxide semiconductor, metal-oxide semiconductor field effect transistor (MOSFET), natural commutation, silicon-controlled rectifier (SCR), source, thyristor

19.1 Introduction

So far, we have seen the most frequently used, yet basic, elements of electronic devices: diodes and transistors. There are many devices and other components in electronics to learn about. The advent of new devices is out of necessity to improve the properties and operational characteristics of their predecessors. The improvements can be reducing the size, better frequency response (acting fast for higher frequency signals), being more stable as temperature changes, being more energy efficient, and so on. This book is limited to the basic content that a technical person dealing with renewable energy needs to know. Other materials fall outside of the scope of this book and, therefore, are not discussed here. An enthusiastic reader should refer to books on the subjects of electronics, solid states, and semiconductors.

In this chapter we discuss a new class of diode and new classes of transistor, with different operational properties. Certain of these devices are employed in today's wind turbines and solar arrays. The simple diode, studied before, does function according to the condition provided by the circuit it is connected to and as far as these conditions correspond to its operational range. In comparison, the new diode (rectifying element) is electronically switchable and can be turned on and off. Also, the output of a bipolar junction transistor is controlled by its base current, whereas the transistors discussed in this chapter are voltage driven. Controlling by voltage is easier and faster.

19.2 Field Effect Transistor

In Chapter 17 we learned that the bipolar junction transistor (BJT) that we studied in Chapters 17 and 18 is a current-driven device. The input current to the device controls the output current. From the point of view of efficiency this type of transistor consumes a relatively high amount of energy due to high current, which converts to heat; thus, the efficiency is low. There are other types of transistor with better energy efficiency. This is due to their higher inherent resistance. A **field effect transistor (FET)** is another type of transistor that, because of its advantages over the junction transistor, is widely used in industrial and domestic electronic appliances. Transistor circuit resistances for a field effect transistor are much larger than their counterparts in BJTs. The current, consequently, is much lower, which leads to much smaller power consumption in a circuit employing this category of transistor.

Field effect transistor (FET): Type of transistor, still made of semiconductor materials, based on a different structure and a different mechanism of operation than the junction transistor. It has a channel or passage around which an electric field can be developed and through which electrons flow. Flow of electrons can be controlled by adjusting the polarity and intensity of the field.

A basic difference between a FET and a junction transistor is that a FET has only one P-N junction. The structure and the manufacturing process of a FET are different from those of a bipolar junction transistor. In a field effect transistor the main body can be N-type or P-type material in which a channel from the opposite type material is diffused. If the channel material is made of N material, then the product is called N-channel FET, and if the channel is made of P-type semiconductor, the FET is called P-channel FET.

A FET is a voltage-driven transistor, dissimilar to BJT, which is a current-driven device. The reason it is called field effect transistor is that the current through the channel is affected by an electrostatic field formed around the channel. The strength of this field can control the flow of the electric carriers in the channel. A schematic of the active area of a FET and its three terminals are shown in Figure 19.1. In practice, a transistor

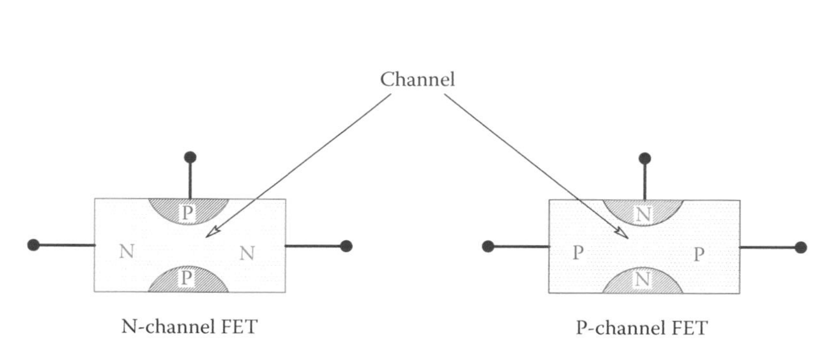

Figure 19.1 Schematic of FET active parts.

is very small, and all its elements are made on a body that itself is a semiconductor and is part of the transistor elements.

The three elements of a field effect transistor have names different than those for BJT. When a FET conducts, electrons move through the channel. One side of the channel is called the **source** (counterpart of emitter) and the other side is called the **drain** (counterpart to collector). The channel is physically surrounded by the opposite type semiconductor material, which can influence the extent (effective size) of the channel, thus introducing less or more resistance for the flow of electrons. This control action is carried out by the **gate**. The channel size (effect size not the physical size) is governed by the *voltage* applied to the gate (see Section 19.3) and *not* the current through the gate. This is why a FET is a voltage-driven device.

Source: (in semiconductors) One of the three terminals in a field effect transistor, comparable with the emitter in a junction transistor.

Drain: One of the three terminals in a field effect transistor (FET) counterpart of collector in a junction transistor.

Gate: Special connection in some semiconductor devices that upon receiving an appropriate signal (in the form of a voltage or pulse) allows controlling of the device, including turning it on and off.

There are many types of FETs and in some (with a symmetrical geometry for the channel) the designation of source and drain is governed by biasing (see Chapter 14 for biasing). In some others the channel does not have a symmetrical shape (structured different in the process of manufacturing) and one side of the channel is designated as the source and the other side is the drain. Figure 19.2 illustrates the schematic of the structure of a FET.

In comparison with a bipolar junction transistor the source, drain, and gate are the counterparts for emitter, collector, and base, respectively. Also, the N-channel FET is the counterpart of NPN transistor and the P-channel FET corresponds to a PNP transistor. Physically, FETs are made similar to BJTs, and there is no way to distinguish one from the other just by looking. One must pay attention to the name tag and the manufacturer's data sheets.

Apart from a number of differences between these two categories of transistors, all that can be done by bipolar transistors, such as amplification, can also be done by FETs. In bipolar transistors we have common emitter, common base, and common collector amplifier. Likewise, in FETs we have common-source, common-gate, and common-drain amplifiers, respectively.

FETs can be used for amplification. Analogous to BJTs, it is possible to have common-source, common-drain, and common-gate amplifiers.

In the same way that all discussions in Chapters 17 and 18 for an NPN transistor is equally applicable to a PNP transistor, whatever applies to

Channel

Gate

N P N

P

N-channel FET

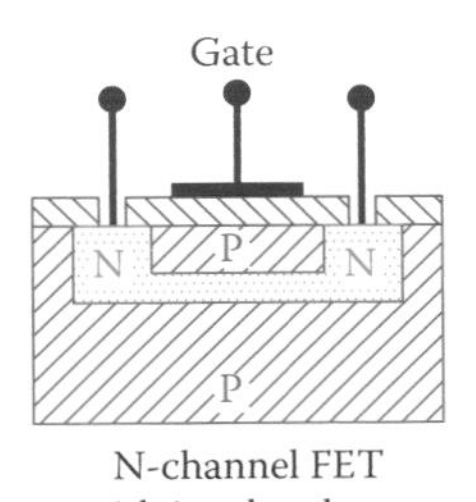

N-channel FET
with insulated gate

Figure 19.2 Schematic structure of FET.

an N-channel FET is true for a P-channel FET, too. The polarity of biasing and operating voltages, accordingly, are the reverse of each other for the N-channel and P-channel FETs. In what follows, we consider an N-channel field effect transistor.

FETs are voltage-controlled transistors.

19.3 FET Types

Junction field effect transistor: One type of field effect transistor (FET) in which the gate is not insulated. Those with insulated gates are more advanced and more common.

Insulated gate field effect transistor: Type of FET with insulated gate, like MOSFET.

Metal-oxide semiconductor: Type of semiconductor device in which metal oxide is used for insulation layers.

Metal-oxide semiconductor field effect transistor (MOSFET): Type of field effect transistor in which metal oxide (like SiO_2) is employed for insulation of the gate.

Field effect transistors are initially categorized into two types, **junction field effect transistor** and **insulated gate field effect transistor**. The first category is referred to by its abbreviation **JFET**. The second category is known as **MOSFET** because of the material used for its gate insulation being **metal-oxide semiconductor**. Thus, a more common name for the second category is **metal-oxide semiconductor field effect transistor**.

The two types are different in structure and there is a difference between the way they can function. In JFET the isolation between gate and channel is a reverse biased PN junction. In MOSFET a layer of metal oxide is used for this purpose. This is schematically depicted in Figure 19.2.

In addition to the final categorization of each type to N-channel and P-channel, MOSFETs have subcategories, as discussed later. Figure 19.3 indicates the possible categories of field effect transistors. As can be seen, JFET family is associated with depletion, whereas MOSFET family is associated with depletion and enhancement. These (depletion and enhancement) are the modes of operation for each family; that is, the mechanism under which each category internally functions. They are described below.

Recall *depletion* from Chapter 14. In the depletion region around a P-N joint the electrons that have moved to this region reject and prevent other electrons to enter or pass through the depletion layer. This is what can happen in the channel of a FET. Suppose that the two sides of a channel formed by an N-type semiconductor are connected to a voltage, as shown in Figure 19.4. Electrons are attracted to the + side (from left to right). The channel material is a semiconductor, and it exhibits some resistance, behaving like a resistor. Consider three cases: (1) there is no voltage on

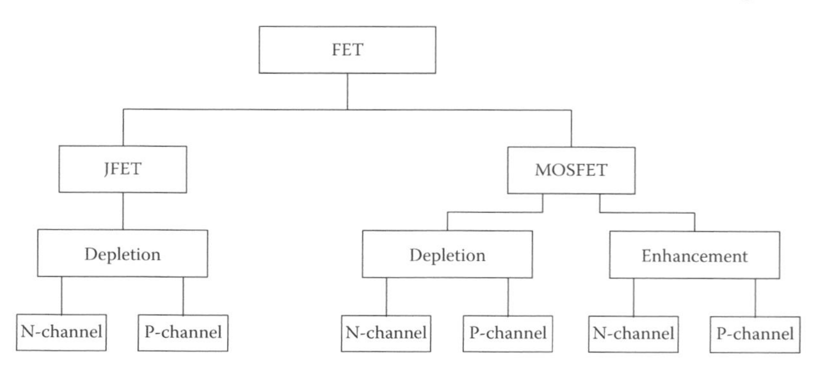

Figure 19.3 Tree of various families of field effect transistors.

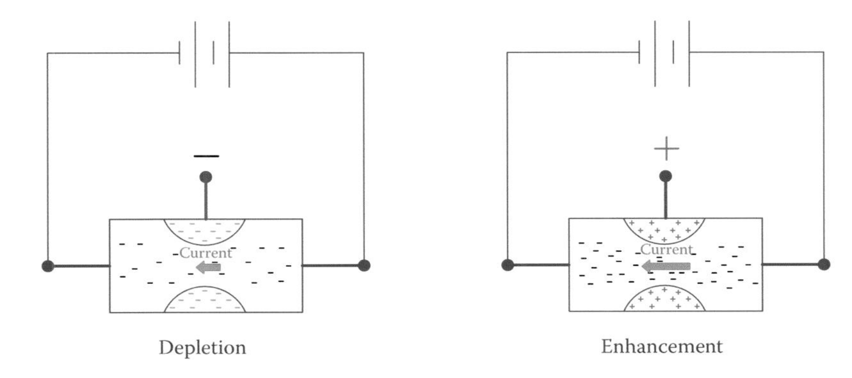

Figure 19.4 Depletion and enhancement in field effect transistor.

the gate, (2) gate is connected to a negative voltage, and (3) gate is connected to a positive voltage.

When there is no voltage on the gate, a current can flow in the channel due to the voltage difference between the two sides of the channel and based on the resistance of the channel. If now a negative voltage is connected to the gate, a negative field will be formed around the channel, as shown in Figure 19.4. The effect of this field is to reject electrons and decrease the current. This is the case called depletion because the negative field depletes the electrons in the channel. The stronger the field is, the lower the number of electrons passing.

When the gate is connected to a positive voltage, a positive field is formed around the channel. The effect on the electrons is to attract them and, as a result, help them move through the channel. This is called **enhancement**; the motion of electrons is enhanced. In both cases (depletion and enhancement) the stronger the field is the larger is the effect on the electron flow. As depicted in Figure 19.3, JFETs work only based on depletion, but MOSFETs can work either in the depletion mode or the enhancement mode.

Enhancement: One of the two methods to control the flow of electrons in the channel of a field effect transistor (FET).

In a FET, by changing the voltage on the gate the current between the source and drain can be controlled.

19.4 JFET

Although JFETs and MOSFETs are both field effect transistors, they have different symbols in the circuit schematics. The symbols for JFET and a comparison with their counterparts in bipolar junction transistors are shown in Figure 19.5.

For an N-channel JFET the drain voltage must be more positive than the source voltage; that is, the drain must be at a higher voltage with respect to the source. This is similar to the polarity for an NPN bipolar junction transistor where the collector-emitter junction must be forward biased. The gate for this type FET, nevertheless, must be negative with respect to the source; that is, the gate-source junction (P-type material in the gate and N type for the channel) must have reverse bias. The gate must

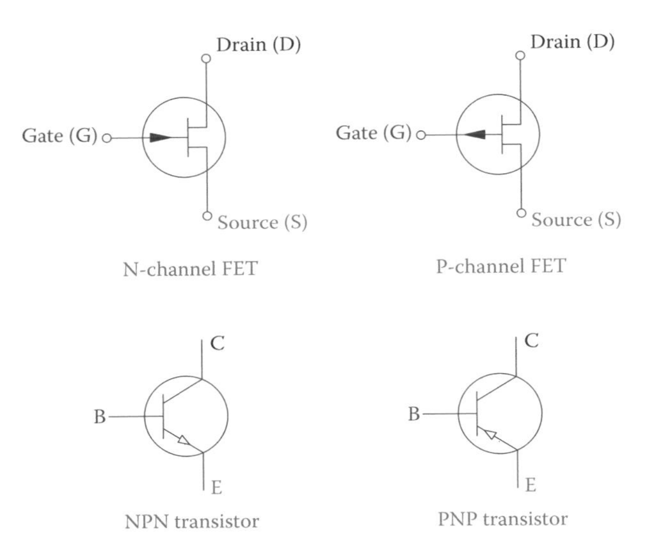

Figure 19.5 Symbols for N-channel and P-channel field effect transistors.

have a *lower* voltage than the source. For this type of FET at the most the gate can be shorted to the source and have a voltage equal to that of the source. Gate voltage going more positive than the source voltage is harmful for the JFET and can damage it. For a P-channel JFET the reverse is true. Typical acceptable voltages are shown in Figure 19.6.

> For N-channel JFET the gate must be negative with respect to the source.

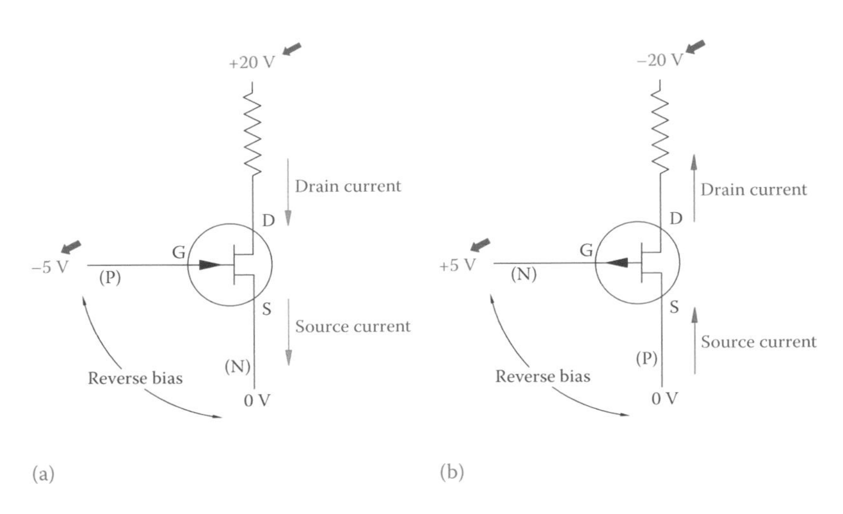

Figure 19.6 Voltage and bias relationships for JFET: (a) N channel and (b) P channel.

In an N-channel JFET when a negative voltage is applied to the gate the effect is to introduce more resistance to the source-drain electric flow, as if the channel becomes narrower. This is due to the growth of the depletion region where the N-type and P-type materials meet. If the gate negative voltage is sufficiently high, then the channel entirely blocks and the current drops to zero. (The transistor goes to cutoff state.) In this respect, the highest source current corresponds to the case when the gate and the source have the same voltage.

Note that the gate-source junction is reverse biased. Consequently, there is no current flowing between the source and the gate. Unlike the case of a BJT, therefore, the source current and the drain current are the same. The voltage at the gate is only to maintain the electrostatic field around the channel. Similar to a BJT, various voltages and currents in a FET are denoted by their corresponding element. These are shown in Figure 19.7. These definitions are good for MOSFET, too.

The required bias condition exemplified in Figure 19.6 necessitates that both positive and negative voltages be available from a power supply. Having two power supplies is undesirable, and normally one power supply is used to power a transistor. In the same way as it was described in Chapter 18 for BJTs, appropriate biasing for FETs can be obtained by fix bias, self-bias or voltage divider bias. Their descriptions will not be repeated in this chapter, but they are shown in Figure 19.8. Example 19.1 helps to clarify the matter.

> In a field effect transistor the source current and the drain current are equal.

Note that in the voltage divider biasing method the voltage at G is determined by the values of R_G and R'_G. This value is positive for a positive

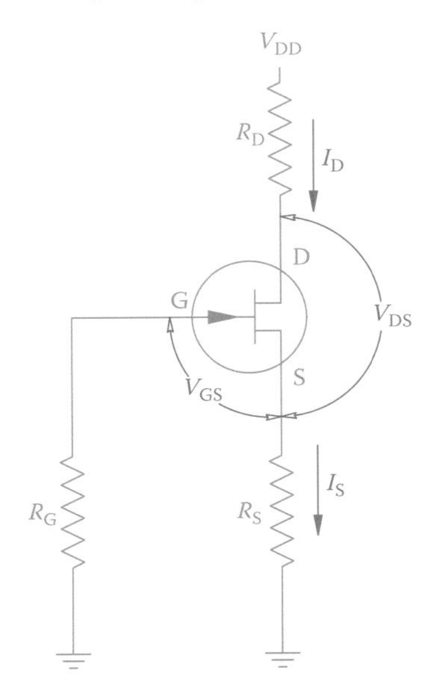

Figure 19.7 Definition of FET circuit components.

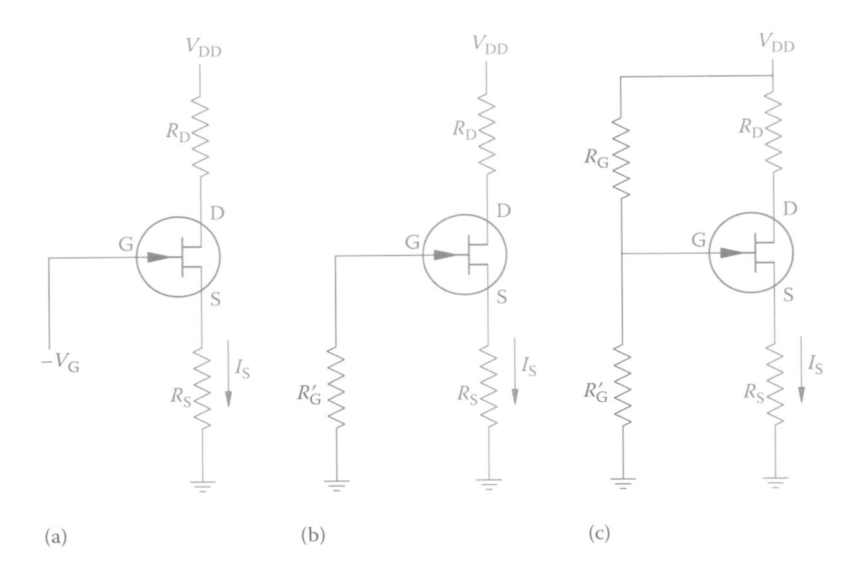

Figure 19.8 Biasing methods for FET: (a) fixed bias, (b) self-bias, and (c) voltage divider bias.

V_{DD}. The voltage at S, however, is determined by the values of R_S and I_S. Correct gate-source bias can be obtained by choosing R_S appropriately. Incorrect magnitude of R_S can make the voltage at S smaller than the gate voltage (in an N-channel transistor), which can cutoff the FET; so, it is important that R_S is correctly selected. When dealing with AC signals that cause I_S to vary between a minimum and a maximum value, the transistor can go to cutoff for some values of I_S.

Example 19.1

In the transistor shown in Figure 19.9, what is the voltage difference between the gate and the source (V_{GS})?

Solution

The supply voltage is 20 V and the voltage at ground is 0. Because the current in the R_S (470 Ω resistor) is 15 mA, there is a voltage drop of

$$(470)(0.015) = 7.05 \cong 7 \text{ V}$$

in the resistor. Thus, the voltage at S is 7 V.

Because the gate is connected to ground and there is no current in the 5.1 K resistor (because of the gate-source junction being reverse biased), the voltage at the gate remains zero. The source, therefore, is at +7 V with respect to the gate.

Example 19.2

In the circuit shown in Figure 19.8c the values of the various resistances are as follows: R_D = 1.5 K, R_S = 470 Ω, R_G = 16 K,

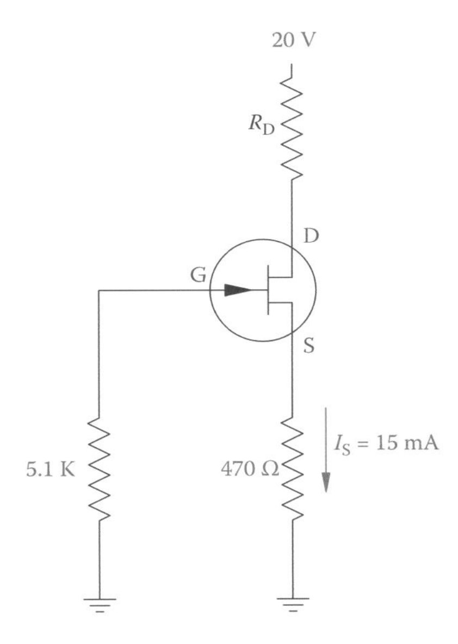

Figure 19.9 Circuit of Example 19.1.

and $R'_G = 1.2$ K. If the supply voltage is 24 V and the drain current is 10 mA, how much are the voltages V_{GS} and V_{DS}?

Solution

Voltage at gate can be found from voltage divider relationship

$$V_G = \left(\frac{1.2}{1.2+16}\right)(24) = 1.67 \text{ V} \cong 1.7 \text{ V}$$

Source current is the same as the drain current. The voltage at S can then be found from the voltage drop in R_S

$$V_S = (470)(0.010) = 4.7 \text{ V}$$

The source-gate voltage, thus, is

$$V_{GS} = V_G - V_S = 1.7 - 4.7 = -3 \text{ V}$$

The drains-source voltage can be found from the difference between the supply voltage and the voltage drops in R_D and R_S:

$$V_{DS} = 24 - (1500 + 470)(0.010) = 4.3 \text{ V}$$

19.5 MOSFET

The metal oxide semiconductor field effect transistor (MOSFET) is a FET whose gate is insulated from the main body of the transistor by a layer of metal oxide semiconductor (such as SiO_2). This layer is very thin, and for this reason MOSFETs are prone to get damaged easily if they are subject to voltages higher than their ratings. They are also very sensitive to static electricity, and they must be handled with care and sufficient protection from receiving static electricity in order not to get damaged.

In addition to the gate being insulated, one more difference between JFET and MOSFET is the channel. While in a JFET, the channel is made of the same material as the source and drain, a MOSFET has no channel, or has a smaller channel made of a semiconductor material that is much less doped than those of the source and drain.

As pointed out earlier, there are two types of MOSFET based on the mechanism that they function, depletion and enhancement. The first type is called D-MOSFET. It has a channel, and it can work both in depletion and enhancement mode (for this reason sometime it is called DE-MOSFET). The second type, called E-MOSFET, can only work in the enhancement mode. Figure 19.10 schematically depicts the structural difference between the two types.

An E-MOSFET is an off device, meaning that when no voltage is applied to the gate ($V_{GS} = 0$), the transistor is in the cutoff mode. If then a positive voltage is applied to the gate, it develops a current between the source and the drain. This current can be controlled by changing the gate voltage. In the D-MOSFET, however, there is a current I_{DS} between the drain and source even when $V_{GS} = 0$ (when a V_{DS} exists between the drain and the source). Then I_{DS} can be lowered (working in the depletion mode) by having a negative voltage applied to the gate or it can be increased (working in the enhancement mode) by applying a positive voltage to the gate.

Symbols for the main categories of MOSFETs are shown in Figure 19.11. There are more varieties of MOSFET, but we do not consider them here.

A MOSFET has a high input impedance (see Section 18.8 for input impedance). This is a definite advantage over the BJT and JFET because it implies low input current and very low power dissipation. Also, it is more appropriate for connecting to devices with a high output impedance. Other advantages that have made the use of MOSFETs more popular than BJTs are (1) insensitivity to temperature change and (2) high-frequency capability (fast switching).

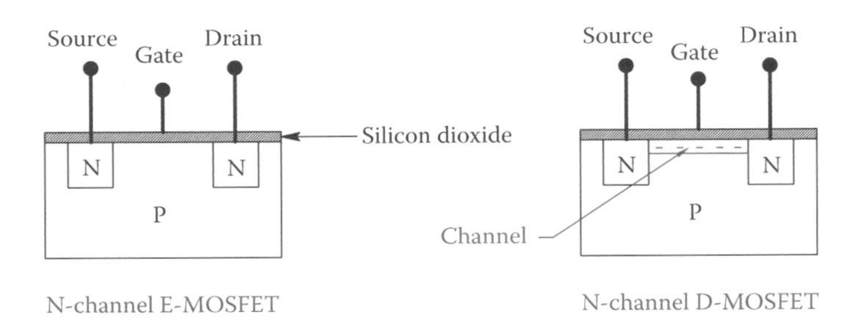

Figure 19.10 Structural difference between D-MOSFET and E-MOSFET.

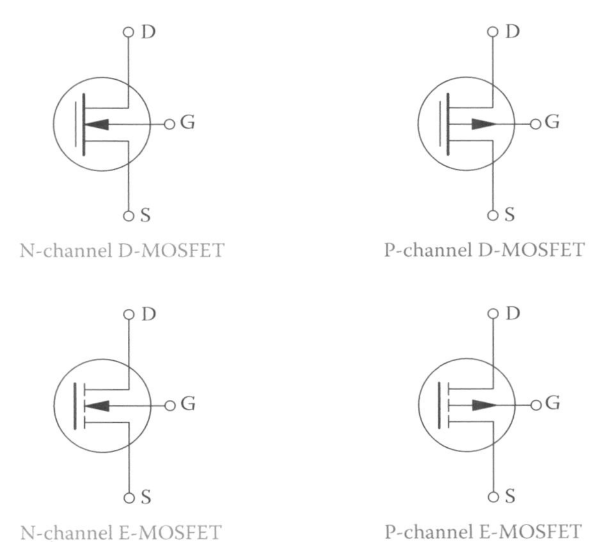

Figure 19.11 MOSFET symbols.

Considering the first issue as an advantage is very obvious. Transistors can become unstable when the operating temperature rises in a device. The second issue is very important for operation on high-frequency signals (e.g., in telecommunications). It is required that a device turns on and off with a desired (normally very high) frequency. A bipolar junction transistor cannot perform this switching at very high frequencies. Despite these advantages, MOSFETs have limitations on the voltage and power that they can handle. Typical maximum values can be said to be 250 V and 500 W.

19.6 Testing a MOSFET

A MOSFET can get damaged in two ways, the insulation of the gate breaks or the gate shorts to the channel. In either case the MOSFET cannot function as expected. In handling a MOSFET, extra care must always be taken not to cause any voltage contact to the terminals and that the terminals do not touch any material that can generate or contain static electricity. In testing a MOSFET the same precaution is necessary.

Testing a MOSFET mainly consists of measuring the resistance between its source and drain. When a MOSFET is conducting, this resistance is lower than when it is in cutoff. In this sense, for the conducting condition, its resistance must be much smaller than for no conducting condition, for which the resistance is very high.

A MOSFET can be tested either by an ohmmeter or by the diode test setting of a digital multimeter (DMM). If an analog ohmmeter is used, it must be put on the 100 K range. Two tests are done, one with the gate field existing and one with the gate field absent. In the presence of the gate field the meter reading indicates some value for the channel resistance, and in the absence of this field the channel resistance is very high. It is necessary first to identify the drain, gate, and source for a MOSFET to be tested.

To test an N-channel MOSFET, connect the black lead of the ohmmeter or DMM to the source of the MOSFET without touching the terminal pins by hand. Then with the red lead touch the gate. This creates the gate field. Move the red lead to the drain, and the meter should read some small value (the needle of an analog meter should move to the midrange).

Now, if without disconnecting any of the two leads touch a finger between the gate and source pins (to discharge the field through your fingers), the MOSFET goes to nonconducting condition and the meter reading goes high; that is, in an analog meter the needle stays on the left, and a DMM shows OL. The above observation is for a good MOSFET. If in both cases the needle moves to the right end (zero ohm) or the DMM shows a near zero value, the device is shorted and damaged.

19.7 Insulated Gate Bipolar Transistor

Insulated gate bipolar transistor (IGBT): Type of gated transistor used in power converters. Its gate is insulated and is connected to a much smaller voltage to turn the transistor on.

An **insulated gate bipolar transistor (IGBT)** is a transistor similar to a BJT but with more and improved capabilities such as fast switching. It is now widely used in many switching devices and vastly used in power converters (particularly in wind turbines), as we discuss in Chapter 20. As the name implies, IGBT has a gate and the gate is insulated from the body. This allows high voltage and high current in the IGBT to be isolated from the control signal to the gate.

Whereas a BJT is a current-driven device, an IGBT has a gate and it is a voltage-controlled device. This is much desirable and preferred, similar to a MOSFET. This means it has the advantages of high-current handling capability of a bipolar transistor with the ease of control of a MOSFET.

In comparison with a BJT, an IGBT has a collector and an emitter, but a *gate* instead of a base. The structure of an IGBT is a combination of a BJT and a MOSFET. From the structure viewpoint, there are two types of IGBTs, PT, and NPT, which stand for punch through and nonpunch through. These are schematically shown in Figure 19.12. The difference is in the additional layer in the PT type IGBT. In fact, an IGBT works as a combination of a BJT and a FET, but we do not go to its details.

The symbol for an IGBT is shown in Figure 19.13. There are two different symbols. Either symbol can be used. Figure 19.14 depicts two 1200 V,

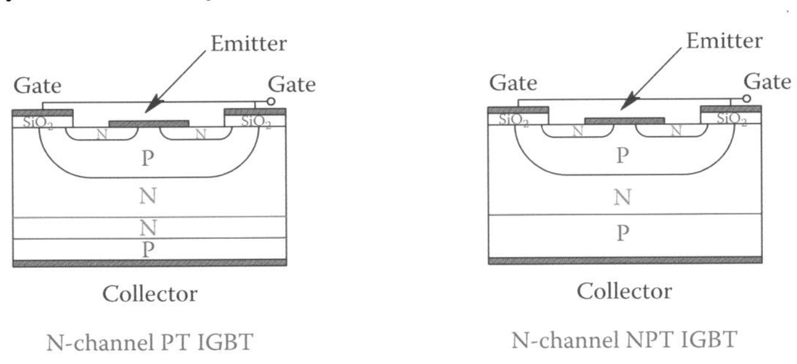

Figure 19.12 Structure schematics of IGBT.

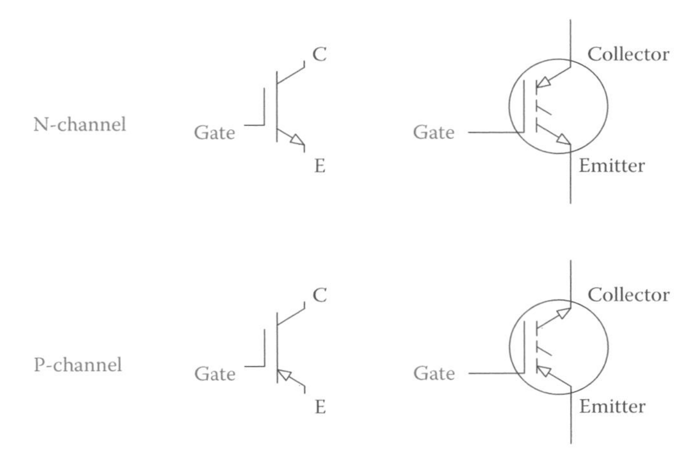

Figure 19.13 IGBT symbols. Two symbols for either N-channel or P-channel IGBT.

600 A IGBTs. They have an embedded controller for overcurrent and overtemperature protection.

As will be discussed in Chapter 21, IGBTs are used as switching devices. For an N-channel IGBT the collector-emitter current can be turned on and off by placing a positive voltage on the gate. As long as there is a forward bias potential difference between the collector and the emitter, a positive voltage on the gate turns the transistor on.

After an IGBT is turned on, it then can be turned off by either removing the positive charge on the gate (introducing a negative or zero voltage to it) or removing the V_{CE} (supply voltage). When connected to AC, if the IGBT is turned on, after half a cycle it will turn off, because this includes reverse biasing the collector-emitter junction.

In operation an IGBT becomes very warm. Cooling by water may become necessary. Figure 19.15 shows an IGBT in the control system of a 1.5 MW wind turbine. The two hoses on the right are for circulating water to take the heat away.

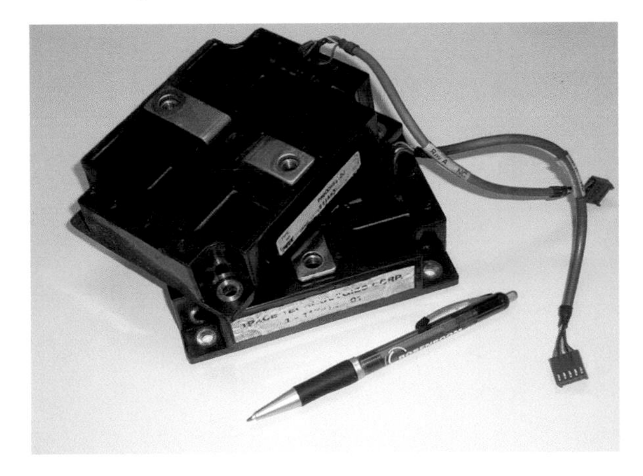

Figure 19.14 Examples of IGBTs.

Figure 19.15 Example of IGBTs inside a circuit, cooled by water circulation.

19.8 Thyristor

Thyristor: Gated semiconductor device whose conduction can be controlled by a pulse sent to its gate. It is used in power converters.

Silicon-controlled rectifier (SCR): A family of gated diodes or transistors that can be controlled (turned on and off) through their gates.

Similar to other electronic devices discussed so far, a **thyristor** is frequently used in electronic appliances, including the electronics for control devices in renewable energy technology. Thyristors are, in fact, a family of about 24 devices, but we are more interested in those that behave like a diode (rectifies electric current). The advantage of a thyristor over a diode is that its function can be turned on and off. Thus, it functions as a switchable diode. For this reason it is also called **silicon-controlled rectifier (SCR)**. It is a rectifier whose functioning is electronically controlled.

Structurally, a thyristor is made of two internally connected transistors, but it can be regarded as a transistor with an extra junction, as shown in Figure 19.16. However, it is employed as a diode and the extra junctions are for its control. Thus, it has three elements, the anode, the cathode, and the gate. Figure 19.16 depicts also the symbol of a thyristor.

In practice, a thyristor is comparable with an IGBT when used for switching. They are interchangeable when the current and voltage rating are compatible, and their required switching frequency allows. A main difference in their operation, however, is as follows.

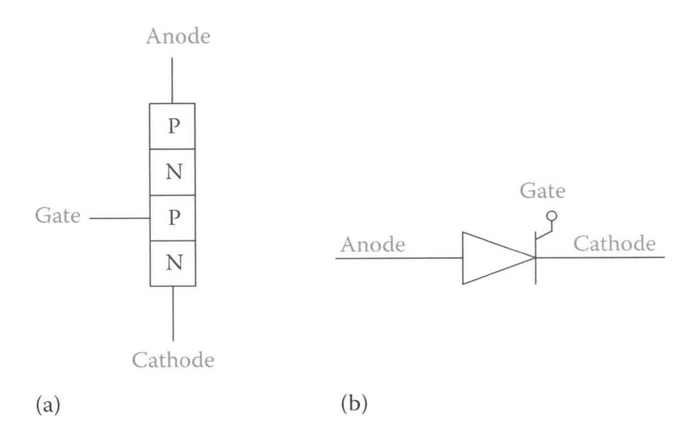

Figure 19.16 (a) Thyristor construction and (b) thyristor symbol.

An IGBT is turned on by applying a voltage to its gate. As long as the voltage is on the gate, provided that correct bias exists between collector and emitter, the IGBT keeps conducting.

A thyristor is turned on by sending a positive pulse (short duration voltage) to its gate. It keeps conducting as long as the forward bias voltage difference between its anode and cathode exists (it is not necessary to repeatedly update the pulse), and there is a current. If the current ceases for any reason, the thyristor turns off, and to turn it back on, another pulse to the gate is necessary. A thyristor can also be turned off by reversing the polarity of the voltage applied between its anode and cathode. This is what happens naturally when a thyristor is used as a rectifier. After a positive half cycle current is followed by a negative current, a thyristor stops conducting. This is further discussed in Chapter 20 (see Section 20.3).

The action that takes place in a thyristor, that when it is turned on it keeps conduction though the activating signal (to the gate) has ceased, is called "latching." This is a desirable property in some applications. A pulse with the correct voltage can start a process, and it is not necessary to maintain the voltage (like in the case of an IGBT). For some other applications, as we will see in Chapter 20, this is an undesirable feature because there is no control on stopping the device. For this reason, some thyristors are equipped with additional capability that they can be turned off when desired by sending a negative pulse to their gate. Among these we only name two, the **gate turn off (GTO) thyristor** and **integrated gate-commutated thyristor (IGCT)**.

The action of turning off a thyristor or a similar device in a circuit is referred to as commutation. If this is done by the circuit itself, as when current direction can reverse, it is called **natural commutation**, and when it is done by an external circuitry to send a signal to the gate or turn the device off by other means, it is called **forced commutation**.

A thyristor is turned on by sending a positive pulse to its gate.

Gate turn off (GTO) thyristor: Type of thyristor with the capability that can be turned off by sending a negative voltage pulse to its gate, also called GTO thyristor, as opposed to those without this extra capability (they turn off as the applied voltage becomes negative).

Integrated gate-commutated thyristor (IGCT): Class of "gate turn off thyristor," in which its gate and antiparallel diodes are integrated with it in a package.

Natural commutation: Act of automatically turning off in a gated diode or transistor (a thyristor) when an alternating current applied across its main terminals changes direction (goes to negative half cycle). This is the opposite of "forced commutation" where the device is turned off through a control signal to its gate.

Forced commutation: Turning off a thyristor by applying a negative voltage to its gate, as opposed to natural commutation in which a thyristor turns off as the alternating current applied to it reaches its negative half cycle.

19.9 Chapter Summary

- The principal structure of a field effect transistor is a channel made of a semiconductor material surrounded by the opposite semiconductor material.
- The three terminals of a FET are called source, drain, and gate.
- In operation, electrons move inside the channel between the source and the drain while affected by an electrostatic field generated and controlled by the gate voltage.
- There are two families of field effect transistors, JFET and MOSFET.
- JFET stands for junction field effect transistor. MOSFET stands for metal oxide semiconductor field effect transistor.
- FETs are voltage-driven devices. The gate voltage controls the drain-source current.
- FETs can be either N channel or P channel.

- A bipolar junction transistor (BJT) can handle higher currents and higher voltages than a FET.
- The source, drain, and gate in a FET correspond to the emitter, collector, and base in a BJT.
- A FET is faster than a BJT and is more suitable for higher-frequency switching.
- A BJT consumes more power than a FET.
- There are two types of MOSFET: D-MOSFET and E-MOSFET.
- A FET like a BJT can be used for amplification and switching.
- IGBT stands for insulated gate bipolar transistor.
- IGBT is a bipolar junction transistor with a gate. It has a collector, an emitter, and a gate.
- SCR stands for silicon-controlled rectifier.
- SCR is the most common member of thyristor family. It is common to use the name thyristor when addressing SCRs.
- A thyristor behaves as a diode with a control gate.
- For switching, an IGBT and a thyristor can be interchanged (if their operating properties permit).
- An IGBT needs a continuous voltage on the gate to keep conducting.
- A thyristor turns on with a positive pulse; then it latches and keeps conducting.
- There are many members in the thyristor family; some can be turned off by a negative pulse sent to the gate, addressed as gate turn off or GTO.

Review Questions

1. What does FET stand for?
2. What is the reason for the name FET?
3. What are the three elements of a FET?
4. What are the counterparts of a BJT base, collector, and emitter in a FET?
5. What is the structural difference between a JFET and a MOSFET?
6. What does a gate do in a FET?
7. Describe two differences between the structures of a BJT and a FET.
8. Describe two differences between the characteristics of a BJT and a FET.
9. Name three ways that a FET can be biased.

10. What is the difference between an N-channel and a P-channel FET?

11. What is/are the difference(s) between D-MOSFET and E-MOSFET?

12. What does SCR stand for?

13. What does IGBT stand for?

14. What is a thyristor?

15. What is the relationship between a thyristor and a SCR?

16. What is an IGBT used for?

17. What is a SCR used for?

18. What is the difference between an IGBT and an IGCT?

Problems

1. In Figure P19.1, find the voltage V_{GS}.

2. If the source current varies in Figure P19.2, what is the minimum acceptable current in the source that does not violate the bias condition?

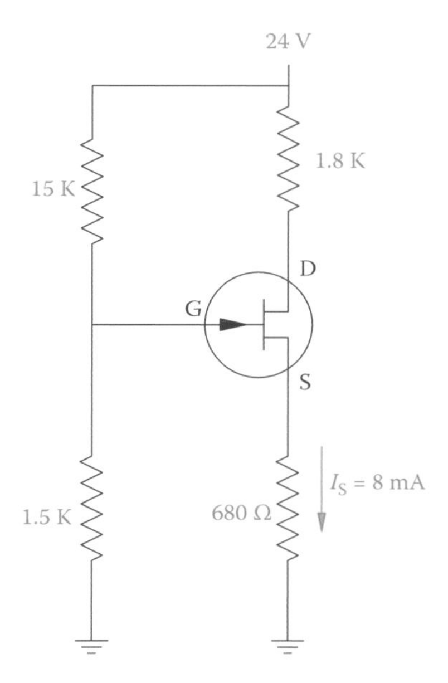

Figure P19.1 Problem 1 circuit.

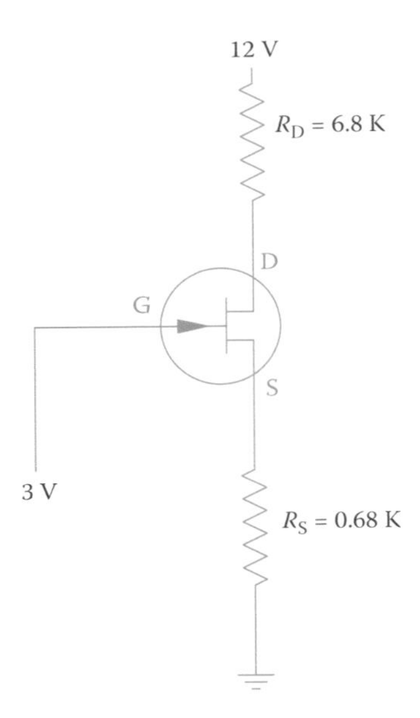

Figure P19.2 Problem 2 circuit.

20

Power Converters

OBJECTIVES: After studying this chapter, you will be able to

- Describe what an inverter is
- Make a single-phase square-wave inverter
- Understand a three-phase inverter
- Define terms such as firing angle and freewheel diode and their purposes
- Describe how output of a rectifier is controlled
- Describe how output of an inverter is controlled
- Explain what a power converter is
- Describe the structure of a thyristor-based converter
- Explain how a converter can be used as a rectifier or as an inverter
- Define pulse width modulation
- Define what pulse width modulation can be used for
- Understand single-phase and three-phase use of inverters based on PWM
- Define amplitude and frequency modulation ratios

New terms: Amplitude modulation ratio, carrier, converter, driver, firing angle, freewheel diode, freewheeling diode, frequency modulation ratio, inverter, overmodulation, photovoltaic (PV) system, pulse width modulation (PWM)

20.1 Introduction

In Chapter 16 we discussed rectifiers, both single phase and three phase: they convert AC to DC. We also learned that diodes are the principal element of any rectifier and that we can use them as half-wave, full-wave, and bridge rectifiers.

The material in this chapter is a continuation of the same framework of devices. First, an inverter is introduced. An inverter does the opposite task of a rectifier and provides an AC waveform from a DC source. Then we continue with replacing a diode in a rectifier by a switchable device; we discussed this in Chapter 19. This allows controlling the output voltage of a rectifier. For rectifiers we focus only on three-phase systems, as in industry three-phase devices are used more frequently than single-phase devices, which are normally small.

Next we discuss three-phase inverters, and we see how the same structure can be used for both rectification and inversion. This is why the term **converter** is employed, referring to both a rectifier and an inverter.

Converter: General term for a rectifier (a device that converts AC to DC) and an inverter (a device that converts DC to AC).

We recall that the output of a rectifier contains ripples, and, to eliminate or reduce the ripples, filters are employed. The same thing is true for the converters, and filters are used in order to smooth the output of a rectifier constructed from switching devices. The discussion about filters is not repeated, and one can use all that was already discussed (see Section 16.5) for the devices explained in this chapter.

Power convertor refers to both rectifier and inverter.

20.2 Inverter

Inverter: Electronic device that generates AC electricity from DC. In the simplest form the generated AC is square wave.

The name **inverter** is used for a device that converts DC to AC; that is, the opposite function of a rectifier. This is the way to provide AC from a battery for the devices that work with AC. For instance, if one wants to use a TV or a razor that works with 110 V AC, and the only available electricity source is a 12 V battery, an inverter becomes necessary. It is necessary that the generated AC electricity has both the required voltage and the required frequency. In the example of using a 12 V battery to generate 110 V AC, thus, using a transformer to increase the AC voltage becomes necessary. The reverse operation (changing from a higher voltage to a lower voltage) does not need a transformer.

Inverters can provide both single-phase and three-phase AC. In both cases, there is only one DC power supply (it is not necessary to have three batteries for three-phase AC conversion). First, we consider the single-phase inverter.

20.2.1 Single-Phase Inverter

Obtaining AC from DC is based on switching the direction of current in a load. This is only possible with power electronic devices that can be controlled, i.e., those whose function can be turned on and off. These are devices such as a thyristor and an insulated gate bipolar transistor (IGBT) that we discussed in Chapter 19. The controls must also allow choosing the voltage and frequency of the generated AC signal as desired. The timing for the switching action, therefore, is quite important. In the simplest form, the output of an inverter is a square-wave AC signal, and not a sinusoidal waveform.

Figure 20.1 shows the schematics of the arrangement for a single-phase inverter. There are four switching devices to control the direction of current at each instant. These can be thyristors or gated transistors (e.g., IGBTs).

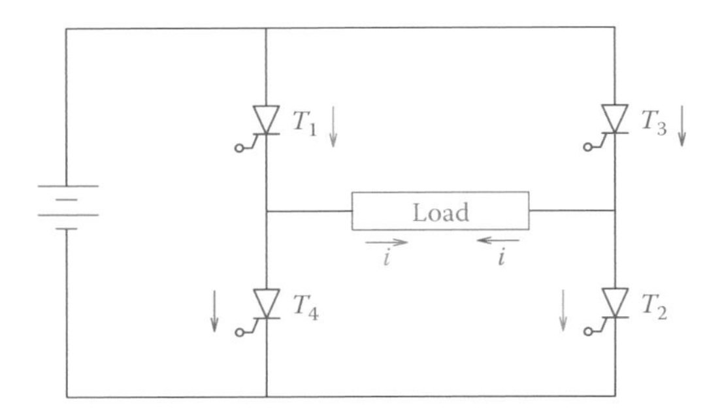

Figure 20.1 Arrangement for a single-phase inverter.

For the sake of simplicity in the rest of this chapter they are only referred to as thyristors. Each of the four thyristors in the circuit allows current in only one direction and only when it is turned on through its gate. At all other times it blocks the current in either direction because of being reverse biased or turned off. These thyristors are turned on and off in an orderly manner and with an appropriate sequence and a desired frequency. The thyristors are divided into two pairs connected between the positive and negative terminals of the power supply, and the load is connected between the two pairs. The thyristors are numbered 1 to 4 for ease of reference.

Consider if with a frequency of 100 Hz sequentially thyristors T_1 and T_2 are turned on together, then turned off, immediately followed by turning on thyristors T_3 and T_4 together and then off together, with the same sequence repeated. The duration each time a pair of thyristors is on, is therefore 1/100th of a second. The result, as observed by the load, is a square-wave alternating current with 50 Hz frequency and a voltage very near the voltage of the DC power supply (three is a small voltage drop in the electronic parts). This is the way that an inverter works.

Note that the two thyristors on the same leg (e.g., T_1 and T_4 or T_2 and T_3) should never be on at the same time because they short the power supply and damage it. Also, in operation, each pair that is on must be turned off *before* the other pair is turned on. For the action of turning the thyristors on and off, a separate electronic circuit is necessary. This circuit is normally called **driver**. The driver circuit must take care of the timing and frequency of turning the thyristors on and off.

Driver: Circuit to turn on and off the gates of gated transistors and similar devices in applications that require an orderly action of one or more of these devices.

> In an inverter, the two thyristors in the same leg should never be in the on state at the same time.

If the load is purely resistive, then the current in the load will have the same pattern of the AC voltage, which is a square wave. For inductive and capacitive loads, as we discussed in Chapter 8, because of the charging effect of capacitor and the Lenz's law in the inductor, the current and voltage are not in phase and some large voltages may develop in part of the circuit. To prevent damage because of these voltages, usually parallel with each thyristor there is a diode allowing current in the opposite direction, as shown in Figure 20.2. The role of the diode is to make a path for current to flow and prevent a voltage buildup. These diodes are called **freewheel diode**, or **freewheeling diode**. Note that these diodes are reverse biased.

Freewheel diode: Diodes put antiparallel to a gated diode (thyristor) in a converter to make a reverse path for current and prevent buildup voltage across the thyristor.

Freewheeling diode: Same as freewheel diode.

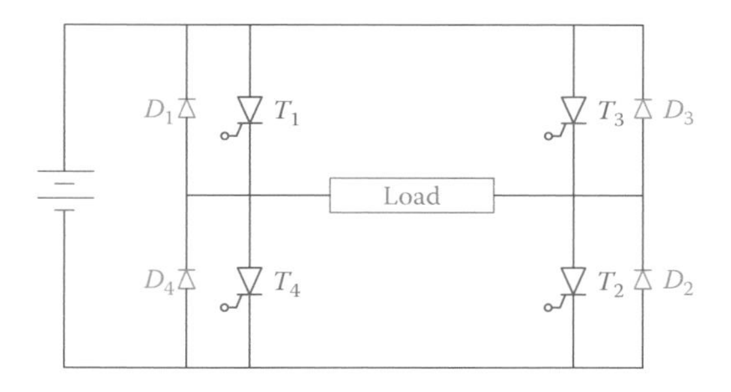

Figure 20.2 Freewheel diodes to protect the circuit from overvoltage.

Thus, they do not short the power supply and do not conduct in the normal conditions.

The inverter as stated, having zero interval between the time a thyristor is turned off and the associated thyristor in the same leg is turned on, generates a square AC waveform as shown in Figure 20.3a. However, if a short interval exists between the two actions (turning off a thyristor and turning on its pair) the generated waveform assumes a different shape, as illustrated in Figure 20.3c. This figure corresponds to the case where there exists a delay of T_2 between the instant one set of thyristors are turned off (after being in the on state for a period of T_1), and the instant the other set are turned on. Thus, for a period of T_2 all the thyristors are turned off.

As can be seen the total period for one cycle of the AC waveform is $2(T_1 + T_2)$. Thus, the frequency of the resulting alternating current is $= \frac{1}{2(T_1 + T_2)}$. For the square waveform of Figure 20.3a, $T_2 = 0$.

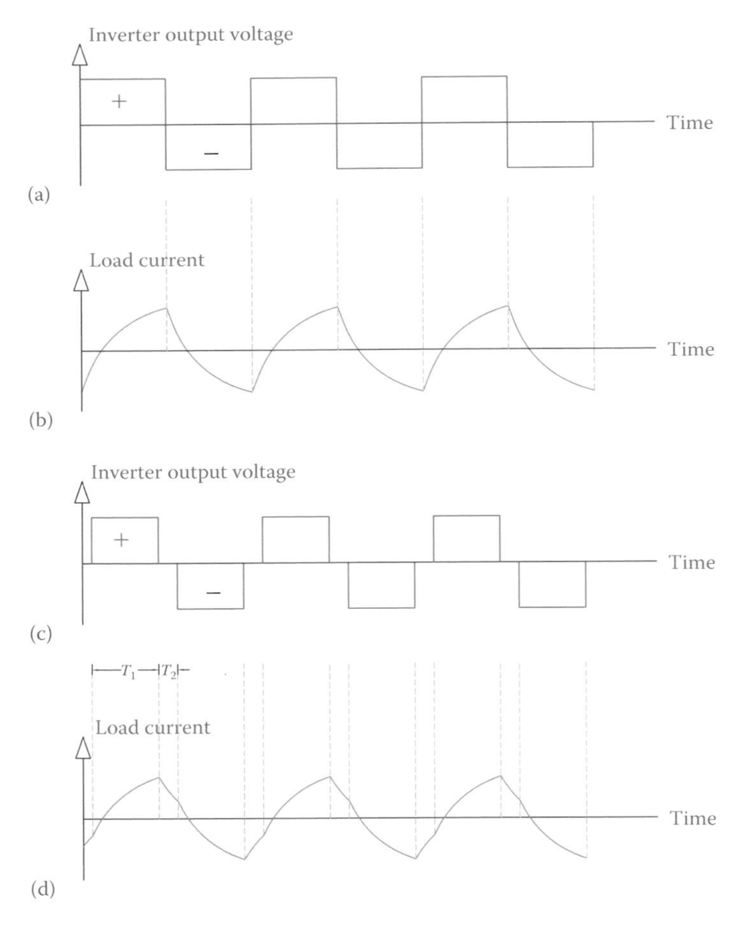

Figure 20.3 Current waveforms for an inductive load based on output voltage. (a) Voltage pattern in a square waveform. (b) Current in an inductive load with a square wave. (c) Square waveform with delay. (d) Current in an inductive load subject to voltage shown in (c).

Although the current variation in a pure resistive load is the same as the voltage, for a different type of load the current variation is different than that of the voltage, as dictated by the load. For instance, for an inductive load the current does not have a square-wave pattern, It rather looks like the waveforms in Figure 20.3b and 20.3d for the voltage waveforms in Figure 20.3a and 20.3c, respectively. As can be seen, the current in an inductive load has a form closer to a sine wave than a square wave.

By changing the time intervals T_1 and T_2 it is possible to change the current waveform to some extent, as can be seen from Figure 20.3d. It is customary to represent T_1 and T_2 in terms of angles. In such a case they do not associate with frequency anymore but can indicate the percentage of time in each cycle that a thyristor is on. For this, $T_1 + T_2 = 180°$, and if, for instance, $T_1 = 120°$ and $T_2 = 60°$, it implies that each set of thyristors are on for 120° and stay off for the rest of a cycle (i.e., for 60° + 180° = 240°). In practice, the driver circuit (software and/or hardware) that controls the thyristors takes care of frequency by repeating the cycle after an elapsed time.

The best result, as far as having an output as close as possible to a sinusoidal waveform is concerned, is when $T_1 = 133.5°$ and $T_2 = 46.5°$.

To improve the quality of the output waveform from an inverter, so that it better resembles a sinusoidal waveform, various techniques are available. One of these is pulse width modulation, which is separately discussed later (see Section 20.6). Another technique is to increase the switching steps and accordingly the necessary controls to generate stepping signals as illustrated in Figure 20.4.

Obviously, the more steps there are, the more costly the device. The steps in the configuration in Figure 20.4 have the same width (time duration). Figure 20.5 shows how a step waveform can be obtained from combining inverters with different outputs of the form in Figure 20.3c.

Practically, a waveform shown in Figure 20.5b can be made by adding the outputs of two inverters together. One way for this is indicated in Figure 20.6, where the outputs of the inverters are added together through a transformer.

The term firing is normally used for turning a thyristor on. Also, **firing angle** represents the delay from the start of a cycle, normally in terms of degrees, when a thyristor must be turned on. For instance, in Figure 20.5 the firing angle for the inverter to generate signal A is 22.5° and for the

Firing angle: Delay from the beginning of a cycle of an AC signal for voltage control in a converter. Delay determines when the gated devices in the converter become active and, thus, the percentage of the 360° cycle that the AC waveform is ineffective.

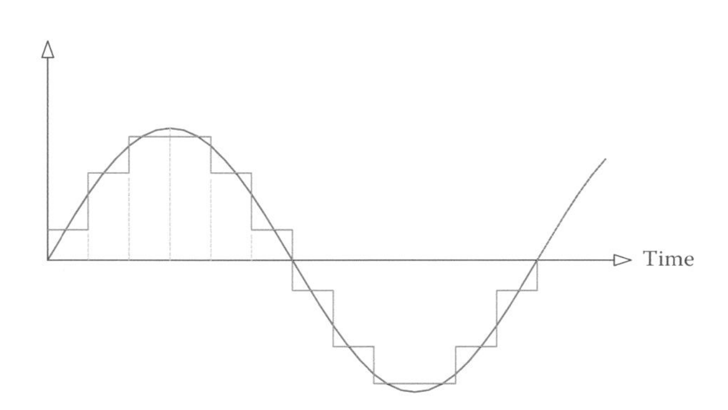

Figure 20.4 A 12-step waveform.

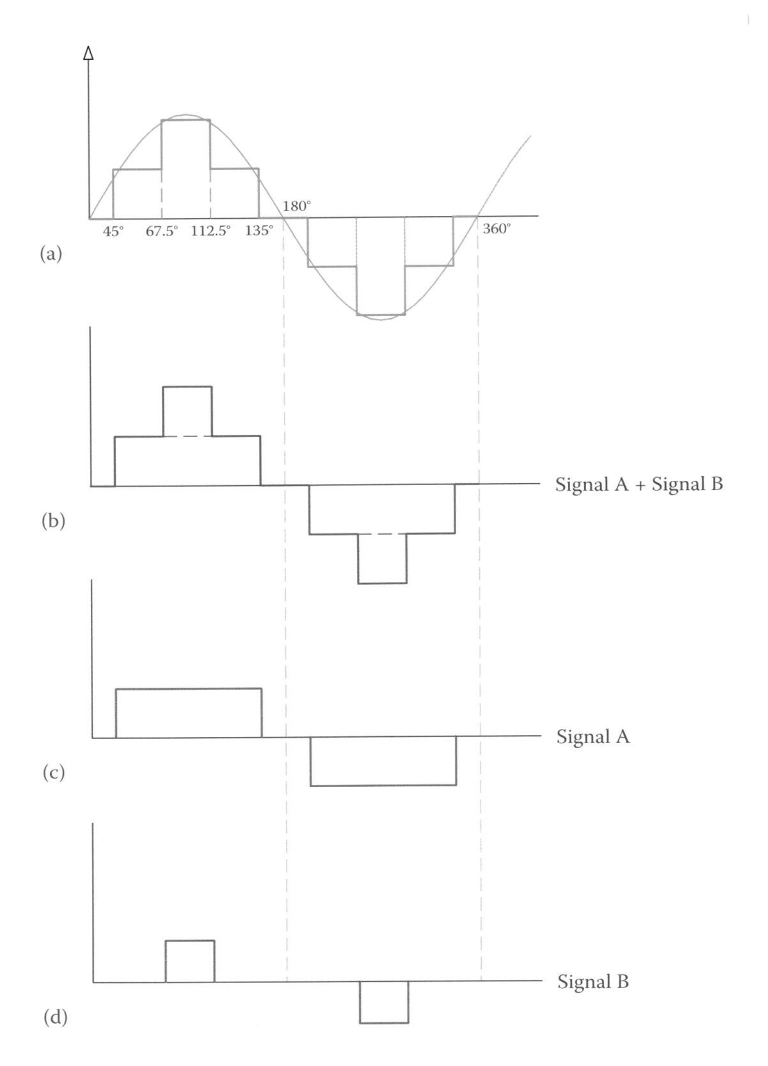

Figure 20.5 A step waveform made up of combining simpler waveforms. (a) Two-step sine form approximation. (b) Two-step waveform made up of two square waves A and B. (c) Square wave A. (d) Square wave B.

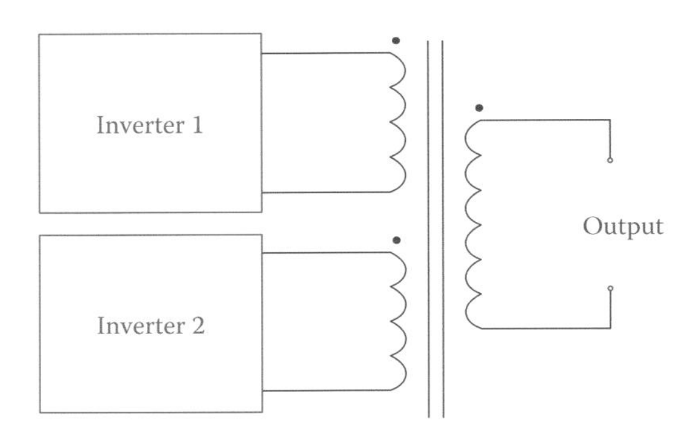

Figure 20.6 Adding two waveforms together through a transformer. A practical way of adding two square waves.

inverter to generate signal B is 67.5°. In this sense, the two inverters in Figure 20.6 can be similar but have different firing schemes.

20.3 Thyristor-Based Three-Phase Rectifier

In Chapter 16 we discussed three-phase rectifiers constructed with diodes (see Section 16.6). The rectifier shown in Figure 16.11 was a three-phase bridge rectifier that uses six diodes. This is sometimes called full-wave bridge rectifier as compared to a three-phase half-wave rectifier with only three diodes. Because at the industrial level a six-diode rectifier is used more commonly, we leave out the description of three-phase half-wave rectifier.

The discussion in this chapter for three-phase converters heavily relates to the diode-based rectifiers already discussed. Thus, we first restate certain details that have been left out for this chapter.

20.3.1 Revisiting Diode-Based Three-Phase Rectifiers

A three-phase diode rectifier is shown in Figure 20.7a. This is the same as was depicted in Figure 16.11a, except that the diodes have been numbered. Diodes 1 and 4 are on phase A, 3 and 6 are on phase B, and 5 and 2 are on

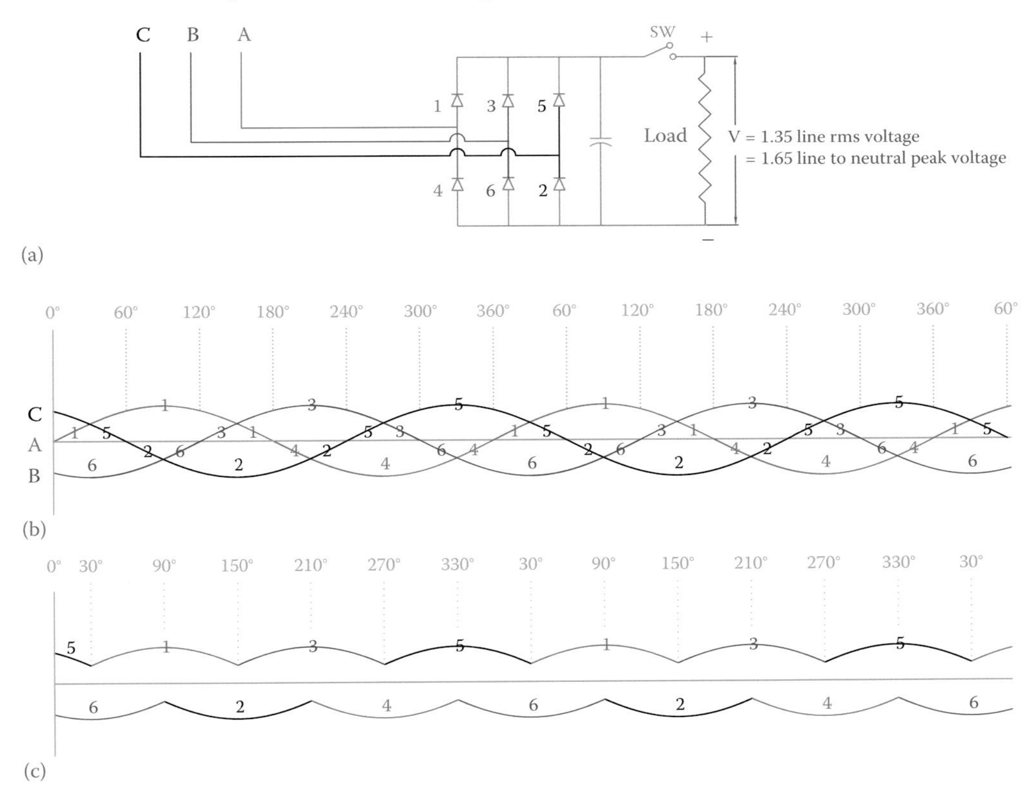

Figure 20.7 A three-phase bridge rectifier employing six diodes. (a) Schematics of a 3-phase rectifier. (b) Waveforms of the individual three phases. (c) Voltage of the DC from rectifier based on conduction of diodes.

phase C. The upper diodes (1, 3, and 5) conduct when the corresponding phase is going through its positive half cycle, and the lower diodes (4, 6, and 2) conduct during the negative half cycle.

Figure 20.7b illustrates the diode numbers corresponding to each phase for each 60° interval, based on their phase and its polarity. As mentioned earlier (see Section 16.6), a diode does not conduct for the whole duration of a half cycle it is forward biased by its line voltage. At each instant, only the two diodes corresponding to the most positive voltage and the most negative voltage conduct and supply the highest possible voltage difference across the load. For example, during the interval between 0 and 30°, in each 360° cycle as represented in Figure 20.7b, diode 5 has a higher voltage than diode 1, while both A and C phases are at their positive cycle. During this period, on the basis of the voltages present at its terminals, diode 1 is reverse biased and does not conduct. The condition reverses for the period between 30° and 60°, where phase A dominates and diode 5 does not conduct.

Figure 20.7c shows the most positive and most negative voltages for two cycles of the AC line as well as the numbers for the conducting diodes at various instants during these cycles. This figure clearly shows which diodes are conducting at each interval. Take notice of the important fact that each diode conducts for 120° during its line cycle and turns off for the rest 240°.

It is customary to use a chart similar to Figure 20.8 to show the on state of each diode with respect to other diodes. This chart is more useful when switchable devices are used instead of diodes; in particular, it shows the instant when each device is turned on and starts conducting. For each diode the hatched area shows the start and end of conduction. The chart also shows which two diodes are on at each instant.

Notice the sequence the diodes start conducting, being 1, 2, 3, 4, 5, and 6. It can be seen from both Figures 20.7c and 20.8 that the functions of diodes switch after an interval of 60°, and one pair of diodes conduct for each interval. Switching action takes place automatically based on the voltage variation in the three lines, and the conducting pairs and their switching order are 1&2, 2&3, 3&4, 4&5, 5&6, and 6&1, repeated. This

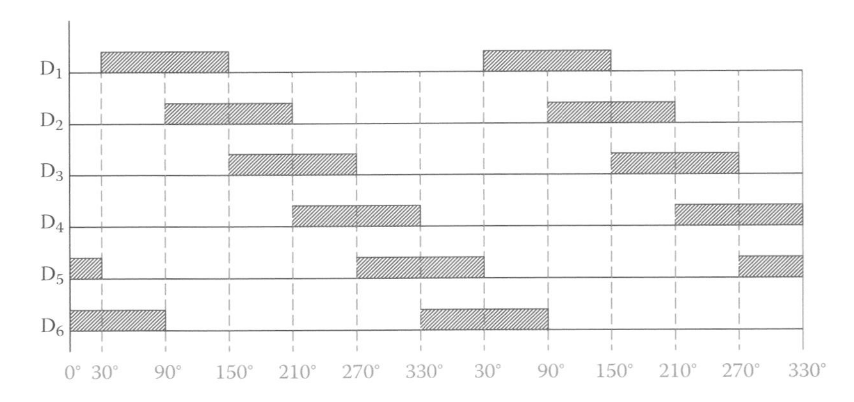

Figure 20.8 Timing chart for conducting diodes in the three-phase rectifier of Figure 20.7.

order also illustrates the reason why the diodes in Figure 20.7 are numbered this way.

20.3.2 Substituting Diodes by Switchable Devices for Three-Phase Rectifiers

As we have discussed, it is possible to use switchable devices such as thyristors and IGBTs instead of diodes. When switchable devices are employed, it becomes necessary to fire each device repeatedly and continuously as far as it is expected to operate. If a thyristor is not fired, then no rectification takes place, and there is no output. This firing or triggering must be based on a proper triggering scheme. For a rectifier, firing of each thyristor must take place in a timely manner and only during the half cycle when it is forward biased. In the rest of this chapter for simplicity we only use the word *thyristor* where it is meant to refer to a switchable device, which can be a thyristor or an IGBT.

Figure 20.9 shows the performance of a thyristor when subject to an alternating current (two cycles are shown). If triggering occurs exactly at the zero crossing point (the instant of zero voltage) for negative to positive voltage, the thyristor starts conducting at that moment and continues to conduct as far as it is forward biased. After a current is established, the device turns off by itself as the applied AC proceeds to its negative half cycle. At the moment of zero crossing from positive voltage to negative,

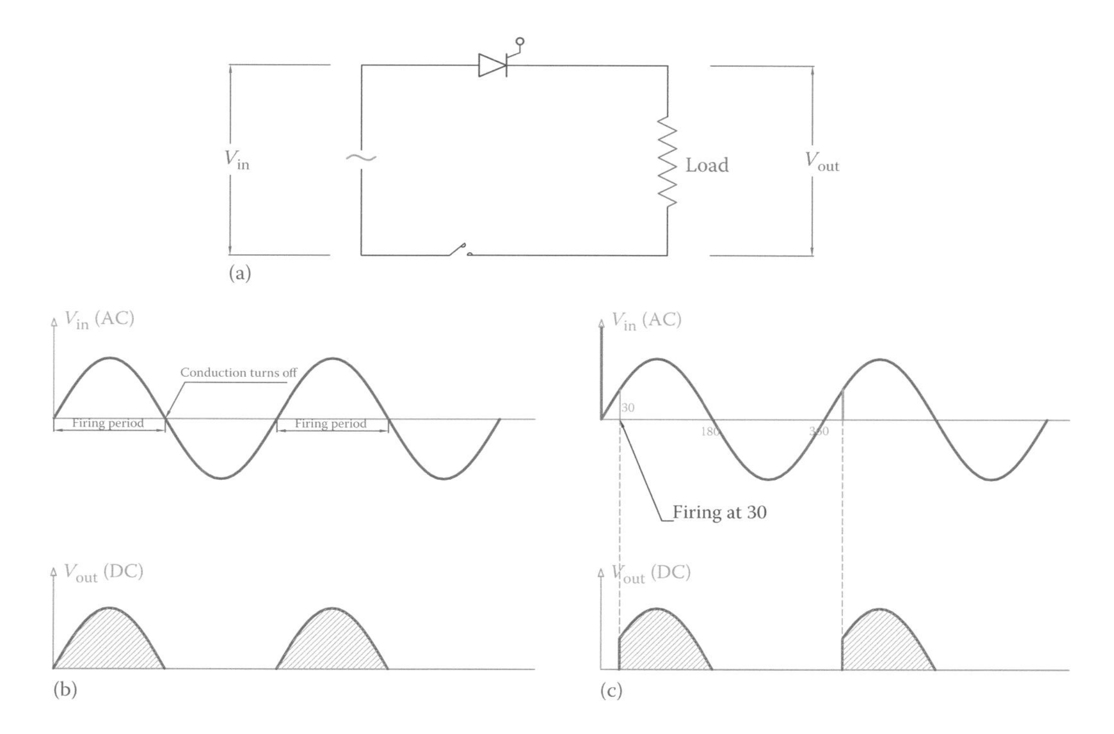

Figure 20.9 Output of a thyristor subject to alternating current and triggered with different firing angles. (a) Thyristor circuit. (b) Firing angle = 0. (c) Firing angle = 30°.

the voltage across the thyristor changes sign and the device becomes reverse biased. Consequently, the current drops to zero. The resulting DC output is shown by the hatched area in Figure 20.9b.

Figure 20.9, furthermore, illustrates what happens if thyristor firing does not take place exactly at the zero crossing point but with a delay. In this case the rectified signal has duration of less than half cycle of the applied AC signal. The case is depicted by the hatched area in Figure 20.9c for a 30° firing angle.

Figure 20.10 illustrates a full-wave three-phase rectifier employing six thyristors numbered T_1 to T_6, following the same numbering method as in Figure 20.7.

This rectifier has a structure and performance similar to that in Figure 20.7, except that it needs a driver to send firing signals to the gate of each thyristor to turn it on. The triggering must be sequential and based on a scheme like that shown in Figure 20.8. The thyristor driver timing must be synchronized by the time of zero crossing of one of the three phases, which is also the instant that the other two phases have the same magnitude, but opposite sign.

Figure 20.11 shows the firing sequence of thyristors, starting from T_1 in the upper leg of phase A. After each 60° of the AC sine wave a thyristor

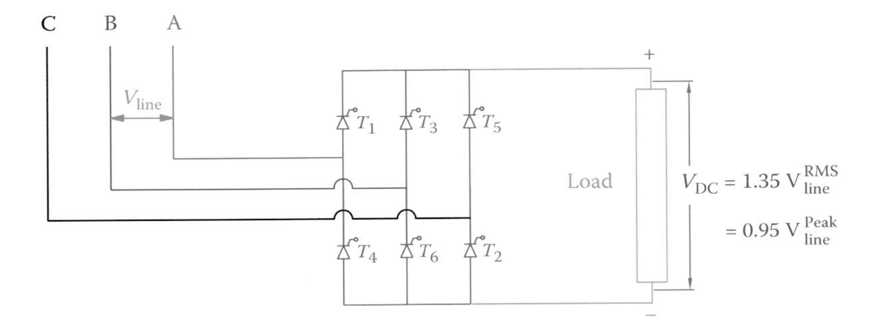

Figure 20.10 A three-phase bridge rectifier employing six thyristors.

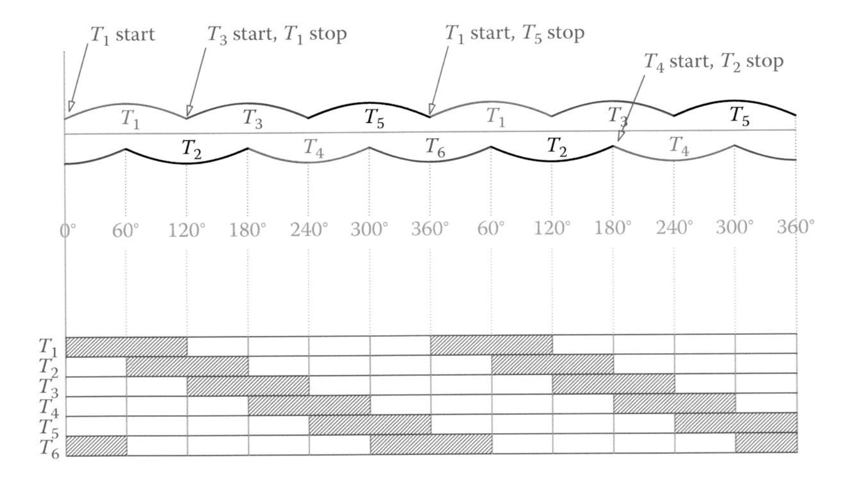

Figure 20.11 Gate firing scheme for a three-phase rectifier.

that is already on must turn off and another one start. This is shown in Figure 20.11 for some instances. Turning off is done naturally by the voltage change across each thyristor creating a reverse bias condition.

DC voltage obtained from the above three-phase rectifier is

$$V_{DC} = 0.95\ (V_{line})_{Peak} = 1.35\ V_{line} \tag{20.1}$$

Thus, for 208 V, line voltage V_{DC} = 280 V, and for 380 V line voltage, V_{DC} = 513 V.

Note that this value is the maximum voltage that can be obtained. In practice, we may want a lower voltage. In such a case the output voltage can be controlled, as discussed next.

20.3.3 Voltage Control in Thyristor-Based Rectifier

One of the advantages of using switchable devices instead of diodes in a rectifier is the possibility to control the voltage of the DC output. This is done by introducing a delay in the firing of all thyristors. This delay, normally expressed in terms of an angle and measured in degrees, is defined as the firing angle (the same term we have already seen for the single-phase inverter). All thyristors must have the same delay in their triggering.

When there is a delay in triggering thyristors, those ones that should turn off were there no delays (because of receiving a voltage at their cathodes that changes their bias condition) continue to conduct. For instance, in Figure 20.12, assuming no triggering delay, T_1 starts conducting at point L and continues until point M when T_3 is triggered and caused T_1 to turn off. But, if T_3 is not triggered, then T_1 stays on. During the interval between M and N, if T_3 is triggered, it turns off T_1; otherwise, T_1 conducts until N. At point N it turns off naturally (natural commutation; see Section 19.8) and stops conducting. It starts conducting again afterward, whenever its gate receives a positive pulse, while it is forward biased. Likewise, if thyristor T_2 is not started at point P, then T_6 does not turn off and conducts until the time point Q is reached.

Note that all preceding statements are true when the load is purely resistive. In cases of inductive and capacitive loads the conditions change. A detailed study of what happens in various conditions is outside the scope of this book.

Figure 20.13 illustrates the case when the firing angle is not zero. The firing times and thyristor states (on or off) can be compared for three cases when the firing angle is 0, 30°, and 50°. Figure 20.13a corresponds to zero firing angle. It also shows the two thyristors that conduct for each

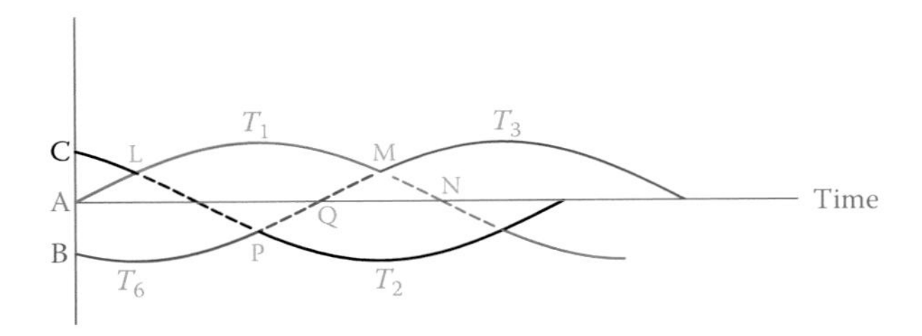

Figure 20.12 On and off conditions of a thyristor.

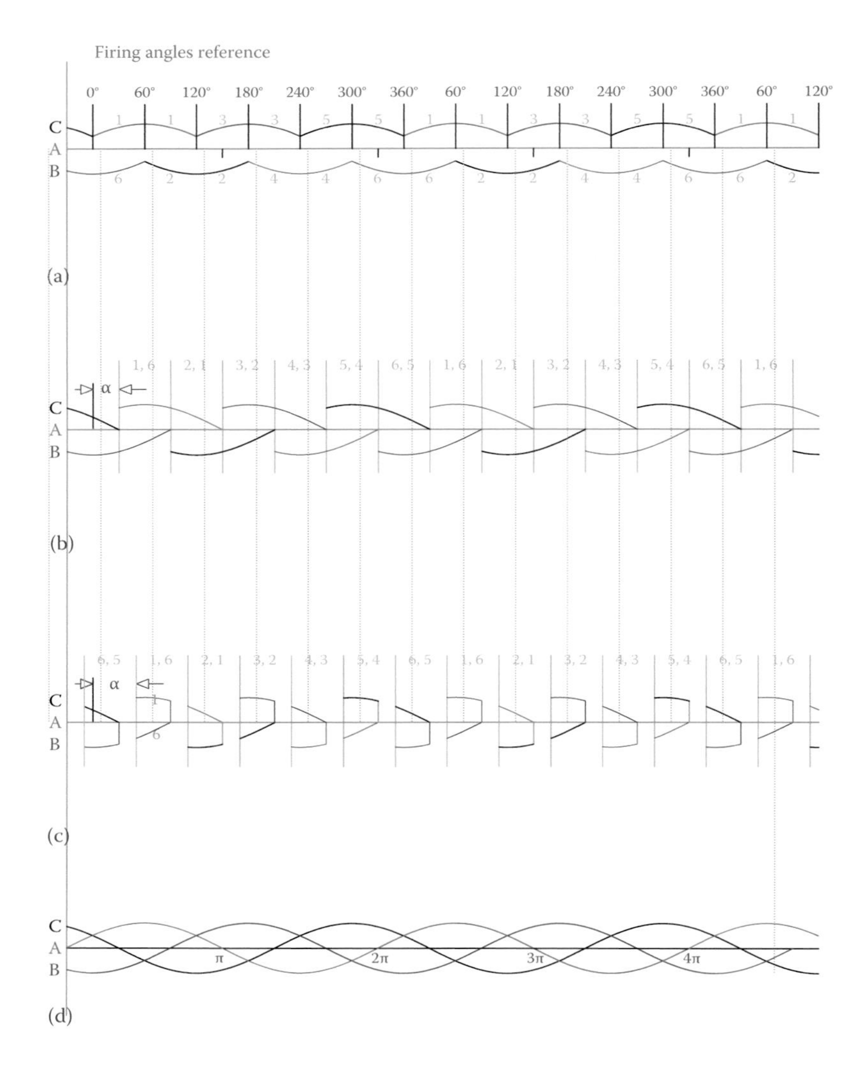

Figure 20.13 Comparison of DC output with (a) 0, (b) 30°, and (c) 50° firing angle and (d) input AC.

60° interval. For example, at the start of the period between 0 and 60°, T_1 was triggered and T_6 was already on and continued to conduct. When there is a delay in firing, then all thyristors are triggered by the same nonzero firing angle, meaning that still the same pair of thyristors as in the case of no delay conduct simultaneously. In other words, the conducting interval that each pair of thyristors share is shifted in time. In Figure 20.13b and 20.13c the delay is represented by α (alpha), which is 30° and 50°, respectively.

For certain firing angles at time intervals when one of the thyristors in a pair turns off by natural commutation, only one thyristor conducts, and, therefore, there is no voltage across the load and no current in it. This is especially true for resistive loads because electricity cannot be stored in them. This has happened for the 50° firing angle in Figure 20.13. It is the

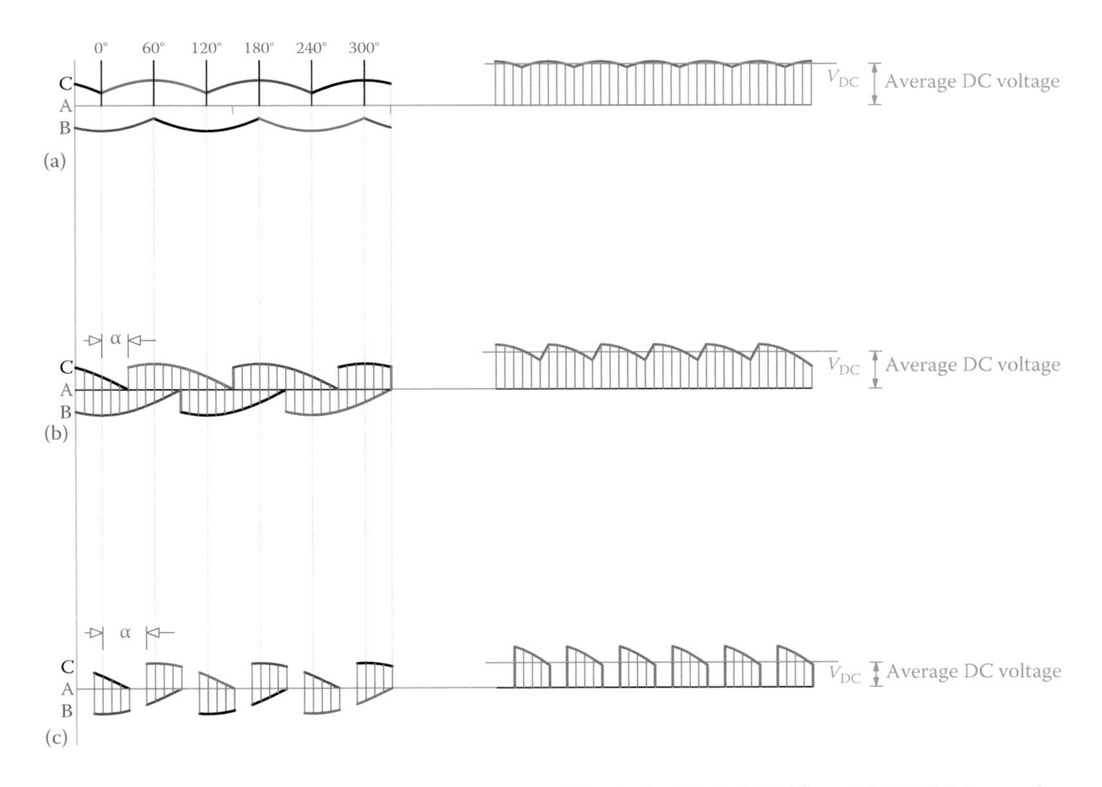

Figure 20.14 Showing examples of the DC output and its ripple. (a) 0, (b) 30°, and (c) 50° firing angle.

reason for the blank intervals in the output DC voltage, shown in Figure 20.14c. Figure 20.14 is a complement to Figure 20.13, showing the DC output voltages and their average values for the three different firing angles.

The result of a delay in triggering thyristors in a rectifier is the reduction of the output voltage. As can be seen from Figures 20.13 and 20.14, the average output voltage becomes smaller as the firing angle is increased. In this way the output of a thyristor-based rectifier can be controlled, as desired, by adjusting the firing angle. The output voltage is proportional to the cosine of the firing angle. If the firing angle is represented by α, then we can write

$$V_{DC} = V_{DC0} \cos \alpha = 1.35\, V_{line} \cos \alpha \tag{20.2}$$

In Equation 20.2, V_{DC0} represents the DC voltage from the rectifier when $\alpha = 0$. Equation 20.2 also implies that if $\alpha = 90°$, there is no DC output ($V_{DC} = 0$).

Figure 20.14 also depicts the DC output of the two aforementioned firing angles compared with that of zero firing angle. It is evident from the figure that the ripple frequency is the same (6 times the line frequency), but the ripple percentage gets larger as the firing angle increases.

> Output voltage from a rectifier can be adjusted by altering the firing angle of its thyristors.

20.4 Thyristor-Based Three-Phase Inverter

Photovoltaic (PV) system: A setup of photovoltaic panels to absorb solar energy. Photovoltaic panels are made of an array of solar cells connected together (in series and parallel) for a desired voltage and a desired current. A solar cell is a semiconductor that converts light to voltage, that is, generates a voltage across its terminals proportional to the light it receives.

Single-phase inverters are good only for small appliances. At the industrial level, such as for motor drives, HVDC, solar energy from **photovoltaic (PV) systems** variable speed wind turbines and turbines equipped with doubly fed induction generator (DFIG), three-phase inverters are widely used. These inverters use thyristors and IGBTs. As mentioned before, thyristor in our discussion implies an electronically controllable switchable device.

The simplest three-phase inverter uses six thyristors and has a structure as shown in Figure 20.15. This configuration is called six-step inverter. The drivers for turning the thyristors on and off are not shown. Each thyristor has a freewheel diode in parallel with it but with the opposite polarity (direction) to protect it from high voltages that can develop in the circuit. As in the case of single-phase inverter, the two thyristors on the same leg cannot conduct at the same time; otherwise, they make a short circuit across the power supply.

20.4.1 180° Conduction

Thyristors must be triggered sequentially on the basis of a specific scheme so that the voltages appearing in A, B, and C have a good resemblance to a three-phase alternating current with 120° phase difference between each two phases. In the triggering scheme in Figure 20.16, each thyristor is in the on state and conducts for 180°, followed by 180° in the off state. The time for this period (i.e., the time for retriggering the same thyristor) defines the frequency of the generated alternating current. For instance, if each thyristor is fired every 20 msec, then the frequency of the output AC is $1 \div 0.020 = 50$ Hz. In this respect, for a frequency of 60 Hz, if this firing scheme is used, then each thyristor must be fired every $1 \div 60 = 0.01667$ sec, or 16.67 msec.

A closer look at the firing scheme in Figure 20.16 reveals that for each interval of 60°, three out of six switches are in the on state, but they alter; one turns off and another is turned on. However, the sequence of firing thyristors is always 1, 2, 3, 4, 5, and 6. Examples of on-off alteration are shown for 120° and 180° angles. At 120°, for instance, first T_6 must be turned off and then T_3 be turned on (not the reverse). Thus, for the period between 120° and 180° the thyristors T_3, T_2, and T_1 conduct, until T_1 (the last shown in sequence T_3,

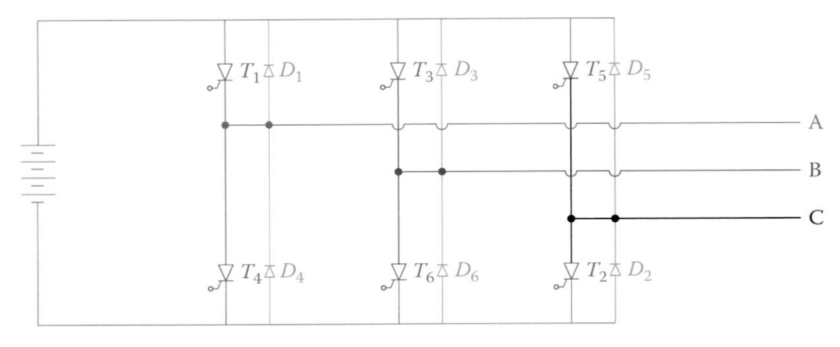

Figure 20.15 Schematic of a three-phase six-step inverter circuit.

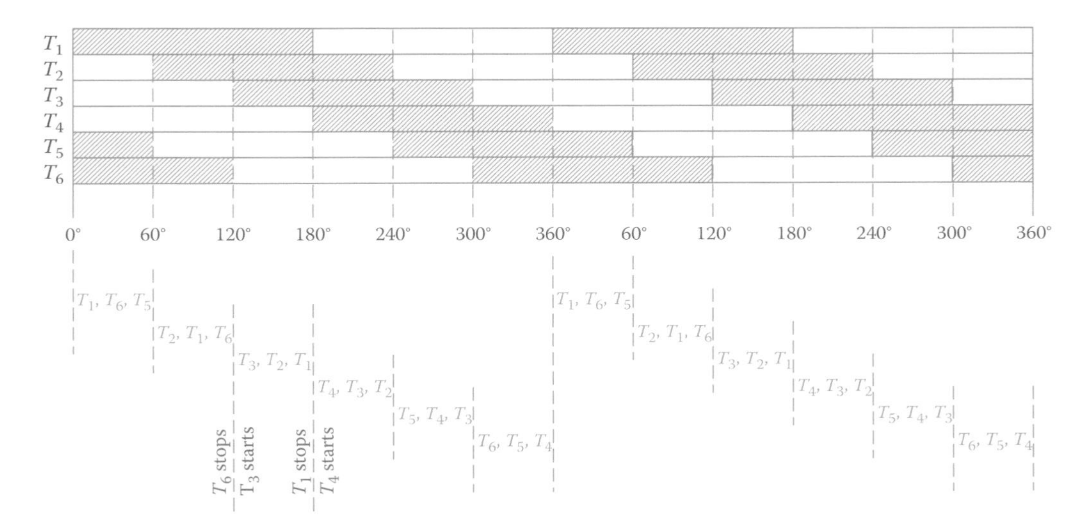

Figure 20.16 Firing scheme for 180° conduction.

T_2, T_1) is turned off and T_4 (the first shown in the next sequence) is fired. For the next period (180° to 240°) the conducting thyristors are T_4, T_3, and T_2 (as shown between the dotted lines in Figure 20.16).

If the firing scheme in Figure 20.16 is employed, then the voltages on all the three lines A, B, and C continuously jump from the positive side of the power supply to the negative side (between $+V_{DC}$ and zero) and vice versa. The power supply voltage is always assumed constant. Table 20.1 summarizes the connections for terminals A, B, and C. In this table V and 0 represent the DC voltage and 0, as a result of connection to the positive side or negative side of the power supply, respectively.

A three-phase load is either star connected or delta connected to them, as shown in Figure 20.17. (For simplicity, the freewheeling diodes are not shown.) Figure 20.18 depicts the states of voltages at A, B, and C for the six different intervals of a 360° cycle and the current direction in the three branches of a Y- or Δ-connected load. In columns 3, 4, and 5 a + sign denotes connection to the positive side of the DC power supply and a – sign denotes connection to the negative side. In column 6 showing the currents a 0 indicates that there is no current between the two terminals for that 60° interval.

When the load is star connected, a neutral point is practically defined for the resulting three-phase system. The phase voltage can be found from the voltage difference of the neutral point N and any of the A, B, or C

TABLE 20.1 Thyristors that are on and Voltages at Terminals A, B and C at 60° Intervals for 180° Conduction

	1, 6, 5	2, 1, 6	3, 2, 1	4, 3, 2	5, 4, 3	6, 5, 4
A	V	V	V	0	0	0
B	0	0	V	V	V	0
C	V	0	0	0	V	V

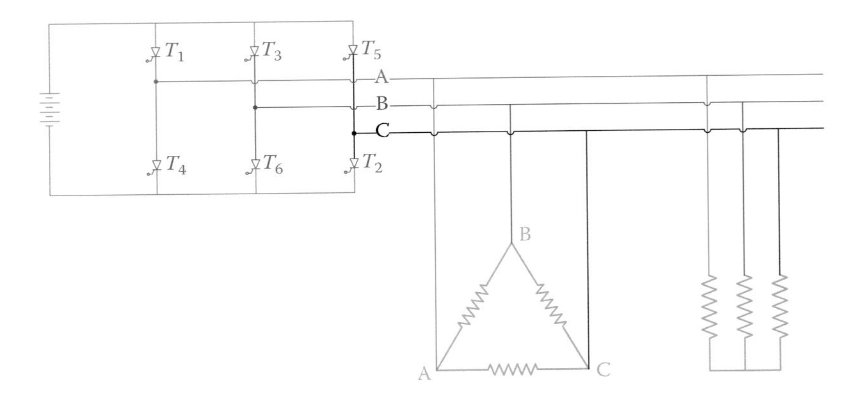

Figure 20.17 Δ- and star-connected loads to an inverter output.

terminals. Figure 20.19 can help understand the voltage at point N with respect to the power supply at each interval of a cycle. The principle of a voltage divider is used for this purpose based on a balanced three-phase resistive load connected to the inverter output.

Figure 20.19 also shows the currents in the three lines for a resistive load, for Y and Δ connection. These currents are proportional to the voltages for this load and have the same waveform similar to the voltage variation. For inductive and capacitive loads, current variation is affected by the load reactance and is smoothed out, as was seen for the single-phase inverter.

The line (phase-to-phase) voltages $V_A - V_B$, $V_B - V_C$, and $V_C - V_A$, shown as V_{A-B}, V_{B-C}, and V_{C-A}, respectively, are presented in Figure 20.20. As can be seen, the three alternating currents of square waveform are 120° out of phase with respect to each other, and the voltage varies between plus and minus V_{DC}. This figure also illustrates the line to neutral voltage (phase voltage) variation denoted by V_{A-N}, V_{B-N}, and V_{C-N}. On the basis of the facts depicted in Figures 20.19 and 20.20 the phase voltage fluctuates between plus and minus $2V_{DC}/3$, and the neutral line can be connected to the negative side of the power supply.

In a simple three-phase inverter, at each instant the line voltage is the difference between a terminal connected to the positive side and one connected to the negative side; thus, the line voltage is equal to 2× the supply (battery) voltage V_{DC}.

20.4.2 120° Conduction

In the preceding firing scheme each thyristor was on for 180° period and off for 180°. It is possible to use an alternative scheme with 120°conduction instead of 180°. In this case each thyristor conducts for 120° followed by an interval of 240° silence. At each interval of 60°, thus, only two thyristors (from two different legs) conduct. When a thyristor is on for a shorter time, less energy is consumed in it. This scheme is shown in Figure 20.21.

Period	Conducting elements	A	B	C	Currents
0°–60°	1, 6, 5	+	–	+	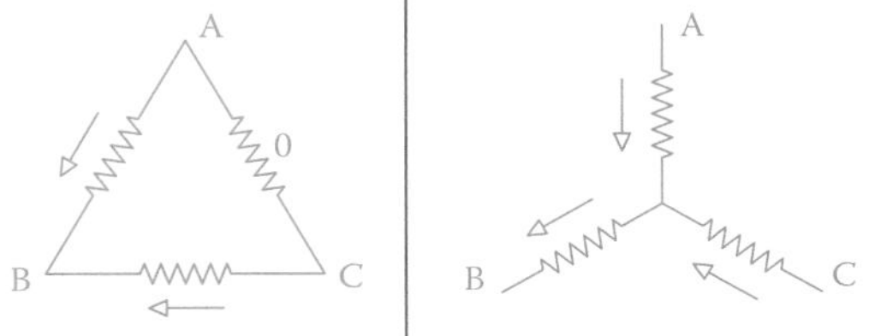
60°–120°	2, 1, 6	+	–	–	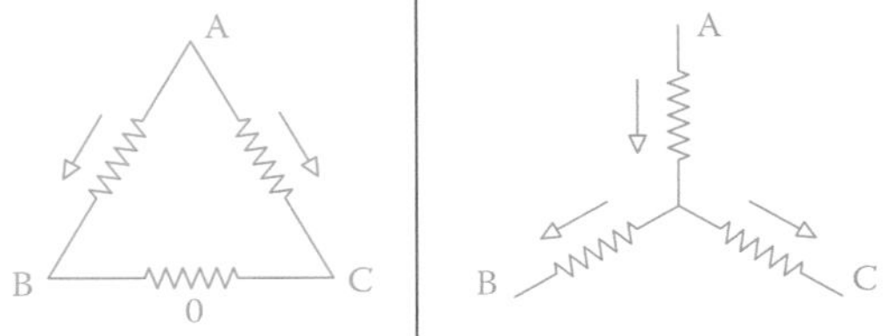
120°–180°	3, 2, 1	+	+	–	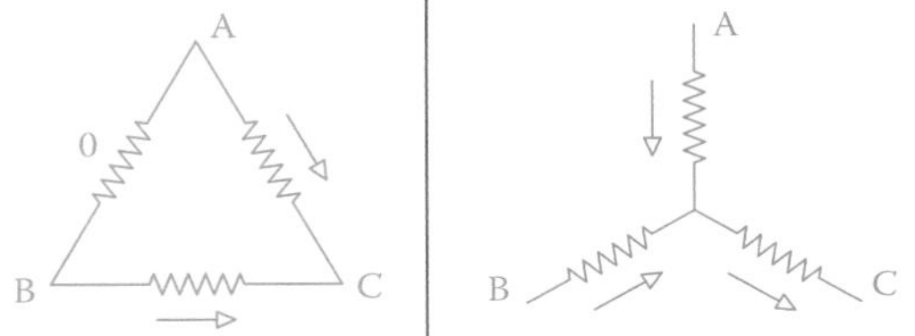
180°–240°	4, 3, 2	–	+	–	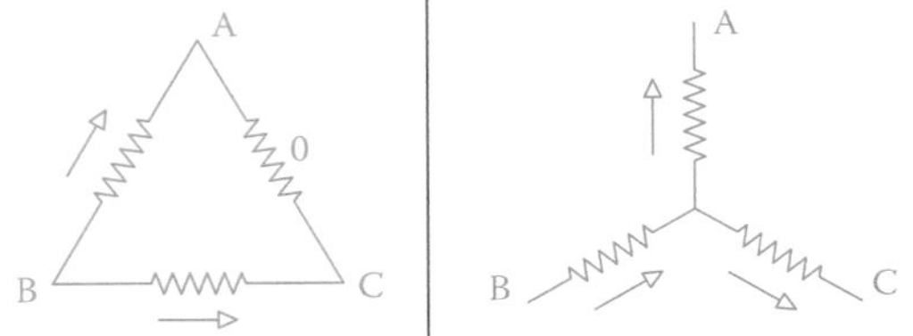
240°–300°	5, 4, 3	–	+	+	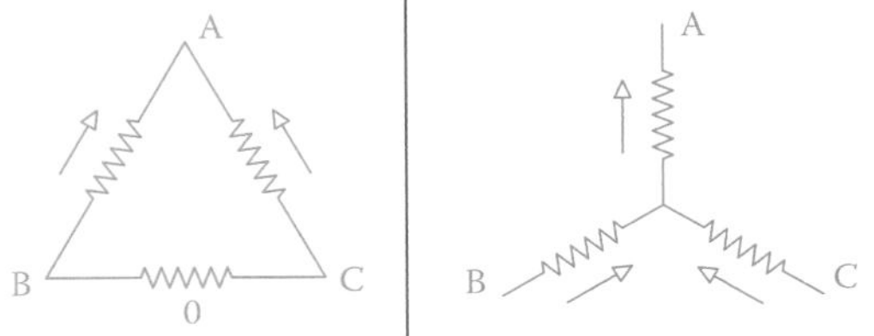
300°–360°	6, 5, 4	–	–	+	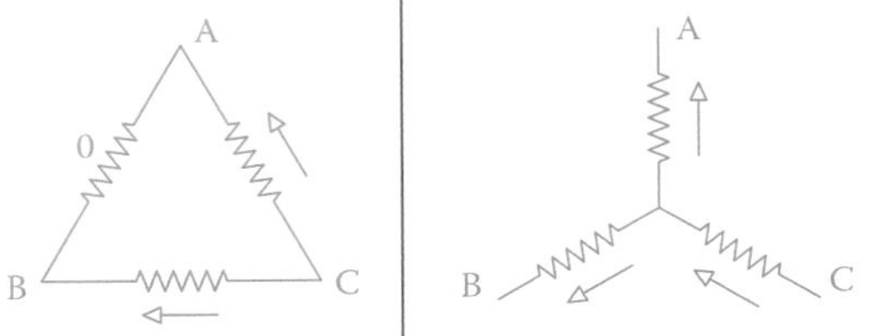

Figure 20.18 Details of voltages and currents for 180° conduction firing scheme in Δ- and Y-connected resistive loads.

Period	Conducting elements	Star connection				Delta connection		
		I_A	I_B	I_C		I_A	I_B	I_C
0°–60°	T_1, T_6, T_5	$\frac{V_{DC}}{3R}$	$\frac{2V_{DC}}{3R}$	$\frac{V_{DC}}{3R}$	B − R ($\frac{2V_{DC}}{3}$) N $\frac{R}{2}$ ($\frac{V_{DC}}{3}$) + A, C	$\frac{V_{DC}}{R}$	$\frac{2V_{DC}}{R}$	$\frac{V_{DC}}{R}$
60°–120°	T_2, T_1, T_6	$\frac{2V_{DC}}{3R}$	$\frac{V_{DC}}{3R}$	$\frac{V_{DC}}{3R}$	B, C − R ($\frac{2V_{DC}}{3}$) N $\frac{R}{2}$ ($\frac{V_{DC}}{3}$) + A	$\frac{2V_{DC}}{R}$	$\frac{V_{DC}}{R}$	$\frac{V_{DC}}{R}$
120°–180°	T_3, T_2, T_1	$\frac{V_{DC}}{3R}$	$\frac{V_{DC}}{3R}$	$\frac{2V_{DC}}{3R}$	C − R ($\frac{2V_{DC}}{3}$) N $\frac{R}{2}$ ($\frac{V_{DC}}{3}$) + A, B	$\frac{V_{DC}}{R}$	$\frac{V_{DC}}{R}$	$\frac{2V_{DC}}{R}$
180°–240°	T_4, T_3, T_2	$\frac{V_{DC}}{3R}$	$\frac{2V_{DC}}{3R}$	$\frac{V_{DC}}{3R}$	A, C − R ($\frac{2V_{DC}}{3}$) N $\frac{R}{2}$ ($\frac{V_{DC}}{3}$) + B	$\frac{V_{DC}}{R}$	$\frac{2V_{DC}}{R}$	$\frac{V_{DC}}{R}$
240°–300°	T_5, T_4, T_3	$\frac{2V_{DC}}{3R}$	$\frac{V_{DC}}{3R}$	$\frac{V_{DC}}{3R}$	A − R ($\frac{2V_{DC}}{3}$) N $\frac{R}{2}$ ($\frac{V_{DC}}{3}$) + B, C	$\frac{2V_{DC}}{R}$	$\frac{V_{DC}}{R}$	$\frac{V_{DC}}{R}$
300°–360°	T_6, T_5, T_4	$\frac{V_{DC}}{3R}$	$\frac{V_{DC}}{3R}$	$\frac{2V_{DC}}{3R}$	A, B − R ($\frac{2V_{DC}}{3}$) N $\frac{R}{2}$ ($\frac{V_{DC}}{3}$) + C	$\frac{V_{DC}}{R}$	$\frac{V_{DC}}{R}$	$\frac{2V_{DC}}{R}$

Figure 20.19 Understanding the voltage at the neutral (reference) point in an inverter with 180° firing scheme.

For this firing scheme at each 60° interval of a cycle the two points that are connected to the power supply positive and negative terminals are shown in Table 20.2. Consider a Y-connected load to the resulting three-phase electricity. The neutral point N is at equal voltages for the two connecting terminals at a time. The voltages on each of the A, B, and C terminals vary between $+V_{DC}$ and zero. Thus, a point at the midrange of the power supply can be considered the neutral for the three-phase system. (If this point is available, it can be physically used as the neutral.)

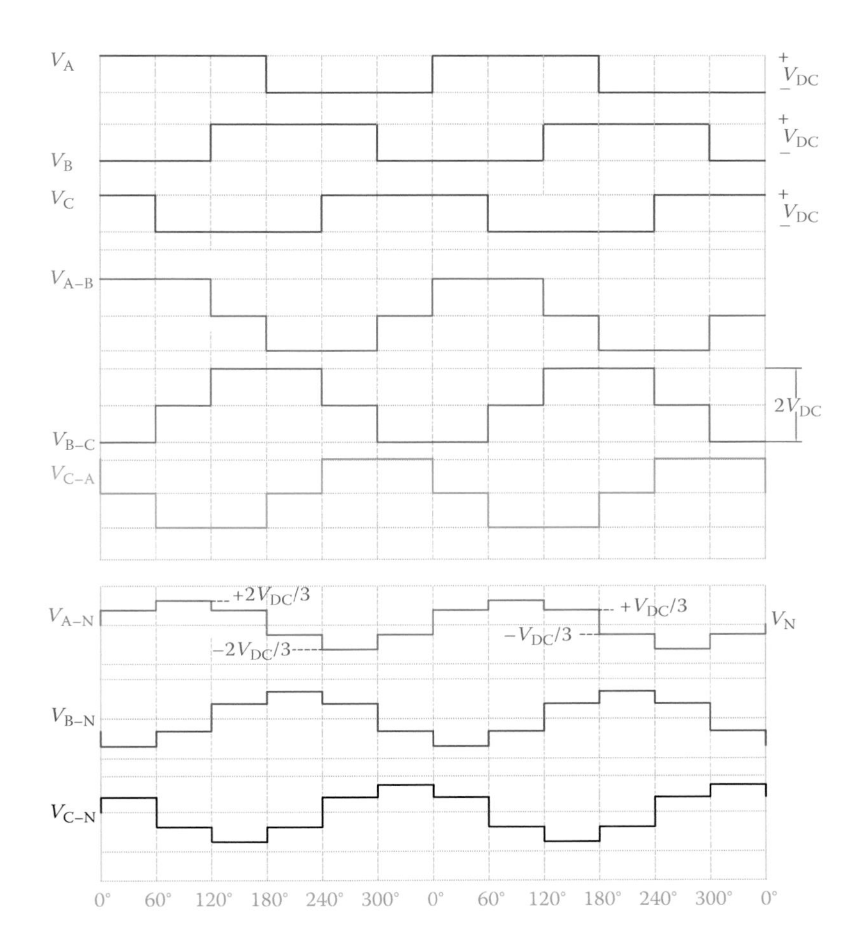

Figure 20.20 Voltage relationships in the generated three-phase waveforms from a six-step inverter with 180° firing.

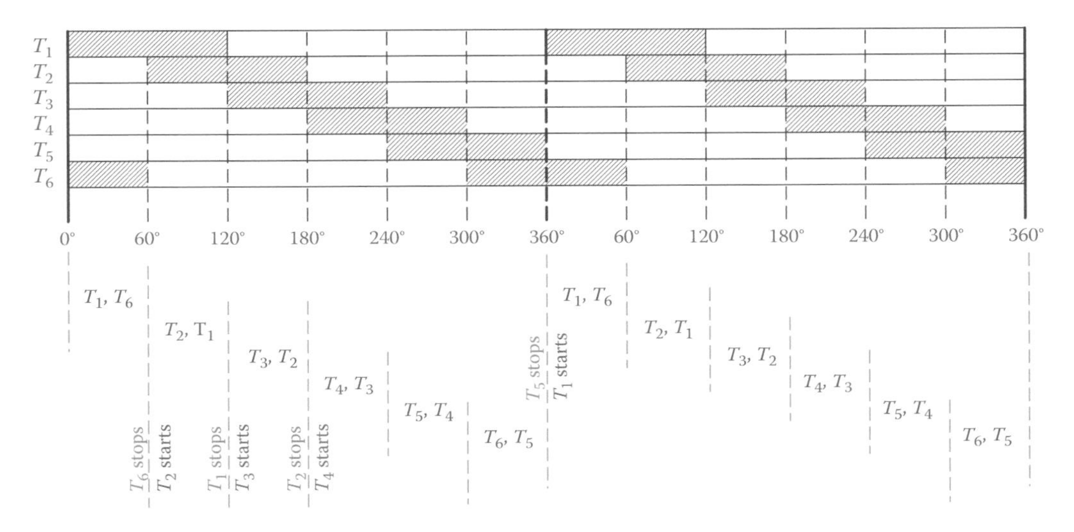

Figure 20.21 A 120° conduction firing scheme.

TABLE 20.2 Thyristors that are on (two) and Voltages at Terminals A, B and C at 60° Intervals for 120° Conduction

	1, 6	2, 1	3, 2	4, 3	5, 4	6, 5
A	V	V	–	0	0	–
B	0	–	V	V	–	0
C	–	0	0	–	V	V

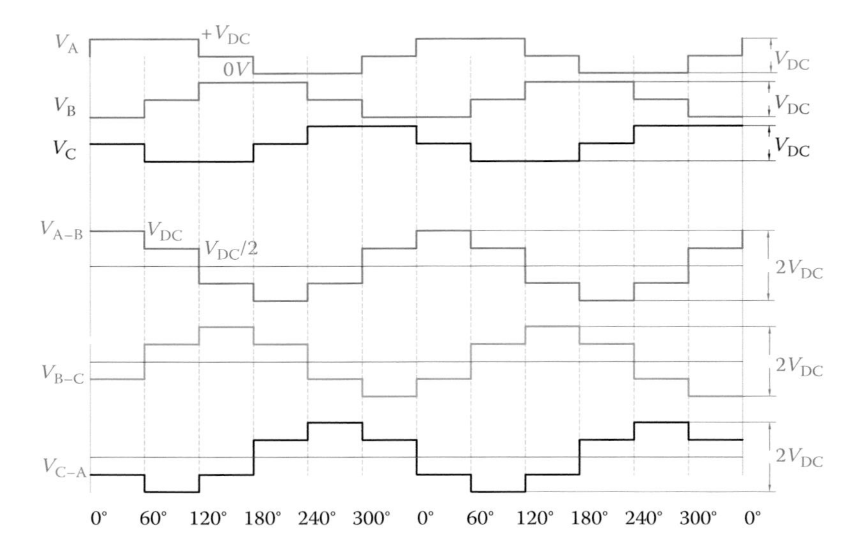

Figure 20.22 Waveform of a six-step inverter with 120° firing scheme.

Figure 20.22 illustrates the variation of phase voltages, based on which the line voltages V_{A-B}, V_{B-C}, and V_{C-A} can be found and are also shown. As depicted in Figures 20.20 and 20.22, in both 180° and 120° conduction the maximum–minimum range for the output waveforms from an inverter is two times the DC supply voltage.

As can be understood from Table 20.2 or Figure 20.21, for this triggering scheme it is not necessary to turn off a thyristor *before* its counterpart is turned on. This is because the two thyristors to be switched (on or off) at the end of a 60° interval do not belong to the same leg.

Figure 20.23 depicts what happens during one cycle (360°) as far as the voltage and current change in Δ-connected and star-connected loads are concerned. In columns 3, 4, and 5 of this figure, + denotes connection to the positive side of DC power supply, – represents connection to the negative side of the power supply and, NC stands for "not conducting" (see Table 20.2). This is for a pure resistive load. The corresponding currents are shown in Figure 20.24, accordingly.

20.5 Power Converter

You have noticed by now that (compare Figures 20.10 and 20.15 together and Figures 20.11 and 20.21 together) a three-phase rectifier has the same

Period	Conducting elements	A	B	C	Currents
0°–60°	1, 6	+	–	NC	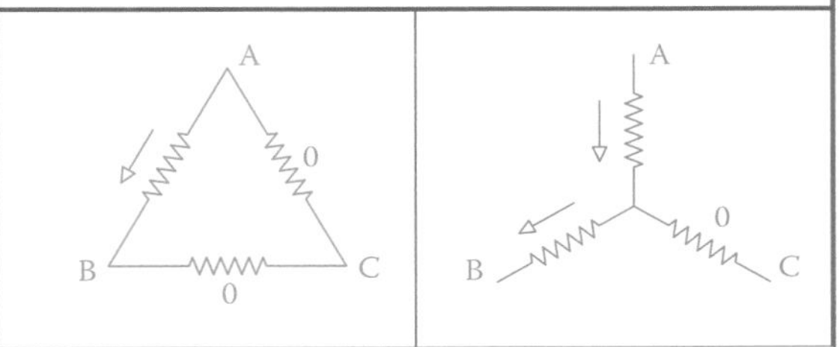
60°–120°	2, 1	+	NC	–	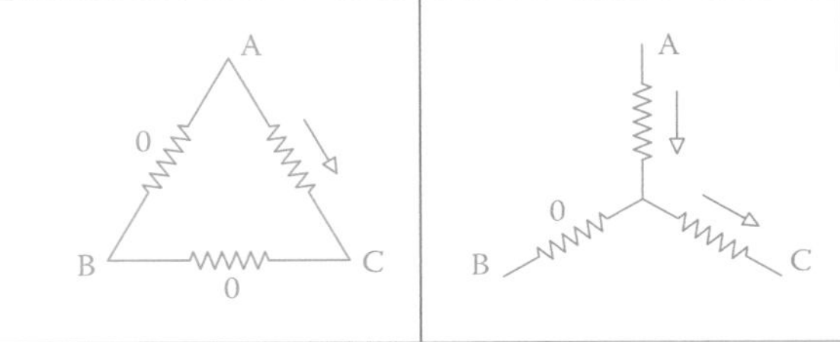
120°–180°	3, 2	NC	+	–	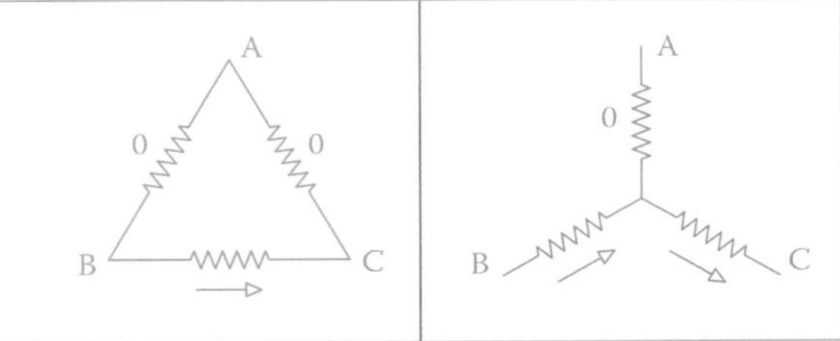
180°–240°	4, 3	–	+	NC	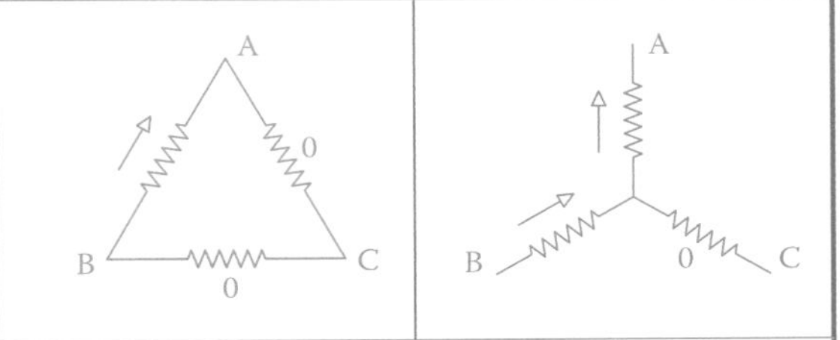
240°–300°	5, 4	–	NC	+	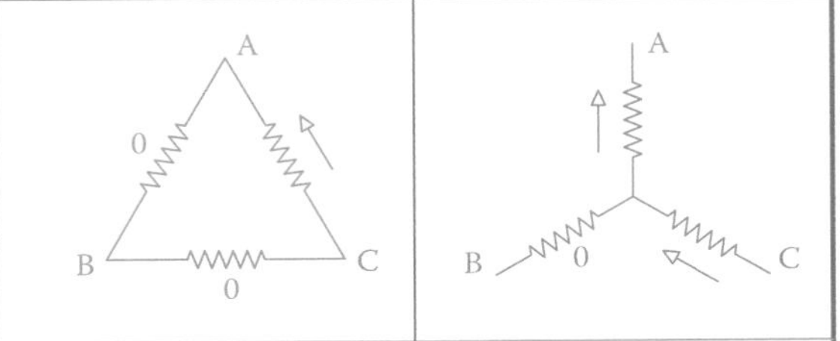
300°–360°	6, 5	NC	–	+	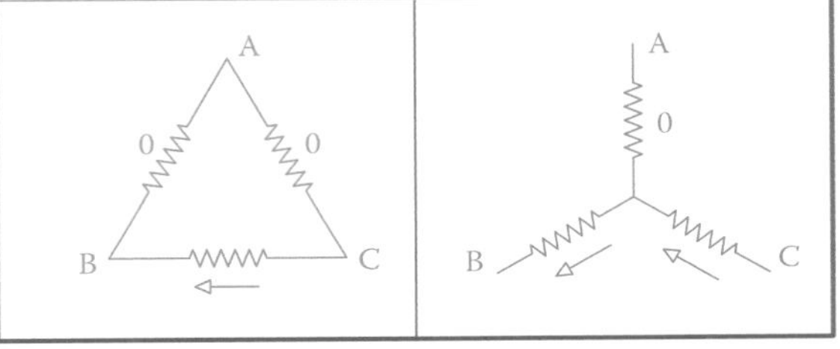

Figure 20.23 Voltages and currents for 120° conduction in Δ- and star-connected loads.

Period	Conducting elements	Star connection			Delta connection		
		I_A	I_B	I_C	I_A	I_B	I_C
0°–60°	T_1, T_6	$\frac{V_{DC}}{2R}$	$\frac{V_{DC}}{2R}$	0	$\frac{V_{DC}}{R}$	$\frac{V_{DC}}{R}$	0
60°–120°	T_2, T_1	$\frac{V_{DC}}{2R}$	0	$\frac{V_{DC}}{2R}$	$\frac{V_{DC}}{R}$	0	$\frac{V_{DC}}{R}$
120°–180°	T_3, T_2	0	$\frac{V_{DC}}{2R}$	$\frac{V_{DC}}{2R}$	0	$\frac{V_{DC}}{R}$	$\frac{V_{DC}}{R}$
180°–240°	T_4, T_3	$\frac{V_{DC}}{2R}$	$\frac{V_{DC}}{2R}$	0	$\frac{V_{DC}}{R}$	$\frac{V_{DC}}{R}$	0
240°–300°	T_5, T_4	$\frac{V_{DC}}{2R}$	0	$\frac{V_{DC}}{2R}$	$\frac{V_{DC}}{R}$	0	$\frac{V_{DC}}{R}$
300°–360°	T_6, T_5	0	$\frac{V_{DC}}{2R}$	$\frac{V_{DC}}{2R}$	0	$\frac{V_{DC}}{R}$	$\frac{V_{DC}}{R}$

Figure 20.24 Understanding the currents in the three phases for a six-step inverter with 120° firing scheme.

structure and the same elements as a three-phase inverter. For this reason both are referred to as three-phase converters, and because they can be built for carrying medium to large power, they are called power converters.

In fact, Equation 20.2 indicates that in a rectifier if the firing angle α is 90°, then the output has zero voltage. This can be observed from Figure 20.25 that shows how for $\alpha = 90°$ out of the pair of thyristors that must be on at the same time only one is conducting, and, therefore, no voltage can be passed to the output. A firing angle of 0° is also included in Figure 20.25, for comparison. It also implies that if $90° < \alpha < 180°$, then the value of DC voltage is negative. Negative DC voltage can be interpreted that if DC power is available in the output, then it can flow in the reverse direction through the switching devices and be converted to AC. This makes it convenient, in practice, because the same device (shown in Figure 20.10 or 20.15) can be used for two different purposes. That is, by changing the triggering times (firing angle) of thyristors the same device (or the same circuit) can be used as a rectifier or as an inverter. If the firing angle is smaller than 90°, the circuit functions as a rectifier, and if the firing angle is greater than 90°, it functions as an inverter.

By changing the firing angle a three-phase converter can behave as a rectifier or can function as an inverter. To act as a rectifier, the firing angle must be smaller than 90°, and to serve as an inverter, the firing angle must be greater than 90°.

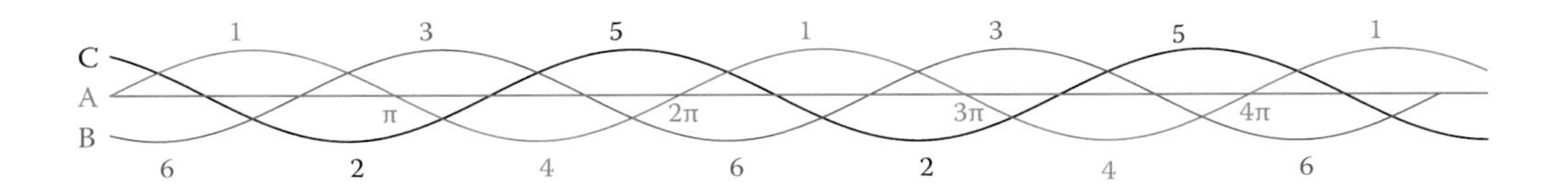

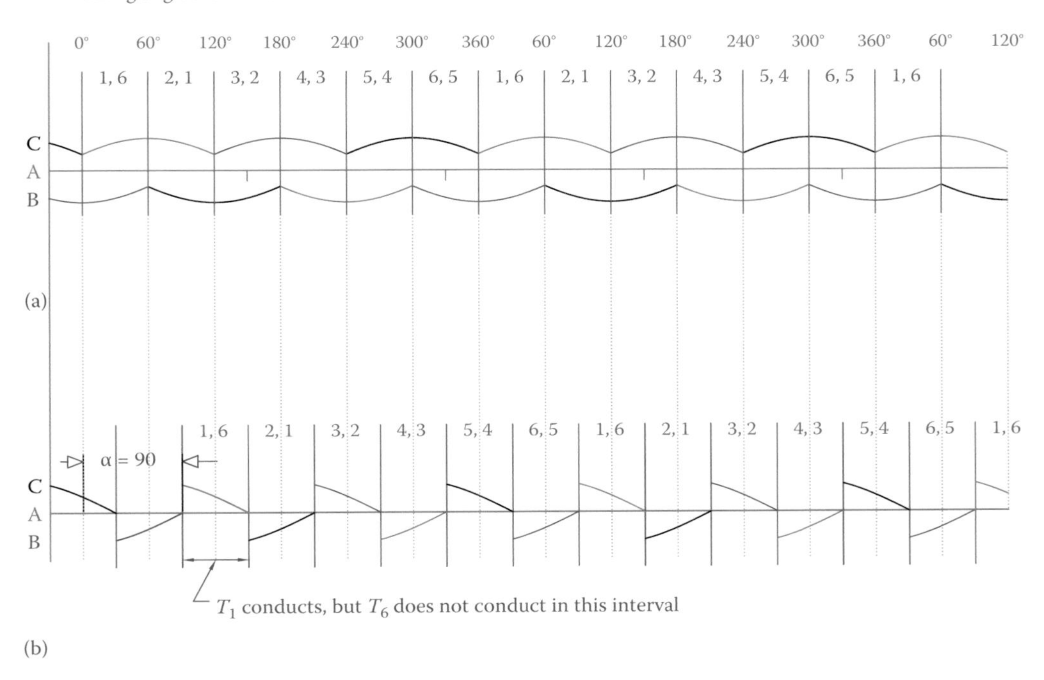

Figure 20.25 A firing angle of (a) 0 and (b) 90° for a rectifier. 90° firing angle leads to zero output.

20.6 Pulse Width Modulation

Pulse width modulation (PWM) is frequently used in electronic operations and circuits for both DC and AC electricity. Among its applications in DC is voltage control. Among its AC applications in addition to voltage control is shaping the output of an inverter. As we have seen before, the output of an inverter is not a sine wave. Using PWM allows generation of an AC signal with a better resemblance to a sinusoidal waveform.

Pulse width modulation (PWM): One way of adjusting/controlling the value of a variable by repeated pulses of appropriate width at a proper frequency. These pulses can directly denote a controlled variable, thus their average defines the value of that variable; or they can be part of the control system.

20.6.1 DC Voltage Regulation by Pulse Width Modulation

The principle of pulse width modulation is shown in Figure 20.26 for a simple case of obtaining a smaller desired DC voltage from a DC power supply. In Section 20.3 (see Figure 20.14) we learned that a DC output of a rectifier may have ripples and even blank intervals. Average DC voltage then depends on the variation of the voltage level, taking into account zero values for the blank intervals.

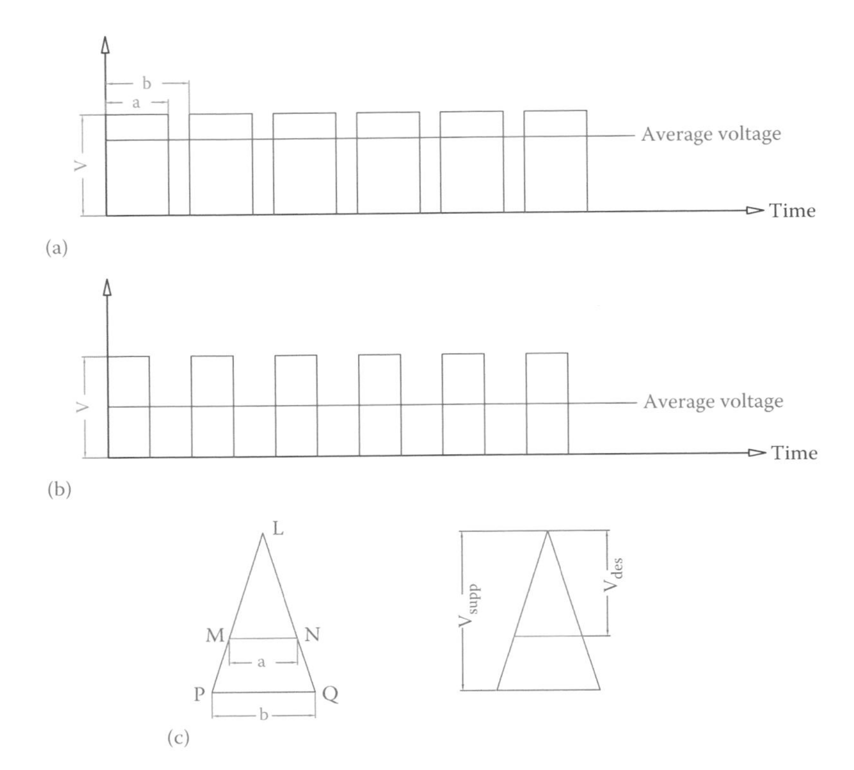

Figure 20.26 Principle of pulse width modulation. (a) Wider pulses lead to higher average compared to (b) with narrower pulses and lower average. (c) Pulse width can be found from position of MN in triangle LPQ.

In Figure 20.26a a repeated series of pulses of constant voltage and constant width (duration), all with the same polarity, are sent across a wire. The width of pulses and the frequency of sending them are important and define the properties of their effects. Pulse width represents its duration and is denoted by a in the figure. Frequency defines the elapsed time between sending the next pulse, which is denoted by b in the figure. Thus, the pulses are sent every b second, and each pulse has a second duration. By this definition, $a < b$, and its maximum value can be b. If the voltage of each pulse is V volts, then the average voltage in the wire is determined by a, b, and V, i.e.,

$$V_{Av} = \frac{a}{b} V \tag{20.3}$$

For example, if a is half of b and $V = 15$ V, then the average voltage is 7.5 V. In practice, a and b can be very small, say, a few milliseconds. Suppose that $b = 5$ msec, $a = 2$ msec, and $V = 15$ V. Then the frequency of pulses is 200 (1/0.005) Hz, and the voltage in the line is 6 V. If a is 0, then no pulses are sent and the (average) voltage is 0.

Figure 20.26c shows the relationship in Equation 20.3. Line MN represents a and can assume any position inside the triangle. When it coincides with PQ, $a = b$ and the available voltage is equal to the supply voltage;

when it is at point L, $a = 0$ and the output voltage is zero. Voltage can be represented by the distance of line PQ from point L.

In practice, the width of an electric pulse and its frequency can be controlled by electronic circuits. These pulses from a low-power function generator can trigger high-power devices to turn on and off and generate high-power pulses for the required applications. For instance, the low-power pulses can be applied to the gates of thyristors and control their outputs. Any thyristor to be used for this purpose must be of the type that can be turned on and off, as necessary.

20.6.2 PWM for AC Generation

Both single-phase and three-phase inverters can use pulse width modulation technique for their operation. There are many varieties of employing pulse width modulation for the operation of inverters. The basic principle is the same as was described for DC voltage control. In an inverter, as far as the output voltage is concerned, the thyristors conduct for short periods, thus generating a series of pulses, instead of conducting for 180° or 120° duration (remember the two schemes for inverter functioning). In this way, during any given period of time, thyristors conduct for much shorter intervals but more frequently. For example, in every 20 msec (corresponding to 50 Hz) instead of switching on and then off only once, they turn on and off 40 times or more.

For AC electricity, in addition to the voltage that must be controlled as required, the waveform and frequency need also to be looked after. The average output voltage variation must be as close to a sinusoidal waveform as possible, and the voltage and frequency must be as desired. Note that all the pulses have the same voltage intensity; only their width can vary.

Figure 20.27 shows two possible scenarios for pulses in a single-phase inverter intended for a sine wave output. The difference between them is in the plus and minus polarity change for the pulse train in Figure 20.27a, whereas in Figure 20.27b all pulses have the same polarity. In other words, in Figure 20.27a the neutral point of the resulting AC waveform is defined, whereas in Figure 20.27b it is floating. Note that in both cases the number of pulses per cycle is odd and they are symmetric with respect to the zero crossing point of the intended waveform.

The frequency of the pulses and width of each pulse need to be known for pulse width modulation. Undoubtedly, frequency of sending pulses must be considerably higher than the frequency of the desired output waveform to be generated. Switching action is required for each pulse, and there are energy losses for each switching. Therefore, the number of pulses must be compromised. For a frequency of 60 Hz for the waveform the pulse train can have a frequency of several times like 1000–2000 Hz.

Carrier: A waveform of relatively high frequency in pulse width modulation and other applications (especially in communications) that behaves as a carrier to transfer or treat signals of low-frequency such as voice (in telecommunication) and an electric alternating current waveform (50–60 Hz).

20.6.3 PWM-Controlled Inverter

From the above discussion it can be understood that for pulse width modulation it is essential to define the width of the pulses to be generated. To determine the pulse width at each instant of time for generating a sinusoidal waveform with a desired frequency and voltage a so-called **carrier**, which is normally a triangular wave, is employed. This carrier

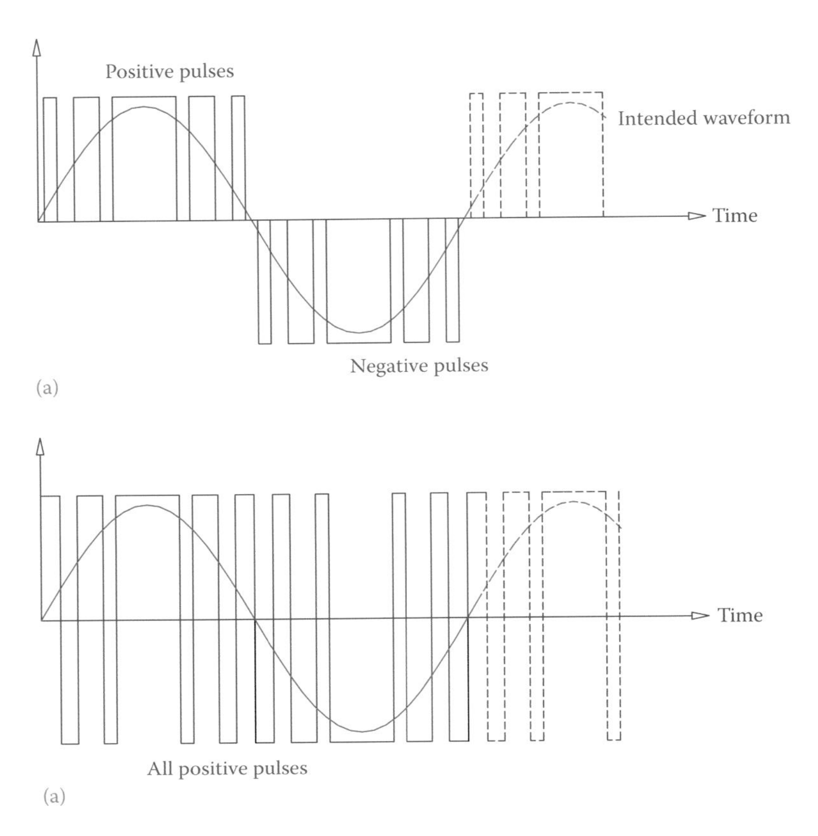

Figure 20.27 Pulse trains for generation of AC waveforms. (a) With positive and negative power supply, (b) with only one power supply.

is compared to the desired waveform with the required frequency. Then based on this comparison the duration of pulses are determined.

In practice, all this process can be done offline by a computer. The results are, then, transferred to the actual device. Figure 20.28 can help to understand the modulation process. For better clarity the triangular waveform shown has a much smaller frequency than what happens in practice.

In Figure 20.28a a triangular carrier wave, to be modulated by a reference sine wave, and the modulating wave are plotted together. This reference wave is the desired sinusoidal AC to be generated by the inverter. The points of intersection of the two waves are numbered 1 to 10. As can be observed, for the interval between 1 and 2 the carrier wave is larger (greater magnitude) than the modulating wave, but for the interval 2 to 3 the modulating function has a larger magnitude. The latter is true for the intervals between 4&5, 6&7, and 8&9. Pulses to be generates are all defined by the points of intersection of the two waves, as follows. For those intervals that the carrier is larger than the modulating wave, there is no pulse, and for the intervals that the modulating wave is larger than the carrier, there is a pulse. For each pulse, width is defined by the horizontal spacing (time interval) between the two associated intersection points. On the basis of the scenario shown in Figure 20.28a the corresponding pulses are shown in Figure 20.28b.

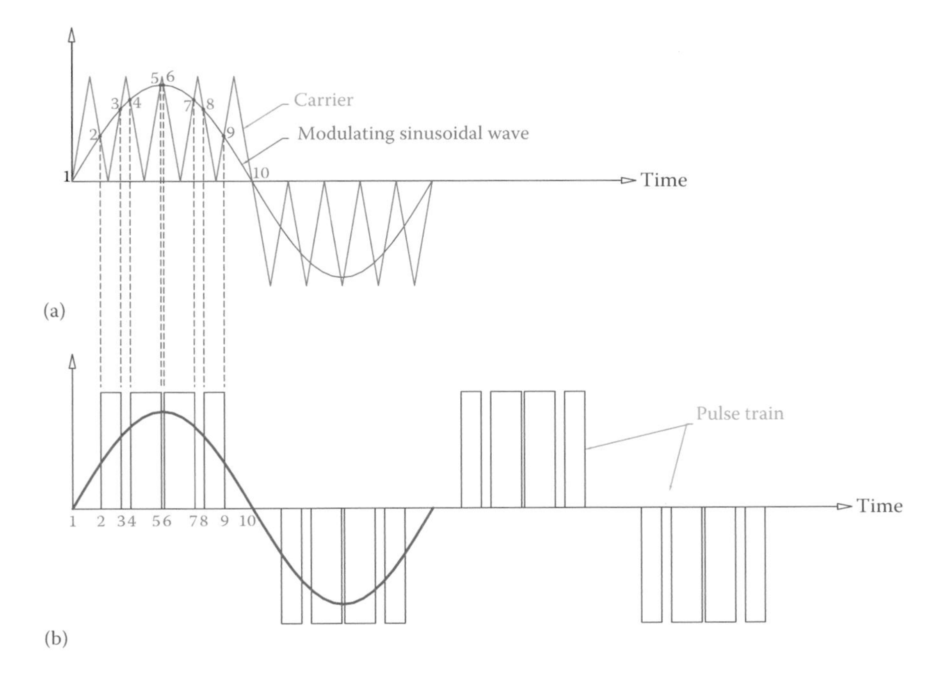

Figure 20.28 Modulating a triangular wave to determine the pulse width for sinusoidal wave generation. (a) Determining the width for pulses. (b) The required pulses.

> In PWM, there is no pulse for those intervals when the carrier is larger than the modulating wave. For the intervals when the modulating wave is larger than the carrier, there is a pulse.

The pulses will be used to trigger the thyristors in an inverter. For a single-phase inverter it is easy to see which thyristors must be turned on or off. At each instant, one thyristor from one leg and its counterpart from the other leg must be turned on at the same time (see Figure 20.1). If the scheme in Figure 20.27b is used (no polarity change for pulses), it is possible to reduce the number of switching action. For this scheme it is possible to keep one of the thyristors on for the whole half cycle and only send pulses to its counterpart. This is shown in Figure 20.29. This method is not practicable for the scheme with polarity change pulses (Figure 20.27a).

Note that to generate the pulse train in Figure 20.29 (without polarity change) the same principle is used and the pulse widths are determined based on comparison of the modulating wave and a carrier wave without polarity. The carrier wave has a larger amplitude than the reference waveform (compare Figures 20.28a and 20.29); otherwise, not sufficient intersections can be found. For thyristor numbering, see Figure 20.1.

In practice, the two waves (the carrier and the modulator) can be compared (by a comparator circuit) to determine which one has a larger magnitude. If digital systems (or a digital computer) are used, then it is possible to put the results of the comparison in memory and send the pulses using a memory look-up table. Among other alternative ways to determine the

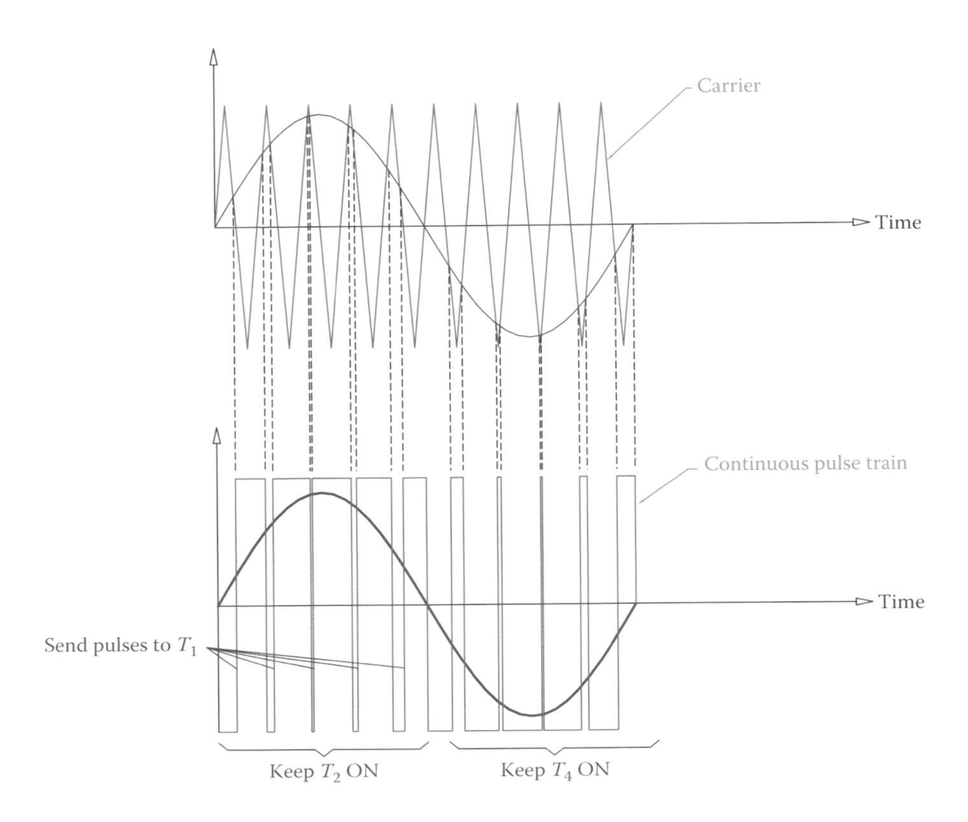

Figure 20.29 Continuous pulses without polarity and with less switching (refer to Figure 20.1).

pulse widths is employing mathematical functions that result in a waveform that better resembles a sine wave. Again, the results can be read from memory-stored data.

Note that in using pulse width modulation both the amplitude ratio and the frequency ratio of the carrier wave to the reference (desired) wave affect the number and width of the pulses. These two parameters are called **frequency modulation ratio** and **amplitude modulation ratio** and are denoted by m_f and m_a, as

Frequency modulation ratio: Ratio of the frequency of the carrier signal to that of the reference signal in pulse width modulation, used in power converters, a number larger than 1 and represented by m_f.

$$m_f = \frac{f_c}{f_r} = \frac{\text{carrier frequency}}{\text{reference frequency}} = \text{frequency modulation ratio} \tag{20.4}$$

Amplitude modulation ratio: Ratio of the amplitude of the reference signal to that of the carrier signal in pulse width modulation, used in power converters.

$$m_a = \frac{V_r}{V_c} = \frac{\text{reference voltage}}{\text{carrier voltage}} = \text{amplitude modulation ratio} \tag{20.5}$$

To see the effect of changing amplitude modulation ratio, see Figure 20.30 and compare the pulses in the two cases shown (focus on width of the pulses). For the effect of changing frequency modulation ratio, compare Figures 20.29 and 20.30. Count the numbers of pulses and the

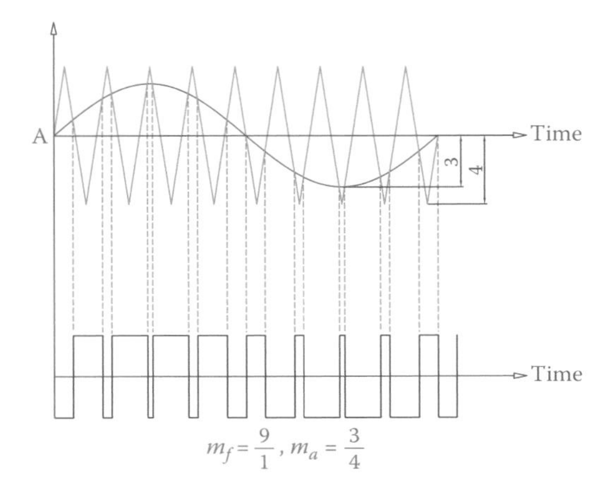

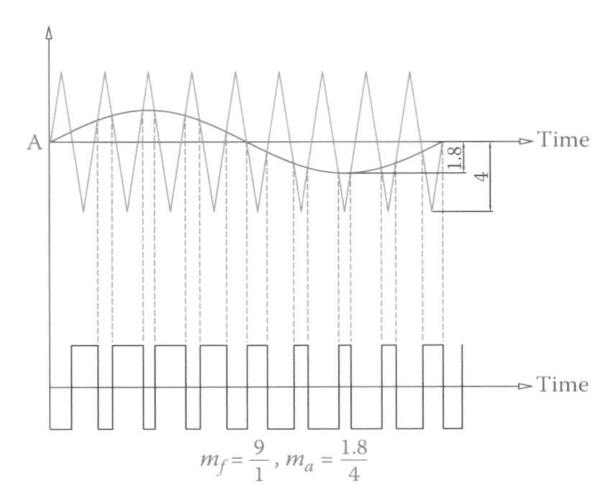

Figure 20.30 Effect of amplitude modulation ratio.

frequency of the carrier for one cycle of the sinusoidal waveform (10 in Figure 20.29 and 9 in Figure 20.30. Also, observe the symmetry). The value of m_f is always greater than 1; it is an odd multiplier of 3 for technical reasons, and its range of values can be between 21 and 35, for instance. The value of m_a is normally less than 1. If it has a magnitude larger than 1, it represents **overmodulation**, which causes distortion in the resulting signal generated from the pulses.

Overmodulation: Having a value larger than 1 for amplitude modulation ratio in pulse width modulation controlled inverters. This happens when the amplitude of the modulating signal is greater than that of the carrier wave.

20.6.4 Pulse Width Modulation for Three-Phase Inverter

The PWM scheme depicted in Figures 20.27 and 20.28 for a single-phase inverter can be extended to three-phase systems. Because there is a 120° phase difference between the three phases, the same pattern for one phase repeats, only with a shift in time. In other words, the same scheme for pulse widths and timing can be used for each individual phase; only they must have a timing shift corresponding to 120° with respect to each other. Figure 20.31 shows the case.

The three pulse trains shown in Figure 20.31 can be used to trigger the gates of the six thyristors in a three-phase inverter. For three-phase system it is not possible to keep any of the thyristors turned on (like the case for a single-phase inverter); thus, a lot of switching is necessary. At each instant, one thyristor of each leg is on and the other is off. At each edge of every pulse (i.e., for both rising edge and falling edge) one thyristor in a leg of the inverter is turned off and the other is turned on, as shown for phase C in Figure 20.32. The thyristor numbers correspond to the arrangement in Figure 20.15.

For a three-phase converter it is not possible to keep any of the thyristors on for a period longer than the associated pulse width.

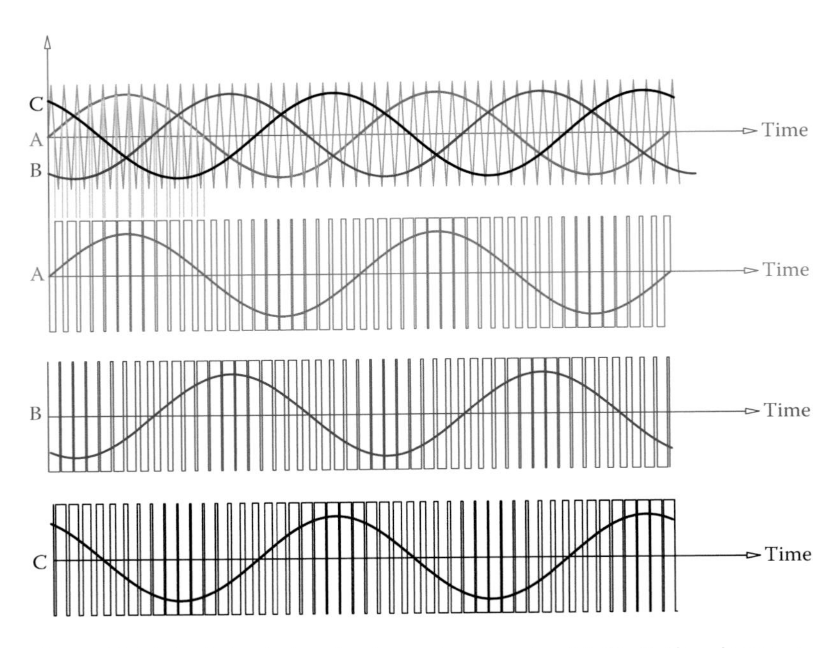

Figure 20.31 A, Pulse trains for three phases using PWM; B, time intersection of pulses for phase A and their counterpart pulses for B and C.

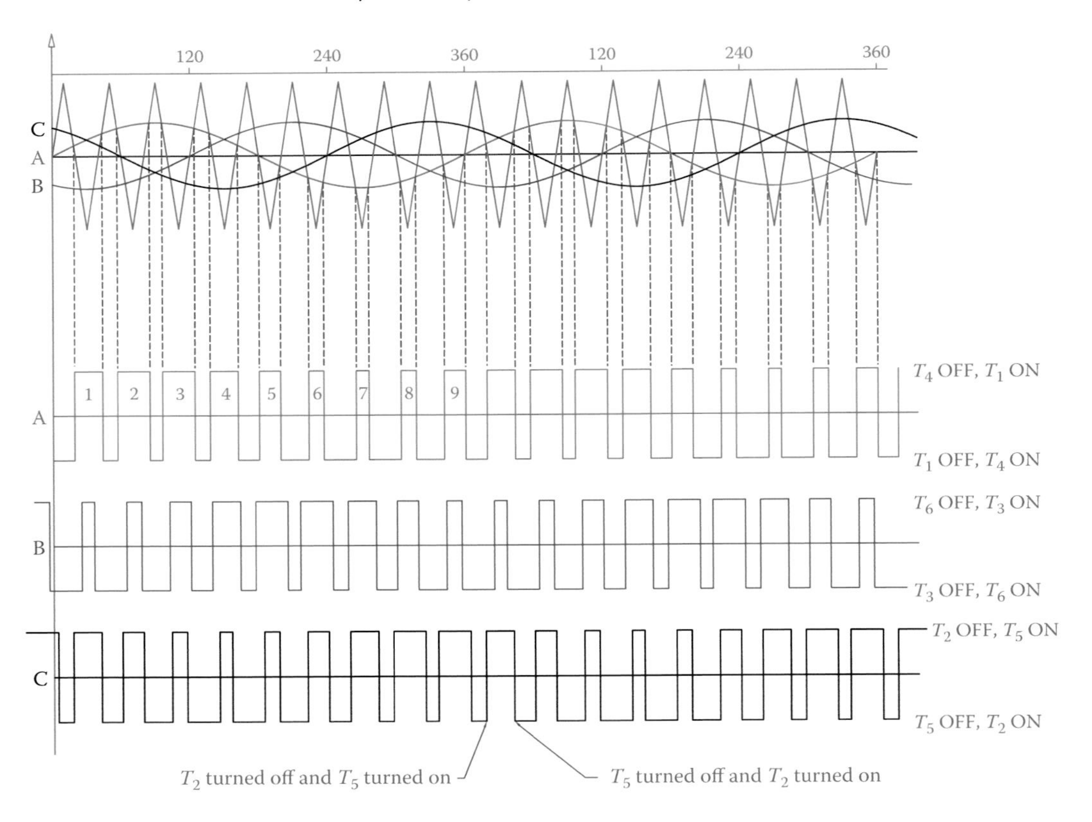

Figure 20.32 Triggering thyristors for three-phase system inverters using PWM.

For a pulse width modulation inverter to work properly the frequency of the carrier wave must be sufficiently larger than the modulating wave frequency; that is, the modulation frequency ratio needs to be greater than 20. Also, overmodulation is not desirable; thus, the amplitude modulation ratio must be smaller than 1.

20.7 Application in Motor Drives

From the knowledge gained in Chapter 11 on AC electric motors and in this chapter about power converters you observe that

1. The speed of an AC motor depends on the line frequency.
2. A converter can change DC to AC with a desired frequency.
3. A converter can also change AC to DC.

Putting together all these facts and the available technology has led to the generation of a family of devices that can be used today for driving motors. Noting that at the industry level the main supply of electricity is in the form of three-phase AC, we can see that

1. DC electricity can be obtained from AC.
2. Single-phase or three-phase electricity with a desired frequency can be obtained from DC.

Also,

3. The frequency of a converter output can be controlled.
4. AC-to-DC and DC-to-AC converters can be put back to back. In such a case, a frequency converter is devised that can provide a variable frequency – variable voltage AC from a fixed frequency AC supply.

As a result, by employing appropriate converters nowadays, (1) DC motors can be run from AC sources and (2) an AC motor speed can be varied as desired. This includes the starting of motors in such a way that they start smoothly without putting too much stress on the machine or on a circuit.

For alternating current machines these devices are called variable speed drives (VSD), or variable frequency drive (VFD), mentioned in Chapter 9. They appear in various forms and at various powers with a variety of options, such as single-phase output from three-phase input, or vice versa in addition to DC output for DC machines.

20.8 Chapter Summary

- An inverter is used to generate AC from DC electricity. An inverter is the opposite of a rectifier.
- An inverter works based on repeatedly altering the direction of current in its load.
- Inverters use switchable diodes or transistors.
- Both single-phase and three-phase inverters are possible.
- Simpler inverters generate square wave. Inverters with an output of stepping waveform have better resemblance to sine wave.

- Single-phase inverters need at least four switching elements; three-phase inverters need six.
- Firing angle is the delay in turning a thyristor or a transistor on, expressed in terms of degrees (referring to a cycle of an AC waveform; that is, 360°). Firing angle is with respect to zero crossing (0° of the AC waveform).
- If rectifiers are made with switchable diodes, their output voltage can be controlled.
- In a rectifier, firing angle can be altered between 0 and 90°.
- In inverters, each switching device has a freewheel diode in parallel with it, but with the opposite polarity (antiparallel). The freewheel diode is for safety and protection of a device.
- In switching devices, there is always some power loss associated with switching.
- Maximum output voltage of an inverter depends on the DC power supply.
- A converter can refer to a rectifier or an inverter because both have the same structure.
- In a three-phase converter if the firing angle is smaller than 90°, the circuit functions as a rectifier, and if the firing angle is greater than 90°, it functions as an inverter.
- Pulse width modulation (PWM) implies defining a pulse train with specified widths of pulses (in terms of time) and frequency to be used for a desired application, such as voltage regulation.
- PWM can be used in single-phase and three-phase inverters and in DC electricity.
- For PWM a carrier wave with much higher frequency is modulated by a reference wave to determine the rate and width of the pulses in a pulse train.
- Both amplitude and frequency of the carrier wave are involved in defining the pulse train.
- The higher the frequency of the pulse train, the closer is the form of an inverter output to the reference (desired) wave but also the more are the switching actions and the energy loss due to switching.
- In inverters the *amplitude modulation ratio* is the ratio of reference wave voltage to that of the carrier wave. Also, the *frequency modulation ratio* is the ratio of carrier wave frequency to that of the reference wave.
- If the amplitude modulation ratio is larger than 1, overmodulation occurs, which is undesirable.

Review Questions

1. What is an inverter for?

2. What is the functioning principle of an inverter?

3. What are the main components of an inverter?

4. Can an inverter work with ordinary diodes? Why or why not?

5. Is the output of an inverter a sinusoidal AC wave? Why or why not?

6. If for single-phase inverters four thyristors are required, how many does one need for a three-phase inverter?

7. How can the quality of the output waveform of an inverter be improved to better resemble a sine wave?

8. Can the output voltage of an inverter be controlled?

9. How can the output voltage of an inverter be changed?

10. What is the firing angle in a rectifier?

11. What is the firing angle in a diode rectifier?

12. If a single-phase rectifier is used for a 120 V, 50 Hz line, what is the DC voltage?

13. If the same rectifier as in the previous question is used on a 60 Hz line (same voltage), what is the DC voltage?

14. What is a freewheel diode for, and where is it used?

15. With a 12 V battery, what is the peak-to peak of the AC electricity obtained from a single-phase inverter?

16. What is the difference between180°- and120°- conducting three-phase inverters?

17. How many thyristors are necessary for a three-phase rectifier?

18. What does PWM stand for?

19. Name two applications of PWM.

20. Can one use a three-phase rectifier as a three-phase inverter? Why or why not?

21. How can frequency of the AC from an inverter be changed?

22. How can output voltage of an inverter be increased?

23. Explain what a power converter is.

24. What is frequency modulation ratio?

Problems

1. A three-phase diode-based rectifier is used on a line with 120 V line-to-neutral voltage. What is the DC voltage, ignoring the losses?

2. If the firing angle in a three-phase rectifier used on 380 V line is 25°, what is the DC voltage?

3. What is the effective voltage of the output of a converter with sufficiently accurate sine waveform if the DC input is 11,050 V DC?

21

Electronic Power Converters in Wind Turbines and Solar Photovoltaic Systems

OBJECTIVES: After studying this chapter, you will be able to

- Name the principal components of a wind turbine
- Understand that wind turbines can work in different ways based on the type of their generator
- Describe how many ways solar energy can be utilized
- Describe how solar energy is converted to electric energy
- Understand that, like batteries, the electricity from solar panels is DC in nature
- Explain what type of electric generators are used in wind turbines
- Explain why most wind turbine generators are not similar to those in other power plants
- Describe how frequency can be changed in AC electricity
- Define DFIG
- Describe how a variable speed wind turbine works
- Understand the use of power converters in wind turbines and photovoltaic systems
- Understand and use the terms direct drive, subsynchronous, and supersynchronous

New terms:
Current source inverter, cut-in speed, cut-out speed, DC link, direct-drive turbine, doubly fed induction generator (DFIG), fill factor, hub, pitch, pitch control, power curve, solar array, solar cell, solar irradiance, solar panel, subsynchronous, supersynchronous, variable speed turbine, voltage source inverter, wound rotor induction machine (WRIM), yaw motion

21.1 Introduction

We discussed DC motors and generators in Chapter 7 and AC motors and generators in Chapter 11. They were addressed as electric machines. It is advisable that you refresh your knowledge about certain parts of those chapters, particularly the induction machines in Chapter 11; these machines are used in the majority of wind turbines. All thermal and hydro power plants use three-phase synchronous machines for AC generation. Wind turbines, however, use induction generators because wind is not under our control and wind speed is not constant, and, therefore, the turbine speed cannot be kept constant in the same way as in any other power plant. An induction generator (the same as induction motor, as we discussed in Chapter 11) has a specific feature that it can work as a generator within a range of speeds above the synchronous speed. Thus, it better suits the way a wind turbine works and is not tied to run at a fixed speed. For this reason, until recently, all wind turbines used induction generators.

The simplest type of induction machine with the property described above is the squirrel cage generator (see Chapter 11). To harness more energy from wind, with the help of electronics, today's wind turbines can use a synchronous machine or a **wound rotor induction machine (WRIM)**. More recently, an increasing number of turbines are made this way because of the benefits and despite the additional cost. This is only possible by using electronic power converters that we studied in Chapter 20. In this chapter, we discuss the mechanism that is used.

Wound rotor induction machine (WRIM): Type of induction machine for AC in which the rotor has wire winding. The windings are accessible through slip rings. Another type is the (squirrel) cage machine that has no wire winding and has no slip rings.

A solar energy system using photovoltaic (PV) technology does not have any generator at all. Energy from the sun is directly turned into electricity, but this is DC electricity. Except for small isolated power generation, to be useful at the industrial and commercial level, DC electricity must be converted to three-phase AC. Again, a DC to AC converter (see Section 20.2) is necessary to carry out this task. In this sense, solar systems employing PV arrays are much simpler than wind turbines because they do not have as many moving components (if any, for tracking the sun) and their converter is also simpler.

Most of this chapter is, necessarily, devoted to the systems employed in wind turbines. We provide only a short overview of wind turbines and photovoltaic systems in this chapter.

21.2 Wind Turbine Basics

A wind turbine consists of five major and many auxiliary parts. The major parts are the tower, rotor, nacelle, generator, and foundation or base. Without all of these a wind turbine cannot function.

The foundation is under the ground for the onshore turbines; it cannot be seen because it is covered by soil. It is a large and heavy structured block of concrete that must hold the whole turbine and the forces that affect it. For offshore turbines the base is under the water and cannot be seen. In offshore turbines that are well into the sea the base is floating, but it is of sufficient mass to support the turbine weight and all the forces exerted on it, and to hold it upright.

The tower in most modern turbines is a round tubular steel of a diameter of 3–4 m (10–13 ft), with a height of 75–110 m (250–370 ft), depending on the size of the turbine and its location. The rule of thumb for a turbine tower is that it has the same height as the diameter of the circle its blades make when rotating. Normally, the taller a turbine is, it is subject to more of the wind with higher speed. This is because the farther we are from the ground, the faster the wind (wind does not have the same speed at various distances from the ground). To increase the height of the tower, turbines with a hybrid tower (the lower section made out of concrete) can become more common in the future.

Rotor is the rotating part of a turbine; it consists of (mostly) three blades and the central part that the blades are attached to, the **hub**. A turbine does not necessarily have to have three blades; it can have two, four, or another number of blades. But the three-blade rotor has the best efficiency and other advantages. Blades are not solid; they are hollow and are made of composite material to be light and strong. The trend is to make them larger (for more power), lighter, and stronger. The blades have the form of an airfoil (same as the wings of an airplane) to be aerodynamic. As well, they are not flat and have a twist between their root and their tip. The blades can rotate up to 90° about their axes. This motion is called blade **pitch**. The function of the hub is to hold the blades and make it possible for them to rotate with respect to the rest of the turbine body.

Hub: In a wind turbine, the hub is the piece to which the blades are attached (thus holding the blades) and all rotate with respect to the rest of the turbine.

Pitch: (wind turbine) Angle of rotation with respect to a reference point of a wind turbine blade about its longitudinal axis. Changing this angle alters the performance of a blade and, thus, the power capacity of a wind turbine.

The nacelle is a housing on top of the tower that accommodates all the components that need to be on a turbine top. There are quite a number of components for the proper and healthy operation of a complicated electromechanical system that a turbine is. A major turbine part among these components is the generator and the turbine shaft that transfers the harvested power from wind to the generator through a gearbox. The gearbox is a vital component of wind turbines; it resides in the nacelle. A gearbox increases the main shaft speed from around 12–25 rpm* (for most of today's turbines) to a speed suitable for its generator. For this reason the shaft on the generator side is called "high speed shaft." Because a turbine must follow the wind and adjust its orientation to the wind direction, its rotor needs to rotate with respect to the tower. This rotation is called **yaw motion** in which the nacelle and the rotor revolve about the tower axis.

Yaw motion: Motion in a wind turbine that the whole rotating part (rotor) and the nacelle orient toward wind to catch the most power from wind. In new turbines, there are sensors and motors for this motion.

The generator is the component that converts the mechanical energy of the rotor, harnessed from wind, to electrical energy. A generator has the same structure of an electric motor. At the commercial production level all electricity generation is in three-phase alternative current. In general, the choice of generator, therefore, is synchronous or asynchronous (induction) generator. Nevertheless, the generator associated with wind turbines, thus far, is the induction generator because a synchronous generator must turn at a tightly controlled constant speed (to maintain a constant frequency). Some of a wind turbine's principal components are depicted in Figure 21.1.

Because a generator must be rotated at a speed corresponding to the frequency of the electric network (50 or 60 Hz in most countries), it must be rotated faster than the turbine rotor. Most generators need to be turned at 1500 rpm (for 50 Hz) and 1800 rpm (for 60 Hz). In no way it is feasible for a turbine rotor to move that fast. A gearbox, therefore, must *increase* the turbine rotor (*main shaft*) rotational speed to a speed that can be used by the generator.

Experience has shown that the gearbox in a turbine is a problematic component. This is due to the fact that the energy in the wind does not remain constant for a relatively acceptable length of time. It continuously fluctuates, because of the nature of wind. This causes the gear teeth to undergo overload and hammering stress that lead to fatigue and failure. In addition, the gearbox is a heavy item in the nacelle on the top of a turbine.

* In some newer turbines, called direct drive, the gearbox is eliminated. Instead, the mismatch for generator speed is performed by a different mechanism through electronics.

Figure 21.1 Principal components of a wind turbine: tower, rotor, nacelle, and foundation (underground).

It is envisaged that in future turbines the gearbox will be eliminated (this does not mean that turbines turn faster). The use of power converters allows this. In the future generation of wind turbines, those with synchronous generators and with permanent magnet generators can be envisaged.

A wind turbine is a complex system to control because the source of power (wind) is not in our control. Wind speed can continuously change, even from one second to the next. The power output from a turbine, therefore, must be adjusted to variation of wind at all times. All new turbines are equipped with **pitch control**, which implies that their blades' pitch angle can be adjusted so that the power output from a turbine is maximized at all times, while it does not overload the generator and mechanical structures of the blades, tower, and rotor shaft.

Pitch control: Changing the pitch angle of blades in a wind turbine in order to alter the power catch capacity or adapt to the wind speed.

The way this adjustment of power to wind speed is carried out is that each turbine has a performance curve called **power curve**. It is used as a power schedule by a computer in the turbine control system to adjust all components that need to be controlled, based on this curve and according to wind speed. A typical sample of this curve is shown in Figure 21.2 (to obtain wind speed values in miles per hour multiply the values in m/sec by 2.2).

Power curve: Curve depicting the maximum power that can be generated from a wind turbine for various wind speeds. This takes into account the mechanical and electrical capacities of the turbine components.

Figure 21.3 depicts part of the components inside the control panel in a wind turbine. This control panel is normally at the bottom and inside the tower.

21.3 Solar Energy Systems

Solar rays reach the Earth at various frequencies and contain light and heat. Solar energy, thus, may basically be utilized in two different ways: absorbing

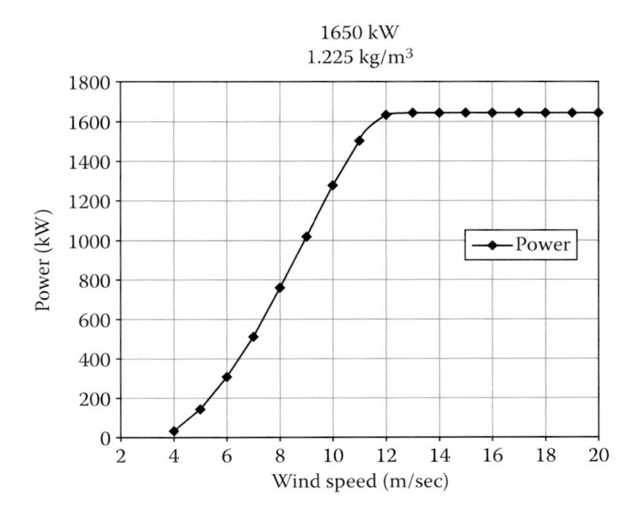

Figure 21.2 Typical power curve for a wind turbine.

Figure 21.3 Part of the control circuitry for a wind turbine.

light and absorbing heat. Heat can be directly used for warming and heating, in addition to generating electricity, whereas absorbing light can only be useful when the contained energy is converted to electricity. This allows both transmission and storing the energy contained in the light from sun. (Storage of electric energy at very small scale is what we do with rechargeable batteries. At large scale, however, work is currently being done on ways/devices to store electrical energy at tens or hundreds of megawatt-hours.)

Both light and heat are parts of the general electromagnetic wave spectrum. Figure 21.4 shows this spectrum, which depicts the standing points of the visible light and the infrared (heat) waves.

At industrial and commercial levels, to be efficient and cost-effective, the electrical power to be transmitted from one point to another must be in megawatts. Because energy from the sun is at low grade (meaning only a small amount of energy per square foot or square meter per second), a large area of coverage is needed. Also, it is obvious that solar energy is

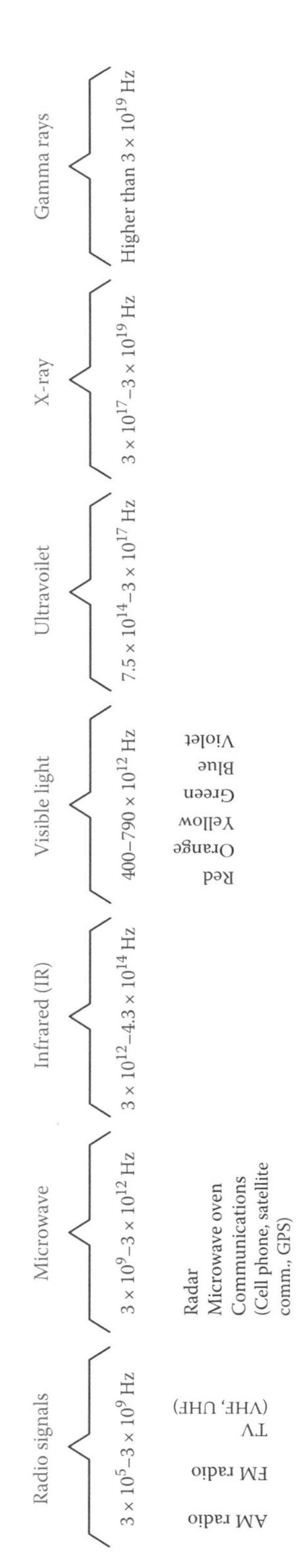

Figure 21.4 General electromagnetic wave spectrum.

available only during the daytime and, moreover, this energy is subject to the weather condition as well as the time of the day.

The average power from the sun outside the Earth's atmosphere is about 1365 W/m^2 (127.0 W/ft^2). This is called **solar irradiance**, which is, indeed, the rate of available solar energy before reaching the Earth's surface. To determine the available energy in any length of time, which can also be converted to calories or any other unit, the solar irradiance must be multiplied by the time duration. For instance, the available energy in 1 hour (3600 s) is 4914 kJ/m^2 per hour or 1174.2 kcal/m^2 per hour (457 kJ/ft^2 or approximately 433 BTU/ft^2 per hour).

Solar irradiance: Rate of available solar energy on Earth per square meter (or per square foot). It is, thus, measured in W/m^2, or W/ft^2.

This energy, however, is decreased when arriving at the Earth's surface owing to various reasons, mostly clouds and particles in the air and partly because of reflection and absorption (air, water vapor, and air pollutants in the air absorb heat and become warmer. They also reflect part of the sun rays back to the atmosphere.) On a very clear day up to 30 percent of the solar energy can be lost this way. This loss can reach 90 percent for a cloudy day.

Despite the fact that a great percentage of the energy from the sun can be reduced by the atmosphere, total energy reaching the Earth's surface is tremendous because of the available area. A study by the US Department of Energy has shown that the total US electrical energy consumption in 2001 (3.24 TW = 3.24×10^{12} W) could be provided by sun if the energy from the sun is collected in an area approximately the size of North Dakota or South Dakota.

> Out of the tremendous amount of energy from the sun reaching the Earth, a good percentage is lost in the atmosphere owing to clouds and air pollution.

21.3.1 Photovoltaic Systems

Photovoltaic systems are based on the reaction of certain semiconductor materials to light. A **solar cell** is the smallest unit in a photovoltaic panel that, on the basis of its material, is sensitive to one or more of the rays in the sunlight spectrum. A small voltage is generated by a solar cell depending on the sensitivity of the material and the intensity of the light. A solar cell is, in fact, a diode with a function opposite to that of an LED; thus, it generates a voltage when subjected to light. Figure 21.5 shows the symbol for a photovoltaic cell.

Solar cell: Semiconductor diode that converts light to voltage (the opposite of what an LED does).

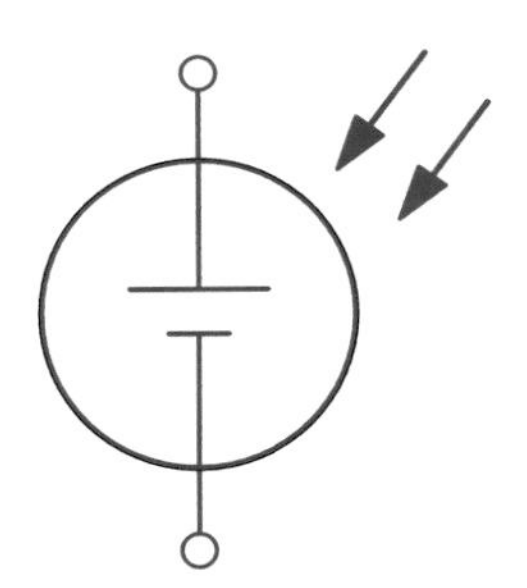

Figure 21.5 Symbol for a photovoltaic cell.

Solar panel: Set of solar cells put together in a frame and connected together in series, parallel combination to deliver a desired voltage at a desired current.

Solar array: Set of solar panels put together in the form of a matrix (rows and columns) and connected together in series, parallel, or a combination to deliver a desired voltage at a desired current.

A solar cell can be regarded as a small battery. The voltage of this battery is small, between 0.5 and 1 V, depending on the material used for the cell construction. In this respect, to obtain a higher voltage, a number of cells must be joined together in series, like inside a car battery. Moreover, current that can be delivered by an individual solar cell is small, on the order of milliamps. For this reason, the cells must be put in parallel with each other so as to deliver more current. Consequently, to use solar energy at a desired voltage and capable of delivering a desired power, a number of solar cells are put together in rows and columns in a **solar panel**, which is a unit that can be practically and physically meaningful to handle. Then, many solar panels can be put together, again in rows and columns, in a **solar array** to provide higher voltage and higher current.

Current that a typical solar cell can deliver is not constant and depends on the voltage at which this current is delivered. In other words, a solar cell works based on its characteristic curve that delineates the relationship between its current and voltage. Figure 21.6 depicts the form of a typical curve for all solar cells. The voltage is shown on the horizontal axis and the current on the vertical axis.

As can be seen, current drops abruptly when the cell voltage approaches its maximum value for a given type of cell. In the graph, point A corresponds to zero voltage and highest current (short circuit of the cell) and point B corresponds to zero current and highest voltage (open circuit). Because power is the product of current and voltage, at both these points the power output of a cell is zero. This implies that a solar cell (and correspondingly solar panels) cannot be used at these extreme points. For any point on the curve the deliverable power is represented by the area of a rectangle formed from that point being a corner. This is shown for two arbitrary points H and K in Figure 21.6. For a point M this deliverable power by a cell is the highest. Thus, to use a solar panel at its maximum power (in other words for gaining the maximum power from sun light), the operating point of a solar panel must be around point M.

Obviously, the I-V curve (characteristic curve), and consequently the point of maximum power, of a solar cell cannot be invariable, because this depends on the intensity of light. If such a curve is drawn for various light conditions, a series of curves are obtained that look like those shown in Figure 21.7. Under different light conditions the variation of the maximum

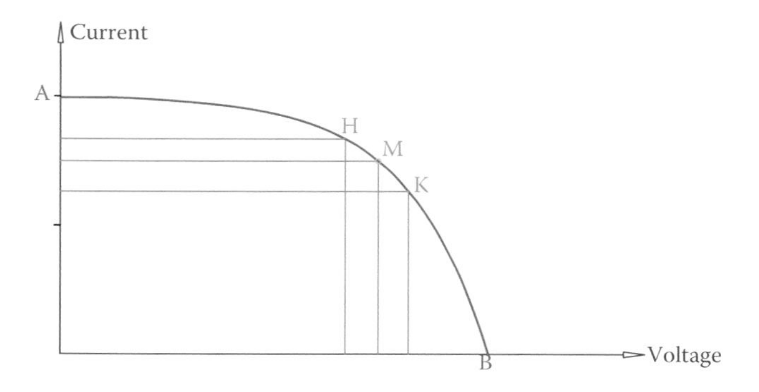

Figure 21.6 Typical I-V curve for the performance of a solar cell.

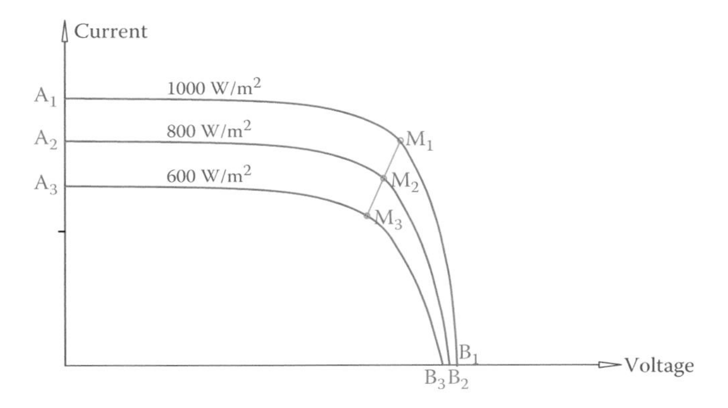

Figure 21.7 Set of I-V characteristic curves for a solar cell under different light conditions.

(open circuit) voltage V_{OC} (point B) is less affected, but the deliverable maximum current I_{SC} (point A) has significant displacement. The position of the point of maximum power (point M) also moves.

To compare the performance of two solar cells under similar conditions, the term **fill factor** is employed. Fill factor is a percentage value, obtained by the ratio of the maximum power possible from a solar cell to the product of the open circuit voltage and short circuit current. This product is shown by the area of the rectangle in Figure 21.8 for two cells that have the same values for V_{OC} and I_{SC}. Observation of this figure depicts that the solar cell with the characteristic curve AM_2B produces more power than that with the curve AM_1B.

Fill factor: Ratio of the maximum power possible from a solar cell to the product of its maximum voltage and maximum current (i.e., the product of the open circuit voltage and the short circuit current).

Also, to have a common criterion for comparing various PV cells, it is customary to plot these curves for a standard condition defined by an irradiation of 1000 W/m^2 (93 W/ft^2) and at 25°C (77°F).

The reason to involve the temperature is the fact that the performance of PV cells is not independent of the operating temperature. The performance of a PV cell deteriorates (and as a result, the characteristic curve alters) as the temperature increases. This change mostly affects the maximum voltage (and the maximum power, as a result). Figure 21.9 illustrates the typical change for the I-V curve. The short circuit current does not remain the same, but its change is negligible.

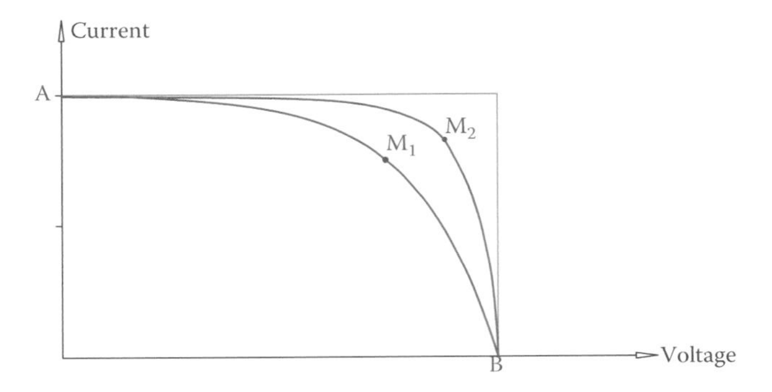

Figure 21.8 Representation of the fill factor concept.

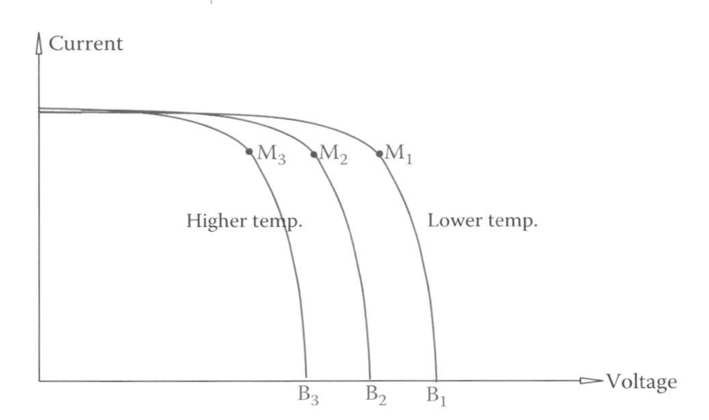

Figure 21.9 Effect of temperature on I-V characteristics of a solar cell.

Another measure for comparing different units of solar panels is their efficiency. Efficiency is always the ratio of output power to the input power, and it is expressed in percentage. For a PV unit, efficiency can be defined as the ratio of the maximum power to the product of cell area and irradiance. Table 21.1 shows the voltages and other information for some solar cells made from different materials.

Fill factor can be used for comparison of two different types of solar cells. Efficiency can be used to evaluate the performance of a solar array or a solar farm.

For an installed PV system consisting of a large number of solar arrays it often becomes necessary to evaluate the efficiency of the system for financial matters. The same formulation for the efficiency of a single unit can be extended for this purpose. The real efficiency for this system, however, can be lower than the theoretical efficiency because the panels may not always work at their maximum power condition. Furthermore, some units can be faulty or damaged. The solar cell area is always known for the number of units in use. The average irradiance for a region can always be found based on statistics of weather conditions. The average power output may be found based on the data from production.

TABLE 21.1 Specifications of Some Solar Cells

Material	V_{OC} (V)	I_{SC} (mA/cm^2)	FF (%)	Eff. (%)
Monocrystalline silicon	0.706	42.2	82.8	24.7
Polycrystalline silicon	0.654	38.1	79.5	19.8
Amorphous silicon (a-Si)	0.887	19.4	74.1	12.7
Gallium arsenide (GaAs)	1.022	28.2	87.1	25.1
Copper indium gallium selenide (CIGS)	0.669	35.7	77.0	18.4
Cadmium telluride (CdTe)	0.848	25.9	74.5	16.4
Dye-sensitized (Graetzel)	0.721	20.53	70.41	10.4

21.3.2 Solar Heating Systems

The heat from the sun can be directly used for heating space or water in a building, at a small scale, or it can be collected and concentrated to run industrial size steam turbines for generation of electricity, at a large scale.

Heating water in a residential building is the simplest case of using solar energy. In such a system, water moves through a number of pipes that are placed on the roof of a building and absorb the heat directly from the sunshine. For maximum performance, these pipes, called collectors, are black colored. The heated water can be moved to a tank by natural convection or by assistance from a pump. For natural convection the reservoir tank must be physically higher than the position of the collector pipes because warm water moves up. In this way, water continuously circulates between the tank and the collector. At each circulation it gets warmer. The warmed water can be used and replaced by fresh cold water, as shown in Figure 21.10.

If the reservoir is not at a higher level than the collectors, a pump is needed to circulate the water to warm up. For this application the size of collectors can be determined for each individual case based on available sunshine, desired temperature, and water consumption.

The above arrangement can also be used for space heating. Instead of consuming the heated water, it can be circulated in radiators inside a building. A reservoir tank always helps to store the heat and smooth the heat flow, just as a capacitor does for electricity, but in particular cases it can be omitted.

Another way that solar energy can be used for space heating at a small scale is shown in Figure 21.11. This approach has been practiced in special cases in manufacturing and process plants in which the air inside the building must be continuously replaced by fresh air owing to fumes or other reasons. For this, blowers must suck the fresh air from outside, heat it, and send it to the inside of a building. In this arrangement, a black glass wall just in front of the main wall on the sunny side of the building absorbs heat from sun and heats the air between the glass wall and the structural wall of the building. Holes on the top of the wall let the heated

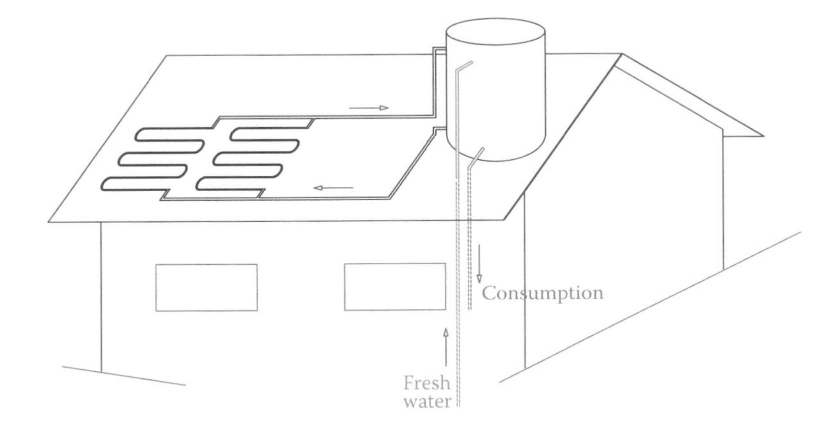

Figure 21.10 Example of solar heating in a residential building.

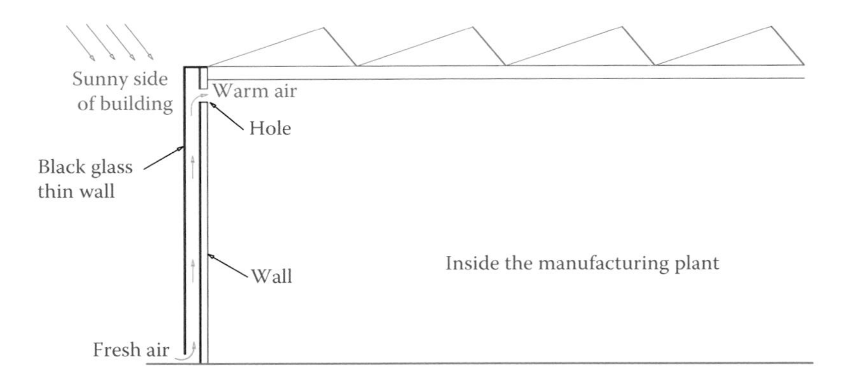

Figure 21.11 Way of utilizing solar heating in an industrial building.

air enter the building. Blowers can help as necessary and the temperature can be adjusted by auxiliary heating when required. If a good candidate for this arrangement, an industrial building can save hundreds of thousands of dollars in a year, for instance, on heating costs.

Solar heating at large scale (megawatts of power) is more involved. This is like a power plant with a steam turbine except that instead of gas or other fuel heating of water and generation of steam must be done through the energy from the sun. In an ordinary steam turbine power plant, water temperature is raised in a boiler for conversion to steam. It is technically necessary, as well as more efficient, that only superheated steam enters a turbine. Thus, steam is further heated to increase its temperature and pressure. This requires temperatures somewhat higher than the boiling temperature for water (unless some other substance is used, which introduces its own technical concerns and adds some other complexities).

To duplicate the same process of producing superheated steam, the solar energy is concentrated by a great number of collectors, consisting of concave mirrors that must reflect their absorbed heat to a common point. For better performance, moreover, these mirrors must track the position of the sun during the daytime.

21.4 AC-to-AC Power Converters

In Chapter 20 we discussed how to convert three-phase AC to DC by a rectifier and how to convert DC to three-phase AC by an inverter. In practice, more types of converters are needed, such as DC-to-DC converters and AC-to-AC converters. These are made by combining the two basic aforementioned converters. For DC-to-DC conversion, voltage is the only parameter that is required to be changed. For AC-to-AC conversion the voltage and/or the frequency are the parameters that might need to be increased or decreased. For voltage only, a transformer can be used. But, for frequency change, and for both frequency and voltage change, use of electronic converters becomes necessary. Here the discussion is confined to three-phase AC-to-AC converters only.

For photovoltaic systems it is necessary to convert the DC output of a system to three-phase AC. Thus, an inverter, as previously studied, suffices for the operation, and the frequency of the alternating current electricity can be controlled as desired. For wind turbines, first an AC to DC is carried out; then, a DC to AC conversion completes the task. However, the operation is more involved, as described in the following sections.

21.4.1 Unidirectional Back-to-Back Converters

The key issue in AC-to-AC converters is to convert AC to DC first and then DC to AC by putting together a rectifier and an inverter. Then, there is the question of if the device is required to work in one direction only, all the time, or if it must work both ways. In the former case one side is always the input and the other side is always the output, and power always is transferred from the input to the output. In the latter case the input and output can switch functions, and power transfer can be either way. The first case is simpler and is discussed in this section.

Figure 21.12 illustrates a unidirectional AC-to-AC converter. AC is first converted to DC through a rectifier that can be based on diodes or thyristors. Then, the DC electricity is converted to AC with the required frequency and voltage. Voltage control can be done through the rectifier if it is thyristor based; otherwise, both voltage control and frequency control is done through the inverter. The rectifier and the inverter are coupled together by a capacitor. The capacitor serves as a voltage source. In other words, there is no need for a battery to store the rectifier output. The capacitor behaves as a short-term storing device. Alternatively, the capacitor can be regarded as a filter to smooth the output of the rectifier. The link between the two parts of the system, containing the capacitor, is referred to as the **DC link**.

DC link: Two-line electric link between the two parts of a back-to-back converter. Voltage in this link is only DC.

The input can be connected to a three-phase system with a frequency f_i and voltage V_i. This alternative electricity is rectified and keeps the capacitor at a voltage determined by V_i. The DC voltage in the DC link then is converted to AC with the required frequency f_o and a desired voltage V_o by using pulse width modulation or other technique described in Chapter 20. Whereas f_o can be greater or smaller than f_i, in general, V_o and V_i are bound through the DC link voltage according to the relationships we have seen before in Chapter 20. Thus, there is a limit for the maximum value that V_o can reach.

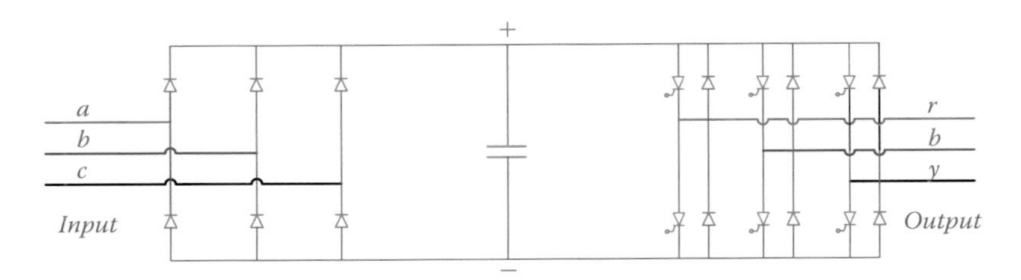

Figure 21.12 Unidirectional diode-based AC-to-AC converter.

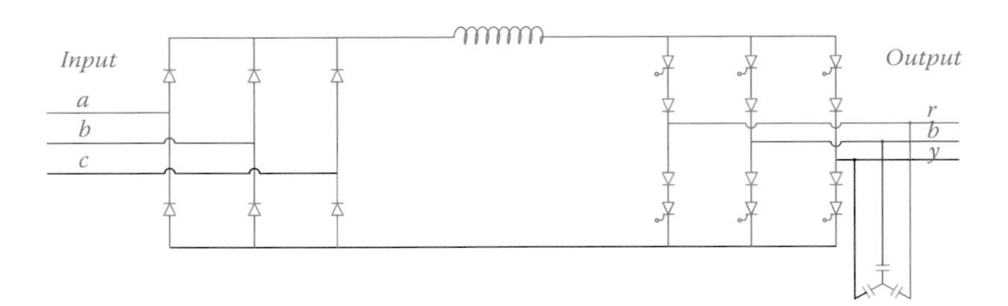

Figure 21.13 Diode-based unidirectional AC-to-AC converter with current source inverter.

Voltage source inverter: (in a back-to-back converter) A back-to-back converter arrangement in which a capacitor is used between the rectifier and inverter for instantaneous electricity storage. In a current source inverter an inductor is used for this purpose.

Current source inverter: (in a back-to-back converter) Back-to-back converter arrangement in which an inductor is used between the rectifier and inverter for instantaneous electricity storage. In a voltage source inverter a capacitor is used for this purpose.

It is possible to use an inductor instead of the capacitor. In such a case the inductor behaves as a short-term current source. The two terms **voltage source inverter** and **current source inverter**, used in industry, stem from whether a voltage source or a current source is used for an inverter (i.e., if a capacitor or an inductor is used in the DC link). Figure 21.13 illustrates the arrangement for a current source inverter. The three capacitors in the output provide a path for the currents even if there is no load. Also, the inclusion of series diodes in series with the thyristors is to protect them (like free wheel diodes) and prevent unwanted discharge or higher voltage at the thyristor terminals during switching.

21.4.2 Bidirectional Back-to-Back Converters

In many applications, including the wind turbines equipped with wound rotor induction machine (the same as DFIG), a bidirectional back-to-back converter is employed. Compared to the unidirectional converter, power can be controlled to transfer from either end to the other. For this arrangement it is necessary to use switchable devices for both converters. We consider only the voltage source inverters in this case because that is the type used in many of the modern wind turbines. The schematic of such a converter is shown in Figure 21.14.

In the circuit of Figure 21.14 if A, B, and C are connected to the mains with a given voltage and frequency the diodes D_1 to D_6 work as a rectifier to provide a DC voltage keeping the DC link capacitor charged. Then, by controlling the thyristors T_1' to T_6' a three-phase alternating current electricity with the required voltage and frequency can be obtained from the DC link. This AC electricity is available at A′, B′, and C′ terminals.

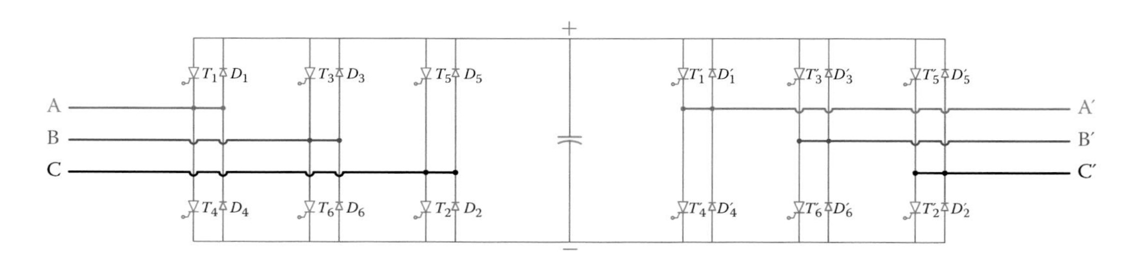

Figure 21.14 Schematic of a bidirectional back-to-back converter.

Figure 21.15 A back-to-back, water-cooled converter in a wind turbine (partly shown).

Likewise, if alternative electricity is available at A′, B′, and C′, then this can be rectified by the three-phase rectifier consisting of diodes D_1' to D_6', which feeds the DC link and then a three-phase alternating current can be obtained at A, B, and C.

As can be seen, the switching actions and the voltages available at A, B, and C terminals and A′, B′, and C′ terminals (which one is larger) determine in what direction the power can transfer. Recall that the function of each converter could be defined by the firing angle of its thyristors. Thus, the operation is not automatic and must be controlled by precise timing for the signals to the gates of the thyristors (IGBTs).

Figure 21.15 shows some of the IGBTs and the DC link capacitors for a 1.5 MW turbine. These IGBTs are water cooled. This turbine is equipped with a DFIG; hence, it has two back-to-back converters with 12 IGBTs.

21.5 Synchronous Generator for a Wind Turbine

If you recall from Chapter 11 (see Section 11.4.2), frequency of the generated AC from a synchronous generator depends on the number of poles in the stator and the speed at which its rotor is turned. All the conventional production of electricity (steam turbine, gas turbine, diesel engines, and hydro) is with a synchronous generator. A salient property of a synchronous generator is the possibility to control the power factor of the grid (electric network), which the generator feeds, through adjusting the rotor magnetic field. This is performed by adjusting the excitation current of the DC machine providing the magnetic field.

To keep the frequency constant, moreover, a generator must run at a constant speed. For a conventional power plant, the speed of the prime mover is controlled such that the frequency variation is within its tight tolerance (e.g., 59.9 to 60.1 Hz). This is performed through adjusting the flow of fuel, steam, or water (in a hydro turbine), based on the load requirement. With a wind turbine, if using a synchronous generator, this is not possible because the generator speed depends on the wind speed (and its load), and wind is not under our control. As a result, keeping the frequency constant is almost impossible.

It is possible to have a small isolated wind turbine equipped with a synchronous generator, provided that the generated energy is used only for applications for which the line frequency has no significant effect on their performance. Examples are heating devices. For instance, if such an isolated turbine is used for heating water, even the voltage does not need to be regulated. Otherwise, voltage needs to be controlled (e.g., in the case of lights). If the output of a turbine is to be connected to the electric network (grid), at any voltage level, its frequency and voltage must be the same as those of the network. As well, it is necessary to synchronize the turbine output with the grid to which it is to be connected.

> With a synchronous generator it is possible to control the power factor of the grid through adjusting the excitation current of an embedded DC generator supplying the rotor magnetic field.

Commercial wind turbines can be equipped with a synchronous generator provided that they are connected to the electric network through a power converter. For such a turbine a unidirectional converter can be employed. The turbine is turned at a speed governed by the wind speed and the load on the turbine. This generates a three-phase AC the frequency of which is variable, say between 50 and 65 Hz, depending on the turbine speed. The back-to-back AC-DC-AC converter, similar to that in Figure 21.12, delivers the required voltage at 50 or 60 Hz. All the power generated this way passes through the converter, but 2 to 3 percent of it is consumed by the converter and is lost.

The aforementioned scenario has the advantage that wind energy is harnessed at a wider range of wind speeds. Thus, such a wind turbine can still remain productive at those wind speeds at which a turbine with a squirrel cage induction generator must be stopped. A turbine working this way is called a **variable speed turbine**, because the rotor (and generator) speed is not constant and is variable. This turbine still has a gearbox. The generator with a synchronous speed of 1200 rpm can be turned at speeds between say 800 and 1600 rpm. Figure 21.16 represents the mode of operation of such a turbine.

Variable speed turbine: Wind turbine whose rotational speed is not constant. Such a turbine's speed can vary between 10 and 25 rpm.

The back-to-back converter must handle all the turbine power. For example, for a 1.5 MW turbine the electronics in the converter must be rated for 1.5 MW. A considerably additional expense, because of the cost

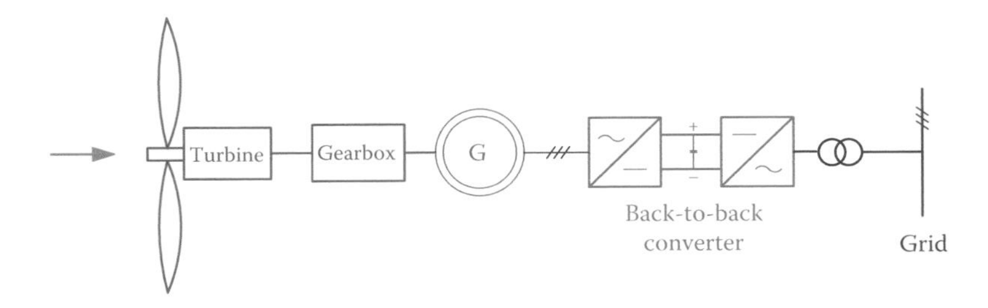

Figure 21.16 Operation scheme of a variable-speed turbine.

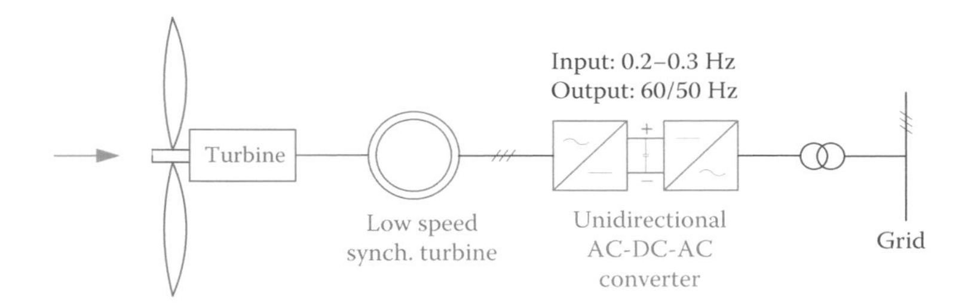

Figure 21.17 Schematic of a direct-drive turbine with electronic converter.

of electronic components in the converter and their maintenance, can be prohibitive. An alternative choice is using a doubly fed induction generator, where only part of the power goes through the converter. In this way, the electronic components are rated only for a fraction of the total power (normally 30 percent) and, therefore, the cost is much less. This is discussed in Section 21.7.

More recently, a version of the variable speed turbine has been introduced in which the gearbox is eliminated. The turbine generator has the the same low speed as the turbine rotor. For instance, if the rotor rotates with a speed of 10 to 20 rpm, the generator has the same rpm instead of 800 to 1600. The generator must have a design more in line with this speed. Permanent magnet generators are introduced for this application. Although a turbine made this way is still a variable-speed turbine, the manufacturers refer to it as a **direct-drive turbine** because there is no gearbox. This is schematically depicted in Figure 21.17.

Direct-drive turbine: Wind turbine with no gearbox.

Direct-drive turbine is a term used for turbines with no gearbox; thus, it also can address certain turbine designs in which a generator's synchronous speed has been reduced by increasing the number of poles. At least one manufacturer has introduced such a turbine. To accommodate a large number of poles in a generator, the diameter must be relatively very large.

21.6 Squirrel Cage Induction Generator

For many years the standard generator for wind turbines has been the squirrel cage induction generator. It is least costly, reliable, and has little maintenance cost mainly because it has no brushes. It matches the requirement for adaptability to wind speed. Nevertheless, in more and more newer turbines it has been replaced by a doubly fed induction generator; and more recently, direct-drive turbines with a synchronous generator have been introduced to the market. The main reason for replacing this generator by other designs is its limitation in capturing the maximum wind power in a broader range as far as the wind speed is concerned.

As we discussed in Chapter 11, an induction generator works around its synchronous speed, which is defined by the number of poles of the stator. The characteristic curve of an induction machine was shown in Figure 11.13; the generator part of it is depicted here in Figure 21.18. Below the synchronous speed the machine becomes a motor and above the synchronous speed, it behaves as a generator. The energy from wind must always

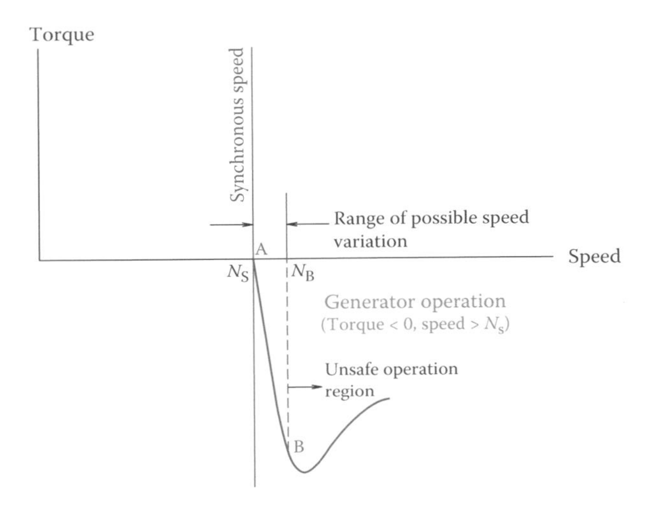

Figure 21.18 Typical characteristic curve (torque versus speed) of an induction machine.

turn the turbine generator above its synchronous speed; otherwise, the generator switches function and becomes a motor, consuming energy. That is, taking energy from the grid instead of injecting power to the grid.

For a wind turbine already working, if the wind speed drops and power grasp from wind cannot keep the rotor turning above the speed that corresponds to the generator synchronous speed, the generator, and thus the turbine, becomes a consumer of electricity. This corresponds to point A on the characteristic curve in Figure 21.18. The turbine controller detects the situation and disconnects the generator from the grid. If, however, the wind speed increases, a turbine speed can increase by a limited percentage (between approximately 2.5 and 3 percent of the synchronous speed, based on the generator design). The generator can work only as far as the operating condition lies between points A and B (the linear part of the characteristic curve). So, the speed of a turbine generator can vary only between the speeds corresponding to points A and B. That is, between N_S and N_B, where N_S is the synchronous speed.

It would be desirable if the slope of the linear part of the curve in Figure 21.18 could be smaller, so that the window between the lower and upper limits of generator speed (the horizontal axis) could be wider and embrace a larger range of corresponding wind speeds. For example, for a sample generator the speed limits are 1200 to 1230 rpm. If this could be expanded to a range of 1200 to 1400 rpm, winds with higher speed could be captured without running the generator into its unstable unsafe zone (see Figure 21.18). In a doubly fed induction generator this expansion is achieved in a smart way.

Doubly fed induction generator (DFIG): Type of induction generator with a wound rotor. Both the stator winding and the rotor winding need to be connected to electricity (thus the term doubly fed) as opposed to a squirrel cage induction generator.

21.7 Doubly Fed Induction Generator

A **doubly fed induction generator (DFIG)** is the same electric machine formerly called *wound rotor induction machine* (or *motor*, when used as a motor), when used as a generator. The reason it is named DFIG is because it has two series of connection to the outside, one for the stator winding and one for the rotor winding. It is more expensive than a squirrel cage induction generator and also, because it has brushes, it is more costly to maintain. Nevertheless, the advantages that it can offer outweigh the additional cost and more of the new wind turbines are equipped with DFIG.

As mentioned in Section 21.5, with a synchronous generator it is possible to correct the power factor. As we have discussed in Chapter 8, this is very important, and a low power factor implies loss of energy and excess voltage drop in the lines. This power factor correction is not possible with a squirrel cage induction generator. In a doubly fed induction generator one can separately control the active and reactive power, which also means controlling the power factor.

In addition to the above advantage, in a wind turbine a doubly fed induction generator allows that wind energy to be harnessed in a wider range of wind speeds. For example, many wind turbines cannot work with a wind speed of less than 4 m/s (9 mph) and they cannot benefit from the winds blowing at a speed more than 20 m/s (44 mph). A turbine with a DFIG can work in the range of 3 to 24 m/s (6.6 to 53 mph) wind speeds. Although the extra power harness at the lower end (3 m/s) can be small, that on the upper end is quite significant.

A DFIG can operate below its synchronous speed and still work as a generator. Also, it can continue to work as a generator at speeds up to 30 to 40 percent higher than its synchronous speed. Speeds below the synchronous speed are called **subsynchronous**, and those above the synchronous speed are called **supersynchronous** speed. By this definition one can say that a DFIG can work in both subsynchronous and supersynchronous modes. This is true also when such a machine works as a motor, but this case is out of the scope of the subject in this book.

Subsynchronous: Speeds below the synchronous speed in an induction machine.

Supersynchronous: Operating speeds higher than the synchronous speed in an induction machine.

With a doubly fed induction generator it is possible to control the power factor.

Here we study, first, the arrangement with which a doubly fed induction generator is employed in a wind turbine. Then we discuss how from the turbine behavior viewpoint this arrangement works.

21.7.1 Wind Turbines with DFIG

In a doubly fed induction generator, both stator and rotor participate in generation. The stator winding is directly connected to the step-up transformer that stands between the turbine and the grid. Rotor winding is connected to the step-up transformer through a bidirectional back-to-back AC-DC-AC converter. This is indicated in Figure 21.19. The two converters are referred to as the rotor side converter and the line side converter (or grid side converter).

When an induction generator works, a low-frequency alternating current flows through the rotor windings. The frequency of this AC electricity is proportional to the machine slip (see Chapter 11), which is the ratio

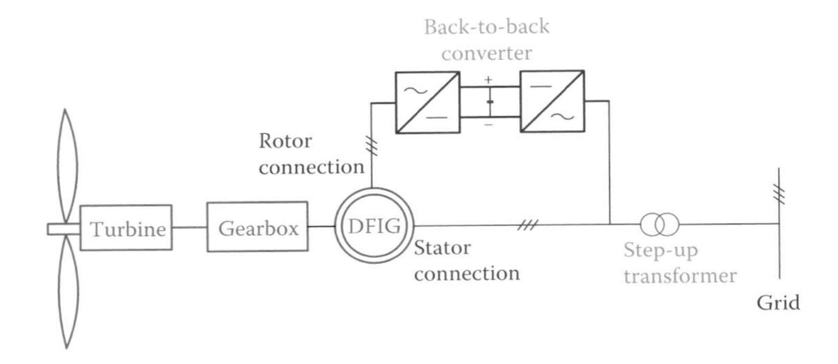

Figure 21.19 Configuration of a wind turbine with a DFIG.

of the difference between the rotor speed and the machine synchronous speed divided by the synchronous speed. It is customary to consider the slip of a motor as positive and that of a generator as negative. The rotor current frequency f_{slip} can be defined as

$$f_{\text{slip}} = f_s - f_r = f_s\left(1 - \frac{f_r}{f_s}\right) = sf_s \text{ Hz} \tag{21.1}$$

where f_s is the synchronous frequency (50 or 60 Hz), f_r is the rotor rotational frequency proportional to the rotor speed such that $\frac{f_r}{f_s} = \frac{N}{N_s}$, with N being the rotor speed. N_s is the synchronous speed, and s is the slip as defined by Equation 11.4. The sign of the frequency is not the focus of the matter at this point and should not drag your attention. It is important, however, to realize the existence of such a low-frequency alternating current in the rotor. For example, if for a machine the synchronous speed is 1200 rpm (such a machine corresponds to 60 Hz frequency) and the rotor rotates at 1230 rpm, the rotor slip is −30/1200 = −0.025. Then the rotor current has a frequency of 0.025 × 60 or 1.5 Hz. If the machine runs at exactly 1200 rpm, the frequency of the rotor current is zero and the machine at this speed neither delivers energy nor does it function as a motor (the torque on its shaft is zero).

As long as the generator speed is sufficiently higher than the synchronous speed, the alternating current in the rotor winding is rectified to DC and charges the DC link in the back-to-back converter. The DC link discharges through the grid side converter, where this DC electricity is converted back into AC and delivered to the step-up transformer. Depending on the number of turns in the rotor winding the rotor voltage might need to be raised to the same voltage as the stator (through another transformer). In the generators for wind turbines this add-on transformer is omitted by having the stator and rotor windings with an equal number of turns. Note that if the rotor of a wound rotor induction machine is prevented from turning, theoretically, the machine behaves as a transformer. In this way, part of the turbine energy is delivered to the grid through the stator winding and the other part through the rotor winding.

When the turbine speed drops so that the generator speed falls below its synchronous speed, the reverse action happens and the grid side converter acts as a rectifier. The power in the rectified DC then converts back to AC through the rotor side converter that now acts as an inverter. As a result, the generator rotor receives power from the grid. The net output delivery of the generator in this case is the power produced by the stator minus the power received by the rotor.

The performance of a turbine is controlled based on the wind speed. At higher wind speeds it must be taken out of production if the wind speed surpasses a value that otherwise can overload and damage a turbine (the **cut-out speed**). Also, it is taken out of production by its controller if the wind speed drops below a certain predetermined value (the **cut-in speed**). Normally, in a turbine with DFIG the converter is designed for 30 percent of the total power from a turbine.

A comparison of Figures 21.16 and 21.19 reveals that the electronic components when a DFIG is utilized are to be rated only for a portion of the power, which is taken from the rotor only. In practice, the electronic components for a DFIG are rated up to 30 percent of the turbine power. This makes a significant difference in the cost associated with the electronic parts. And this is a reason why the design in Figure 21.19 is preferred and more of the newer turbines are equipped with doubly fed induction generators.

Cut-out speed: Wind speed at which a wind turbine has reached its maximum design capacity and must be stopped. If it continues to work, the power in the wind surpasses the mechanical and electrical capacities; the generator or a mechanical component becomes overloaded and gets damaged.

Cut-in speed: Wind speed at which a wind turbine starts generating electricity and below which there is not sufficient power in the wind. Also, a wind turbine stops generating electricity if the wind speed drops below cut-in speed.

> In supersynchronous mode of a DFIG, electrical energy is extracted from its rotor. In a subsynchronous mode, electrical energy is injected to the rotor.

21.7.2 How a DFIG Works

The torque-speed characteristic curve of a three-phase induction machine is always as shown in Figure 11.13 or Figure 21.18 when only the generator operation is considered. The alternative current that flows in the rotor when the machine works has a frequency, as was mentioned earlier and according to Equation 21.1, proportional to the difference between the rotor speed and the synchronous speed of the machine. The direction of this current at a given instant depends if the machine works as a generator or if it works as a motor. Because in the case of alternating current *direction of current* does not give a clear definition, we may say that the two currents, corresponding to when a machine works as a motor and when it works as a generator, have 180° phase difference (or phase shift) with respect to each other. This is shown in Figure 21.20.

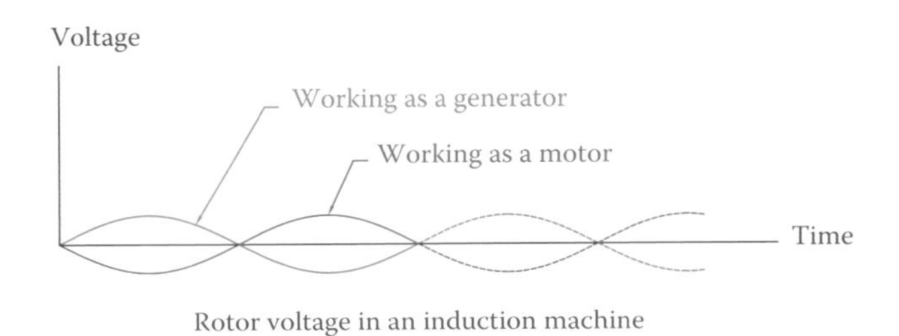

Figure 21.20 Rotor alternating current in an induction machine.

Figure 21.20 corresponds to a specific machine and at a particular condition. Consider, for example, an induction machine whose synchronous speed is 1200 rpm. If this machine runs at 1170 rpm, it is working as a motor and the slip frequency is 1.5 Hz (see Equation 21.1). If the machine is driven at 1230 rpm, it functions as a generator and the slip frequency is –1.5 Hz.

When an induction machine works as a motor, the rotor magnetic field follows the stator rotating magnetic field (see Chapter 11), and when it works as a generator the stator magnetic field follows that of the rotor. Figure 21.21 illustrates schematically the two magnetic fields and their resultant (which is the magnetic field in the air gap between the stator and the rotor) for the two cases. In each case the normal situation and the small load condition are depicted. In transition from generator to motor, and vice versa, a machine passes through this small load condition. The transition takes place when the rotor runs at the synchronous speed, where no torque exists on the shaft and no power is handled by a machine.

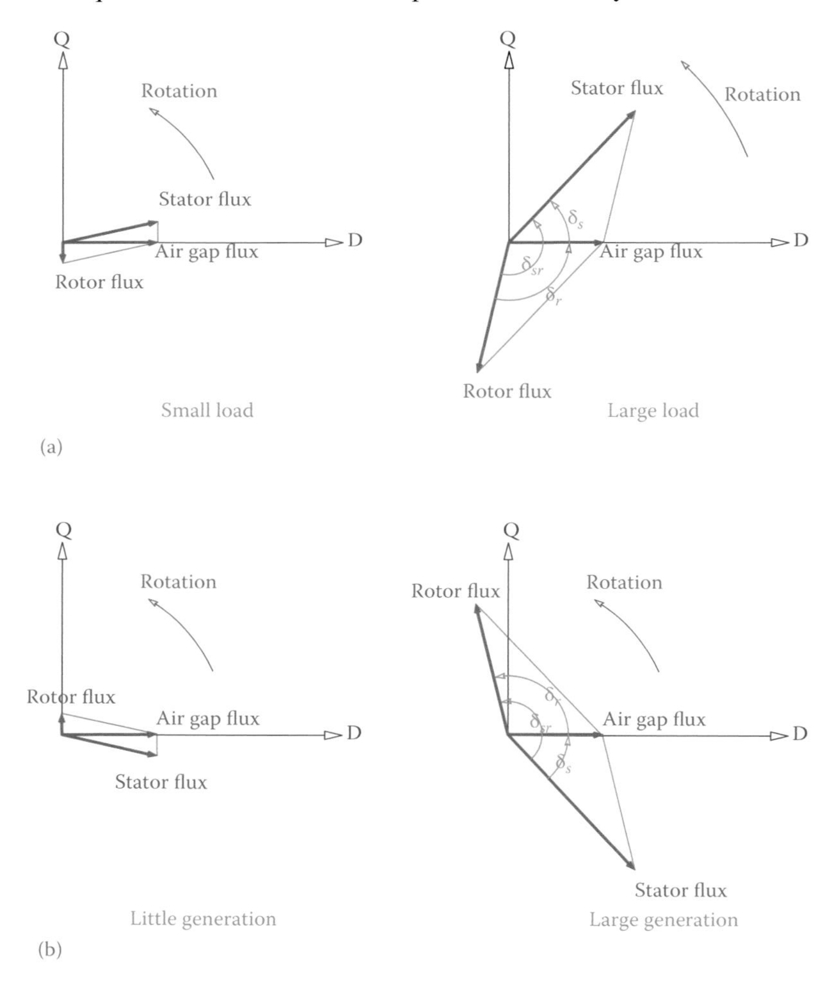

Figure 21.21 Magnetic flux variation in the operation of an induction machine: (a) motor action and (b) generator action.

To understand what happens in a doubly fed induction generator, we study the two cases of sub- and supersynchronous modes separately.

21.7.2.1 Subsynchronous Operation

Consider a DIFG working as a wind turbine generator. If the input power decreases (wind speed drops), then the rotor speed drops. If the power taken from the turbine is not adjusted, then eventually f_{slip} becomes zero and no electric power is generated. Any further decrease in the rotor speed turns the generator into a motor, that is, a load.

Suppose that when f_{slip} is near to becoming zero, an AC current is injected to the rotor winding by connecting it to an AC source, so that the rotor current stays above zero, and the working condition as a generator is maintained (see Figure 21.20). The result is that the rotor sees a magnetic field is maintained, leading the magnetic field of the stator; thus, the machine will continue working as a generator, even if the rotor speed is actually below the synchronous speed. In other words, the generator still generates electricity at speeds less than the synchronous speed. By injecting electricity to the rotor winding electrical energy is transferred to the rotor, which is equivalent to adding mechanical energy to the generator.

It is imperative to understand that the polarity of the AC current injected to the rotor winding must be opposite to that as if a motor. Otherwise, it only just adds to the rotor current while working as a motor. This injected signal, in fact, cancels the current that could flow in the rotor if working as a motor, as well as forcing a current of opposite polarity, as shown in Figure 21.22.

In Figure 21.22 a voltage (curve C) is introduced (injected) to the rotor of a DFIG. Part of this voltage (curve B) is used to cancel the current that at the given speed (as a motor) would be flowing in the rotor (curve A). The rest of that (curve D) is the resultant and corresponds to the consequential working condition as a generator.

This process of injecting a low-frequency alternating current to the rotor winding in fact sets a new characteristic curve for the machine. Whereas at a speed $N < N_s$, a machine should act as a motor and its *corresponding* torque should be positive, but now it is acting as a generator and the torque is negative. This is shown in Figure 21.23. Compare point M on curve A (motor) with the same point denoted as G in curve B (generator).

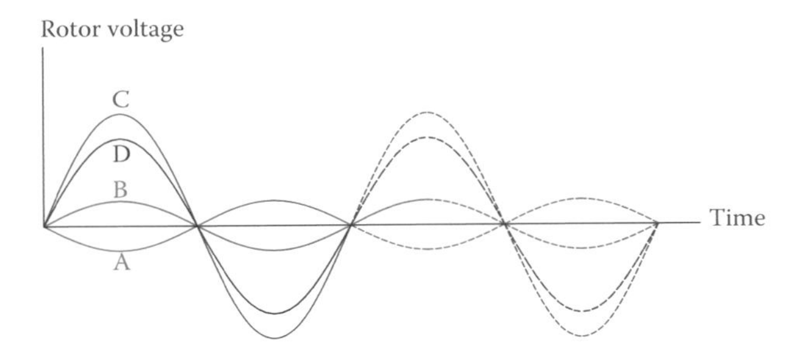

Figure 21.22 Injecting an alternating current into the rotor winding of a DFIG.

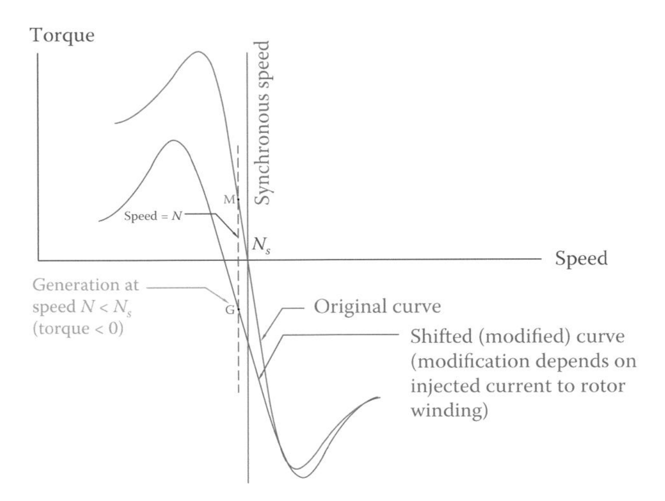

Figure 21.23 Effect of injecting AC to the rotor of a DFIG on the speed-torque curve.

21.7.2.2 Supersynchronous Operation

In supersynchronous operation, current is taken from the rotor winding through a set of slip rings. This is the current feeding the rotor side converter. The action of taking current from the rotor can be interpreted as injecting a negative current in the rotor winding. Thus, by the same token, the effect is shifting the characteristic curve in the reverse direction of the previous case. This is true for the case a wound rotor induction motor works in the supersynchronous mode and electric power is extracted from its rotor winding.

In the case of a DFIG, when power is extracted from the rotor current the slope of the characteristic curve changes. This change is due to reduction in the rotor current, which leads to reduction in torque. In reality part of the current that otherwise would flow in the rotor winding is removed. Consider if the winding was shorted at the points that the rotor side converter is connected. For the same rotor speed, therefore, the torque on the rotor shaft is smaller (compare points H and K in Figure 21.24). Consequently, the characteristic curve is more stretched along the horizontal axis and toward right.

Note that we are interested in the linear part of the curve around the synchronous speed. The result of both effects, injecting current to the rotor winding and extracting power from it, is shown in Figure 21.24, which also indicates the range of speeds for a DFIG. This range is significantly larger compared to the curve for a squirrel cage machine.

In a DFIG by controlling the rotor current a number of objectives can be achieved. The active and reactive power from the generator can be independently controlled or the torque on the generator shaft can be adjusted. This implies that the torque can be kept constant while the speed increases beyond what normally would be available with a squirrel cage generator.

We can summarize the advantages of a doubly fed induction generator as follows. This is due to the fact that the rotor winding is accessible

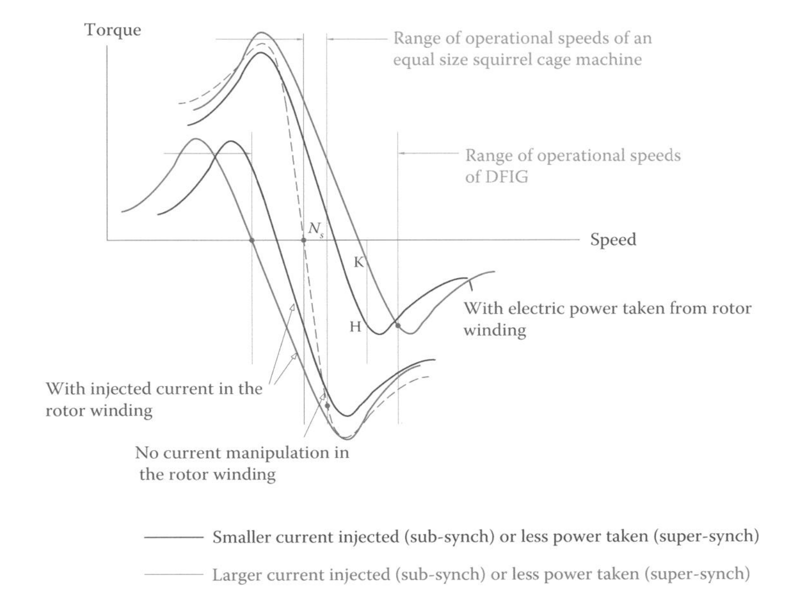

Figure 21.24 Extension in the range of operational speeds in a DFIG.

and can be connected to the external devices (such as the back-to-back converter), as required.

1. Active and reactive powers can be independently controlled.
2. Power factor can be set as desired.
3. Generator output torque can be adjusted.

The first two items are all that can be achieved with a synchronous generator. We can say that a doubly fed induction generator has the performance of a synchronous generator, while it does not need to work at a fixed speed. As for wind turbines, it can perform as a variable speed synchronous generator without the need to utilize a full rate power converter. In the subsynchronous mode some energy is injected to the machine through the rotor in the form of electrical energy. The result is that the generator receives part of its energy in the form of mechanical energy (from wind) on its shaft and part in the form of electrical energy. Consequently, it can continue to work and convert the mechanical energy to electricity. In the supersynchronous mode it receives all its energy in the form of mechanical energy on its shaft, converts this energy to electricity and delivers a major portion of that through the stator and a smaller portion of it through the rotor. Figure 21.25 shows this.

The power that must be given to the rotor winding in the electrical form when working in the subsynchronous mode is approximately proportional to the power output from the stator, that is,

$$P_r \approx sP_s \tag{21.2}$$

where P_r is the rotor power, P_s is the stator power, and s is the rotor slip.

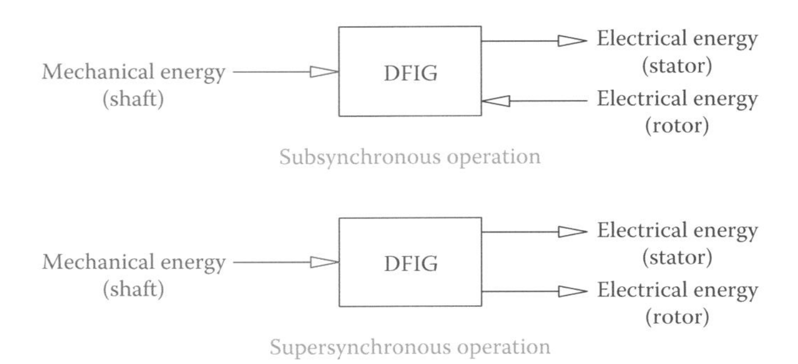

Figure 21.25 Summary of operation of a DFIG.

21.8 Chapter Summary

- A wind turbine consists of a foundation, a tower, a nacelle, a rotor, and a generator, plus many auxiliary devices.
- The majority of wind turbines have a gearbox to increase the rotor speed for the generator.
- The rotor consists of the blades and a hub.
- The nacelle houses the generator and gearbox on top of the tower.
- Solar irradiance is the amount of power per unit of area reaching the Earth above the atmosphere, which is about 1365 W/m^2 (127.0 W/ft^2).
- Part of the solar irradiance is lost before reaching the Earth's surface owing to clouds and particles in the air.
- A solar panel converts solar light/heat to DC electricity.
- A solar panel consists of many solar cells in series and parallel combination.
- A solar cell is a diode with a function opposite to that of an LED.
- The output of a solar cell depends on sun irradiance, weather condition, air pollution, and characteristic of the cell, which depends on the cell material. The fill factor is a percentage value, obtained by the ratio of the maximum power possible from a solar cell to the product of the open circuit voltage and short circuit current.
- Most wind turbines have an induction generator.
- A back-to-back converter consists of two power converters put together. It can be used for DC-AC-DC conversion or AC-DC-AC conversion.
- In wind turbines using a squirrel cage generator the rotor speed is almost constant.
- In a variable speed turbine the rotor speed can vary within a large range, for instance, between 10 and 20 rpm.
- A wind turbine can use a synchronous generator with or without a gearbox, if it has an AC-DC-AC converter.
- Power converters add significantly to the cost of a turbine.
- A wind turbine equipped with a doubly fed induction generator (DFIG) can be regarded as a variable speed turbine with a

synchronous generator, but only part of the power is transferred through the converter.
- A DFIG is the same as a WRIM, working as a generator.
- A turbine with a DFIG captures more wind power than a turbine with a squirrel cage generator.
- A DFIG can work in a wider range of wind speeds. It can work in subsynchronous mode and supersynchronous mode.
- Up to 30 percent of the generator power can be drawn from the rotor and the rest from the stator of a DFIG.
- In the subsynchronous operation of a DFIG, electrical power is given to the rotor winding through the converter.
- In the supersynchronous operation of a DFIG, electrical power is taken from the rotor winding through the converter.
- DFIG with a controlled bidirectional converter has the same performance as a synchronous generator.

Review Questions

1. What are the principal components of a wind turbine?
2. In what way is a wind turbine different than other turbines?
3. What does a solar cell do?
4. What is the average voltage of a single solar cell?
5. Why does a solar array need a converter to connect to the electric grid?
6. In how many ways can you say the energy from the sun can be used?
7. Define open circuit voltage and short circuit current of a solar cell.
8. Describe the characteristic curve of a typical solar cell.
9. What is the fill factor in a solar cell?
10. In what way(s) can a solar cell be compared to a small battery?
11. What is the most common generator for a wind turbine?
12. Do wind turbines work at a constant speed? Why or why not?

13. What does DFIG stand for?

14. What are the advantages of a DFIG over a squirrel cage induction generator in a wind turbine?

15. What are the disadvantages of DFIG over a squirrel cage induction generator in a wind turbine?

16. What is the relationship between DFIG and WRIM? Explain.

17. In what way can a synchronous generator work in a wind turbine?

18. What is a back-to-back converter and what is its purpose?

19. Why is it called a back-to-back converter?

20. What is the difference in design between a unidirectional and a bidirectional converter?

21. What type of converter (unidirectional, bidirectional) is used in a doubly fed induction generator? Why?

22. What is DC link?

23. What are the advantages of a DFIG in a wind turbine compared to a synchronous generator?

24. What is meant by subsynchronous speed?

25. How can a DFIG work in a subsynchronous speed? Explain.

22

Digital Electronics

OBJECTIVES: After studying this chapter, you will be able to

- Explain binary, hexadecimal, and octal number systems
- Explain the practical importance of all those number systems
- Write decimal numbers in binary, octal, and hexadecimal systems and convert those to decimal numbers
- Define bit, byte, word, LSB, MSB, overflow, flag, microprocessor, and microcontroller
- Understand conversion from an analog measurement to a digital number
- Describe the relationship between precision in measurement and the number of bits of digital representation of the measured value
- Define and use DAC, ADC, A/D, and D/A
- Define signed and unsigned numbers and their application in measurement of physical systems
- Do basic arithmetic operations in binary numbers
- Describe one's complement and two's complement of binary numbers
- Give definitions for terms sampling, sampling time, and sampling rate
- Explain how an analog value is updated for being used in a computer-controlled device
- Understand a digital signal representation and a clock signal
- Understand how microprocessor control is performed in many small or large devices and equipment
- Define what binary-coded decimal is and what it is used for

New terms:
Actuator, analog-to-digital converter, binary, binary-coded decimal (BCD), decimal system, digital-to-analog converter, falling edge, flag, hexadecimal, leading edge, least significant bit (LSB), microcontroller, microprocessor, most significant bit (MSB), negative going edge, one's complement, overflow, positive going edge, register, rising edge, rotation, sample and hold, sampling, sampling frequency, sampling rate, sampling time, shift right, signed number, trailing edge, two's complement, unsigned number

22.1 Introduction

Many of the technological advancements we observe today, both at domestic level and industrial applications, are due to the progress in digital electronics. Examples are laptops, cellular phones, TV broadcasting, and many processes that are now controlled by a digital computer. In fact, all that started from a digital calculator leading to a digital computer (replacing an analog computer) and the rest.

Part of this technological advancement is due to progress in electronics, computer science, material science, miniaturization in manufacturing, and research and development on all the pertinent subjects. In this chapter and the next we cover the fundamentals of digital electronics without addressing the materials or the manufacturing processes. All digital electronics are based on a few analog devices that we have learned so far, on top of them the *transistor*. A computer chip, an integrated circuit that is made in a miniature scale and can be only 1.5 × 1.5 inch (4 cm × 4 cm), can contain 1 million components, most of them transistors.

22.2 Number Systems

Decimal system: Most common numbering system with ten numbers (0 to 9) as the base for making all other numbers.

The numbers we are familiar with and use every day are based on 10 symbols and rules for making and displaying all numbers, both positive and negative, from them. This system of numbers is the **decimal system**. We learned about most of these rules in elementary school. At this time these rules may look primitive, and we do not even think about them. But they are the same for other number systems used in digital computers that have revolutionized the world, which we will discuss in this chapter.

Note that in the decimal system the **base** is 10; we have 10 symbols, and the first number is 0. After counting to 9, the next number is composed of two digits: a 1 followed by the very first number, 0. Consider an ordinary number such as 543. Because we are used to ordinary numbers like 543, it is not necessary to indicate its base. Otherwise, we would have to write it as 543_{10}, showing that its base is 10, or it is a number written in a decimal system.

In the rest of this chapter, unless it is required, we do not refer to a number as decimal or write it with the base. We follow the ordinary way of writing numbers.

Hexadecimal: Numbering system with 16 as the base. Thus, there are 16 basic numbers (0 to 15) in order to make all other numbers. Furthermore, there are 16 symbols 10 of which are 0 to 9 and for the rest the capital letters A to F are used (A = 10, B = 11, C = 12, D = 13, E = 14, and F = 15).

Binary: Having two states: 0 and 1, true and false, high and low, on and off, yes and no.

The meaning of the number 543 is as follows:

$$543 = 5 \times 100 + 4 \times 10 + 3 \times 1 = 5 \times 10^2 + 4 \times 10^1 + 3 \times 10^0$$

(Remember that any number raised to power 0 is equal to 1.)

Also, note that any number multiplied by 10 receives an additional zero at its right, or if divided by 10 loses its rightmost digit (The rightmost digit is the least significant digit. We may need to round a number):

$$543 \times 10 = 5430$$

$$543 \div 10 = 54\ (+0.3)$$

All these and other rules are applicable to any other number system. We are going to see only 3 other number systems: **hexadecimal**, **octal**, and **binary** numbers. These are widely employed in the development or usage of industrial and domestic digital products. All that is done in calculators, computers, and the like are practically based on binary system, but for presentations, analysis, discussions, and theorem formulations, hexadecimal and octal systems are used. The reason for using hexadecimal and octal numbers is the convenience in converting from binary to those and vice versa. Conversion between decimal and binary numbers is not as convenient.

22.2.1 Hexadecimal Number System

Hexadecimal represents a number system whose base is 16 (made out of the Greek word hex for 6 and decimal denoting association with 10). Initially, therefore, 16 symbols are necessary to represent all numbers. Also, in writing hexadecimal numbers it is necessary to somehow indicate that its base is 16 and not 10.

The 16 numbers for the base of the hexadecimal system are numbers 0 to 15. For numbers 0 to 9 the same symbols used in the decimal system are employed. For numbers 10 to 15 the letters A to F are utilized, without any name. The base for hexadecimal system is then

0 1 2 3 4 5 6 7 8 9 A B C D E F

and their corresponding decimal values are

0 1 2 3 4 5 6 7 8 9 10 11 12 13 14 15

The following number after F in hexadecimal system is 10, in the same way that in decimal numbers after 9 we have 10. To show that a number is written in the hexadecimal system (or its base is 16), we can put the base as a subscript for the number, like 543_{16}. A more common way is to put an X before the number (e.g., X543) or add an H after the number, like 543H. Thus,

$$\text{X543} = 543_{16} = \text{543H} = 5 \times 16^2 + 4 \times 16^1 + 3 \times 16^0 = (5)(256) + (4)(16) + (3)(1) = 1347$$

The following shows examples of hexadecimal numbers (without the preceding X):

48, 312, A9, C3, 2AA, ABCD, 3FF, 6A2B, 12DF

To have a better feeling for the values of these numbers, we calculate their decimal equivalents. You notice that if we do not use the preceding X, it is not clear if numbers such as 48 and 312 are decimal or hexadecimal, and these are quite different. The decimal magnitudes of hexadecimal numbers are calculated in the same way shown earlier for 543 in both cases of having 10 or 16 as for base. Because $16^0 = 1$, we refrain from its repetition and use 1 for the multiplier of the rightmost (least significant) digit.

$$\text{X48} = 4 \times 16^1 + 8 \times 1 = 72$$

$$\text{X312} = 3 \times 16^2 + 1 \times 16^1 + 2 \times 1 = 786$$

$$\text{XA9} = 10 \times 16^1 + 9 \times 1 = 169$$

$$\text{XC3} = 12 \times 16^1 + 3 \times 1 = 195$$

$$\text{X2AA} = 2 \times 16^2 + 10 \times 16^1 + 10 \times 1 = 682$$

$$\text{XABCD} = 10 \times 16^3 + 11 \times 16^2 + 12 \times 16^1 + 13 \times 1 = 43{,}981$$

$$\text{X3FF} = 3 \times 16^2 + 15 \times 16^1 + 15 \times 1 = 1023$$

$$\text{X6A2B} = 6 \times 16^3 + 10 \times 16^2 + 2 \times 16^1 + 11 \times 1 = 27{,}179$$

$$\text{X12DF} = 1 \times 16^3 + 2 \times 16^2 + 13 \times 16^1 + 15 \times 1 = 4831$$

Also, notice the following

$$X99 = 9 \times 16^1 + 9 \times 1 = 153$$

$$XFF = 15 \times 16^1 + 15 \times 1 = 255$$

$$X100 = 1 \times 16^2 + 0 \times 16^1 + 0 \times 1 = 256$$

$$XFF + 1 = X100$$

Note that equivalent to 99, 999, and similar decimal numbers (i.e., those figures before the number of digits increases) in hexadecimal system we have FF, FFF, and so on. Thus,

$$X1000 - 1 = XFFF$$

$$XFFFF + 1 = X10'000$$

(Note that it is not necessary to write X1 instead of 1 because there is no ambiguity.)

Example 22.1

Find the decimal equivalent of XABCD0.

Solution

In the same way that in the decimal number system a number like 1230 is 10 times the number 123, we can use the analogy to say the number XABCD0 is 16 times the number XABCD. The decimal equivalent of the latter has already been found to be 43,981. Thus, for the decimal value of XABCD0, we have

$$XABCD0 = (16)(43{,}981) = 703{,}696$$

The reverse process of finding the hexadecimal expression of decimal numbers can be done by successive divisions by 16. The remainder of a division determines the digit for the number. Consider, for example, the small figure 23. Dividing this figure by 16, we can write

$$23 = 1 \times 16 + 7 \quad \leftarrow \text{The first digit from right is 7}$$

$$\swarrow$$

$$1 = 0 \times 16 + 1 \quad \leftarrow \text{The second digit from right is 1}$$

$\uparrow$ Stop dividing when this quotient has reached 0

The arrows on the left show the procedure, and the arrows on the right show the answer. The answer is 17, as defined by the numbers 7 and 1 denoted by arrows. The oblique arrow starting from 1 on the first row toward the second row denotes the transfer of the quotient to a new line for continuation of the same process of dividing by 16. The vertical arrow shows that when the quotient becomes 0 the process should stop.

For this small number, only one division is sufficient (the last one is not really necessary). But for larger numbers the dividing process must be repeated a number of times until no more division is possible (or no extra results are obtained). The following examples show the same procedure for larger numbers. It is not necessary to show the arrows. Here they help understanding.

Example 22.2

Find the hexadecimal representation of 2002.

Solution

We carry out dividing by 16 showing the process in the same format as above.

$$2002 = 125 \times 16 + 2 \leftarrow$$
$$\swarrow$$
$$125 = 7 \times 16 + 13 \leftarrow$$
$$\swarrow$$
$$7 = 0 \times 16 + 7 \leftarrow$$
$$\uparrow$$

The corresponding hexadecimal number, therefore, is X7D2 (D is for the decimal 13).

Example 22.3

Find the hexadecimal representation of decimal 52,778.

Solution

$$52{,}778 = 3298 \times 16 + 10 \rightarrow \mathrm{A}$$
$$3298 = 206 \times 16 + 2 \rightarrow 2$$
$$206 = 12 \times 16 + 14 \rightarrow \mathrm{E}$$
$$12 = 0 \times 16 + 12 \rightarrow \mathrm{C}$$

The answer is XCE2A.

22.2.2 Binary Number System

The binary number system has only two base numbers, 0 and 1. The reason for developing such a number system is its applicability to many physical processes that have two states, and thus a binary number is sufficient to describe them. These physical states can be in the form of

- On and off
- True and false
- High and low
- Yes and no
- Positive and negative

depending on the application. For example, for switches and switching devices in an electrical circuit the on-off defines the state of one switch. This can be the basis of storing data, which can consist of the many switches or devices where each one assumes one of the two states at a time.

A number in the binary system, thus, can have only 0's and 1's because there is no other symbol. Note that in the same way that in the decimal system the number after 9 (the last number element) is 10, in the binary system the number after 1 is 10. The following lines show a few examples of some decimal numbers when written in the binary system. A subscript b or 2 is used to indicate that a number is binary.

$$\begin{aligned}
0 &= 0 \\
1 &= 1 \\
2 &= 10_b \\
3 &= 11_b \\
4 &= 100_b \\
10 &= 1010_b \\
11 &= 1011_b \\
12 &= 1100_b \\
15 &= 1111_b \\
16 &= 10000_b
\end{aligned}$$

Before we learn how to find the binary equivalent of a number, note that in the same way that 10 + 1 = 11, in the given binary numbers the same thing can be verified:

$$1010_b + 1 = 1011_b$$

and after 1111_b (the last four digit binary number) is 10000 (the first five-digit number), in the same way that after 9999 in decimal numbers we have 10000.

The decimal equivalent of any binary number can be found by the same procedure we have used earlier for the number 543. We use the above figures to verify that they really are equivalent to each other:

$$\begin{aligned}
10_b &= 1\times 2^1 + 0\times 2^0 = 2+0 = 2 \\
11_b &= 1\times 2^1 + 1\times 2^0 = 2+1 = 3 \\
1111_b &= 1\times 2^3 + 1\times 2^2 + 1\times 2^1 + 1\times 2^0 = 8+4+2+1 = 15 \\
10000_b &= 1\times 2^4 + 0\times 2^3 + 0\times 2^2 + 0\times 2^1 + 0\times 2^0 = 16+0+0+0+0 = 16
\end{aligned}$$

Example 22.4

Find the decimal number for the binary 1011010110_b.

Solution

For clarity, it is convenient to show the number as $10{,}1101{,}0110_b$, that is, separating each four digits from right. This number has 10 digits.

$$1 \times 2^9 + 0 \times 2^8 + 1 \times 2^7 + 1 \times 2^6 + 0 \times 2^5 + 1 \times 2^4 + 0 \times 2^3 + 1 \times 2^2 + 1 \times 2^1 + 0 \times 2^0 = 512 + 128 + 64 + 16 + 4 + 2 = 726$$

But, note that it is not necessary to write the multipliers because they are either 1 or 0. Moreover, 0 multiplied by any number results in 0; hence, it is not necessary to write. The above mathematical line can be reduced to

$$2^9 + 2^7 + 2^6 + 2^4 + 2^2 + 2^1 = 512 + 128 + 64 + 16 + 4 + 2 = 726$$

To find the binary number equivalent of a decimal number, use the following procedure:

1. Divide the number by 2 and write the quotient and the remainder.
2. If the remainder is 0, the rightmost digit is zero; otherwise, it is 1.
3. Substitute the quotient for the number and repeat the division by 2.
4. If the remainder is 0, the second digit is 0; otherwise, it is 1.
5. Continue the same procedure until the quotient reaches 0.

The following example shows this procedure.

Example 22.5

Find the binary equivalent number for 86.

Solution

The sequence of divisions by 2 can be arranged in the following format:

	Quotient				Remainder	Digit from right	
86 =	43	×	2	+	0	0	1st
43 =	21	×	2	+	1	1	2nd
21 =	10	×	2	+	1	1	3rd
10 =	5	×	2	+	0	0	4th
5 =	2	×	2	+	1	1	5th
2 =	1	×	2	+	0	0	6th
1 =	0	×	2	+	1	1	7th

The binary representation of 86, thus, is 1010110.

22.2.3 Octal Number System

Another number system that is more utilized in industry is the octal numbers, having number 8 as the base. The rules for conversion from octal to decimal and vice versa are the same as the other number systems. Numbers 0 to 7 are used for the eight main symbols. Thus, the positive numbers start from 0 and are

0 1 2 3 4 5 6 7 10 11 12 13 14 15 16 17 20 …

Again, each octal number must be followed by a subscript 8 to indicate its base. The following examples show calculation of the decimal equivalent of an octal number and the octal equivalent of a decimal number.

Example 22.6

Find the decimal value of 2012_8.

Solution

$$2012_8 = 2 \times 8^3 + 0 \times 8^2 + 1 \times 8^1 + 2 \times 1 = 2 \times 512 + 8 + 2 = 1034$$

Example 22.7

Find the octal representation of number 7777.

Solution

$7777 = 972 \times 8 + 1$	→	1	first digit from right
$972 = 121 \times 8 + 4$	→	4	second digit
$121 = 15 \times 8 + 1$	→	1	third digit
$15 = 1 \times 8 + 7$	→	7	fourth digit
$1 = 0 \times 8 + 1$	→	1	fifth digit

The octal number is 17141_8.

22.2.4 Conversion between Binary and Hexadecimal Numbers

The fact that hexadecimal numbers are based on 16, and 16 is the fourth power of 2 makes the conversion between hexadecimal and binary numbers easy. Consider a single digit hexadecimal number. The values expressed by this digit are between 0 and 15. Now consider the numbers that may be represented by four digits in a binary number. Again, they are between 0 and 15. All these numbers are compared together below (the prefix X for hexadecimal numbers is omitted for simplicity).

$$0000_b = 0$$
$$0001_b = 1$$
$$0010_b = 2$$
$$0011_b = 3$$
$$0100_b = 4$$
$$0101_b = 5$$
$$0110_b = 6$$
$$0111_b = 7$$
$$1000_b = 8$$
$$1001_b = 9$$
$$1010_b = A$$
$$1011_b = B$$
$$1100_b = C$$
$$1101_b = D$$
$$1110_b = E$$
$$1111_b = F$$

The above equivalence can be directly carried to all other digits of a hexadecimal number and the set of four digits of a binary number. The following are some examples.

X123 = $0001{,}0010{,}0011_b$
X47A = $0100{,}0111{,}1010_b$
XBCD1 = $1011{,}1100{,}1101{,}0001_b$
XA00F5 = 1010,0000,0000,1111,0101b
1111,1100,0011,0000,1110 = XFC30E
0000,1001,0000,1111 = X90F

Note that in the last number the leftmost 0 is not shown in the hexadecimal number, whereas in a binary number it is very common to have 0's on the left, as you will see in the next section.

Example 22.8

Find the hexagonal equivalent of 1001,1010,1100,1111,1000, 0001.

Solution

Each digit can be calculated or looked up in the preceding list. Thus,

$$1001{,}1010{,}1100{,}1111{,}1000{,}0001 = 9ACF81_{16}$$

22.2.5 Direct Conversion between Binary and Octal Numbers

On the basis of the discussion in Section 22.2.4, a direct conversion between octal and binary numbers is possible. A binary number can be divided in groups of three, each of which corresponds to one digit of its associated octal number. Only the first 8 out of 16 in the preceding list correspond to octal numbers, and the 0 on the left must be omitted so that there are only three digits for each binary number. Some examples are as follows.

$$000{,}101{,}000{,}100\text{b} = 504_8$$
$$000{,}101{,}111{,}100\text{b} = 574_8$$
$$001{,}001{,}111{,}101\text{b} = 1175_8$$
$$010{,}111{,}110{,}000\text{b} = 2760_8$$
$$100{,}111{,}110{,}000\text{b} = 4760_8$$
$$111{,}101{,}011{,}001\text{b} = 7531_8$$

As said before, the 0's on the left side of a binary number do not need to be included in the converted number.

Microprocessor: Integrated circuit for all arithmetic and logic operations in a computer.

22.3 Number Conversion and Precision

Analog-to-digital converter: Electronic device that generates a digital signal from an analog signal so that the signal can be treated with digital devices. This is normally used for measured values. In such a case converted signals must be updated at a desired frequency.

Digital-to-analog converter: Electronic device that translates the value of a digital signal generated by a computer or a digital device into an analog value that can be used in an electric circuit.

Actuator: Any physical device that can generate a mechanical action (e.g., rotation, force exertion, moving) based on an electric, hydraulic, or pneumatic excitation.

Most of the physical parameters that we deal with every day have an analog nature, such as temperature, wind speed, line voltage, pressure in a tube, flow rate in a pipe, sun's position with respect to a solar farm, angle of a solar panel, and so forth. For using a computer or a **microprocessor** for measurement and control, and for other purposes such as display and data storage, their values can be converted to digital numbers. This is carried out through converters. These converters must not be confused with power converters that we have studied before.

For altering the physical values that come out of measurements to the form that can be used by a microprocessor an **analog-to-digital converter** (ADC) is used. Also, for the reverse action a **digital-to-analog converter** (DAC) is employed. The latter is used to send an analogue value to an **actuator**, for instance. An actuator is any device that acts on a mechanical system, for example, a motor or hydraulic piston that turns the blades of a wind turbine to a desired pitch angle, or the adjusts the angle of a solar panel toward the sun's rays. The desired angle is obtained from some calculations that the system microprocessor performs.

A microprocessor is the brain of a computer that controls all the numerical operations and peripheral actions of the computer. It is not necessarily used only in a computer. There are special purpose microprocessors for specific purposes. Also, in many devices, microprocessors are employed for adjusting the performance of the device based on given conditions and desired objectives. Examples can vary from a huge turbine to a camera, a car, a washing machine, a coffeemaker, and many more. For each purpose an appropriate microprocessor is used. Figure 22.1 illustrates the schematic of what happens in any microprocessor controlled device. In this figure, A/D and D/A stand for analog-to-digital (ADC)

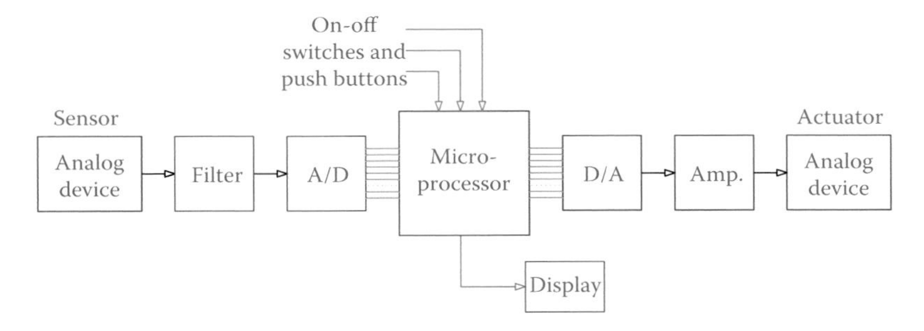

Figure 22.1 Schematic of a microprocessor control system.

and digital-to-analog (DAC) converter, respectively. For control systems a small microprocessor, specifically designed for control purposes, built together with other necessary components such as memory and input/output devices, is referred to as a **microcontroller**.

Microcontroller: A one chip integrated circuit (IC) comprising a main processor plus all components and input/output terminals that can be used for control purposes (process control and control of machinery).

In Figure 22.1, you can see that output of a measurement device may need to be filtered for noise and unwanted signals. Also, the output from a microprocessor may be too weak to be able to drive a system. Thus, an amplifier is used to amplify the signal.

Not all the inputs to a microprocessor (or microcontroller) are initialized from a parameter with an analogue quantity. Certain measurements are naturally in a binary form and can be readily used. The output of a switch in most cases is *on* and *off*, which is already in the binary form. These parameters do not need any conversion. A switch is either on or off, a key is either pressed or not pressed and a pushbutton either is or is not pushed. Binary numbers serve as the best way to show these *on* and *off* states because the associated data can be easily stored and accessed in the form of 0's and 1's.

The best example for conversion from analog to digital is a digital watch in which the time, an analog quantity, is converted to a digital variable and is digitally displayed. Certain variables have a digital nature, such as the number of people entering a library, or the number of cars crossing a bridge. In fact, anything that can be counted, has a digital nature and anything that can be measured has an analog nature.

No matter if the numbers correspond to an analog quantity, come directly from a digital counter, or are already in a binary form, binary numbers must be used for their processing. Processing implies measuring, recording, performing arithmetic operations on numbers, displaying numbers, and similar actions that are performed by a digital computer. For recording and storing binary numbers, at the base level, an electronic device is used that is capable of holding a value of 0 or 1. For example, a charged capacitor can represent 1 and a discharged capacitor 0. This is for a single one digit binary number.

For larger numbers, more than one unit of the 0 and 1 storing device is essential. Then the size of the physical device in terms of how many 0's and 1's can be allocated for a number becomes important.

22.3.1 Bits and Bytes

In dealing with binary numbers, the terms **bit** and **byte** come into picture. A bit is the unit for the digits that a binary number can have. Thus, the

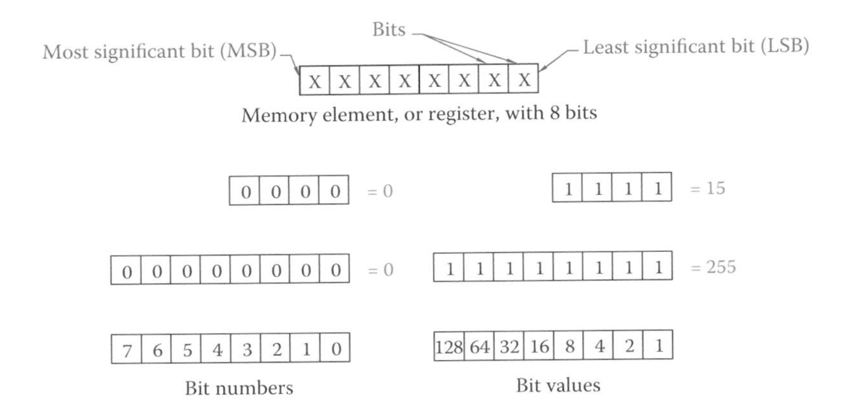

Figure 22.2 Representation of bits, bit numbers, and their values.

smallest binary number has 1 bit, and its value can be 0 or 1. A byte is a group of 8 bits together. The memory and the digital processor for many devices may contain 8 bits (one byte), but computers have larger number of bits for their number processing. Today's laptops are made with 32-bit or 64-bit processors and the larger computers can be based on 128 bits. This number of bits is for representing or storing one digital number, only. A computer can handle millions of numbers. Thus, it has many units of the 128-bit device. This defines the computer memory, which is normally measured in bytes; for example, 32 megabytes or 4 gigabytes, etc. Associated with bit and byte in computer terminology a **word** is a group of 16 bits or two bytes.

Register: Electronic device to store digital data (in the form of binary numbers), made of a number of flip-flops, one for each bit.

Least significant bit (LSB): The bit on the far right in a binary number.

Most significant bit (MSB): Leftmost bit in a binary number.

Figure 22.2 shows the representation of the electronic circuits that hold binary numbers. We refer to this electronic device as a **register**. At this time we are only concerned about the bits and bytes, without caring about what circuits constitute the register. When there are a number of bits, like in a byte, the rightmost bit is called the **least significant bit (LSB)**, and the leftmost bit is called the **most significant bit (MSB)**. Each bit is numbered for ease of reference. The rightmost bit is bit 0, followed by bit 1, bit 2 and so forth. The last bit in a byte, therefore, is bit 7. There is a value associated with each bit. This value is obtained by raising "2" to the power defined by the bit number. For instance, the first one from the right has a value $2^0 = 1$, and the most significant figure in a byte has a value of $2^7 = 128$. These numbers are also shown in Figure 22.2 for a byte.

As can be understood, with 4 bits only, the smallest number that can be addressed or stored is zero, when all bits are 0, and the largest number is 15, when all bits are 1. Altogether, 16 values can be recorded. Similarly, with 8 bits, the smallest number is zero and the largest number is 255. This implies that only 256 values are possible. These are 255 positive integer values plus zero.

Example 22.9

Determine with 64 bits how many numbers can be stored.

Solution

The smallest number (negative numbers are not yet considered) is 0, when all bits are 0. The problem is finding the number when all bits are 1. We can use the same procedure used to find the decimal equivalent of a binary number, that is,

$$11...11_b = 1 \times 2^{63} + 1 \times 2^{62} + ... + 1 \times 2^1 + 1 \times 2^0$$

But, this is tedious and time-consuming. Consider what was said earlier about 1111_b and 10000_b. That is, the number with all bits equal to 1 is one less than the next number with only the most significant bit being 1 and all the other bits being 0.

Thus, we have

$$\begin{aligned} 11...11_b = 1 \times 2^{64} - 1 &= 18446744073709551616 - 1 \\ &= 18{,}446{,}744{,}073{,}709{,}551{,}615 \end{aligned}$$

You may ask where such a large number is used. Remember that negative and fractional numbers need to be handled, too. Computers use 64 bits, but for industrial use sometimes even 4 bits may suffice for very simple applications.

In practice, we often have negative values, particularly in measurements. We continue this discussion to see how negative binary numbers can be stored. For dealing with negative binary numbers it is crucial to adopt a fixed number of bits. If not, numbers cannot be interpreted correctly. Normally, a commercial device comes with a fixed bit size, and the greater the number of bits, the more expensive the device is.

22.3.2 Representation of Negative Binary Numbers

For simplicity and better clarity, in what follows only 8 bits are used. With 8 bits the total numbers that can be addressed is limited. Dealing with larger numbers requires larger bit size. Nevertheless, in many small applications, 8 bits is sufficient.

When an 8-bit device is in use we need to fill all the places in the register with 0's and 1's. In other words, for each bit a value must be defined and it cannot be left without a value. For instance, for storing the decimal number 86 in a binary form it was found that (see Section 22.2) 86 = 1010110_b. This number has only 7 digits and occupies 7 bits in a register. To represent this number with 8 bits the remaining bit on the left is filled with a 0 and thus 86 is represented by $0101{,}0110_b$. For any other number in the same way all remaining bits from left are filled with 0's.

In computer terminology, positive numbers are referred to as **unsigned numbers**, and when we have both positive and negative numbers, then we need to use **signed numbers**. With 8 bits there are 256 possible numbers. For any application an initial decision must be made if these 256 possibilities are to be used for the range of 0 to 255 unsigned numbers, or half are negative and half positive and, therefore, the 256 positions must accommodate 256 signed numbers.

Unsigned number: Binary number representing a positive value, as opposed to a signed number, which can represent a positive or negative value.

Signed number: Binary number that can be positive or negative. Sign (of the number) must be embedded with the bits representing the number.

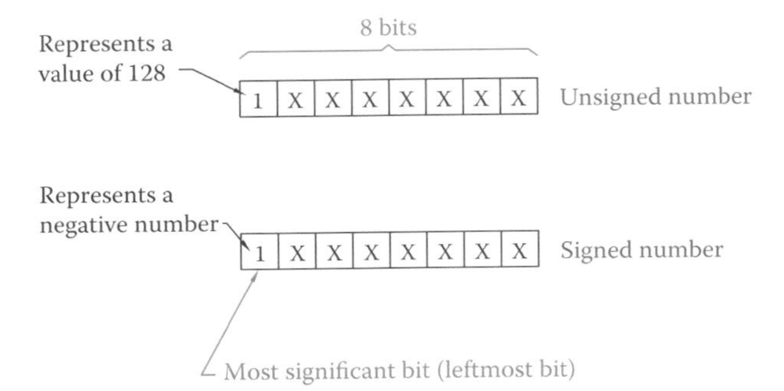

Figure 22.3 Signed and unsigned 8-bit numbers.

In unsigned numbers (for 8 bits) the most significant bit has a value of $2^7 = 128$, whereas in signed numbers this bit represents the negative sign. Figure 22.3 shows this fact. The unsigned numbers start from 0 and end with 255, and are as follows:

$$
\begin{aligned}
&0000,0000 = 0\\
&0000,0001 = 1\\
&0000,0010 = 2\\
&\ldots\\
&1111,1110 = 254\\
&1111,1111 = 225
\end{aligned}
$$

and the signed numbers are from −128 to +127, as follows:

$$
\begin{aligned}
&1000,0000 = -128\\
&1000,0001 = -127\\
&1000,0010 = -126\\
&\ldots\\
&1111,1111 = -1\\
&0000,0000 = 0\\
&0000,0001 = +1\\
&\ldots\\
&0111,1111 = +127
\end{aligned}
$$

Below we see some examples of signed numbers. Pay attention to the fact that if a positive number and a negative number of the same quantity are added, the resultant is zero. Note and compare the numbers having similar bits except for the most significant bit.

Unsigned:

0000,1100 = 12

0010,1100 = 44 (32 + 12)

0110,1100 = 108 (64 + 44)

$$1110{,}1100 = 236 \quad (128 + 108)$$

$$1111{,}1111 = 255$$

Signed:

$$0010{,}1100 = 44 \quad \text{(same as above)}$$

$$0110{,}1100 = 108 \quad \text{(same as above)}$$

$$1\boxed{110{,}1100} = -20 \quad (-128 + 108)$$

$$1\boxed{101{,}0100} = -44 \quad (-128 + 68)$$

$$1\boxed{111{,}0100} = -12 \quad (-128 + 116)$$

$$1\boxed{111{,}1111} = -1 \quad (-128 + 127)$$

Although a way to find the value of a negative number is shown above (in the brackets), this is not the best way. Also, we want to find the negative of a given number. There is a rule for finding the negative of a binary number. This is discussed under one's and two's complement in Section 22.4.

22.3.3 Precision in Analog-to-Digital Conversion

Today, in many industries, microprocessors are used for control of processes and plants. Examples of that are plenty. Today's cars are equipped with a microprocessor that controls the performance of the engine by measuring a number of parameters, such as air temperature, car speed, position of the gas pedal, and other pertinent parameters. Likewise, the operation of a wind turbine, steam turbine, and gas turbine is supervised and controlled by a microprocessor based on many parameters that need to be measured.

A microprocessor is the brain of any computer and any equipment with digital control. There are many different microprocessors with different bit size and different capabilities. In all microprocessor-based control systems, when an analog quantity is measured, it must be converted to its digital equivalent, so that it can be read or stored digitally or better, say, by a binary number.

The output of most measurement devices is in the form of a voltage. For some variables like wind speed this value is always positive (we do not have negative wind speed) and for some other ones, like temperature this can be positive or negative. According to what was said before, a signed or unsigned number can be decided on based on the requirement.

Suppose that the output of a measurement device is a voltage that can vary between 0 and 10 V. Considering an 8-bit converter, there are 256 possibilities for the values that can be recorded, processed, and displayed. In this case the 256 numbers correspond to 10 V range. Because there are 255 intervals (between 0 and 255), each digital number corresponds to 10/255 V, which is approximately 0.039 V. This implies that within the converted digital values the difference between any number and the number next to it is 0.039 V. This value of 0.039 V is the resolution of the representation of 10 V by 8 bits. It determines the precision of the operation. Samples of the binary values and their corresponding voltages for this example are shown below:

0000,0000	0 V	
0000,0001	0.039 V	(1 × 0.039)
0000,0011	0.117 V	(3 × 0.039)
0011,1100	1.95 V	(50 × 0.039)
1000,0000	5.0 V	(128 × 0.039)

The above numbers show that the difference between any two consecutive numbers that can be displayed through a digital display is 0.039 V. For instance, considering the two first numbers, a measured value can be 0 or 0.039 V, and nothing in between. With 8 bits it is not possible to have a better precision. To have a higher precision, either the voltage range must be smaller or the number of bits (bit size) must be increased, as shown in the following examples.

Example 22.10

Determine how much is the precision in the conversion of 5 V range by an 8-bit system.

Solution

Divide the 5 V range by 255 possible graduations.

$$5\ \text{V} \div 255 = 0.0196\ \text{V} = 19.6\ \text{mV}$$

The precision in this case is twice that of the previous case.

In analog-to-digital conversion the higher the number of bits, the higher the precision of conversion.

Example 22.11

Determine the precision if a 10-bit converter is used for conversion of numbers between −5 and +5 V.

Solution

With 10 bits the possible number of binary values is $2^{10} = 1024$ (0 to 1023); that is, 4 times more than with 8 bits. Also, note that for the range of values between −5 and +5 V, we have 10 V.

Following the same procedure as in the previous example, we come up with

$$10 \div 1023 = 0.00977\ \text{V} = 9.77\ \text{mV}$$

Note that for this example it is necessary to use signed numbers to be able to accommodate and display the negative values.

Precision determines the maximum error that can exist due to conversion to digital values. Only those values that are a multiplier of 0.00977 can be represented by their real value. For other numbers, there exists an error. For instance, consider the real value of a voltage to be 2 V. Also, consider the following binary numbers and their corresponding decimal values

$$00{,}1100{,}1100_b \quad (204) \to 204 \times 0.00977 = 1.99308$$
$$00{,}1100{,}1101_b \quad (205) \to 205 \times 0.00977 = 2.00285$$
$$00{,}1100{,}1110_b \quad (206) \to 206 \times 0.00977 = 2.01262$$

The nearest binary number corresponding to 2 V is the middle one, $00{,}1100{,}1101_b$ but is not exact (the exact value is 204.8 × 0.00977, which cannot be addressed by a binary number here). The error in this case is 0.00285 V (2.85 mV); the error may be larger than this, but not greater than 9.77 mV.

Conversion from analog to digital values contains inevitable error. Only some of the digital values that will be displayed represent the true measured value without error.

22.4 One's and Two's Complement

For the analysis of practical applications it is necessary to do mathematical operations on binary numbers. This is the basis of all that a microprocessor does. In this section we study how two binary numbers are added, how one is subtracted from the other. We also discuss how the negative of a binary number is formed.

22.4.1 Addition

The addition of two binary numbers is performed in the same way as we add two decimal numbers with the exception that the only nonzero element in the set of basic elements of the binary number system is the number 1. Thus, in the same way that in decimal numbers if we add 1 to 9, the answer is 10, in binary numbers if we add 1 to 1, the answer is 10. That is, the addition of 1 and 1 is a 0 with a 1 carried over to the next digit on the left. Also, practically, which is the way this operation is performed in microprocessors, only two numbers are added together at a time. We consider the following three examples. Assume unsigned numbers.

$$\begin{array}{r} 10100011\,+ \\ 00101010 \\ \hline 11001101 \end{array} \qquad \begin{array}{r} 10100011\,+ \\ 10100011 \\ \hline 101000110 \end{array} \qquad \begin{array}{r} 01010101\,+ \\ 11111110 \\ \hline 101010011 \end{array}$$

Because any binary numbers is practically the contents of a register, only a limited number of bits are available. This is because of the hardware, which determines the bit size to be 8, 10, 12, or some other. In the result of adding two numbers you see that the very last digit on the left is shown in red, when it happens to exist. This digit (bit) is ignored. There is physically no place for this bit to go to. It is either not significant (in a subtraction operation, as you will see next), or it is an indication of **overflow** (as shown above). Overflow is a situation where a number is larger than the capacity of the machine. When there is an overflow situation, the last bit is ignored, but it signals out a flag. A **flag** is an indicator of an undesirable or fault situation in digital systems. A message will be issued when a flag is signaled. An example is when you have a division by zero in your calculator.

Overflow: State of reaching a value beyond the capacity of a digital device in terms of the number of bits.

Flag: Representation of the occurrence of of some unusual or unwanted situation in digital circuits and computer terminology.

22.4.2 Subtraction

To subtract a number B from number A the common method is to add A to the negative of B.

$$A - B = A + (-B)$$

The subtraction, then, changes to an addition operation. Then, in the same way as discussed above the two numbers are added.

Consider for example subtraction of 44 ($0010,1100_b$) from 108 ($0110,1100_b$) in binary operation. Instead, we add −44 ($1101,0100_b$) to 108

```
 01101100 +
 11010100
---------
101000000
```

You can easily check that 108 − 44 = 64, and the resultant binary number ($0100,0000_b$) is 64. Also, you can verify the number for −44 to be correct. If it is added to 44, the resultant is 0.

```
 01101100 +  (44)
 11010100   (−44)
---------
100000000
```

Note that in both cases the very last digit, shown in red, does not play any role and is ignored.

22.4.3 One's Complement of a Binary Number

The **one's complement** of a binary number is used to find the **two's complement** of a number, which is used to find the negative of that number. Note that negatives are valid with signed numbers. The one's complement of a number is obtained by switching each bit to its opposite value, 1's to 0's and 0's to 1's. For example, the one's complement of 0110,1010 (106) is 1001,0101.

One's complement: Result of swapping each bit value to its complement in a binary number.

Two's complement: A binary number obtained by adding 1 to the one's complement of a binary number.

22.4.4 Two's Complement of a Binary Number

Two's complement of a binary number is obtained by adding 1 to the one's complement of that number. For instance,

if one's complement of a number is	10010101
then the two's complement of it is	10010101+
	1
	10010110

The two's complement defines the negative of a number. Therefore, the negative of $0110{,}1010_b$ is $1001{,}0110_b$. We can check this by adding the two numbers together.

$$\begin{array}{r} 01101010+ \\ 10010110 \\ \hline 100000000 \end{array}$$

Example 22.12

Find the negative of the signed binary number 1110,0111.

Solution

First, note that this number is already negative because its most significant bit is 1. The negative of a negative number is positive. No matter if the number is initially positive or negative, the two steps to be taken are

Step 1. Find the one's complement of the number
One's complement of 1110,0111 is 0001,1000.
Step 2. Find the two's complement of the number (or, add 1 to the one's complement)

$$0001{,}1000_b + 1 = 0001{,}1001_b$$

The answer, thus, is 0001,1001.
Verifying the result:

$$\begin{array}{r} 11100111+ \\ 00011001 \\ \hline 100000000 \end{array}$$

The given numbers was –25. You cannot say that easily just from the number directly by looking at the digits, except that one has a look-up table showing all the numbers in the range. Figure 22.4 shows some numbers between –128 and –1. You may notice a pattern to determine the numbers for 8 bits as shown in the table. On the basis of this pattern, find

−127	1	0	0	0	0	0	0	1	1−128
−126	1	0	0	0	0	0	1	0	2−128
−125	1	0	0	0	0	0	1	1	3−128
−120	1	0	0	0	1	0	0	0	8−128
−64	1	1	0	0	0	0	0	0	64−128
−32	1	1	1	0	0	0	0	0	96−128
−28	1	1	1	0	0	1	0	0	100−128
−20	1	1	1	0	1	1	0	0	108−128
−10	1	1	1	1	0	1	1	0	118−128
−3	1	1	1	1	1	1	0	1	125−128
−2	1	1	1	1	1	1	1	0	126−128
−1	1	1	1	1	1	1	1	1	127−128

Represents a negative number

Figure 22.4 Negative binary numbers.

the decimal magnitude of the binary number leaving out the leftmost bit, and subtract 128 from that number. For other bit numbers the value to subtract (128 in this case) must be changed accordingly.

> The two's complement of a binary number defines the negative of that number.

22.4.5 Multiplication and Division of Unsigned Binary Numbers

Multiplication of binary numbers is very easy. This is the way multiplication is performed in a microprocessor. First, consider multiplication by 10_b and 100_b, which is, in fact, multiplication by 2 and 4. Consider $0011{,}1011_b$ (always assume a limited number of bits, here 8). Assume an unsigned number.

$$0011{,}1011_b \times 10_b = 0111{,}0110_b \quad (59 \times 2 = 118)$$
$$0011{,}1011_b \times 100_b = 1110{,}1100_b \quad (59 \times 4 = 236)$$

As in the case of decimal numbers 1 or 2, 0's are added to the right side of the number for multiplication by 10_b and 100_b, respectively. Because the number of bits must remain the same, this operation is like bringing the 0's from the left hand to the right hand. Figure 22.5 depicts this. This operation of moving 0's to the right is called **rotation**.

Rotation: (in binary number operations) Operation of moving one or more leftmost zeros in a binary number with a limited number of digits to the far right. This implies multiplying that number by 10_b, 100_b, etc.

Rotation to the right is performed for multiplication, 1 bit for multiplication by 10_b, 2 bits for multiplication by 100_b, and so on. Note that here if multiplication is continued, then eventually overflow happens when the result becomes larger than a machine can handle (here because there are only 8 bits, not many numbers can be handled, but in a machine with 128 bits the operation can go on much further).

For multiplication by numbers like 3 (11_b) and 6 (110_b) and so forth, these numbers are broken into their components. For example,

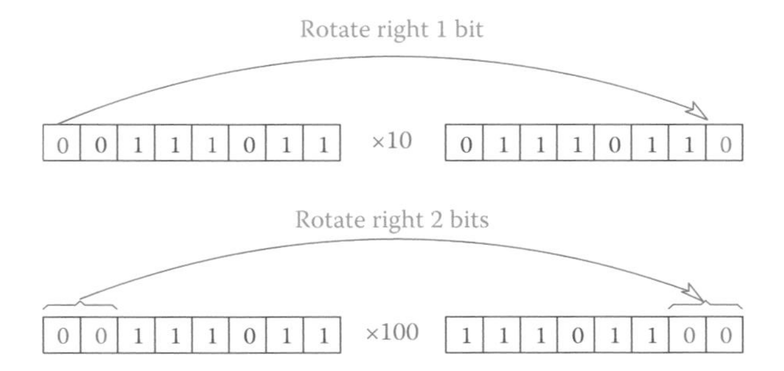

Figure 22.5 Multiplication of a number by 10_b and 100_b.

$$11_b = 10_b + 1$$

Thus, multiplication of a binary number by 11_b includes multiplication of that number by 10_b, as above, and multiplication of the number by 1(multiplication by 1 is trivial and does not change the number), and adding the results. Similarly, multiplication by 110_b includes multiplication by 100_b plus multiplication by 10_b,

$$110_b = 100_b + 10_b$$

both of which we have already seen. The results of the two multiplications then must be added together. Multiplication by all other numbers can be treated the same way.

Example 22.13

Multiply the 10-bit binary number $00{,}0110{,}0111_b$ by 7.

Solution

Multiplication of the number by 7 consists of multiplication by 4 + 2 + 1. Because $7 = 111_b$, then the number must be multiplied by 100_b, multiplied by 10_b, and multiplied by 1_b followed by adding all the results.

$$00{,}0110{,}0111_b \times 100_b = 01{,}1001{,}1100_b$$
$$00{,}0110{,}0111_b \times 10_b = 00{,}1100{,}1110_b$$
$$00{,}0110{,}0111_b \times 1_b = 00{,}0110{,}0111_b$$

The final result is

$01{,}1001{,}1100_b + 00{,}1100{,}1110_b + 00{,}0110{,}0111_b = 10{,}1101{,}0001_b$

You may want to check the results with decimal numbers. The initial number was 103, and the final answer is 721, which is 103 × 7.

In conjunction with multiplication is division. For division, numbers are shifted right. **Shift right** is shown in Figure 22.6. In dividing integer numbers, round-off errors can occur, as can be seen in the figure.

Shift right: (in binary number operations) Operation of moving all digits of a binary number (with a limited number of digits) to the right by one or more places. The left-most places are then filled with zeros. This implies dividing that number by 10_b, 100_b, etc. The rightmost digits are moved out and introduce the round-off error that occurs in division.

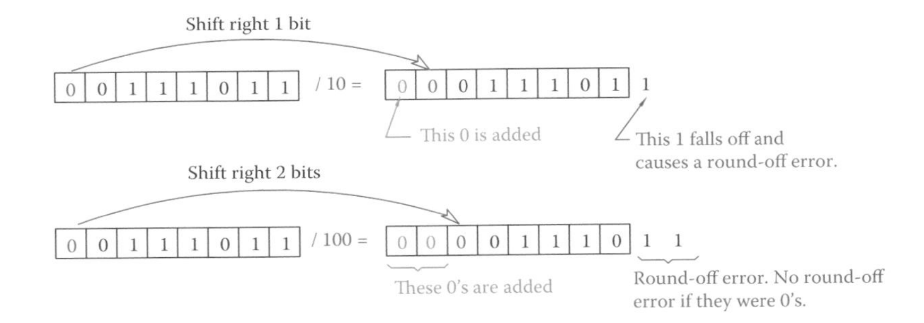

Figure 22.6 Division in binary unsigned numbers.

Example 22.14

Multiply the 12-bit number $0000{,}0110{,}1101_b$ by 1011_b.

Solution

We see that the multiplier consists 4 bits as follows:

$$1 \times 2^3 + 1 \times 2^1 + 1 \times 2^0 = 1000_b + 10_b + 1_b$$

Therefore, the given number must be multiplied by 1000_b and 10_b and 1, and the results be added together.

For multiplication by 1000_b, rotate right 3 bits, for multiplication by 10_b, rotate right 1 bit and for multiplication by 1 do not change the number:

Rotate right 3 bits	0011, 0110, 1000 +	
Rotate right 1 bit	0000, 1101, 1010	
Add together	0100, 0100, 0010	(intermediate result)

Multiplication by 1	0000, 0110, 1101	
Add to the previous result	0100, 0100, 0010	
Result	0100, 1010, 1111	$(2^{10}+2^7+2^5+15=1199)$

We can verify the result. The original number is 109 and the multiplier is 11. The result of multiplication must be 109 × 11 = 1199, which is true.

22.5 Number Display and BCD

For digital display of numbers, for example, in a digital watch, a DMM, and many other devices a seven-segment display, matrix display, or other display types are used. In all of these, every digit in a number is handled independent of the other digits. For instance, 216 is not treated as two hundred and sixteen, but as 2, 1, and 6. This is completely acceptable and widely practiced because no mathematical operation is involved in displaying a number.

In displays, all digits are not always needed to be present. In a digital clock, for instance, the leftmost digit must be able to show only 1 (and 2 if a 24 hours format is to be used), and the left digits for minutes and seconds must only show figures 0 to 6; other digits must be able to display 0 to 9. In each particular case, if some unnecessary components can be removed, this is normally done by manufacturers.

For handling numbers 1 and 2 by a binary device, only 2 bits are enough (2 bits can represent four different states or four entities), and for handling 0 to 6 only 3 bits suffice. Three bits can handle up to 8 values (0 to 7). This is one reason for using octal numbers for some practical applications. In a more general case where all digits vary between 0 and 9, 4 bits are necessary for their binary representation. Although up to 16 numbers can be addressed by 4 bits, only 10 of them are utilized (0 to 9) and the rest are unused but are always available. It is clear that for addressing 10 values, 3 bits are definitely not sufficient.

Binary-coded decimal (BCD) refers to treating a decimal number as a set of independent binary numbers corresponding to each individual digit in the number. As an example, the number 192 consists of a 1, a 9, and a 2. It is not before 193 and it is not after 191. Each individual digit, in this way, needs 4 bits to be addressed. For a three-digit decimal number, then, 12 bits are necessary. The binary-coded decimal numbers corresponding to 1, 9, and 2 in 192 are

Binary-coded decimal (BCD): Special representation of decimal values when each individual digit is coded separately in binary form and can be displayed by digital circuits.

$$0001,\ 1001,\ 0010 \quad (1, 9, 2)$$

Compare the above number with that representing (unsigned) 192, which requires only 8 bits.

$$192 = 1100{,}0000_b$$

If we add 8 to 192, the result is $200 = 1100{,}1000_b$, but adding 8 to the BCD number has no meaning (0001,1001,1010 BCD has no magnitude content). The binary-coded decimal for 200 is

$$\begin{array}{ccc} 0010, & 0000, & 0000 \\ \uparrow & \uparrow & \uparrow \\ 2 & 0 & 0 \end{array}$$

Example 22.15

Find the BCD number for 1376.

Solution

We need to put the binary equivalents of the digits 1, 3, 7, and 6 together and in the same order.

The answer is

$$\begin{array}{cccc} 0001, & 0011, & 0111, & 0110 \\ \uparrow & \uparrow & \uparrow & \uparrow \\ 1 & 3 & 7 & 6 \end{array}$$

22.6 Sampling Analog Variables

In the processes or machinery that are controlled by a microcontroller or microprocessor, as we learned earlier in this chapter, all analog data must be converted to digital, so that it can be represented by binary numbers and be usable by the microprocessor. We learned about precision in converting analog numbers to their digital representation and its relationship with the number of bits.

When a variable entity such as wind speed, car speed, or electric current in a motor is measured, the value obtained is for only the moment the measurement is performed. In a continuous process the measured value is used for one or more necessary actions to be taken. For example, when a car is driven, the position of the gas pedal depressed by the driver is used to adjust car speed.

Sampling: Action of measuring a variable in a process or machine repeatedly at a fixed frequency, for control purposes.

Sampling rate: Same as the sampling frequency. The (normally fixed) frequency at which a sampling action is performed.

Sampling frequency: The number of samples taken (or measurements made) in 1 second in a sampled data system. This is for converting analog values to digital values for digital control, computer control, display of values, and so on.

Sampling time: Time interval between each two consecutive measurements in a sampling action.

22.6.1 Sampling

In a continuous process any measurement must be frequently repeated for updated values. Therefore, for microprocessor control of a device another factor comes into picture: the rate at which the data are updated.

This updating of measured values is called **sampling** and how often a variable is updated is determined by the **sampling rate**. The sampling rate can also be called **sampling frequency**, which implies how many times in one second the variable is measured. The time interval between each two measurements is called the **sampling time**.

For example, if in each second a variable is measured 1000 times, the sampling rate is 1000 per second, and the sampling time is 0.001 second. Depending on how important a role a parameter has in a process, this updating of values must be performed at a higher sampling rate.

In Figure 22.7 a variable is sampled between times A and B. The variable is assumed to remain constant at the measured value for the total period of the sampling time until the next measurement is performed. In reality, it is more likely that the value of a variable changes even if the time interval is very small. For example, the value of the parameter at M is changing to that at N, whereas in the microcontroller it is kept constant until the new reading comes in, which implies a jump from N′ to N. It is thus obvious that more precision in measurement and control can be

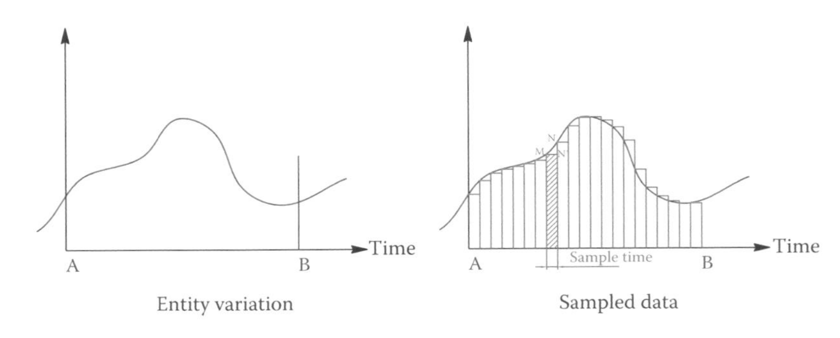

Figure 22.7 Sampling a variable parameter.

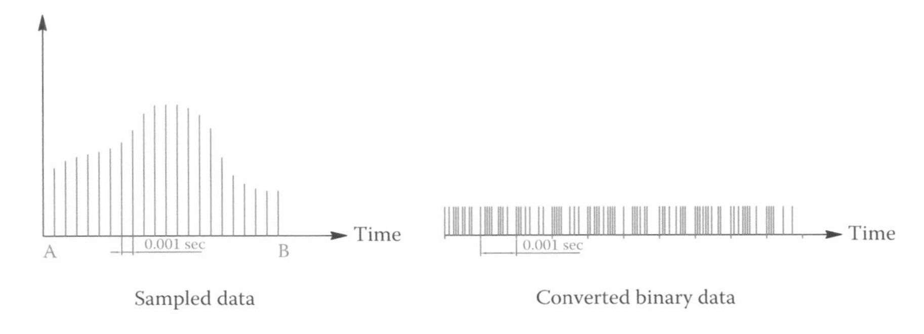

Figure 22.8 Sampled data and their binary conversion.

achieved by having a smaller sampling time. The measured value of any variable is stored by a circuit called **sample and hold**.

Sample and hold: Action of measuring a variable and storing the measured value for use. This is common in computer control systems where measurements are performed repeatedly (but not continuously; say at 1 msec intervals) at a proper fixed frequency.

Nowadays, fast sampling, for instance, one million samples per second (a sampling time of 1 microsecond) is possible for many measurements. This implies that the necessary time for measuring the variable under question, converting its value to digital and storing the result can be less than 1 millionth of a second. It must be noted that, in general, the higher the sampling rate is, the higher the price of the sampling equipment.

Increasing the sampling rate is not always necessarily essential, useful, or worth the cost even if possible. For example, consider the pitch angle of a wind turbine. If it takes 1 sec for the pitch angle to change (through its mechanism), any measurement faster than 1 per second does not have any useful effect. Also, if the measurement itself takes a long time, or the process is slow, such as heating a liquid, there is no point in sampling at a high rate. It can only add to the cost.

After sampling is performed at any instant, the measured value is converted to its binary form. This is repeated after each sampling time interval. Figure 22.8 shows a typical example of sampled data and its conversion to binary data (the binary conversion is only for demonstration).

22.6.2 Clock Signal

Each microprocessor has a clock to regulate its operation. Each and every single operation, such as adding two bits of data, is performed with the clock pulse. Each clock pulse is a step for synchronized work of all components (for reading new values, conversions, mathematical operations, etc.) and all operations progress step by step. A clock signal is a square pulse train with a fixed frequency. The clock rate can be decreased, and slower pulse trains be made from it for slower processes.

Rising edge: Same as positive going edge and leading edge in a square waveform.

Falling edge: Edge in a square waveform signal where the voltage drops from a positive value to a negative value (or to zero), as opposed to the rising edge.

Figure 22.9 shows the pulse train of the clock signal and two other signals with half and one quarter of the clock frequency. Each pulse has a **rising edge**, a **falling edge**, and a level part. A device can be sensitive to the rising edge or the falling edge, and, consequently, its function takes place with that part of a pulse. Some devices are level sensitive.

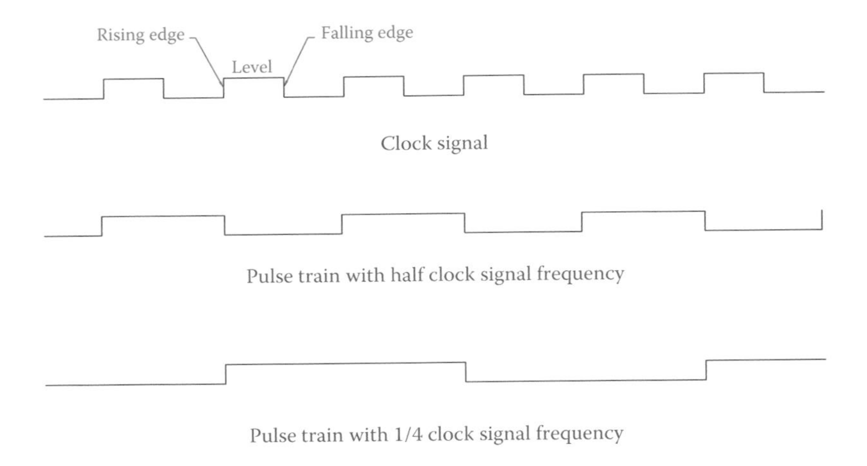

Figure 22.9 Representation of fixed frequency pulse signals.

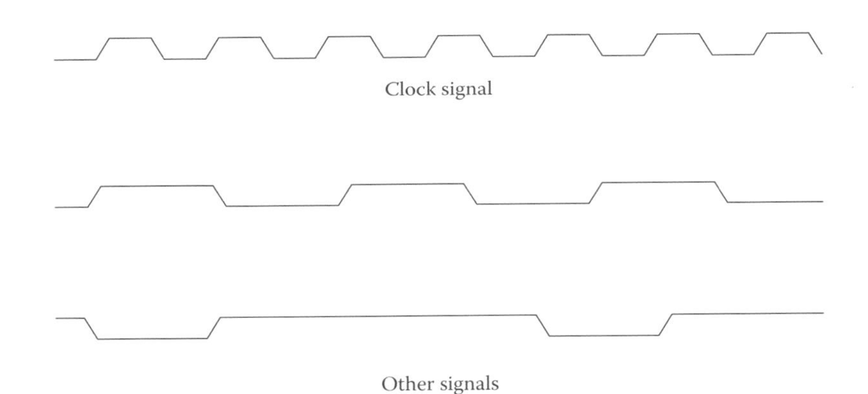

Figure 22.10 Representation of clock and other signals in a microprocessor.

Leading edge: Same as rising edge in a square waveform.

Positive going edge: Rising edge in a square wave where the magnitude sharply goes positive from negative or zero.

Trailing edge: Same as negative going edge, which is that edge of a rectangular waveform that the value drops from positive to negative or zero.

Negative going edge: Same as falling edge or trailing edge (changing from a positive value to a negative or zero value in a square wave).

A rising edge is also called a **leading edge** or **positive going edge**. A falling edge is also called a **trailing edge** or a **negative going edge**. A pulse train continuously switches between zero and a positive value. We refer to these values as low and high.

Although microprocessors are very fast and their clock frequencies are in the range of MHz (million hertz), the change from a low level to a high level and vice versa is not instant, and, thus the time for switching (duration of the rising edge and falling edge) is not zero. In this sense, the actual clock signal and other signals are represented as shown in Figure 22.10.

22.7 Chapter Summary

- In addition to the decimal number system that we are accustomed to, there are binary, hexadecimal, and octal number systems for use in digital technology.

- The base in decimal numbers is 10, the base for binary, hexadecimal, and octal systems are 2, 16, and 8, respectively.
- In the same way that in the decimal system the symbols for base numbers are from 0 to 9, in the binary, hexadecimal, and octal systems the base numbers are 0 to 1, 0 to 15, and 0 to 7, respectively.
- In the hexadecimal number system for showing numbers 10 to 15 the symbols A, B, C, D, E, and F are used. F is the last base number; the number after that is 10_{16}. After FF is 100_{16}, after FFF is 1000_{16}, and so forth.
- The binary number system contains only 0 and 1. The number after 1 is 10_b, after 11_b is 100_b, and so on.
- The octal number system has 8 digits, from 0 to 7. The number after 7 is 10_8, after 77_8 is 100_8, and after 777_8 is 1000_8, and so on.
- Any single digit in a binary number is called a bit. An ensemble of 8 bits is called a *byte*, and two bytes make a *word*.
- In a binary number the leftmost bit is called the most significant bit (MSB) and the rightmost bit is called the least significant bit (LSB).
- A microprocessor is the brain of a digital computer. Customized microprocessors for small control systems are called microcontrollers.
- An analog-to-digital converter (ADC or A/D converter) translates an analog value to a digital value.
- A digital value is made out of a finite number of 0's and 1's. This finite number is based on the number of bits in a device. For a very simple application it can be 4, but in today's laptops it is 64 and for larger computers it is at least 128.
- A digital-to-analog converter (DAC or D/A converter) translates a digital value to an analog value.
- In conversion from an analog (measured) value to digital the number of digits is limited by the physical electronic device. The number of digits defines the precision of conversion.
- All positive integer numbers in a computer application are referred to as unsigned numbers.
- All integer numbers (negative and positive) fall in the category of signed numbers.
- One's complement of a binary number is obtained by changing all 0's to 1's and vice versa.
- Two's complement of a binary number is obtained by adding 1 to the one's complement of that number.
- In the binary system the negative of a number is represented by its two's complement.
- In practice, because the number of bits are fixed, the most significant bit determines if a signed number is positive or negative. A 1 as the MSB signifies a negative number.
- For digital control of a machine or a process the entities that need measurement are sampled at a constant sampling rate. Sampling rate is the frequency of repeated measurement at "sampling time" intervals.

- Any microprocessor or microcontroller works in steps. All mathematical, logical, and conversion operations are performed one step at a time. The steps are synchronized by a clock signal.
- All signals in a microprocessor are in the form of square wave or pulses, where each pulse has a rising edge, a level and a falling edge.

Review Questions

1. What are hexadecimal and octal number systems?
2. How many basic symbols are there in the binary number system?
3. How many base numbers are required for hexadecimal numbers?
4. What is the importance of binary numbers?
5. What is the use of octal numbers?
6. What do LSB and MSB stand for?
7. What is a microcontroller?
8. What are meant by overflow and flag in microprocessor terminology?
9. What does BCD stand for and what does it mean?
10. What do DAC and ADC mean?
11. What is the difference between signed and unsigned numbers?
12. If the most significant bit in a binary number is 1, what does it imply?
13. What is meant by two's complement of a number?
14. What is a D/A converter used for?
15. What is the result of $1_b + 1_b$?

Problems

1. Convert the following unsigned binary numbers to their decimal values:
   a. 001011110001
   b. 101011010110
   c. 001100111011
   d. 100010110011
   e. 111100011101
   f. 100111000110
   g. 110110010110
   h. 100000100001

2. Find the decimal values of the following signed binary numbers:
   a. 111000111001
   b. 100010111010
   c. 111100011011
   d. 110001011001
   e. 100001000011

3. Convert all the unsigned numbers given in Problem 1 to their hexadecimal representation.

4. Write the unsigned numbers in Problem 1 in their octal equivalents.

5. Convert the following numbers to binary. Use 12 digits.
   a. 18
   b. 25
   c. 132
   d. 1456
   e. 2333
   f. 4000

6. Write the one's complement and two's complement of the following numbers:
   a. 0001110011
   b. 0010110111
   c. 0010100011
   d. 0000010110
   e. 0001001101
   f. 0011001100

7. Find the binary numbers for the negative of those in Problem 5.

8. Find the octal values of the following decimal numbers:
   a. 22
   b. 33
   c. 65
   d. 178
   e. 4355

9. Find the decimal values of the following octal numbers:
   a. 105_8
   b. 231_8
   c. 427_8
   d. 177_8
   e. 7777_8
   f. 1000_8
   g. 12345_8

10. Write the following numbers in hexadecimal:
   a. 13
   b. 198
   c. 522
   d. 1292
   e. 2020
   f. 10000
   g. 99999

11. Convert the following hexadecimal numbers to decimal:
   a. 155
   b. 2A3B
   c. AABB
   d. 10001
   e. CDA00
   f. 1A5FF
   g. FFFF

12. Convert the numbers in Problem 11 directly from hexadecimal to binary.

13. Find the BCD number for the following decimal numbers:
   a. 16
   b. 415
   c. 840
   d. 2233

14. Write the decimal numbers corresponding to the following BCD numbers:
   a. 1001,0110,0101,1000
   b. 0001,0011,0001,1001

15. If an 8-digit converter is used for a sensor with 2.5 V range, what is the smallest nonzero voltage that can be displayed?

16. For measurement in the range of 0 to 10 V, how many bits are necessary to have a minimum resolution (or precision) of 10 mV?

17. Add the numbers given in Problem 6. Hint: Add 2 numbers at a time.

23

Logic Circuits and Applications

OBJECTIVES: After studying this chapter, you will be able to

- Describe logic gates
- Define the meaning of the logic gates AND, OR, NOT, NAND, NOR, and XOR
- Understand the symbols for the logic gates
- Explain the use of logic gates in digital processes
- Demonstrate the use of logic gates with timing diagrams
- Define the logic functions for AND, OR, NOT, NAND, NOR, and XOR gates
- Put logic gates together to form logic circuits
- Simplify logic circuits and determine the logic function of a given circuit
- Describe the laws of Boolean algebra and employ them for logic circuits
- Use truth tables
- Explain what the product of sums and sum of products are and what their purposes are
- Define what K-maps are and how they can be used
- Use K-maps to define logic function of practical problems

New terms:
Boolean algebra, complement, Karnough map, K-map, logic gates, product of sums, sum of products, timing diagram, transistor-transistor logic (TTL), truth table

23.1 Introduction

In Chapter 22 we learned about the mathematics behind the microprocessor and digital circuits and digital control. In this chapter we study the way all these are implemented and put to work. In many applications a number of conditions must be satisfied before an action is taken. For instance, in a camera, for a picture to be taken all the following conditions must be satisfied: (1) sufficient light, (2) picture in focus, (3) film in camera (for digital cameras a memory card instead of film), (4) flash ready, if it is called on, and finally (5) the button is pressed to take picture. A similar set of conditions must be met for any other application such as starting a car, starting a controlled electric motor, and starting a washing machine.

In the above examples it is necessary that all required conditions are true. There are, furthermore, other examples in which at least one of multiple conditions must be present. For example, in a more advanced camera, the right exposure is achieved by various combinations of shutter speed, aperture opening, and film speed (or its equivalent in digital cameras). In such an example, only one of many combinations is required, not all of them (various combinations are predefined in the form of a table look-up).

In a wind turbine, as another example, the adaptation with wind speed together with many other required conditions for the safe operation of the machine creates a complex situation. Consider decisions that must be reached in terms of when to stop or restart the turbine when wind speed is around the cut-in speed or change the blade pitch angle during the normal operation so that the output voltage is within its specified values while the maximum power is captured. All these and other pertinent conditions that must be satisfied define a great number of binary input parameters that interact with each other.

Logic circuits define the ways the essential conditions can be put together in the required manner. Also, if some conditions conflict with each other, are redundant, or can be represented in a different way, they can be identified by the rules of the logic circuits and their associated mathematics called **Boolean algebra**.

Boolean algebra: Number of rules (different from regular algebra rules) for logic gates and logic circuits.

23.2 Logic Gates

Logic gates: Number of special electronic devices exhibiting switching properties according to Boolean algebra.

A number of basic circuits, called **logic gates**, define the fundamental elements of logic circuits. These are AND, OR, NOR, NOT, XOR, and NAND. Pay attention to the fact that often a required condition can be expressed in two ways using opposite statements but leading to the same meaning. For instance, if for the correct execution of a process the temperature needs to be above 60°, the statement for this condition can be one of the following:

The temperature must be above 60°

or equivalently

The temperature must NOT be below 60°.

As mentioned in Chapter 22, for all signals used for binary applications and bit setting the terms *high* and *low* are used. These two refer to two values that are opposite to each other. On the basis of the application they can represent *true* and *false*. The result of action of a logic gate is always expressed in the form of true or false. The practical application of it is almost always in the form of a voltage.

The output of a logic gate is represented by two voltages that are used to define a distinct difference between 0 and 1. On the basis of the hardware used (e.g., type of transistor), *high* can be represented by zero voltage and *low* by a nonzero voltage. Throughout the chapter, when necessary, we assume true to be high and represented by 1, and false to be low and denoted by 0.

When using transistors, the voltage difference between *high* and *low* is nominally 5 V. But, in practice, anything below 1 V is acceptable as 0 V and anything above 3 V is considered to represent 5 V. The value of 0 and

5 V for the lower and upper ends of the spectrum is an accepted standard value for **transistor-transistor logic** interface and is normally addressed **TTL**.

Transistor-transistor logic (TTL): Standard voltage for connecting electronic circuits together.

Output of a logic gate is categorized as either true or false.

23.2.1 AND Gate

Imagine two conditions that must both be satisfied before some action takes place. This can be demonstrated by the two switches in series with each other in Figure 23.1a. The two conditions for this simple example can be (1) a photocell switch that turns on only when the ambient is dark and (2) a manual switch that is to be turned on.

An AND gate has two inputs and one output, which is the result of AND'ing action on the two inputs. For one gate, there are only two inputs, but if there are more conditions, one can combine AND gates. The inputs considered all through this chapter are only binary values that can be represented in the form of 0 and 1. Also, the output assumes only two values either 1 or 0. Through the chapter the two inputs for this and other gates are addressed, in general, as A and B, and the output is denoted by Q. These are the names for the inputs and output; hence, A, B, and Q can be 0 and 1 only. The output of an AND gate is 1 only if both inputs are 1. In other words, $Q = 1$ if $A = B = 1$.

Figure 23.1 also shows the symbol for an AND gate. There is another symbol illustrated. This is an ANSI (American National Standards Institute) standard symbol. Either of these symbols can be used in North America, but the first one is more common.

Figure 23.2 shows an application of the AND gate with two signals. The square wave diagrams represent the variation of an electric signal versus time. In the case shown, both A and B have constant frequency, the frequency of A being twice as that of B. At each instant of time each signal can have a value of 1 or 0, for high and low, respectively, and alters between these values. The gate output is 1 only when both signals are 1. Representation of signals as shown in Figure 23.2 are called a **timing diagram**. A signal can be periodic

Timing diagram: Diagrams that indicate the variation of logic parameter (change of state between 1 and 0) versus time in a logic circuit.

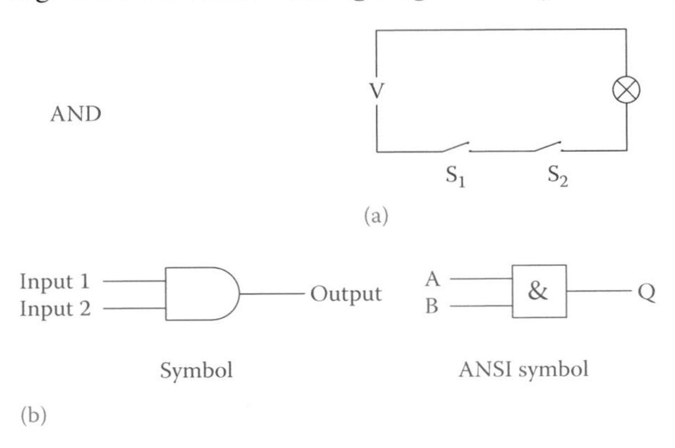

Figure 23.1 AND gate. (a) Representation of AND gate action. (b) Symbol for AND gate. (c) AND ANSI symbol.

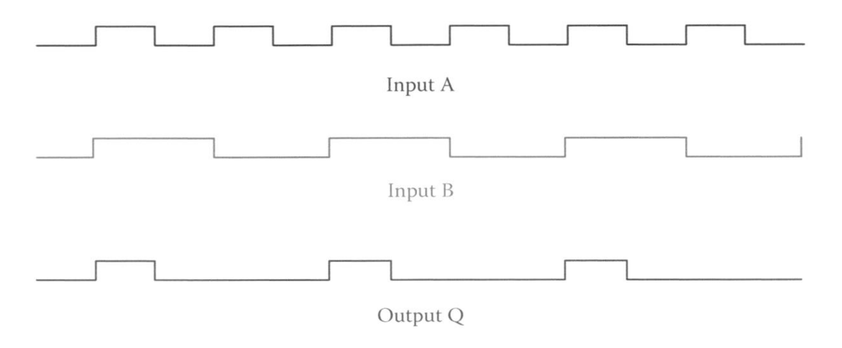

Figure 23.2 Applying AND gate to inputs A and B.

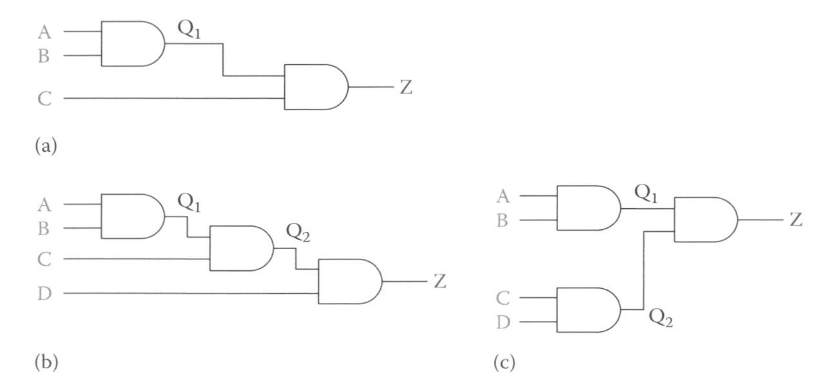

Figure 23.3 Combining AND gates for more inputs. (a) 3 inputs to be AND'ed together. (b) AND'ing 4 inputs. (c) Alternative way for AND'ing 4 inputs.

or random. The signal from a clock (a function generator) is periodic, but a signal from a sensor counting people that enter a library is random.

In Figure 23.3a two AND gates are combined to accommodate three conditions that must be simultaneously satisfied. More conditions can be accommodated by adding more gates. Nevertheless, as depicted in Figure 23.3b, and c, alternative ways are possible. In this case the number of gates in both arrangements is 3, but cases can be found where one arrangement requires fewer components than the other, and thus it is preferred. We learn about these cases later in this chapter.

The inputs in Figure 23.3 are denoted as A, B, C, and so forth and the final output is shown as Z. The value of *Z* depends on the value of each input. It is either 1 or 0. It is 1 when all the inputs are 1, and is 0 otherwise.

> The final output of AND'ing a number of inputs is 1 if all the inputs are 1.

23.2.2 OR Gate

Now we consider two conditions where if any of them is true an action takes place. Suppose that there are two generators feeding a manufacturing

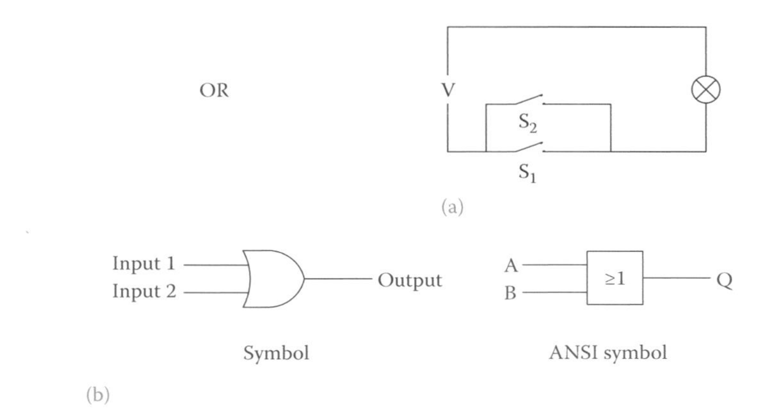

Figure 23.4 OR gate. (a) Representing application of OR gate. (b) Symbols for OR gate.

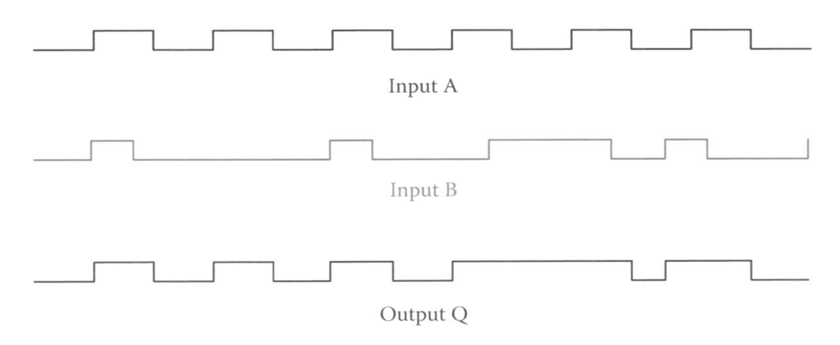

Figure 23.5 Applying OR gate to two inputs.

plant. If any of the two is in service, no matter which, then the plant has electricity. Of course, it is completely normal that both be in service. Nonetheless, if none is working, then the plant is without electricity.

The representation of this scenario can be seen in Figure 23.4a, where the two switches denote two conditional situations that are independent and only one is sufficient to be true for a desired purpose. The common symbol and the ANSI symbol for OR gate are shown in Figure 23.4b. The output of an OR gate is 1 if at least one of the inputs is 1.

The result of OR'ing two inputs is illustrated in Figure 23.5. For more than two inputs, OR gates can be combined together in the same way that was depicted for AND gates.

> The output of OR'ing a number of inputs is false if all inputs are false.

Complement: Opposite of a binary entity. It is shown by an overbar or a prime. For example, if A is a binary variable and its value is 1, its complement is represented by $\bar{A}$ or A' and its value is 0.

23.2.3 NOT Gate

The NOT gate is for swapping a quantity to its **complement**. Then 1 becomes 0 and 0 becomes 1 after going through a NOT gate. The NOT gate has only one input. It has one output, which is the inverse or complement of the input. The NOT gate and its symbols are shown in Figure 23.6.

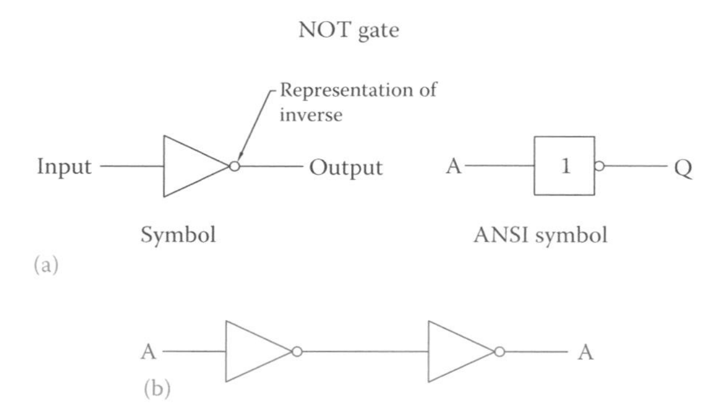

Figure 23.6 NOT gate. (a) Symbols for NOT gate. (b) Result of two consecutive NOT gates.

Note that the NOT of an already NOT gated entity is the entity itself (Figure 23.6b). The application of a NOT gate is obvious. A high signal becomes low, and a low signal becomes high.

There are two gates that result from combining AND and OR gates with a NOT gate. These are NAND and NOR.

23.2.4 NAND Gate

NAND is a separate gate but can be composed of two gates, one AND and one NOT gate. It implies Not AND; that is, the output of an AND gate is inverted. Figure 23.7 depicts the NAND gate and its symbols. The output of a NAND gate with two given inputs is shown in Figure 23.8. The output

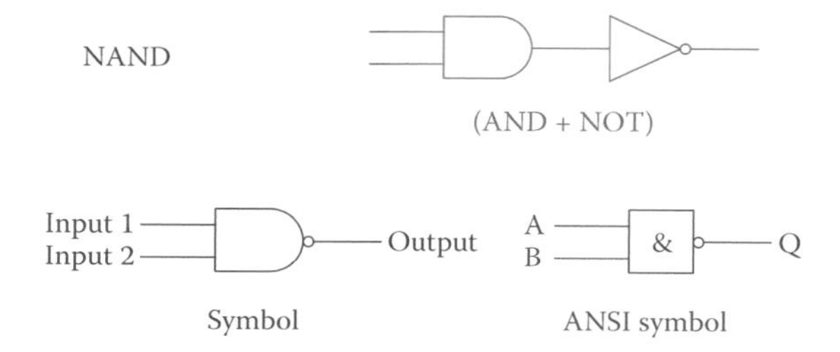

Figure 23.7 NAND gate.

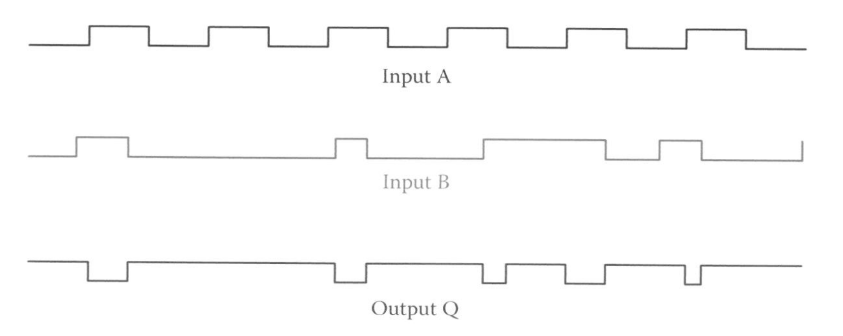

Figure 23.8 Output of a NAND gate.

is low when both inputs are high, which implies that the output is high when any input is low.

23.2.5 NOR Gate

In the same way that a NAND gate was the combination of NOT and AND gates, the NOR gate is made up of an OR gate and a NOT gate. It is shown in Figure 23.9.

The outcome of applying NOR to two signals is shown in Figure 23.10. In this figure the inputs *A* and *B* are the same signal, but with a delay (or phase shift) between them. Thus, they have the same frequency. A new signal with the same frequency is generated by NOR'ing the two signals.

23.2.6 XOR Gate

XOR stands for eXclusive OR, which is a special gate with a special property. The output for this gate is true (1) if only one of the inputs is true. In other words, the output is 1, if the inputs are *not similar*. The common symbol and the ANSI symbol for the XOR are shown in Figure 23.11. An application of XOR gate is depicted in Figure 23.12. A signal with

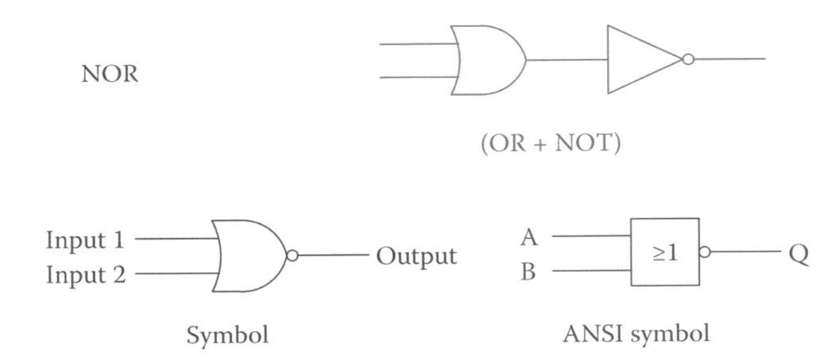

Figure 23.9 NOR gate.

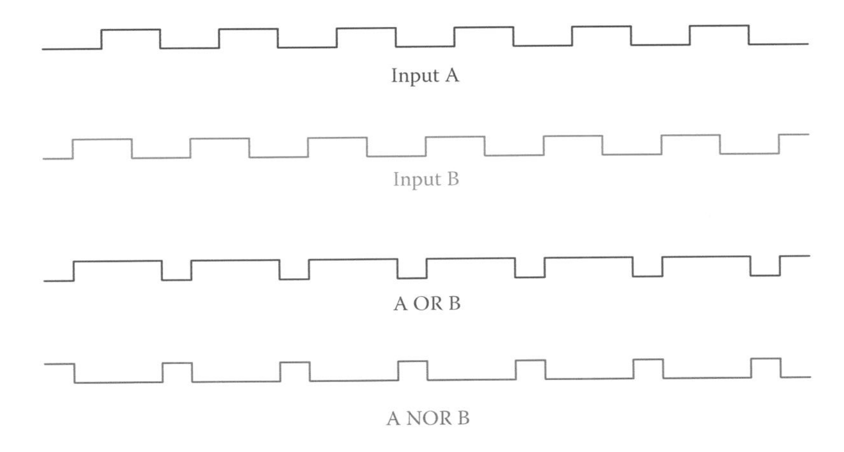

Figure 23.10 NOR output of two similar signals with a phase difference.

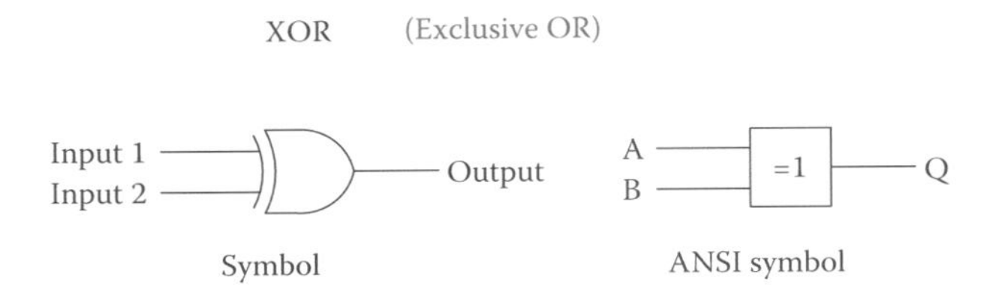

Figure 23.11 Symbols for XOR gate.

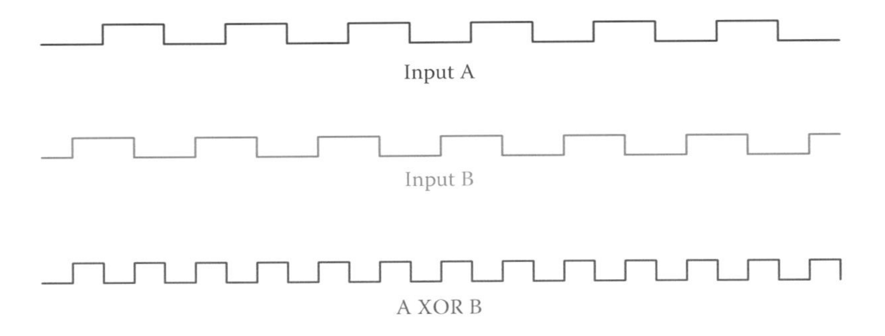

Figure 23.12 Generating a waveform with double frequency using XOR gate.

double frequency is generated with two similar waveforms with a phase difference.

> The output of a XOR gate is true if only one of the entries is true, but not both.

23.3 Logic Gate Math Functions

In conjunction with the definition of various gates introduced in Section 23.2, one needs to mathematically express the relationship between a gate output and its input(s). This is particularly important when complicated processes with many parameters are involved. Consider the work of an elevator. Even with only a few floors to serve, the number of entries becomes large because of many combinations that can exist defined by pressing the up and down keys from inside or outside of the elevator and the position of the elevator at the time a key is pressed. Many of these conditions can be simplified or eliminated. For instance, consider if a key is pressed again and again or if more keys from different locations are pressed for the same floor. The mathematics associated with the logic gates allows formal and systematic handling of the matter.

23.3.1 Gate Functions by Definition

Consider an AND gate (or any other gate except the NOT gate). Each input can be either 1 or 0. Consequently, there are four possibilities as follows (the inputs are represented as *A* and *B*):

A is 0 and *B* is 0
A is 1 and *B* is 0
A is 0 and *B* is 1
A is 1 and *B* is 1

Because the output must be either 0 or 1, the mathematical relationship must convey such a relationship. Normally, we show the summary of the possible cases for *A* and *B*, and the results *Q* in the following form

A	*B*	*Q*
0	0	0
1	0	0
0	1	0
1	1	1

This representation is called **truth table**. The mathematical expression for *Q* in terms of *A* and *B* must be such that the correct result is obtained for all the four case with two inputs. We study this for all the gates.

Truth table: Table for a logic circuit showing the output based on various input entries.

AND Gate

The best mathematical relationship describing the function of an AND gate is multiplication because we can observe that

$$0 \times 0 = 0$$
$$1 \times 0 = 0$$
$$0 \times 1 = 0$$
$$1 \times 1 = 1$$

That is, the output is equal to 1 only when both inputs are 1. In general terms, with *A* and *B* as the two inputs and *Q* as the output, we may write

$$Q = A \times B \quad \text{or} \quad Q = A \cdot B$$

The truth table for AND gate is shown above, and it is not repeated.

OR Gate

On the basis of the function of the OR gate, we expect that the output be 0 only when both inputs are 0 and in all the other cases be equal to 1. The mathematical expression to define the OR function is addition. But, as can be seen below, this is not a regular addition; it follows the Boolean algebra rules. We observe that

$0 + 0 = 0$ (output is 0 when both inputs are 0)
$1 + 0 = 1$ (output is 1 when one input is 1)
$0 + 1 = 1$ (output is 1 when one input is 1)
$1 + 1 = 1$ (this is the difference between Boolean algebra and the regular addition)

The truth table for an OR gate is

A	B	Q
0	0	0
1	0	1
0	1	1
1	1	1

The mathematical expression for an OR gate is

$Q = A + B$ (following the Boolean algebra, described later)

NOT Gate

The NOT gate has only one input; therefore, there is no need for a truth table. To show the relationship between the output and the input, a symbol is used to show the inverse (or complement) of a 1 or 0 entry. A bar on top of a letter name is used for this purpose. For example,

A NOT is shown as $\overline{A}$ and B NOT is shown as $\overline{B}$. For an input A to a NOT gate, thus, the output can be written

$$Q = \overline{A}$$

On the basis of this definition

$$\overline{1} = 0$$

$$\overline{0} = 1$$

Alternatively, instead of an overbar, the complement of an entity A is shown as A'. In this book we follow the first convention.

NAND Gate

The truth table for a NAND gate is based on the one for the AND gate. It is

A	B	$A \cdot B$	$Q = \overline{A \cdot B}$
0	0	0	1
1	0	0	1
0	1	0	1
1	1	1	0

As shown in the table, the expression for the output is

$$Q = \overline{A \cdot B}$$

NOR Gate

The truth table for a NOR gate is

A	B	$A + B$	$Q = \overline{A + B}$
0	0	0	1
1	0	1	0
0	1	1	0
1	1	1	0

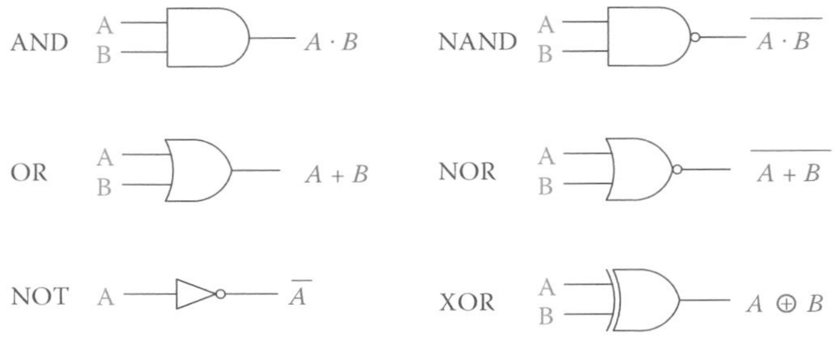

AND $A \cdot B$ NAND $\overline{A \cdot B}$

OR $A + B$ NOR $\overline{A + B}$

NOT $\overline{A}$ XOR $A \oplus B$

Figure 23.13 Summary of all logic gates.

The output is represented by

$$Q = \overline{A + B}$$

XOR Gate

On the basis of the definition for an XOR gate, its truth table is

A	B	Q
0	0	0
1	0	1
0	1	1
1	1	0

A special symbol is used for the mathematical expression of the XOR gate, as follows

$$Q = A \oplus B$$

A summary of the logic gates and their output expressions are shown in Figure 23.13.

23.4 Boolean Algebra

The Boolean algebra is a set of specific rules that governs the mathematical relationships corresponding to the logic gates and their combinations. Their application is limited to two-valued (0 and 1) entries such as the inputs and outputs of logic gates. Dealing with one single gate and a pair of inputs is a trivial task. When there are many parameters that are combined together through gates of various types, rules of Boolean algebra help to simplify and analyze the problem. Boolean algebra was developed by George Boole in 1854.

23.4.1 Laws of Boolean Algebra

There are a number of laws for Boolean algebra. Here we study 10 of these laws considered to be more important, together with some examples for them. These laws govern the relationships that exist between two or more inputs to logic gates. They can be used to simplify circuits.

First Law

Any entity OR'ed with itself is equal to itself:

$$A + A = A$$

Example

$$1 + 1 = 1$$
$$0 + 0 = 0$$

Second Law

Any entity AND'ed with itself is equal to itself:

$$A \cdot A = A$$

Example

$$1 \times 1 = 1$$
$$0 \times 0 = 0$$

Third Law

OR and AND are commutative (the order does not matter):

$$A + B = B + A$$

$$A \cdot B = B \cdot A$$

Fourth Law

OR and AND are associative (the meaning is clear from the mathematical expression):

$$A \cdot (B + C) = A \cdot B + A \cdot C$$

$$A + (B \cdot C) = (A + B) \cdot (A + C)$$

This law is very important and useful, though can be confusing. The first one of these relationships follows the rules of algebra, but because the second one is not so clear the following truth table verifies its validity for various values of the three inputs. Compare the fifth and the last columns.

A	B	C	BC	$A + BC$	$A + B$	$A + C$	$(A + B) \cdot (A + C)$
0	0	0	0	0	0	0	0
0	0	1	0	0	0	1	0
0	1	0	0	0	1	0	0
0	1	1	1	1	1	1	1
1	0	0	0	1	1	1	1
1	0	1	0	1	1	1	1
1	1	0	0	1	1	1	1
1	1	1	1	1	1	1	1

Fifth Law

The result of any entity OR'ed with its own complement is 1:

$$A + \bar{A} = 1$$

Example

$$1+\overline{1}=1+0=1$$
$$0+\overline{0}=0+1=1$$

Sixth Law

Anything AND'ed with its own complement is 0:

$$A\cdot\overline{A}=0$$

Example

$$1\times\overline{1}=1\times 0=0$$
$$0\times\overline{0}=0\times 1=0$$

Seventh Law

Anything OR'ed with 0 is equal to itself; anything AND'ed with 0 equals 0:

$$A+0=A \qquad (A \text{ can be 0 or 1})$$
$$A\times 0=0$$

Eighth Law

Anything OR'ed with 1 is equal to 1; anything AND'ed with 1 is equal to itself.

$$A+1=1 \quad A\times 1=A$$

Example

$$0+1=1 \quad 0\times 1=0$$
$$1+1=1 \quad 1\times 1=1$$

The following two laws are called De Morgan laws:

Ninth Law (De Morgan First Law)

The inverse of the result of OR'ing two entities A and B is the same as if the inverse of those entities are AND'ed.

$$\overline{A+B}=\overline{A}\cdot\overline{B}$$

Example

Because there are four different cases for two inputs all are shown in a truth table.

TRUTH TABLE FOR THE FIRST DE MORGAN LAW

A	B	$A+B$	$\overline{A+B}$	$\overline{A}$	$\overline{B}$	$\overline{A}\cdot\overline{B}$
0	0	0	1	1	1	1
0	1	1	0	1	0	0
1	0	1	0	0	1	0
1	1	1	0	0	0	0

Note that $\overline{A+B}$ is the output of a NOR gate.

Tenth Law (De Morgan Second Law)

The inverse of the result of AND'ing two entities *A* and *B* is the same as if the inverse of those entities are OR'ed.

$$\overline{A \times B} = \overline{A} + \overline{B}$$

Example

The following truth table illustrates the four possibilities.

TRUTH TABLE FOR THE SECOND DE MORGAN LAW

A	B	$A \times B$	$\overline{A \times B}$	$\overline{A}$	$\overline{B}$	$\overline{A} + \overline{B}$
0	0	0	1	1	1	1
0	1	0	1	1	0	1
1	0	0	1	0	1	1
1	1	1	0	0	0	0

The term $\overline{A \times B}$ represents the output of a NAND circuit.

The implementation of De Morgan laws is converting AND and OR gates and vice versa when they are combined with a NOT gate. Consider the equivalence between the two expressions $\overline{A + B}$ and $\overline{A} \cdot \overline{B}$ and between $\overline{A \times B}$ and $\overline{A} + \overline{B}$ based on De Morgan's laws. The associated gates combinations are shown in Figure 23.14a. Note that an alternative way when a NOT gate is involved to invert an input signal is shown in Figure 23.14b.

23.4.2 Application of Boolean Algebra Laws

Boolean algebra is employed to simplify logic circuits. Simplification often leads to having fewer components. Moreover, many cases can be found where two logic circuits lead to the same results. The two circuits in this case are equivalent to each other. Boolean algebra can help to verify and identify these circuits. This helps to reduce the number of gates in a circuit or synthesize a logic gate by some other gates, when necessary. In this section we see a number of examples where the laws of Boolean algebra are implemented.

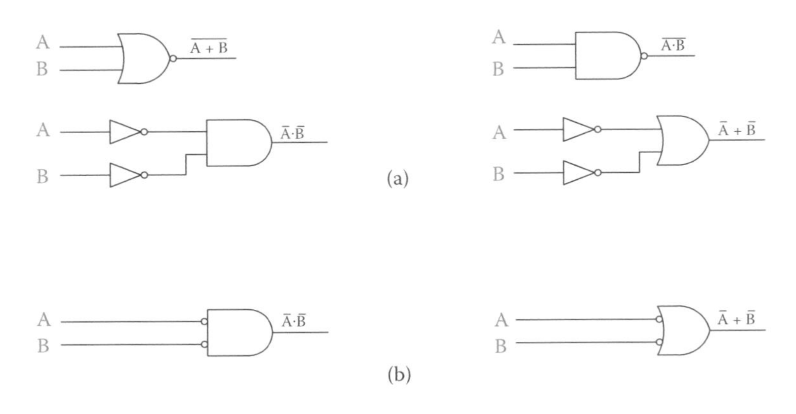

Figure 23.14 (a) Showing the first and second De Morgan law equivalent circuits. (b) Another way to show inputs to AND and OR gates if complements of signals A and B are the entries.

Example 23.1

Simplify the expression $Z = (A + B) \cdot \overline{C} + AC$.

Solution

This expression implies three inputs, two of which are OR'ed together and the result is AND'ed to the inverse of the third one. The result then is OR'ed to the outcome of AND'ing the first and third inputs. We first look at the logic circuit for it, shown in Figure 23.15. This circuit uses six gates.

Using the preceding laws, we can write

$$Z = (A + B) \cdot \overline{C} + AC = A\overline{C} + B\overline{C} + AC = A(\overline{C} + C) + B\overline{C}$$

On the basis of the fifth law $\overline{C} + C = 1$. Thus,

$$Z = A + B\overline{C}$$

This expression is simpler. Now let see its logic circuit in Figure 23.16.

The circuit in Figure 23.16 has fewer components (only three gates) compared to that in Figure 23.15. Thus, it is preferable and more economical.

Example 23.2

Simplify the expression $Z = (A + B) \cdot \overline{BC} + A \cdot C$. Show the logic circuit before and after it is simplified.

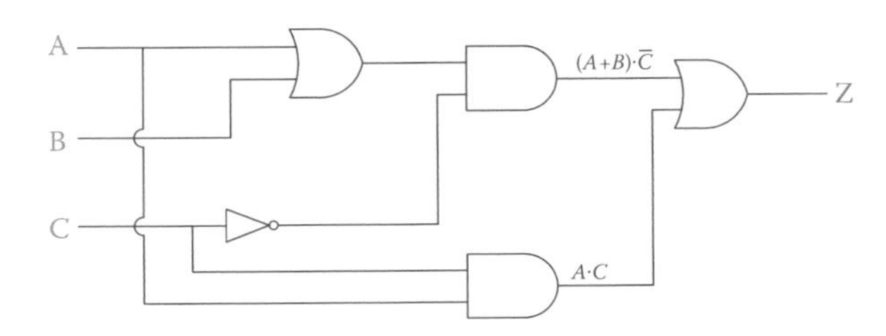

Figure 23.15 Logic circuit for $Z = (A + B) \cdot \overline{C} + AC$.

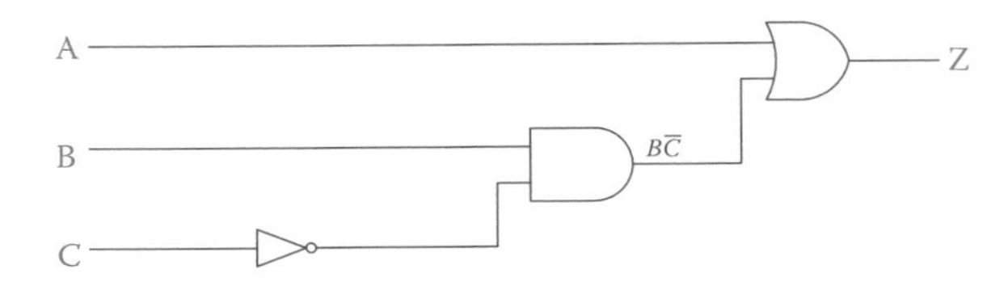

Figure 23.16 Simplified circuit $(A + B\overline{C})$.

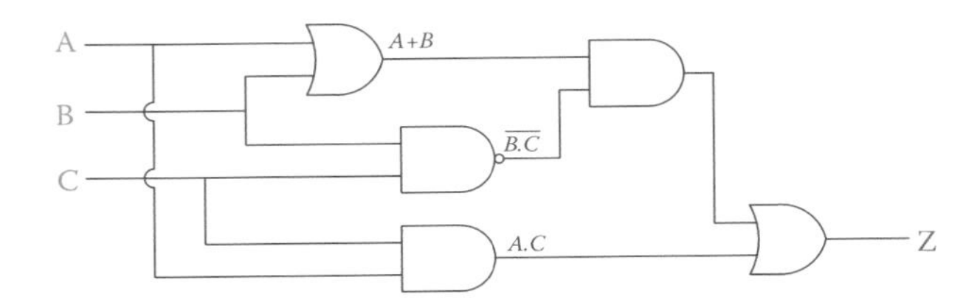

Figure 23.17 Circuit for Example 23.2.

Solution

The circuit for original expression is shown in Figure 23.17.

The expression for Z can be simplified by first using the second De Morgan law, as follows:

$$Z = (A + B) \cdot \overline{BC} + AC = (A + B) \cdot (\overline{B} + \overline{C}) + A \cdot C$$
$$= (A + B) \cdot \overline{B} + (A + B) \cdot \overline{C} + A \cdot C$$
$$= A \cdot \overline{B} + B \cdot \overline{B} + A \cdot \overline{C} + B \cdot \overline{C} + A \cdot C$$

But $B \cdot \overline{B} = 0$ (sixth law), and by rearrangement

$$Z = A \cdot \overline{B} + B \cdot \overline{C} + A \cdot \overline{C} + A \cdot C$$
$$= A \cdot \overline{B} + B \cdot \overline{C} + A \cdot (\overline{C} + C) \qquad (\overline{C} + C = 1.\ \text{Fifth law})$$
$$= A \cdot \overline{B} + B \cdot \overline{C} + A \cdot 1 = A \cdot \overline{B} + B \cdot \overline{C} + A \qquad (A \cdot 1 = A.\ \text{Eighth law})$$
$$= A \cdot (\overline{B} + 1) + B \cdot \overline{C} = A + B \cdot \overline{C} \qquad (\overline{B} + 1 = 1.\ \text{Eighth law})$$

The simplified expression for Z is the same as in the previous example. Its circuit is illustrated in Figure 23.16.

It is possible to have the two inputs for a gate to be the same signal. The output then is the same as the input. This is used when a gate is synthesized from other gates. Because commercial logic gates come as packages, more of one type of gate may be available, which can be employed for the purpose of synthesizing other gates.

Example 23.3

By using Boolean algebra show the equivalent circuit of the three NOR gates, as shown in Figure 23.18. Verify the results by a truth table.

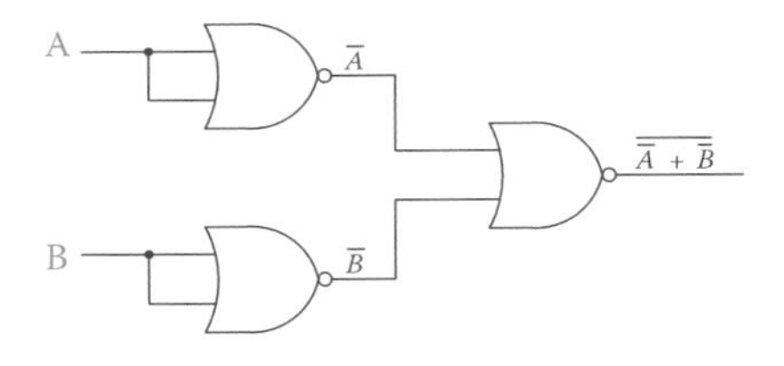

Figure 23.18 Output of three NOR gates.

Solution

First, we need to point out that the inverse of the inverse of any entity is the entity itself; that is, $\overline{\overline{A}} = A$.

The solution follows by the application of De Morgan first law. It can be seen that the output of the three gates is $\overline{\overline{A} + \overline{B}}$. This expression can be replaced by its equivalent $\overline{\overline{A}} \cdot \overline{\overline{B}}$, which is $A \cdot B$. That is, the three NOR gates are equivalent to an AND gate.

The following truth table verifies this.

A	B	$\overline{A}$	$\overline{B}$	$\overline{A}+\overline{B}$	$\overline{\overline{A}+\overline{B}}$	$A\cdot B$
0	0	1	1	1	0	0
0	1	1	0	1	0	0
1	0	0	1	1	0	0
1	1	0	0	0	1	1

Example 23.4

Find the simplified form of the circuit shown in Figure 23.19 by finding the equivalent expression for the output. Verify the result by a truth table.

Solution

On the basis of Figure 23.19, the expression for Z is

$$\overline{[(A\cdot B)+\overline{C}]\cdot\overline{C}}$$

This expression looks complicated, but it can be simplified. First, let see how it was obtained, because for some it may look uneasy, especially if there are more components involved. To reduce the problem to smaller segments, one may always name the outputs of each gate or write the output of each gate on the figure, as in Figure 23.20, you see both a given name and the expression for it.

Simplification of expression for Z is carried out in the following steps:

1. Use De Morgan second law to separate the parts under the overbar:

$$\overline{\left[(A\cdot B)+\overline{C}\right]\cdot\overline{C}} = \overline{\left[(A\cdot B)+\overline{C}\right]}+\overline{\overline{C}}$$

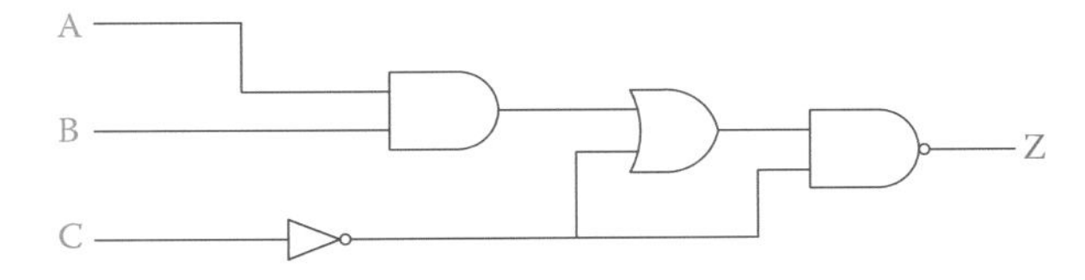

Figure 23.19 Circuit of Example 23.4.

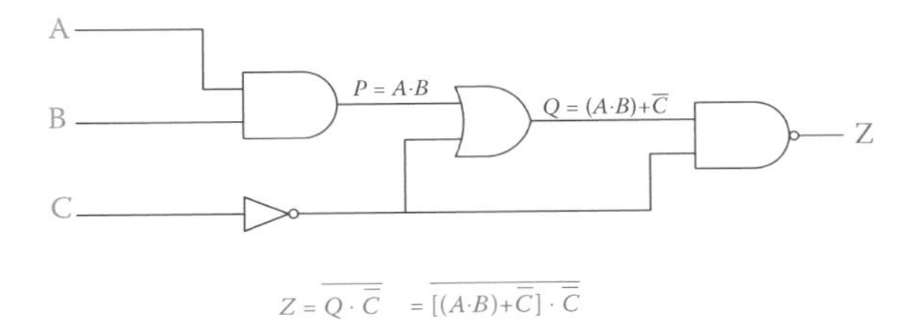

Figure 23.20 Details of local outputs of all gates.

2. Use De Morgan first law, again to separate the parts under the overbar:

$$\overline{[(A \cdot B) + \overline{C}]} = \overline{(A \cdot B)} \cdot \overline{\overline{C}}$$

3. Now Z is reduced to

$$Z = \overline{(A \cdot B)} \cdot C + C$$

because the inverse of the inverse of C is equal to $C(\overline{\overline{C}} = C)$.

4. Finally, we simplify Z as follows:

$$Z = \overline{(A \cdot B)} \cdot C + C = (\overline{A} + \overline{B}) \cdot C + C = [(\overline{A} + \overline{B}) + 1] \cdot C$$

But, according to the eighth law anything OR'ed with 1 equal 1. The term inside the parentheses is OR'ed with 1, and the result is 1, no matter what A and B are. The final result is $Z = C$.

The following truth table shows the results.

A	B	C	$\overline{C}$	$P = A \cdot B$	$Q = P + \overline{C}$	$Q \cdot \overline{C}$	Z
0	0	0	1	0	1	1	0
0	0	1	0	0	0	0	1
0	1	0	1	0	1	1	0
0	1	1	0	0	0	0	1
1	0	0	1	0	1	1	0
1	0	1	0	0	0	0	1
1	1	0	1	1	1	1	0
1	1	1	0	1	0	0	1

23.4.3 Sum of Products and Product of Sums

It is always possible to write the Boolean expression for an application based on a truth table. Suppose that for a process application there are three inputs that determine what action must be taken. For example, in a small office it is expected that during the working hours the door is unlocked and some people are in the office. For this example the three parameters and their associated binary states are

1. Office time (denoted by A): outside office time $A = 0$, inside office time $A = 1$;
2. Door (denoted by B): door opened $B = 0$, door locked $B = 1$;
3. Workers (denoted by C): no worker in $C = 0$, some workers present $C = 1$.

The condition is monitored by a security system, and in three cases an alarm sounds and notifies the security personnel: (1) when outside the office hours nobody is in the building and the door is opened, (2) when outside the office hours the door is locked, but some motion is sensed in the office, and (3) when inside the office hours, the door is open, but nobody is in the office.

You have noticed so far the pattern to represent three conditions, as seen in the previous truth tables with three inputs A, B, and C. The leftmost column starts with 0's and for all other combinations of the other inputs has repeated 0's, followed by the same number of 1's. To the right of that, the other parameters are shown each one with repeated 0's and 1's covering all the possible variations of the parameters to its right. Similar patterns can be formed for four, five, and more parameters. In the following truth table the three sensitive cases for the alarm to sound are identified by asterisks.

A	*B*	*C*	*Output*	
0	0	0	*1	1 (outside of office hours, door
0	0	1		opened, no person in)
0	1	0		2 (outside of office hours, door
0	1	1	*2	locked, motion is observed)
1	0	0	*3	3 (inside the office hours, door
1	0	1		open, but nobody in the office)
1	1	0		
1	1	1		

From this table we want to write the logic expression for the scenario.

Note that for each case that the alarm must sound three conditions must simultaneously exist. This calls for AND'ing the three outputs corresponding to the cases. In this respect the expressions for the three cases are, respectively, $\bar{A}\cdot\bar{B}\cdot\bar{C}$ (all inputs have 0 values), $\bar{A}\cdot B\cdot C$, and $A\cdot\bar{B}\cdot\bar{C}$. Because each of these combinations must trigger the alarm, they must be OR'ed together. The expression for the output, therefore, is defined by

$$\bar{A}\cdot\bar{B}\cdot\bar{C}+\bar{A}\cdot B\cdot C+A\cdot\bar{B}\cdot\bar{C}$$

We are going to see the associated circuit for the above logic expression without simplification. The purpose is to show the way more

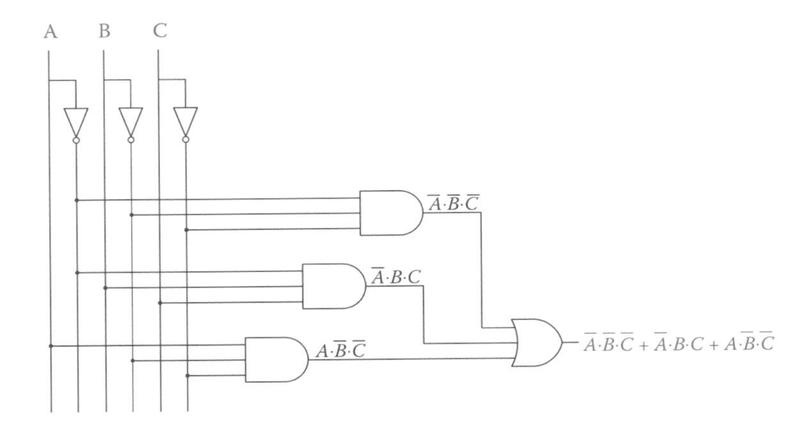

Figure 23.21 Method for representing inputs to complex circuits.

Sum of products: A logic function represented in the form like (AB) + (AC) + ($\overline{A}\cdot\overline{B}\cdot\overline{C}$), for example, which is the sum of items obtained by multiplication of logic variables. This happens when the results of some AND gates are OR'ed together.

Product of sums: Logic function represented in the form like $(A + B) \cdot (A + C) \cdot (\overline{A} + \overline{B} + \overline{C})$, for example, which is the multiplication of items obtained by addition of logic variables. This happens when the results of some OR gates are AND'ed together.

involved circuits are presented. In Figure 23.21 each parameter is shown with two vertical lines, one for its 1 value and one for its complement value through a NOT gate. Then, all connections for gates are drawn with horizontal lines while the points of connection are denoted by a black dot (there is no connection point between vertical and horizontal lines if there is no dot).

The representation of the circuit for the preceding problem is in the form of three AND gates driving an OR gate. This representation is often called **sum of products** because the final expression is in the form of a number of multiplication products adding together.

In the same way that sum of products consists of a number of AND gates driving an OR gate, it is possible to have a number of OR gates driving an AND gate. Such a representation is called **product of sums** because the final output function consists of multiplication of a number of components each of which is the sum of some parameters.

De Morgan laws are the key to convert from multiplication to sum and vice versa. We have already seen examples of employing these laws for two parameters. For the current subject, first we see the expansion for more parameters.

Considering the product $A \cdot B \cdot C$ to be written as $A \cdot (B \cdot C)$, by De Morgan law

$$\overline{A \cdot B \cdot C} = \overline{A \cdot (B \cdot C)} = \overline{A} + \overline{B \cdot C} = \overline{A} + \overline{B} + \overline{C}$$

Similarly,

$$\overline{A + B + C} = \overline{(A + B) + C} = \overline{A + B} \cdot \overline{C} = \overline{A} \cdot \overline{B} \cdot \overline{C}$$

The sum of products in question is

$$Z = \overline{A} \cdot \overline{B} \cdot \overline{C} + \overline{A} \cdot B \cdot C + A \cdot \overline{B} \cdot \overline{C}$$

It has three components that add together. But,

$$\overline{A} \cdot \overline{B} \cdot \overline{C} = \overline{A + B + C}, \quad \overline{A} \cdot B \cdot C = \overline{A + \overline{B} + \overline{C}}, \quad \text{and} \quad A \cdot \overline{B} \cdot \overline{C} = \overline{\overline{A} + B + C}$$

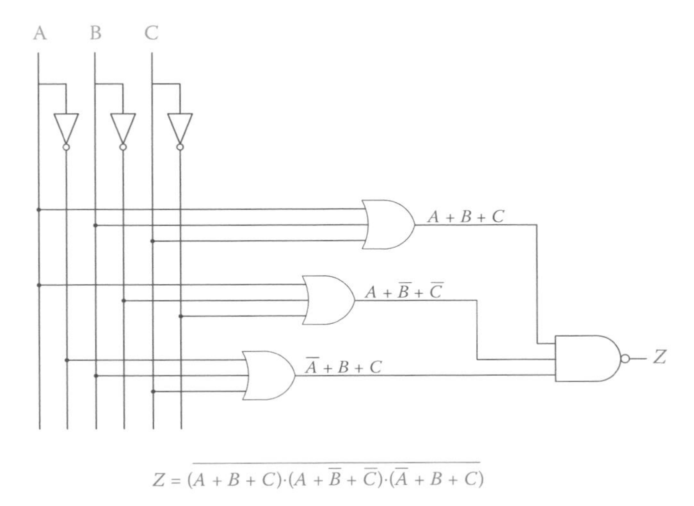

Figure 23.22 Alternative circuit for that in Figure 23.21.

Because

$$\overline{Z} = \overline{\overline{\overline{A+B+C} + \overline{A+\overline{B}+\overline{C}} + \overline{\overline{A}+B+C}}}$$

$$= \left(\overline{\overline{A+B+C}}\right)\cdot\left(\overline{\overline{A+\overline{B}+\overline{C}}}\right)\cdot\left(\overline{\overline{\overline{A}+B+C}}\right)$$

$$= (A+B+C)\cdot(A+\overline{B}+\overline{C})\cdot(\overline{A}+B+C) \quad (\text{Remember } \overline{\overline{x}} = x)$$

Therefore,

$$Z = \overline{(A+B+C)\cdot(A+\overline{B}+\overline{C})\cdot(\overline{A}+B+C)}$$

which is in the form of product of sums; however, the result must be inversed by a NOT gate. The corresponding circuit is depicted in Figure 23.22.

Example 23.5

Consider the starting operation of a wind turbine. A number of safety issue and technical conditions must be satisfied before the brakes are released from the rotor. We summarize the similar conditions in sets of inputs to AND gates, as depicted in Figure 23.23. All the hatches must be closed; otherwise, the turbine will not start (hatches are the doorways between each two sections of the tower and on the nacelle roof; each one has a microswitch that relays whether the hatch is open or closed).

The line voltage, oil temperature, oil level, and similar measurements are checked, and we assume they are put together as the inputs of two other logic gates, one for the electrical conditions and one for the mechanical conditions. Two more gates bring the condition of the wind speed and the position of a key if it is set for *run* or *stop*. Note that this is a very simplified version of the real system. It is just to serve as an example.

There are two ways of looking at the switch scenarios. For instance, all the hatches must be closed. This can be interpreted in two different ways for implementation in the logic circuit.

1. We want to check that the outputs from all the first set of gates are 1, indicating that all conditions are satisfied. So they must be AND'ed together, and the final gate output is sending a command for the brake to release.
2. We say that if any of the outputs from the first set of gates is not as desired, then the brakes must not

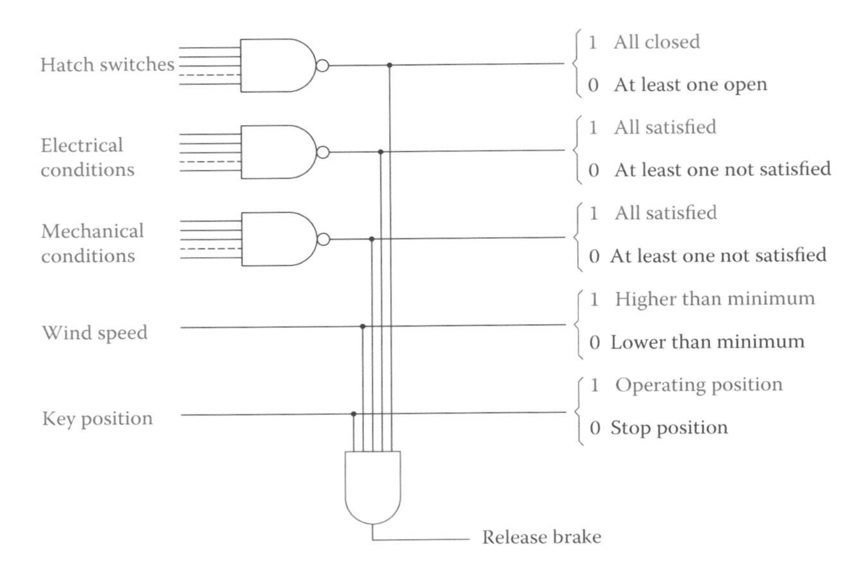

Figure 23.23 Example for part of process to start a turbine.

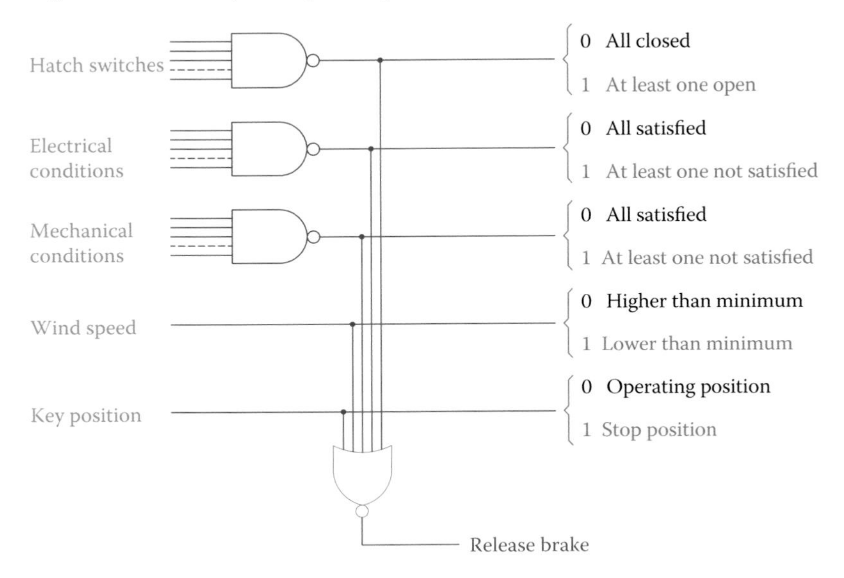

Figure 23.24 Alternative circuit for Figure 23.22.

release. In this case the desired output of each gate is 0, and outputs must be OR'ed together. An action takes place based on the complement of the OR gate output.

This latter case is shown in Figure 23.24. This example confirms that there is not a unique solution to any problem and alternatives are possible. In using logic circuits it is always necessary to know and check which value is considered *true*. It is not always 1 that is considered *true*. In many circuits, *true* is the case when the voltage (or the output of a gate) is 0.

Note that in Figures 23.23 and 23.24 the multiple entries to a gate implies that it is not a single gate and consists of various gates (as many as necessary) where their outputs are combined together, as illustrated in Figure 23.3 for AND gates.

23.5 K-Maps

An alternative to the truth table to determine and simplify the logic function for an application is **Karnaugh map**, named after its originator Karnaugh. Karnaugh map abbreviates to **K-map** offers a simpler solution to find the logic function for applications with two, three, and four inputs. Its application to cases with a higher number of inputs is possible but difficult to tackle.

Karnaugh map: Special tables by the use of which logic circuits can be simplified and their output can be determined.

K-map: Abbreviation for Karnaugh map.

23.5.1 K-Map for Two Inputs

Applications with only two inputs A and B are easy to handle by any method. For easier understanding of how a K-map works, nevertheless, we start with two inputs. As you have noticed, with two inputs there are four possible states, as shown in the truth table depicted in Figure 23.25, which also shows the corresponding K-map.

For a two-input problem the K-map has four cells. The horizontal cells represent the two states of one of the inputs (B in Figure 23.25, starting from $\bar{B}$ on the left), and the vertical cells exhibit the states of the other input starting from the NOT value. The correspondence between the truth table lines and the K-map cells are shown by arrows for two cells.

The required values for the output are entered in the corresponding cells (Figure 23.25b). For example, in the figure shown, it is required that the output be 1 in the two cases: $\bar{A}\bar{B}$ and AB; that is when both inputs have their complement values, and when both inputs have their 1 values, or in other words, when the two inputs have similar values. (Note that for simplicity the multiplication symbol is not shown; thus, AB denotes $A{\cdot}B$.) The arrows show the corresponding transfers. The other cells are required to be 0's in this case. Thus, they are filled with 0's.

Each cell represents one of the four combinations of the two inputs. For example, the top left cell represents $\bar{A}\bar{B}$. This is the logic expression for that cell. When the cell values in the K-map are identified based on the output requirement and the 1's and 0's are assigned, the logic expression can be derived from the cells containing ones. The sum of the expressions

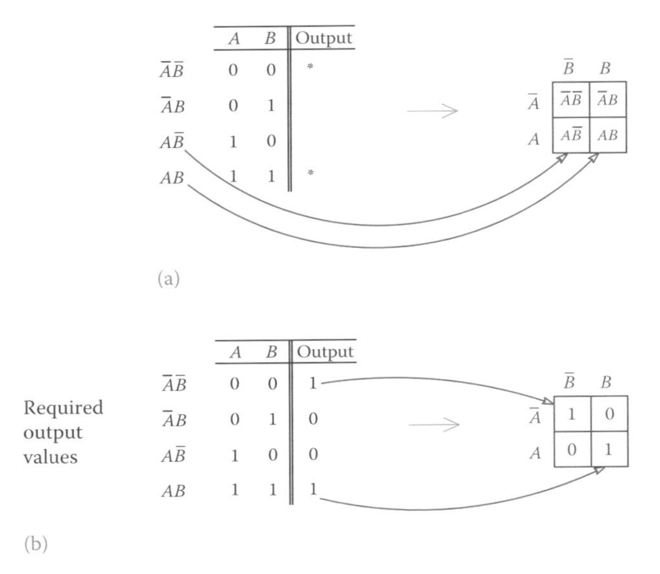

Figure 23.25 Truth table and K-map for a two-input problem. (a) Structure of the K-map. (b) Example of implementation of requirements.

for those cells with 1 inside defines the logic function for the application. In the present two-input example the expression for the output Z is

$$Z = \overline{A} \cdot \overline{B} + A \cdot B$$

This is the expression for the complement of a XOR gate. So, for this application the required circuit consists of a XOR gate and a NOT gate. For this problem it was not necessary to use a K-map because it is a trivial case of two inputs. But, for three and four inputs the problem is more involved.

One important property of a K-map is that moving from any cell to a neighboring cell implies only one change (and never more than one). One change means that only one of the inputs switches to its complement. You can see that for two-input K-map moving horizontally changes B, only, and moving vertically changes A, only. This property helps to select the correct entry for extension of the two-input K-map to three and four inputs, as can be seen next.

Also, take note that in K-map notation *all* cells have two neighbors vertically and two neighbors horizontally. In other words, the cells are assumed to be seamlessly organized in circles and not in rows and columns. This matter is more important and clearer for three and four inputs and will be visited again.

23.5.2 K-Map for Three Inputs

The K-map for two inputs can be extended to three inputs by combining the third input either in the horizontal or vertical direction with the input already placed there. Here we do that horizontally, and the third variable C is combined with B, as it is shown in Figure 23.26. The K-map for three variables has eight cells, each one of which represents one of the possible eight combinations of three inputs. In Figure 23.26 the cells are numbered from 1 to 8 for referencing.

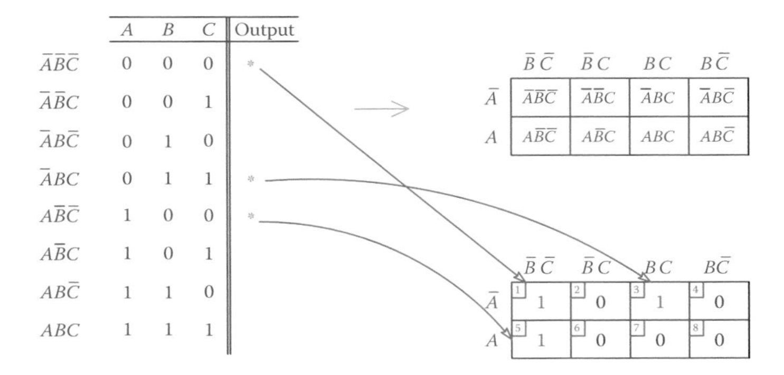

Figure 23.26 K-map for a three-input case.

Note the following important points:

1. Each cell represents the combination of three inputs. Thus, the expression of the function for each cell is the product of inputs or their complements like $\bar{A}B\bar{C}$.
2. Variation of two inputs is reflected in the horizontal direction.
3. In the horizontal direction moving from each cell to its right or left neighbor implies a change only either in B or in C but not both. This is why in the third column from left BC appears and not $B\bar{C}$.

On the basis of the requirement for output, as shown by the asterisks in the truth table, 1's are entered in the corresponding cells in the K-map. The rest of cells are filled with 0's.

The essence of a K-map now can be shown based on the pattern that appears after all cells are filled. Both cells 1 and 5 (in Figure 23.26) must be selected as part of the output. Cell 1 contains $\bar{A}$, and cell 5 contains A. This implies that changing A to $\bar{A}$ (and vice versa) does not have any effect on the output, meaning that no matter what input A is the output is independent of it. In such a case the output is defined by the common part of these two cells, plus any other cell that contains 1. For this case,

$$Z = \bar{B}\cdot\bar{C} + \bar{A}\cdot B\cdot C$$

Note from the asterisks in the truth table that they represent the output of the logic circuit in Figure 23.21. The above expression for Z after simplification could also be reached if the logic function of the circuit in Figure 23.21 was simplified, using Boolean algebra rules.

The process of definition and simplification of the logic function by using a K-map is as follows.

1. Encircle all occurrences of 1's in the cells putting together all those in the neighboring cells in groups of two and four. The neighboring cells can occur in a row, column or rectangle when more than one column (or row) is involved.
2. Write the corresponding expression for each encircled group.
3. Add together the logic expressions for all groups.

For step 1, one should pay attention to the fact that moving from cell 4 to cell 1 (or vice versa) and from cell 8 to cell 5 implies only one change; it is a switch between B and its complement. In this respect, cells 1 and 4 are considered neighbors, and cells 5 and 8 are also considered neighbors. Thus, they must be encircled together if they contain 1's.

For step 2, note that a row change implies a switch between A and its complement, a change between the first and fourth column implies a switch between B and its complement, and a change between columns 1 and 2 and between columns 3 and 4 implies a switch between C and its complement (see Figure 23.27).

We try to form groups of two or four, whenever possible. For each case the variable(s) that do not change within the group describe the logic function for that group. Forming groups is the process of simplification of the logic function for an application. Examples of forming groups and

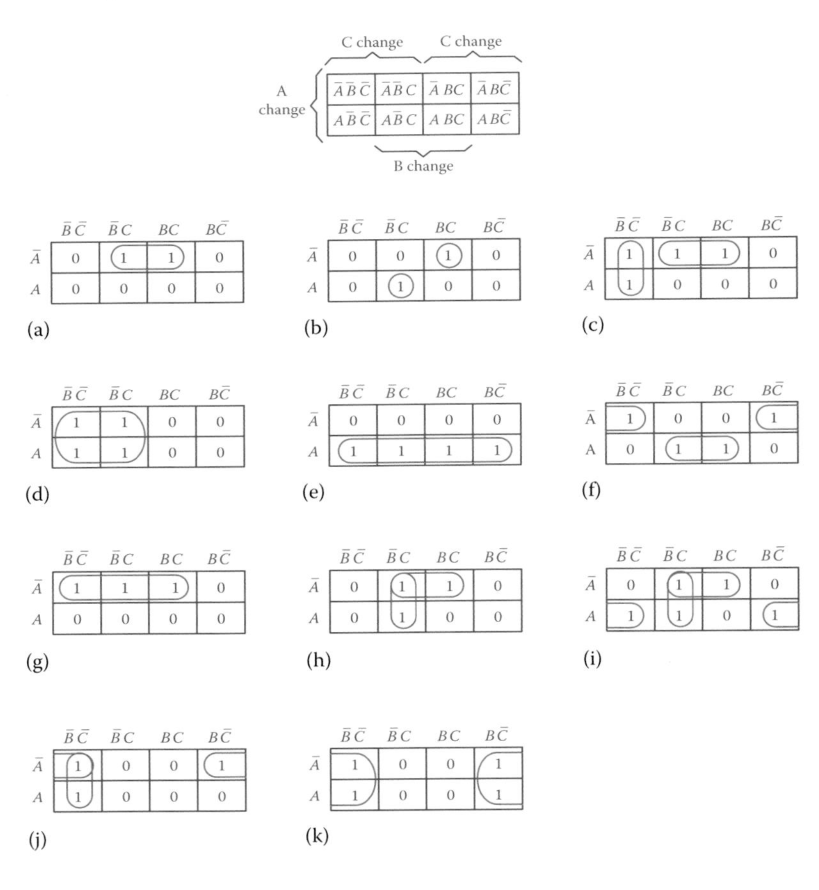

Figure 23.27 Examples of encircling neighboring cells containing 1's in a three-input scenario. (a) Only one group. (b) No grouping is possible. (c) One vertical and one horizontal group. (d) One group of four cells in two rows. (e) One group of four cells in a row. (f) Extreme cells in a row make a group. (g) In this case, there are one pair plus one cell, or two pairs with a shared cell. (h) Two groups, one horizontal and one vertical. (i) Three groups. (j) Two groups. (k) One group of four cells.

encircling neighboring cells are shown in Figure 23.27. The corresponding logic function for the cases shown is described below. More examples are shown in the next section for four-input problems.

The logic functions for cases a to k illustrated in Figure 23.27 are as follows:

a. The encircled cells belong to $\bar{A}\bar{B}C$ and $\bar{A}BC$. That is, the output is defined by the sum of $\bar{A}\bar{B}C$ and $\bar{A}BC$. Because a change of $\bar{B}$ to B has no effect on the output (which is shown by having 1 in both cells), then they are removed and the output is defined by

$$Z = \bar{A}C$$

b. The individual encircled cells do not have any common element. The output is, thus, the sum of these two cells.

$$Z = \bar{A}BC + A\bar{B}C$$

c. Although a cell can be shared between the two encircled areas, it is better to divide the cells containing 1's to two groups each one containing a pair of cells.

$$Z = \bar{B}\bar{C} + \bar{A}C$$

d. All the encircled cells have $\bar{B}$ in common and change in A and C do not have any effect. The output is only defined by $\bar{B}$.

$$Z = \bar{B}$$

e. Similar to case d, all encircled cells contain A, and changes for B and C have no effect on the output.

$$Z = A$$

f. There are two encircled pairs in this case; one corresponds to no change in $\bar{C}$ (it exists in both cells) and the other one implies no change in C.

$$Z = \bar{A}\bar{C} + AC$$

g. When three cells are involved, as encircled, we can consider two pairs with one shared cell. For the first pair, $\bar{B}$ does not change, and for the second pair, C does not change.

$$Z = \bar{A}\bar{B} + \bar{A}C$$

(We could consider the third cell independently and say $Z = \bar{A}\bar{B} + \bar{A}BC$, but the final result is the same.)

h. For this case, again, we consider a horizontal pair and a vertical pair (one cell is shared between the two). For the horizontal pair, B changes (thus, its function excludes B) and for the vertical pair A changes (so, A is omitted).

$$Z = \bar{A}C + \bar{B}C$$

i. This case can be considered a combination of that in case h plus an additional term for the cells 5 and 8. Thus,

$$Z = \bar{A}C + \bar{B}C + A\bar{C}$$

j. The cells 1 and 4 lead to $\bar{A}\bar{C}$, and the cells 1 and 5 lead to $\bar{B}\bar{C}$; therefore,

$$Z = \bar{A}\bar{C} + \bar{B}\bar{C}$$

k. All the encircled cells contain $\bar{C}$, and changes in A and B do not play a role.

$$Z = \bar{C}$$

23.5.3 K-Map for Four Inputs

The K-map for four variables has 16 cells; that is, twice as many as that for three variables. As was done for the horizontal extension from the two-input case, both horizontal and vertical extensions are implemented. The result is shown in Figure 23.28. Each cell in this case reflects the value corresponding to one of 16 possible states for combination of four inputs. The logic function of each cell is composed by multiplying (AND'ing) the logic expression of the row and the column in which it lies. For example, $ABCD$ corresponds to the third row (AB) and the third column (CD).

Simplification is done in the same manner that was described for the three-input problem. We try to find groups of two, four, and eight neighbor cells that share the same variable(s). Those variables then define the logic function for the group. Figure 23.29 illustrates a number of examples, whose logic functions are as follows.

a. The four encircled cells are independent of each other, and they cannot be grouped. The output, thus, is the sum of all four of them.

$$Z = \bar{A}\bar{B}\bar{C}\bar{D} + \bar{A}\bar{B}CD + \bar{A}B\bar{C}D + \bar{A}\bar{B}C\bar{D}$$

AB ↓ \ CD →	$\bar{C}\bar{D}$	$\bar{C}D$	CD	$C\bar{D}$
$\bar{A}\bar{B}$	$\bar{A}\bar{B}\bar{C}\bar{D}$	$\bar{A}\bar{B}\bar{C}D$	$\bar{A}\bar{B}CD$	$\bar{A}\bar{B}C\bar{D}$
$\bar{A}B$	$\bar{A}B\bar{C}\bar{D}$	$\bar{A}B\bar{C}D$	$\bar{A}BCD$	$\bar{A}BC\bar{D}$
AB	$AB\bar{C}\bar{D}$	$AB\bar{C}D$	$ABCD$	$ABC\bar{D}$
$A\bar{B}$	$A\bar{B}\bar{C}\bar{D}$	$A\bar{B}\bar{C}D$	$\bar{A}\bar{B}CD$	$A\bar{B}C\bar{D}$

Figure 23.28 K-map for four-input problem.

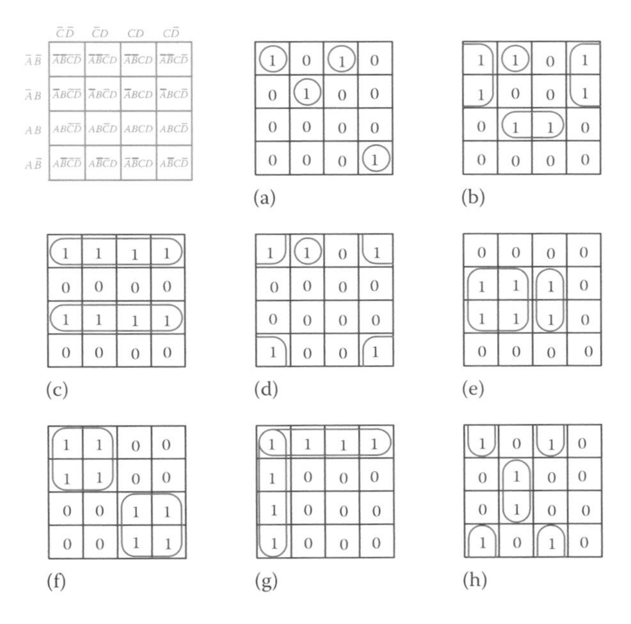

Figure 23.29 Examples of four-input K-maps.

b. There are three groups, part of column 1 and 4, the individual circle, and the two in the middle.

$$Z = \overline{A}\overline{D} + ABD + \overline{A}\overline{B}\overline{C}D$$

c. The function of a row or column is always that written in front of the row or on the top of column. Thus,

$$Z = \overline{A}\overline{B} + AB$$

d. All the 1's in the four corners can be grouped together. The only two variables that do not change are $\overline{B}$ and $\overline{D}$. The individual cell is $\overline{A}\overline{B}\overline{C}D$.

$$Z = \overline{B}\overline{D} + \overline{A}\overline{B}\overline{C}D$$

e. A group of four and a pair can be identified here.

$$Z = B\overline{C} + BCD$$

Note that an alternative function can be obtained if the four cells in the middle are grouped together. Then $Z = B\overline{C}\overline{D} + BD$. The final result, nevertheless, is the same. The degree of complexity or simplicity in terms of the required number of gates is also the same.

f. Two independent groups exist. In the top group, change of B and D does not play any role; in the bottom one, again, change of B and D with no effect is involved. Therefore,

$$Z = \overline{A}\overline{C} + AC$$

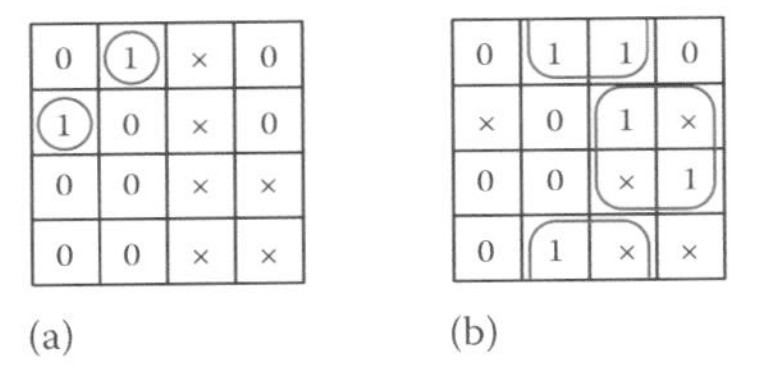

Figure 23.30 Benefitting from do not care entries. (a) All X's are considered 0 and do not appear in the output. (b) In order to group with 1's, some X's are considered 1.

g. This case is similar to case c with the exception that the two groups share one cell, but that does not affect the result.

$$Z = \bar{A}\bar{B} + \bar{C}\bar{D}$$

h. The answer based on the three pairs of cells forming three groups is

$$Z = \bar{B}\bar{C}\bar{D} + B\bar{C}D + \bar{B}CD$$

23.5.4 Don't Care Entries

The entries of a K-map are not always 1's and 0's. The application requirement determines those cells that must contain 1. Some of the cells also must contain 0 as a requirement, but, also, it can happen that for certain cells the content either is not defined or its value is immaterial. In such a case, it does not matter if a value of 1 or 0 is assigned to the cell. For a case like this an X is entered in the cell instead of 1 or 0. An example is shown in Figure 23.30a. These cells are called "don't care."

One can benefit from the "don't care" entries in simplifying the logic function. This advantage is not available in the example in Figure 23.30a. On the contrary, in the example shown in Figure 23.30b the logic function can be quite simplified if certain of these "X" cells are assumed to contain 1. One option, based on the encircled groups as shown and according to the definition of cells in Figure 23.28, is

$$Z = \bar{B}D + BC$$

23.6 Chapter Summary

- Logic gates are electronic devices that govern the digital process in computers, microprocessors, and digital control systems.
- The most common logic gates are AND, OR, NOT, NAND, NOR, and XOR.
- All gates except NOT have two inputs and one output. The output is the result of the action of the gate on its inputs.

- Each gate has a specific function that applies on its inputs.
- The function of each gate can be defined by a mathematical expression.
- *True* in logic circuits implies that a condition is satisfied.
- AND implies that both its inputs must be true so that the output becomes true.
- OR gate output implies that out of two alternative conditions at least one is satisfied; that is, at least one is true.
- NOT gate reverses its input. The output is the complement of the input.
- NAND is a combination of AND gate followed by a NOT gate.
- NOR is a combination of OR gate followed by a NOT gate.
- The output of a XOR (eXclusive OR) gates is true when ONLY one of its inputs is true, or in other words when the inputs are not alike.
- A logic circuit is made of a number of interactive gates. It may have a number of inputs to some gates. It has at least one output.
- The mathematical function of an AND gate is the product of the inputs.
- The mathematical function of an OR gate is the sum of the inputs.
- A truth table is used to define the output of a logic circuit and form the logic function. It can also be used to find the logic function of a logic circuit.
- Boolean algebra deals with logic (true and false) entries. It has a number of rules governing the relationships between logic values.
- Boolean algebra is used to simplify logic circuits.
- Simplified logic circuits are mainly in the form of sum of products and product of sums. In this way a logic circuit can be constructed by a number of AND and OR gates.
- Karnaugh map or K-map is another way of defining and simplifying a logic circuit.

Review Questions

1. What is a logic gate?
2. What is a logic gate used for?
3. What is meant by complement of a logic value?
4. Name the common gates.
5. How many inputs do logic gates normally have, and how many outputs?
6. Can gates be combined with each other? For what purpose?

7. When does the output of the AND gate go *true*?

8. What are the conditions that make the output of a NAND gate true?

9. If two inputs are simultaneously running an AND and a NAND gate, what is the relationship between the gate outputs?

10. In what condition is the output of an OR gate false?

11. In what condition is the output of an XOR gate false?

12. Which gate do you use if you need to multiply two logic values?

13. Can a gate have only one input? How?

14. What is the meaning of a truth table?

15. What is a truth table used for?

16. What is the result of AND'ing a logic variable with 1?

17. What is the result of OR'ing a logic variable with 1?

18. What mathematical operation defines the action of the OR gate?

19. Why do we need to simplify logic functions?

20. What is meant by sum of products?

21. What is meant by product of sums?

22. What is special about sum of products and product of sums?

23. What is a K-map used for?

24. How many cells does a K-map for four inputs have?

25. What is the main property of a K-map?

Problems

1. Simplify the following logic expressions:

$$Z_1 = \bar{A}\bar{B}C + \bar{A}BC$$

$$Z_2 = (\bar{A}\bar{B} + C)\cdot(\bar{A}B + \bar{C})$$

$$Z_3 = (\bar{A}\bar{B} + C)\cdot C$$

$$Z_4 = (\overline{AB} + C)\cdot(\bar{A} + B)$$

$$Z_5 = (\overline{AB} + B + C)\cdot(\overline{AB} + D)$$

$$Z_6 = (\overline{ABC} + B + C)\cdot(\overline{AB} + D)$$

$$Z_7 = (\overline{ABCD} + C)\cdot(\overline{ABC} + CD + \bar{C})$$

2. For all the expressions in Problem 1, draw the logic circuit (1) for the original problem and (2) for the simplified problem.

3. For all the logic expressions in Problem 1, make the truth table (1) for the original problem and (2) for the simplified problem. Verify your answers to Problem 1 by comparing the results of the truth tables.

4. Derive the logic expressions for the outputs of the circuits in Figure P23.1.

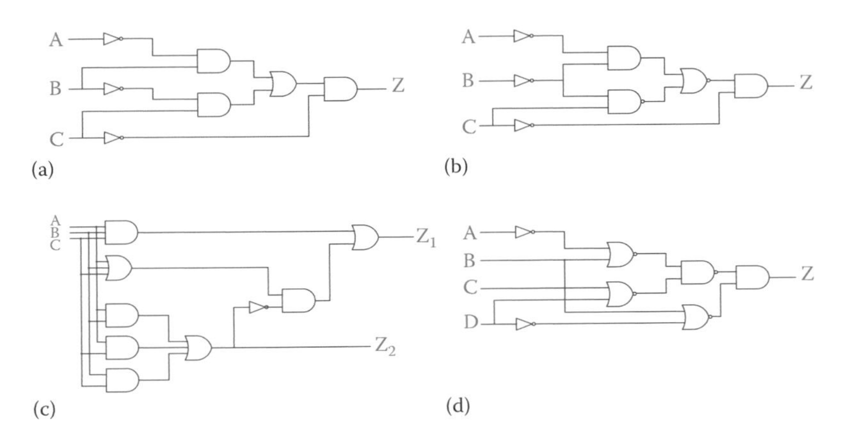

Figure P23.1 Circuits for Problem 4.

5. Derive the logic expression from the following K-maps (Figure P23.2). Use the notation in Figure 23.28.

6. Make the truth table for Problems 5a and 5c.

7. Simplify the output Z of the logic circuit in Figure P23.3 and draw the simplified circuit.

8. Convert the expression $Z = A\overline{B}\overline{C}D + ABCD + AB\overline{C}\overline{D} + A\overline{B}\overline{C}D + \overline{A}B\overline{C}\overline{D}$ to a Karnaugh map. By working on the K-map get a simple expression for Z.

9. Draw the waveform for the output Z of the logic circuit in Figure P23.4 based on the given input waveforms. Hint: Draw the waveforms for each individual gate first.

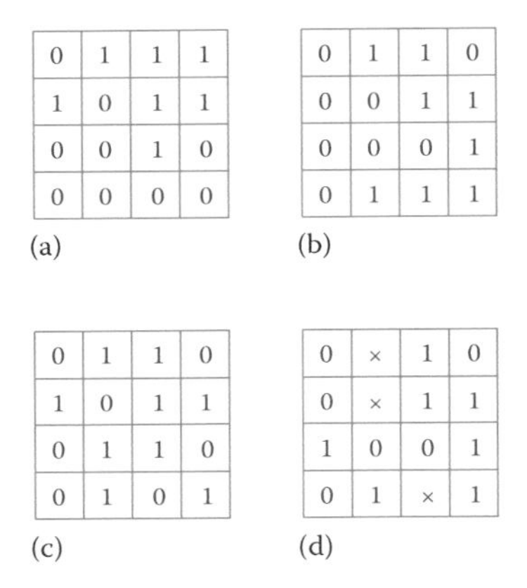

0	1	1	1
1	0	1	1
0	0	1	0
0	0	0	0

(a)

0	1	1	0
0	0	1	1
0	0	0	1
0	1	1	1

(b)

0	1	1	0
1	0	1	1
0	1	1	0
0	1	0	1

(c)

0	×	1	0
0	×	1	1
1	0	0	1
0	1	×	1

(d)

Figure P23.2 Circuits for Problem 5.

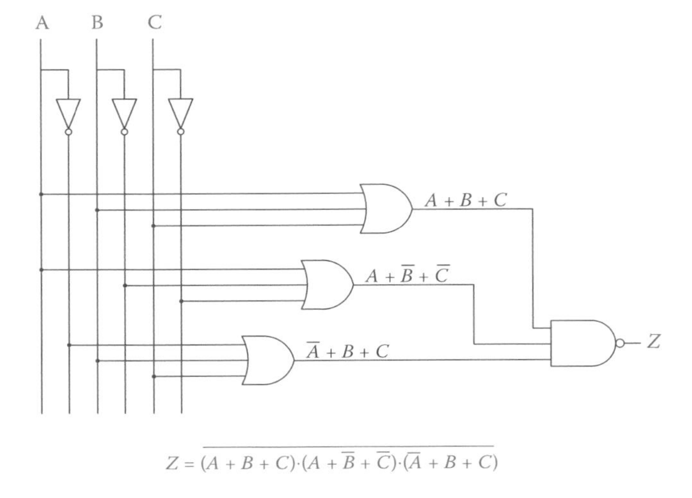

Figure P23.3 Circuit for Problem 7.

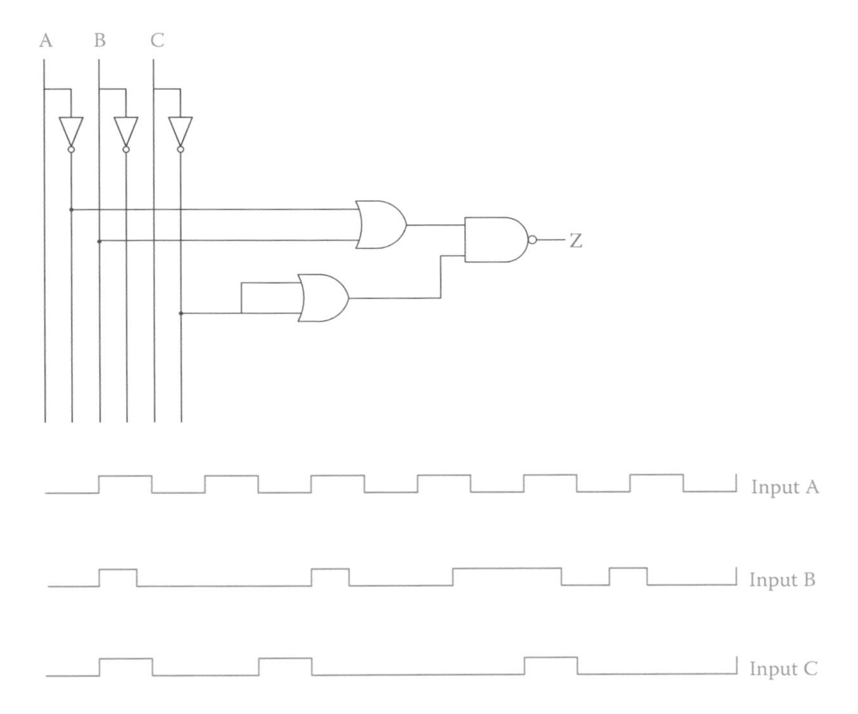

Figure P23.4 Circuit for Problem 9.

10. Make the Karnaugh map for the following truth table.

A	B	C	D	Output
0	0	0	0	*
0	0	0	1	
0	0	1	0	*
0	0	1	1	
0	1	0	0	*
0	1	0	1	
0	1	1	0	
0	1	1	1	
1	0	0	0	*
1	0	0	1	*
1	0	1	0	
1	0	1	1	*
1	1	0	0	
1	1	0	1	Undefined
1	1	1	0	Undefined
1	1	1	1	Undefined

24

Encoders and Decoders

OBJECTIVES: After studying this chapter, you will be able to

- Explain what a flip-flop is and what it is used for
- Define various types of flip-flops
- Understand toggle action in a flip-flop
- Describe how a flip-flop can be used for counting
- Describe how a flip-flop can be used for storing a *bit* of data
- Explain how a digital clock works
- Explain the difference between a latch and a flip-flop
- Describe the function and use of an encoder
- Describe the function and use of a decoder
- Understand the precision in a shaft encoder
- Explain what a multiplexer is and why we need it
- Describe the difference between a multiplexer and a demultiplexer

New terms:
Absolute encoder, active high, active level, active low, anemometer, asynchronous counter, bistable, combinational logic, D flip-flop, feedback, frequency divider, incremental encoder, J-K flip-flop, latch, reset, sequential logic, set, synchronous counter, wind vane

24.1 Introduction

In Chapter 23 we discussed logic gates, logic circuits, their mathematical functions, and how they can be used to formulate digital systems. Digital systems are the backbone of digital computers and digital industrial control.

In this chapter we discuss how data are handled and stored; for example, how an 8-bit system holds 8-bits and how codes from one form are converted to the other form. An example of the latter is how digital displays work. Considering a seven-segment display, the information in a binary form must light up different segments of the display. This requires a conversion of code from one form to another (binary to BCD).

All gates that we studied and their circuits lead to an output as a result of a combination of the actions of the elements in a circuit. For this reason, those circuits are referred to as **combinational logic**. In the new material an element of time plays a role, as well. Any stored data have to be maintained until they are altered. Stored data are in the form of 0 and 1. How these data can be controlled, in terms of whether and when to switch to a new value, is the subject of **sequential logic**, which we study in this chapter.

We start with the study of flip-flop, which is the basic memory element.

Combinational logic: Logic circuits made of a combination of logic gates but without any memory element, as opposed to a sequential circuit.

Sequential logic: Kind of logic circuit with timing sequence, memory, and synchronization.

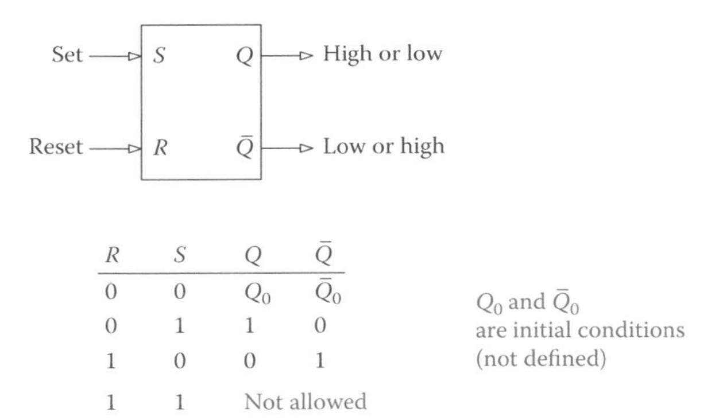

Figure 24.1 Basic flip-flop.

Bistable: Having two stable (controllable) states.

Set: (for a flip-flop) Putting a device to a desired state. Some devices have only two states on and off, or high and low. Set implies putting the device in a desired (usually nondefault) state.

Reset: Action of putting or setting a device to its default state.

Active level: The (high or low) level in which a logic device functions.

Active low: Implying that the active level for a device is low, as opposed to active high.

Active high: Two-state logic device (e.g., true and false, on and off, high and low) that acts (executes its function) when it is in a high state, as opposed to being active when in the low state.

24.2 Flip-Flop

A flip-flop is the basic memory element for storing a *bit* of information. It is an edge-triggered device. That is, it reacts to the edge of a pulse. A simple flip-flop has two stable states (remember, for instance, that a capacitor has two states: charged and discharged). States are represented by 1 and 0. Stable state implies that the device remains in one state (e.g., 0) until an input changes its condition to the alternative state. Because a flip-flop has two states only, it is called a **bistable** device.

A basic flip-flop has two inputs and two outputs, as shown in Figure 24.1. The two inputs are used to **set** or **reset** the device and are denoted by *R* and *S*. The outputs are denoted by Q and $\bar{Q}$; they reflect the state of the flip-flop either 1 or 0. If $Q = 1$, then $\bar{Q} = 0$, and if $Q = 0$, then and $\bar{Q} = 1$. Figure 24.1 illustrates also the truth table for the states of a flip-flop.

A flip-flop acts like a single-pole double throw (SPDT) switch. When activated, its output toggles. This property can be used for many operations performed in digital electronics, as we will see in this chapter. The toggle action is performed through the inputs to a flip-flop. There are a few types of flip-flops, as will be described shortly, in terms of their operational properties. The one shown in Figure 24.1 is the basic flip-flop or the simplest type.

Before continuing further we need to categorize flip-flops based on their structure, which defines the active level for a flip-flop. **Active level** is the logic level (high or low) that activates a device. For flip-flops it implies if a flip-flop is set, or its state toggles, with a logical high or a logical low applied to its *set* input. In this sense, there are two categories of flip-flops, those whose active level is 1 and those whose active level is 0, i.e., those that are **active high** and those that are **active low**.

24.2.1 Basic Flip-Flop

The basic flip-flop is set by applying an active level signal to its *set* input and the inverse signal to its *reset* input. When the unit is set, the Q output is high and the $\bar{Q}$ output is low. If, instead, the active level signal is applied to the reset terminal and its complement is applied to the set terminal, then the $\bar{Q}$ output is high and the Q output is low.

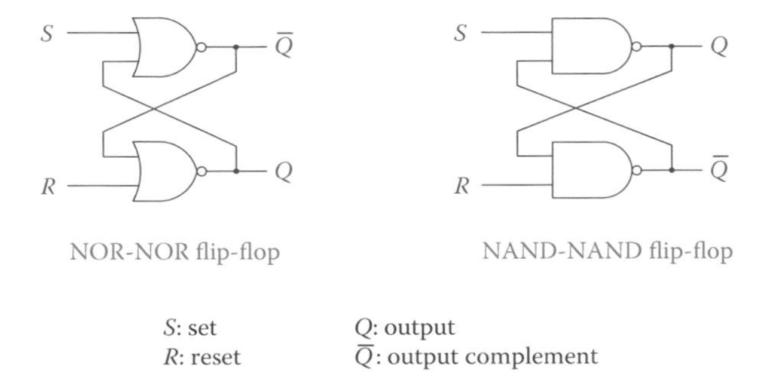

S: set
R: reset
Q: output
$\overline{Q}$: output complement

Figure 24.2 Internal structure of flip-flops.

Also, as shown in Figure 24.1, one cannot apply a logical high to both set and reset of a basic flip-flop at the same time. It is like simultaneously pressing the start and stop button of an electric motor.

Structurally, a flip-flop can be made of two NOR gates or two NAND gates. This is shown in Figure 24.2. The gates are interconnected to each other or, in other words, they have **feedback**. The output of each gate is one of the two inputs to the other gate.

Feedback: Taking the output of a device to compare with its input for the purpose of regulation and adjustment.

Note the output of the gate designated as S. For the NOR-NOR flip-flop this is $\overline{Q}$; whereas for the NAND-NAND flip-flop, it is Q. Also, the active level for the former is 1, and for the latter is 0. In both types a *set* input sets the content of output Q to high. To change the value of the output to low, the *reset* must be used. In other words, when the flip-flop is set, if the value of the input *set* changes to its complement it does not have any effect on the output.

> In a NOR-NOR flip-flop the active level is high, but for a NAND-NAND flip-flop it is low.

A flip-flop is either set or it is reset. If it is set, it implies that the *set* button is activated. In this case it is meaningless to set it again. Also, it is meaningless to make *reset* active, too. Thus, to reset a flip-flop, first, the *set* must be deactivated, and then the *reset* must be activated. Deactivation takes place by removing (turning off) the input signal.

Likewise, to set a flip-flop after it has been reset, first, the *reset* must be deactivated (this by itself has no effect and does not change in the status of the output), and then, if *set* is activated, it sets the flip-flop (it is meaningless to activate *set* if the *reset* is still active).

> A flip-flop is set by the *set* input when the *reset* is not activated. It is reset by *reset* input when *set* is not activated.

> **Note:** In a pushbutton the activation is performed by pushing, but immediately a spring brings the button to deactivated position after being released. A flip-flop does not have the spring action; thus, the deactivation has to be separately performed.

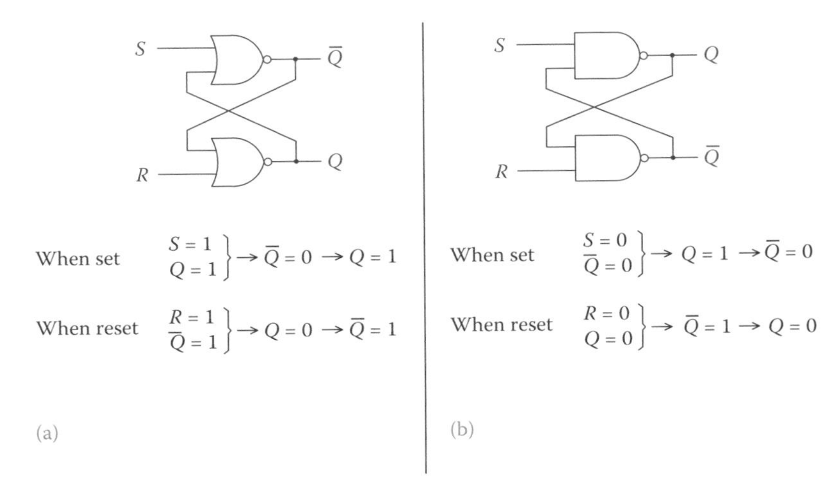

Figure 24.3 Stability reinforcement in (a) active high and (b) active low flip-flops.

What was said above can be observed in Figure 24.3 for both a NOR-NOR gate and for a NAND-NAND gate. Each gate has two inputs. One of the inputs, being either $\bar{Q}$ or Q, reinforces the stability of the output.

Figure 24.3 also shows that when a flip-flop is set, if S changes, nothing happens, and when it is reset, if R changes, nothing happens.

24.2.2 Clock-Controlled Flip-Flops

Clock-controlled flip-flops are more appropriate for synchronized operations, which is the case in microprocessor-controlled systems. The action of a flip-flop takes place with the clock pulse; therefore, it is in synchronism with all other operations (by other flip-flops). Setting, resetting, or toggling are the three actions of a flip-flop; depending on the inputs one is carried out at either the rising edge of the clock pulse or the falling edge of the clock pulse. This depends on whether a flip-flop is positive edge sensitive or negative edge sensitive. The symbols for the two are depicted in Figure 24.4.

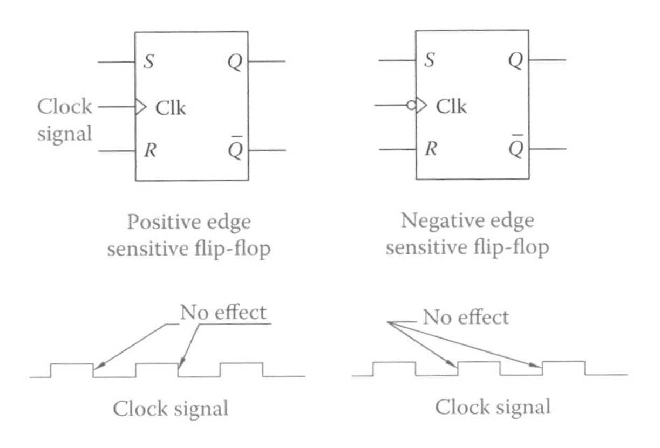

Figure 24.4 Symbols for positive and negative edge triggering flip-flops.

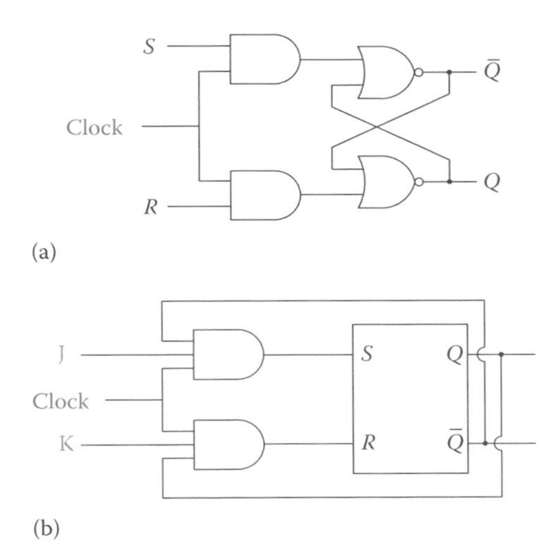

Figure 24.5 (a) Adding a clock input to a flip-flop and (b) structure of a J-K flip-flop.

For a desired action the data for the flip-flop inputs (0 and 1 values) are applied to them before the clock pulse enables the action. The clock input, shown as *clk* in Figure 24.4, is sometimes referred to as *enable* because it enables an action to take place. With the clock pulse, if the set input is activated, the flip-flop sets, and if the reset input is activated, it resets.

Figure 24.5a indicates how a clock or an enable signal is added to a flip-flop. This is shown only for a NOR-NOR flip-flop, but a NAND-NAND flip flop has a similar structure. Figure 24.5 also illustrates how a basic flip-flop can be used with two AND gates in forming a J-K flip-flop.

A **J-K flip-flop** is clock controlled and has the advantage that unlike the basic flip-flop it can accept both inputs (now called J and K inputs) to have an active level signal. This combination is prohibited for a basic flip-flop. The effect of having both inputs activated is a toggle of the output state. In a J-K flip-flop, J input is used as *set* and K input as *reset*. When both J and K inputs are active, the flip-flop toggles its output (each output switched to its complementary value, no matter what its current value is). The symbol and the truth table of a J-K flip-flop are shown in Figure 24.6.

J-K flip-flop: Flip-flop with two data entries, one for set and one for reset the output. It is also possible to toggle the output by activating both inputs.

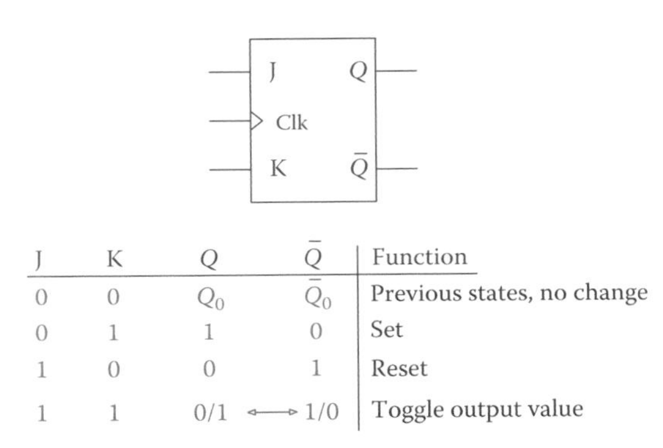

J	K	Q	$\bar{Q}$	Function
0	0	Q_0	$\bar{Q}_0$	Previous states, no change
0	1	1	0	Set
1	0	0	1	Reset
1	1	0/1 ↔	1/0	Toggle output value

Figure 24.6 J-K flip-flop.

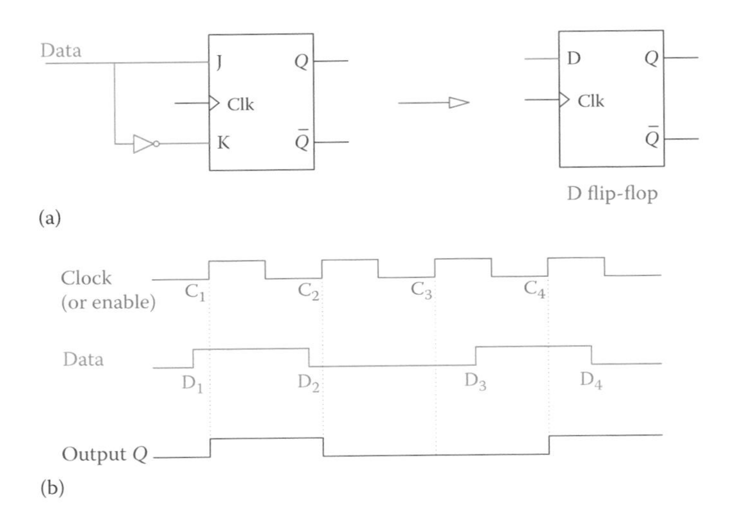

Figure 24.7 (a) Structure and symbol for D flip-flop. (b) Example of a D flip-flop timing diagram.

D flip-flop: Data flip-flop: a flip-flop with only one input whose output follows the input after the enable or clock signal.

T flip-flop: A flip-flop that can toggle. It has one data input and a clock input. It toggles its output when data input is 1 after clock input allows.

A data or **D flip-flop** is a clocked J-K flip flop that has only one input. The arrangement is as shown in Figure 24.7a, which indicates how the one output relates to the J and K terminals. The property of a D flip-flop is that the output follows the input. That is, with the clock pulse, the output will be whatever (0 or 1) has been placed on the input. Figure 24.7b shows this; even if the input (data) changes at points D_1, D_2, and so on, the output values update does not occur until the rising edge of the clock pulse, at C_1, C_2, and C_4 (assuming the device is rising edge sensitive).

Another product based on a J-K flip-flop is a **T flip-flop**. Similar to D flip-flop it has a clock input and only one data input. If the input is logic 1, then at the active clock pulse edge the output is toggled. Figure 24.8 depicts the

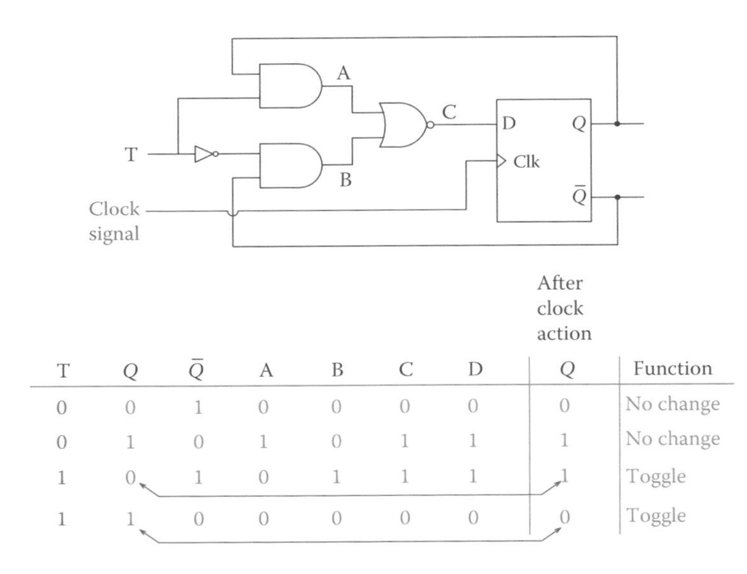

T	Q	$\bar{Q}$	A	B	C	D	After clock action Q	Function
0	0	1	0	0	0	0	0	No change
0	1	0	1	0	1	1	1	No change
1	0	1	0	1	1	1	1	Toggle
1	1	0	0	0	0	0	0	Toggle

Figure 24.8 Structure of a T flip-flop.

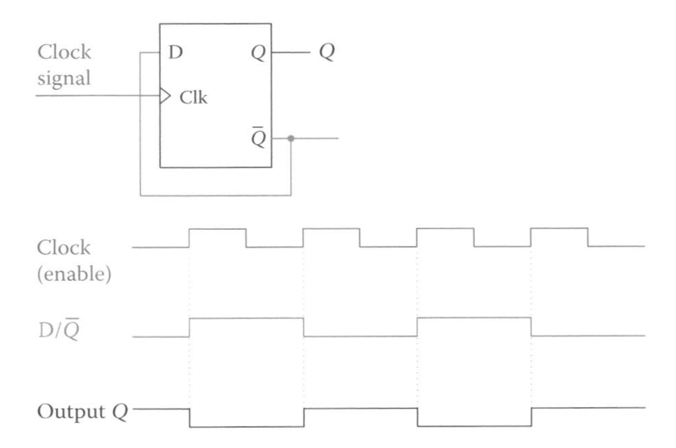

Figure 24.9 Toggle action with D flip-flop.

structure for a T flip-flop and the truth table indicating what happens after the clock signal enable the flip-flop action based on the input and the output state. Note that a T flip-flop is based on a D flip-flop in which Q follows D.

As we have observed, a J-K flip-flop toggles its output when both J and K inputs are set high. The toggle action takes place with the active edge of the clock pulse. Also, the T flip-flop toggles its output when the T input is high. Again, the action is carried out at the clock pulse edge. There is another arrangement where a flip-flop toggles with the clock pulse. This is with a D flip-flop as shown in Figure 24.9, in which the complementary output is fed back to the data input. Because the output must follow the input, as the clock enables the action Q takes the value of $\bar{Q}$, thus toggling. Notice that when a flip-flop toggles the frequency of its output pulse train is 1/2 of the clock frequency.

24.2.3 Initialization

Normally, when a logic circuit becomes active, for example, at power up, the initial condition of a flip-flop is not defined or not known. We may need to make sure in what state it is before proceeding with operations. For this reason, two other inputs are added to flip-flops, a **preset** and a **clear**. If the *preset* is activated it forces the output to go high ($Q = 1$), and if the *clear* is activated it forces the output to low ($Q = 0$). *Clear* and *preset* override the clock action. Figure 24.10 shows the symbol for a D flip-flop with these two entries and a sample timing diagram indicating their effect.

24.3 Latch

Flip-flops are devices that trigger with the edge of the *enable* (or clock) signal. In other words, they are edge-sensitive devices. Similar devices exist that are level sensitive instead of being edge sensitive. These are called **latch**. Giving this name stems from the fact that they hold the status after the input is removed. This is in contrast to a switch that when its contact is broken the corresponding effect is terminated.

Latch: Term in electronics when a device retains the effect of an input after the input has stopped; like the case of a pushbutton that is pressed to make a circuit, but when it is released, the circuit is not broken because another device (like a magnetic switch) maintains the connection.

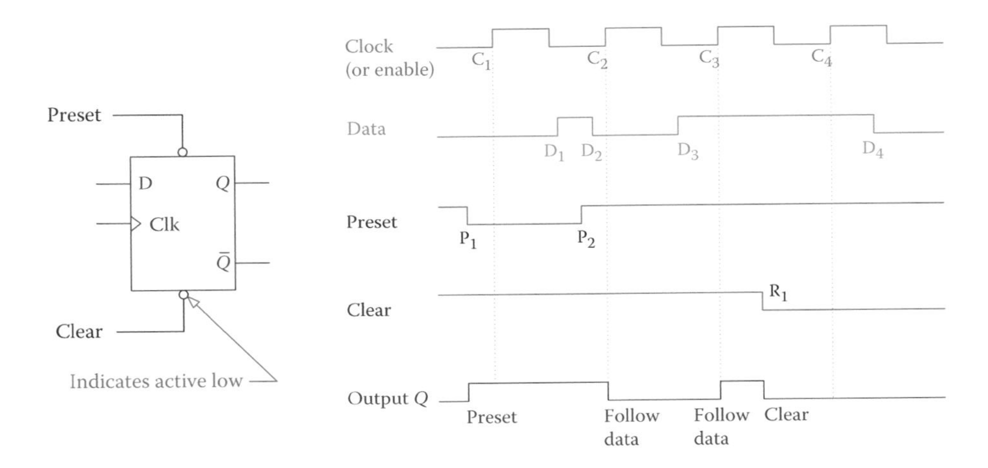

Figure 24.10 Flip-flop with *preset* and *clear.*

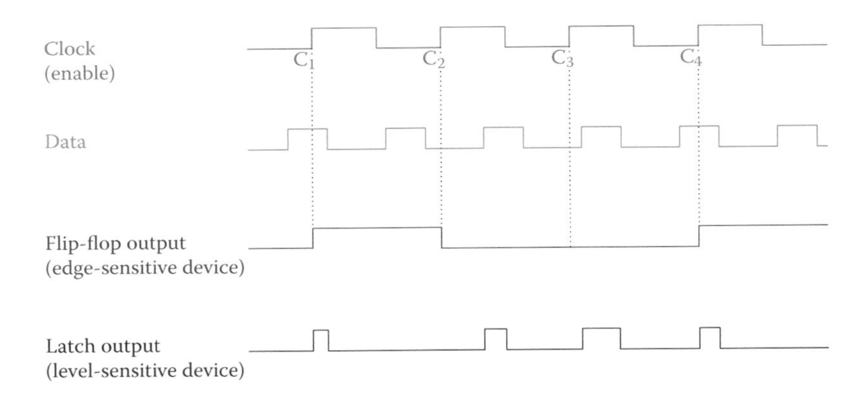

Figure 24.11 Comparison of the outputs of a latch and a flip-flop.

Figure 24.11 compares the actions of a flip-flop and that of a latch. Whereas for a flip-flop the change of the output state occurs when the clock signal edge enables it, for a latch the output is high as long as the clock has a high level and the input is high. For a flip-flop the clock pulse edge allows the output to go high if the input is already high. A rising edge sensitive flip-flop has been assumed in Figure 24.11.

24.4 Flip-Flop Applications

We can now look at a few examples of the applications of flip-flops as a storing device to hold data. A flip-flop can hold 1 bit of information. For N number of bits, thus, N numbers of flip-flops are needed.

Figure 24.12 indicates a register to hold 8 bits. As an example, the binary number 0110,1011 has been entered by some means, such as a switch set, or received from another device. The register holds the number for further processing.

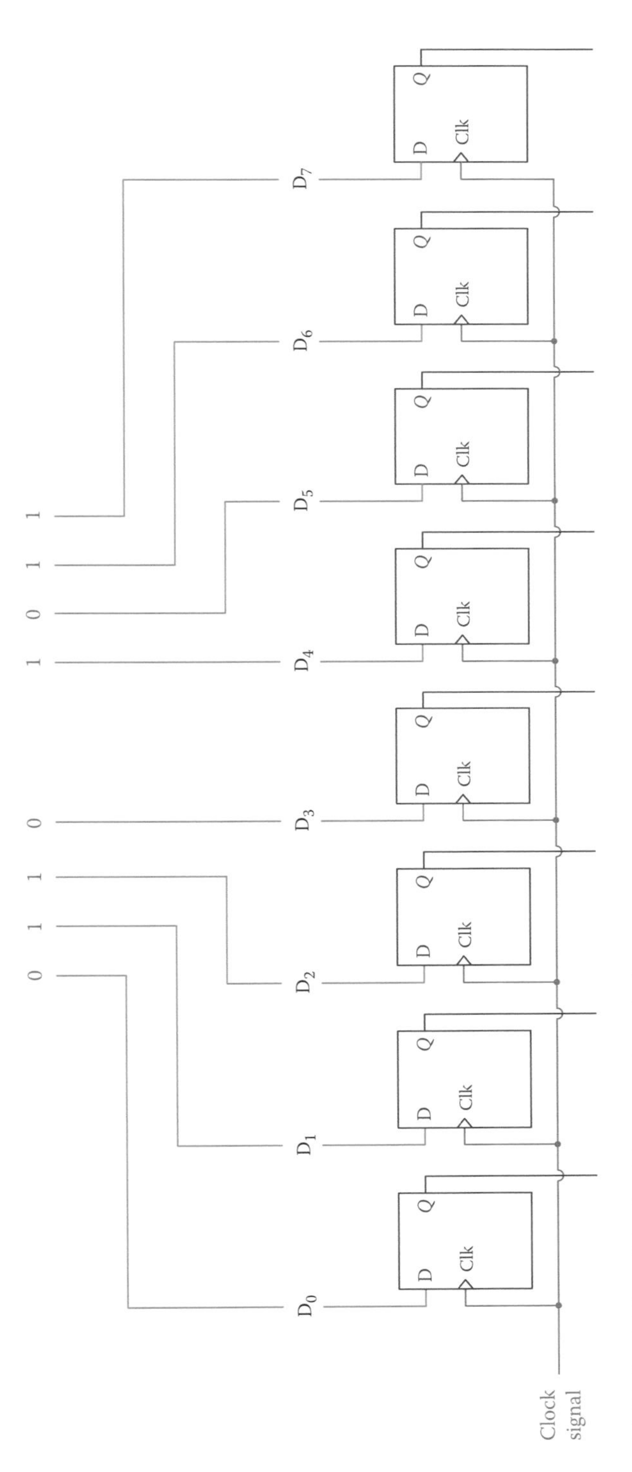

Figure 24.12 An 8-bit register for binary number storing.

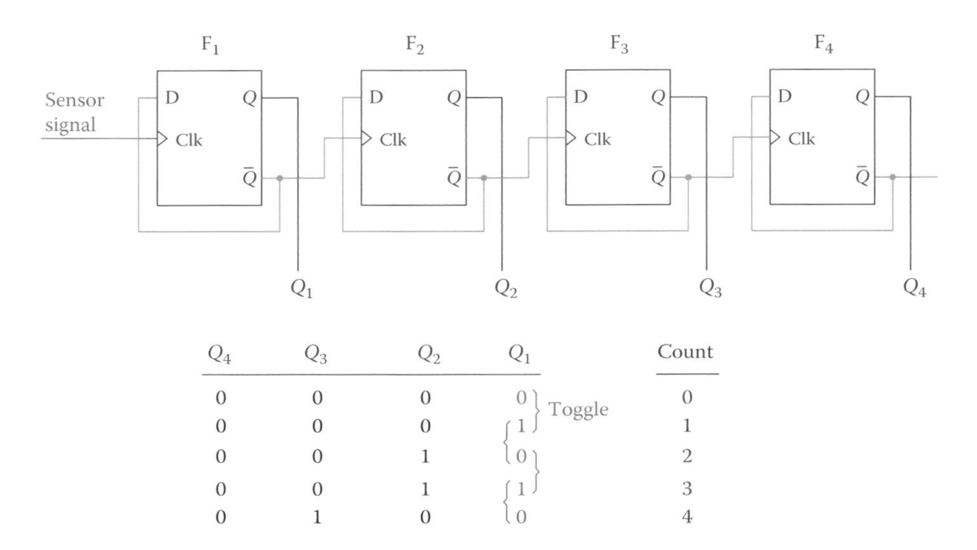

Q_4	Q_3	Q_2	Q_1	Count
0	0	0	0	0
0	0	0	1	1
0	0	1	0	2
0	0	1	1	3
0	1	0	0	4

Figure 24.13 Flip-flops used as a random event counter.

As a second example, consider the cascade of four D flip-flops in which the complementary output from each is the data or the clock input to the next. This arrangement can be used for counting. Two types of counting may be addressed, random event counting and clock counting. Figure 24.13 illustrates how a random event counter works. This can be used for counting people entering a building, for instance.

In the arrangement in Figure 24.13 all flip-flops toggle when their clock input receives an active signal. The flip-flops are numbered for ease of referencing. The complementary output of each D flip-flop is fed back to the data input as well as serving as the clock signal of the following flip-flop. In this way the stored value of each flip-flop is transferred to the next when it toggles from 1 to 0.

F_1 toggles with each entry from the external sensor. F_2 toggles with each change of state Q_1 of F_1 from 1 to 0; that is, for each two times toggle of F_1 it only toggles once. Likewise, F_3 toggles with each change of state of Q_2 from 1 to 0. Thus, it toggles once after F_1 has toggled 4 times and F_2 has toggled twice; and so forth. In this way the numbers stored in the output of the flip-flops is a binary number that represent the total count of events picked by the sensor. This set of four flip-flops can count up to 16. After 16 counts ($1111_b = 15$ is reached) the number goes back to 0. The total count can be extended by adding more flip-flops in the same way. Adding one more flip-flop extends the maximum count to 32. The count capacity is, therefore 2^N, where N is the number of flip-flops.

The flip-flops do not need to be D type. Any flip-flop that can toggle its states with the clock pulse can be employed for this purpose. Figure 24.14 depicts the circuit with gated *set-reset* flip-flops (gated implies it has clock input). Equivalently, a J-K flip-flop can be used, as shown in Figure 24.14b. In this case the two inputs for J and K must be held high, so that the flip-flop toggles with the clock pulse. Figure 24.14c indicates the contents of the flip-flop outputs and the count number starting with a *clear* signal that has set all the flip-flops to 0.

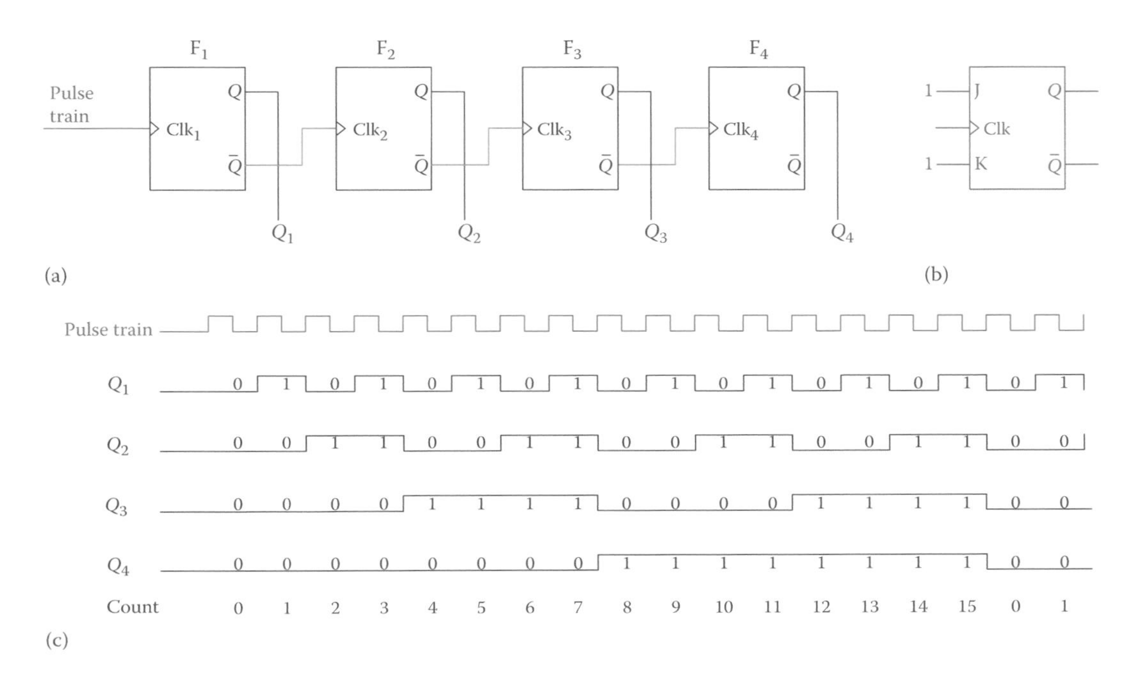

Figure 24.14 (a) Counting up circuit, (b) if J-K flip-flop is used for counter, and (c) toggle action at various flip-flops.

> Any flip-flop that has a clock enable input and can toggle its output with the clock pulse may be used for counting.

In Figure 24.14 a pulse train is shown as the main input. This implies that in addition to random count this circuit can have other applications. For example, a **frequency divider** uses this circuit. Observe the frequency of the outputs from the succeeding flip-flops. Each one has a frequency one half that of the previous one. Note that alternative of this circuit is also possible, for example, if the output (instead of the complementary output) of each flip-flop drives the clock signal of the next device.

Frequency divider: Electronic device (normally made up of flip-flops in digital electronics) that generates waveforms of lower frequency from a higher-frequency waveform.

With this circuit it is possible to count down instead, as well. One way is to initially preset all the flip-flops to high, instead of *clear*ing them. Another way is to use the complements of the output for counting (that is, if the readings are taken from the complementary outputs instead of Q_1 to Q_4). In such a case, the initial count is 15, and with each active edge of the clock it is reduced by 1 until 0 is reached. The next clock pulse then changes all the states to 1 (thus, number 15 again) and the process repeats.

Another application of the preceding circuit is the construction of a digital clock/watch. If the input to the first flip-flop is a pulse train with a constant frequency, the counting can be decoded to hours, minutes and seconds, and displayed. Decoding and conversion to display is discussed in Section 24.5. The clock pulse generator can be a circuit containing a quartz crystal, as that in digital watches, or the mains frequency (50 or 60 Hz) can be used for this purpose, as in many digital clocks. The former is more precise because it uses a quartz crystal whose frequency is in megahertz, but more divisions are necessary to bring the frequency to around 1 Hz, for the measure of 1 sec.

As can be observed from Figure 24.15, at some point it becomes necessary that the counting be interrupted and reset to 0. This is because the result of all counting is in binary; thus, the natural count goes up to 32 or 64, if we have 5 or 6 flip-flops, respectively. For example, using the 60 Hz line frequency in North America corresponds to a count of 60 for 1 sec. Thus, after counting up to 60, the seconds display is advanced by 1 and the counting process is reset to start from zero again. The same must be done after 60 sec are counted for 1 min, and after 60 min are counted for 1 hour.

For the hours, assuming 12-hour display, after the number count has reached 12, it cannot go to 13 and must start from 0 again. This sort of action is possible by adding a logic circuit so that when the count has reached its maximum allowed, a clear signal is activated to clear all flip-flops corresponding to that part of the circuit. This is shown in Figure 24.16.

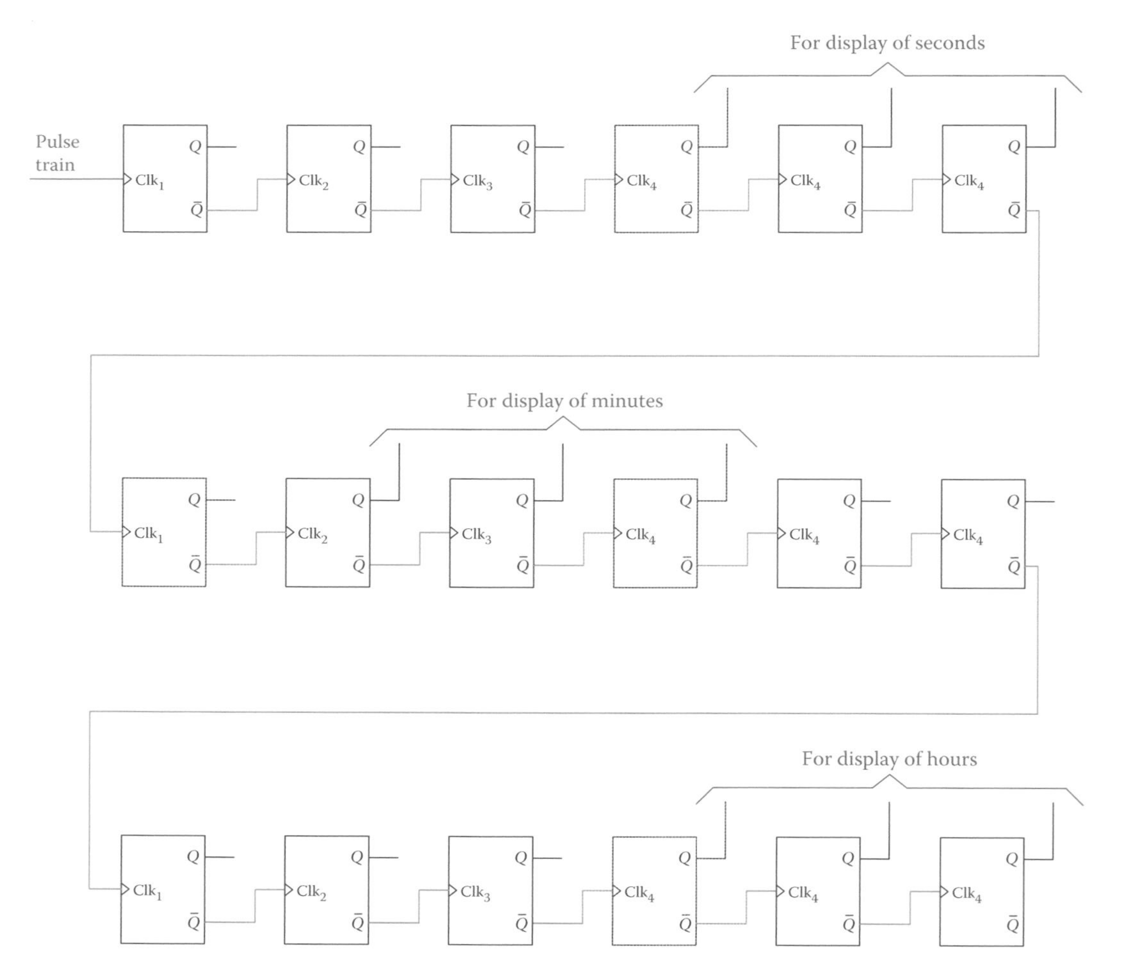

Figure 24.15 Basic circuit of a digital clock.

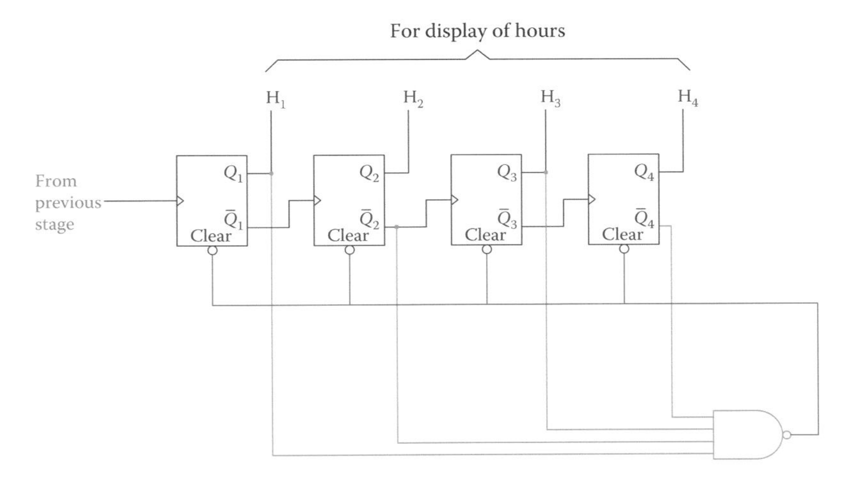

Figure 24.16 Interruption of counting by sending a "clear" signal to all flip-flops after a desired count is reached.

In Figure 24.16 the binary number reached by four flip-flops for the display of hours is 1010_2. We need to stop this number from increasing to 13. The solution is that after this condition is reached an enable signal is generated and sent to the clear terminal for all the four flip-flops. The clear input was not shown in the basic configuration in Figure 24.15 for more clarity, but as you expect a clock has more circuits than shown, for example, to set the clock to a desired hour.

Note that in Figure 24.16 the inputs to the AND gate are Q_1, Q_3, $\bar{Q}_2$, and $\bar{Q}_4$, corresponding to $12 = 1010_b$.

The arrangement shown in Figure 24.16 is also used in a frequency divider for dividing a frequency by a number that is not a power of 2. For instance, in Figure 24.14 it was shown that a waveform of pulses can be converted to another waveform with a frequency 16 times smaller. However, if we need to divide a frequency by 12, then we need to modify a circuit, similar to what was shown in Figure 24.16. A counter that counts up to N is referred to as MOD-N counter. For example, the counter in Figure 24.16 is a MOD-12 counter.

Figure 24.17 illustrates the difference between the output waveforms for a MOD-16 and a MOD-12 counter. Comparing the period (one cycle) of the square wave outputs, the first one (Figure 24.17a) corresponds to 16 counts and the second one corresponds to 12 counts.

The type of counter that we have studied is called **asynchronous counter**. The reason is that the toggling operations in various flip-flops do not take place at the same time; each one acts after receiving an enable signal from the previous flip-flop. In reality, there is a slight time delay in each operation, and, as a result, if the pulse train frequency is high and there are many flip-flops involved, an ambiguity in states can occur. In such a case a **synchronous counter** is used in which all the transactions are synchronized to happen at the same time governed by a clock pulse.

In a synchronous counter depending on how many flip-flops are involved, the outputs of certain flip-flops can be AND'ed together so that

Asynchronous counter: Electronic circuitry that can store binary numbers (used, for instance, for counting random events) consisting of a number of cascaded flip-flops. The change of state in various flip-flops does not occur at the same time, as opposed to synchronous counter.

Synchronous counter: Binary counter made up of flip-flops arranged such that all flip flops change state at the same time. Synchronization is governed by the clock pulse.

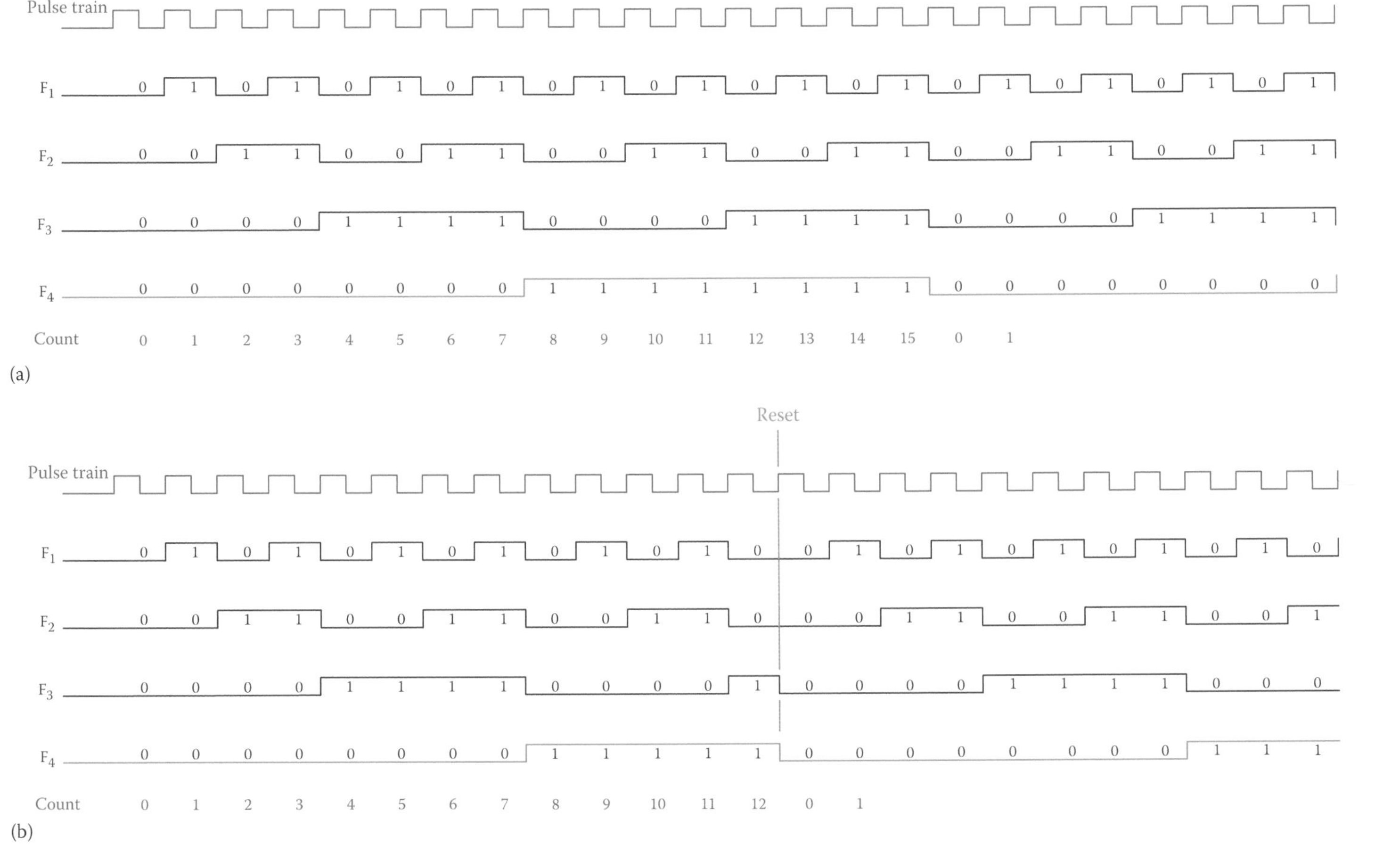

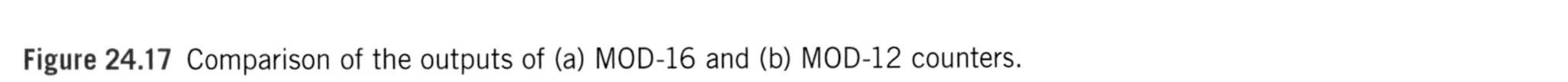

Figure 24.17 Comparison of the outputs of (a) MOD-16 and (b) MOD-12 counters.

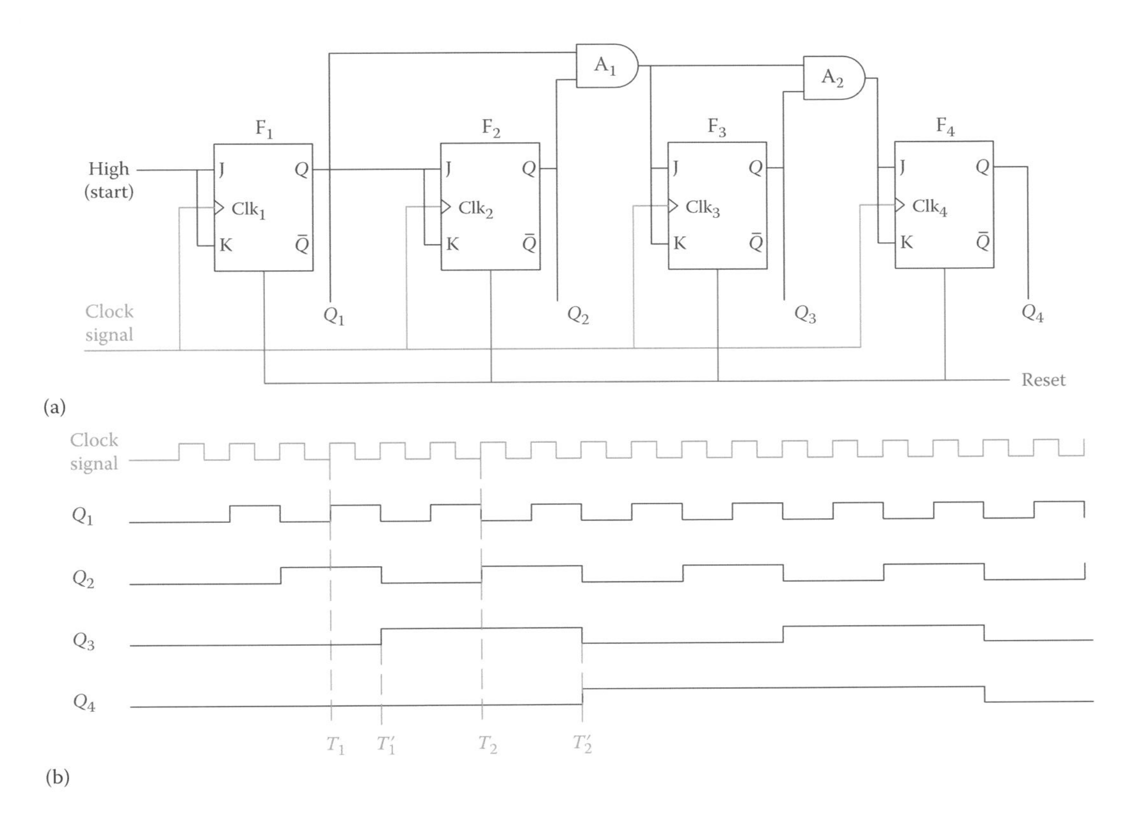

Figure 24.18 Synchronous counter. (a) Circuit and (b) timing diagram of a synchronous counter.

data transfer from one to the next is delayed and then performed when the AND gate allows. Otherwise, some flip-flops toggle before they are expected to do that. An example is shown in Figure 24.18 for a counter with four J-K flip-flops.

In this arrangement the outputs of flip-flops F_2 and F_3 are not directly taken to their succeeding flip-flop. Instead, they are routed through two AND gates. As can be observed from Figure 24.18b, at the time shown as T_1 the output Q_2 is high, and therefore flip-flop F_3 would toggle if the AND gate A_1 was not there. Also, at time T_2 the output Q_3 is high, and flip-flop F_4 would toggle if the AND gate A_2 was not there. But, AND'ing the appropriate outputs delays T_1 until T_1' for when both Q_1 and Q_2 are high. The same is true for the second AND gate.

24.5 Code Conversion

As we have learned so far, data can be stored in the form of binary numbers. Depending on the number of digits available, the magnitude of information stored changes, but more information can always be stored with less data. Having 4 bits of storage (say, four flip-flops), for example, offers the possibility of $2^4 = 16$ possible cases, not 4. The binary data are called *code*. Code conversion refers to the translation between the data stored in bits and the data for an application, in other words, conversion of data from one form to another.

A good example is the conversion from four bits of information that are available on four lines lighting up 10 decimal numbers (from 0 to 9) on a seven-segment LED. Although four bits (lines) can hold up to 16 numbers, all 16 do not necessarily need to be utilized. A seven-segment LED has seven LED bars, but not all must be turned on. The code converter receives the inputs for a number to be displayed and depending on the input states (corresponding to the number) determines which bars must be lit.

24.5.1 Decoder

In general, a binary code of N bits can represent 2^N distinct values. A decoder is a combination circuit that converts the binary information from N inputs to a maximum of 2^N distinct outputs. In a way it extracts information from a packaged data.

A decoder is a combination circuit meaning that it consists of various gates that put together a number of conditions carried by the input code. Figure 24.19 depicts the circuit for a decoder, which translates from BCD (binary-coded decimal) to decimal numbers. This circuit is good for lighting up one light out of 10 or opening a door out of 10 doors, for example, corresponding to each number. Comparing with 10 switches and 10 lines, the same function is performed with only four lines.

As another example consider a decoder that converts from binary digits to a seven-segment display to illuminate numbers 0 to 9, that is, for lighting up a one digit decimal number.

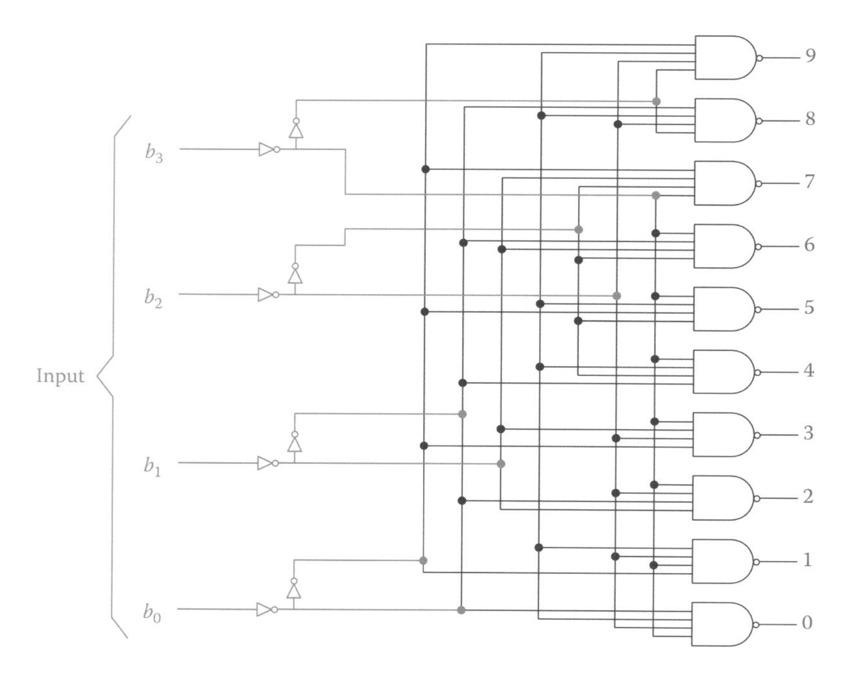

Figure 24.19 Circuit for BCD to decimal decoder.

Input				Output							Number displayed
b_3	b_2	b_1	b_0	a	b	c	d	e	f	g	
0	0	0	0	1	1	1	1	1	1	0	0
0	0	0	1	0	1	1	0	0	0	0	1
0	0	1	0	1	1	0	1	1	0	1	2
0	0	1	1	1	1	1	1	0	0	1	3
0	1	0	0	0	1	1	0	0	1	1	4
0	1	0	1	1	0	1	1	0	1	1	5
0	1	1	0	1	0	1	1	1	1	1	6
0	1	1	1	1	1	1	0	0	0	0	7
1	0	0	0	1	1	1	1	1	1	1	8
1	0	0	1	1	1	1	1	0	1	1	9

Figure 24.20 Truth table for BCD to seven-segment display.

The decoder logic circuit is defined based on writing the requirement for each output element and simplifying the resulting logic expression. For example, the requirement for a decoder to light up a seven-segment display is as shown in Figure 24.20. For each single digit number, certain segments in the display must be lit. The segments are named a to g, as shown in the figure. Those that must be lit are tabulated for all digits from 0 to 9.

Study of the table in Figure 24.20 reveals that it shows the truth table of the decoder circuit. Each segment in the display (a to g) can be expressed as a function of the four inputs. A Karnaugh map can be used to simplify the logic expression. This is done for segment a as an example.

Example 24.1

Find the logic expression for segment a in a seven segment display.

Solution

From the truth table we see that segment a must be lit in eight cases. These are shown in the Karnaugh map in Figure 24.21a. Before filling the rest of the cells in the K-map we notice that the combinations starting with 11 do not exist. Also, the combinations 1010 and 1011 do not exist (the binary combinations for numbers 10 to 15 do not exist). All the corresponding cells, therefore, can be filled with "don't care" X. But before that, we are going to alter the arrangement (swapping b_1 and b_2) in the K-map and draw it as shown in Figure 24.21b. This is a rearrangement of input combinations to have the two 0's on the same row, for simplification. This figure also shows the logic values (0 and 1) of the inputs on the right and bottom of the K-map. The don't care cells are now filled with X's in Figure 24.21c. The next step is simplification by grouping. The cells

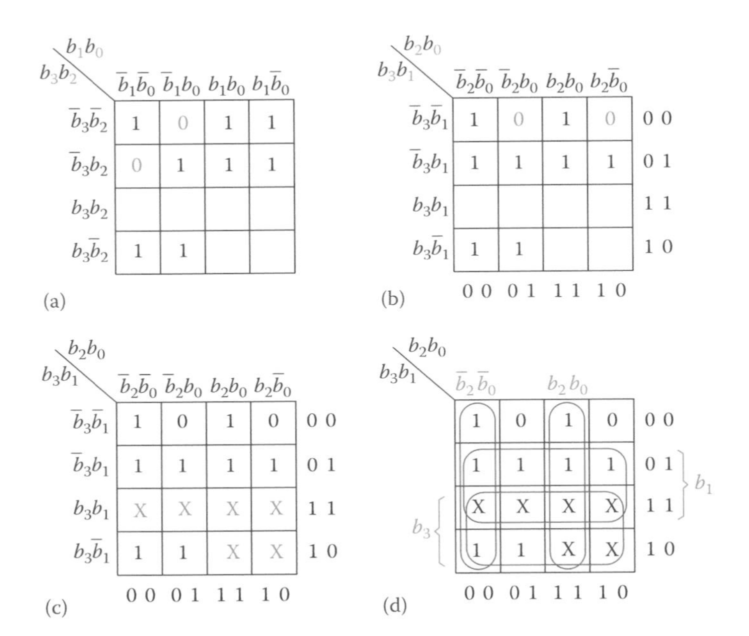

Figure 24.21 K-map for segment *a* of seven-segment display. (a) Information from Figure 24.20. (b) Swapping b_1 and b_2 for bringing both 0's to row 1. (c) Entering X in cells with "don't care" values. (d) Grouping cells for simplification.

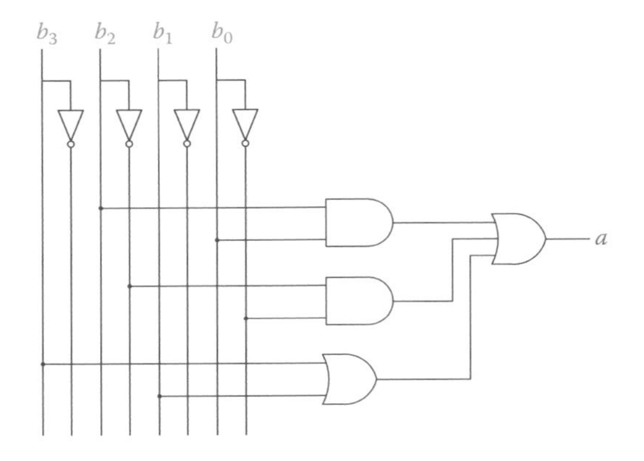

Figure 24.22 Logic circuit for segment *a*.

are grouped as in Figure 24.21d, which leads to the following expression:

$$\text{seg } a = b_3 + b_1 + b_0 b_2 + \overline{b}_0 \overline{b}_2$$

The logic circuit for segment *a* is illustrated in Figure 24.22.

24.5.2 Encoder

An encoder does the reverse of a decoder. It has 2^N or less inputs containing information, which are converted to be held by N bits of output.

The best example of an encoder is what is used to measure the rpm of a rotating shaft or to find the angle position of a shaft in one revolution. The first one, called **incremental encoder**, can be used in an **anemometer**, a device that measures the wind speed. The second one, called **absolute encoder**, can be used in a **wind vane** to detect the wind direction. There are plenty applications of encoders in industry and home devices. Without going into much detail we briefly study the encoder part of the second type.

This encoder generates a four- or five-digit binary code that can be transmitted with the same number (4 or 5) of wires carrying 0 and 1 values, which then can be decoded to find the angle position of a shaft within 360°. The resolution of measurement depends on whether four or five (or other) number of digits are used. The more the number of digits (and the wires) the more accurate the reading is. For four digits, each reading corresponds to $360 \div 2^4 = 22.5°$. For five digits, each reading corresponds to $360 \div 2^5 = 11.25°$.

The physical part of the encoder consists of a flat round disk on which four or five tracks are painted. Each track is divided into a number of black and white segments and a sensor can detect if a segment is black or white, which stand for 0 and 1. The arrangement of segments is such that each position of the disk (within the possible resolution) can be defined by a unique combination of 0's and 1's.

Figure 24.23 shows a disk with four tracks. Two types of segments are shown, one called binary code and the other gray code. With this type of encoder the gray code (which uses a different code than the binary numbers) offers an advantage that makes it more desirable in application. The advantage of the gray code over the binary code is in that moving from one number to the next implies only one bit change and not more than one. This allows detecting of any error if takes place in data reading and number processing.

For better clarity and comparison of the binary code and gray code the tracks containing black and white segments are shown straight (instead of round) in Figures 24.24 and 24.25.

Table 24.1 shows the difference between the binary code and the gray code.

Incremental encoder: Type of encoder (for a rotating shaft) that adds the number of revolutions and determines the total displacement of a shaft during a given time. This is as opposed to an absolute encoder that determines the displacement (and position) of a shaft within only one revolution.

Anemometer: Device to measure the wind speed.

Absolute encoder: Determines the angular position of a shaft or rotating body in its 360° angular displacement as opposed to incremental encoder.

Wind vane: Instrument for detecting the direction of wind, consisting of an aerodynamic body that rotates about an axis and can orient itself with the wind direction.

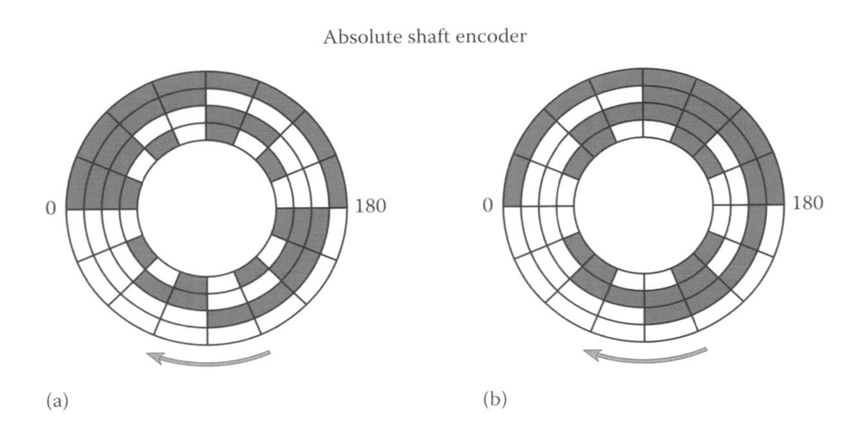

Figure 24.23 Tracks on a shaft encoder: (a) binary code and (b) gray code.

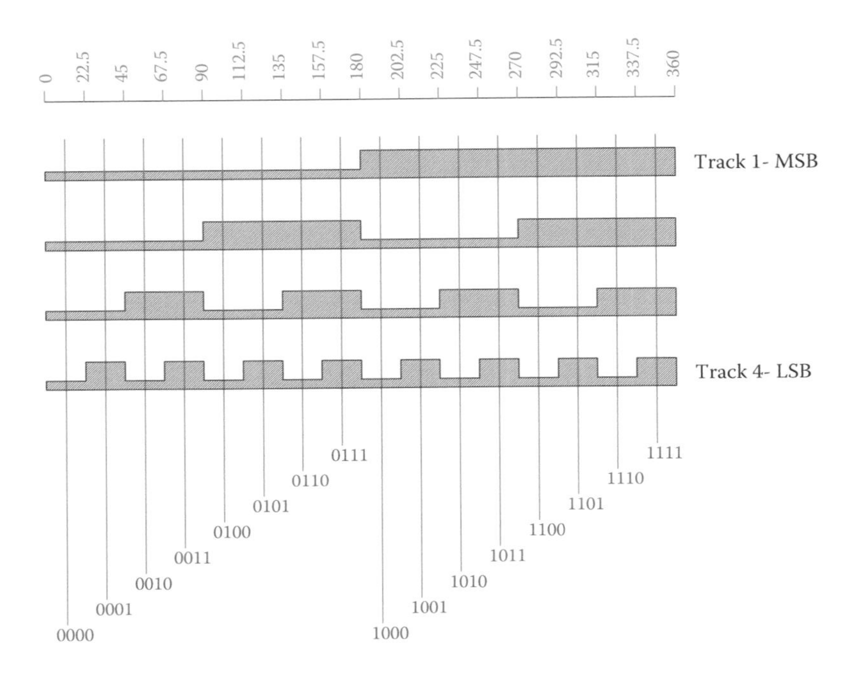

Figure 24.24 Representation of binary code tracks on the disk of a shaft encoder.

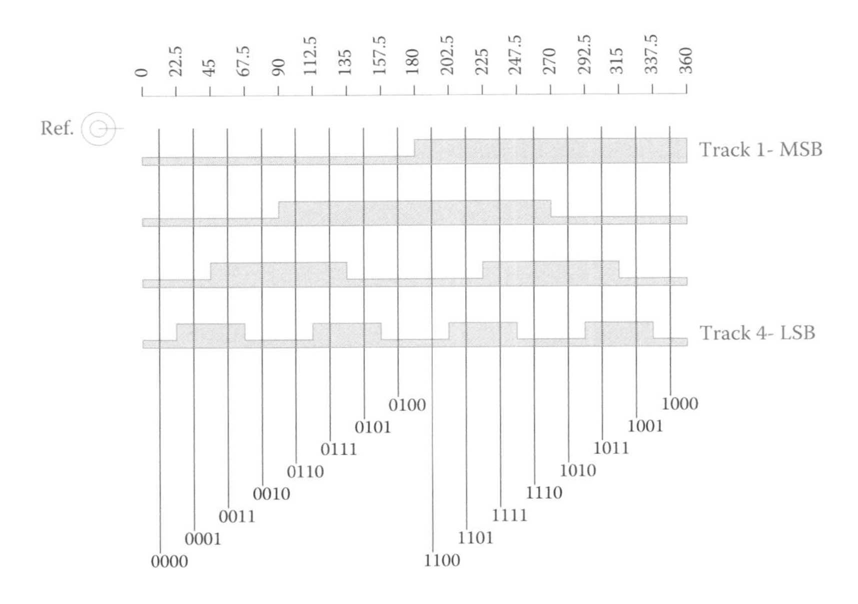

Figure 24.25 Representation of gray code tracks on the disk of a shaft encoder.

TABLE 24.1 Four-Digit Binary and Gray Code Comparison

Decimal	Binary Code	Gray Code
0	0000	0000
1	0001	0001
2	0010	0011
3	0011	0010
4	0100	0110
5	0101	0111
6	0110	0101
7	0111	0100
8	1000	1100
9	1001	1101
10	1010	1111
11	1011	1110
12	1100	1010
13	1101	1011
14	1110	1001
15	1111	1000

24.6 Multiplexer and Demultiplexer

Multiplexer and demultiplexer are two devices very important in data communications. As the name implies, their functions are opposite to each other (similar to encoder and decoder). These devices are used for sharing a device between two or more applications. Consider, for instance, a decoder for seven-segment display as discussed before. If we have a four-digit decimal number to display then we need four seven-segment displays and each one requires a decoder. However, by using a multiplexer and a demultiplexer, one unit of a decoder can be shared between the four displays.

The role of a multiplexer is to receive the information from different senders and deliver it to its output. By receiving a *select* data, a multiplexer is instructed from which sender to receive data. We may say, a single multiplexer (we may have multiple units of multiplexer) at a time selects information from one of many input lines and directs it to a single output. The *select* data then can successively address other inputs. N *select* lines can address 2^N data input lines as senders of information.

The role of the demultiplexer is to receive information from one line input and deliver it to one of many output lines or devices as specified by the *select* data, which is also received by the demultiplexer. By receiving a *select* signal a demultiplexer channels the main data to a particular address. In other words, a simple demultiplexer receives one channel of input data and directs it to one of the many output channels, as selected by the *select* data. Depending on the *select* data bit size a maximum of 2, 4, 8, 16, 32, and so forth can share the main data.

Figure 24.26 shows the schematic of a pair of multiplexer/demultiplexer for transferring data over a single line.

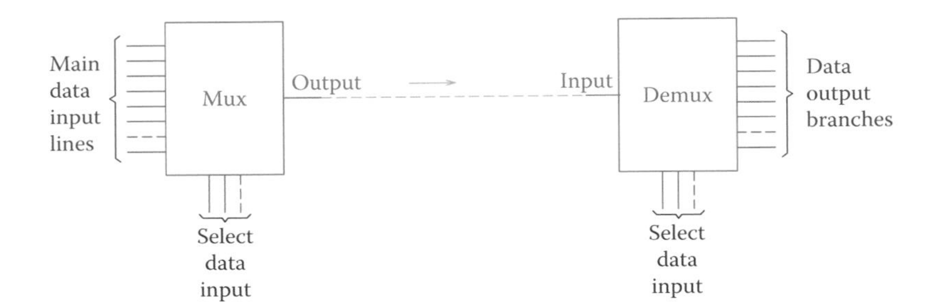

Figure 24.26 Schematic of a multiplexer and a demultiplexer.

In the arrangement in Figure 24.26, if the select data in the multiplexer have a pattern based on which the data lines are continuously selected in an orderly manner, the same pattern can be given to the demultiplexer, and consequently after synchronization all data are sent and received over the line between the two. This is what happens in communication devices.

Figure 24.27 illustrates a very simplified circuit for the example of the driver for a seven-segment display. The multiplexer in this example directs a set of four incoming data to its output, which goes to the decoder (driver). After decoding is done, the information for *a* to *g* segments are put on a data bus (a bus is a main line in an electric or electronic device that provides voltage or data to all components) that is shared between the four display units. Only one of the units will light up at a time because the select data are also sent to the demultiplexer, based on which only one of the display units is connected to the positive voltage. In here it is assumed that the display segments have a common anode that needs to be connected to the positive terminal. In general, the LED's in a seven-segment display can be connected in two ways. They have either a common anode or a common cathode. Their other sides are then connected to the driver data bus.

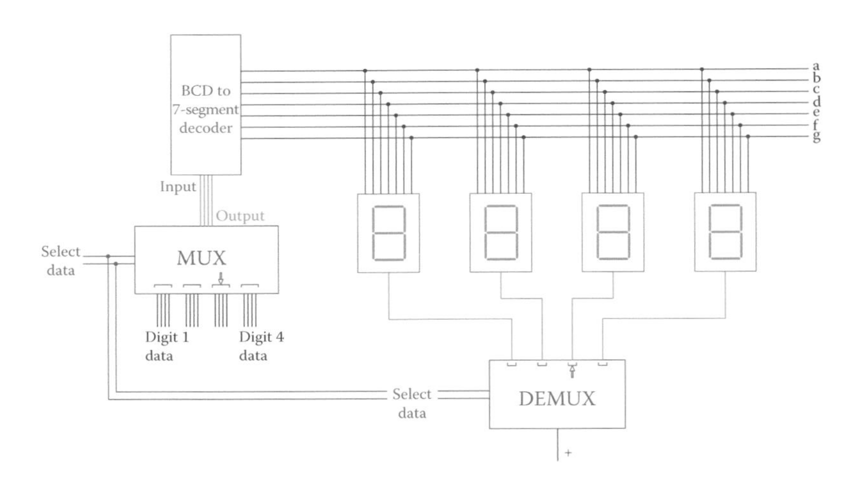

Figure 24.27 Simplified circuit showing multiplexer and demultiplexer for seven-segment display units sharing a driver.

24.7 Chapter Summary

- A flip-flop is a device to hold (store) one bit of information in the form of 1 and 0.
- When a simple flip-flop is set, its output is 1; when it is reset, its output is 0. The output is available for connection to another device.
- The output complement is also accessible for reading or using.
- A simple flip-flop is a set-reset device and does not have a provision for a clock signal.
- A clocked flip-flop has an input for clock signal or enable signal. Its function takes place at the active clock (enable) signal either at the rising edge or the falling edge of the clock signal.
- The active level is the logic level (high or low) that activates functioning of a device.
- A data flip-flop, also called D flip-flop, has only one input. Output takes the value of the input as the clock signal activates it.
- Other flip-flops are J-K and T flip-flops. A J-K flip-flop toggles when both of its inputs are logic high. A T flip-flop toggles when its only input is logic high.
- A significant property of the above flip-flops is that they toggle their outputs (swap its value). Flip-flops can be used for counting, time keeping, and frequency dividing (generating another signal with smaller frequency).
- A latch is a device similar to a flip-flop, but it is level sensitive. Its function can be performed when the clock signal has an active level.
- An encoder is a device that places up to 2^N binary data on N data lines.
- A decoder extracts up to 2^N binary information (made by an encoder) from N received data.
- A shaft encoder has a round disk with a number of tracks shaded differently in black and white segments. An incremental encoder is used for rpm measurement and an absolute encoder is used to determine the angular position of a shaft during one revolution.
- A multiplexer is a device to select one set out of several sets of received input data and to send it to its output. A particular input is selected based on the *select* data also received by the multiplexer.
- The function of a demultiplexer is opposite to that of a multiplexer. It receives one set of input data and directs it to one of the many outputs, as specified by the *select* data.

Review Questions

1. What is the main function of a flip-flop?

2. What is the principal difference between a flip-flop and a switch?

3. What is meant by toggle?
4. How many inputs does a J-K flip-flop have?
5. What is *preset* in a flip-flop used for?
6. What is the role of a *clear* input in a flip-flop?
7. What is the difference between functions of a D flip-flop and a T flip-flop?
8. How many inputs does a D flip-flop have, and how many does a T flip-flop have?
9. What are flip-flops used for?
10. What is the difference between a flip-flop and a latch?
11. What is the active level of a signal?
12. What is the advantage of using an encoder?
13. Why does one need a decoder?
14. What does an encoder do?
15. What is the function of a decoder?
16. Give an example of a decoder.
17. What is the advantage of a multiplexer?
18. What does a multiplexer do?

Glossary

Absolute encoder: Determines the angular position of a shaft or rotating body in its 360° angular displacement as opposed to incremental encoder.

AC circuit: Any electrical circuit with alternating current as the energy source.

Active high: Two-state logic device (e.g., true and false, on and off, high and low) that acts (executes its function) when it is in a high state, as opposed to being active when in the low state.

Active level: The (high or low) level in which a logic device functions.

Active low: Implying that the active level for a device is low, as opposed to active high.

Active power: The portion of electric power in AC circuits that converts to heat or mechanical work (like the work by a motor). It is also called true power and real power.

Active region: An area in its characteristic curve that a transistor can function. It represents the range of voltage and current values for a transistor collector. If the collector voltage or current is very low, a transistor falls in saturation region or cutoff region and cannot function.

Actuator: Any physical device that can generate a mechanical action (e.g., rotation, force exertion, moving) based on an electric, hydraulic, or pneumatic excitation.

All-day efficiency: Average efficiency of a transformer during a 24-hour period.

Alternating current: Type of electricity in which current continuously and regularly changes direction (i.e., electrons in the wires move back and forth, as opposed to direct current in which electrons move only in one direction).

Alternative current: Same as alternating current.

Alternator: Generator of alternating current (a machine that produces alternating current electricity when turned).

Ammeter: Device to measure electric current.

Ampacity: A word (name) made of ampere and capacity. It is used to define the current capacity of standard conductors (wires) in different working conditions for safe operation. Ampacity is determined based on the heat generated in a conductor due to the current through it.

Ampere: Unit to measure electric current.

Amplification: Enlarging the intensity of a signal. For an electric signal, intensity can be the power, through the voltage, the current or both.

Amplifier: A (normally electronic) device that amplifies (enlarges the strength of) an electric signal.

Amplitude modulation: A way of embedding a signal to be transmitted (called the modulating signal; for example, music for radio broadcast) to a signal called carrier (for example, radio frequency electromagnetic waves) that can be transmitted through antennas across the air (or empty space) to a distant destination. In amplitude modulation, the amplitude of the carrier signal is modified by the effect of the modulating signal. AM radio uses this type of modulation (see also FM: *frequency modulation*).

Amplitude modulation ratio: Ratio of the amplitude of the reference signal to that of the carrier signal in pulse width modulation, used in power converters.

Analog: Any natural phenomenon, which has a continuous nature and is not discrete, like the output of a sensor and a measurement device (e.g., temperature, current, voltage).

Analog-to-digital converter: Electronic device that generates a digital signal with a desired frequency from an analog signal so that the signal can be treated with digital devices.

Anemometer: Device to measure the wind speed.

Angle of flow: Measure of percentage of time (based on 100 percent for 360°) for one cycle of a sinusoidal waveform during which an amplifier is in an operating state (transistor is conducting in a transistor amplifier.)

Anode: Positive side terminal in a diode and a thyristor.

Apparent power: Power that must be supplied by an AC source for powering its loads.

Armature: Rotating part in a DC machine, especially the windings of this rotating part.

Asynchronous counter: Electronic circuitry that can store binary numbers (used, for instance, for counting random events) consisting of a number of cascaded flip-flops. The change of state in various flip-flops does not occur at the same time, as opposed to synchronous counter.

Asynchronous motor: Another name for induction motor (see induction machines).

Atom: Smallest particle of an element having the same properties of the element, as comparable to a molecule that is the smallest particle of a compound material.

Atomic mass: The mass of an atom or molecule measured in atomic mass unit. The unit is 1/12 of the mass of carbon atom (atomic mass of carbon = 12). For the elements with isotopes, the atomic mass is the average value based on the atomic mass for each isotope and its abundance in nature. Atomic mass depends on the number of electrons, protons, and neutrons in an element.

Atomic number: Unique number for any element representing the number of protons in the element.

Atomic weight: Atomic weight is the ratio of the atomic mass of an atom or molecule to the atomic mass of carbon-12. Therefore, it does not have a unit since it is a ratio. The number for atomic weight is the same as that for atomic mass (without a unit).

Attenuation: Reducing the intensity of an electric signal to match measurement and other devices.

Autotransformer: Transformer with only one winding per phase (instead of two). Primary and secondary windings are part of the same wire winding.

Avalanche current: Relatively higher current in a semiconductor device after breakdown occurs.

Back electromotive force (BEMF): Electromotive force that is generated in an electric motor due to its rotation (as what happens in a generator) and is in the opposite direction to the applied voltage and tends to stop the motor from rotation. It acts as a brake to the operation of a motor.

Balanced load: Three identical loads in a three-phase system that come together as a three-phase load.

Band pass filter: Type of electric filter that lets signals in a range of frequencies pass from the input side to the output. All other signals with frequencies outside this range are blocked.

Band reject filter: Opposite of band pass filter. All the signals with frequencies in a defined range are blocked from reaching the output.

Band stop filter: Another name for band reject filter.

Bandwidth: Range of frequencies in a band pass or band reject filter.

Base: One of the three principal terminals of a junction transistor (ordinary transistor).

BEMF: Back electromotive force.

Biasing a transistor: Connecting the three terminals of a transistor to the appropriate voltage to adjust/select the operating region of a transistor.

Binary: Having two states: 0 and 1, true and false, high and low, on and off, yes and no.

Binary-coded decimal (BCD): Special representation of decimal values when each individual digit is coded separately in binary form and can be displayed by digital circuits.

Bipolar junction transistor: Another name for junction transistor.

Bistable: Having two stable (controllable) states.

Bolt short circuit: Short circuit by direct contact between two lines of different voltages.

Boolean algebra: Number of rules (different from regular algebra rules) for logic gates and logic circuits.

Breakdown point: Point in the characteristic curve of a semiconductor device where the applied voltage is beyond the limit that the device can stand without a sudden change in behavior or without getting damaged.

Breakdown torque: The same as the pull-out torque. The maximum torque on an induction motor speed–torque characteristic curve that can be provided by the motor before reaching the operating speed, where the motor continues to work at full-load torque (see the characteristic curve of an induction motor).

Breakdown voltage: Voltage at which a semiconductor device changes behavior or gets damaged.

Breaker: Normally physical device that breaks an electric current and disconnects a device from electricity in order to protect it from damage.

Breakover voltage: Voltage below which a diac or a triac is not conducting, but when the voltage surpasses that conduction starts.

Bridge rectifier: Full-wave rectifier of AC using four diodes (for single phase) or six diodes (for three phase) to obtain direct current from alternating current.

Brush: Solid block component made mainly from carbon (can have other substances such as metal powder) for transferring electricity from a rotating part to the stationary part in a motor and generator. Brush is stationary and pressed by springs on the rotating rings/segments.

Byte: Binary number consisting of 8 bits; 8 bits of information form a byte.

Capacitance: Measurable property of a capacitor for storing electricity, measured in Farad.

Capacitive reactance: Reactive effect of a capacitor when connected to AC electricity, measured in ohms. Capacitive reactance determines the current in a circuit containing a capacitor.

Capacitor: Electric device mainly made of two conductors separated by an insulator. It has the property of storing electric energy in the form of an electric field. Energy stored in a capacitor can only be discharged instantly, not like a battery in which the stored energy can be gradually discharged.

Capacitor-run motor: Alternative current type of motor that needs a capacitor in parallel with it in order to work.

Carrier: A waveform of relatively high frequency in pulse width modulation and other applications (especially in communications) that behaves as a carrier to transfer or treat signals of low-frequency such as voice (in telecommunication) and an electric alternating current waveform (50–60 Hz).

Cascade: Way of putting devices of the same functionality together in succession to each other for enhancing their effect, such that the output from one is the input to the next, like two or more amplifiers.

Cathode: Negative side terminal of a diode or similar device, as opposed to anode.

Cathode ray tube: Vacuum tube for showing images in devices such as TV and oscilloscope, nowadays replaced by LCD and other types of screens.

Characteristic curve: Line or curve that shows the relationship between two or more parameters affecting the functionality of a device, like the relationship between voltage and speed in an electric motor.

Circuit: Any combination of electric and electronic components connected together by wires to be connected to an electric source.

Circuit breaker: Same as a breaker, to disconnect a device or a circuit from power source to prevent damage.

Circular mil: Unit for measuring the thickness (cross section) of wires. A CM is the area of a circle whose diameter is one mil (1/1000 of an inch).

Coil: Any wire wound in circles to form a helix, usually with a core to enhance the inductance.

Collector: One of the three main terminals of an ordinary transistor.

Combinational logic: Logic circuits made of a combination of logic gates but without any memory element, as opposed to sequential circuit.

Combined circuit: Circuit having a number of the same electric components in both series and parallel at the same time.

Common base: One of the three configurations for using transistors in a circuit, where the base is a part of both the input and the output.

Common collector: One of the three configurations for using transistors in a circuit, where the collector is a part of both the transistor input and the output.

Common emitter: One of the three configurations for using transistors in a circuit, where the emitter is a part of both the transistor input and its output.

Commutator: Part of the armature of a direct current motor or generator consisting of many small metallic (normally copper or brass) segments in the form of a ring which can slide under a brush. The brushes are spring pressed on the commutator and transfer the current from the rotating commutator to the outside of the armature. The commutator has the role of rectifying the otherwise AC current in the armature.

Complement: Opposite of a binary entity. It is shown by an overbar or a prime. For example, if A is a binary variable and its value is 1, its complement is represented by $\bar{A}$ or A' and its value is 0.

Compound: Any material that is not a simple element and is made of at least two elements.

Conduction angle: The same as "angle of flow."

Converter: General term for a rectifier (a device that converts AC to DC) and an inverter (a device that converts DC to AC).

Copper loss: Amount of power loss in a transformer that corresponds to the resistance of the wire winding and it depends on the load current (the percentage of loading).

Core loss: Amount of power loss in a transformer that corresponds to the quality of design and core material and is independent of the load current.

Core loss current: Current in the primary winding of a transformer that corresponds to the core losses; that is, it is equal to the core loss power divided by the primary voltage. It exists even if there is no load connected to a transformer.

Core-type transformer: Category for the construction of a transformer in which the core has the shape of a rectangular frame, and the windings are normally on the opposite sides of the rectangle.

Corona: Corona is a local electric discharge around the conductors of high-voltage transmission lines, when the (especially humid) air is ionized. It is manifested by a bluish glow and a small humming sound. This phenomenon is accompanied with a slight loss of power used for making the sound and the glow.

Cosine (of an angle): Trigonometric function for an angle represented by the ratio of the length of the side adjacent to the angle to the length of the hypotenuse in a right triangle.

Cotangent (of an angle): Trigonometric function for an angle represented by the ratio of the length of the side adjacent to the angle to the length of the side opposite to the angle in a right triangle.

Coulomb: Measure of the amount of electricity equal to the electric charge of 6.241×10^{18} number of electrons.

Counter electromotive force (EMF): Same as back electromotive force (BEMF).

Coupling (in amplifier): Act of properly connecting an amplifier to its input or output or connecting two amplifiers in cascade.

Cumulative compound wound: Type of DC electric machine with both parallel and series field windings where the magnetic effects of the parallel and series windings add together, as opposed to the differential compound wound.

Current gain: Ratio of the output current to the input current in an amplifier.

Current source inverter (in back-to-back converter): Back-to-back converter arrangement in which an inductor is used between the rectifier and inverter for instantaneous electricity storage. In a voltage source inverter a capacitor is used for this purpose.

Cut-in speed: Wind speed at which a wind turbine starts generating electricity and below which a there is not sufficient power in the wind. Also, a wind turbine stops generating electricity if the wind speed drops below cut-in speed.

Cutoff: Status of a transistor when it is not operating because of very low current.

Cut-out speed: Wind speed at which a wind turbine has reached its maximum design capacity and must be stopped. If it continues to work, the power in the wind surpasses the mechanical and electrical capacities; the generator or a mechanical component becomes overloaded and gets damaged.

Cycle: Pattern of all values that will be continuously repeated in a cyclic variable. For example, in alternative current electricity a cycle is the complete set of positive and negative values based on a sine function of an angle when it varies from zero to 360°.

Cyclic waveforms: Waveform corresponding to a variable entity whose variation has a repeat of the same pattern over and over.

D flip-flop: Data flip-flop: a flip-flop with only one input whose output follows the input after the enable or clock signal.

DC circuit: Electric circuit with only DC source(s).

DC link: Two-line electric link between the two parts of a back-to-back converter. Voltage in this link is only DC.

Decimal system: Most common numbering system with ten numbers (0 to 9) as the base for making all other numbers.

Delta connection: Way of connecting loads and sources to a three-phase system, where the components of a load or a source form a triangle (or delta, Δ) each corner of which is connected to the three lines.

Δ connection: Same as delta connection.

Demodulation: Opposite action to modulation. Action of separating the original modulating signal from a modulated signal by filters.

Depletion region: Narrow width area in a semiconductor junction just between the N-type and P-type materials.

Deuterium: Isotope of hydrogen. When composed with oxygen, it produces heavy water.

Diac: Type of semiconductor device. It acts as a diode for alternative current, restricting the current until certain voltage is reached and then it conducts normally.

Diamagnetic: Materials without any significant magnetism property, such as wood and clay. They can never become a magnet.

Differential compound wound: Type of DC electric machine with both parallel and series field winding where the magnetic effects of the parallel and series windings are opposite to each other (the magnetic field from the parallel winding is reduced by the smaller magnetic field of the serial winding).

Digital (1): Variable that has discrete (not continuous) values, as opposed to analog. It could be the result of digitization of an analog value.

Digital (2): Representation by readable numbers.

Digital multimeter (DMM): A device to measure resistance, current, voltage, and other electrical parameters, in which the reading is automatically adjusted and displayed by 3 or 4 digits, as compared to an "analog multimeter" in which the position where a moving needle stops indicates a reading.

Digital-to-analog converter: Electronic device that translates the value of a digital signal generated by a computer or a digital device into an analog value that can be used in an electric circuit.

Direct current: Type of electricity in which there are positive and negative poles (or sides) and electric current is always in the same direction between the two poles, as opposed to alternative current in which the current direction continuously changes (at a fixed frequency).

Direct-drive turbine: Wind turbine with no gearbox.

Dopant: Very small impurity (from elements with three or five valence electrons) added to pure silicon or germanium to form a semiconductor material.

Doped material: Pure silicon or germanium to which a very small impurity has been added.

Doping: Act of adding a very small amount of dopant to pure silicon or germanium.

Doubly fed induction generator (DFIG): Type of induction generator with a wound rotor. Both the stator winding and the rotor winding need to be connected to electricity (thus the term doubly fed) as opposed to a squirrel cage induction generator.

Drain: One of the three terminals in a field effect transistor (FET).

Driver: Circuit to turn on and off the gates of gated transistors and similar devices in applications that require an orderly action of one or more of these devices.

Dynamic braking: Using the back EMF in a motor for the purpose of braking.

Effective value: Equivalent DC value of an AC value based on producing/consuming the same power. This is mathematically equal to the root mean square (RMS) of the AC waveform. In AC electricity, effective value can be used for voltage and current.

Effective voltage: Because the voltage continuously fluctuates in alternating electricity, the effective voltage is the value that if a DC system was used instead, it produced the same power. For sinusoidal waveform the effective voltage is 0.707 of the peak voltage.

Electric power: Power in the form of electricity and measured by electrical units (power is the amount of work in 1 sec).

Electromagnet: A (not permanent) magnet made by a wire coil wrapped around a ferromagnetic core when carrying an electric current. The magnet can be turned on and off or its strength can be adjusted.

Electromotive force: Electrical potential difference causing electrical current between two points. A battery or an electric generator is a source of electromotive force, measured in volts.

Electrons: Main carrier of negative electric charge. One of the main components in the structure of an atom or molecule that orbits the nucleus at high speed.

Elements: Material that cannot be decomposed to other simpler materials, like oxygen and copper. All elements are listed in the Mendeleev table.

Emitter: One of the three main terminals of a common (not FET) transistor.

Emitter follower: Name given to common collector amplifier.

Energy efficiency: Overall efficiency of a transformer in a period of time, for example, in one year; That is, the ratio of the energy provided by a transformer in a length of time divided by the energy consumed by the transformer in the same period.

Enhancement: One of the two methods to control the flow of electrons in the channel of a field effect transistor (FET).

Equivalent resistor: Resistor that can replace two or more resistors in series, parallel, or in any combination, that is, having the same effect on the circuit.

Falling edge: Edge in a square waveform signal where the voltage drops from a positive value to a negative value (or to zero), as opposed to the rising edge.

Farad: Unit for measurement of capacitance of capacitors. Farad is a relatively large unit and fractions of it such as microfarad and picofarad are commonly used.

Feedback: Taking the output of a device to compare with its input for the purpose of regulation and adjustment.

Ferromagnetic: Type of material from the iron family that is suitable for magnetization.

FET: Field effect transistor.

Field effect transistor (FET): Type of transistor, still made of semiconductor materials, based on a different structure and a different mechanism of operation than the junction transistor. It has a channel or passage around which an electric field can be developed and through which electrons flow. Flow of electrons can be controlled by adjusting the polarity and intensity of the field.

Fill factor: Ratio of the maximum power possible from a solar cell to the product of its maximum voltage and maximum current (i.e., the product of the open circuit voltage and the short circuit current).

Filter: Device to separate, for selection or rejection, electric signals with a certain property. Selection criteria can be frequency or sudden deviation in voltage and current.

Filtering: Act of separating unwanted electric signals (normally with various frequencies) or reducing the effect of sudden and abrupt changes in voltage or current (smoothing) when there is only one frequency.

Firing angle: Delay from the beginning of a cycle of an AC signal for voltage control in a converter. Delay determines when the gated devices in the converter become active and, thus, the percentage of the 360° cycle that the AC waveform is ineffective.

Fixed bias: Type of biasing a transistor in which a fixed resistor is connected between the base terminal and the power supply, thus providing a fixed voltage at the base.

Flag: Representation of the occurrence of some unusual or unwanted situation in digital circuits and computer terminology.

FLT: Full load torque (in a motor).

Flux lines: Imaginary lines around a magnet indicating the direction of magnetic effect. Stronger magnetism effect implies more flux lines through the same area.

Forced commutation: Turning off a thyristor by applying a negative voltage to its gate, as opposed to natural commutation in which a thyristor turns off as the alternative current applied to it reaches its negative half cycle.

Forward bias: Applying a more positive voltage to the P-type side of a PN-junction than that to the N-type.

Freewheel diode: Diodes put antiparallel to a gated diode (thyristor) in a converter to make a reverse path for current and prevent buildup voltage across the thyristor.

Freewheeling diode: Same as freewheel diode.

Frequency: The number of repetitions per second of any cyclic phenomenon. In AC electricity, the number of cycles per second for alternating current.

Frequency divider: Electronic device (normally made up of flip-flops in digital electronics) that generates waveforms of lower frequency from a higher-frequency waveform.

Frequency modulation: A way of loading the information contained in a signal (like electric waveforms for music and voice) to another type of signal, with a much higher frequency, which is capable of being transmitted through antennas (like electromagnetic signal). In frequency modulation, the frequency of the latter is changed according to the variations in the former signal. FM radio uses this type of modulation (see also amplitude modulation).

Frequency modulation ratio: Ratio of the frequency of the carrier signal to that of the reference signal in pulse width modulation, used in power converters, a number larger than 1 and represented by m_f.

Full load torque: The torque provided by an induction motor at its operating speed. This is the point of intersection of the load line and the motor characteristic curve, and must be sufficiently smaller than the pull-out torque to avoid motor stall.

Full-wave bridge rectifier: Same as bridge rectifier.

Full-wave rectifier: Rectifier in which both half cycles of an alternating current waveform are rectified and delivered to the output as DC, as opposed to a half-wave rectifier in which only one half of each cycle reaches the output.

Fuse: Protective device for an electric device or circuit consisting of a piece of wire that melts and disconnects the circuit if current surpasses a specific value.

Galvanometer: A device consisting of a needle attached to a coil that can rotate around a pin shaft as a result of an electric current flowing through the coil. The coil behaves also as a spring, limiting the motion of the needle. A galvanometer is used to measure electric current, but the needle position can be graduated for other electric entities, like voltage.

Gate: Special connection in some semiconductor devices that upon receiving an appropriate signal (in the form of a voltage or pulse) allows the device to turn on and performs its function.

Gate turn off (GTO) thyristor: Type of thyristor with the capability to be turned off by sending a negative voltage pulse to its gate, also called GTO thyristor, as opposed to those without this extra capability (they turn off as the applied voltage becomes negative).

Generator step up (GSU) transformer: Transformer just after a three-phase generator in power generation plants that increases the voltage from the generator to match the voltage of the power line to the plant.

Geometrical Mean Distance (GMD): A term in power transmission lines corresponding to the spacing between the two (in single phase) and three or six (in three phase) power lines used to determine the inductance and capacitance associated with lines.

Geometrical mean radius (GMR): Term in power transmission lines corresponding to the size (radius) of power lines used to determine their associated inductance and capacitance.

GMD: Geometrical mean distance.

Grounding transformer: Transformer used only for providing a reference point for connection to ground, in order for detection of faults in power generation and distribution.
GTO: See gate turn off thyristor.

Half-wave rectifier: Simplest type of rectifier for alternating current, consisting of only one diode (for single phase) and three diodes (for three phase) that block the negative half cycle of AC, so that only the positive half cycles are passing to the output.
Heat sink: Piece of aluminum attached to an electronic component to provide a larger area of contact with air, so that the heat produced in the component is removed from it better and faster and overheating and damage is prevented.
Heavy water: Special water made out of deuterium (an isotope of hydrogen), thus with the chemical structure D_2O, which is used for storing fuel rods in a nuclear power plant.
Henries: Plural of Henry. Henry is the unit of measurement of inductance of an inductor.
Hertz: Cycles per second. The unit for measurement of frequency.
Hexadecimal: Numbering system with 16 as the base. Thus, there are 16 basic numbers (0 to 15) in order to make all other numbers. Furthermore, there are 16 symbols 10 of which are 0 to 9 and for the rest the capital letters A to F are used (A = 10, B = 11, C = 12, D = 13, E = 14, and F = 15).
High pass filter: Type of electric filter that blocks the lower frequencies in a signal consisting of various frequencies and allows the higher frequencies to pass to the output.
Hole: State of missing electrons in an atom in a semiconductor material, equivalent to having a positive charge. In the same way that electric current can be attributed to flow of electrons, in a semiconductor device one may use flow of holes, when appropriate.
Hub: In a wind turbine, hub is the piece to which the blades are attached (thus holding the blades) and all rotate with respect to the rest of the turbine.
Hysteresis: Property of keeping an effect after its cause has ceased. This is particularly noticeable in electromagnets and electric machines that maintain the effect of magnetization.

Ideal transformer: When in a transformer all the losses are assumed to be zero and, as a result, input power equals output power.
Impedance: Load in AC circuits consisting of resistive and reactive components. Impedance is measured in ohms and can be represented by a vector. It is normally denoted by Z.
Impedance matching: Selecting suitable values for resistors, capacitors and/or other components so that the output from one device matches the input of the following device for a desired purpose such as maximum power transfer or other criteria.
Incremental encoder: Type of encoder (for a rotating shaft) that adds the number of revolutions and determines the total displacement of a shaft during a given time. This is as opposed to an absolute

encoder that determines the displacement (and position) of a shaft within only one revolution.

Inductance: A property associated with the magnetic effect of electricity, exhibited by a winding carrying electric current. The magnetic field generated by a winding is proportional to the winding inductance.

Induction: Generation of electricity in a wire when the magnetic flux is cut by the wire (e.g., when wire moves with respect to a magnetic field or the strength and/or direction of the magnetic field varies).

Inductive reactance: The apparent resistance to flow of electricity exhibited by an inductor in an AC circuit. It is measured in ohms and determines the current in the inductor based on the applied voltage.

Inductor: A winding (a coiled wire) with only magnetizing property and without any electric resistance. Physically it is not possible to have a pure inductor, but at certain conditions, particularly high-frequency electric signals, the resistance of the coiled wire can be ignored in comparison with its inductance.

In phase: When two cyclic waveforms of the same shape and frequency have their pairwise maximum and minimum points occurring at the exact same time.

Input impedance: Value of impedance an amplifier (or a circuit, in general) exhibits to its input device, so that the device sees the amplifier as an impedance of that value.

Inrush current: Relatively high current that a motor initially experiences when connected to electricity (at zero speed). Current decreases as the motor speeds up.

Insulated gate bipolar transistor (IGBT): Type of gated transistor used in power converters. Its gate is insulated and is connected to a much smaller voltage to turn the transistor on.

Insulated gate field effect transistor: Type of FET with insulated gate, like MOSFET.

Integrated gate-commutated thyristor (IGCT): Class of "gate turn off thyristor," in which its gates and antiparallel diodes are integrated with it in a package.

Inverter: Electronic device that generates AC electricity from DC. In the simplest form the generated AC is a square wave.

Isolation transformer: Transformer with the same number of turns in the primary and secondary windings, thus not changing voltage. Its purpose is only to protect a device from possible high currents in the primary side by isolating it from being directly connected to the mains.

Isotope: Variation of a chemical element in nature, having different number of neutrons in the nucleus.

J-K flip-flop: Flip-flop with two data entries, one for set and one for reset the output. It is also possible to toggle the output by activating both inputs.

Junction field effect transistor: One type of field effect transistor (FET) in which the gate is not insulated. Those with insulated gates are more advanced and more common.

Junction transistor: Same as bipolar junction transistor, which includes any ordinary transistor.

Karnaugh map: Special tables by the use of which logic circuits can be simplified and their output can to determined.
K-map: Abbreviation for Karnaugh map.

Lagging: Implying that a cyclic waveform reaches its maximum (minimum) *after* another cyclic waveform of the same frequency. More commonly used in AC circuits to indicate the timing relationship of current with respect to voltage.
Latch: Term in electronics when a device retains the effect of an input after the input has stopped; like the case of a pushbutton that is pressed to make a circuit, but when it is released, the circuit is not broken because another device (like a magnetic switch) maintains the connection.
Leading: Opposite of lagging, implying that a cyclic waveform reaches its maximum (minimum) *before* another waveform of the same pattern and same frequency. Most commonly employed in AC electricity (see also lagging).
Leading edge: Same as rising edge in a square waveform.
Leakage (in transformer): Part of the (magnetic) flux produced by the primary winding that is lost and does not reach (link) the secondary of a transformer, due to structural imperfection.
Least significant bit (LSB): The bit on the far right in a binary number.
Light-emitting diode (LED): Diode that when subject to a DC voltage with correct polarity (positive to anode) and magnitude (around 5 V) emits light. The color of the light depends on the material used; it can be red, orange, green yellow, blue, and so on.
Line current (in three-phase systems): Current in lines bringing electricity to the loads in a three-phase system.
Line voltage: Voltage between each pair of (three) lines bringing electricity to the loads in a three-phase system.
Liquid crystal display (LCD): A kind of technology for display in calculators, TV, monitors, and so on, based on blocking a backlight from passing through to a viewer according to the electric voltage applied to each unit in the display.
Load line (in transistor): Line showing the operating points of a transistor on a transistor characteristic curve, based on the supply voltage and the resistors in the transistor circuit. This line is between the points corresponding to the maximum collector voltage and maximum collector current.
Locked rotor current (LRC): Current in a motor corresponding to when the rotor speed is zero (the current at starting a motor or if it is prevented from turning); this is the highest current that the winding in a motor experiences.
Locked rotor torque (LRT): Starting torque in a motor (corresponding to when the rotor speed is zero) or the torque if the rotor is prevented from turning or cannot turn the load.

Logic gates: Number of special electronic devices exhibiting switching properties according to Boolean algebra.

Long shunt compound: A kind of compound wound DC electric machine in which the shunt winding receives the entire terminal voltage, as opposed to the short shunt compound machine. The shunt winding current is independent of the series winding current.

Low pass filter: Type of electric filter that blocks the higher frequencies in a signal consisting of various frequencies and allows the lower frequencies to pass to the output.

Magnetic field: Limited part of the space around a magnet where the magnetic effect exists.

Magnetizing current: Part of the current in a wire winding that corresponds to creating a magnetic field.

Metal-oxide semiconductor: Type of semiconductor device in which metal oxide is used for insulation layers.

Metal-oxide semiconductor field effect transistor (MOSFET): Type of field effect transistor in which metal oxide (like SiO_2) is employed for insulation of the gate.

Microcontroller: A one chip integrated circuit (IC) comprising a main processor plus all components and input/output terminals that can be used for control purposes (process control and control of machinery).

Microprocessor: Integrated circuit for all arithmetic and logic operations in a computer.

Modulation: Act of combining the information contained in a waveform (like voice) to a carrier waveform with much higher frequency (like radio frequency) for transmission.

Molecule: Smallest piece in a combined substance that still has the properties of the substance and cannot be broken into smaller pieces.

MOSFET: Metal oxide semiconductor field effect transistor.

Most significant bit (MSB): Leftmost bit in a binary number.

Multimeter: Device for electrical measurements with selectable switches to function as voltmeter, ohmmeter, and ammeter, and some more capabilities (all in the same unit).

Natural commutation: Act of automatically turning off in a gated diode or transistor (a thyristor) when an alternative current applied across its main terminals changes direction (goes to negative half cycle). This is the opposite of "forced commutation" where the device can be turned off through a control signal to its gate.

Negative going edge: Same as falling edge or trailing edge (changing from a positive value to a negative or zero value in a square wave).

Negative sequence: One of three sets of balanced components (represented by equal size vectors, 120° apart from each other) of any unbalanced condition in a three-phase system. The unbalance condition is more noticeable when a fault happens. Thus, decomposition to balanced components is used for detection of faults in three-phase generators, loads, and transformers. The sense of direction for negative sequence is negative, as opposed to the direction for positive sequence.

Neutral grounding resistance (NGR): An added resistance to the grounding strap to limit the (otherwise high) current flow through a device or a circuit in case of a fault.

Neutral line: Fourth line in a three-phase system, which ideally does not carry current and can be grounded; also called null line.

Neutron: One of the two main particles in the nucleus of an atom. Existence of neutrons adds the atomic mass and defines isotopes of a chemical element.

Node: Any point in an electric circuit where more than two lines (branches) join each other.

N-type material: Semiconductor material carrying extra electrons, thus having negative charge an attracting positive charge, as opposed to P-type semiconductor material.

Nucleus: Part of an atom in the middle, consisting of protons (and neutrons when they exist), around which electrons orbit. The nucleus holds almost the entire mass of an atom.

Null line: Same as neutral line in a three-phase system.

Ohm: The unit for measurement of the electric resistance and electric reactance.

Ohm meter: Device for measurement of electric resistance.

Ohm's law: One of the most important laws of electric circuits: relationship between voltage across a component, the current in the component and the electric resistance exhibited by the component to the flow of electricity. For a simple resistor it is $V = RI$.

Omega: The name for the Greek letter Ω/ω (corresponding to W/w). The upper case omega (Ω) is used to represent ohm (for example, 5 Ω, 20 Ω, 1000 Ω, etc.).

One's complement: Result of swapping each bit value to its complement in a binary number.

Open-circuit test: Test for measuring current in the primary of a transformer when the secondary winding is not connected to any load. This measurement determines the core loss (the losses for magnetization, leaks, and Eddy current, which are load independent) of a transformer.

Out of phase: Fact of having a finite time difference between the instances that two waveforms of the same shape and frequency reach their maximum or minimum values. This can be extended for logic signals by comparing their rising or falling edge timing.

Output impedance: Equivalent impedance an electronic device (such as an amplifier) exhibits to the device connected to its output.

Overflow: State of reaching a value beyond the capacity of a digital device in terms of the number of bits.

Overmodulation: Having a value larger than 1 for amplitude modulation ratio in pulse width modulation controlled inverters. This happens when the amplitude of the modulating signal is greater than that of the carrier wave.

Pad mount (transformer): Transformer that is mounted on a flat surface (a flat slab) as opposed to those which are designed to be mounted on an electric pole (pole mount).

Parallel circuit: Electric circuit having two or more electric components with their terminals connected to the same point in the circuit, so that current divides between the branches formed by those components.

Paramagnetic: Metals with very little (negligible) magnetic property that never can retain any magnetism after being in a magnetic field, as opposed to ferromagnetic materials.

Peak inverse voltage (PIV): It is the maximum reverse bias voltage that a diode can withstand without getting damaged (breaking down). This is important in selection of diodes in the design of rectifiers. It is also called Peak Reverse Voltage (PRV).

Peak reverse voltage (PRV): The same as peak inverse voltage (PIV).

Peak-to-peak: Magnitude between the minimum and the maximum values in a cyclic (periodic) waveform. Frequently used for voltage and current in AC electricity.

Peak voltage: Highest magnitude of a sinusoidal waveform (like in AC electricity).

Performance curve: Same as characteristic curve: A line or curve that shows the relationship between two or more parameters related to the performance of a device.

Period: The duration of time for one cycle of a cyclic phenomenon, like the period for going through a complete cycle of positive followed by negative and back to zero value for variation of voltage and current in AC electricity.

Permanent magnet: Magnet with permanent magnetic property that cannot be turned on and off or altered.

Phase current: Current inside each electric component connected to a three-phase system, as opposed to the line current representing the current in the lines carrying electricity to that component.

Phase voltage: Voltage between each line and the common reference point (ground) in a three-phase system.

Photovoltaic (PV) system: A setup of photovoltaic panels to absorb solar energy. Photovoltaic panels are made of an array of solar cells connected together (in series and parallel) for a desired voltage and a desired current. A solar cell is a semiconductor that converts light to voltage, that is, generates a voltage across its terminals proportional to the light it receives.

Pitch (wind turbine): Angle of rotation with respect to a reference point of a wind turbine blade about its longitudinal axis. Changing this angle alters the performance of a blade and, thus, the power capacity of a wind turbine.

Pitch control: Changing the pitch angle of blades in a wind turbine in order to alter the power catch capacity or adapt to the wind speed.

Pole mount: Type of transformer which is suitable for mounting on an electric pole.

Positive going edge: Rising edge in a square wave where the magnitude sharply goes positive from negative or zero.

Positive sequence: One of three balanced components of any unbalanced condition in a three-phase system. The positive sequence comprises three vectors with a positive sense of direction of rotation. Balanced components are used for detection of faults, particularly ground fault, in three-phase generators, loads and transformers (see also negative sequence).

Potential difference: The difference in intensity in an electric field or various points in an electric circuit, also called voltage. The potential difference between the + and – terminals in a car battery is around 12 V.

Power curve: Curve depicting the maximum power that can be generated from a wind turbine for various wind speeds. This takes into account the mechanical and electrical capacities of the turbine components.

Power gain: Ratio of the output power to the input power in an amplifier.

Prime mover: Mechanical power source (e.g., steam turbine, gas turbine, diesel engine) that turns an electric generator or any other device requiring mechanical energy.

Product of sums: Logic function represented in the form like $(A+B)(A+C)$ $(\bar{A}+\bar{B}+\bar{C})$, for example, which is the multiplication of items obtained by addition of logic variables. This happens when the results of some OR gates are AND'ed together.

Proton: One of the two main particles in the nucleus of an atom. A proton holds a positive charge.

P-type material: Semiconductor material missing some electrons, thus carrying *holes* and having positive charge, as opposed to N-type semiconductor material having negative charge.

Pull-out torque: The same as the breakdown torque. The maximum torque that an induction motor can provide after being started before reaching its operating speed. The operating speed is within a region where the motor continues to work at full-load torque and the speed variation is proportional to the motor torque (see the characteristic curve of an induction motor).

Pull-up torque: The minimum torque, after starting, in an induction motor that can be provided by the motor. If this torque is not sufficiently higher than that of a load, the motor cannot accelerate the load fast enough (see the characteristic curve of a three-phase induction motor).

Pulse width modulation (PWM): One way of adjusting/controlling the value of a variable by repeated pulses of appropriate width at a proper frequency. These pulses can directly denote a controlled variable, thus their average defines the value of that variable; or they can be part of the control system.

Quiescent point: Point on the load line of a transistor in a circuit. This point corresponds to a condition that the circuit input does not change (stays quiescent).

Rated value: Value(s) affecting the operation of a device that the device is designed for and must operate under conditions within the neighborhood of those values. For example, an electric device has a rated voltage and current. Operating the device outside of the rated conditions is either inefficient or harmful to the device.

Reactance: The apparent resistance (measured in ohms) that a capacitor or an inductor when connected in an AC circuit exhibits to the flow of electricity. Reactance depends on AC frequency and unlike a resistor the energy involved does not convert to heat.

Reactive power: Power in an AC circuit corresponding to an inductor or a capacitor. Reactive power is not consumed but is momentarily stored in the component (inductor or a capacitor) and is returned to the circuit in the next half cycle (when the AC polarity alters).

Reactor: Wire winding with or without a core, dry or oil immersed, which is specifically designed for a number of purposes, including grounding in a three-phase system.

Real power: Another name for active power.

Rectification: Converting AC to DC.

Rectifier: Electrical device that converts AC to DC.

Register: Electronic device to store digital data (in the form of binary numbers), made of a number of flip-flops, one for each bit.

Reluctance: Equivalence of resistance in an electric circuit in a magnetic circuit; resistance to magnetic flux.

Reluctance force: Force on a ferromagnetic material caused by a magnetic field to minimize the path of magnetic flux.

Reluctance motor: Single-phase AC motor that works based on reluctance force.

Reset: Action of putting or setting a device to its default state.

Resistance: (in electricity) (a) The property of resisting (but not blocking) the flow of electric current, leading to limiting the flow and absorbing (consuming) electric energy. Resistance is measured in ohms (Ω). (b) One of the three basic components that any electric load (a device using electricity) can be assumed to be composed of.

Resistivity: Same as specific resistance: the electric resistance of a specific size (based on the measurement system) of a metal or material.

Resistor: (a) An electric component that only exhibits resistance (and no other reaction) to the flow of electricity. (b) Any of the standard components, made in different physical sizes, that are used in electrical and electronic circuits to absorb electrical energy.

Resonance: Special condition in AC circuits where all the energy stored by inductive components is provided by capacitive components, and vice versa. This occurs in a particular frequency. This condition implies other facts such as the net reactive power to be zero, the power factor to be unity, and a number of other correlations based on the circuit layout.

Resonance frequency: Frequency at which resonance happens in an AC circuit.

Resonant frequency: A unique frequency for each AC circuit containing both reactive components (inductors and capacitors) at which the resultant reactance of all capacitive components is equal to the resultant reactance of all the inductive components. As a result, the two types of components cancel the effect of each other, and the total reactive power of the circuit is zero.

Reverse bias: Opposite of forward bias: Connecting the P-type side of a PN-junction to a less positive or more negative voltage than the voltage connected to the N-type side.

Ripple: Fluctuations in a rectified AC waveform. Fast fluctuations in an electrical value, such as voltage, that is supposed to be constant.

Ripple factor: The ratio of the AC content (peak-to-peak value in volts) of a rectified voltage to its DC content (which is the average value of the rectified voltage). The smaller this ratio becomes, the better the rectified value is.

Rising edge: Same as positive going edge and leading edge in a square waveform.

Rotating magnetic field: Magnetic field with constant strength but rotating about an axis, formed in the stator of three-phase machines when connected to electricity. Such a rotating filed is developed because the three windings are physically apart by 120°.

Rotation (in binary number operations): Operation of moving one or more leftmost zeros in a binary number with limited number of digits to the far right. This implies multiplying that number by 10_b, 100_b, etc.

Rotor: The rotating part of any motor, generator, turbine, and so on. In a propeller-type turbine, the rotor consists of the hub and the blades.

Safety rules: Sets of rules that deal with safe operation of a machine or completion of a task, thus minimizing the possible harm for an operator and others.

Safety standards: Safety rules issued by authorities.

Sample and hold: Action of measuring a variable and storing the measured value for use. This is common in computer control systems where measurements are performed repeatedly (but not continuously; say at 1msec intervals) at a proper fixed frequency.

Sampling: Action of measuring a variable in a process or machine repeatedly at a fixed frequency, for control purposes.

Sampling frequency: The number of samples taken (or measurements made) in 1 second in a sampled data system. This is for converting analog values to digital values for digital control, computer control, display of values, and so on.

Sampling rate: The (normally fixed) frequency at which a sampling action is performed.

Sampling time: Time interval between each two consecutive measurements in a sampling action.

Saturation (in a transistor): The state of a transistor at which the collector current has reached its maximum value for the present collector-emitter voltage, and cannot increase further by only increasing the base current I_B.

Self-bias: Type of biasing a transistor in which a fixed resistor is connected between the base terminal and the connector, thus providing a voltage at the base that varies with the current through the collector.

Semiconductor: Term referring to a semiconductor device or a semiconductor material.

Semiconductor device: Any simple or complex device made up of semiconductor material, like diode, transistor, and integrated circuit.

Semiconductor material: Man-made material by adding elements with five covalent electrons or three covalent electrons to silicon or germanium (with four covalent electrons). The electric properties of such a material can be controlled. It is the base material for all semiconductor devices.

Sequential logic: Kind of logic circuit with timing sequence, memory, and synchronization.

Series circuit: Type of electric circuit in which the loads form a single loop between the two terminals of the power source.

Servomotor: Motor whose speed or displacement can be controlled (must be capable of being controlled).

Set (for a flip-flop): Putting a device to a desired state. Some devices have only two states on and off, or high and low. Set implies putting the device in a desired (usually nondefault) state.

Shaded pole: Type of single-phase AC motor (usually small motors of a fraction of horsepower) based on the creation of an initial (starting) torque by magnetic shading of its poles.

Shell-type transformer: Category for the construction of a transformer in which the core has the shape of a figure 8 (i.e., with a bridge in the middle of a rectangular frame).

Shift right (in binary number operations): Operation of moving all digits of a binary number (with a limited number of digits) to the right by one or more places. The leftmost places are then filled with zeros. This implies dividing that number by 10_b, 100_b, etc. The rightmost digits are moved out and introduce the round-off error that occurs in division.

Short-circuit test: One of the two important tests for transformers. Short-circuit test measures the copper losses of a transformer. The secondary winding is shorted for this test.

Short shunt compound: A kind of compound wound DC electric machine in which the field shunt winding receives only a part of the terminal voltage (the field has a shunt winding plus a series winding). The shunt winding is parallel with the armature winding, only, and the current in the shunt winding is a fraction of the field series winding current.

Signed number: Binary number that can be positive or negative. The sign (of the number) must be embedded with the bits representing the number.

Silicon-controlled rectifier (SCR): A family of gated diodes or transistors that can be controlled (turned on and off) through their gates.

Sine (of an angle): Trigonometric function for an angle represented by the ratio of the length of the side opposite to the angle to the length of the hypotenuse in a right triangle setting.

Sinusoidal: The most common and preferred waveform for alternating current electricity. The variation of a sinusoidal waveform is according to the values of sine (or cosine) of an angle for 360° (one revolution) change.

Slip: The fact that the rotor of an induction machine does not rotate with the same speed as the rotating magnetic field (turning faster in a generator and slower in a motor).

Slip rings: Set of two (in single-phase) and three rings (in three-phase) in some types of AC electric machines that are connected to windings that rotate with the rotor. Brushes pressed by springs on the slip rings can transfer electricity from or to the winding from the external circuit.

Slip speed: Difference between the speeds of the rotor and the rotating magnetic field in an AC induction machine.

Solar array: Set of solar panels put together in the form of a matrix, and connected together in series, parallel, or a combination to deliver a desired voltage at a desired current.

Solar cell: Semiconductor diode that converts light to voltage (the opposite of what an LED does).

Solar irradiance: Rate of available solar energy on earth per square meter (or per square foot). It is, thus, measured in w/m^2, or w/ft^2.

Solar panel: Set of solar cells put together in a frame and connected together in series, parallel, or other combination to deliver a desired voltage at a desired current.

Solid-state: Electronic circuits and devices being made of semiconductors, as opposed to those made out of vacuum tubes.

Source (in electricity): Active component powering an electric circuit.

Source (in semiconductors): One of the three terminals in a field effect transistor, comparable with the collector in a junction transistor.

Specific resistance: Same as resistivity: the electric resistance of a specific size (based on the measurement system) of a metal or material.

Split-phase motor: Type of single-phase AC motor in which a starting torque is generated by an additional winding in the rotor.

Squirrel cage machine: Type of alternative current induction machine in which the rotor winding has little resistance and thus carries very high current. To withstand high currents, the rotor structure is modified and more resembles a cage than a winding.

Star connection: Same as wye connection. One of the two ways for connection of loads and sources in a three-phase system.

Static electricity: Type of electricity where charged bodies instantly discharge (when conditions allow). There is no long time duration for controlled flow of electrons, as in electric devices.

Stator: Stationary part of AC electric machines, as opposed to rotor that rotates.

Step-down transformer: Transformer with secondary winding having fewer turns than the primary winding, thus decreasing voltage.

Step-up transformer: Transformer with secondary winding having more turns than the primary winding, thus increasing voltage.

Stray loss: Losses of various origins, including that in the air gap, in an electric machine. These losses cannot be included in the other categories for a machine.

Substation: Point in the electric distribution network where transformers and switchgears are installed.

Subsynchronous: Speeds below the synchronous speed in an induction machine.

Sum of products: A logic function represented in the form $(AB)+(AC)+(\bar{A}\cdot\bar{B}\cdot\bar{C})$, for example, which is the sum of items obtained by multiplication of logic variables. This happens when the results of some AND gates are OR'ed together.

Superposition: A way of analysis of problems containing multiple parameters or conditions (inputs for example) by supposing that each parameter exists without the presence of the others, and then the different outcomes (of applying each parameter) are added together.

Supersynchronous: Operating speeds higher than the synchronous speed in an induction machine.

Surge: Sudden and short-term increase of voltage in a line, for instance, during a few seconds after a large load is taken off the line.

Surge suppressor: Device or arrangement to stop or reduce the effect of a surge.

Swing: Total change (the difference between the maximum and minimum values) of the input or the output in an amplifier.

Synchronous counter: Binary counter made up of flip-flops arranged such that all flip flops change state at the same time. Synchronization is governed by the clock pulse.

Synchronous motor: Type of AC motor (usually very large size) that is insensitive to load change and always runs at a fixed (synchronous) speed, unless the load becomes larger than the motor capacity (when it stops working).

Synchronous speed: Special speed for an AC machine depending on its number of poles and the frequency of the AC line.

Tangent (of an angle): Trigonometric function for an angle represented by the ratio of the length of the side opposite to the angle to the length of the side adjacent to the angle in a right triangle representation.

Tank circuit: Part of an electric circuit consisting of an inductor and a capacitor in parallel with each other, used in filters.

Temperature coefficient: Numerical value (positive for metals) representing how much the specific resistance of a material changes with temperature.

Tertiary: The same as tertiary winding.

Tertiary winding: A third winding in large transformers to serve for different purposes such as reducing the unbalance effect of loads and fault currents.

Thyristor: Gated semiconductor device whose conduction can be controlled by a pulse sent to its gate. It is used in power converters.

Time constant: Duration of time for a circuit containing a capacitor or inductor and resistors to reach 63 percent of a new value after a change has happened to the circuit (e.g., its power has been turned on or off).

Timing diagram: Diagrams that indicate the variation of a logic parameter (change of state between 1 and 0) versus time in a logic circuit.

Trailing edge: Same as negative going edge, which is that edge of a rectangular waveform that the value drops from positive to negative or zero.

Transformer percent impedance: Percent ratio of the voltage in the short-circuit test of a transformer that causes the rated current to the rated voltage.

Transistor-transistor logic (TTL): Standard voltage for connecting electronic circuits together.
Transposition: Special arrangement in the transmission of three-phase power, in which the wires are rotated such that the location of wires for each phase is physically interchanged so that each wire equally occupies the same location as the other wires within the length of a line.
Triac: Type of gated diode for alternative current. Its main terminals do not connect until the gate allows connection or the voltage across it has reached certain value.
Trigonometric functions: Four functions sine (sin), cosine (cos), tangent (tan), and cotangent (cotan) for an angle.
True power: Same as active power.
Truth table: Table for a logic circuit showing the output based various input entries.
TTL: See transistor-transistor logic.
Turns ratio: Ratio of number of turns in the secondary and primary windings of a transformer.
Two's complement: A binary number obtained by adding 1 to the one's complement of a binary number.

Unbalanced load: When in a three-phase system, the load consists of three separate and not identical single-phase devices.
Universal motor: Type of motor that can work with both AC and DC. A series-wound DC motor has this capability.
Unsigned number: Binary number representing a positive value, as opposed to a signed number, which can represent a positive or negative value.

Valence: Outermost orbit of electrons in the structure of an atom. Valence electrons move on this orbit.
Valence electrons: The electrons on the outermost orbit around the nucleus of an atom.
Variable speed turbine: Wind turbine whose rotational speed is not constant. Such a turbine's speed can vary between 10 and 25 rpm.
Volt: The unit for measurement of voltage (or potential difference) in an electric source or load or between two points in an electric circuit. The voltage in a pen size battery is 1.5 V.
Volt-Ampere-Reactive (VAR): Unit of measurement for reactive power.
Voltage: A main property of electricity representing the intensity of electric charges based on accumulation of electrons. Normally a voltage difference between two points defines the potential for a discharge. Voltage is measured in volts.
Voltage divider: Electric circuit made up of components in series that can deliver two or more smaller voltages from a voltage applied to the circuit.
Voltage gain: Ratio of output voltage variation (output swing) to the input voltage variation (input swing) in an amplifier.
Voltage regulation: Percent value representing how good a transformer is, obtained from the ratio of voltage loss to the transformer nominal voltage. The smaller this number is, the better the transformer.

Voltage source inverter (in back-to-back converter): A back-to-back converter arrangement in which a capacitor is used between the rectifier and inverter for the instantaneous electricity storage. In a current source inverter an inductor is used for this purpose.

Voltmeter: Electrical instrument to measure electric voltage.

Volts per turn: Number indicating the value of volts for each turn of winding in a transformer. This value can be obtained from either the primary side values or the secondary values by dividing the voltage by the number of turns.

Waveform: The form (shape) of variation of a signal (electric or other) that varies with time. The shape of a rectangular waveform, for example, is made up of succeeding rectangles.

Wind vane: Instrument for detecting the direction of wind, consisting of an aerodynamic body that rotates about an axis and can orient itself with the wind direction.

Windage: Aerodynamic losses associated with ventilation fans and rotating parts in an electric motor.

Word: Binary number consisting of 16 bits; 16 bits of information form a word.

Wound rotor induction machine (WRIM): Type of induction machine for AC in which the rotor has wire winding. The windings are accessible through slip rings. Another type is (squirrel) cage machine that has no wire winding and has no slip rings.

WRIM: Wound rotor induction machine (or motor, when used as a motor).

Wye connection: A way of connecting loads and sources to a three-phase system, where the components of a load or a source form a letter Y (that is one side of each component are tied together) and the free sides of the Y are connected to the three lines.

Y connection: Same as wye connection.

Yaw motion: Motion in a wind turbine where the whole rotating part (rotor) and the nacelle orient toward wind to catch the most power from the wind. In new turbines, there are sensors and motors for this motion.

Zener current: Same as the avalanche current (the higher current after breakdown occurs in a semiconductor device) in a zener diode.

Zener knee: Sharp curve in the characteristic curve of a zener diode where the breakdown voltage is reached and breakdown occurs.

Zener region: Region in the characteristic curve of a zener diode where the breakdown has occurred because of the reverse voltage surpassing a certain value.

Zero sequence: One of the three balanced components when any unbalanced condition (or load) in a three-phase system can be decomposed into balanced components. The zero sequence comprises three equal vectors that together with positive sequence and negative sequence vectors can represent the unbalanced condition. See also the positive sequence and negative sequence.

Zigzag transformer: Special three-phase transformer in which the primary and secondary windings have the same number of turns and are combined in a particular way such that they cancel the magnetic effect of each other. It can be used for various purposes including fault detection.

Appendix A: Internal Resistance of Batteries and Battery Aging

Ideally for any source of electricity, it is desirable that the voltage is constant, irrespective of how much power is taken from the source (provided that it is within the capacity of the source). This is usually not the case, particularly for batteries. Available voltage at the terminals of a battery does not stay the same if the current (and consequently the power) taken from the battery varies. As current increases, voltage decreases; in other words, there is more voltage drop in the source. As long as this voltage drop due to an increase in the delivered current is small and acceptable, everything is fine.

Figure A.1 illustrates the decrease in voltage versus the increase in current for a battery. The graph for an ideal source is a horizontal line, and for a new battery this line has a small slope. For example, if the case of a car battery is considered, measurement by a voltmeter can show 14 V, but when starting a car if 200 A is taken from the battery, the voltage can only be 12 V. An old battery has a steep slope, and although it can look good for turning the lights on, when cranking a car, its voltage can drop to 5 to 6 V, which is not high enough to deliver the required torque to the engine.

We can always think of an internal resistance in the battery where when a battery is new, its value is low, but it gradually increases as the battery becomes older. In this sense, part of the battery's power is consumed in this resistance, and as expected, it causes a voltage drop because it is in series with a load to the battery. This resistance, nevertheless, is very small. For instance, the value for this resistance can be on the order of 10 to 40 milliohm (mΩ). Figure A.2 shows the schematic for what can be represented as the internal resistance of a battery. Note that it is not accessible from outside a battery because it is an intrinsic part of the battery.

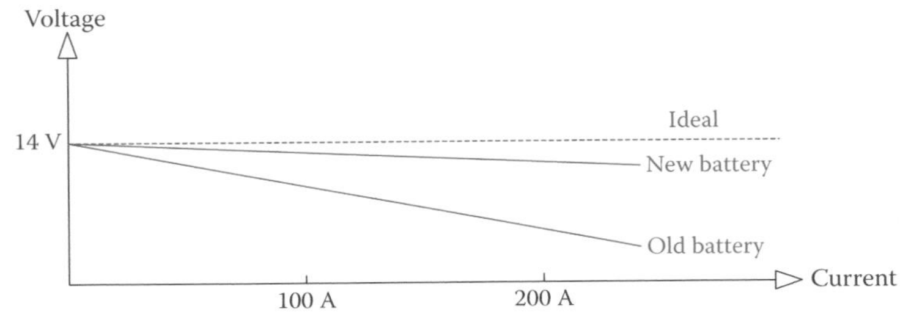

Figure A.1 Voltage drop in a battery due to the current drawn from it.

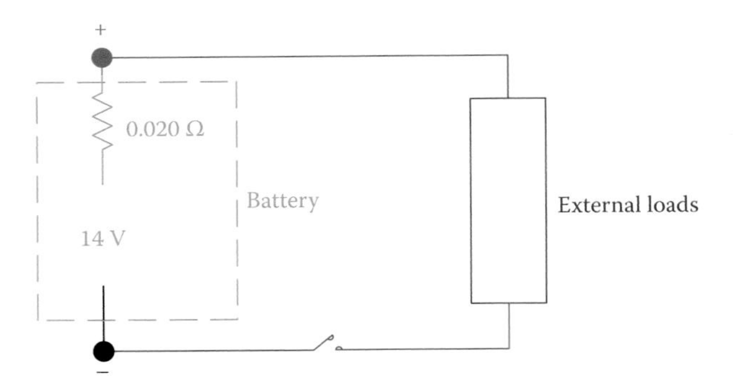

Figure A.2 Representation of the internal resistance of a battery.

Appendix B: Switches

A switch is a principal element in any device or circuit because we need to have a control to turn a device on and off. Like any other component, to be used in a circuit, a switch must be rated for the circuit. For example, in buildings in North America, most of the switches that are used for lighting are 15 A and good for 120 V. Thus, for the circuit of a dryer, for instance, which takes more current than the other appliances, such a switch should not be used.

In addition to the ordinary switches you are more familiar with, there are other switches that you should know about. The ordinary switch has two terminals and has two states: *on* and *off*. Thus, its function is to connect or disconnect the two terminals. This type of switch is called *single-pole single throw* (SPST) switch.

A *single-pole double throw* (SPDT) switch is a type of switch that has three terminals. By toggling the switch, one terminal alters connection between the other two terminals (see Figure B.1b). Thus, with this switch, one line connected to the main or common terminal is connected either to terminal 1 or to terminal 2 (but not both at the same time). If one of the terminals is not used, then this switch can function as a SPST.

In a single throw of a switch it is possible to simultaneously connect two independent pairs of lines. This switch is called *double-pole single throw* (DPST). Its schematic is shown in Figure B.1c. Such a switch acts as two simultaneous single-pole single throw switches.

By combining two single-pole double throw switches, a *double-pole double throw* (DPDT) switch is obtained. This switch has six terminals. By one action, two separate terminals are independently connected either to position 1 or to position 2, as shown in Figure B.1d.

There is also another switch as shown in Figure B.1e. This rotary switch has four terminals. By turning the switch knob, either A is connected to A′ or B is connected to B′. This switch can be used in conjunction with two SPDT switches when a device must be turned on from multiple points. For instance, for a stairway light (accessed from two points) a pair of SPDT switches are employed. To add a third, fourth, etc., point of action, this type of switch must be added in between the SPDTs.

In smaller-scale switches for low current, particularly for electronic circuits, you can find sliding switches that function as a DPDT, but with more terminals; for instance, there are five pairs of terminals, A through E and A′ through E′. When in the off position, no terminals are connected to each other, but in the on position, A is connected to A′, B′ is connected to B, and so on.

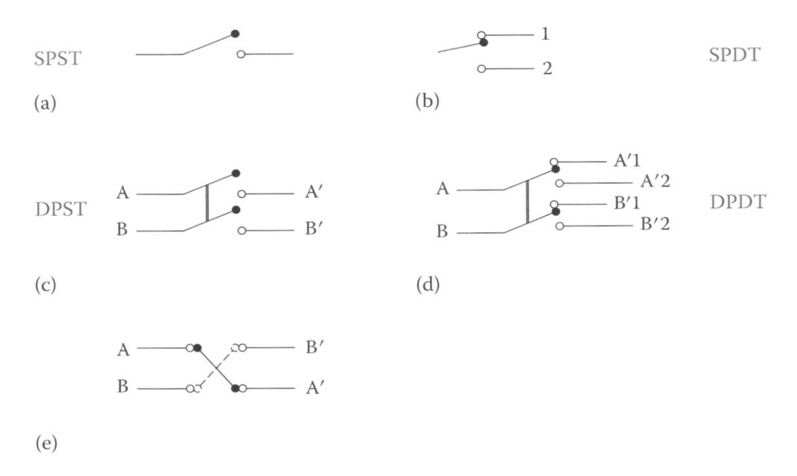

Figure B.1 Schematics of various switches. (a) Single-pole single throw. (b) Single-pole double throw. (c) Double-pole single throw. (d) Double-pole double throw. (e) Double-cross switch.

Appendix C: Special Transformers

C1.1 Zigzag Transformer

A zigzag transformer, alternatively known as an interconnected-star (or wye) transformer, is a three-phase special transformer for the purpose of creating a neutral point for grounding, where there is not such a point available. For example, in three-phase lines coming out of a delta connection, there is no neutral reference point. Also, in transformers where for a technical reason a delta connection is compulsory and for fault protection it becomes necessary to have a ground reference, a zigzag transformer can provide the neutral point.

In a zigzag transformer each phase has two identical windings wound in opposite directions to each other, similar to that in Figure 10.4. In other words, for each phase, there is one winding that is divided into two halves. Because the two half-phase windings are similar, they can be made by a double winding process when constructed. Connections are made in such a way that each half winding of each phase is in the opposite sense of its other half, hence cancelling the effect of each other. In this sense, one could say that this transformer has only a primary winding and no secondary is necessary, as depicted in Figure C.1a, where all the windings are shown along each other.

The interconnection between the windings of the three phases is done in such a way that the three windings are star connected to each other (follow the windings shown in Figure C.1). An alternative way to show the interconnection is as illustrated in Figure C.1b, where it is assumed that each pair of windings is wound as in Figure 10.4.

Figure C.2 illustrates the phase relationships for the voltages through the windings of a zigzag transformer and how they generate a zero voltage point.

The outstanding feature of the zigzag transformer is that while for normal circumstances it has large impedance, it can exhibit little impedance in case of a short to ground fault.

A zigzag transformer may also be used as an autotransformer with a fixed turns ratio of 2:1. For this the two half-phase windings are joined together with the same polarity (rearrangement of the interconnection). Furthermore, it can be used as a three-phase isolation transformer. Nevertheless, its main purpose is for creating a neutral reference point.

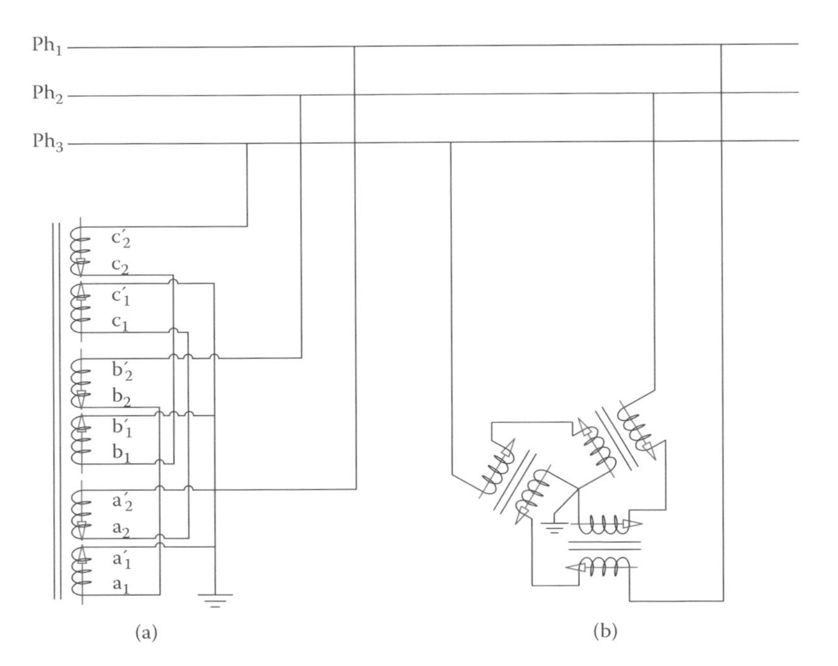

Figure C.1 Interconnected-wye (zigzag) transformer connections. The schematics of a zigzag transformer are shown: (a) only a common core or (b) three cores.

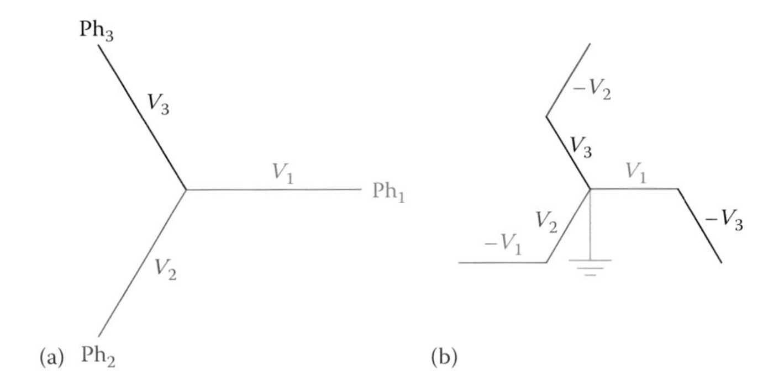

Figure C.2 Phase relationships for voltages in a zigzag transformer. (a) Voltages in the primary winding. (b) Voltages in the secondary windings.

C1.2 Current Transformer

A current transformer is a single-phase transformer with only one or only a few turns for the secondary winding. The purpose of such a transformer is to measure current in a circuit (1) without opening the circuit and (2) without the need to have provision and special tool for high currents. In fact, by having only a few loops in the secondary, the voltage is tremendously reduced, and, instead of measuring the current in the

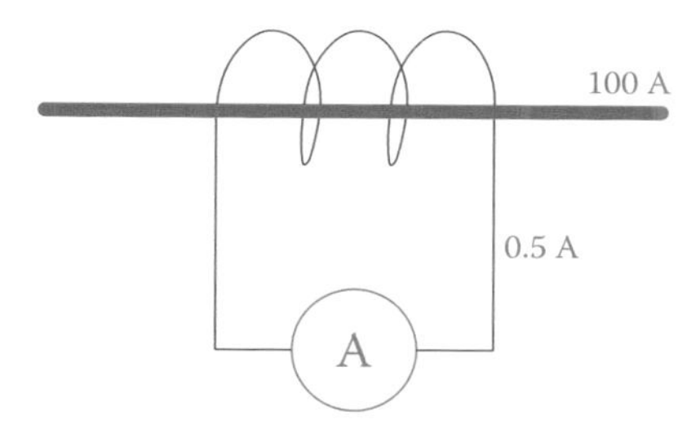

Figure C.3 Principle of current transformer.

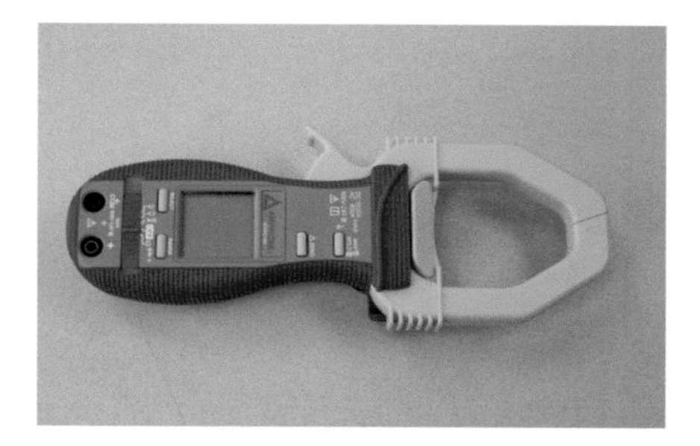

Figure C.4 A clip on ammeter.

main circuit, a much smaller current in the secondary of a transformer is measured. Nevertheless, the graduating of the needle, or the reading of a digital meter, is accordingly adjusted for the main current. Figure C.3 illustrates the principle of a current transformer, and Figure C.4 shows a familiar hand-held clamp-on meter with one loop.

Appendix D: Mil and Circular Mil

In the English system for measuring small lengths such as paper thickness and precision machining dimensions 1/1000 of an inch is used as the unit. It is called *mil*. For items that are round and have a circular cross section, like most wires, another small unit is used. This unit, called *circular mil*, is a unit of measure for area. One circular mil is the area of a circle whose diameter is 1 mil. In this respect, the area called a circular mil is

$$1\,\text{CM} = \frac{1}{4}(1)^2(\pi) = \frac{\pi}{4}\,\text{mil}^2 = \frac{\pi}{4} \times 10^{-6}\,\text{in}^2$$

$$= \frac{\pi}{4} \times 0.00064516\,\text{mm}^2 = 0.00050671\,\text{mm}^2$$

If the diameter of a wire is d mils, then its area is

$$\frac{1}{4}(d)^2(\pi) = d^2\text{CM}$$

By defining the circular mil, in fact, the multiplication by the factor $\pi/4$ is included in the unit, making the expression for measurement easier.

1000× of a circular mil is also used for larger sizes and is represented by kcmil or MCM.

Example D.1

If the diameter of a wire is 10 mil (0.01 in), what is the area in circular mil?

Solution

The area of this wire is 100 CM.

Note: Do not confuse CM (uppercase) with centimeter, cm (lowercase).

Appendix E: Standard Resistors

Resistors for electronic and small electrical circuits are made from various materials such as carbon film or metal film (a thin layer covering a supporting body) mostly in a cylindrical form with two leads coming out from the two ends of the cylinder for connection to other parts in a circuit. They come in different values and with a few power values (the maximum power they can dissipate). Resistance values can be as small as a fraction of 1 Ω to large values such as 10 M Ω (10 million ohm). There are three issues in understanding how to select a resistor: tolerance, magnitude, and power.

E1.1 Resistor Value Tolerance

The manufacturing process does not guarantee that a resistor has an exact value as desired. More precision implies better machines and processes and a higher price, as a result. For this reason, there is always fluctuation in the value of a resistor. A resistor can have a value higher or lower than the expected (nominal) value. Nevertheless, this difference with the nominal value can be controlled within certain limits. In other words, this error in value can be limited in the manufacturing process not to exceed a certain percentage.

The percentage difference from a nominal value for a resistor is called *tolerance*. For any resistor in addition to its resistance value one needs to know its tolerance, which can define the variation in its value. For example, if the nominal value of a resistor is 100 Ω and its tolerance is 10 percent, its real resistance can be any value from 100 – 10 percent to 100 + 10 percent; that is, from 90 to 110 Ω. The actual value of a 100 Ω resistor with 1 percent tolerance is somewhere between 99 and 101 Ω.

Resistors are made with 20, 10, 5, and 1 percent and even smaller tolerances. Obviously, those with 1 percent or smaller tolerance are more precise, and also more expensive, than the resistor with 5 and 10 percent tolerance. For many applications, resistors with 10 percent tolerance are sufficiently good. For very precise circuits, smaller tolerances (0.5, 0.25, and 0.1 percent) are available. Also, resistors with a 20 percent tolerance are very common.

E1.2 Standard Resistors

It is economically and technically inefficient and not feasible to manufacture resistors of all possible values. For this reason, resistor sizes are standardized and only resistors of certain values are manufactured. Thus, a resistor of an arbitrary value cannot be found to purchase. Consequently, in selecting resistors, one has to select the nearest standard size or arrange the required resistance value by combining standard size resistors. The

TABLE E.1 Standard Resistor Values for 10, 5, and 1 Percent Tolerances

10% Standard Values											
10	12	15	18	22	27	33	39	47	56	68	82
5% Standard Values											
10	11	12	13	15	16	18	20	22	24	27	30
33	36	39	43	47	51	56	62	68	75	82	91
1% Standard Values											
10.0	10.2	10.5	10.7	11.0	11.3	11.5	11.8	12.1	12.4	12.7	13.0
13.3	13.7	14.0	14.3	14.7	15.0	15.4	15.8	16.2	16.5	16.9	17.4
17.8	18.2	18.7	19.1	19.6	20.0	20.5	21.0	21.5	22.1	22.6	23.2
23.7	24.3	24.9	25.5	26.1	26.7	27.4	28.0	28.7	29.4	30.1	30.9
31.6	32.4	33.2	34.0	34.8	35.7	36.5	37.4	38.3	39.2	40.2	41.2
42.2	43.2	44.2	45.3	46.4	47.5	48.7	49.9	51.1	52.3	53.6	54.9
56.2	57.6	59.0	60.4	61.9	63.4	64.9	66.5	68.1	69.8	71.5	73.2
75.0	76.8	78.7	80.6	82.5	84.5	86.6	88.7	90.9	93.1	95.3	97.6

standard values for resistors with different tolerance are not the same. The most commonly used resistor values are shown in Table E.1.

Table E.1 shows that all values are between 10 and 100. This does not mean that all standard resistors are below 100 Ω. Numbers given in the table are base values and can be multiplied by any power of 10. For instance, a value of 22 implies that the following standard resistors are available: 2.2, 22, 220, 2200, 22,000 Ω, and so forth.

E1.3 Resistor Color Code

A number of resistors are color coded by carrying a number of color bars on their body, which define their value and tolerance. Examples are shown in Figures E.1 and E.2. The difference between the two figures is the number of color bars. That in Figure E.2 carries five color bars, which implies more digits and more precision. Thus, it is used for lower tolerances such as 1 percent.

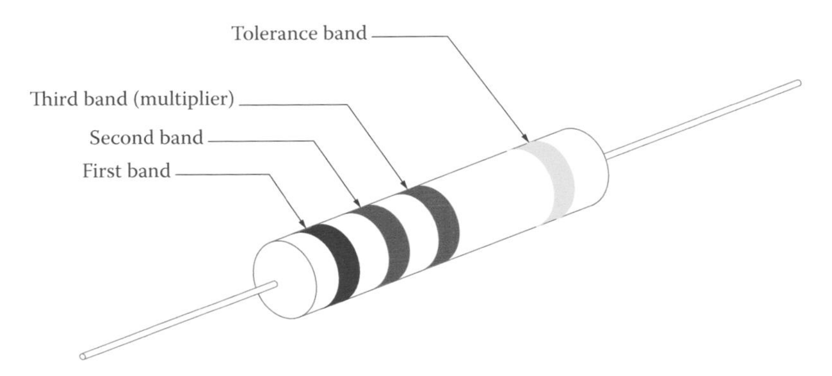

Figure E.1 Color-coded resistor with four color bars.

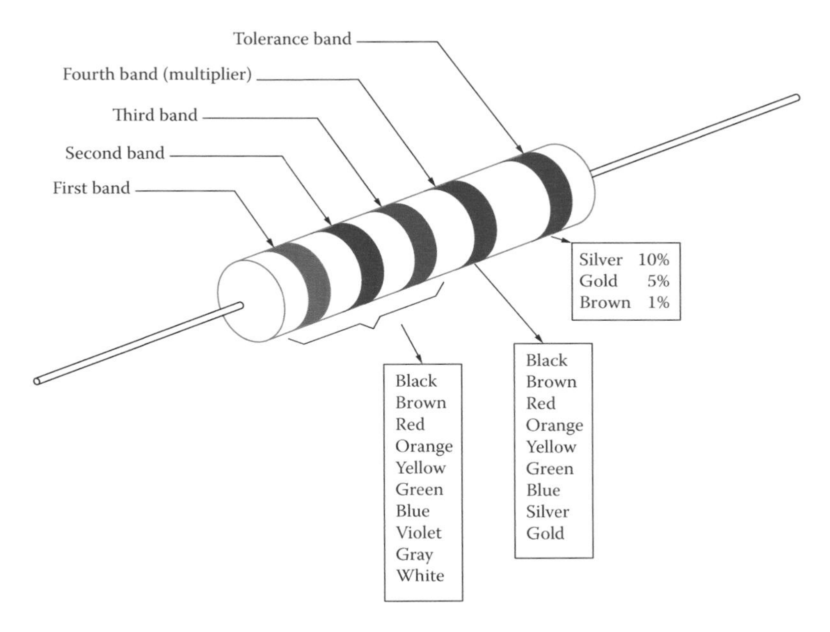

Figure E.2 Color-coded resistor with five color bars.

The way to use the color code is based on the numbers that are assigned to colors. These numbers are shown in Table E.2. As you see from the figures, for both four color bars and five color bars one of the bands is separated from the others. This bar determines the resistor tolerance. The rest of the color bars define the value of the resistor. The bar numbers are shown in Figures E.1 and E.2. The first two (three) bands in a four color (five color) bar resistor show a two-digit (three-digit) number. The remaining bar is the multiplier, which determines the number of zeros that must be added in front of the number just obtained (or, alternatively, a multiplier number obtained by raising 10 to the power defined by the number corresponding to the color). The tolerance colors are shown in Figure E.2.

TABLE E.2 Color Table for Color-Coded Resistors

Digit Bars		Multiplier Bar	
Band Color	**Digit Value**	**Color**	**Multiplier Value**
Black	0	Black	1
Brown	1	Brown	10
Red	2	Red	100
Orange	3	Orange	1000
Yellow	4	Yellow	10,000
Green	5	Green	100,000
Blue	6	Blue	1,000,000
Violet	7	Silver	0.01
Gray	8	Gold	0.1
White	9		

The following examples illustrate the matter more clearly.

Four color bars:

Brown, black, black	10 Ω
Brown, black, brown	100 Ω
Brown, black, red	1000 Ω
Red, red, black	22 Ω
Red, red, red	2200 Ω
Green, brown, orange	51,000 Ω
Green, brown, gold	5.1 Ω

Five color bars:

Brown, brown, red, red	11,200 Ω
Yellow, violet, green, black	475 Ω
Yellow, violet, green, gold	47.5 Ω
Yellow, violet, green, silver	4.75 Ω
Green, violet, blue, brown	5.76 Ω

E1.4 Resistor Power Values

In addition to the tolerance and the nominal value, when selecting a resistance, one needs to specify its power. As discussed in Chapter 6, standard resistors are available in 1/4, 1/2, 1, 1.5, and 2 W rating. They are made from materials such as metal film, carbon film, and composition carbon. Resistors of larger power value are wire wound (see Figures 4.1 and 4.2).

It is important to understand that the power value of a resistor represents the maximum power that a resistor can dissipate without getting damaged. It does not represent the power consumption of a resistor when it is used in a circuit. When inside a circuit, power consumption of a resistor must be determined by Ohm's law. This value must not surpass the power rating.

Appendix F: Quality Factor of *RLC* Circuits

This appendix is added for those readers who might need it for other application than electric circuits, such as in communications. For parallel and series circuits a quality factor Q can be defined using the values of the three components. Q is a ratio; so it is a number without any unit. Here R is the resistance of a resistor that is part of the RLC and must not be confused with the internal resistance of a coil as an inductor.

For a parallel RLC circuit the quality is

$$Q = R\sqrt{\frac{C}{L}} \tag{F.1}$$

At the resonance frequency, for which $X_C = X_L$, C can be substituted in terms of L and Equation F.1 assumes the form

$$Q = \frac{R}{X_L} \tag{F.2}$$

Because $X_C = X_L$, Q can also be written as $Q = \dfrac{R}{X_C}$.

For a series RLC the quality is

$$Q = \frac{1}{R}\sqrt{\frac{L}{C}} \tag{F.3}$$

and for the resonance frequency ($X_C = X_L$) substituting for C in terms of L leads to

$$Q = \frac{X_L}{R} = \frac{X_C}{R} \tag{F.4}$$

Appendix G: Structure of Logic Gates

It was mentioned in Chapter 22 that all the digital electronics components are made out of analog electronic circuits, such as transistors. Construction of some of the logic gates are shown in Figures G.1 and G.2. These are the simplest designs, which are based on transistor-transistor logic. Other designs are also possible. Furthermore, the basic designs can be enhanced by amplification.

Figure G.3a depicts a simple transistor working as a NOT circuit. An enhancement of the same circuit with another transistor to amplify the input is shown in Figure G.3b. Also, another version of the NOR gate is illustrated in Figure G.3c. However, the previous one in Figure G.2b is preferable to this in the sense that the two inputs cannot have interaction with each other, so that one influences the other (when one is high and the other is low). A XOR gate is more involved and can be synthesized from other gates.

In practice, logic gates come as packages in the form of integrated circuit (IC). For example, the family of 74xx IC's is a series of logic gates and other digital devices all defined by the number 74 and some other characters and numbers. Figure G.4 shows the inside structure and the pin-out of a sample called 7486, which contains four two-input XOR gates.

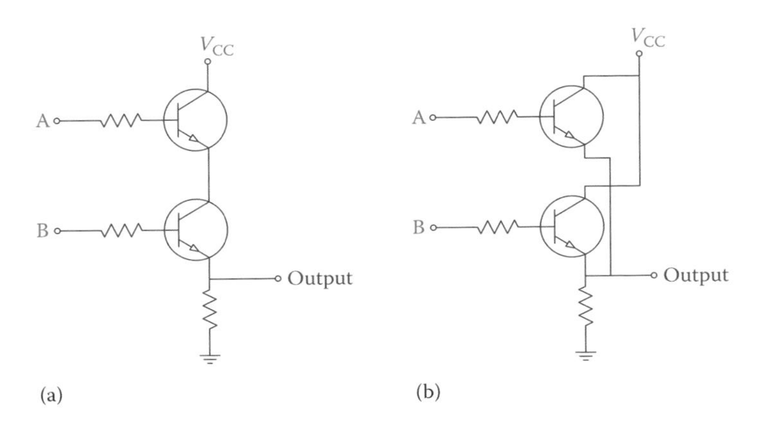

Figure G.1 Structure of (a) AND and (b) OR gates from transistors.

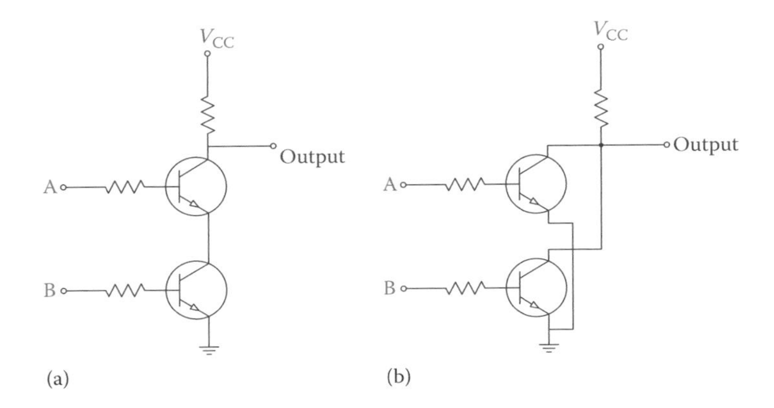

Figure G.2 Structure of (a) NAND and (b) NOR gates from transistors.

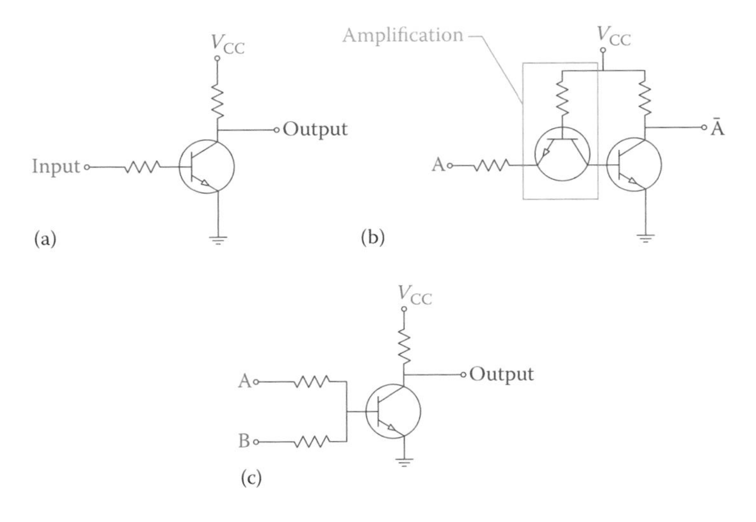

Figure G.3 (a) NOT gate, (b) NOT gate with input amplification, and (c) alternative NOR gate.

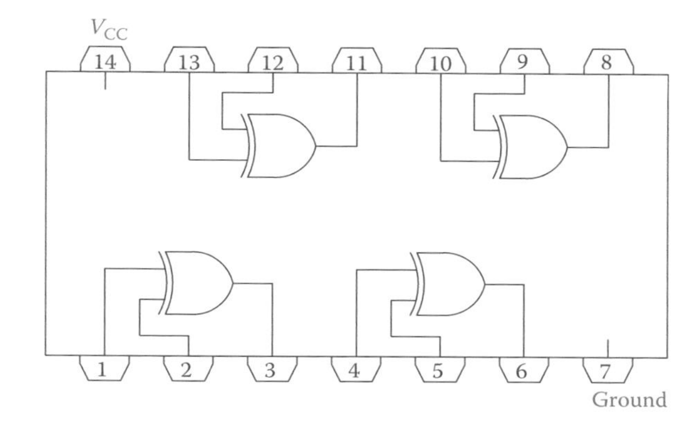

Figure G.4 Layout of four XOR gates in 7486 IC.

Index

Page numbers followed by *f* and *t* indicate figures and tables, respectively.

C

D

E

J

K

L

M

Q

R

S

U

V